Tectonics of Sedimentary Basins

Tectonics of Sedimentary Basins

Edited By

CATHY J. BUSBY
University of California, Santa Barbara

RAYMOND V. INGERSOLL
University of California, Los Angeles

b

Blackwell
Science

Blackwell Science

Editorial offices:
238 Main Street, Cambridge, Massachusetts 02142, USA
Osney Mead, Oxford OX2 0EL, England
25 John Street, London, WC1N 2BL, England
23 Ainslie Place, Edinburgh EH3 6AJ, Scotland
54 University Street, Carlton, Victoria 3053, Australia
Arnette Blackwell SA, 1 rue de Lille, 75007 Paris, France
Blackwell Wissenschafts-Verlag GmbH, Kurfürstendamm 57,
    10707 Berlin, Germany
Blackwell MZV, Feldgasse 13, A-1238 Vienna, Austria

DISTRIBUTORS:

USA
    Blackwell Science, Inc.
    238 Main Street
    Cambridge, Massachusetts 02142
    (Telephone orders: 800-215-1000 or 617-876-7000)

Canada
    Oxford University Press
    70 Wynford Drive
    Don Mills, Ontario M3C 1J9
    (Telephone orders: 416-441-2941)

Australia
    Blackwell Science Pty Ltd
    54 University Street
    Carlton, Victoria 3053
    (Telephone orders: 03-347-5552)

Outside North America and Australia
    Blackwell Science, Ltd.
    c/o Marston Book Services, Ltd.
    P.O. Box 87
    Oxford OX2 0DT
    England
    (Telephone orders: 44-865-791155)

Acquisitions: Jane Humphreys

Development: Debra Lance

Production: Maria Hight

Manufacturing: Kathleen Grimes

Typeset by A-R Editions, Inc.

Printed and bound by Braun-Brumfield, Inc.

©1995 by Blackwell Science, Inc.
Printed in the United States of America

Library of Congress Cataloging-in-Publication Data

Tectonics of sedimentary basins / [compiled by] Cathy J.
    Busby, Raymond V. Ingersoll.
        p.   cm.
    Includes bibliographical references and index.
    ISBN 0-86542-245-1
        1. Sedimentary basins. 2. Geology, Structural. I. Busby,
    Cathy J. II. Ingersoll, Raymond V.
    QE571.T355   1995
    551.4'4—dc20   94-39440
                                                              CIP

Contents

Contributors

GERARD C. BOND
Lamont-Doherty Earth Observatory, Palisades, New York

CATHY J. BUSBY
Department of Geological Sciences, University of California, Santa Barbara, California

WILLIAM R. DICKINSON
Department of Geosciences, University of Arizona, Tucson, Arizona

STEPHAN A. GRAHAM
Department of Geological and Environmental Sciences, Stanford University, Stanford, California

RAYMOND V. INGERSOLL
Department of Earth and Space Sciences, University of California, Los Angeles, California

TERESA E. JORDAN
Department of Geological Sciences, Cornell University, Ithaca, New York

GEORGE D. KLEIN
New Jersey Marine Sciences Consortium, Fort Hancock, New Jersey

MICHELLE A. KOMINZ
Department of Geological Sciences, University of Texas at Austin, Austin, Texas

CHARLES A. LANDIS
Geology Department, Otago University, Dunedin, New Zealand

MICHAEL R. LEEDER
Department of Earth Sciences, University of Leeds, Leeds, United Kingdom

KATHLEEN M. MARSAGLIA
Department of Geological Sciences, University of Texas at El Paso, El Paso, Texas

ANDREW D. MIALL
Department of Geology, University of Toronto, Toronto, Ontario, Canada

GREGORY F. MOORE
Hawaii Institute of Geophysics, University of Hawaii, Honolulu, Hawaii

TOR H. NILSEN
Consulting Geologist, San Carlos, California

A.M.C. ŞENGÖR
ITÜ Maden Fakültesi, Jeoloji Bölümü, Ayazağa, Istanbul, Turkey

ROBERT E. SHERIDAN
Department of Geological Sciences, Rutgers, The State University, New Brunswick, New Jersey

GARY A. SMITH
Department of Earth and Planetary Sciences, University of New Mexico, Albuquerque, New Mexico

ARTHUR G. SYLVESTER
Department of Geological Sciences, University of California, Santa Barbara, California

MICHAEL B. UNDERWOOD
Department of Geology, University of Missouri, Columbia, Missouri

Preface

For at least a decade, courses in sedimentation and tectonics have been taught at many universities, but there has been no textbook available on this subject. Instructors have, instead, cultivated long reading lists of journal articles that they attempt to synthesize in lecture, typically covering a different basin type in each lecture. Students have been able to read only a few of these articles, resulting in uneven treatment of each basin type.

The plate-tectonic revolution some twenty-five years ago provided an extremely useful paradigm for interpreting basin stratigraphy and structure. This interpretive process has been underway "in a piecemeal manner" (Miall, 1990), involving literally thousands of investigators the world over. The data base has now grown to the point where sophisticated basin models can be linked to specific plate-tectonic settings, and used for interpreting regional plate-tectonic histories. These basin models integrate sedimentology, structural geology, geophysical theory, and petrology to characterize the architecture of basins in most tectonic settings. The authors of the chapters in this book have all taken a broad-based, multidisciplinary approach to their research, as demonstrated by the numerous important papers they have published on the topics they have synthesized.

As outlined in the introductory chapter, the table of contents of this book follows the basin classification of Ingersoll (1988b). Thus, the organization and subject matter do not overlap substantially with previously published books. We provided the authors with an outline of topics to be used in every chapter, to encourage uniform treatment. Emphasis was placed on pointing out distinctive characteristics of each basin type. We also worked to eliminate gaps or overlaps in subject matter among chapters by involving authors of related chapters in our review process, as well as by reviewing all chapter drafts ourselves. Complementary readings may be found in *Basin Analysis* by Allen and Allen (1990), which provides more complete discussions of subsidence mechanisms and quantitative models.

Both of us "test-drove" drafts of the chapters in our courses at the University of California, and gave the authors feedback from these trial runs, in order to make sure that the material would be understandable to a college senior majoring in Earth sciences. This book will be equally useful for graduate students and professional researchers. Because of the multidisciplinary nature of the topic and the exhaustive reference list, we believe that even professors teaching from this book will learn a great deal; we know we did!

We thank the following formal and student reviewers for their thoughtful comments and suggestions: B. F. Adams, S. B. Bachman, G. Bates, B. Benumoff, E. Cann, P. A. Craig, S. Critelli, T. A. Cross, W. R. Dickinson, S. Dougherty, L. Hasbargen, T. E. Jordan, G. D. Klein, P. B. Kokelaar, M. R. Leeder, G. H. Mack, K. M. Marsaglia, A. D. Miall, P. Miloe, J. A. Nunn, M. Parris, P. E. Rumelhart, D. W. Scholl, A.M.C. Şengör, D. Smith, G. A. Smith, F. Surlyk and M. B. Underwood.

We also thank the staff of Blackwell Science for their help in organizing this book. In particular, we thank Simon Rallison for encouraging us to propose it in 1989, Jane Humphreys, who took over his job as acquisitions editor in 1991, and Gail Segal and Debra Lance, developmental editors, and Maria Hight, assistant production editor.

Last but not least, we thank our families and friends for putting up with our hauling around huge piles of manuscripts and hunching endlessly over our computers for the past five years.

Santa Barbara, August 1994   Cathy J. Busby
Los Angeles, August 1994   Raymond V. Ingersoll

Tectonics of Sedimentary Basins

Tectonics of Sedimentary Basins

1

Raymond V. Ingersoll and Cathy J. Busby

Introduction

This chapter has three primary purposes. First, it serves as an overview of the general subject of tectonics of sedimentary basins. Second, it complements the following twelve chapters, which inevitably, do not cover all aspects of all types of sedimentary basins. Third, it highlights controversies and/or disagreements both within and among these chapters. As a result of these three complementary, but somewhat conflicting purposes, the coverage of topics in this chapter is variable. Subjects that are covered in great depth in other chapters are treated lightly herein (e.g., most basins in convergent settings); we expand upon subjects that are covered lightly or that are controversial (e.g., modes of continental extension). Thus, the entire book should be read for a complete treatment of tectonics of sedimentary basins.

The chapter authors were invited to contribute to this volume because of their expertise concerning at least one class of basin; each chapter is an independent contribution that can stand alone. Nonetheless, many of the chapters were reviewed by authors of other chapters, as well as by the editors and additional reviewers, and we all have attempted to cross-reference each other wherever possible. As a result, we hope that there is coherence to the book, a common lack of multiply authored texts. We, as editors, have attempted to enforce uniform nomenclature in all of the chapters, with some authors agreeing more readily than others. This introductory chapter illustrates where some of our disagreements over both nomenclature and process remain. The authors are a diverse lot, and open discussion of disagreements is a sign of healthy science!

Ingersoll (1988b) summarized parallel advances in our understanding of processes and development of models at both large and small scales during the last two decades; these advances have occurred in several disciplines central to basin analysis. They include the plate-tectonic revolution itself (e.g., Cox and Hart, 1986), as well as a revolution in our understanding of modern depositional systems, and consequent major advances in sophistication of actualistic depositional models (e.g., Davis, 1983; Walker, 1984; Reading, 1986). Actualistic petrologic models relating sediment composition, especially sand and sandstone, to plate-tectonic settings have been developed (e.g., Dickinson and Suczek, 1979). Exploration techniques, especially seismic stratigraphy and detailed mapping of ocean floors, have provided new avenues for investigation of both ancient and modern basins, including development of sequence-stratigraphic concepts (e.g., Van Wagoner et al., 1990). In addition, refinement of chronostratigraphic methods, subsidence analysis, microanalytical investigation of thermal history, paleomagnetism, paleoclimatology and other fields have revolutionized basin analysis, as summarized by Miall (1990) in a landmark textbook. Miall pointed out that a "new stratigraphy" has arrived. The last 25 years have been the beginning of the "golden age of the new stratigraphy" due to concurrent revolutions affecting the Earth sciences. In fact, stratigraphy has become a true science only with the development of testable actualistic models based on combinations of theory, observation and experiment. These actualistic models are the key to basin analysis from the micro- to the megascale. Dott (1978) reviewed the development of pre-plate-tectonic (primarily nonactualistic) models for megascale basins (geosynclines). Windley (1993) reviewed the application of actualistic (uniformitarianistic) models to the interpretation of Precambrian crustal evolution.

The emphasis of this chapter and the book is on actualistic plate-tectonic basin models that are derived from the study of modern basins and applied to the interpretation of ancient basins. To paraphrase R.G. Walker (1984, p. 6), a basin model should act as: 1. a norm, for purposes of comparison; 2. a framework and guide for future observations; 3. a predictor in other situations; and 4. an integrated basis for interpretation of the class of basin it represents. As demonstrated below, individual basins are far more complex than any model (this is true of all natural systems), but testing and refinement of models based on observations provide the primary way that progress occurs. Our understanding improves by the clear statement of a model (whether quantitative or qualitative), and the subsequent refutation of the model and its replacement with an improved model. "Truth emerges more readily from error than from confusion" (Francis Bacon) (Kuhn, 1970, p. 18).

Classification of Sedimentary Basins

Our classification and conceptualization of sedimentary basins have been guided primarily by Dickinson (1974b, 1976a) and Ingersoll (1988b). The following discussion draws heavily on these articles, and is essentially a progress report. Some of the basin types are understood in great detail, including sophisticated quantitative models. Other types remain almost unstudied; in fact, some basin types are virtually ignored by modelers, either because they are too difficult to model quantitatively or because the modelers are unaware of their existence as a class of sedimentary basin. Ingersoll's (1988b) 23 basin types are a daunting menu from which to choose, but they accurately reflect the complexity of the real world. In fact, Table 1.1 lists 26 basin types because, during the course of editing this book, we realized that a few unusual basins could not be classified, and even some very common basins needed to be distinguished. We do *not* apologize for "creating additional basin types." Plate tectonics has done this; we are only attempting to classify, and thereby better understand, actual sedimentary basins of the world. As S.J. Gould (1989, p.

98) stated, "Classifications are theories about the basis of natural order, not dull catalogues compiled only to avoid chaos."

W.R. Dickinson (1993, personal communication) (also, see Dickinson, 1993b) wrote the following in defense of our subdivision of basins into so many types, and as a reason why "geophysicists need to know about basin classification": "It is witless to emulate the geosynclinal theorists of yore (who dismissed the variability of geosyncline types, the 'menagerie of Marshall Kay', in favor of a focus on supposedly ruling 'orthogeosynclines'); similarly, modern geodynamic analysis cannot gain ground with subsidence models uninformed by the variant conditions imposed by the diverse tectonic settings of sedimentary basins (plowing ahead with 'ideal' models, except as devised to mimic actual variants, would just insure that none would truly fit real cases well)." Furthermore, he states, "Basin modelers who bypass the constraints supplied by insight into the tectonic setting of different types of sedimentary basins run the risk of inventing a virtual world that has no crossover with reality." Thus, we urge modelers to delve into the complexity of processes acting in the very messy tectonic setting of Earth's crust, modern and ancient, and we hope that this book helps them.

Dickinson (1974b, 1976a) originally provided the most comprehensive actualistic classification of basin types related to plate-tectonic processes. This classification is modified and updated, following Ingersoll (1988b). To paraphrase Dickinson (1974b, 1976a), plate tectonics emphasizes horizontal movements of the lithosphere, which induce vertical movements due to changes in crustal thickness, thermal character and isostatic adjustment. These vertical movements cause the formation of sedimentary basins, uplift of sediment source areas, and reorganization of dispersal paths. Primary controls on basin evolution (and the basis for classification) are: 1. type of substratum; 2. proximity to plate boundary(s); and 3. type of nearest plate boundary(s). Types of substratum include continental crust, oceanic crust, transitional crust and anomalous crust. Primary types of plate boundaries are divergent, convergent and transform; intraplate and hybrid settings also are common (Tables 1.1 and 1.2).

Table 1.1   Basin Classification (modified after Dickinson, 1974b,1976a, and Ingersoll,1988b)

Divergent Settings		
	Terrestrial Rift Valleys:	rifts within continental crust, commonly associated with bimodal volcanism.
	Proto-Oceanic Rift Troughs:	incipient oceanic basins floored by new oceanic crust and flanked by young rifted continental margins.
Intraplate Settings		
	Continental Rises and Terraces:	mature rifted continental margins in intraplate settings at continental-oceanic interfaces.
	Continental Embankments:	progradational sediment wedges constructed off edges of rifted continental margins.
	Intracratonic Basins:	broad cratonic basins floored by fossil rifts in axial zones.
	Continental Platforms:	stable cratons covered with thin and laterally extensive sedimentary strata.
	Active Ocean Basins:	basins floored by oceanic crust formed at divergent plate boundaries unrelated to arc-trench systems (spreading still active).
	Oceanic Islands, Aseismic Ridges and Plateaus:	sedimentary aprons and platforms formed in intraoceanic settings other than magmatic arcs.
	Dormant Ocean Basins:	basins floored by oceanic crust, which is neither spreading nor subducting (no active plate boundaries within or adjoining basin).
Convergent Settings		
	Trenches:	deep troughs formed by subduction of oceanic lithosphere.
	Trench-Slope Basins:	local structural depressions developed on subduction complexes.
	Forearc Basins:	basins within arc-trench gaps.
	Intra-Arc Basins:	basins along arc platform, which includes superposed and overlapping volcanoes.
	Backarc Basins:	oceanic basins behind intraoceanic magmatic arcs (including interarc basins between active and remnant arcs), and continental basins behind continental-margin magmatic arcs without foreland foldthrust belts.
	Retroarc Foreland Basins:	foreland basins on continental sides of continental-margin arc-trench systems (formed by subduction-generated compression and/or collision).
	Remnant Ocean Basins:	shrinking ocean basins caught between colliding continental margins and/or arc-trench systems, and ultimately subducted or deformed within suture belts.
	Peripheral Foreland Basins:	foreland basins above rifted continental margins that have been pulled into subduction zones during crustal collisions (primary type of collision-related forelands).
	Piggyback Basins:	basins formed and carried atop moving thrust sheets.
	Foreland Intermontane Basins (Broken Forelands):	basins formed among basement-cored uplifts in foreland settings.

Table 1.1 Cont'd.
Transform Settings (and transcurrent-
fault-related basins)

	Transtensional Basins:	basins formed by extension along strike-slip fault systems.
	Transpressional Basins:	basins formed by compression along strike-slip fault systems.
	Transrotational Basins:	basins formed by rotation of crustal blocks about vertical axes within strike-slip fault systems.
Hybrid Settings		
	Intracontinental Wrench Basins:	diverse basins formed within and on continental crust due to distant collisional processes.
	Aulacogens:	former failed rifts at high angles to continental margins, which have been reactivated during convergent tectonics, so that they are at high angles to orogenic belts.
	Impactogens:	rifts formed at high angles to orogenic belts, without preorogenic history (in contrast with aulacogens).
	Successor Basins:	basins formed in intermontane settings following cessation of local orogenic or taphrogenic activity.

Rifts are associated with all three types of plate boundaries, and for this reason, a general overview of rift tectonics is presented first in this book (Chapter 2). The Wilson Cycle is followed in the organization of most of the rest of the book, that is, divergent and derivative intraplate settings are discussed first (Chapters 3–4), followed by convergent settings (Chapters 5–11). Chapter 12 on strike-slip basins deals with basins both along transform plate boundaries and along transcurrent faults within plates. Finally, intracratonic basins, most of which probably originated as fossil rifts, are discussed in Chapter 13.

Terminology

Throughout this discussion, the distinction between continental margins and plate boundaries is important; they may or may not correspond. For this reason, usage such as "collision of plates" should be avoided. Plates "converge"; only nonsubductable (buoyant) crustal components can "collide" (e.g., Cloos, 1993). We use terms such as "extensional," "compressional," "transtensional" and "transrotational" to refer to crustal deformation, whereas "divergent," "convergent," "transform" and "intraplate" refer to plate-tectonic settings.

The literature is full of confusing or ambiguous terminology, especially in situations where geosynclinal or physiographic terminology has been retained. One example of ambiguity is the use of "miogeoclinal" and "eugeoclinal" as synonyms for "miogeosyncline" and "eugeosyncline." Dietz and Holden (1966) demonstrated how geosynclines could be reinterpreted using actualistic models based on modern continental margins. In doing so, they changed "geosynclines" to "geoclines" (thus, reflecting the nonsynclinal nature of intraplate-margin sequences), and pointed the way to developing actualistic plate-tectonic models. The "geocline" terminology, however, is inherently confusing, and we recommend abandonment of these terms.

A second example of terminological confusion involves the terms "foredeep," "foreland" and "hinterland" (pre-plate-tectonic terms), which have been adapted to plate tectonics in different ways by different workers. We recommend that these terms retain their physiographic and/or bathymetric definitions, namely deeps or lands before or behind related compressional mountain fronts. As used without modifiers, these are geographically relative terms, without plate-tectonic significance. Thus, the Po Valley of northern Italy is a foreland relative to the Apennines, a foreland relative to the south side of the Alps, and a hinterland relative

Table 1.2 Modern and Ancient Examples of Basin Types

Basin Type	Modern Example	Ancient Example
Terrestrial Rift Valleys	Rio Grande rift (New Mexico)	Proterozoic Keweenawan rift
Proto-Oceanic Rift Troughs	Red Sea	Jurassic of East Greenland
Continental Rises and Terraces	East Coast of USA	Early Paleozoic of USA and Canadian Cordillera
Continental Embankments	Mississippi Gulf Coast	Early Paleozoic Meguma terrane of Canadian Appalachians (?)
Intracratonic Basins	Chad Basin (Africa)	Paleozoic Michigan basin
Continental Platforms	Barents Sea (Asia)	Middle Paleozoic, North American midcontinent
Active Ocean Basins	Pacific ocean	miscellaneous ophiolite complexes (?)
Oceanic Islands, Aseismic Ridges and Plateaus	Emperor-Hawaii seamounts	Mesozoic Snow Mountain Volcanic Complex (Franciscan) (California)
Dormant Ocean Basins	Gulf of Mexico	Phanerozoic Tarim basin (China) (?)
Trenches	Chile Trench	Cretaceous, Shumagin Island (Alaska)
Trench-Slope Basins	Central America Trench	Cretaceous Cambria slab (California)
Forearc Basins	Sumatra	Cretaceous Great Valley (California)
Intra-Arc Basins	Lago de Nicaragua	Early Jurassic Sierra Nevada (California)
Backarc Basins	Marianas	Jurassic Josephine ophiolite (California)
Retroarc Foreland Basins	Andes foothills	Cretaceous Sevier foreland (Wyoming)
Remnant Ocean Basins	Bay of Bengal	Pennsylvanian-Permian Ouachita basin
Peripheral Foreland Basins	Persian Gulf	Mid-Cenozoic Swiss Molasse basin
Piggyback Basins	Peshawar basin (Pakistan)	Neogene, Apennines (Italy)
Foreland Intermontane Basins	Sierras Pampeanas basins (Argentina)	Laramide basins (USA)
Transtensional Basins	Salton Sea (California)	Carboniferous, Magdalen basin (Gulf of St.Lawrence)
Transpressional Basins	Santa Barbara basin (California) (foreland)	Miocene Ridge basin (California) (fault-bend)
Transrotational Basins	Western Aleutian forearc (?)	Miocene Los Angeles basin (California)
Intracontinental Wrench Basins	Qaidam Basin (China)	Pennsylvanian-Permian Taos trough (New Mexico)
Aulacogens	Mississippi Embayment	Paleozoic Anadarko aulacogen (Oklahoma)
Impactogens	Baikal rift (Siberia) (distal)	Rhine Graben (Europe) (proximal)
Successor Basins	Southern Basin and Range (Arizona)	Paleogene Sustut basin (?) (British Columbia)

to the Molasse basin of Switzerland (see Fig. 11.2). It is only when plate-tectonic setting is known that the more refined classification into peripheral, retroarc or transpressional foreland basins can be attempted. The Po Valley is an Apennine peripheral foreland basin that has been superposed on an Alpine retroarc foreland basin. "Hinterland" is such a general term that we recommend its abandonment, as it basically is synonymous with "mountains" and areas "behind" them. It is also important to point out that some workers use "foredeep" when they really mean "forethick" (W.R. Dickinson, 1994, personal communication). Thick foreland sediments may accumulate in the absence of any bathymetric basin; a bathymetric basin "deep" should not be confused with thick strata.

Most terminology for convergent-margin settings, such as trenches, forearcs, backarcs and intra-arcs, has been defined actualistically and is unambiguous, although recognition of ancient examples is seldom unambiguous. Most terminology for other orogenic basins has evolved from study of ancient collisional orogens involving intraplate continental margins and subduction zones; the resulting hybrid terms can be confusing if not carefully defined. Some workers use "foredeep" as synonymous with "foreland," whereas a more refined usage is that the former term refers to the bathymetrically deepest part of the foreland, immediately in front of the foldthrust belt, whereas the latter refers to the shallower, more distal part. During collision orogeny, a trench evolves into a peripheral foredeep as the edge of a continent is pulled into a subduction zone and a broader peripheral foreland forms on top of the former intraplate continental margin (see Chapters 10 and 11).

We use "terrane" as a nongenetic term, which refers to any assemblage of rocks. We eschew "terranology" with its complex mixing of genetic and descriptive terms into a nonactualistic stew (see also Şengör and Dewey, 1991). "Terrane" is a useful general word, which we refuse to give up, in spite of its misappropriation by the "terranologists." "Terrain" should be restricted to descriptions of surficial features, as in "rugged terrain" or "desert terrain."

We do not follow Şengör's (Chapter 2) restriction of the word "rift" to structures that penetrate the lithosphere (as opposed to "grabens," which do not) because this distinction commonly cannot be made (due to insufficient data), and because rifts commonly initiate in thick continental crust, where lower-crustal flow may transfer extension laterally.

It is essential to remember that a basin should be classified according to its tectonic setting at the time of deposition of a given stratigraphic interval; thus, a "basin" (whose stratigraphic "fill" may actually be a wedge, a slab, a prism, or another shape) may change its tectonic setting rapidly and often. "The evolution of a sedimentary basin thus can be viewed as the result of a succession of discrete plate-tectonic settings and plate interactions whose effects blend into a continuum of development" (Dickinson, 1974b, p. 1). Therefore, a complete basin analysis must incorporate all phases of development and must consider both proximal and distal tectonic influences. Also, it is important to remember that "basin," as used herein, refers to any sedimentary (and/or volcanic) stratigraphic succession. Some of these successions accumulate due to subsidence of shallow substrate ("sinking substratum"), whereas others result from filling of space below base level (commonly sea level) ("filling hole"). Many are hybrid sinking, tilting and filling holes. Some basins are considerably above sea level (commonly surrounded by mountainous terrain that may rise faster than the basins), resulting in relative basin subsidence concurrent with absolute uplift of the substrate. In contrast, all mature oceanic basins are "holes" that fill with sediment, independent of tectonic subsidence. Precise paleobathymetry and paleoelevation data are critical to any quantitative analysis in either case (e.g., Dickinson et al., 1987); unfortunately, these data are extremely difficult to acquire, especially for deep-marine and nonmarine paleoenvironments.

Subsidence Mechanisms

Subsidence of the upper surface of the crust is induced by the following processes (Dickinson, 1974b, 1976a, 1993b) (Table 1.3): 1. thinning of crust due to stretching, erosion and magmatic withdrawal; 2. thickening of mantle lithosphere during cooling;

3. sedimentary and volcanic loading of both crust and lithosphere; 4. tectonic loading of both crust and lithosphere; 5. subcrustal loading of both crust and lithosphere; 6. dynamic effects of asthenospheric flow; and 7. crustal densification. Crustal thinning dominates in extensional settings, whereas lithospheric thickening is most important in intraplate settings that originated at divergent plate boundaries (Fig. 1.1). Sediment loading is most important in areas of high sediment flux, especially where oceanic crust is adjacent to major deltas (e.g., continental embankments and remnant ocean basins). Tectonic loading dominates in foreland (including transpressional) settings; lithospheric flexure can result in both subsidence and uplift far from applied loads. Virtually all settings experience complex combinations of processes. Ancillary effects such as paleolatitude, paleogeography and eustatic changes provide modifying influences.

Gross subdivision of basins into compressional and extensional types, or active versus passive margins is even less useful than geosynclinal classification, which if nothing else, recognized the diversity and complexity of basin types (e.g., Kay, 1951). Likewise, simple subdivision into stretching models and flexural models (e.g., Allen and Allen, 1990) ignores many sedimentary basins, including most of Earth's largest sediment accumulations (e.g., Bengal and Indus fans), which are deposited in oceanic basins experiencing neither stretching nor tectonic loading (see Chapter 10).

Preservation Potential and Paleotectonic Reconstruction

Some modern basin types are common and volumetrically important, whereas others are rare and volumetrically minor. In addition, even some common modern basin types are rarely found in the very ancient record because they are prone to uplift and erosion, and/or deformation and destruction (e.g., remnant ocean, trench-slope and backarc basins; Fig. 1.2); their rarity in ancient orogenic belts is predicted by their susceptibility to erosion and deformation. Thus, their absence is not a valid test of plate-tectonic models.

The preservability of tectonostratigraphic assemblages is an important but seldom discussed factor in basin analysis and paleotectonic reconstruction. An important distinction must be made between the preservation of a basin (e.g., the Bay of Bengal) versus preservation of the basin fill (e.g., the Bengal Fan). In this example, the basin is destined to be destroyed during subduction and continental collision (see Chapter 10), whereas much of the sedimentary fill will be accreted into an orogenic belt to form continental crust (e.g., Şengör and Okurogullari, 1991). Thus, basin preservation potential is low, although stratal preservation is moderately high. Generally, any basin formed on oceanic crust has low preservation potential because of the probability that subduction will destroy the basin.

Table 1.3 Subsidence Mechanisms

Crustal Thinning:	extensional stretching, erosion during uplift, and magmatic withdrawal.
Mantle-Lithospheric Thickening:	cooling of lithosphere following either cessation of stretching or heating due to adiabatic melting or rise of asthenospheric melts.
Sedimentary and Volcanic Loading:	local isostatic compensation of crust and regional lithospheric flexure, dependent on flexural rigidity of lithosphere, during sedimentation and volcanism.
Tectonic Loading:	local isostatic compensation of crust and regional lithospheric flexure, dependent on flexural rigidity of underlying lithosphere, during overthrusting and/or underpulling.
Subcrustal Loading:	lithospheric flexure during underthrusting of dense lithosphere.
Asthenospheric Flow:	dynamic effects of asthenospheric flow, commonly due to descent or delamination of subducted lithosphere.
Crustal Densification:	increased density of crust due to changing pressure/temperature conditions and/or emplacement of higher-density melts into lower-density crust.

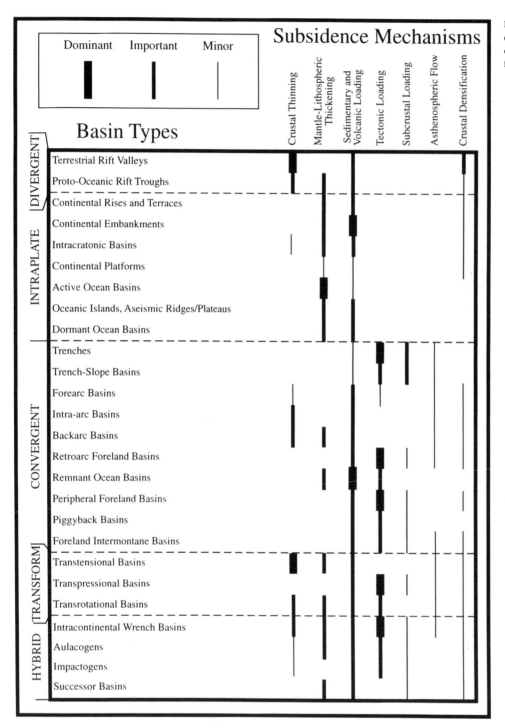

Figure 1.1 Suggested subsidence mechanisms for all types of sedimentary basins. See text for discussion.

Veizer and Jansen (1979, 1985) provided an empirical method of determining half lives of tectonostratigraphic elements. They estimated the half lives of "active-margin basins" at 30 my, oceanic sediments at 40 my, oceanic crust at 55 my, "passive margins" at 80 my, "immature orogenic belts" at 100 my, and "mature orogenic belts" and platforms at 380 my. The application of Veizer and Jansen's type of analysis to all of the basin types discussed herein should provide additional quantitative constraints on paleotectonic reconstructions.

Certain tectonic settings (e.g., continental rises and terraces, and foreland basins) have received much attention in the last decade, so that actualistic models are becoming quantitatively sophisticated. In contrast, other settings (e.g., successor basins, intra-arc basins and remnant ocean basins) have received relatively little attention. As all of these models improve, it is likely that integration of successive orogenic processes into a continuum of models will provide strong predictive capabilities for paleotectonic reconstructions.

The first step in identifying essential components controlling basin development is the construction of stratigraphic sections, maps and cross sections (preferably at true scale) of modern plate-tectonic systems. Figure 1.3 illustrates a courageous attempt to summarize almost every basin type in one diagram.

Each of these basin types is discussed below and in the following chapters.

Divergent Settings

Sequential Rift Development and Continental Separation

Most modern intraplate ("passive") continental margins originated as divergent plate margins formed during the breakup of the supercontinent Pangea in the early Mesozoic (e.g., Dietz and Holden, 1970). Many Paleozoic intraplate margins (e.g., Cordilleran and Appalachian) likewise originated during the latest Proterozoic breakup of the supercontinent Rodinia (Dewey and Bird, 1970; Stewart, 1972; Bond et al., 1984; McMenamin and McMenamin, 1990; Dalziel, 1991; Moores, 1991). A supercontinent cycle, with a period of 350–400 my, may be the result of the random motions of continental blocks around Earth's surface. Alternatively, some workers (e.g., Anderson, 1982, 1994; Fischer, 1984; LePichon and Huchon, 1984; Worsley et al., 1984) have attributed this cycle to convection within the mantle, including the focusing of heat beneath supercontinents to initiate breakup (also, see Chapters 2 and 13 for discussion of rifting and continental breakup). This model attributes

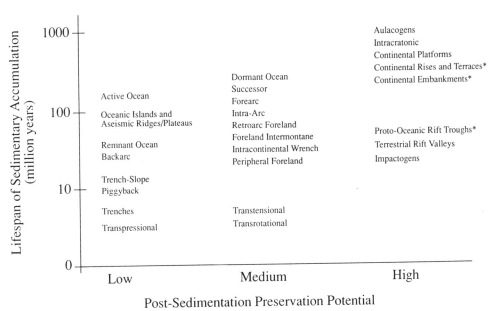

Figure 1.2 Typical life spans for sedimentary basins versus their post-sedimentation preservation potential. "Preservation potential" refers to average amount of time during which basins will not be uplifted and eroded, or be tectonically destroyed during and following sedimentation. Sedimentary or volcanic fill may be preserved as accretionary complexes during and after basin destruction (true of all strata deposited on oceanic crust). Basins with asterisks (intraplate continental margins) are "preserved" in the sense of retaining their basement, but they are likely to be subcreted beneath or within suture belts, and are difficult to recognize in the ancient record in such settings.

continental rifting to active-mantle processes (i.e., Şengör and Burke, 1978; Chapter 2); in contrast, most well studied rifts (modern and ancient) seem to have formed by passive-mantle processes (i.e., Şengör and Burke, 1978), where plate-tectonic processes initiate rifting (e.g., Ingersoll et al., 1990). An alternative mechanism for supercontinent breakup is suggested by the observation that density-driven slab subduction is one of the primary driving forces of plates (Jarrard, 1986a; see below). As a result, large old oceanic plates tend to subduct rapidly, with significant slab rollback, which induces extension in overriding plates and rapid sea-floor spreading. This would have been the dominant mode of oceanic plate tectonics during the time of supercontinents (e.g., in the Paleopacific during the Triassic), and any continents attached to such large old oceanic plates would have been prone to extension, either in immediate backarc areas or within plate interiors. We speculate that this scenario is the most likely way in which supercontinents inevitably break up to form new ocean basins and eventually, intraplate continental margins. (For a summary of the competing hypothesis, see Chapter 13.)

Intraplate continental margins contain evidence of two phases in their evolution (Fig. 1.4): a rift phase, which occurs before the break up of the continent, and a drift or post-rift phase, which occurs after the onset of sea-floor spreading adjacent to the rifted continental margins (Keen and Beaumont, 1990). Well studied modern examples of the rift phase include the East African rift system, and the Rio Grande rift of the western United States. The Red Sea and the Gulf of California are examples of young post-rift (i.e., "early-drift") margins, herein called "proto-oceanic troughs". The rift phase is a tectonically active one, with normal faulting, thinning of crust, elevation changes, volcanism and high heat flow, with locally high rates of basin subsidence and sediment accumulation. The post-rift phase is one of lithospheric cooling, thermal subsidence, and development of broad flexural basins. Over time, subsidence by sediment loading becomes more important than thermal subsidence, especially at continental embankments (see below).

Kinsman (1975) presented a model for the evolution of terrestrial rift valleys into juvenile ocean

basins, and ultimately, mature continental margins. He postulated that early domal uplifts preceded crustal stretching, and that subaerial erosion of the thermal dome thinned the upper crust enough to result in major subsidence once the margin was rafted away from the heat source, during the drift stage (Sleep, 1971). Recent studies of modern continental margins have shown that thinning cannot be caused mainly by erosion because continental crust has been thinned by more than 15 km, and remnants of older sedimentary rocks and other near-surface crustal rocks that predate rifting are found above the thinned regions (Keen and Beaumont, 1990). Furthermore, Kinsman's hypothesis did not take into consideration the possibility that domal uplifts are responses to lithospheric stretching, rather than its

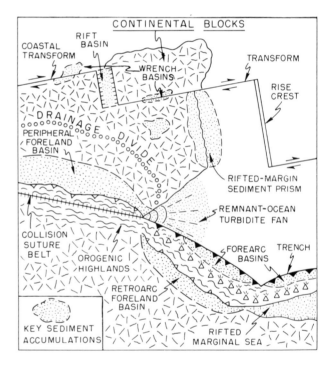

Figure 1.3 Sketch map showing sites of key sediment accumulations (basins) in relation to plate boundaries, continental margins and associated sources of detrital sediment. Unpatterned areas represent oceanic crust. Continental margins and tectonic features are indicated by solid lines; other basin margins are indicated by dashed lines. Stipples indicate areas of sediment accumulation; smoking triangles represent magmatic arc. Solid barbs represent subduction zones; open barbs represent foreland foldthrust belts. An intracratonic basin is shown between "drainage divide" (peripheral bulge) and "rifted-margin sediment prism" (continental rise and terrace). Modified from Dickinson (1980); from Ingersoll (1988b).

cause (e.g., Şengör and Burke, 1978; Morgan and Baker, 1983b). In this case, referred to as "passive rifting" (Şengör and Burke, 1978), tensional stresses within the lithosphere result in lithospheric thinning, causing upwelling of asthenosphere beneath the thinned region (McKenzie, 1978a; Royden and Keen, 1980; Beaumont et al., 1982b; Steckler and Watts, 1982). In the "passive-rifting" model, magmatism results from decompressive partial melting of the asthenosphere as it rises (Foucher et al., 1981).

Processes of Rifting

"Active rifting" requires the presence of an upwelling convective plume at the base of the lithosphere prior to crustal extension (Şengör and Burke, 1978; Spohn and Schubert, 1983). It is postulated that this process may result in convective thinning of the lithosphere from below, either thermally (due to increased heat input displacing the solidus upward) or mechanically (due to removal of material from the base of the lithosphere). Lithospheric thinning should cause uplift of several kilometers but little extension of near-surface rocks; thus, the crust will be significantly thinned tectonically only if the base of the lithosphere rises above the crust-mantle boundary (Keen and Beaumont, 1990).

Other mechanisms of rifting have been proposed, but stretching of the lithosphere has received the most attention, and models of lithospheric stretching appear to satisfy most of the observed properties of continental margins; what produces these extensional forces, however, is still unclear, and it has not been proven that lithospheric extension (i.e., "passive rifting") is the driving force for all continental rifting (Keen and Beaumont, 1990).

Kinsman (1975) assumed "normal" continental crust prior to rifting, which is at odds with theoretical studies showing that thickened and heated crust is weaker, and thus, is more likely to be the locus of extension when tensile stresses are applied (Kusznir and Park, 1987; Lynch and Morgan, 1987). In any case, elevated rift shoulders ("arch rims" of Veevers, 1981) form on either side of a terrestrial rift and persist through the proto-oceanic phase (also, see Hellinger and Sclater, 1983; Steckler and Omar, 1994) (Fig. 1.4C). Erosion thins the crust while these rims

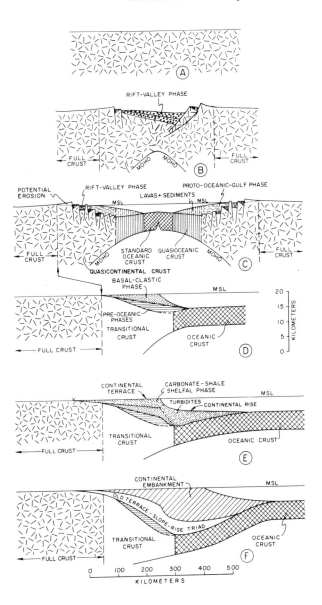

Figure 1.4 Schematic diagrams (vertical exaggeration 10X) to illustrate continental separation (that may "fail" at any time), with evolution from "normal" continental crust (A); to terrestrial-rift-valley phase (B); to proto-oceanic phase, showing terrestrial-rift-valley deposits on top of attenuated continental (quasicontinental) crust, adjacent to thickened basaltic (quasioceanic) crust (C); to end of proto-oceanic phase, when thermal subsidence is nearing completion; quasicontinental and quasioceanic crust are combined into "transitional crust," underlying subsiding continental margin (D); to continental terrace-slope-rise configuration during open-ocean phase, during which sediment loading is dominant subsidence mechanism (E); to continental-embankment phase, reached only where sediment delivery is voluminous enough to cause progradation of shoreline over oceanic crust (in areas of major deltas, usually at open ends of fossil rifts) (F). Only left side of ocean is shown in D through F. Modified from Dickinson (1976a) and Ingersoll (1988b).

are elevated, resulting in subsidence below sea level during subsequent cooling as the continental edges move away from the spreading ridge.

Kinsman (1975) suggested that attenuated continental crust seaward of the rim is relatively narrow (60–80 km), although other workers suggest that wider attenuated margins are common (e.g., Cochran, 1983a; Bohannon, 1986b; Lister et al., 1986, 1991). Rifting and continental separation commonly occur along alternating divergent and transform margins linking hot spots (Burke and Dewey, 1973), resulting in interplume, transcurrent and plume margins, respectively.

Kinsman's (1975) model emphasizes simple isostatic compensation at a depth of 100 km, whereas regional lithospheric flexure (e.g., Walcott, 1972) is a major control on subsidence, especially in terms of the seaward tilting of the shallow-marine (shelf) prism (e.g., Pitman, 1978). Nonetheless, Kinsman's model successfully predicts the maximum stratigraphic thickness found in continental embankments (16–18 km) based on the loading capacity of oceanic lithosphere.

As summarized by Frostick and Steel (1993), "active" and "passive" rifts should be distinguishable on the basis of their sedimentary history (also, see Şengör and Burke, 1978), but many rifts have features diagnostic of both types. "Active" rifts should be floored by erosional unconformities; initial centrifugal drainage patterns should result in clastic sediment starvation, although volcanogenic sedimentation may compensate for this. By contrast, in "passive" rifts, unconformities do not form and early sagging diverts drainages into centripetal or axial patterns. Voluminous basaltic volcanism should occur early in "active" rifts, and later in "passive" rifts, following significant crustal extension.

We agree with Şengör (Chapter 2) that the proposal to call "passive" rifting "closed-system rifting" and "active" rifting "open-system rifting" (Gans, 1987; also used in Chapter 3) is misleading because it is not obvious to which parameter the system is open or closed. Admittedly, a similar ambiguity exists with "passive" and "active" terminology, as it is not clear to many people that these terms refer to the absence or presence of asthenospheric rift-*initiating* processes,

respectively. "Passive" rifts are certainly not closed systems with respect to magmas; the important role of magmatism in them has long been recognized (e.g., Wernicke, 1985). Dunbar and Sawyer (1988, 1989) proposed that both "active" and "passive" rifts result from regional horizontal stresses and that differences reflect contrasts in pre-existing weaknesses in the continental lithosphere, but this distinction also is simplistic.

Aside from the "passive-active" debate, modelers of extension disagree on the relative importance and potential interdependence of three main effects (Wernicke, 1992): 1. the consequences of adding magma to the crust during extension; 2. the effect of rheological layering of the lithosphere, with discrete brittle faulting in the upper crust being accompanied by uniform stretching of the lower crust, and 3. the importance of lateral relaying of simple shear on detachment faults through the lower crust into upper-mantle shear zones.

Continental crust is inherently weaker in extension than is oceanic crust (Vink et al., 1984; Steckler and tenBrink, 1986). Preferential rifting of continents results, given lithospheric plate interactions that favor extension. This observation helps explain the common occurrence of thin continental slivers along complex plate margins (e.g., Baja California and the Levant) (e.g., Steckler and tenBrink, 1986; Dunbar and Sawyer, 1989). It also is consistent with the observation that late Cenozoic extension along the complex continental margin of the western United States has been concentrated in areas of pre-existing weaknesses within continental crust, especially thick crust (e.g., Axen et al., 1993), in preference to adjoining oceanic crust (i.e., Ingersoll, 1982a; Ingersoll et al., 1990).

Models for continental rifting must consider all of the following aspects of lithospheric behavior: 1. rheology at different horizons; 2. contrasts in crust and mantle, both in composition and structure; 3. contrasts in oceanic and continental crust; 4. "active" (asthenospherically driven) processes versus "passive" (lithospherically driven) processes; 5. the effects of pre-existing heterogeneities, especially in continental crust; and 6. the temporal context. The following characteristics tend to favor the development of low-angle detachment faults (in preference to high-angle planar faults): previously thickened crust, high heat flow and

rapid extension. No existing model considers all these aspects, and there is great variability in natural rifts, both modern and ancient (see Chapters 2 and 3).

Earliest models of rifting invoked symmetrical, pure-shear extension (Fig. 1.5A) to explain crustal thickness, subsidence histories and gravity profiles of extended terranes (e.g., McKenzie, 1978a; Sclater and Christie, 1980; LePichon and Sibuet, 1981). Symmetric-extension models, however, do not predict the wide variability in continental-margin architecture (e.g., Lister et al., 1986, 1991; Etheridge et al., 1989; Mutter et al. 1989). Also, models involving symmetric pure shear (e.g., Miller et al., 1983) provide no mechanism for bringing mid-crustal rocks to shallow levels (e.g., Wernicke, 1981, 1985).

Prior to the 1980s, most continental rifts were viewed as symmetrical features (grabens) bounded by high-angle normal faults. Based primarily on field work in the Basin and Range Province of the western United States, several models involving low-angle detachment faults have now been proposed (Fig. 1.5). Wernicke (1981, 1985) originally depicted detachment faults as low-angle faults that cut the entire lithosphere (Fig. 1.5B; also see Fig. 3.16); more recently, he has depicted them as intracrustal (Fig. 1.6), at least during early stages of extension. An alternative model (Fig. 1.5C) includes complex delamination of the lithosphere, with crustal detachment (at the brittle-ductile transition) connecting with the Moho (Lister et al., 1986).

Lister et al. (1986), while stating that the concept of a single lithospheric dislocation may be overly simple, presented a detachment model for the evolution of intraplate margins that predicts asymmetry at all scales (Fig. 1.7). "Upper-plate margins," which originate in hangingwalls of detachments, are characterized by thick continental crust with narrow continental shelves and thin sedimentary cover, and are structurally simple, with only weakly rotational normal faults (Fig. 1.7B). This is the "simple passive margin" of Rosendahl (1987). "Lower-plate margins," which originate in footwalls of detachments, consist of thin continental crust with broad shelves and thick sedimentary cover; the basement consists of exhumed middle- to lower-crustal rocks of the lower plate, commonly overlain by remnants of the upper plate in tilt blocks (Fig. 1.7A). This is the "complex passive margin" of Rosendahl (1987). Where the detachment fault changes dip direction across a transfer fault, the passive margin changes along its length from upper-plate to lower-plate (Fig. 1.8). Where more than one detachment system is involved, marginal plateaus and continental ribbons may result (Fig. 1.9).

Many modern and ancient intraplate margins have been interpreted using the upper- and lower-plate model of Figure 1.7 (e.g., Mutter et al., 1989; Wernicke and Tilke, 1989; Hansen et al., 1993), but it remains unclear how central the connectivity/co-linearity between upper-crustal and upper-mantle faults is to this model. Lower-plate margins are hypothesized to undergo greater subsidence than upper-plate margins, due to isostatic compensation resulting from thinning of the crust). No translitho-spheric detachments have been documented, although inclined reflections imaged in the upper mantle beneath extended terranes around the British Isles can be interpreted as extensional shear zones that flatten upward toward the Moho (Reston, 1993). These appear to be laterally offset from structures in the upper crust, however, requiring a considerable

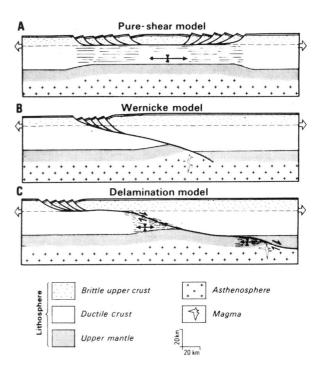

Figure 1.5 Three end-member models for continental extension. From Lister et al. (1986).

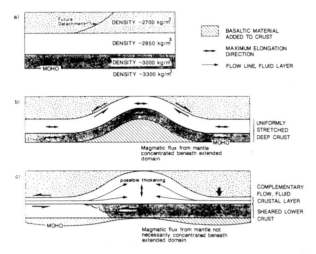

Figure 1.6 Contrasting modes of deep-crustal response to extension. a) Unstrained crust. b) Deep crust stretches uniformly, causing displacement of stretched layer from beneath stable blocks and into extended domains. Mantle magmatic flux is concentrated beneath extended domain (possibly mixing with and remobilizing old deep crust), thereby smoothing Moho. c) Deep crust is divided into an upper quartzose, fluid layer and a lower mafic, relatively viscous layer. Fluid-layer flow is primary mechanism for keeping relief on Moho low, and is governed, in large part, by flotational equilibrium of upper-crustal stable blocks as they separate. Thus, fluid layer may thicken substantially beneath extended domain. Mantle magmatic flux need bear no particular relation to position of extended domain or stable blocks on either side, but flotational flow may serve as a mechanism to concentrate either magmatic additions or remelted crust in extended domains relative to stable blocks via simple shear. Situation in c) may typify early extensional history of thickened crust (e.g., northern Basin and Range), whereas latest stages of extension prior to seafloor spreading may resemble b) (e.g., Salton trough). From Wernicke (1992).

component of simple shear in the lower crust (Reston, 1993).

Şengör (Chapter 2) provides an indepth analysis of continental extensional settings on Earth, both modern and ancient. As he demonstrates, rifts form in almost any type of plate-tectonic setting. "Successful" rifting leads to continental separation and sea-floor spreading; "failure" of rifting creates fossil rifts, which are crustal flaws prone to reactivation. Many of the diverse types of sedimentary basins discussed below are underlain by fossil rifts.

Terrestrial Rift Valleys

Modes of Rifting

The past decade has seen an enormous increase in our understanding of the tectonic development of terrestrial rifts, with important implications for basin models. One of the most important recent concepts is that the lower crust may flow during continental extension. The other is that rifting may be asymmetric with regard to structure, thermal evolution, uplift/subsidence patterns, and stratigraphic development. The basic structural element of a continental rift is now thought to be a half graben (e.g., Gibbs, 1984; Leeder and Gawthorpe, 1987; Rosendahl, 1987; Frostick and Reid, 1990; Frostick and Steel, 1993). The single, major fault zone that controls the asymmetrical basin is called the border fault, and border faults may step to the right or left and/or "flip" dip direction across transfer faults (see Fig. 3.22).

As summarized by Wernicke (1992 and references cited therein), surface observations in the Basin and Range show that upper-crustal extension is spatially highly variable, resulting in localized tectonic domains, where the upper one-third to one-half of the crust has been removed. If isostatic compensation of this extension had taken place in the mantle, then one would expect upwarping of the Moho; yet a subhorizontal Moho has been imaged across several of the boundaries between extended domains and stable blocks (e.g., Klemperer et al., 1986). Furthermore, the lower crust shows pervasive, strong reflectors that also occur near metamorphic core complexes (c.f., Wernicke, 1992). Heterogeneous upper-crustal strain and uniform deep-crustal structure across extensional domain boundaries may be accomplished by flow of deep crust from beneath stable blocks and into extended domains (e.g., Spencer and Reynolds, 1984; Wernicke, 1985; Gans, 1987; Block and Royden, 1990) (Fig. 1.6c). The concept of intracrustal isostasy (e.g., Block and Royden, 1990) is supported by the lack of Moho relief across domain boundaries and by the observation that topographic depressions over extended domains are not as great as they would be if the isostatic compensation took place in the mantle rather than in the lower crust. An important implication is that, for this situation, extension would not cause differential loading of the upper mantle, thus eliminating the mechanism for additional decompression melting beneath extended domains.

On the other hand, smoothing of the Moho may result if the mantle magmatic flux is concentrated

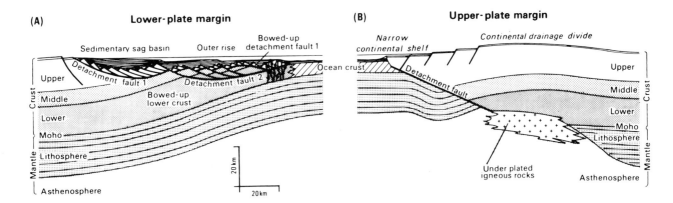

Figure 1.7 Detachment-fault model of intraplate continental margins with lower-plate and upper-plate characteristics. "Lower-plate" and "upper-plate" terminology refers to position during terrestrial-rift stage; both types of margins evolve into intraplate continental margins during seafloor spreading. Lower-plate margin (A) has complex structure; tilt blocks are remnants from upper plate, above bowed-up detachment faults. Multiple detachment has led to two generations of tilt blocks. Upper-plate margin (B) is relatively coherent. Rise of asthenosphere during detachment faulting would cause uplift of adjacent continent, but "underplating of igneous rocks" would result in subsidence as asthenosphere converts to lithosphere during cooling as an intraplate margin. Diagrams shown are for proto-oceanic stage (Fig. 1.4C); as either type of margin moves away from spreading ridge, it subsides in an intraplate setting (Fig. 1.4D–F). From Lister et al. (1986).

preferentially beneath extended domains (McKenzie, 1984; Gans et al., 1989) (Fig. 1.6b). Magmatic additions to the crust are clearly important during extension, and a broad correlation between magmatism and extension is apparent (Coney, 1980a; Lipman, 1980). It appears, however, that syn-rift magma centers occur over broad regions, across stable blocks as well as extended domains (R.E. Anderson, 1989; Wernicke, 1992), thus obscuring causal relations between extension and magmatism. The degree to which mantle-derived magma underlies strongly extended domains (relative to stable blocks) may even be controlled by midcrustal fluid flow (Fig. 1.6c), if mantle-derived magmas can be transported laterally into highly extended regions (Wernicke, 1992). This model of midcrustal fluid flow may be more applicable where rapid extension of heated and/or thickened crust has occurred (e.g., early Basin and Range extension), whereas the model of uniform stretching (Fig. 1.6b) may be more applicable in other situations, including late stages of extension prior to sea-floor spreading (Wernicke, 1992). It is difficult to distinguish arc magmatism from decompression melting because arc magmatism swept southwestward along the length of two active N–S-trending extensional belts in the western United States from the Oligocene to the Early Miocene (Axen et al., 1993), so that Basin and Range extension has occurred in intra-arc, backarc and transtensional settings at different times, unlike any other modern or known ancient rift belt (Ingersoll et al., 1990).

A temporal progression from core-complex mode to wide-rift mode to narrow-rift mode is predicted by a simplified model of crustal extension that includes lower-crustal flow (Buck, 1991) (Fig. 1.10). Buck's numerical results indicate that there is no clear difference in the rate of extension inferred for each of the modes of extension; instead, crustal thickness and thermal condition at the time of rifting appear to control the mode of extension. In core-complex mode, upper-crustal extension is concentrated over narrow regions, while the lower crust thins over a broad area. As extension thins the crust within a core complex, the lower crust "pours in" from surrounding regions. This occurs where crust is thick and heat flow is high. When the lower crust does not flow fast enough, then the mode of extension depends on the strength of the lithosphere: if it is initially hot and weak, then a wide rift forms and if it is initially cold and strong, then a narrow rift forms. The southern Basin and Range may be an example of a region that has passed through all three stages, from the core-complex mode (between 20 and 10 Ma), through the wide-rift mode to the narrow-rift mode (e.g., Gulf of California). For this reason, and also because it is more descriptive, we prefer to retain Buck's terminology of modes, rather

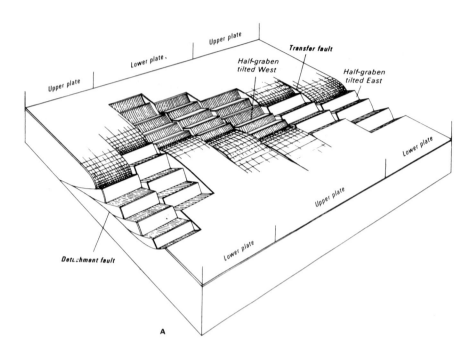

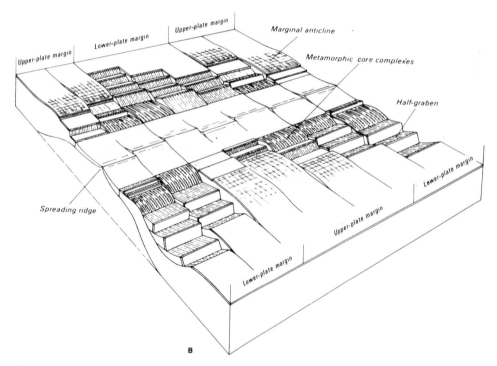

Figure 1.8 Changes from upper plate to lower plate occur across transfer faults. A: Half-graben complexes. Where detachment faults change dip across transfer faults, sense of rotation of tilt blocks also changes. B: If extension continues until ocean basin forms, then transfer faults mark changes from upper-plate margins to lower-plate margins. Architecture of intraplate margins is determined by where final continental separation began relative to detachment systems. From Lister et al. (1986).

than using the "Amerotype," "Aegeotype" and "Afrotype" terminology introduced in Chapter 2.

Another important implication of lower-crustal flow is that voluminous silicic volcanism (i.e., the "ignimbrite flareup"; Coney and Reynolds, 1977; Keith, 1978; Lipman, 1992) that may accompany core-complex-mode extension may be produced by decompressive melting of the lower crust as it flows into the extending upper crust, in a manner analogous to the flow of mantle asthenosphere under a

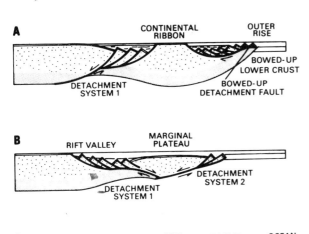

Figure 1.9 More than one detachment system may be involved in continental separation. Paired detachment systems may lead to formation of marginal plateaus, internal rift valleys, and/or isolated ribbons of continental crust. From Lister et al. (1986).

mid-ocean ridge (Buck, 1991). Mantle heat sources (i.e., from slab rollback following the Laramide orogeny; Dickinson and Snyder, 1978; Keith, 1978) would provide additional thermal energy. Wernicke (1992) made the insightful observation that a silicic fluid middle crust would prevent the rise of higher-density basaltic magma until crustal occlusion occurred (i.e., after the fluid part of the crust had been removed by lateral flow). In contrast, no fluid middle crust forms in narrow rifts (e.g., Rio Grande rift), and basaltic magmatism characterizes all of the rift history (e.g., Baldridge et al., 1984; Ingersoll et al., 1990).

Structural Style

Several structural models have been proposed for half-graben development. "Domino" faulting (Fig. 1.11A) involves high-angle, normal faults that extend deep into the upper crust with nearly constant dip. Both the faults and the intervening fault blocks rotate during extension. Seismic data from currently active faults are consistent with this model (e.g., Jackson and MacKenzie, 1983; see Chapter 3). The problem with the domino model is that it predicts gaps beneath

tilt blocks and at the ends of the array (Wernicke and Burchfiel, 1982); this is not a problem, however, if magma fills these gaps or if the corners are abraided.

In many highly extended terranes, listric normal faults terminate downward into subhorizontal detachment faults of regional extent, and fault blocks are highly rotated (Fig. 1.11B; Wernicke and Burchfiel, 1982; Gibbs, 1983, 1984). In a variation on the detachment-fault model (Jackson, 1987; Groshong, 1989), the fault system may consist of two planar segments (Fig. 1.12A).

In both domino and listric models of extension, rotated blocks form highlands separated by basins filled with strata with dips that fan, although in the domino model, oldest strata in a basin are less tilted and angular unconformities are rarer. Domino faulting ideally should result in simultaneous rotation of all blocks by the same degree, so that strata of similar age should have similar dip in any basin, and unconformities produced by changes in strain rate should be the same age in all basins. The locus of listric faults, in contrast, migrates with time, producing strata and unconformities of different ages in separate basins (Fig. 1.11B).

The flexural-rotation or "rolling-hinge" model of Buck (1988), Hamilton (1988b) and Wernicke and Axen (1988) (Fig. 1.11C) predicts that basin formation should young in one direction. In this model, an initially high-angle normal fault is progressively rotated to lower dips by isostatic uplift resulting from tectonic denudation (e.g., Spencer, 1984). This results in migration of an active hinge, where the detachment may be breached and the lower plate commonly exposed. If heterogeneous lower-plate rocks are exposed by this mechanism, then the provenance of sediments derived from the footwall should track migration of the active hinge in both space and time (e.g., Miller and John, 1988; Yarnold, 1994) (Fig. 1.13).

A three-dimensional model of half-graben development, the fault-growth model (Fig. 1.12C), utilizes faults that show greatest displacement in their centers and no displacement at their ends (e.g., Gibson et al., 1989; Schlische, 1991). Extensional basins thereby grow both in length and width through time and resemble elongate synclines in longitudinal section.

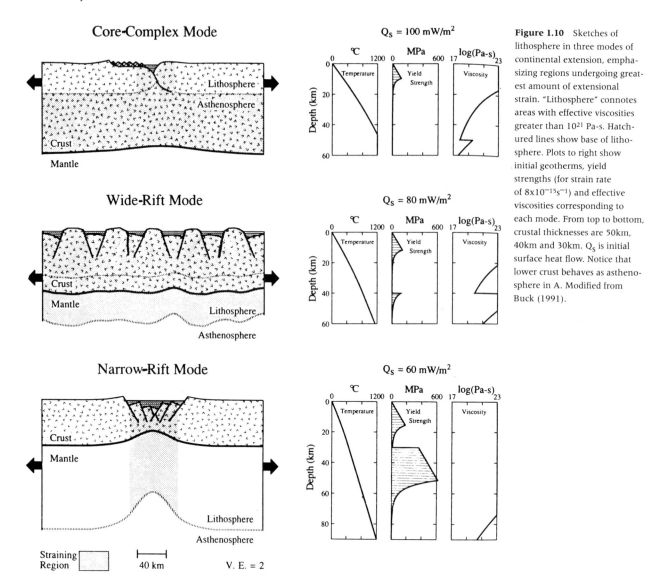

Figure 1.10 Sketches of lithosphere in three modes of continental extension, emphasizing regions undergoing greatest amount of extensional strain. "Lithosphere" connotes areas with effective viscosities greater than 10^{21} Pa-s. Hatchured lines show base of lithosphere. Plots to right show initial geotherms, yield strengths (for strain rate of $8 \times 10^{-15} s^{-1}$) and effective viscosities corresponding to each mode. From top to bottom, crustal thicknesses are 50km, 40km and 30km. Q_s is initial surface heat flow. Notice that lower crust behaves as asthenosphere in A. Modified from Buck (1991).

For example, early Mesozoic half-graben basins of eastern North America appear to be deepest adjacent to centers of border-fault systems and shallow toward all edges (Fig. 1.12); thicknesses of fixed-period (Milankovitch) lacustrine cycles reflect this along-strike variation in fault displacement (Schlische and Olsen, 1990; Schlische, 1993). The fault-growth model may also apply to many parts of the Basin and Range, where the ranges are higher in their centers due to greater fault displacement and footwall uplift (Schlische, 1991). Change in the rate of increase in basin volume predicted for the fault-growth model is positive, in contrast with that predicted for the detachment-fault model (zero) or the domino-fault model (negative) (Fig. 1.12D). If the rate of increase of basin capacity is faster than the rate of sediment infilling, then there may be an upward transition from fluvial sedimentation to lacustrine sedimentation (Fig. 1.14), such as that seen in all 11 Mesozoic rift basins of eastern North America (Schlische, 1993).

In the fault-growth model, linkage of originally isolated basins may produce basins composed of multiple synclinal subbasins separated by transverse anticlines (Fig. 1.15). Basins may widen both by onlap of the hangingwall and through development of younger faults within footwalls, forming step-faulted margins with numerous "rider blocks"; thus, most intrabasinal faults are synthetic. Most of the border

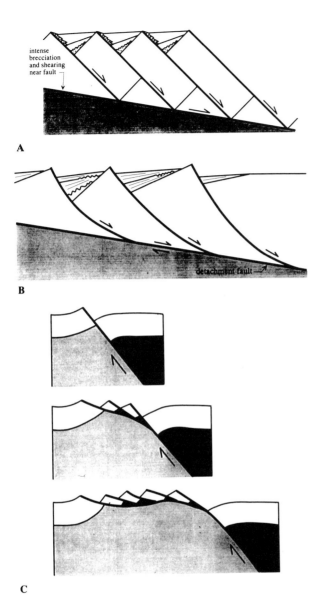

Figure 1.11 Styles of upper-plate faulting. A. Domino faulting, in which movement initially occurs on planar, high-angle, normal faults that rotate to lower angles with continued extension. At some critical angle, these faults may be abandoned, and new high-angle faults formed. Faults do not merge with detachment fault, and zone of intersection is extremely brecciated and sheared. Ideally, ages of unconformities and tilt of similar-aged strata in individual basins are the same. B. Listric faulting, in which movement occurs on curviplanar faults that flatten with depth and merge with detachment fault. This mode predicts that ages of unconformities and degree of stratal tilting may be different among basins. C. Rolling-hinge model, in which an initially high-angle normal fault is progressively rotated to lower dips by isostatic uplift resulting from tectonic and erosional denudation. New high-angle normal faults are produced when flattened (rotated) fault can no longer accommodate extension. Ideally, basin formation and unconformities are younger in one direction (left to right). From Lucchitta and Suneson (1993).

faults in eastern North America are known or inferred to be reactivated Paleozoic thrust faults (Schlische, 1993). In these basins, when a border fault became an intrabasinal fault and began propagating upward through basin fill, it may have changed its strike to become normal to the extension direction (Fig. 1.14). Although the sense of asymmetry varies along the eastern North American rift system, individual rifts exhibit no polarity reversals, in marked contrast to the East African rift system. This contrast probably reflects the fact that pre-existing structures dipped the same over large regions in eastern North America. Furthermore, where polarity reversals occur between rift zones in eastern North America, no accommodation zones are required (in contrast to East Africa) because displacement on faults dies out toward their tips.

Basin Fill

Although it is clear that coarse-grained aprons of sediment are spatially associated with boundary faults of rift basins, the interpretation of progradational-retrogradational cycles remains controversial. Some workers propose that progradation of clastic wedges corresponds to times of minimum tectonic activity along the basin margin, and that fine-grained intervals (lacustrine or marine mudrock) correspond to times of high rates of basin subsidence (e.g., Leeder and Gawthorpe, 1987; Blair and Bilodeau, 1988; Heller and Paola, 1992; Steel, 1993). These models assume constant sediment supply, where progradation results from reduction in accommodation during times of decreased subsidence. Surlyk (1978, 1990), in contrast, proposed that sedimentary architecture is controlled by episodicity in footwall-generated sediment discharge into depocenters subject to continuous deepening. Upward fining, deep-water conglomerate-sandstone-shale sequences (up to 300 m thick) are common in mid-Mesozoic rift basins of East Greenland, the North Sea and Scotland (Surlyk, 1978, 1990; Pickering, 1984; Turner et al., 1987; Underhill, 1991). Surlyk (1978) interpreted each upward fining sequence as recording uplift of the footwall block followed by erosional retreat of the scarp.

Leeder's (Chapter 3) discussion of extensional processes to form rifts complements the above discussion. In addition, he provides the most complete

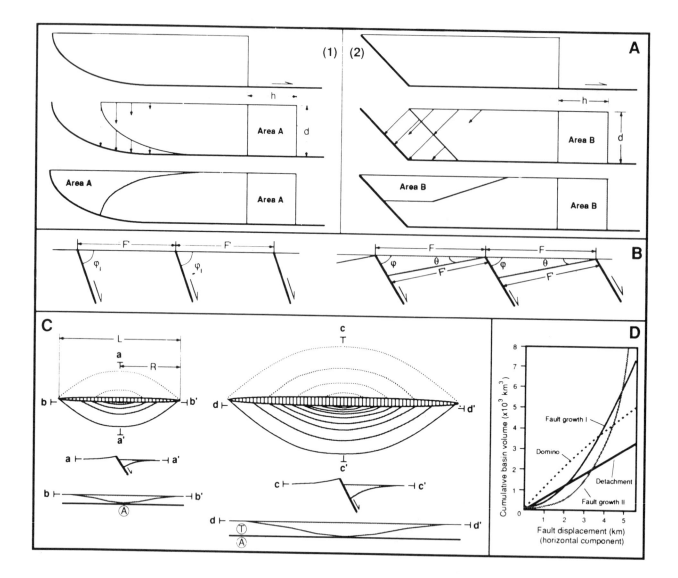

Figure 1.12 Three modes of extensional basin development. A. Linked-fault system, with two end members: (1) listric fault - subhorizontal detachment, and (2) planar kink fault. In both instances, horizontal displacement (h) on detachment creates a potential void between hangingwall and footwall, which is erased by collapse of hangingwall along either vertical faults (shown in (1)) or antithetic faults (shown dipping at 45° degrees in (2)). Deformation is area-balanced. B. Domino fault blocks, with both faults and intervening fault blocks rotating during extension. C. Essential elements of fault-growth model. Ruled "ellipse" is map view of a normal fault, along which displacement is greatest at fault center and decreases to zero at ends. Contours represent elevation change (positive for dotted contours, negative for solid contours) of originally horizontal free surface. Note that footwall uplift is smaller than hangingwall subsidence. L is length of fault; R is radius of fault (L/2); T is fault motion toward reader; A is away. D. Graph of cumulative basin volume versus horizontal component of fault displacement. Change in rate of increase in basin volume is zero for detachment-fault model, negative for domino model and positive for two fault-growth models. From Schlische (1991).

summary available of the sedimentary fill of modern and ancient terrestrial rifts.

Proto-Oceanic Rift Troughs

The Red Sea is the type proto-oceanic rift trough. However, caution is needed in constructing a general model for such settings because the Red Sea (including the Gulf of Aden) is the only modern proto-ocean on Earth. The Gulf of California is primarily a transtensional feature, although it shares many characteristics with the Red Sea. It is not surprising that a summary of the history of the Red Sea is the basis for a proto-oceanic model.

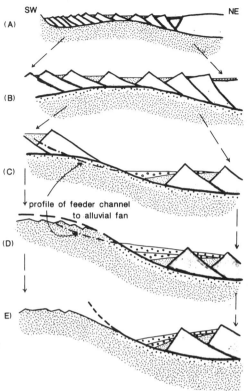

Figure 1.13 Interpretive sketches to show probable evolution of Chemehuevi detachment fault and coeval alluvial-fan sedimentation. Note that thickness of sedimentary section is exaggerated. (A) Extensional fault system after some doming. Sediments derived from hangingwall only. (B) Enlargement of central part of (A). (C) Enlargement of central part of (B); continued doming and extension of hangingwall, and exhumation of Chemehuevi fault; feeder channel to alluvial fan erodes debris from hangingwall, fault zone and footwall. (D) Erosion has breached Chemehuevi fault; alluvial-fan debris now dominated by footwall sources. (E) Continued slip tilts and displaces strata. From Miller and John (1988).

Sedimentary rocks containing dominantly footwall clasts

Sedimentary rocks containing hanging-wall and footwall clasts

Sedimentary rocks containing hanging-wall clasts only

Hanging-wall rocks excluding Tertiary sedimentary rocks

Chloritic brecciated granite in fault zone

Footwall rocks

Cochran (1983a) presented convincing geophysical evidence for active sea-floor spreading in the southern Red Sea, whereas north of 25° N, the center of the trough appears to be underlain by transitional crust. Plate reconstructions dictate that significant divergence has occurred between Africa and Arabia since the end of the Oligocene. This necessitates the presence of quasicontinental and quasioceanic crust under most of the Red Sea (e.g., Fig. 1.4C). Cochran (1983a) argues that the northern and southern Red Sea represent the earlier and later stages, respectively, of the transition from terrestrial rifting to sea-floor spreading. "An initial period of diffuse extension by rotational faulting and dike injection over an area perhaps 100 km wide is followed by concentration of extension at a single axis and the initiation of sea-floor spreading." (Cochran, 1983a, p. 41).

Bohannon (1986a,b) analyzed in more detail the type of crust underlying southern Red Sea margins. He showed that a minimum divergence of about 320 km would be indicated if all the transitional crust were quasicontinental (closer to Cochran's [1983a] assumption), whereas a maximum divergence of about 380 km would be indicated if it were all quasioceanic. The most reasonable assumption undoubtedly is intermediate, and possibly quite variable along the margins; this consideration has significant effects on coastal reconstructions.

An extended phase of rifting and diffuse extension must be accounted for in thermal models for postrifting subsidence (Cochran, 1983b). Horizontal heat flow causes additional cooling within the rift and uplift of the rift shoulders. Subsequent thermal subsidence of the continental margins is less than would be expected following "instantaneous" subsidence (e.g., McKenzie, 1978a). Magnetic quiet zones (subdued magnetic anomalies) along ancient rifted margins may have resulted from high sedimentation rates, which led to intrusion rather than extrusion of basaltic magma (e.g., off Labrador; Keen and Beaumont, 1990). The result is quasioceanic crust.

LePichon and Cochran (1988), Coleman (1993) and Sultan et al. (1993) summarize other aspects of the Red Sea area. Frostick and Steel (1993) and Leeder (Chapter 3) provide additional discussion of proto-oceanic rift troughs.

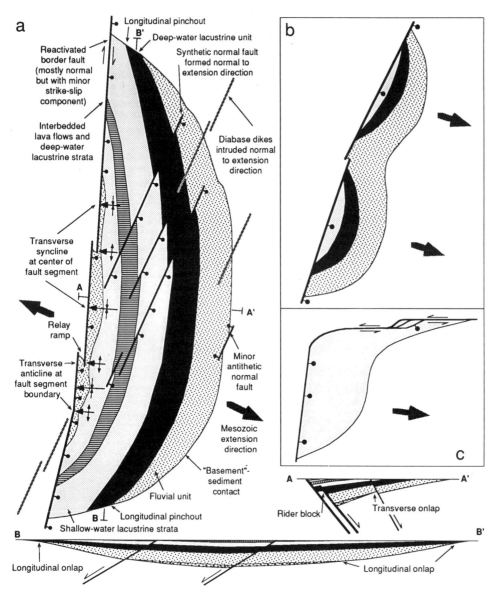

Figure 1.14 a. Geologic map and cross sections of idealized, dip-slip-dominated Mesozoic rift basin of eastern USA. Note that, although trace of prerift-synrift contact defines a large syncline, it is not affected by transverse folds adjacent to border-fault system. b. Geologic map of idealized Mesozoic basin containing multiple subbasins. c. Geologic map of basin with both dip-slip-dominated and strike-slip-dominated margins. From Schlische (1993).

Intraplate Settings

Post-Rift Deposits

Large-scale basins overlying fossil rifts (referred to as "koilogens" in Chapter 2) are believed to result from thermal relaxation in response to declining heat flow, although normal faults may be reactivated with minor displacement during this stage (e.g., Schlee and Hinz, 1987; Badley et al., 1988). If continental separation is successful, then a new ocean basin is formed ("Thalassogeny"; see Chapter 2).

Distinguishing synrift from post-rift (thermal subsidence) basin fill can be difficult. For example, the ca 900 Ma Torridon Group of western Scotland was interpreted as synrift deposits by Stewart (1982), but has more recently been reinterpreted as the fill of part of an originally wider (incompletely preserved) post-rift basin undergoing thermal subsidence (Nicholson, 1993). This interpretation is based on the absence of unequivocal intrabasinal features supporting syndepositional taphrogenesis, the absence of evidence for active basin-bounding fault margins, the relative uniformity of petrography and facies, the absence of vol-

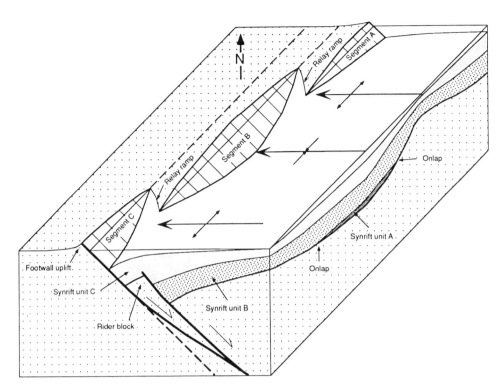

Figure 1.15 Block diagram illustrating geometric relations among border-fault segmentation, relay ramps, transverse folds and rider blocks. Synrift unit A forms a restricted wedge, suggesting that fault segment B lengthened through time. Synrift unit B is absent from hangingwall block of fault segment C, suggesting that segment C is younger than segment B. Segments B and C may merge at depth, forming a kinematically linked fault system, although this is not required. Faults (dashed) may extend from ends of segments B and C if they involve only partial reactivation of preexisting structures. From Schlische (1993).

canic rocks, and scale considerations of the fluvial system, which suggest that the Torridon Group basin was too broad to be a rift graben (Nicholson, 1993).

Continental Rises and Terraces

Pitman (1978) elegantly demonstrated how transgressive/regressive sequences are formed on shelves along intraplate continental margins. He showed that subsidence rates at shelf edges are normally greater than rates at which eustatic sea level changes. Shelves can be modeled as platforms rotating about landward hinge lines. Thus, shorelines seek locations that reflect balance among sea level change, subsidence rate and sediment flux. The net result is that transgressive/regressive sequences reflect changes in rates of sea level change, rather than changes in sea level. All modern intraplate continental margins are younger than Paleozoic (post-breakup of Pangea), and most seismic "sea level curves" (e.g., Vail et al., 1977) are constructed primarily from Cretaceous and Cenozoic sequences along Atlantic margins. Sea level has been generally falling since its high point in the Cretaceous, due to the combination of increase of the

average age of oceanic crust (e.g., Heller and Angevine, 1985) and the initiation of continental glaciation during the middle Cenozoic. The record of Cenozoic transgressive/regressive sequences along Atlantic margins is consistent with continually falling sea level, at varying rates. Diverse tectonic processes provide mechanisms for second-order changes in sea level (e.g., Cloetingh et al., 1985; Karner, 1986).

Pitman's (1978) model shows how the seaward thickening wedge of the continental terrace can form (e.g., Fig. 1.4E) in response to the combined effects of tilting about a landward hingeline, changes in sea level and changes in sedimentation rates. The shoreline would be at the shelf edge only during times of unusually rapid sea level fall (e.g., glaciation or sudden flooding of a formerly isolated dry ocean basin [e.g., the Messinian Mediterranean]) or in areas of very high sedimentation (e.g., continental embankments).

Pitman (1978) also argued that, except for glacial variations and phenomena such as the dessication or flooding of small ocean basins, volumetric changes in ridge systems are the fastest way to change sea level. If this is true, then the rate of subsidence at shelf edges of intraplate margins (which is everywhere

greater than 20 m/my) is always greater than the rate at which sea level may fall or rise during nonglacial times. Under these conditions, sediment is deposited on the shelf, and the shoreline could prograde over the shelf edge only in areas of unusual sediment flux (e.g., major deltas to create continental embankments; see below). However, if rapid and significant glaciation occurs, then the strandline could migrate beyond the shelf edge, and rivers could cut canyons across the shelf, depositing their load at the top of the slope, to be resedimented into abyssal plains.

Cloetingh et al. (1985) argued that tectonically controlled changes in relative sea level occur over such large regions that they may be misinterpreted as global (eustatic) sea-level changes. Furthermore, these may occur with rates and magnitudes similar to those of glacioeustatic events (10–100 m/my, up to 100 m) (Cloetingh, 1988). This suggests that exposure of shelf areas and resedimentation into deep water may have occurred more commonly in the geologic past than Pitman (1978) would predict. Numerical modeling and observations of modern and ancient stress fields have demonstrated the existence of large stress provinces in the interiors of plates, providing evidence for far-field propagation of intraplate stresses from plate boundaries into plate interiors (Cloetingh et al., 1985). This may help explain short-term cycles in sea level, which have traditionally been interpreted in terms of glacioeustatic control, but for which there is no geologic or geochemical evidence. Furthermore, the Vail et al. (1977) "global" coastal onlap/offlap curves, although based on data from basins around the world, are heavily weighted with circum-Atlantic and Gulf Coast data; therefore, at least some of the "global" cycles may reflect rifting and post-rifting events in the north and central Atlantic (Miall, 1986, 1992, 1993).

Bond et al. (Chapter 4) summarize plate-tectonic setting, basin architecture, geophysical modeling, and sedimentary environments and facies of continental terraces and rises.

Continental Embankments

Continental embankments are seldom discussed, as such; most effort has been concentrated on the more common shelf-slope-rise configuration of most intra-plate continental margins (see Chapter 4). Nonetheless, continental embankments form one of the major classes of sediment accumulation today; in contrast, their ancient counterparts are difficult to identify because of the high probability that they will be partially subducted and highly deformed during continental collision. The only possible ancient example we have identified is the Meguma terrane of the Canadian Appalachians (Keppie, 1993), but this interpretation is quite speculative (Table 1.2; additional suggestions are appreciated!). Identification of ancient examples of continental embankments should be a high priority, especially in the search for hydrocarbons.

Continental embankments (e.g., Fig. 1.4F), such as the submarine extensions of the Niger Delta and the Mississippi Delta, form in areas of unusually high sediment flux, primarily at the mouths of failed rifts (Burke and Dewey, 1973) (also, see Audley-Charles et al., 1977). Drainage from large cratons commonly is directed toward the mouths of failed rifts and away from adjacent "normal" rifted continental margins. Similar massive outbuilding of continental margins due to deltaic progradation occurs along sutures associated with remnant ocean basins (e.g., Bengal and Indus fans; see Chapter 10); however, these latter deposits form in tectonically active settings related to continental suturing. Therefore, provenance, dispersal orientations, sediment types, related petrotectonic assemblages and deformational features contrast with continental embankments formed at failed rifts, even though gross structural cross sections of these progradational continental margins are similar. Kinsman (1975) predicted that the maximum possible stratal thickness of such margins is 16–18 km, which agrees with total thickness of both the Niger and Mississippi margins (Burke, 1972; Worrall and Snelson, 1989).

Winker and Edwards (1983) summarized the nature of "unstable progradational clastic shelf margins," herein referred to as "continental embankments." Rapid subsidence at such margins is attributable to at least 4 factors: 1. sedimentary loading induces lithospheric flexure (e.g., Walcott, 1972), with maximum subsidence located near the shelf-slope break; 2. listric normal (growth) faulting concentrates

subsidence on downthrown blocks; 3. salt withdrawal enhances subsidence; and 4. compaction amplifies subsidence. Progradation of upward coarsening sequences results in coarser-grained nonmarine delta-plain, shoreline and shallow-marine facies overlying finer-grained outer-shelf and slope facies. Winker and Edwards (1983) demonstrated Cenozoic seaward migration of the clastic shelf margin in the northwestern Gulf of Mexico and the Niger Delta (Fig. 1.16).

Burke (1972) summarized depositional processes leading to development of the Niger Delta (the surficial expression of a continental embankment). Prevailing southwest winds impinge symmetrically on the Niger Delta and neighboring coasts, thus setting up converging longshore drifts, which meet at the delta corners. Submarine canyons at these corners (and at the center of the delta during low sea level) feed voluminous sediment to submarine fans at the delta foot. The net result is a five-layer structure as the continental margin progrades seaward over oceanic crust and older pelagic sediments. From bottom to top, these layers are: 1. deep-sea-fan sand; 2. transitional sand and mud; 3. slope mud, with abundant diapirs and slumps; 4. transitional shoreline and shelf sand and mud; and 5. continental (primarily fluvial) sand. Continental embankments are gravitationally unstable due to rapid burial of low-density, water-saturated sediments, and the common occurrence of deeply buried evaporite deposits inherited from the proto-oceanic stage. Therefore, salt diapirs, mud diapirs and growth faulting are common.

One of the most comprehensive summaries of a continental embankment has been presented by Worrall and Snelson (1989) (also, see Salvador, 1991). Four impressive aspects of the northern Gulf of Mexico are: 1. the width of continental-margin sedimentary sequences (close to 1000 km from the edge of the coastal plain in Oklahoma to the Sigsbee Escarpment); 2. stratal thicknesses of approximately 16 km over large areas of the margin; 3. progradation of shorelines over quasioceanic crust; and 4. the dynamic nature of salt movement below the growing sedimentary wedge, including lateral flow and formation of the Sigsbee salt nappe (Fig. 1.17). "The immense scale of the Sigsbee salt nappe is unique among described geometries in the world's salt basins, and is one of the largest single structural features of the North American continent" (Worrall and Snelson, 1989, p. 113). The Swiss Alps are dwarfed by the Sigsbee nappe (Fig. 1.17). Pleistocene strata reach thicknesses of over 7.6 km, resulting in some of the highest subsidence rates on Earth. This subsidence is the direct result of the high sediment flux caused by the Mississippi and other rivers; recent subsidence has occurred without significant thinning of crystalline crust or lithospheric thickening or tectonic loading, although the former two mechanisms helped initiate the margin during the mid-Mesozoic. Sedimentary loading and resulting lithospheric flexure accommodate additional sedimentary deposition; loading also results in massive lateral flow of salt beneath the migrating load, which, in turn, encourages the formation of local growth faults with associated basins.

Lateral flow of salt is similar, in some ways, to the lateral flow of a fluid crustal layer (e.g., Fig. 1.6), although first-order tectonic controls are quite different in core-complex extension and continental-embankment loading. The Gulf Coast example of massive lateral salt movement illustrates why sedimentary strata might be considered part of the crust, instead of "overburden" (T.E. Jordan, 1994, personal communication). Our preference is to include lithified strata within the crust, but to exclude unconsolidated sediment. This question of whether sediments are part of the crust illustrates how all definitions of "crust" and "rocks" will inevitably include some transitional elements such as stratigraphic sequences.

Intracratonic Basins

With increased exploration of deeper levels of well known intracratonic basins (e.g., the Michigan basin), it has become clear that most intracratonic basins overlie fossil rifts (see Chapter 13).

DeRito et al. (1983) developed a lithospheric-flexure model with a nonlinear Maxwell viscoelastic rheology that helps explain how intracratonic basins can have long histories of synchronous subsidence over broad areas of continents. They pointed out that predicted subsidence due to stretching and cooling following cessation of rifting (e.g., McKenzie, 1978a; Sclater and Christie, 1980) is insufficient to explain

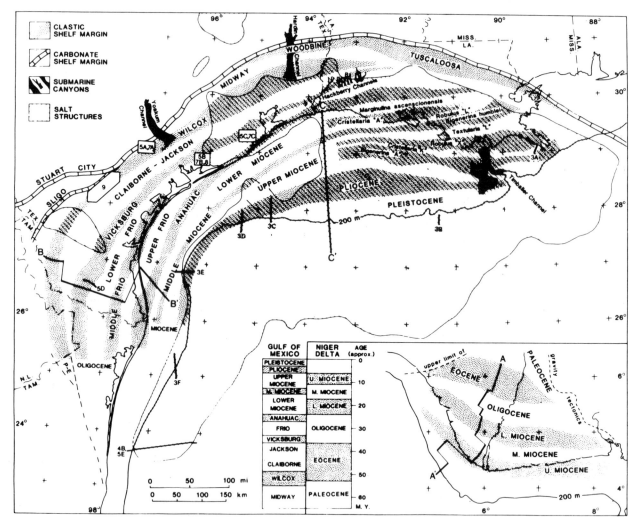

Figure 1.16 Regional shelf-margin trends of northwestern Gulf of Mexico and Niger Delta, based on isopach maxima, flexures (particularly for Niger Delta), timing of maximum growth-fault activity (particularly in Texas and northeast Mexico), and stratigraphic top of geopressure (particularly in Louisiana). Where necessary, shelf-margin trends are extrapolated along regional growth-fault trends (particularly for Niger Delta). In Gulf basin, Midway and Woodbine represent stable progradation; all other clastic sequences represent unstable progradation. From Winker and Edwards (1983).

the magnitude or timing of subsidence for most intracratonic basins. Many intracratonic basins (e.g., Michigan, Illinois and Williston basins) have experienced renewed subsidence during times of orogeny in adjacent orogenic belts. These periods of reactivation may be due to the reduction of effective viscosity of the lithosphere resulting from applied stress (DeRito et al., 1983). Due to flexural rigidity of the lithosphere, dense loads emplaced in the crust during rifting (e.g., basaltic dikes) remain isostatically uncompensated for geologically long periods of time. Any stress applied to

the lithosphere results in a geologically instantaneous relaxation and increased subsidence. Repeated changes of stress due to orogenic activity decrease the effective viscosity beneath the basin, and the basin subsides more rapidly than during nonorogenic periods. Thus, nearby orogenic activity allows the basin (underlain by a fossil rift) to approach isostatic compensation more rapidly than it would without orogenic activity. Cumulative subsidence over hundreds of millions of years approaches that predicted by simple thermal models (e.g., McKenzie, 1978a). The

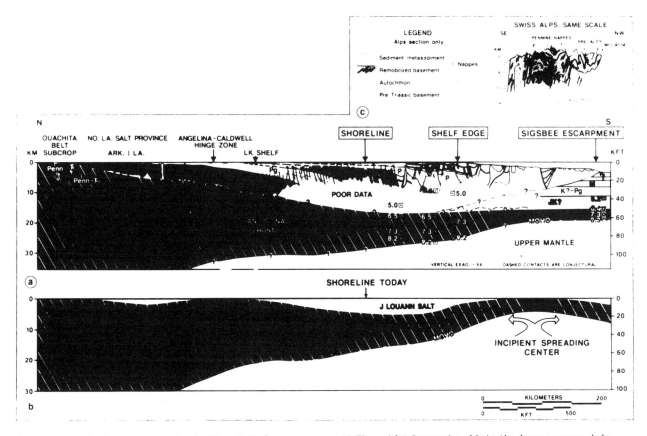

Figure 1.17 (a) North-south cross section, Louisiana. Vertical exaggeration = 5x. Dashed contacts highly conjectural. JK = Jurassic-Cretaceous; Pg = Paleogene; M = Miocene; P = Pliocene; Ps = Pleistocene. (b) Hypothetical reconstruction, mid-Jurassic time (proto-oceanic setting).

(c) (Upper right) Cross section of Swiss Alps drawn to same scale for comparison. Sigsbee salt nappe is one of largest structural features in North America and is apparently unique in scale if not structural style among Earth's salt basins. From Worrall and Snelson (1989).

model of Derito et al. (1983) presents an elegant explanation for continent-wide synchroneity of depositional sequences due to orogenic activity along nearby continental margins.

Quinlan (1987) and Klein (Chapter 13) provide thorough reviews of mechanisms and models for intracratonic basins.

Continental Platforms

Continental platforms include all of Earth's continental cratonal crust. By definition, cratons experience little tectonic activity; therefore, there is little to say about them in the context of this book! The depositional sequences found on cratons primarily reflect global tectonic events and eustasy, as summarized by Bally (1989) and Sloss (1988a, b, c).

Platforms are transitional into continental margins of all types, as well as forelands, rift valleys, intracratonic basins, intracontinental wrench basins and successor basins. Continental lithosphere must behave as if it is homogeneous and strong in order to qualify as "cratonal," in contrast to intracratonic basins, with their underlying heterogeneities, which are prone to reactivation (see above and Chapter 13). Distinction of distal forelands and platforms (e.g., Cretaceous epeiric seaway of North America; Elder et al., 1994) is difficult and in some ways arbitrary; most workers would consider the flexural bulge of a foreland as the boundary, but the lateral flexural effect of lithospheric loading is, in theory, infinite, although with rapid decay (e.g., Allen and Allen, 1990; see Chapter 9). During times of high sea level and broad foreland flexure, sedimentary sequences of forelands and platforms merge imperceptibly.

Cratons are not isolated from tectonic effects (Cloetingh, 1988). Nonetheless, of all tectonic settings, platforms are the most likely locations for isolating eustatic sea-level effects (e.g., Sahagian and Holland, 1991; Sahagian and Jones, 1993), as exemplified by mid-continent cyclothems (e.g., Heckel, 1984; Klein, 1992b; Klein and Kupperman, 1992; Chapter 13). Paleolatitudinal and paleoclimatic signatures also are best isolated within the stratigraphic record of platforms (e.g., Berry and Wilkinson, 1994).

Continental platforms have generally been exposed above sea level during times of continental assemblage into supercontinents (e.g., Neoproterozoic and Permo-Triassic), and maximum flooding has occurred approximately 100 my following supercontinent breakup (e.g., Ordovician and Cretaceous) (e.g., Heller and Angevine, 1985). Supercontinent assemblage and breakup provide first-order control of relative sea level, which is modulated by other processes, such as worldwide averages of sea-floor spreading, glacioeustatic effects and regional tectonics (e.g., Hays and Pitman, 1973; Pitman, 1978; Cloetingh et al., 1985; Karner, 1986) to produce higher-order cycles (e.g., Posamentier et al., 1988). In spite of the theoretically simpler eustatic signal to be found on continental platforms (relative to other tectonic settings), the debate remains intense concerning the relative importance of various controls on mid-continent cyclothems (e.g., Heckel, 1984; Klein, 1992b; Klein and Kupperman, 1992; Chapter 13).

Active Ocean Basins

Heezen et al. (1973) presented a kinematic model to explain the distribution of Cretaceous and Cenozoic sediment in the modern Pacific Ocean. The model is based on the systematic increase in depth of oceanic crust with age, combined with the dependence of sediment type on depth of water (Fig. 1.18). Biogenic oozes are volumetrically dominant away from continental margins, with generally more rapid production of calcareous pelagic detritus than siliceous pelagic detritus. Virtually all of this biogenic material is produced within the photic zone (approximately the upper 150m of water); when pelagic organisms die,

their shells (tests) settle slowly through the water column, dissolving as they fall. The depth below which none of the carbonate tests remain is the carbonate compensation depth (CCD). This depth is determined by the balance of biologic productivity in the photic zone, and the rates at which tests fall and dissolve in the water column (Berger and Winterer, 1974). Changes in any of these rates can change the CCD, which presently averages -3700m in the open ocean (Heezen et al., 1973), and -5000m near the equator (Berger, 1973). The CCD is depressed near the equator (and in other areas of upwelling) due to higher biologic productivity. The silica compensation depth (SCD) is less well defined than the CCD, but it is universally deeper than the CCD due to the slower dissolution of silica tests. Below the SCD, only nonbiogenic clay accumulates (Fig. 1.18).

Oceanic crust formed at the East Pacific Rise south of the equator has the potential to subside below both the CCD and the SCD before crossing the equator; given the right depth (age) of crust and depth of equatorial CCD and SCD, an equatorial sequence of siliceous and calcareous oozes also may be deposited. Ancient oceanic basaltic crust with this history should, therefore, be overlain by the following vertical lithologic sequence: carbonate ooze, siliceous ooze, abyssal clay, siliceous ooze, calcareous ooze, siliceous ooze, abyssal clay. Heezen et al. (1973) developed this model to explain stratigraphic sequences resulting from contrasting equatorial-transit histories. The result is a predictive model that is in excellent agreement with the Cretaceous and Cenozoic stratigraphy of the Pacific plate. They discussed the formation of time-transgressive lithofacies (see the interesting discussion of oceanic stratigraphic principles by Cook [1975]), including pyroclastic material derived from western Pacific magmatic arcs, and turbidite fans derived from the North American margin. Winterer (1973) provided an additional example of paleotectonic reconstructions using Pacific plate stratigraphy.

Heezen et al.'s (1973) kinematic model is a potentially powerful tool for interpreting depositional histories of ophiolite sequences preserved in subduction complexes and suture zones. Caution is advised, however, because most ophiolites probably

formed in backarc (including interarc) basins rather than open-ocean settings, in which case, models for backarc basins are more useful for reconstructing paleotectonic settings (see Chapter 8). In addition, biologic controls on carbonate and siliceous sedimentation have changed markedly due to evolutionary processes (Berger and Winterer, 1974); therefore, pre-Cenozoic oceanic models cannot be truly actualistic.

Oceanic Islands, Aseismic Ridges and Plateaus

The model discussed above for the systematic subsidence of oceanic lithosphere as it travels away from divergent boundaries has important implications for intraoceanic sedimentation in general. All islands, aseismic ridges and plateaus constructed of unusually thick basaltic material (e.g., Hawaiian-Emperor chain) experience regional subsidence equal to the subsidence of oceanic lithosphere on which they ride. During growth due to active basaltic volcanism, only minor fringing sediments are likely to be deposited.

Clague (1981) divided the post-volcanic history of seamounts into three sequential stages: subaerial,

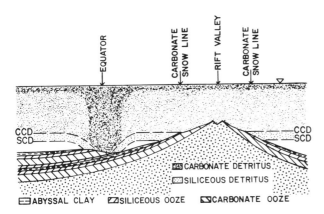

Figure 1.18 Model of axially accreting oceanic sedimentation (extreme vertical exaggeration, especially for stratal thicknesses). Deposition of carbonate ooze dominates on oceanic crust (solid-triangle pattern) shallower than CCD; siliceous ooze accumulates between CCD and SCD, and only abyssal clay accumulates below SCD. Both CCD and SCD are depressed near equator and in other areas of high biologic productivity (zones of upwelling). Predictable stratigraphic sequences result as oceanic crust formed at spreading center (rift valley) cools and subsides as it moves away from spreading ridge (see text). After Heezen et al. (1973); from Ingersoll (1988b).

shallow water and deep water or bathyal. Following cessation of volcanism, seamounts are likely to accumulate pelagic and shallow-marine carbonate sediments, with rate and type of sedimentation controlled largely by latitude (Clague, 1981). If carbonate sedimentation is equal to or greater than thermal subsidence, then fringing reefs and atolls will form atop the submerged islands and plateaus. Some of the depositional and migratory behavior of ancient islands, ridges and plateaus accreted to continental margins at subduction zones may be reconstructed based on this model.

Dormant Ocean Basins

This class of basin is newly defined herein. Dormant ocean basins are floored by oceanic crust, which is neither spreading nor subducting (there are no active plate boundaries within or adjoining the basin). This setting contrasts with active ocean basins, which include at least one active spreading ridge (e.g., Atlantic, Pacific and Indian oceans) and remnant ocean basins, which are small shrinking ocean basins bounded by at least one subduction zone (e.g., Bay of Bengal and Huon Gulf; see Chapter 10). The term "dormant" implies that no orogenic or taphrogenic activity is occurring within or adjacent to the basin; "oceanic" requires that the basin is underlain by oceanic lithosphere, in contrast to intracratonic basins, which are typically underlain by partially rifted continental lithosphere (see Chapter 13).

Dormant ocean basins initiate by two very different processes: 1. spreading ridges of proto-oceanic troughs or small ocean basins cease activity due to plate-margin rearrangement (e.g., Gulf of Mexico; Dickinson and Coney, 1980; Pindell and Dewey, 1982; Pindell, 1985) or 2. remnant ocean basins are not completely closed, but subduction ceases, resulting in an oceanic gap in the suture belt (e.g., Black Sea, North Caspian depression and Tarim basin of western China; S.A. Graham and A.M.C. Şengör, 1994, personnal communications). Once spreading and/or subduction cease, the dominant mode of subsidence is sediment loading, although lithospheric thickening due to residual cooling may be important

if the oceanic lithosphere is young (especially important in the first type [e.g., Gulf of Mexico]). As oceanic lithosphere cools and thickens underneath a dormant ocean basin, flexural rigidity increases. The residual "hole" in the top of the lithosphere (oceanic crust surrounded by continental crust) is a natural sediment trap, with centripetal drainage. Dormant ocean basins are the ultimate example of a "filling-hole" basin, with oceanic sediments overlain by deep-marine turbidites and upward shallowing continental-margin deposits. Relatively rapid filling of dormant ocean basins results where continental embankments prograde across the continental-oceanic boundary (e.g., northern Gulf of Mexico). Even in this case, the life span of a dormant ocean basin may be hundreds of millions of years.

As a dormant ocean basin fills to near sea level, the residual depression might have the appearance of an intracratonic basin, although it would probably react differently to in-plate stresses (i.e., Cloetingh, 1988). Mature oceanic lithosphere would be an unusually rigid part of the plate, in contrast to the relative weakness of a fossil continental rift underlying an intracratonic basin (i.e., DeRito et al., 1983; Chapter 13). The weakest part of the lithosphere related to a mature dormant ocean basin would be along the residual continental margins, where crustal flaws related to initial continental rifting or extinct subduction zones would tend to localize deformation; thus, subsequent orogeny might create intermontane basins corresponding to ancient dormant ocean basins (e.g., modern Tarim basin of western China; A.M.C. Şengör and S.A. Graham, 1994, personal communications), surrounded by uplifted formerly intraplate continental margins.

Hoffman (1988b) proposed that the Proterozoic Belt basin of Montana and the southern Canadian Cordillera formed as "a landlocked remnant ocean basin? (analogous to the South Caspian and Black seas)." We agree with his possible modern analogs, although he has misused "remnant ocean basin," as first defined by Graham et al. (1975) and illucidated in Chapter 10. "Dormant ocean basin" is our suggested term for the Belt basin, if the modern analogs are appropriate. On the other hand, the Belt basin probably had a more complicated billion-year history than simply as a dormant ocean basin (e.g., Winston, 1986).

Convergent Settings

Arc-Trench Systems

Dewey (1980) classified arc-trench systems as extensional, neutral or compressional, analogous to Dickinson and Seely's (1979) migratory-detached, noncontracted-stationary and contracted categories (Fig. 1.19). Extensional systems are commonly intra-oceanic due to formation of oceanic crust within or behind magmatic arcs (e.g., Kanamori, 1986), although continental systems may also be extensional, with normal faults in backarc, intra-arc and forearc positions (Jarrard, 1986a; Busby-Spera, 1988b; Saleeby and Busby-Spera, 1992). Extension is favored where trench rollback is faster than trenchward migration of the overriding plate. The western Pacific intraoceanic arcs are typical modern examples, with steep Wadati-Benioff zones dipping westward and subduction of old oceanic lithosphere (e.g., Molnar and Atwater, 1978). Compressional systems occur where an overriding plate advances trenchward faster than trench rollback. The Andes typify these systems, with shallow Wadati-Benioff zones dipping eastward and subduction of young oceanic lithosphere. Neutral systems result where trench rollback is approximately equal to the trenchward advance of the overriding plate. The Aleutian and Indonesian arcs (both of which include oceanic and continental crust in the overriding plates) typify these arcs, with intermediate-dip Wadati-Benioff zones dipping northward and subduction of intermediate-age oceanic lithosphere.

Dewey's (1980) kinematic model is useful, whatever the dynamic controls on the tectonics of arc-trench systems. Second-order effects include the probability that extensional arc-trench systems experience primarily basaltic magmatism, have low relief, thin sediments and deep trenches. In contrast, compressional arc-trench systems include significant silicic magmatism, have high relief, abundant sediments and shallow trenches. Most arc-trench systems have intermediate characteristics, commonly including transform motion along arc trends. An understanding of arc-trench dynamics improves our ability to interpret tectonic controls on basin devel-

opment; therefore, the following section reviews this subject.

Uyeda and Kanamori (1979) used seismicity of this century to define two types of subduction zones: "strongly coupled," with great earthquakes, and "weakly coupled," with smaller earthquakes. They depicted the former type as continental ("Chilean type"), and the latter type as intraoceanic ("Mariana type"), even though they listed examples of weakly coupled continental subduction zones. Kanamori (1986) examined the inferences that increased convergence rate should result in increased coupling and seismicity, whereas increased age of the subducting slab should result in more rapid sinking and reduced seismicity. Separately plotting each of these parameters against estimated rate of seismic-energy release (per unit time and unit length of the subduction zone), he found a positive correlation with large scatter; combining these parameters, he found a strong correlation (Fig. 1.20). Kanamori (1986) concluded that convergence rate and plate age together control seismicity, although parameters of potentially equal importance, such as velocity of the upper plate relative to the hot-spot frame of reference, were not treated. This study also showed that backarc opening is generally associated with subduction zones of low seismicity (Fig. 1.20).

Jarrard's (1986a) study supports Kanamori's earthquake-magnitude model. In Jarrard's model, convergence rate controls horizontal slab migration, and slab age controls vertical slab migration; together, they define a preferred trajectory for the shallow part of a slab. The difference between this preferred trajectory and the dip of the interface between overriding and underriding plates directly affects coupling and maximum earthquake magnitude.

Jarrard (1986a) used 26 measurable parameters from each of 39 modern subduction zones to isolate causal relationships among parameters. These parameters include arc curvature, geometry of the Wadati-Benioff zone, strain regime of the overriding plate, convergence rate, "absolute" motion (relative to hot-spots), slab age, arc age, and trench depth. Stepwise regression was used to isolate key independent variables. Of these variables, the strain regime of the overriding plate is the most difficult to quantify. Class 1 subduction zones are defined as the most

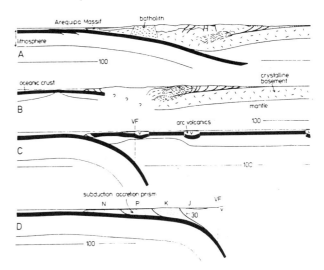

Figure 1.19 True-scale sections across Pacific plate margins (scale in kilometers). A. Central Peru (compressional). B. Western Canada (compressional). C. Marianas (extensional). D. Alaska (neutral) (age of subduction-accretion prism: J - Jurassic, K - Cretaceous, P - Paleogene, N - Neogene). See text for discussion. Modified from Dewey (1980); from Ingersoll (1988b).

extensional, with active backarc spreading. Class 2 are those with very slow, incipient or waning backarc spreading, and class 3 zones are mildly extensional, with normal faults. Class 4 zones include both neutral zones, and zones with mildly compressional forearcs and mildly extensional backarcs. Class 5 zones are mildly compressional, with somewhat more reverse faulting than normal faulting. Class 6 zones are moderately compressional, with consistent reverse faulting and folding, and class 7 zones are strongly compressional, with "substantial" folding and faulting. These strain classes are arbitrary points along a continuum, and perhaps some of them should be combined.

Jarrard's (1986a) study used more subduction zones, and compiled more measurable parameters for each of them, than any previous study; nonetheless, he cautioned that reliable isolation of key variables may be precluded by the restricted number of modern subduction zones. This factor is important in determining the primary controls of strain regime because most proposed causative parameters are intercorrelated. For example, the highly extensional Mariana system has a steeply dipping slab, an old plate, "absolute" motion of the upper plate away from the trench, and high convergence rate. In contrast, the highly compressional Chilean system has a shallowly dipping slab, a young

plate, "absolute" motion of the overriding plate toward the trench, and high convergence rate. Determining which (if any) of these four parameters controls strain regime requires examination of subduction zones with completely different combinations of these parameters. A brief summary of Jarrard's examination of these parameters is given here.

Convergence Rate - Jarrard's study showed a moderate (R = 0.52) correlation between convergence rate and strain class (i.e., faster is more compressive). This observation supports the conclusion of previous workers that additional causative factors must be sought.

Slab Age - Jarrard found a much weaker correlation between slab age and strain regime (R = 0.35) than did Molnar and Atwater (1978). One proposed mechanism for backarc spreading is age-dependent slab rollback due to gravitational instability of old lithosphere. Alternative driving forces include: (1) trench suction due to mantle flow toward the trench and down the upper surface of the descending slab ("ablative subduction"), and (2) "absolute" motion of the overriding plate away from the trench.

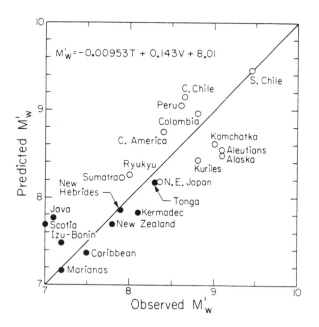

Figure 1.20 Relation between M'_W calculated from T (plate age) and V (convergence rate), using relation $M'_W = -0.00953T + 0.143V + 8.01$, and the observed M'_W. Closed and open symbols indicate subduction zones with and without active backarc spreading, respectively. From Kanamori (1986).

"Absolute" Motion - Of all possible independent variables considered in Jarrard's analysis, "absolute" motion of the upper plate appears to be the best single predictor of strain regime, although it does not account for all modern variations in strain. The correlation between backarc spreading and a retreating overriding plate is strong (R = 0.71). One possible mechanism for this effect is that the downgoing slab acts as an anchor that cannot readily move landward or seaward because of the difficulty in displacing mantle. By this model, "absolute" motion of the overriding plate away from the trench requires an equal amount of extension, and motion toward it, an equal amount of shortening in the overriding plate. Another possible mechanism to explain "absolute" motions of overriding plates may be westward tidal lag due to Earth's rotation (e.g., Moore, 1973; Dickinson, 1978).

"Absolute" Motion and Slab Age - Molnar and Atwater (1978) first suggested that strain regime may depend on both "absolute" motion and slab age. Jarrard (1986a), however, did not find that a combination of these two factors accounts for significantly more of the variance in strain regime than does "absolute" motion alone. Furthermore, Jarrard examined age-dependent slab rollback by plotting the calculated descent angle of subducting slabs (trajectories) against their dip angles (architectures) (Fig. 1.21). This plot shows that slabs descend more steeply than slabs dip; that is, they are sinking vertically, as predicted by the rollback models of Molnar and Atwater (1978) and Dewey (1980). However, the older slabs do not appear to descend more steeply than the younger slabs (Fig. 1.21), suggesting that rollback is not age-dependent. Jarrard (1986a) argued that downward gravitational force, therefore, cannot be the only factor causing the difference between dip and descent angle, and that deep convective mantle flow may cause rollback. It should be noted, however, that estimates of descent angle vary according to which "absolute" motions are used.

Convergence Rate and Slab Age - As mentioned above, these factors successfully predict earthquake magnitude. Jarrard (1986a) also concluded that these two variables together are about as successful as "absolute" motion of the overriding plate in predict-

ing strain regime, but since both slab age and convergence rate are correlated with "absolute" motion, further analysis is needed. A significant improvement is achieved by adding a third variable, slab dip, to the strain-regime equation based on convergence rate and slab age.

Slab Dip - It is clear that low-angle subduction results in strongly compressional arc-trench systems, probably due to increased transmission of compressive stress (see Chapter 9). Jarrard (1986a) concluded, however, that dip alone is a poor predictor of strain regime.

Convergence Rate, Slab Age and Slab Dip - Stepwise multiple regression by Jarrard (1986a) indicated that a linear combination of these three variables probably determines strain regime (R = 0.88 or 0.93, depending on data set used). Shallow-depth slab dip (that is, to 100 km) was used. According to this model, slab age and convergence rate determine shear stress at plate boundaries, whereas dip angle is a measure of the contact area over which this stress acts. Jarrard's multivariate analysis is incapable of distinguishing whether "absolute" motion or slab age is the third variable, although he considers slab age more likely.

In summary, Jarrard's (1986a) analysis of modern subduction zones indicates that strain regime is probably determined by a combination of convergence rate, slab age, and slab dip. Strain regimes of modern subduction zones are indirectly correlated with crustal type in the overriding plate because extension ultimately results in sea-floor spreading and compression commonly closes oceanic basins as well as thickens continental crust. The strong tendency for old slabs in the western Pacific, intermediate-age slabs in the north Pacific and young slabs in the eastern Pacific biases our sample set, or possibly reflects Earth's eastward rotation (Dickinson, 1978). Royden (1993a, b) provides additional insights regarding the dynamics of arc-trench systems, especially in regard to arc-continent collision.

Continental Growth in Extensional Arcs

The importance of extensional continental arcs in the geologic record has only recently been appreciated. For example, the Mesozoic arc of the Southwest

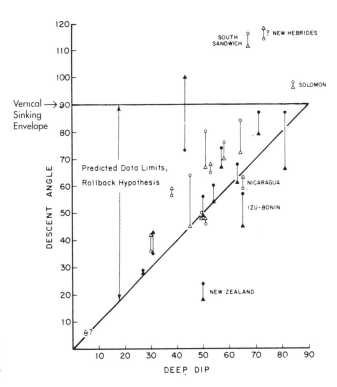

Figure 1.21 Descent angle of deep slab (trajectory) versus dip of deep slab (architecture) for modern subduction zones. Solid symbols indicate slabs older than 70 my; open symbols indicate slabs younger than 70 my. Note that few deep slabs sink at angles parallel to deep dip (diagonal line); more commonly, they descend at angles between deep dip and vertical (90), as predicted by rollback models of Molnar and Atwater (1978) and Dewey (1980). However, some deep slabs sink at angles shallower than deep dip or greater than 90 degrees. Also, no tendency is seen for open symbols to plot near diagonal line and closed symbols to plot near horizontal line. Both observations are inconsistent with a model in which greater slab age gravitationally swings descent angle toward vertical. From Jarrard (1986a).

Cordilleran U.S. was interpreted by Burchfiel and Davis (1972, 1975) as an "Andean-type" arc (i.e., highstanding, with a retroarc foldthrust belt). During earlier phases of subduction (early Mesozoic), however, the overriding plate was undergoing extension along the arc, and possibly in forearc and backarc areas (Busby-Spera, 1988b; Busby-Spera et al., 1990b; Saleeby and Busby-Spera, 1992; Riggs et al., 1993). Busby-Spera (1988b) noted that continuous belts of thick continental-arc volcanic/volcaniclastic sequences are common in the geologic record, and suggested that these are preserved due to extensional tectonics. Busby-Spera et al. (1990a) noted that extension characterized the early history of the South

American Mesozoic arc as well, and suggested that slab age controlled this, inferring that the paleo-Pacific ocean basin was composed of large, relatively old plates at the time of breakup of Pangea.

Effects of Oblique Convergence On Convergent-Margin Basins

Oblique convergence has an important impact on the upper plates of convergent margins because it may result in strike-slip faulting near or within magmatic arcs (Jarrard, 1986b). Coupling appears to be the primary factor controlling the presence of strike-slip faulting, however, so that strongly oblique convergence is not required for strike-slip faults to form in compressional systems (Jarrard, 1986b). There also appears to be a correlation between strike-slip faulting and lithospheric type, since two-thirds of modern continental arcs have strike-slip faults, whereas only one-fifth of oceanic arcs have them. This correlation may be due to the greater strength of oceanic lithosphere (e.g., Vink et al., 1984; Steckler and tenBrink, 1986), or it may only be an indirect correlation because continental overriding plates tend to be more compressional than oceanic ones (Jarrard, 1986a).

An interesting example of this phenomenon is illustrated by a comparison of the Aleutian and Indonesian arc-trench systems. Both systems involve the northward subduction of moderate-aged (e.g., 100 my) oceanic crust, with normal convergence dominating at the eastern ends of the arcs, and highly oblique convergence dominating at the northwestern ends. The primary contrast in the two systems is that oceanic lithosphere dominates in the western Aleutians and continental lithosphere in the eastern Aleutians, whereas the opposite is true of the Indonesian system. The strong oceanic lithosphere of the western Aleutians resists disruption, whereas the relatively weak continental lithosphere of western Indonesia (Sumatra) is broken by a major strike-slip fault (with the associated giant Toba caldera; Hamilton, 1979), resulting in a forearc sliver, which is partly coupled with the obliquely converging Indian plate. Neither the Aleutian nor the Indonesian arc-trench system is compressional or extensional. In summary, forearc slivering seems to result preferentially where the overriding crust is weaker (i.e., continental forearcs).

Glazner (1991) argued that it is only during periods of oblique convergence that plutonism is likely in a continental arc, and that during periods of normal convergence, volcanism predominates, although Ingersoll (1992) challenged his specific California examples. Grocott et al. (1994) suggested that this analysis may be valid for compressional systems, but that it does not apply to extensional continental systems. In any case, in compressional systems, emplacement of plutons may be aided by arc-parallel strike-slip faults. For example, the biggest plutons in the Mesozoic Sierran arc (the Late Cretaceous Whitney series and coeval plutons) were emplaced along the "proto-Kern Canyon fault zone" contemporaneous with dextral strike-slip movement (Busby-Spera and Saleeby, 1990). The arc-trench system was clearly compressional at this time (e.g., Sevier orogeny and retroarc foreland basin; see Chapter 9).

Trenches

Thornburg and Kulm (1987a, b) provided one of the most detailed studies of sedimentation in a modern trench. The Chile Trench is especially interesting because it provides markedly different climatic and sedimentologic conditions along the studied length from 18 to 45 degrees south latitude. Subducted crust is younger to the south; subduction is orthogonal and rapid everywhere. Thus, plate-tectonic processes are relatively constant along the trench, and sedimentary processes can be isolated as causes and effects.

Karig and Sharman (1975) and Schweller and Kulm (1978b), among others, discussed the dynamic nature of sedimentation and accretion at trenches. The sediment wedge of a trench is in dynamic equilibrium when subduction rate and angle, sediment thickness on the oceanic plate, rate of sedimentation, and distribution of sediment within the trench are constant. Schweller and Kulm (1978b) presented an empirical model relating convergence rate and sediment supply to characteristics of trench deposits (see Fig. 1.22). Thornburg and Kulm

(1987a) provided documentation of the dynamic interaction of longitudinally transported material (trench wedge with axial channel) and transversely fed material (trench fan). With increasing transverse supply of sediment to the trench, the axial channel of the trench wedge is forced seaward and the trench wedge widens. Contrasts in dynamic trench-fill processes help determine not only trench bathymetry and depositional systems, but also accretionary architecture (see Fig. 5.29). This dynamic model may be useful for reconstruction of sedimentary and tectonic processes in trenches, as expressed in ancient subduction complexes.

Underwood and Moore (Chapter 5) provide a comprehensive summary of sedimentary and tectonic processes in modern trenches, and models for interpreting ancient subduction complexes.

Trench-Slope Basins

G.F. Moore and Karig (1976) developed a model for sedimentation in small ponded basins along inner trench walls. Deformation within and on the subduction complex results in irregular bathymetry, commonly including ridges subparallel to the trench. Turbidites are ponded behind these ridges, and trench-slope basins form. Average width, sediment thickness and age of basins increase up slope due to the progressive uplift of deformed material and the widening of fault spacing during dewatering and deformation of offscraped sediment. In ancient subduction complexes, trench-slope basins are filled with relatively undeformed, locally derived turbidites surrounded by highly deformed accreted material of variable origin. Contacts between trench-slope basins and accreted material are both depositional and tectonic. G.F. Moore and Karig's (1976) model was developed for Nias Island near Sumatra, an area of rapid accretion of thick sediments. Their model is less useful for sediment-starved forearc areas. Nonetheless, their general principles governing the development of sedimentary basins on the lower trench slope are fundamental to reconstructing ancient subduction settings.

Underwood and Moore (Chapter 5) discuss trench-slope basins, both modern and ancient.

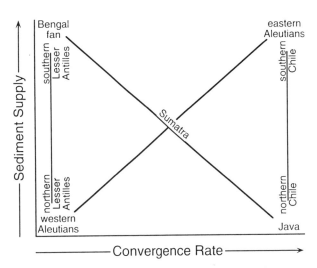

Figure 1.22 Examples of individual trenches with varying sediment supply and/or convergence rate along strike. Convergence rates for the Aleutians and Sunda trench (Bengal-Java system) vary due to increased obliquity of convergence at northwestern ends of both trenches; in the former case, sediment supply decreases in this intraoceanic setting, whereas in the latter, sediment supply increases as the Himalayan suture zone is approached (also, see Fig. 10.5). Convergence rates are fairly constant for the Antilles and Chile trenches, whereas sediment supply varies greatly, with resulting variation in trench facies and accretionary complexes. Inspired by Schweller and Kulm (1978b).

Forearc Basins

Dickinson and Seely (1979) provided a classification of arc-trench systems, similar to Dewey's (1980), and outlined plate-tectonic controls governing subduction initiation and forearc development. Figure 6.4 illustrates the variability of forearc types. Factors controlling forearc geometry include: 1. initial setting; 2. sediment thickness on subducting plate; 3. rate of sediment supply to trench; 4. rate of sediment supply to forearc area; 5. rate and orientation of subduction; and 6. time since initiation of subduction. Arc-trench gaps tend to widen through time (Dickinson, 1973) due to prograde accretion at trenches and retrograde migration of magmatic arcs following subduction initiation. Prograde accretion is especially rapid where thick sequences of sediment are accreted. The net result of widening of the arc-trench gap is the general tendency for forearc basins to enlarge through time (e.g., Great Valley forearc basin; Ingersoll, 1979, 1982b; Chapter 6).

Forearc basins include the following types (Dickinson and Seely, 1979): 1. intramassif (transitional to intra-arc); 2. accretionary (trench-slope); 3. residual

(lying on oceanic or transitional crust trapped behind the trench when subduction initiated); 4. constructed (lying across the boundary of arc massif and subduction complex); and 5. composite (combination of above settings). Residual and constructed basins tend to evolve into composite basins; commonly, this evolutionary trend is accompanied by filling and shallowing of the forearc basins.

Stern and Bloomer (1992) discussed extension along the front edge of the overriding plate at the time of initiation of the Mariana subduction zone (Oligocene). This type of extension is only likely within 10–20 my of initiation of an intraoceanic subduction zone within old (strong) oceanic lithosphere, where slab rollback begins as soon as subduction initiates, and the weakest part of the overriding plate is near the edge. Soon after subduction begins, forearc areas are cooled by the cold subducting oceanic lithosphere; thus, mature intraoceanic forearcs tend to be underlain by cold and strong lithosphere, and resist extension (i.e., Vink et al., 1984; Steckler and tenBrink, 1986). In contrast, arc axes of mature intraoceanic systems tend to be the weakest part of the overriding plate, and extension is accommodated by intra-arc and backarc spreading (see Chapter 8).

Dickinson (Chapter 6) updates Dickinson and Seely (1979), with new information and synthesis of modern and ancient examples.

Intra-Arc Basins

The origin of basins within volcanoplutonic (magmatic) arcs is, in general, poorly understood, largely due to the paucity of studies that integrate volcanology, sedimentology and basin analysis (Ingersoll, 1988b). A further deterent to many sedimentologists is the fact that arcs are characterized by high heat flow with steep geothermal gradients and intense magmatism, and are commonly subjected to crustal shortening at some time in their history; therefore, experience in "seeing through" the overprints of hydrothermal alteration, metamorphism and deformation is required. Recent publications on sedimentation in volcanic settings show that neglect of this topic is ending; these include two textbooks (Fisher and Schmincke, 1984; Cas and Wright, 1987), a spe-

cial issue of Sedimentary Geology (Cas and Busby-Spera, 1991) and an SEPM Special Publication (Fisher and Smith, 1991).

Magmatic arcs vary according to crustal foundation, stress regime and longevity. Primitive intra-oceanic arcs are founded on oceanic crust and are dominated by tholeiitic mafic igneous rocks; the edifice is largely subaqueous. More evolved arcs are dominated by calc-alkaline mafic through intermediate to silicic igneous rocks, and are generally founded upon thicker crust composed of older arc rocks or other convergent-margin assemblages such as accreted terranes. In these arcs, the edifices form broad submarine platforms, and many parts may be emergent (e.g., arcs of the Caribbean). Arcs formed on thick continental crust are silicic to intermediate in composition and are dominantly subaerial (e.g., the Andean arc). As discussed above, however, the overriding plates on which arcs form commonly undergo extensional or transtensional tectonism, in which case, subsidence of arc edifices may be accompanied by marine inundation. With marked extension or transtension, rifting processes result in formation of new oceanic crust (Karig, 1971a, 1972; Dewey, 1980; Hamilton, 1988a, 1989). Thus, extensional arcs form a continuum with backarc basins formed by sea-floor spreading (discussed below).

Volcanic arcs produce large volumes of clastic material that may form much of their arc edifices, in addition to intrusive and extrusive igneous rocks. Clastogenic processes include: (1) magmatic explosion, due to exsolution of volatiles, (2) hydroclastic fragmentation, due to magma-water interactions, (3) autoclastic lava fragmentation, and (4) epiclastic processes of weathering and erosion. These clastic materials move downslope into basins of various types within or on the flanks of arcs.

There are at least three major types of depocenters for volcanic and sedimentary accumulations within arcs. First, depocenters may occur in low regions between volcanoes and along their flanks, although these have high preservation potential only below sea level (i.e., generally in oceanic arcs). Second, depocenters with high preservation potential may form when the axis of arc volcanism shifts to a new position on an oceanic arc platform, thus

creating a low region between the active chain and the abandoned chain. Both the first and second types of intra-arc basin are referred to as "volcano-bounded basins" in Chapter 7. The third type is referred to as "fault-bounded basins" (Chapter 7); these are rapidly subsiding basins where tectonic structures, rather than constructional volcanic features, account for relief along the basin margins. Fault-bounded basins occur in both *continental* arcs (e.g., Fitch, 1972; Carr, 1976; Erlich, 1979; Dalmayrac and Molnar, 1981; Burkart and Self, 1985; Tobisch et al., 1986; Cabrera et al., 1987; Smith et al., 1987; Busby-Spera, 1988b; Kokelaar, 1988; Smith, 1989) and *oceanic* arcs (e.g., Hussong and Fryer, 1982; Geist et al., 1988; Bloomer et al., 1989; Klaus et al., 1992a, b).

The origin of structural depressions within magmatic arcs is poorly understood. The most important mechanisms for accumulating and preserving thick stratigraphic successions in continental arcs appear to be, in descending scale (Busby-Spera et al., 1990a): (1) plate-margin-scale extension or transtension, (2) extension on a more local scale during pluton or batholith emplacement, and (3) localized subsidence of calderas during large-volume ignimbrite eruptions. Plate-margin-scale extension or transtension produces belts of continental-arc sequences that are continuous or semi-continuous over hundreds to thousands of kilometers and record high rates of subsidence over tens of millions of years. The effects of extension in the roofs of plutons or batholiths (e.g., Tobisch et al., 1986) may be difficult to distinguish from plate-margin-scale extension, but the former should operate over shorter time scales (i.e., less than a few million years) and should not, by itself, produce a low-standing arc capable of trapping sediment derived from outside the arc. Continental calderas form small (10–60 km wide) but deep (1–4 km) depocenters for ignimbrite erupted during caldera collapse, as well as for volcanic and sedimentary strata ponded within the caldera after collapse (e.g., Riggs and Busby-Spera, 1991; Lipman, 1992; Schermer and Busby, 1994). This type of depocenter may be short-lived because resurgent magmas commonly invade it within a few-hundred-thousand years.

The standard facies models for clastic sedimentary strata are not applicable to many volcaniclastic sequences. This is largely due to the extremely episodic sediment supply typical of settings with explosive eruptions (Busby-Spera, 1985; Smith, 1991; Kokelaar, 1992). Explosive eruptions instantaneously produce large volumes of unconsolidated or poorly consolidated material; the effects of eruptions on sedimentation in subaerial settings have been relatively well studied (e.g., Kuenzi et al., 1979; Vessel and Davies, 1981; Hackett and Houghton, 1989; Smith, 1991). Far less is known about the response of submarine sedimentary systems to explosive eruptions, making paleoenvironmental interpretations more controversial (e.g., see contrasting paleogeographical interpretations for strata enclosing an Ordovician welded tuff given by Fritz and Howells [1991] and Orton [1991]). An additional complicating factor is major volcanic slope failure (sector collapse), a degradation-dispersal process whose frequency has only recently been recognized. The result may be debris avalanches and associated lahars that travel tens of kilometers or more (e.g., Siebert et al., 1987). Given the importance of sector collapse in generating debris-avalanche and associated debris-flow deposits around subaerial Quaternary volcanoes (e.g., Lipman and Mullineaux, 1981; Palmer et al., 1991), one would expect these deposits to occur in marine sections inferred to be associated with ancient arc volcanoes. Only one marine formation has been interpreted as including debris-avalanche deposits, however (the "Parnell Grit" of Ballance and Gregory [1991]). This paucity is probably due to problems with recognition, rather than scarcity; efficient transformation of debris avalanches into debris flows occurs by ingestion of water on the submarine flanks of volcanoes (Ballance and Gregory, 1991).

As discussed in Chapter 7, the distinction of forearc, intra-arc and backarc basins is not always clear. In this book, intra-arc basins are defined as thick volcanic-volcaniclastic and other sedimentary accumulations along the arc platform, which is formed of overlapping or superposed volcanoes. The presence of vent-proximal volcanic rocks and related intrusions is critical to the recognition of intra-arc basins

in the geologic record, since arc-derived volcaniclastic material may be spread into forearc, backarc and other basins. A more general term, "arc massif," refers to crust generated by arc magmatic processes (Dickinson, 1974a, b), and arc crust may underlie a much broader region than the arc platform. The distinction of forearc and intra-arc basins is also discussed in Chapter 6. Many backarc basins form by rifting within the arc platform (Chapter 8), and are thus intra-arc basins in their early stages. Also, forearc, intra-arc and backarc settings change temporally and are superposed on each other due to both gradual evolution and sudden reorganization of arc-trench systems resulting from collisional events, plate reorganization and changes in plate kinematics.

Smith and Landis (Chapter 7) provide the definitive synthesis of intra-arc basins.

Backarc Basins

Backarc basins are defined in this book as: (1) oceanic basins behind intraoceanic magmatic arcs, and (2) continental basins behind continental-margin arcs that lack foreland foldthrust belts. Many backarc basins are extensional in origin, forming by rifting and sea-floor spreading. These commonly originate through rifting of the arc, either along its axis (intra-arc) or immediately to the front or rear of its axis. The term "interarc basin" (Karig, 1970a) has been widely superseded by the term "backarc basin," and is not used in this chapter, but it is used to describe the specific situation where rifting occurs along or in front of the arc axis, thus eventually producing a remnant arc behind the backarc basin (Fig. 8.2). If rifting occurs to the rear of the arc axis, however, no remnant arc will be formed. The presence or preservation of a remnant arc is thus not a necessary condition for recognition of a backarc basin (Taylor and Karner, 1983).

Many backarc basins are nonextensional (Chapter 8), forming under neutral strain regimes. The most common type of nonextensional backarc basin consists of old ocean basins trapped during plate reorganization (e.g., the Bering Sea). Also, nonextensional backarc basins develop on continental crust (e.g., the Sunda shelf of Indonesia).

Modern oceanic backarc basins may be distinguished from other ocean basins petrologically or by their positions behind active or inactive trench systems (Taylor and Karner, 1983). It is unlikely that such diagnostic features are preserved in ancient backarc basins, which commonly undergo metamorphic and structural modifications during emplacement in orogenic zones. The nature and timing of deposition of sediment on top of ophiolite sections have proven more diagnostic for determining original plate-tectonic settings (e.g., Tanner and Rex, 1979; Sharp, 1980; Hopson et al., 1981; Kimbrough, 1984; Busby-Spera, 1988a; Robertson, 1989).

Many modern backarc basins have been penetrated by the Ocean Drilling Program (formerly the Deep Sea Drilling Project) (Klein and Lee, 1984). Yet, few detailed studies have been carried out on ancient backarc-basin assemblages. An additional gap in our knowledge arises from the fact that arc-proximal facies of modern backarc basins have been studied least, due to coring difficulties and poor recovery in coarse-grained unconsolidated deposits. For example, the excellent model constructed by Sigurdsson et al. (1980) and Carey and Sigurdsson (1984) for the Granada backarc basin (Fig. 8.13) used piston cores up to 150 km from the arc; the most proximal deposits could not be cored. The most detailed study of a backarc volcaniclastic apron and its substrate comes from Middle Jurassic rocks in Mexico (Busby-Spera, 1987, 1988a) (Fig. 8.17). That study supported Karig and Moore's (1975) assertion that oceanic backarc basins isolated from terrigenous sediment influx may show the following simple, uniform sedimentation patterns: (1) Lateral and vertical differentiation of facies due to progradation of a thick volcaniclastic apron into a widening backarc basin; such an apron may extend for more than 100 km from a volcanic island and grow to a thickness of 5 km in 5 my (Lonsdale, 1975). This phase is followed by: (2) Blanketing of the apron with a thin sheet of mud and sand eroded from the arc after volcanism and spreading have ceased. This cycle reflects the temporal episodicity of oceanic backarc basins, which appear to form in 10–15 my or less (Taylor and Karner, 1983).

Backarc basins are depicted in Fig. 1.2 as generally having shorter life spans and lower preservation

potential than intra-arc basins. The shorter life span reflects temporal episodicity of oceanic backarc basins, the most common type of backarc basin. In contrast, arcs may undergo episodic extension for many tens of millions of years, particularly in continental settings. Oceanic backarc basins have lower preservation potential than intra-arc basins because of their susceptibility to subduction. Although backarc basins and their fill make an important contribution to orogenic belts, we assume that most backarc basins are subducted, and that the frontal-arc sides of backarc basins are preferentially preserved in the geologic record (Busby-Spera, 1988a).

Marsaglia (Chapter 8) summarizes present knowledge of interarc and backarc basins, with discussion of modern and ancient examples, both oceanic and continental, and both extensional and neutral settings.

Mesozoic Convergent Margin of Baja California, Mexico

Mesozoic rocks of the Baja California Peninsula (Fig. 1.23) form one of the most areally extensive, best exposed, longest-lived (160 my) and least tectonized and metamorphosed convergent-margin basin complexes in the world. This convergent margin shows an *evolutionary trend* that the junior author (CJB) suggests may be typical of arc systems facing large ocean basins: *a progression from highly extensional through mildly extensional to compressional strain regimes* (Fig. 1.24). In this evolutionary model, subduction is initiated by rapid sinking of very old, cold oceanic lithosphere, but over many tens of millions of years the subducting lithosphere becomes progressively younger. In addition to slab age, Jarrard's (1986a) analysis (discussed above) indicates that slab dip and convergence rate also control strain regime. He showed that slab dip tends to decrease with duration of subduction, resulting in increasingly compressional strain; this effect enhances the evolutionary trend proposed here. A decrease in convergence rate or increase in age of subducted crust, on the other hand, may slow or even reverse this evolutionary trend; in the case of the Baja margin, however, convergence rate increased, thus accelerating the proposed evolutionary trend.

A summary of the types of basins formed during three phases of subduction (highly extensional, mildly extensional and compressional) is given in Figures 1.23 and 1.24, and a brief description is provided here.

PHASE 1 - The oldest depositional systems represented in the Baja California convergent margin are arc-apron deposits, consisting of pyroclastic and lesser volcanic epiclastic debris deposited in Late Triassic to Late Jurassic oceanic intra-arc and backarc basins (Fig. 1.24). These basins were isolated from continental sediment sources for most of their history, although the presence of continentally derived coarse-grained detritus in the Upper Jurassic Eugenia and Coloradito formations suggests that these oceanic terranes were near a continental margin by the Late Jurassic. As discussed above, the fill of the Middle Jurassic backarc basin preserved on Cedros Island (Gran Cañon Formation) (Fig. 1.23) shows a simple, uniform sedimentation pattern that may be typical of backarc basins isolated from terrigenous sediment influx (Busby-Spera, 1988a).

PHASE 2 - The fringing arc of phase 2 is now part of the western wall of the Peninsular Range batholith (Lower Cretaceous Alisitos Group) (Fig. 1.23). The western Peninsular Range has been interpreted as an exotic oceanic arc terrane that was accreted to the continental margin, represented by the eastern Peninsular Ranges, in mid-Cretaceous time (Gastil et al., 1978; Todd et al., 1988). The Julian Schist, however, occurs in both the western and eastern Peninsular Ranges, and is, therefore, interpreted as the fill of a composite (continental to oceanic) extensional forearc basin that developed in front of the Jurassic extensional arc (Saleeby and Busby-Spera, 1992; Thomson et al., 1994). This relationship appears to tie the Early Cretaceous arc of the western Peninsular Ranges to the edge of the continent. The Baja segment of the Early Cretaceous arc was at least mildly extensional, with syndepositional normal faults in the forearc region and high rates of subsidence within the arc (Fig. 1.25) (Busby-Spera and Boles, 1986; Adams and Busby, 1994), although in the California segment, there is evidence for localized shortening within the arc at around 118 to 114 Ma (Thomson et al., 1994).

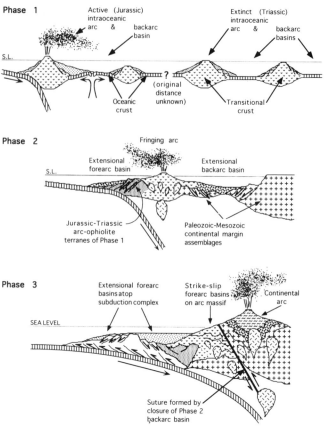

Figure 1.23 Locations of Mesozoic convergent-margin assemblages in Baja California, Mexico, grouped by evolutionary phases (see Fig. 1.24): PHASE 1 - Vizcaino Sur: Upper Triassic La Costa ophiolite and overlying Upper Triassic to Lower Jurassic San Hipolito Formation volcaniclastic rocks (Whalen and Pessagno, 1984; Moore, 1985); Jurassic volcaniclastic strata of Cerro El Calvario (Moore, 1984, 1985). Vizcaino Norte: Upper Triassic Sierra San Andres ophiolite and overlying Upper Triassic Puerto Escondido tuff (Barnes, 1984; Moore, 1985); Upper Jurassic to Lower Cretaceous Eugenia Formation (arc-apron strata: Hickey, 1984; Kimbrough et al., 1987). Cedros Island: Middle Jurassic Cedros Island ophiolite and Choyal oceanic-arc assemblage, overlapped by Middle Jurassic Gran Cañon Formation volcaniclastic rocks (Kimbrough, 1984, 1985; Busby-Spera, 1988a); Upper Jurassic Eugenia Formation (Kilmer, 1984); Upper Jurassic Coloradito Formation sedimentary melange (Kilmer, 1984; Boles and Landis, 1984). PHASE 2 - Vizcaino Sur: Lower Cretaceous Asuncion Formation (forearc strata: Barnes, 1984; Moore, 1984, 1985; Busby-Spera and Boles, 1986). Western Peninsular Ranges, Baja Norte: Alisitos Group (oceanic-arc assemblage: Silver et al., 1963; Gastil et al., 1975; Beggs, 1984; Gastil, 1985; Silver, 1986; Busby-Spera and White, 1987; White and Busby-Spera, 1987; Adams and Busby, 1994). Central Peninsular Ranges, Baja Norte: (backarc strata and continental-margin substrate: Gastil, 1985; Griffith et al, 1986; Goetz, 1989). PHASE 3 - Vizcaino Peninsula and Cedros Island: Upper Cretaceous Valle Formation (forearc strata: Barnes, 1984; Patterson, 1984; Morris et al., 1989; Smith and Busby, 1993, 1994). Western Peninsular Ranges of Baja California Norte and Southern California: Upper Cretaceous Rosario Group (forearc strata: Kilmer, 1963; Gastil and Allison, 1966; Nilsen and Abbott, 1981; Bottjer and Link, 1984; Yeo, 1984; Boehlke and Abbott, 1986; Cunningham and Abbott, 1986; Morris and Busby-Spera, 1988, 1990; Morris et al., 1989; Morris, 1992; Fulford and Busby, 1994). Eastern Peninsular Ranges: continental-arc plutons, which record an eastward-migrating linear locus of magmatism from ca. 100 to 75 Ma (Krummenacher et al., 1975; Silver, 1986).

Figure 1.24 Evolutionary model for arc systems facing large ocean basins: a case study from Baja California, Mexico. (A) Three main phases are recognized (localities and references given in Figure 1.23 caption): PHASE 1 - HIGHLY EXTENSIONAL INTRAOCEANIC-ARC SYSTEM(S) - Earliest stages of subduction (ca. 220-130 Ma) are represented by intra-oceanic arc-ophiolite systems, where small, steep-sided basins received volcaniclastic debris. PHASE 2 - MILDLY EXTENSIONAL FRINGING-ARC SYSTEM - Intermediate stage of subduction (ca. 140-100Ma) is represented by a fringing island arc separated from a continental margin by a narrow backarc basin, partly floored by continental-margin rocks, that received continental as well as island-arc detritus. Forearc-basin complex developed atop amalgamated oceanic arc-ophiolite terranes of phase 1, was dominated by normal faults, and lacked an accretionary complex (Fig. 1.25). PHASE 3 - COMPRESSIONAL CONTINENTAL-ARC SYSTEM - In third phase of subduction (ca. 100-70 Ma), backarc basin of second phase closed, and a highstanding continental arc was established. Strongly coupled subduction led to development of forearc strike-slip basins, as well as a thick subduction complex with extensional basins atop it (Fig. 1.26).

The Early Cretaceous fringing-arc terrane of the western Peninsular Ranges provides an excellent opportunity to develop a better understanding of intra-arc basins because the size (450 X 50 km) and unusually good exposure of the terrane permits comparison of its stratigraphy and structure with those of modern systems. Initial studies in a 100km-long segment of the terrane have resulted in recognition of two stages in its evolution (Adams and Busby, 1994, and unpublished data):

(1) A locally emergent, silicic to intermediate-composition, dominantly explosive stage. Tectonic subsidence of the arc during this stage resulted in rapid accumulation of thick shallow-marine deposits. Pyroclastic flows accumulated in marine as well as nonmarine environments, including a deep-water welded tuff filling a small caldera. Deep-water equivalents to debris-avalanche deposits have also been recognized.

(2) A dominantly mafic-composition stage, including formation of lava flows and hyaloclastites, as well as dike swarms and other hypabyssal intrusions, that records an arc rifting event.

Coarse-grained slope aprons developed in the forearc region of the fringing arc in response to extension (Fig. 1.25). The slope-apron deposits represent scree-cone debris aprons that build directly from coastal fault scarps onto half-graben floors at bathyal marine water depths. They are thick, laterally extensive wedges of texturally and mineralogically immature detritus. Two major facies are recognized: (1) a fault-scarp facies, consisting of *in situ* basement breccia mantled by monomictic, closely packed sedimentary breccia or by bioclastic limestone, and (2) a basin-fill facies that onlaps the scarp with steep depositional dips. Basin-fill deposits include rock-fall/avalanche and debris-flow deposits, as well as turbidites. Although most of the detritus in the slope-apron deposits can be shown to have been derived from immediately adjacent basement horst blocks, the presence of at least a small percentage of fresh volcanic pebbles and sand indicates that arc-derived detritus was able to make its way into each sub-basin, suggesting that the half grabens stepped down toward the trench (Fig. 1.25).

Plate reconstructions suggest that old oceanic lithosphere was subducting under North America during Early Cretaceous time (Engebretson et al., 1985); this may have been a major control on extension in the forearc to backarc region during phase 2.

PHASE 3 - During this phase, a high-standing continental arc was established in the present eastern Peninsular Ranges. Although this was part of an overall trend to a progressively more compressional strain regime, this trend accelerated in mid-Cretaceous time (ca. 105–95 Ma), as shown by: (1) the presence of reverse faults of that age within the arc (Griffith, 1987; Goetz et al., 1988; Todd et al., 1988; George and Dokka, 1994; Thomson et al., 1994), and (2) sudden influx into forearc basins of coarse-grained sediment eroded from relatively deep structural levels of the arc (Valle Formation) (Figs. 1.23 and 1.24). This acceleration of trend may have been due to an increased convergence rate hypothesized at that time (Engebretson et al., 1985), resulting in telescoping of the fringing arc.

Strongly coupled subduction along the compressional arc resulted in: (1) mid-Cretaceous accretion of blueschist metamorphic rocks, and development of an extensional forearc-basin complex atop the subduction complex as these blueschists were exhumed (Smith and Busby, 1993), and (2) Late Cretaceous development of strike-slip basins on arc-massif basement in front of the arc (Morris, 1992); these strike-slip forearc basins record the oblique convergence inferred from plate reconstruction models of Engebretson et al. (1985) and Glazner (1991). This oblique convergence may have resulted in the 10–19 degrees northward displacement of Baja California relative to North America that has been proposed by Hagstrum et al. (1985).

The transition from a mildly extensional arc to a compressional arc is recorded in mid-Cretaceous strata that occur in outboard parts of the forearc complex (Valle Formation) (Figs. 1.23 and 1.24). Detailed studies on Cedros Island show that deep-marine conglomerates that flooded the forearc during arc uplift were deposited at the onset of extensional brittle deformation of the upper crust, concomitant with initial uplift of blueschist. The conglomerates fill a

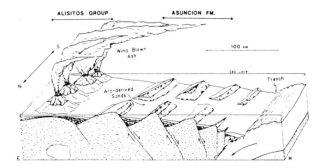

ALISITOS GROUP   ASUNCION FM.

Wind Blown
Ash   100 km

Sea Level

Trench

Arc-derived
Sands

Figure 1.25 Reconstruction of arc-forearc region during phase 2 subduction (mildly extensional fringing-arc system; Fig. 1.24), drawn at latitude of southern Vizcaino Peninsula (Baja California, Mexico). Presence
of normal faults in forearc and lack of a subduction complex make this
region similar to modern extensional forearcs (e.g., the Marianas [Hus-
song and Fryer, 1982], and the Peruvian forearc [Moberly et al., 1982];
see Chapters 5 and 6). From Busby-Spera and Boles (1986).

deep-marine half-graben structure that formed by
reactivation of a Jurassic fault zone (Fig. 1.26A). The
near coincidence of peak blueschist metamorphism
in the lower plate of the convergent margin and the
onset of extension in the upper plate (that is, syndepositional faulting in the forearc basement reported
here at about 95 Ma) support blueschist unroofing
models calling for uplift soon after peak metamorphism (e.g., Platt, 1986; Jayko et al., 1987).

Arc magmatism in the present eastern Peninsular
Ranges gradually migrated eastward during Late
Cretaceous time (Krummenacher et al., 1975; Silver,
1986), so that by Campanian time (or perhaps earlier), forearc basins formed atop the Early Cretaceous arc terrane of the western Peninsular Range.
The Peninsular Ranges forearc-basin complex is thus
an arc-massif type (classification of Dickinson and
Seely, 1979). Numerous authors have proposed that
syndepositional faults controlled basin shape and
bathymetry, distribution of depositional systems,
and cyclic sedimentation (e.g., Gastil and Allison,
1966; Boehlke and Abbott, 1986; Cunningham and
Abbott, 1986; Morris and Busby-Spera, 1987, 1988;
Morris, 1992).

The Rosario embayment (Fig. 1.23) is the most
areally extensive and best-exposed segment of the
Peninsular Ranges forearc-basin complex. Two main
lines of evidence (Morris, 1992) point to strike-slip
deformation of the Peninsular Ranges forearc-basin

complex at a time when plate reconstruction models
(Engebretson et al., 1985) and paleomagnetic studies
(Hagstrum et al., 1985) indicate oblique convergence:
(1) kinematic indicators show a strike-slip component
for basin-bounding normal faults, and (2) the basin
fill shows evidence for rapid alternation of contractional and extensional events, which is typical of
modern strike-slip forearc basins (e.g., Kimura, 1986;
Geist et al., 1988).

Retroarc Foreland Basins

Compressional arc-trench systems commonly
develop foreland basins behind the arcs due to partial subduction of continental crust beneath the arc
orogens (Fig. 1.19A, B). "Foreland basin" is a pre-
plate-tectonic term used to describe a basin between
an orogenic belt and a craton (Allen et al., 1986).
Dickinson (1974b) proposed that the term "retroarc"
be used to describe foreland basins formed behind
compressional arcs, in contrast to peripheral foreland
basins formed on subducting plates during continental collisions. Thus, although "backarc" and
"retroarc" literally are synonymous, the former is
used for extensional and neutral arc-trench systems,
whereas the latter is used for compressional arc-
trench systems.

Jordan (1981) presented an analysis of the asymmetric Cretaceous foreland basin associated with
the Idaho-Wyoming thrust belt. She used a two-
dimensional elastic model to show how thrust loading and sedimentary loading resulted in broad flexure of the lithosphere. The location of maximum
flexure migrated eastward as thrusting migrated
eastward. The area of subsidence was broadened
due to the erosional and depositional redistribution
of part of the thrust load, and possibly enhanced by
high eustatic sea level of the Late Cretaceous. Comparison of modeled basin and basement geometries
with isopach maps provides tests of possible values
of flexural rigidity of the lithosphere. The modern
sub-Andean thrust belt and foreland basin have
similar topography to that proposed for the Cretaceous of the Idaho-Wyoming system. Topography is
controlled by thrust-fault geometry and isostatic
subsidence.

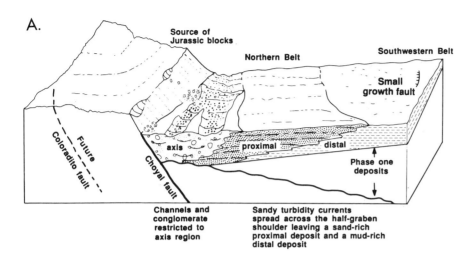

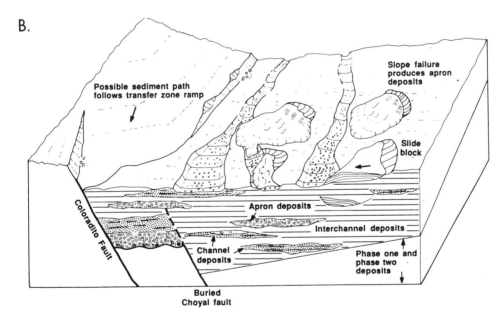

Figure 1.26 Marine mid-Cretaceous forearc strata on Cedros Island (Fig. 1.23) record onset of brittle deformation of upper crust concomitant with uplift of blueschist. (A) In Cenomanian time, a deep-marine half graben formed by reactivation of a Jurassic fault zone (Fig. 1.25A). Axis of half graben acted as a submarine canyon that funneled sediment gravity flows toward southwest while shoulder of half graben was draped by sandy turbidites and mud. (B) By end of early Cenomanian time, a second episode of faulting on a fault parallel to, and synthetic with, the older half-graben fault produced a stress-transfer zone that steepened intervening block, producing a catastrophic slope failure. Older fault was buried, and half graben filled with stacked channel-interchannel deposits interstratified with thick intrabasinal slump deposits. These abundant slope-failure features, including numerous 100m-long slide blocks, resulted from frequently recurring intrabasinal seismicity. From Smith and Busby (1993).

The model presented by Jordan (1981) is broadly applicable to other retroarc foreland basins. It demonstrates that tectonic activity in the foreland foldthrust belt is the primary cause of subsidence in associated foreland basins. Sedimentary redistribution, autocyclic sedimentary processes, and eustatic sea-level changes are important modifying factors in terms of regressive-transgressive sequences, but compressional tectonics behind the arc-trench system is the driving force. The Cretaceous seaway of North America was largely the result of this compressional tectonic activity (combined with high eustatic sea level). Details concerning timing of thrusting and initial sedimentary response to thrusting within the Idaho-Wyoming thrust belt are debated (e.g., Heller et al., 1986), but the essential role of compressional tectonics in creating retroarc foreland basins is clear.

Jordan (Chapter 9) points out that most of the oldest sandstones in the Bermejo retroarc foreland basin of Argentina are volcaniclastic, in contrast to the younger primarily recycled-orogenic sandstones

typical of most forelands (i.e., Dickinson and Suczek, 1979; Dickinson, 1985). Noncompressional backarc basins (either neutral or extensional) are characterized by volcaniclastic sandstones of magmatic-arc provenance (also, see Chapter 8). It is interesting to note that where the Antarctic plate is being subducted beneath southern South America, the subduction rate is quite slow (in contrast to rapid subduction of the Nazca plate farther north), and the Andean arc-trench system here is neutral, with no retroarc foreland foldthrust belt. In fact, the foldthrust belt is also absent more than 1000 km north of the triple junction. The result of this absence is that modern sand is volcaniclastic all the way to the Atlantic intraplate margin (admittedly, not far to the east) (e.g., Blasi and Manassero, 1990; Ingersoll et al., 1993). Sand and sandstone deposited in backarcs with narrow continental crust generally have magmatic-arc provenance, even if continental margins are intraplate (e.g., southern South American Atlantic and western Gulf of Mexico).

Jordan (Chapter 9) updates her analysis of the Cretaceous retroarc foreland of North America, synthesizes her definitive work on the Neogene to Holocene retroarc of South America, and discusses general models for retroarc foreland basins.

Remnant Ocean Basins

Intense deformation occurs in suture belts during the attempted subduction of nonsubductable, buoyant continental or magmatic-arc crust (e.g., Cloos, 1993). Suture belts can involve rifted continental margins and continental-margin magmatic arcs (terminal closing of an ocean basin) or various combinations of arcs and continental margins. Figure 10.3 illustrates stages in the development of suture belts, either in time or space. Colliding continents tend to be irregular, and great variability of timing, structural deformation, sediment dispersal patterns and preservability occurs along strike (Dewey and Burke, 1974).

Graham et al. (1975) used Cenozoic development of the Himalayan-Bengal system as an analog for the late Paleozoic development of the Appalachian-Ouachita system, and proposed a general

model for sediment dispersal related to sequentially suturing orogenic belts (Fig. 10.1). "Most sediment shed from orogenic highlands formed by continental collisions pours longitudinally through deltaic complexes into remnant ocean basins as turbidites that are subsequently deformed and incorporated into the orogenic belts as collision sutures lengthen." (Graham et al., 1975, p. 273). This model provides a general explanation for many syn-orogenic flysch and molasse deposits associated with suture belts, although many units called "flysch" and "molasse" have different tectonic settings (see Chapters 10 and 11).

North American examples of arc-continent collisions, with variable volumes of remnant-ocean-basin flysch, include the Ordovician Taconic orogeny of the Appalachians (Rowley and Kidd, 1981; Stanley and Ratcliffe, 1985; Lash, 1988; Bradley, 1989; Bradley and Kidd, 1991) and the Devonian-Mississippian Antler orogeny of the Cordillera (Speed and Sleep, 1982; Dickinson et al., 1983). In both cases, it is difficult to clearly distinguish remnant ocean basins from incipient peripheral foreland basins as the depositional sites for "flysch" (also, see Chapters 10 and 11).

Ingersoll et al. (Chapter 10) provide a general synthesis of remnant ocean basins, including several case studies, both modern and ancient.

Peripheral Foreland Basins

As continental collision occurs between a rifted continental margin and the subduction zone of an arc-trench system, a tectonic load is placed on the rifted margin, first below sea level, and later subaerially (Fig. 10.3). A peripheral foreland basin forms as the elastic lithosphere flexes under the encroaching dynamic load. The discrimination of ancient peripheral and retroarc forelands associated with suture belts is difficult, but may be possible based on the following characteristics: 1. polarity of magmatic arc; 2. presence of oceanic subduction complex associated with earliest phases of peripheral foreland; 3. greater water depths in peripheral foreland (foredeep stage); 4. asymmetry of suture belt (closer to peripheral foreland); 5. protracted development of retroarc foreland (longterm arc evolution) versus discrete

development of peripheral foreland (terminal ocean closure without precursor); and 6. possible volcaniclastic input to retroarc foreland throughout history versus minimal volcaniclastic input to peripheral foreland.

Stockmal et al. (1986) provided a dynamic 2-dimensional model for the development of peripheral foreland basins, following finite times of rifting. They modified the model of Speed and Sleep (1982), and demonstrated the effects of rifted-margin age and topography on lithospheric flexure and basin development. The primary effect of age shows up as a higher flexural forebulge and thicker trench fill during earlier stages of attempted subduction of an old (120 my) margin. Subsequent development is relatively insensitive to margin age. Foreland-basin subsidence is sensitive to overthrust load, with depths possibly exceeding 10 km. Crustal thickness may reach 70 km during the compressional phase (e.g., Himalayas). Tens of kilometers of uplift and erosion, of both the allochthon and the proximal foreland basin, are predicted during and after deformation. Most of this eroded material is deposited elsewhere due to uplift within the foreland; longitudinal transport into remnant ocean basins results (Graham et al., 1975; see Chapters 10 and 11). Thick overthrusts with low topographic expression are to be expected where broad, attenuated rifted continental margins have been pulled into subduction zones.

Miall (Chapter 11) summarizes all "collision-related foreland basins," which include peripheral foreland basins and collisional retroarc basins formed on the overriding plate during crustal collision. The latter may have noncollisional retroarc foreland basins as precursors.

Piggyback Basins

Ori and Friend (1984) defined "piggyback basins" as basins that form and fill while being carried on moving thrust sheets. They discussed examples from the Apennine and Pyrenean foldthrust systems. Piggyback basins are dynamic settings for sediment accumulation; most sediment, if not all, is derived from associated foldthrust belts. The foldthrust belts can be in peripheral, retroarc or transpressional settings. Piggyback basins share characteristics with foreland basins and trench-slope basins. They have low preservation potential due to their formation on top of growing thrust belts; therefore, they are generally found only in young orogenic systems (e.g., Burbank and Tahirkheli, 1985).

Jordan (Chapter 9), Miall (Chapter 11) and Nilsen and Sylvester (Chapter 12) provide additional insights regarding piggyback basins.

Foreland Intermontane Basins (Broken Forelands)

Low-angle subduction beneath compressional arc-trench systems may result in basement-involved deformation within retroarc foreland basins (Dickinson and Snyder, 1978; see Chapter 9). The Rocky Mountain region of the western United States is the best-known ancient example of this style of deformation; similar modern provinces have been documented in the Andean foreland (e.g., Jordan et al., 1983a, b; Chapter 9). Overthrusting and wrench deformation, similar to processes related to intracontinental wrench basins, are likely.

Chapin and Cather (1981) discussed controls on Eocene sedimentation and basin formation of the Colorado Plateau and Rocky Mountain area. Types of associated uplifts include: 1. Cordilleran thrust-belt uplifts; 2. basement-cored Rocky Mountain uplifts; and 3. monoclinal uplifts of the Colorado Plateau. Resulting basins can be classified into 3 types: 1. Green River types: large equidimensional to elliptical, bounded on three or more sides by uplifts, and commonly containing lakes; 2. Denver type: elongate, open, asymmetric synclinal downwarps with uplift on one side; and 3. Echo Park type: narrow, highly elongate, fault-bounded, with through drainage, and strike-slip origin. Composite uplifts and basins also formed. Green River-type basins have quasiconcentric facies zonation, in contrast to wedge-shaped, unidirectional facies distribution in Denver-type basins. Echo Park-type basins have complex facies, with common sedimentary breccias and sheetwash deposits, associated with active faulting, erosion and stream diversion typical of strike-slip basins. Chapin and Cather (1981) also discussed geomorphic and

climatic effects on basin evolution, and used the occurrence and timing of different basin types to constrain interpretations of the Laramide orogeny. They proposed a two-stage model for basin formation, which can be related to changes in North American plate interactions both in the Atlantic and the Pacific oceanic basins.

Dickinson et al. (1988) independently classified Laramide basins from northern New Mexico to Montana, but came to somewhat different conclusions. Their three comparable, but slightly different basin types are: 1. ponded, 2. perimeter and 3. axial, respectively. They compiled stratigraphic data indicating the following (Dickinson et al., 1988, p. 1023): 1. Laramide structures of varying trend, style and scale reflect heterogeneity of crustal strain caused by shear between the continental lithosphere and an underlying subhorizontal slab of subducted oceanic lithosphere. 2. Maastrichtian initiation of basement deformation was approximately synchronous throughout the Rocky Mountain province, but termination of deformation was systematically diachronous from north to south from early to late Eocene time. 3. Widespread Eocene erosion surfaces truncate syntectonic Laramide sequences and are overlain by largely volcanic and volcaniclastic strata of Eocene age in the north and Oligocene age in the south.

Chapin and Cather's (1981) and Dickinson et al.'s (1988) models for Laramide basin evolution were discussed by Cather and Chapin (1990), Dickinson (1990), Hansen (1990) and Dickinson et al. (1990). The primary disagreements among these workers regard paleodrainage networks, the relative importance of strike-slip deformation along the east side of the Colorado Plateau, and whether the Laramide orogeny occurred in two distinct stages or was a continuum of responses to a generally synchronous strain field. A recently proposed model for Laramide crustal strain and basin evolution (Yin et al., 1992; Yin and Ingersoll, 1993) is consistent with the latter interpretation.

Jordan (Chapter 9) provides a synthesis of the analogous modern retroarc broken foreland of Argentina.

Transform Settings

Strike-Slip Systems

The complexity and variety of sedimentary basins associated with strike-slip faults are almost as great as for all other types of basins. Transform faults in oceanic lithosphere generally behave according to the plate-tectonic model, whereas strike-slip faults in continental lithosphere are extremely complex and difficult to fit into a model involving rigid plates. Simple mechanical models based on homogeneous media have little application to the heterogeneous media of continental crust.

The Reading Cycle (e.g., Reading, 1980) predicts that any strike-slip fault within continental crust is likely to experience alternating periods of extension and compression as slip directions adjust along major crustal faults. Thus, opening and closing of basins along strike-slip faults (the Reading Cycle) is analogous, at smaller scale, to the opening and closing of ocean basins (the Wilson Cycle).

Basins related to strike-slip faults can be classified into end-member types, although most basins are hybrids. Transtensional (including pull-apart) basins form near releasing bends and transpressional basins form at constraining bends (Crowell, 1974b). Basins associated with crustal rotations about vertical axes within the rotating blocks ("transrotational"; Ingersoll, 1988b) may experience any combination of extension, compression and strike slip.

Christie-Blick and Biddle (1985) provided a comprehensive summary of the structural and stratigraphic development of strike-slip basins, based, in large part, on the pioneering work of Crowell (1974a,b). They illustrated the structural complexity likely along strike-slip faults (Fig. 1.27), and implications for associated basins. Primary controls on structural patterns are: 1. degree of convergence and divergence of adjacent blocks; 2. magnitude of displacement; 3. material properties of deformed rocks; and 4. preexisting structures (Christie-Blick and Biddle, 1985, p. 1). Subsidence in sedimentary basins results from crustal attenuation, thermal subsidence during and following extension, flexural loading due to compression, and sedimentary loading. Thermal subsidence is less important than in elongate orthogonal rifts due to lateral heat conduction in narrow pull-apart basins. Distinctive aspects of sedimentary basins associated with strike-slip faults include (Christie-Blick and Biddle, 1985, p. 1): 1. mismatches across basin margins; 2. longitudinal and lateral basin asymmetry; 3. episodic rapid subsidence; 4. abrupt lateral facies changes and local unconformities; and 5. marked contrasts in stratigraphy, facies geometry and unconformities among different basins in the same region.

Nilsen and Sylvester (Chapter 12) provide an additional synthesis of all aspects of strike-slip basins.

Transtensional Basins

Pull-apart basins form at left-stepping sinistral fault junctures and at right-stepping dextral fault junctures (Fig. 12.3a,b). Mann et al. (1983) proposed a model for such basins based on a comparative study of well studied pull-apart basins at various stages of development. Pull-apart basins evolve through the following stages: 1. nucleation of extensional faulting at releasing bends of master faults; 2. formation of spindle-shaped basins defined and commonly bisected by oblique-slip faults; 3. further extension, producing "lazy-S" or "lazy-Z" basins; 4. development into rhombochasms, commonly with two or more sub-circular deeps; and 5. continued extension, resulting in the formation of oceanic crust at short spreading centers offset by long transforms. Basaltic volcanism and intrusion may become important from stages 3 through 5 (e.g., Crowell, 1974b). Most pull-apart basins have low length-to-width ratios, due to their short histories in changing strike-slip regimes (Mann et al., 1983).

Transpressional Basins

Transpressional basins include two types: 1. severely deformed and overthrust margins along sharp restraining bends that result in flexural subsidence due to tectonic load (e.g., south side of modern San Gabriel Mountains, southern California); and 2. fault-wedge basins at gentle restraining bends that result in uplift of one or two margins and downdropping of a basin as one block moves past the restraining bend (e.g., Neogene Ridge basin, southern California) (Crowell, 1974b) (Fig. 12.3d). A basin model for type 1 would involve flexural loading similar to the foreland models discussed above, although at smaller scale.

Ridge basin is one of the most elegantly exposed and carefully studied transpressional basins in the world, as summarized by Crowell and Link (1982). Crowell (1982a) presented a dynamic model for the evolution of Ridge basin (12–5 Ma), a narrow crustal sliver caught between the San Gabriel fault to the southwest, and a set of northwest-trending faults that became active sequentially in a northeast direc-

tion (Figs. 1.28, 12.10 and 12.11). Ridge basin became inactive when motion was transferred completely to the modern San Andreas fault. As a result of movement on the San Gabriel fault, the southwest side of the basin was uplifted and the Violin Breccia was deposited along the basin margin. The depressed floor of the basin moved past this uplifted margin, at the same time it received abundant sediment from the northeast. Previous depocenters moved southeastward past the restraining bend, after receiving sediment in conveyor-belt fashion, with uplift and tilting following deposition. The result is a stratigraphic thickness of over 11 km in outcrop, although vertical thickness of the basin fill is approximately one-third of this. Many extraordinarily thick coarse clastic units in ancient, narrow fault-bounded basins likely were deposited in similar settings.

We reject a newly published model for Ridge basin presented by May et al. (1993) (Fig. 12.11a) on the following basis: 1. Their model is based on a single two-dimensional seismic-reflection line that is oblique to the San Gabriel fault, which they claim to have imaged. As a result, their seismic line shows an apparent dip, less steep than the actual feature. 2. Nowhere does their seismic line cross the San Gabriel fault, which is mapped on the surface as a linear feature with close to vertical dip for tens of kilometers (e.g., Crowell and Link, 1982). 3. The truncation of Miocene dipping strata (shown on their seismic line) is identical to the relationship predicted by Crowell's (1982a) model, namely a buttress unconformity resulting from sequential deposition and rotation of

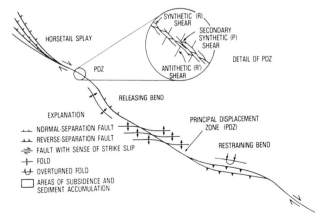

Figure 1.27 Spatial arrangement, in map view, of structures associated with an idealized right-slip fault. From Christie-Blick and Biddle (1985).

strata concurrent with strike-slip motion along the San Gabriel fault. The data presented by May et al. (1993) are consistent with Crowell's (1982a) model, which we, therefore, find more useful than the new model of May et al.

Transrotational Basins

Paleomagnetic data from southern California have documented extensive clockwise rotation of several crustal blocks, beginning in the Miocene and continuing today (Luyendyk et al., 1980; Hornafius et al., 1986) (Figs. 12.3c and 12.15). Luyendyk and Hornafius (1987) further developed their geometric model in order to make testable predictions concerning amount and direction of slip on faults bounding rotated and nonrotated blocks, and areas of gaps (basins) and overlap (overthrusts) among blocks. Within this setting, all of the complexity of transtensional and transpressional basins is likely. The unique aspect of Luyendyk and Hornafius' (1987) geometric model is that it successfully predicts the positions, shapes, areas and ages of most southern California Neogene basins. Similar geometric models may be possible along other complex transform boundaries involving significant crustal rotations.

Crouch and Suppe (1993) proposed that large parts of the southern California and Baja borderland should be included in the late Cenozoic Basin and Range province. They proposed that large-magnitude, core-complex-style extension progressed along this margin from north to south, in response to arrival of the Farallon-Pacific ridge segment at the trench. The Los Angeles basin (Fig. 12.16) - inner borderland is more complex than the southern borderland because it formed in the wake of the rotating western Transverse Ranges (approximately 90 degrees of clockwise rotation) (Luyendyk et al., 1980, 1985; Luyendyk, 1991). The southern borderland basin is floored by highly extended continental crust and by oceanic crust. The Los Angeles basin - inner borderland is floored by the Catalina Schist, interpreted by Crouch and Suppe (1993) as a footwall metamorphic tectonite, tectonically denuded below a detachment. Unroofing is recorded in uppermost Oligocene to mid-Miocene strata, with hangingwall source rocks

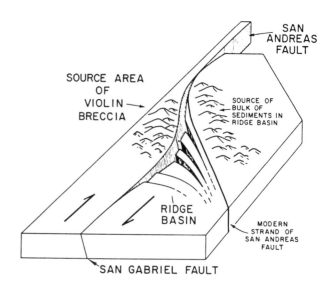

Figure 1.28 Block diagram (looking toward northwest) illustrating origin of Ridge basin at a sigmoidal bend in San Andreas fault. See text for discussion. From Crowell (1982a).

(e.g., Great Valley sequence and its ophiolitic basement) represented at lower stratigraphic levels, and footwall basement (Catalina Schist) represented at higher stratigraphic levels.

A model that successfully explains the extraordinarily complex basins of the Los Angeles area will need to integrate the transrotational model of Luyendyk and Hornafius (1987), the detachment model of Crouch and Suppe (1993), the microplate-capture model of Nicholson et al. (1994) and the detailed history of the Los Angeles basin (e.g., Wright, 1991).

Hybrid Settings

Intracontinental Wrench Basins

The collision of continents of varying shapes and sizes can lead to bewildering complexity in ancient orogenic belts (e.g., Dewey and Burke, 1974; Graham et al., 1975; Molnar and Tapponnier, 1975; Şengör, 1976b; Tapponnier et al., 1982). As Tapponnier et al. (1982) demonstrated through the use of plasticine models, the collision of India and Asia has resulted in major intracontinental strike-slip faults, with associated transtensional, transpressional and transrotational basins, including the formation of impactogens.

Şengör et al. (1978) developed criteria for distinguishing fossil rifts formed during the opening of nearby oceans that are later closed (aulacogens) from intracontinental rifts formed due to crustal collisions (impactogens). Both types of rift valleys trend at high angles to orogenic belts; however, aulacogens have a rifting history coincident with initiation of a neighboring ocean basin prior to collision, whereas impactogens have no precollisional rift history. A third category of rifts at high angles to orogenic belts are rifts unrelated to either ocean formation or collision orogenesis. All of these rift basins are likely to be deformed during suturing in associated orogens. Tests for distinguishing them must come from the stratigraphic record because temporal correlation of initial rifting (or lack thereof) is the primary test for the geodynamic origin for these ancient rifts located at high angles to orogenic belts. One suggestive type of evidence for discrimination is that aulacogens tend to form at reentrants along rifted continental margins (Dewey and Burke, 1974), whereas impactogens are more likely to form opposite coastal promontories, where deformation of colliding continents is more intense (Şengör, 1976b). This criterion must be applied cautiously, however, due to the difficulty of definitively reconstructing precollision geometry (e.g., Thomas, 1983, 1985).

Şengör (Chapter 2) discusses all settings of fossil rifts.

Aulacogens

During continental rifting, three rifts commonly form at approximately 120 degrees, probably because this is a least-work configuration (Burke and Dewey, 1973). Regardless of whether initiating processes are "active" or "passive" (i.e., Şengör and Burke, 1978; Morgan and Baker, 1983b), in the majority of cases, two rift arms proceed through the stages of continental separation, whereas one rift arm tends to fail (Fig. 2.12). Hoffman et al. (1974) discussed the resulting sedimentary accumulations, with emphasis on a Proterozoic example. They outlined 5 stages in the development of the Athapuscow aulacogen, which with slight modification, provide a model applicable to most aulacogens (linear sedimentary troughs at high angles to orogens): 1. rift stage; 2. transitional stage; 3. downwarping stage; 4. reactivation stage; and 5. postorogenic stage.

Although the quasiactualistic model developed by Hoffman et al. (1974) may be applied to several fossil rifts and aulacogens, Hoffman (1987) questioned whether it is the best model for the Athapuscow "aulacogen." Also, Thomas (1983, 1985) discussed the possibility that the Southern Oklahoma "aulacogen" originated as a transform boundary, rather than as a rift. Neither of these reinterpretations discredits Hoffman et al.'s (1974) model; rather, both examples illustrate the difficulty in applying any model to the complexity of the real world.

Successful rifts evolve into shelf-slope-rise margins, with continental embankments at reentrants (e.g., Niger Delta). Failed rifts grade from embankments at their mouths to terrestrial rifts within cratons. As lithospheric extension ceases (for any reason), the rift areas cool and flexurally subside to form intracratonic basins, especially where three failed rifts meet. Upon activation or collision of a continental margin, the rifted-margin sedimentary prisms are intensely deformed, especially at continental promontories (Dewey and Burke, 1974; Graham et al., 1975). As orogeny proceeds, fossil rifts become aulacogens, which may experience compressional, extensional or translational deformation.

Şengör (Chapter 2) provides the most comprehensive review of both aulacogens and impactogens.

Impactogens

Impactogens (Şengör et al., 1978; Chapter 2) resemble aulacogens, but without preorogenic stages. They typically form during the attempted subduction of continental crust (continental collision, either with another continent or a magmatic arc). Two excellent Cenozoic examples, of contrasting style and tectonic setting, are the middle Cenozoic Rhine graben and the late Cenozoic Baikal rift. The Rhine graben formed as a transtensional impactogen that was proximal to the Alpine collision orogen (Şengör, 1976b). It formed on the subducting plate (Europe), in a peripheral-foreland setting. The Baikal rift, which is still active, is also transtensional, but it is distal to the related Himalayan collision. It can also be characterized as an intracontinental wrench basin, which has

formed on the overriding plate (retroarc setting, but far from the pre-collisional arc of the southern Asian margin). Şengör (Chapter 2) discusses these and other examples in detail.

Successor Basins

The original definition of successor basins (King, 1966) as "deeply subsiding troughs with limited volcanism associated with rather narrow uplifts, and overlying deformed and intruded eugeosynclines" (Kay, 1951, p. 107; Eisbacher, 1974) needs modification; "deeply subsiding" and "eugeosynclines" should be replaced by "intermontane" and "terranes," respectively. Within the context of plate tectonics, successor basins form primarily in intermontane settings on top of inactive foldthrust belts, suture belts, transform belts and noncratonal fossil rifts. The presence of successor basins indicates the end of orogenic or taphrogenic activity; therefore, their ages constrain interpretations of timing of suturing, deformation and rifting. Thus, they have special significance in "terrane analysis"; they represent overlap assemblages which provide minimum ages for terrane accretion (Howell et al., 1985). (As mentioned in our discussion of terminology, above, we use "terrane" in the nongeneric sense of any assemblage of rocks.)

Little work has been published on actualistic models for such basins; Eisbacher (1974) summarized models based on work on ancient basins in the Canadian Cordillera. This dearth of work may reflect the diversity of successor basins and their tectonic settings. In a sense, all basins are successor basins because they form following some orogenic or taphrogenic event represented in the basement of the basin. In fact, one of Kay's (1951) examples of epieugeosynclines (successor basins) is the post-Nevadan basin of central California, which is now interpreted as a forearc basin, overprinted in the Tertiary by transform tectonics (Ingersoll, 1982b; Ingersoll and Schweickert, 1986; Graham, 1987; Chapter 6). Modern use of the term "successor basin" should be restricted to post-orogenic and post-taphrogenic basins that do not fall into any other plate-tectonic framework. For example, the southern Basin and

Range Province has been tectonically inactive since the Miocene (Wernicke, 1992). Therefore, modern intermontane basins of this region may be considered successor basins.

Because successor basins develop in the absence of orogenic or taphrogenic activity, the primary mechanisms available to cause subsidence are sediment and water loading. Therefore, one key to their recognition is the absence of any residual (tectonic) subsidence due to either crustal thinning or tectonic loading. The only other basins lacking both crustal-thinning and tectonic-loading subsidence are intraplate settings (Fig. 1.1); successor basins are distinguished from intraplate settings based on their locations at or near plate boundaries.

Discussion

Readers of this chapter, as well as readers of the entire book, undoubtedly will be overwhelmed by the complexity of tectonic processes controlling the evolution of sedimentary basins. The more we know about these processes and their consequences, the more complex become our models, and the more each basin seems unique. This experience is an inevitable consequence of applying the scientific method to the study of complex systems; it is also both exhilarating and frustrating. Exhilaration results from new discoveries of both fact and insight; frustration results from the need to assimilate the overwhelming crush of new information. Fundamental progress results each time new data are used to improve existing models, as well as each time insightful simplifications or generalizations are made. Integration of observation, modeling and experiment is an interative, self-adjusting process.

The ultimate goal of this book is the improvement of paleotectonic and paleogeographic reconstructions through the application of actualistic models for basin evolution. Related features, whose recognition aids paleotectonic reconstruction, include suture belts (e.g., Burke et al., 1977), magmatic arcs (e.g., Şengör et al., 1991a), foldthrust belts (e.g., McClay, 1992) and metamorphic belts (e.g., Miyashiro, 1973). A skilled basin analyst needs to integrate all these topics, as well as geo-

chemistry, geophysics, petrology, paleoecology, and an array of other disciplines. In a complementary manner, workers in these other fields should draw on the insights provided by the sedimentary record to constrain their paleotectonic reconstructions. We hope this book encourages this process of interdisciplinary development and testing of models for Earth's evolution.

Acknowledgments

We thank all of the chapter authors, both for their willingness to participate and for their patience with the extended process of assembling this book. RVI thanks W.R. Dickinson and S.A. Graham for years of rewarding interaction, during which many of these concepts were formulated. Stimulating interactions with A.M.C. Şengör have also been enlightening. CJB thanks R.V. Fisher, R.S. Fiske, B.P. Kokelaar, J.B. Saleeby, and F.B. VanHouten for discussions over the years. Also, we thank students at the University of California (Los Angeles and Santa Barbara) for helping to review the chapters and refine our critical thinking on these diverse basins.

We thank W.R. Dickinson and T.E. Jordan for quickly reviewing a preliminary version of this chapter.

Further Reading

Allen PA, Allen JR, 1990, *Basin analysis: principles and applications:* Blackwell, Boston, 451p.

Bally AW, Snelson S, 1980, *Realms of subsidence:* Canadian Society of Petroleum Geologists Memoir 6, p. 9–75.

Cox A, Hart RB, 1986, *Plate tectonics: how it works*: Blackwell, Palo Alto, 392p.

Dickinson WR, 1974b, *Plate tectonics and sedimentation*: Society of Economic Paleontologists and Mineralogists Special Publication 22, p. 1–27.

Dott RH Jr., 1978, *Tectonics and sedimentation a century later:* Earth-Science Reviews, v. 14, p. 1–34.

Einsele G, 1992, *Sedimentary basins: evolution, facies, and sediment budget*: Springer-Verlag, Berlin, 628p.

Ingersoll RV, 1988b, *Tectonics of sedimentary basins*: Geological Society of America Bulletin, v. 100, p. 1704–1719.

Kay M, 1951, *North American geosynclines*: Geological Society of America Memoir 48, 143p.

Kleinspehn KL, Paola C (eds), 1988, *New perspectives in basin analysis*: Springer-Verlag, New York, 453p.

Miall AD, 1990, *Principles of sedimentary basin analysis (2nd ed.)*: Springer-Verlag, New York, 668p.

Reading HG (ed), 1986, *Sedimentary environments and facies*: Blackwell, Boston, 615p.

Sedimentation and Tectonics of Fossil Rifts 2

A.M.C. Şengör

Introduction

Rifts are fault-bounded elongate troughs, under or near which the entire thickness of the lithosphere has been reduced during their formation (Fig. 2.1; cf., Şengör and Burke, 1978, p. 419). They form in most tectonic settings (Fig. 2.2) and at *all* stages of the Wilson Cycle of ocean opening and closing (Burke, 1978). Because of their morphology, they are commonly convenient sediment receptacles preserving, in various states of completeness, a record of the tectonic environment in which they originate and/or evolve (see Chapter 3). This record may be hidden owing to burial under younger sedimentary (Fig. 2.3A) and/or tectonic units (Fig. 2.3B), but it is rarely completely destroyed. Associated igneous activity is also common, whose products enrich rifts' archive of geological history. Rift-related metamorphism is more modest than that in orogenic belts, the most extreme cases being known from the "metamorphic core complexes" of the southwestern United States (Armstrong, 1982). Owing to their very widespread occurrence, great versatility in terms of the tectonic environment they inhabit, and the geological documentation they preserve, rifts form an ideal object of study for the historian of the earth.

Many rifts do not survive as rifts. Commonly, when the extension factor (ß = extended width/unextended width: cf., McKenzie, 1978a) grows beyond 3 (cf., Le Pichon and Sibuet, 1981), sea-floor spreading starts opening an ocean and destroys the rift. Remnants of most rifts forming continental margins later fall prey to orogeny (i.e., convergent plate phenomena) and become deformed and metamorphosed beyond easy recognition. Some rifts, however, do not become oceans and terminate their tectonic life during the rift stage and get incorporated into the cemetery of fossil structures of our planet as rifts, comparable to an individual who dies at infancy. It is these that are customarily called "failed rifts" in the geological literature, implying that they "failed to generate an ocean." What are generally called "failed rifts" are, in reality, perfectly successful rifts, *as rifts*, but are "failed oceans" and that is why "failed rift" is an inappropriate designation. Thus "fossil rifts" are preferred in the title of this chapter and not "failed rifts".

Before we begin discussing fossil rifts, it is useful to obtain an unambiguous nomenclature and a practical classification of rifts lest we get lost in the richness of the variability of rifts.

Rift, Graben, and Taphrogen

Although "rift" and "graben" commonly are used interchangeably in the geological literature, it is proposed here to confine "graben" to those structures that do not penetrate the lithosphere (i.e., "thin-skinned") and apply rift to those that do (i.e., "thick-skinned"). This proposition is supported by the history of these two terms.

Graben is a German word meaning a ditch or trench. It entered the language of geology early, via mining. In the miners' jargon, "grabens... are depressions or troughs in horizontal beds, which are much longer than they are wide" (Jacobsson, 1781, cited in Rosenfeld and Schickor, 1969). It did not become a common term, however, until Suess' (1883, p. 166) usage for strips of country subsided between two normal faults. The way Suess used it, especially in relation to the East African Rift Valley, "graben" is equivalent to rift.

For Suess' graben, Gregory (1894, p. 295) used the word "rift valley." "Rift" comes from the root "reve," meaning to tear apart or to pull asunder. Thus, whereas the word "graben" is purely descriptive, "rift" involves an interpretation (i.e., extensional rupturing

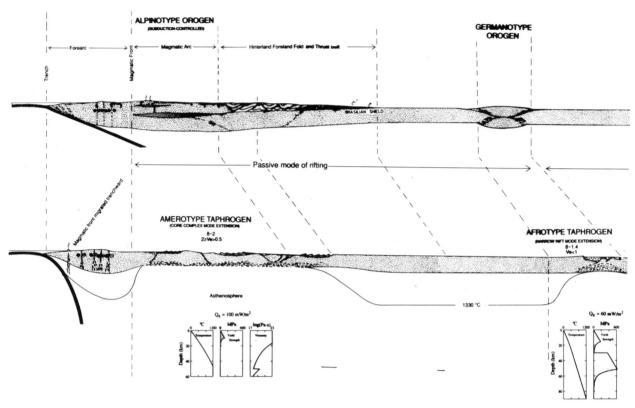

Figure 2.1 Schematic, simplified cross section illustrating concepts of *rift, graben* and *taphrogen*. A hypothetical continent shown in the cross section above, including two *orogenic complexes* (sensu Şengör, 1990a) one of them with both *alpinotype* and *germanotype* parts, located on the margins of a *craton* (cross section at top), is stretched, creating various types of taphrogens including *afrotype* , *amerotype* and *aegeotypes* (cross section at bottom). Notice that quasi-cratonic area, deformed in a germanotype manner during orogeny, leads to an afrotype, largely symmetric rift, immediately under which lithospheric thinning has occurred. By contrast, alpinotype deformed cordilleran-type orogenic area leads to largely asymmetric amerotype rifting, where sub-detachment lithospheric thinning and crustal thinning do not overlie one another at every locality. A similar picture, but this time creating an aegeotype taphrogen, is seen to result from the rifting of a collisional alpinotype orogen (e.g., Aegean Sea rifted atop orogen connecting Hellenides with Turkish ranges). Grabens do not penetrate the lithosphere and form only in areas of superficial deformation, commonly along external margins of taphrogens. For the structural sections, I have consulted papers that reported data on the shallow and deep structure of rifts and grabens, and especially shallow and deep seismic-reflection data. For the amerotype

of a formerly continuous medium). As originally used by the miners, "grabens" referred to smaller and undoubtedly shallower structures than what are called "rifts". It would avoid confusion if one adhered to this distinction in the study of fossil extensional structures. In this chapter, we are concerned almost exclusively with rifts, as defined herein.

Currently, no term is employed to designate *regions of intense extension*, in which many rifts and grabens occur as a result of general lithospheric stretching. For comparable *regions of intense shortening* the term *orogen* has been employed since Kober's (1921, p. 21) suggestion. A corresponding term for zones of intense extension is clearly needed, since some have used the con-

fusing appellation "extending orogens" (e.g., Wernicke, 1981, 1985). *Taphrogen,* derived from Krenkel's (1922, p. 181 and footnote) term *taphrogeny* meaning trough-building (from the Greek τάφροσ = ditch or trench), is suggested to designate the extensional counterparts of orogens. Thus, taphrogens are lithospheric-scale structures, commonly formed from a linked system of rifts and grabens that stretch the lithosphere. Advanced taphrogeny eventually leads to ocean formation (that may be called *thalassogeny,* Kober, 1921, p. 48: from the Greek θαλασσα = sea). If taphrogeny stops before producing ocean (i.e., before leading to thalassogeny), then it leads to subsidence and creates large basins overlying taphrogens (cf., McKenzie, 1978a). In other

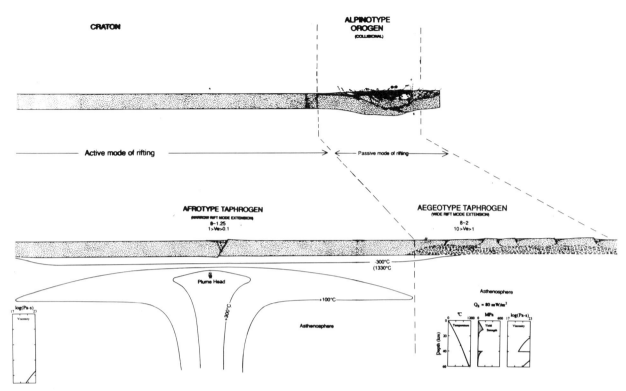

model, I have mainly consulted Allmendinger et al. (1983b, 1986), Miller et al. (1983), Gans et al. (1985), Wernicke (1985), Cheadle et al. (1986) and Hamilton (1987). For the afrotype model, I have used Illies (1970), Rosendahl (1987), and Burgess et al. (1988). For the aegeotype model I utilized Şengör (1982, 1987c) and Eyidoğan and Jackson (1985), plus unpublished seismic-reflection lines in the southern and central Aegean made available to me by courtesy of the late Mr. Ozan Sungurlu of the Turkish Petroleum Co. Owing to the small scale of the diagram such details of normal faulting geometries as those reported by Colletta and Angelier (1982) or Miller et al. (1983) could not be shown.

The three graphs for each type of taphrogen at the bottom of the figure show initial model geotherms, yield strengths (for a strain rate of $8 \times 10^{-15} \ \mathrm{s}^{-1}$), effective viscosities (for dry quartz crust overlying a dry olivine mantle) for different crustal thicknesses (50km for amerotype, 40km for aegeotype and 30km for afrotype) and initial surface heat-flow values (Q_s) (taken from Buck, 1991, fig. 1). The large mantle plume under the right-hand afrotype taphrogen is from White and McKenzie (1989). The width of the magmatic arc in the upper cross section is about 200km; there is no vertical exaggeration in either section.

words, intracontinental taphrogeny leads to *koilogeny* (Spizaharsky and Borovikov, 1966, p. 113ff: from the Greek ϗοῖλος = hollow) (Fig. 2.4).

As we need to recognize in the past not only individual extensional structures, but also *patterns of structures* (i.e., whole taphrogens), in the following section, I present a hierarchical classification of rifts in the framework of taphrogens, which goes from pure geometry to dynamics, to enable us in our study of old rifts to be able to use data not only from individual rift fragments preserved in the record, but also from parts of larger patterns of rifts in relation to one another and/or to other structures such as koilogens, thalassogens and orogens.

Classification of Rifts

The classification of rifts offered here is concerned more with *groups of rifts* (i.e., taphrogens), than with individual rifts. It has three different categories that do not completely overlap, namely *geometric, kinematic* and *dynamic*. In the following, the three different categories are identified with their initials (i.e., *g*, *k*, and *d*, respectively). However, it is helpful to review, in passing, the three main kinds of individual rift types proposed by Buck (1991) that form in response to changes in crustal thickness, heat flow, and extension rate, to show how they relate to the classification here proposed.

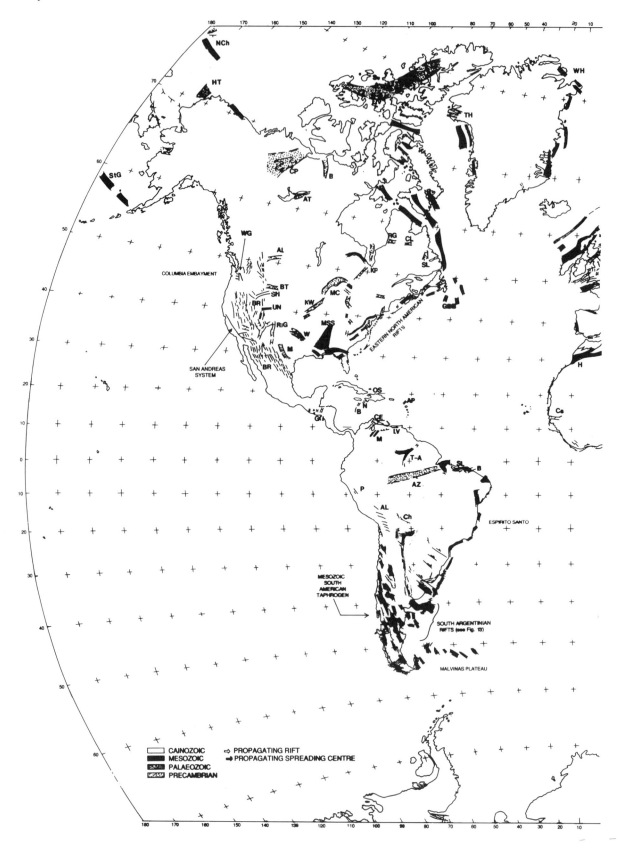

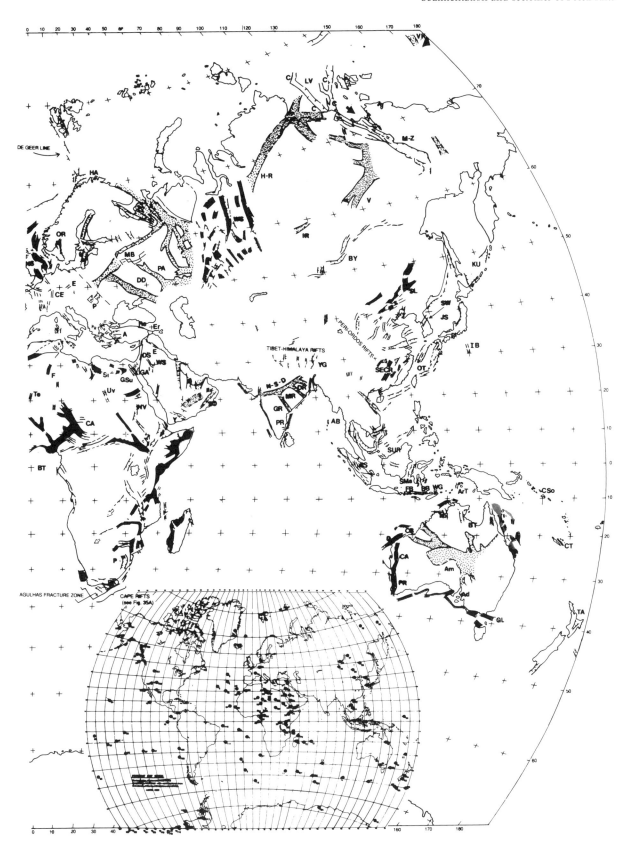

Figure 2.2 World distribution of rifts, showing only those rifts that did not undergo enough later compression and/or strike-slip so as to obliterate their rift signatures. Owing to our heterogeneous knowledge of rifts of the world and to the scale, this map is more representative than exhaustive, and is intended to show that rifts indeed occur in all tectonic settings and that this has been so since the Archean! The map was updated using numerous sources, from a similar one in Burke et al. (1978). A few notable features of the map are the following:

1. Characteristics of taphrogens:

1.1. The map shows that taphrogens come in all scales ranging from singular, insignificant rifts, such as the Tesoffi rift in North Africa (Te) to immense rift complexes, such as the huge Mesozoic taphrogen of Argentina and Chile.

1.2. Taphrogens share plan-view characteristics with orogens. They exhibit curves in trend called *deflections* (e.g., pronounced deflection displayed by the Mid-Continent rift in Proterozoic basement of the United States). They can splay into horsetail-shaped *virgations*, as seen in the case of the East European rifts under the Moscow basin (MB); the virgations may be *one-sided*, as in the case of the Moscow virgation or they may be *two-sided*, as in the case of the smaller Central African rift system (CA). In East Africa, virgations are locally termed *divergences* (e.g., Daly et al., 1989). *Linkages* between two rift systems also occur, as the Kapuskasing rift (KP) meeting the Mid-Continent rift system under the Great Lakes area. Both linkages and virgations give rise to rift stars. *Taphroclines* are sharp bends in rift chains induced by bending after the formation of the rift chain, analogous to oroclines in orogenic belts. The best documented taphrocline is possibly formed by rifts that led to the opening of the Japan Sea, that became bent by nearly 90° as a result of the trap-door opening mechanism of the marginal basin. This involved an along-strike elongation as well, that thus makes the taphrocline really a *taphroclinotath*!

Unlike orogens, taphrogens may acquire roughly *equant* shapes, as exhibited by the West Siberian Rift System (WS) or the North Sea Mesozoic taphrogen (NS), or the Aegean rifts (A). Equant taphrogens also come in various sizes, as these examples show, and they usually result from rift clusters.

1.3. Rift clusters do not commonly seem to lead to ocean generation (i.e., to thalassogeny) because nearly all examples exhibited on this map have remained intact (e.g., West Siberian taphrogen (WS), Basin and Range taphrogen (BR), North Sea taphrogen (NS) or at least the major part of it, and Aegean taphrogen (A). This is true also for some of the better known rift clusters that have later fallen prey to orogeny (e.g., the early Paleozoic rift cluster in southern Tasman orogen in eastern Australia: B.C. Burchfiel, personal communication, 1977; Burke et al., 1978; Veevers, 1984). The immense Mesozoic taphrogen of South America is, in this regard, an exception, but here we may be witnessing the superposition of a rift chain of Late Jurassic - Early Cretaceous age on an older Mesozoic rift cluster!

1.4. More so than in orogens, in taphrogens, spatially and temporally persistent trends are easily discernible. In the Mesozoic Central African rift system, for example, northwesterly trends dominate. A similar trend dominates the large South American taphrogen. In both cases, these trends are a reflection of the grain of the basement fabric. Because rocks are weaker under tension than under compression, basement trends control taphrogeny more easily than they can orogeny.

1.5. Cenozoic collision-related rifts are almost invariably narrower than intra-plate or constructive-margin rifts.

1.6. If this map is compared with one showing age of crust, it will be seen that the younger a piece of crust, the narrower the rifts that form in it. I interpret this in terms of lithospheric thickness. Younger crust indicates thinner lithosphere and the thinner a piece of lithosphere is, the narrower are the rifts that disrupt it.

2. Characteristics of ocean margins related to taphrogens:

2.1. Most ocean margins of Atlantic type today appear to have grown out of Mesozoic taphrogens. In some cases, the parental taphrogen was a rift chain split down its axis (e.g., eastern African margin, central Atlantic margins of North America and Africa). In others, a rift cluster became severed along a line oblique to its dominant rift grain (e.g., in South Africa and in the Falkland spur of South America). This latter is either a result of shear separation (e.g., South Africa/Falkland Spur separation along the Agulhas fracture zone) or the superposition of a rift chain on an older rift cluster (as in the case of Argentina). We may thus distinguish two sub-groups within Atlantic-type continental margins, namely a *Central Atlantic type* (rift chain split down its axis) and a *South Atlantic type* (thalassogeny obliquely cuts older taphrogenic grain).

2.2. A consideration of the geometry of the major taphrogens (e.g., in Eastern Europe or in East Africa) shows how microcontinents might have been produced if all depicted rifts had formed oceans. Rift chains or rift nets commonly lead to roughly equant or at best "diamond-shaped" microcontinents. By contrast, rift clusters create, at the thalassogenic stage, more ribbon continents than anything else (e.g., Baja California).

2.3. Present continental margins of Atlantic type are overwhelmingly of Mesozoic age. Almost no Atlantic-type margin of Paleozoic or older age survives.

Key to abbreviations not mentioned above:

ASIA: AB = Andaman Basin; ArT = Aru trough; BB = Bone basin; BY = Baikal rift; CS = Central Sumatra basin; DR = Damodar rift; Er = Erzincan pull-apart rift; FB = Flores back-arc rift; GR = Godavari rift; H-R = Hantaj-Rbyninsk rift; I-B = Izu-Bonin rifts; Ir = Irkineev rift; KU = Kurile basin; MR = Mahanadi Basin; MW = Gulf of Mannowar rift; M-Z = Moma-Zoryansk rifts; N-S-D = Narmada-Son-Damodar rift lineament; OT = Okinawa Trough; PR = Pondicherry rift; Re = Reşadiye pull-apart rift; SECR = Southeast China rifts; SL = Song Liao rifts; SMa = South Makassar rift; SUR = Sundaland rifts; V = Vilyuy rift; WG = Wetar "graben" rift; Y-G = Yadong-Gulu rift.

EUROPE: CE = Central European "pack-ice-type" rifts and impactogens; DD = Dnyepr-Donetz aulacogen; E = Eger rift; HA = Hammersfest basin; OR = Oslo rift; P = Pannonian basin rifts; PA = Pachelma aulacogen.

AFRICA: BT = Benue Trough; Cs = Casamance rift; H = Haha rift; NV = Nile Valley faults; P = Pongola basin (world's oldest preserved rift: see Burke et al., 1985); Te = Tessofi rift.

ARABIA: (The gently compressed Syrian rifts are not shown) DS = Dead Sea pull-apart rift; E = Euphrates "graben" rift; GA = Gulf of Aqaba rift; GSu = Gulf of Suez rift; RA = Rub-al-Khali rift; SO = South Oman rift, WS= Wadi Sirhan rift, Y = South Yemen rift.

AUSTRALIA, NEW ZEALAND, OCEANIA: Ad = Adelaide aulacogen; Am = Amadeus basin; BT = Batten trough rift; CA = Carnarvon rift; CB = Canning basin rift (with the Fitzroy rift); CSo = Central Solomons trough (Russel Basin); CT = Coriolis trough; DR = Dampier rift; GL = Gippsland basin rift; PR = Perth rift; TA = Taupa rift.

NORTH AMERICA: AL = Alberta; AT = Athapuscow rift; B = Bathurst rift; BT = Belt rift; CL = Cambrian Lake; CP = Coppermine; GBB = Grand Banks basins; GG = Guatemalan rifts; KW = Keweenawan rifts (part of the Mid-Continent rift system); M = Marathon rift; MSS = Mississippi embayment rift (including the Reelfoot rift); RiG = Rio Grande rift; RG= Richmond Gulf rift; SL = Seal Lake; SN = Snake River; UN = Uinta rift; W = Wichita rift.

SOUTH AMERICA: AL = Altiplano-Puna rifts; AZ = Amazon rift; B = Barreirinhas rift; Ch = Chiquitos rift; M = Maracaibo rifts; P = Cordillera Blanca rift; SL = San Luis; T-A = Takutu-Apoteri rift.

Inset: A liberal interpretation of the active hot spots of the world (defined as "non-plate margin-magmatism": see Burke et al., 1978, Fig. 6.2.14), after the unpublished work of W.S.F. Kidd, S. Anderson, K. Burke, C. Chatelain, and C. White (Albany Global Tectonics Group, 1980). Other and somewhat simplified versions of this map were published in Burke et al. (1978, Fig. 6.2.4) and Turcotte and Schubert (1982, fig. 1-49). For a more conservative view of hot-spot distribution, see Crough and Jurdy (1980). Another map published by Vogt (1981) is only a slightly modified version of the Kidd et al. map. I follow here the view of Kidd et al., with the reservations expressed below.

Hot spots on or near divergent plate boundaries:

No.	Name	Coordinates	Plates
1	Jan Mayen	71N, 9W	EUR/NAM
2	Iceland	65N, 16W	EUR/NAM
3	45°	44N, 28W	EUR/NAM
4	Azores	38N, 27W	EUR/AFR/ NAM
5	Colorado Seamount	34N, 38W	AFR/NAM
6	Ascension	8S, 14W	SAM/AFR
7	St. Helena	16S, 6W	AFR/SAM
8	Tristan/ Gough	38S, 11W	AFR/SAM
9	Discovery Seamounts	46S, 8W	AFR/SAM
10	Bouvet	54S, 3E	ANT/AFR/ SAM
11	Prince Edward/ Marion	47S, 38E	ANT/AFR
12	Rodriguez	20S, 62E	AFR/IND
13	St. Paul/New Amsterdam	39S, 78E	ANT/IND
14	Naturaliste **Q**	48S, 105E	ANT/IND
15	Kangaroo **Q**	49S, 135E	ANT/IND
16	Balleny	67S, 163E	ANT/PAC/ IND
17	Easter	27S, 109W	NAZ/PAC
18	Galapagos	0, 91W	NAZ/COC
19	Islas Revillagigedo	19N, 111W	PAC/NAM/ COC
20	Explorer Seamount	49N, 132W	PAC/GOR
21	Bowie Seamount	53N, 136W	PAC/NAM/ GOR
22	Harrat Er-Raha **Q**	27N, 36E	ARA/AFR
23	Aden	14N, 46E	ARA/AFR
24	Danakil	15N, 41E	AFR/ARA
25	Awash	10N, 41E	AFR/ARA

Hot spots on major active taphrogens:

No.	Name	Coordinates	Plates
26	S. Wonji	7N, 38E	AFR
27	Nakuru	0, 36E	AFR
28	Kilimanjaro	3S, 36E	AFR
29	Nyirangongo	1S, 29E	AFR
30	Rungwe	9S, 34E	AFR
31	NE Ordos Plateau **Q**	40N, 114E	CHI
32	Balagan Tas Indigirsky **Q**	67N, 144E	CHI
33	Baikal/ Irkutsk **Q**	52N, 104E	EUR/CHI

Hot spots within plates:

No.	Name	Coordinates	Plates
34	Massif Central	46N, 3E	EUR
35	Vogelsberg **Q**	50.5N, 7.5 E	EUR
36	Ara Hangay		
37	**Q**	48N, 100E	CHI

38	Vitim Plateau	54N, 113E	CHI
39	Nen-Chiang	49N, 125E	CHI
40	**Q**		
	SW Hsing-An	45N, 115E	
41	**Q**		CHI
42	Soeul	39N, 127E	
43	Wonsan **Q**	33N, 126E	CHI
44	Cheju **Q**	21N, 110E	CHI
45	Hainan **Q**	16N, 107E	CHI
46	SE Laos **Q**	13N, 108E	CHI
	SE Vietnam **Q**	1N, 110E	CHI
47	W. Borneo **Q**		CHI
	Central	1N, 110E	
48	Borneo **Q**		CHI
49	Vema	33S, 8E	
50	Seamount	33N, 17W	AFR
51	Madeira	28N, 15W	AFR
52	Canary	31N, 7W	AFR
53	Anti-Atlas	16N, 24W	AFR
54	Cape Verde	15N, 18W	AFR
55	Dakar	18N, 9E	AFR
56	Air	24N, 6E	AFR
57	Ahaggar	32N, 13E	AFR
58	Tripoli	29N, 15E	AFR
59	Jebel Sawda	28N, 17E	AFR
60	Haroudj	23N, 11E	AFR
61	Ih Ezzane	23N, 18E	AFR
62	Tibesti	23N, 25E	AFR
63	Jebel Uweinat	18N, 33E	AFR
64	Bayuda	14N, 25E	AFR
65	Jebel Marra	11N, 12E	AFR
66	Biu	10N, 9E	AFR
67	Jos Plateau	6N, 10E	AFR
68	Adamawa	6N, 13E	AFR
	Ngaoundere	4N, 9E	AFR
69	Cameroon		AFR
70	Sao Tome/	0, 6E	
71	Annobon	12S, 44E	AFR
	Comores	14E, 49E	AFR
72			AFR
		20S, 47E	
73			AFR
		21S, 56E	
74			AFR
75		31N, 29W	
76		22N, 42E	AFR
77		26N, 41E	ARA
78		33N, 37E	ARA
			ARA
79		37N, 39E	ARA
		20S, 145E	
80	Cap dÕAmbre	38S, 143E	IND
	Central Madagascar		IND
81	Reunion/	39S, 155E	
	Mauritius		IND
82	Great Meteor Seamount	36S, 160E	
83	Mecca **Q**	29S, 168E	IND
84	Medina **Q**	46S, 173E	IND
	Damascus **Q**		PAC
85	Karacada◻ **Q**	44S, 174E	
86	W. **Q**ueensland	44S, 176W	PAC
	W. Victoria		PAC
87	Tasmantid Seamounts	52S, 169E	

	Lord Howe Rise		PAC
88	Norfolk	50S, 166E	
	E. Otago **T**		PAC
89	Banks Peninsula **T**	50S, 179E	
90	Chatham Id. **T**	14S, 172 W	PAC
91	Campbell Id. **T**	5N, 164E	PAC
92	Auckland Id. **T**	20N, 155W	PAC
93	Antipodes Id. **T**	29N, 118W	PAC
94	Samoa	22S, 158W	PAC
95	Caroline	18S, 148W	PAC
96	Hawaii	11S, 138W	PAC
	Guadalupe		PAC
97	Rarotonga	29S, 140W	
	Tahiti		PAC
98	Marquesas	27S, 120W	
99	Macdonald Seamount	67N, 165E	PAC
100	Pitcairn/	64N, 170W	NAM
101	Gambier	61N, 165W	NAM
	Anjuisky		NAM
102	St. Lawrence	66N, 160W	
103	Nunivak	58N, 131W	NAM
	Seward Peninsula		NAM
104	Mt. Edziza **T**	44N, 112W	
105	Yellowstone/Snake River **S**	35N, 112W	NAM
106	Flagstaff	36N, 106W	NAM
	Santa Fe **S**		NAM
	Rio Grande/Big Bend **Q**, **S**		
107	Cocos	30N, 105W	
108	Isla San Felix	5N, 87W	NAM
109	Isla Juan Fernandez	26S, 80W	COC
	Fernando de Noronha		NAZ
110	Trinidade/	34S, 81W	
	Martin Vaz		NAZ
111	Peter 1st Id.	4S, 32W	
	Merick Mts.		SAM
112	Alexander	21S, 28W	
113	Hudson/	69S, 91W	SAM
114	Jones Mts.	75S, 72W	ANT
115	Murphy/	72S, 70W	ANT
	Crary Mts		ANT
116	Mt. Siple	74S, 97W	
	Ames/Flood Ranges		ANT
117	Executive Committee Range	77S, 115W	
118	Fosdick Mts.	73S, 126W	ANT
	Mt. Early		ANT
119	Mt. Erebus	76S, 135W	
	Mt. Melbourne/		ANT
	Adare Pen.		
120	Gaussberg	78S, 126W	
121	Heard	77S, 144W	ANT
122	Kerguelen	87S, 153W	ANT
123	Crozet	78S, 167E	ANT
			ANT
124		73S, 170E	
125		67S, 89E	ANT
126		53S, 74E	ANT
127		49S, 70E	ANT
		47S, 52E	ANT
			ANT

Notes after names
Q: **Q**uestionable existence of hot spot; **S**: Associated with small taphrogen; **T**: Secondary extension in wide zone associated with major transform fault.
Plate abbreviations: EUR: Eurasia; CHI: China; AFR: Africa; ARA: Arabia; IND: India; PAC: Pacific; NAM: North America; COC: Cocos; NAZ: Nazca; SAM: South America; ANT: Antarctica.

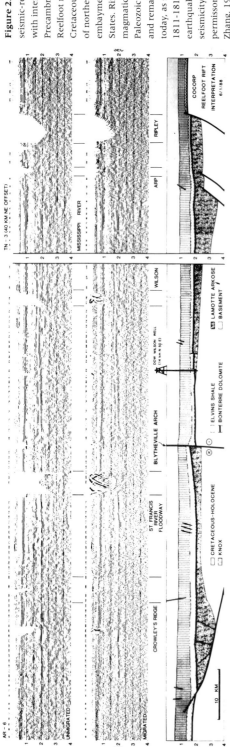

Figure 2.3A Deep seismic-reflection profile with interpretation across late Precambrian(?)/early Paleozoic Reelfoot rift buried beneath Cretaceous to Holocene strata of northern Mississippi embayment in southern United States. Rift was structurally and magmatically reactivated in Late Paleozoic and late Cretaceous, and late Cretaceous and remains seismically active today, as shown by destructive 1811–1812 New Madrid earthquakes and contemporary seismicity. (Reproduced with permisson from Nelson and Zhang, 1991, Fig. 4.)

Figure 2.3B Deep seismic-reflection profile with interpretation across southern Appalachians along Appalachian Ultra-Deep Core Hole (ADCOH), line 3 showing Late Proterozoic/early Paleozoic rifts (labeled as "LPB", i.e., Late Proterozoic basin) now tectonically buried beneath Appalachian allochthons and parautochthon. Abbreviations are BRT: Blue Ridge thrust; HVT: Hayesvile fault; CGMT: Chunky Gal Mountain fault; SFF: Shope Fork fault; SCW: Shooting Creek window; BRA: Blue Ridge allochthon; LPSS: lower Paleozoic shell strata (parautochthon); I(?): intrusion; DS: duplex struccture; MCR: mid-crustal reflections. (Reproduced with permisson from Hubbard et al., 1991, Fig. 4.)

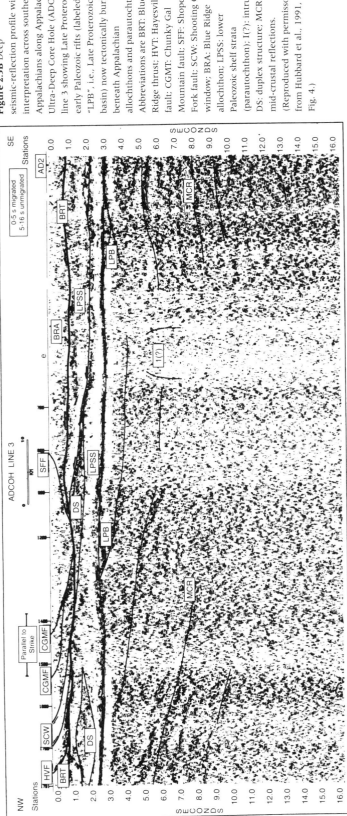

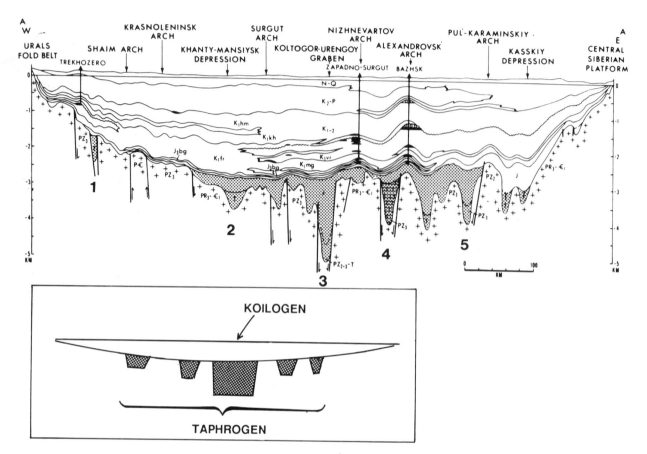

Figure 2.4 E-W schematic cross section across southern part of West Siberian basin, showing middle to late Mesozoic-Cenozoic *koilogen* nested on a late Paleozoic - early Mesozoic *taphrogen*, consisting of several rifts (major ones labeled 1 through 5). Sedimentary rocks filling rift basins (i.e., those deposited during taphrogen stage) are stippled. Location of the section (A-A') is shown in Fig. 2.8B. PC: Precambrian; PR$_3$: Vendian (Neoproterozoic); PZ$_2$: middle Paleozoic; PZ$_3$: upper Paleozoic; C$_1$: lower Carboniferous; T: Triassic (may include some Permian); J: Jurassic;

J$_3$bg: Upper Jurassic Bazhenov and Georgiev formations, forming important source rocks; Kfr: Fedorov Formation; KKl: Kulomzin Formation; Kmg: Megion Formation; Kvr: Nizhnevartov Formation; Kkh: Kharosoimsk Formation; Khm: Khantimansisnsk Formation; K$_1$: Lower Cretaceous; K$_2$: Upper Cretaceous; P: Paleogene; N: Neogene; Q: Quaternary. (Modified with permisson from Meyerhoff, 1982, Fig. 23.) Inset shows the "bovine head" model for the formation of koilogens through taphrogeny.

Structural Classification of Rifts (Fig. 2.1)

When buoyancy forces resulting from crustal-thickness differences created by extension can overcome tensional stresses, rifting loci migrate about a taphrogen; if not, then rifting occurs along a single locus. When, in addition, lower crust can flow faster than extension can augment crustal-thickness differences (see esp. Block and Royden, 1990, and Wernicke, 1991, for lower-crustal flow under extension), extension remains localized in the upper (i.e., non-flowing) crust, but is diffused in the lower (i.e., in the readily flowing) crust. For a discussion on the pos-

sible precise location and nature of these parts, see Wernicke (1991). These differences lead to the following three modes of rifting (Buck, 1991):

1) Briefly, when heat flow is high (about 100 mW/m^2) and the crust thick (about 60 km), fairly rapid extension (about 1 to 2 cm/y) leads to localized upper-crustal extension, compensated at depth by crustal flow distributing thinning in an area larger than the surficial extension. Buck (1991) calls this "core-complex mode." A taphrogen extending in a core-complex mode is herein termed an *amerotype taphrogen* from its type locality

in the western United States (Fig. 2.1: Coney, 1979, 1980a; Armstrong, 1982). Amerotype rifts are poor sediment receptacles, for they do not create large depocenters, associated localized extension being confined to the upper crust, and the local changes in crustal thickness being small (leading to small isostatic subsidence). For the same reason, such taphrogens must lead to very wide, but shallow koilogens.

2) When heat flow is lower (about 80 mW/m^2) and the crust thinner (about 45 km or so), rapid to very rapid stretching (1 to almost 10 cm/y) leads to rifting in a wide area (a few hundreds of km. wide). Buck (1991) calls this "wide-rift mode" extension, which leads to what is here called *aegeotype* taphrogens (Fig. 2.1), after its best example in the Aegean Sea and its surroundings. Aegeotype taphrogens create rifts deeper and wider than the amerotype ones, but still not nearly as deep as the afrotype ones (see below). By contrast, the aegeotype taphrogens are followed by large and deep koilogens, such as the West Siberian Lowlands, the "younger internal basins" of eastern Australia, the North Sea, and the Song Liao basin in northeastern China (Fig. 2.2).

3) Normal-thickness crust (about 30 to 35 km) and heat flow (about 60 mW/m^2) lead to the generation of a narrow locus of rifting, when low extension rates (generally less than 1 cm/y) are applied to the lithosphere (Buck, 1991). I call this type of rifting *afrotype*, after its most spectacular representative in East Africa (Fig. 2.1). Afrotype taphrogens create deep and wide rifts, but small koilogens afterwards.

Geometric Classification of Rifts (Fig. 2.5)

Rifts display five kinds of patterns in map view (Şengör, 1983). From simplest to more complex, these are:

g1) Solitary rifts

Solitary rifts form small, fairly insignificant and rare taphrogens and are extremely difficult to ascertain in the geological record because it is often impossible to tell whether a given rift fragment is isolated or part of a larger area of rifting. The Cambrian Tesoffi Rift (Fig. 2.2; Kampunzu and Popoff, 1991) may be one such solitary rift.

g2) Rift stars

Rift stars (Fig. 2.6) form when more than two rifts radiate away from a common center, building a fairly equant taphrogen. Rift stars are common features of the structural repertoire of our planet today (e.g., the numerous rift stars called triple junctions by Burke and Whiteman, 1973, and Burke and Dewey, 1973; also, see Burke et al., 1972) and they seem to have been so in the past as well (see Burke and Dewey, 1973; Burke, 1976, 1977a; Şengör, 1987a).

g3) Rift chains

When several rifts are aligned end-to-end along linear/arcuate belts of rifting, they form *rift chains*. The East African Rift System constitutes the best known active rift chain in the world (Fig. 2.7A). A similar rift system appears to have been responsible for the opening of the Atlantic Ocean, which constitutes the best-studied fossil example of a rift chain (Fig. 2.7B: Burke, 1976). In general, major rift chains do not fail in toto to form ocean, as the example of the rift chain associated with the Atlantic Ocean shows. An ancient rift chain that did not lead to ocean opening, on the other hand, is represented by the Keweenawan (or the "Midcontinent") rift located mainly in the basement of the U.S. Interior Lowlands (Fig. 2.2), but its length is considerably shorter than either the Atlantic rift chain or the extant East African one (Van Schmus and Hinze, 1985; Gordon and Hempton, 1986).

Solitary rifts, rift stars and rift chains commonly form from afrotype rifts, although aegeotypes are also known to house rift stars (e.g., the rifts possibly generated by migration of the Yellowstone hot spot in the Columbia basalt province plus the Snake River plain: see Davis, 1977, esp. Fig. 2R C–6). A fossil example may be the Jurassic-Cretaceous aulacogen-generating rifts superposed on an earlier Triassic-Jurassic rift cluster in southern South America (Fig. 2.2; for some possible west Siberian cases, see Milanovsky, 1987, esp. Fig. 12).

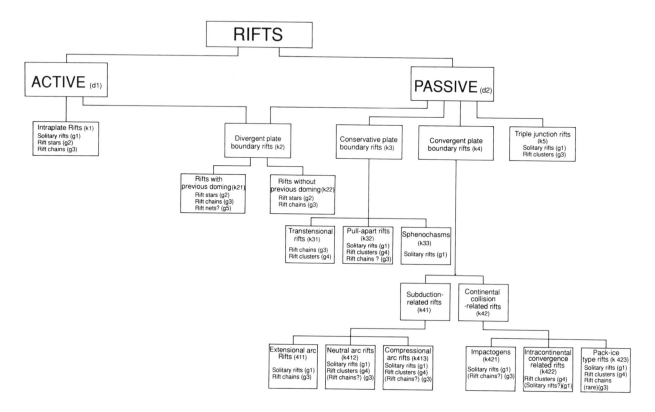

Figure 2.5 Classification of rifts

g4) Rift clusters

When several sub-parallel rifts occur in roughly equant areas, they are said to form a *rift cluster* (Şengör and Burke, 1978, p. 419). The two best-known active rift clusters in the world are the Basin and Range extensional area in the Great Basin (Fig. 2.8A: Gilbert, 1928; Becker, 1934; King, 1977; Stewart, 1978) and the Aegean Sea and surrounding regions ("Aegea" of Le Pichon and Angelier, 1981; McKenzie, 1978b; Şengör, 1978, 1982; Dewey and Şengör, 1979; Şengör et al., 1985; Şengör, 1987c). An equally impressive, but less well-known active rift cluster is the one in Tibet (Fig. 2.2; Tapponnier et al., 1986). The rift cluster underlying the West Siberian Lowlands is the most impressive fossil example (Fig. 2.8B: Rudkevich, 1976; Surkov and Djero, 1981; Meyerhoff, 1982). The rift clusters commonly form from aegeotype rifts, but some amerotype rifts may also lead to rift-cluster formation (e.g., the early Basin and Range).

g5) Rift nets

Rift nets constitute a rare pattern, which comes about when rifts form a roughly checkered pattern, as in the Neoproterozoic basement of the East European platform (Fig. 2.2) or in the late Mesozoic in central North Africa (Fig. 2.2). They resemble chocolate-bar boudinage and may have a similar origin, but more commonly rift nets form in complex and rapidly shifting stress environments in which dominant extension directions change rapidly. Many rift nets, in fact, may represent two superposed rift clusters. All three types of rifts may generate rift nets, but genuine rift nets (i.e., those not formed by the crossing of two rift clusters) are generally formed from afrotype members.

Kinematic Classification of Rifts (Fig. 2.5)

As rifts occur during all stages of the Wilson Cycle, the kinematic characteristics of the plate boundaries may be taken as bases for classifying them according

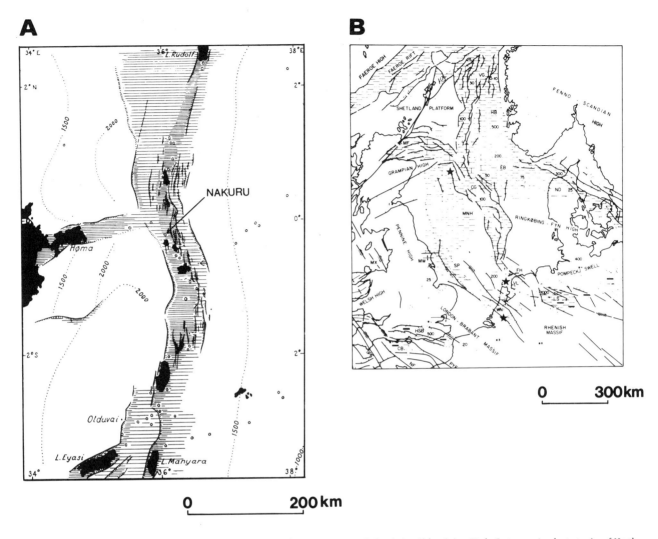

Figure 2.6 Two rift stars that have formed through different mechanisms:

A. Nakuru rrr junction (see also Burke and Whiteman, 1973, p. 743 and Fig. 5), at which two active arms forming Gregory rift, and Kavirondo rift, which ceased activity in Pliocene, meet. This junction formed through doming-rifting mechanism (path d1-k2-k21-g3). (Modified with permission from Shackleton, 1955, Fig. 1.)

B. Berriasian-Valanginian (Early Cretaceous) paleotectonics of North Sea rift star, whose arms were formed from Viking Graben (VG), Central Graben (CG), and Moray Firth rift (MF). This rift star formed mainly in latest Permian - Early Triassic through superposition of a pull-apart system on a previously formed intracontinental-convergence-related rift system (path d2-k4-K42-k422-?g1 superposed by the path d2-k3-k32-g4). (Reproduced with permission from Ziegler, 1990, Fig. 1.13.)

to the environment of the overall displacement and strain field in which they form. There are three types of plate boundaries, plus the plate interiors, with which four types of rift families correspond. In addition, problems of compatibility arise around some unstable triple junctions, commonly owing to involvement of hard-to-subduct buoyant lithosphere. Some of these problems lead to complex rifting that should be treated separately from the other four classes, thus creating a fifth kinematic class, herein called triple-junction rifts.

k1) Intraplate rifts (exclusively afrotype)
Rifts surrounded entirely by undeformed lithosphere occupy this category. Such rifts are usually solitary, small and rare, and are difficult to detect in the geological record. Some active examples are found in the northeastern United States in the Lake George and

A

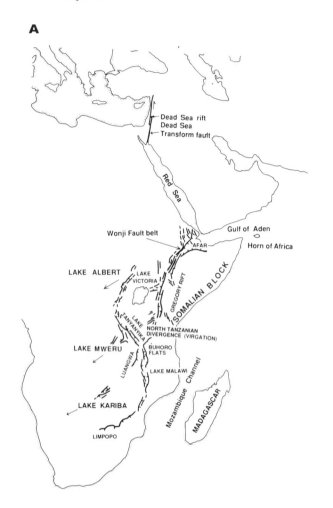

B

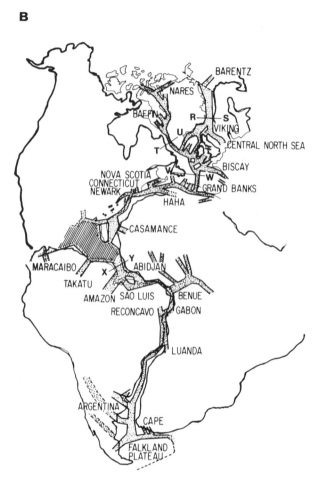

B. Sketch map illustrating dominantly Mesozoic taphrogenic complex in western Pangea, forming a large, diachronous rift chain that eventually led to thalassogeny creating the present Atlantic Ocean. Many small rift basins and horsts could not be shown at this scale. Rifts developed between 210 and 170 Ma in area between lines V-W and X-Y. Oblique shading marks place from which continental material was removed prior to deposition of basal Jurassic salts on young ocean floor of Gulf of Mexico (see Burke et al., 1984, for a review of identity and placement of these continental fragments). Rifts south of line X-Y formed between about 145 and 125 Ma. Rifts north of line T-U formed about 80 Ma, and rifts north of R-S, 80-60 Ma. North of line R-S, Devonian, Permian, Triassic, and Jurassic taphrogenic episodes preceded Cretaceous taphrogeny that finally led to thalassogeny. (Modified from Burke, 1976, Fig.1.)

Figure 2.7   Rift chains:
A. Taphrogen of Great Rift Valley of East Africa as the best example of an active rift chain. Modified from Dixey (1956, Fig. 1) using more recent sources. Note how various branches of the rift chain are connected through rift stars. Compare this map with Fig. 2.11B. Also, compare it with one showing the distribution of sutures in a major collisional orogen (e.g., in the Tethysides: see , Şengör et al., 1988, Fig. 3) in order to have a better idea of the origin of various map views of former continental margins and microcontinents now caught up in an orogenic collage (see also Dewey and Burke, 1974; Scrutton, 1976). Arrows in central Africa indicate directions of rift-tip propagation after Girdler (1991, Fig. 1b).

Lake Champlain rift structures (e.g., Burke, 1977). The Tesoffi rift in North Africa may be one fossil example (Fig. 2.2; Kampunzu and Popoff, 1991).

k2) Rifts associated with divergent plate boundaries (commonly afrotype, but may evolve from amero- or aegeotypes to afrotype just before the onset of thalassogeny)

These rifts form as a direct consequence of plate separation along nascent divergent boundaries. All rifts along the East African Rift system (Figs. 2.2 and 2.7A) belong to this category. In fact, along the Wonji fault belt in the Ethiopian rift segment, sea-floor spreading may have just commenced by axial dike injection (Mohr, 1971, Fig. 6). The latest Paleozoic-early Mesozoic grabens in eastern Canada and the

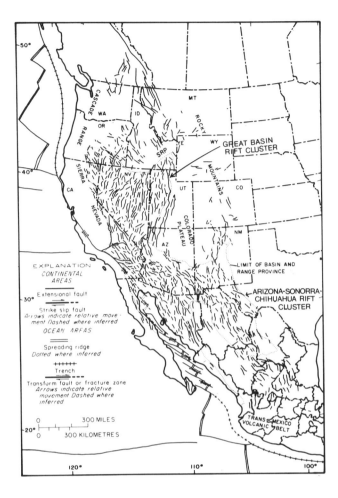

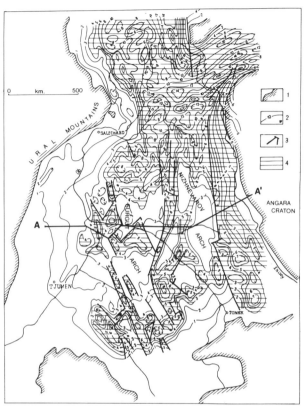

B. Sketch map of major structures of West Siberian basin at the top of the basement (i.e., pre-Triassic). Legend: 1: boundary of basin; 2: iso-baths; 3: faults of the taphrogenic stage (i.e., latest Permian to medial Jurassic, forming a rift cluster); 4: basement areas formed during the middle Paleozoic through (?) Early Triassic (reproduced with permission from Aleinikov et al., 1980, Fig. 7). Note that West Siberian rift cluster is similar in size to Great Basin cluster. For cross section AA', see Fig. 2.4.

Figure 2.8   Rift clusters:
A. Distribution of late Cenozoic extensional faults and a few strike-slip faults in western North America defining Basin and Range province *sensu lato*. The whole area consists of one immense rift cluster, but it is possible to define two smaller clusters, namely the Great Basin cluster (Basin and Range Province *sensu stricto*: Gilbert, 1928; see also, King, 1977, Fig. 100) and the Arizona-Sonora-Chihuahua cluster. Key to abbreviations: WA, Washington; OR, Oregon; CA, California; CO, Colorado; ID, Idaho; MT, Montana; WY, Wyoming; NV, Nevada; UT, Utah; AZ, Arizona; NM, New Mexico; SRP, Snake River Plain; YE, Yellowstone. (Reproduced with permission from Stewart, 1978, Fig. 7.)

eastern United States constitute perhaps the best-known fossil examples of this category of rift structures (Fig. 2.7B) (Benson and Doyle, 1988; Hutchinson and Klitgord, 1988; Manspeizer, 1988; but see also Swanson, 1982). This category of rifts may be further subdivided into two classes as follows:

k21) Rifts that form following an episode of doming. The divergent boundary along which rifts form is, in this case, preceded by an episode of lithospheric doming. The East African Rift Valleys are outstanding extant examples of such a situation (cf., Thiessen et al., 1979). Rifts of Mesozoic age on the Atlantic margins of the Iberian peninsula yield evidence for a comparable situation (Wilson, 1975).

k22) Rifts that form with no pre-rift doming. The late Cenozoic Laptev-Moma-Zoryansk rift system probably represents a recent example (Fujita and Cook, 1990). A good fossil example is the rifting of the Alpine Neo-Tethys in the earlier Mesozoic.

k3) Rifts that form in association with conservative plate boundaries

Conservative boundaries are, by definition, those along which neither extension nor shortening takes place. However, various reasons conspire to induce both extension and shortening along considerable stretches of these boundaries (Christie-Blick and Biddle, 1985; Sylvester, 1988). Rifts along conservative plate boundaries form in three different settings:

k31) Transtensional conservative boundaries. If a conservative boundary is leaking because of a component of extension, it is called transtensional (Harland, 1971, p. 30 and esp. Fig. 2). Many active rifts have a transtensional component (e.g., the Gulf of Suez: Fairhead and Stuart, 1982, Fig. 1; the East African Rift System: Rosendahl, 1987, Fig. 1d) and fossil examples of such rifts may be recognized largely through the structures they contain (e.g., see Chorowicz et al., 1989, for an active example and Beauchamp, 1988, for a fossil one) or from their former bounding-transform-fault endings (e.g., see Tiercelin et al., 1988; Chorowicz, 1989; Daly et al., 1989; Chorowicz and Sorlien, 1992, for active examples; Şengör, 1990b, for a fossil example).

k32) Pull-apart basins along conservative boundaries. Major strike-slip faults, the main structural expression of conservative plate boundaries, commonly have bends along them that either facilitate ("releasing bends": Crowell, 1974b, Fig. 3) or obstruct ("restraining bends": Crowell, 1974b, Fig. 3) movement along the strike of the fault. These bends may be primary, related to the initial nucleation of the fault (e.g., Tchalenko, 1970; Merzer and Freund, 1974) or secondary, formed through structural modifications imposed on a preexisting fault and/or system of faults (e.g., Merzer and Freund, 1975; Mann et al., 1983, esp. Fig. 3; Şengör et al., 1985, esp. Figs. 10–12). In both cases, extensional basins form along the releasing bends, in which the magnitude of extension equals the magnitude of cumulative strike-slip offset along the strike-slip fault since the formation of the releasing bend. Such basins are called "pull-apart basins" after Burchfiel and Stewart's (1966) apposite appellation, but the concept is much older. These basins come in all forms and shapes, notwithstanding the claim by Aydin and Nur

(1982) that they display a constant aspect ratio at all scales. Some, especially the small ones, display a thermal regime colder than other kinds of rifts owing to the proximity of cold bounding-fault walls (e.g., Pitman and Andrews, 1985) that may also contaminate their magmatic signature (e.g., Baş, 1979). One of the sedimentologically best-studied examples is the Dead Sea rift in the Middle East (Fig. 2.2; Manspeizer, 1985; for a recent alternative view, see: Ben-Avraham and Zoback, 1992; Ben-Avraham, in press; Ben-Avraham and Lyakhovsky, in press).

k33) Sphenochasms. Not all basins created by secondary extension associated with strike-slip faults are pull-apart basins. Some represent tears caused by either an asperity or differential drag along the strike-slip fault in one of the fault walls, in which the amount of extension changes from a maximum along the fault to zero at the pole of opening of the tear basin. Carey (1958, p. 193) called such wedge-shaped rifts that open towards a major strike-slip fault *sphenochasms* (from the Greek σφεν = corner , and χαω = to yawn). Highly seismic active examples of sphenochasm generation are seen in the Balkan Peninsula, between Albania and Thrace, as a consequence of the North Anatolian fault's attempt to tear Albania away from the rest of the peninsula (cf., Dewey and Şengör, 1979, esp. Fig. 1A). Another active example is constituted by the grabens around the Ordos block (Fig. 2.2; see Molnar and Tapponnier, 1975; Ma et al., 1982) and the Rio Grande rift in the western United States (Fig. 2.2; e.g., Eaton, 1979, esp. Fig. 7; for a recent overview see Ingersoll et al., 1990; also see Eardley, 1962, Fig. 31.22).

k4) Rifts that form in association with convergent plate boundaries

A large family of rifts forms in association with convergent plate boundaries. In this group, a first-order subdivision is between rifts associated with subduction zones and rifts associated with continental collision, although this may artificially split some genetic groups, such as those rifts that presumably form due to tension generated by excessive crustal thickening (e.g., Dewey, 1988; England and Houseman, 1988).[1] The usefulness of the present grouping is that it

enables a rapid overview of the presently active rift environments and comparison with past ones.

k41) Rifts associated with subduction zones. Three separate environments of rifting associated with subduction zones correspond with three different types of arc behavior, namely, extensional, neutral, and compressional arcs (Dewey, 1980; Şengör, 1990a).[2] In these environments, an enormous variety of rifts forms and many evolve into oceans. In the following discussion, I consider only those that fail to generate oceans and get preserved as fossil rifts.

k411) Rifts associated with extensional arcs. Once an arc begins extending, it generally splits along the magmatic axis (if such an axis is already in existence) and forms a small rift chain (Karig, 1972, 1974; Klaus et al., 1992a) (see Chapters 7 and 8). Such a situation is today known from both the Okinawa rift (Fig. 2.2; Lee et al., 1980; Kimura, 1985; Letouzey and Kimura, 1985; Viallon et al., 1986; Kimura et al., 1988), and the Izu-Bonin arc system (Fig. 2.2; Klaus et al., 1992a), where marginal basins are in the process of rifting. Such rifts generally do not get preserved intact in the geological record, both because of the vicissitudes of the tectonic evolution of arcs involving common changes of behavior (see esp. Dewey, 1980; Şengör, 1990a), and because of later collisions with other arcs or continents. Preservation of rifts associated with extensional arcs in an uncompressed state takes place commonly when the associated arc switches from extensional behavior to neutral behavior.

Extensional arcs may also generate strike-slip faults parallel with the direction of extension, separating arc segments with different rates of extension. In such cases, some rifts in the less extended parts may be left behind and end up trending into fully developed marginal basins. The Neogene to Recent Seoul-Wansan graben in North Korea (Fig. 2.2) may be such a rift related to the opening of the Japan Sea marginal basin (Burke, 1977b).

k412) Rifts associated with neutral arcs. Neutral arcs are defined to have neither shortening nor extension across them. Therefore, the only rifts that form in neutral arcs are those associated with arc-parallel strike-slip faults and may be classified in exactly the same way as the rifts that form along conservative plate boundaries. More complex rift basins may form

along such arc-parallel strike-slip faults, if the sliver plate in the forearc area (Jarrard, 1986b, p. 235; "fore-arc plate" of Woodcock, 1986) disintegrates and its various pieces rotate about vertical axes (e.g., Jesinkey et al., 1987; Beck, 1988).

Pull-apart basins in arcs are difficult to recognize. None of the major active strike-slip faults located in arcs has well-developed pull-apart basins along them (e.g., Median Tectonic Line in Japan, the Atacama fault in the Andes, or the Philippine fault in the Philippine Archipelago), except the Andaman Basin that connects the right-lateral Sagaing and the Semangko faults and that is likely floored by oceanic crust (cf., Hamilton, 1979). Fossil and relatively undeformed examples of such basins have been inferred and mapped, however (e.g., the Chuckanut, Puget, and Swauk basins in Washington, U.S.A.: Fig. 2.2; Johnson, 1985).

Sphenochasms along strike-slip faults in arcs are rarer still. Davis et al. (1978) have discussed two possible examples, the more recent of which may have created the "Columbia Embayment" (Fig. 2.2; Rogers and Novitsky-Evans, 1977a,b) by motion along the Straight Creek fault in the latest Cretaceous and the earliest Cenozoic.

K413) Rifts associated with compressional arcs. In compressional arcs, crust commonly thickens and lithosphere thins, both by heating and by eventual delamination (Isacks, 1988; Şengör, 1990a). The arc becomes shortened across, and elongated along, its trend (see esp. Dewey and Lamb, 1991). This elongation commonly generates rifts at high angles to the trend of the arc (e.g., in the Altiplano: Mercier, 1981; Mercier et al., 1991; in Crete: Le Pichon and Angelier, 1979).

In the Andes of Peru, however, Dalmayrac and Molnar (1981) have observed normal faulting parallel with the trend of the arc and ascribed this to gravitational stresses set up by the potential energy of the thickened crust. Later, Jordan et al. (1983a,b) indicated that this phenomenon was more widespread in the Andes than documented by Dalmayrac and Molnar (1981). Coney and Harms (1984) ascribed the onset of Basin and Range extension in the Cenozoic to such gravitational stresses generated by a thickened crust in an Andean setting over a thinned lithosphere. The extensional structures described by Dalmayrac

and Molnar (1981) seem to have at least an order of magnitude less extension on them than the rifts of the Basin and Range province and may be related to left-lateral strike-slip faulting along the Peruvian Andes (cf., Dewey and Lamb, 1991, Fig. 9 and Mercier et al., 1991, Fig. 3) rather than to gravity spreading.

k42) Rifts associated with zones of continental collision. Three different environments of rifting are associated with the collision of continents: 1)*Lines of extension* that radiate from points at which collision commences; 2) *regions of extension* abutting against sutures, and 3) *nodes of extension* in areas of complex deformation in fore- and hinterlands shattered by collisions. Impactogens (k421), rifts forming along intracontinental convergence belts (k422), and pack-ice-type rifts (k423) correspond with these three environments, respectively.

k421) Impactogens. Impactogens are rifts that originate as a result of tensional stresses set up in a continent when it is hit by a pointed promontory of another continent (Şengör, 1976b, 1987b; Şengör et al., 1978). The best example today is the Upper Rhine graben between Germany and France (Fig. 2.9), which formed in the medial Eocene upon collision in the Alps (Şengör, 1976a,b, 1987b; Şengör et al., 1978). Impactogens are commonly solitary rifts, but several impactogens may form along a long front of collision, if more than one promontory collide with the opposing continent (e.g., Oslo Graben and the Viking/Central Graben in the North Sea along the Variscan collision front in northern Europe (Şengör, 1976b; see Fig. 2.6B).

k422) Rifts forming along intracontinental convergence belts. These rifts are similar in principle to those described under k413 (rifts associated with compressional arcs) and indicate the elongation of the orogen along its trend during post-collisional convergence. The north-south grabens in southern and central Tibet (Fig. 2.2), which formed as a consequence of the shortening and east-west elongation of the Tibetan high plateau following collision along the Indus-Yarlung suture represent the best active examples of these (Fig. 2.18; also see Tapponnier et al., 1986, Figs. 1 and 2).

k423) Pack-ice-type rift basins. When a continental collision generates first impactogens and then rifts

related to ongoing intracontinental convergence, along with conjugate strike-slip faults that help the sideways elongation of the shortening region along the orogen, the whole deformed area becomes divided into rigid and semi-rigid blocks, in central Europe termed *Schollen* (see Dewey and Şengör, 1979, footnote 1), that move with respect to one another along compressional, extensional, and strike-slip boundaries similar to drifting pack-ice. In such a setting, rifts and grabens form in diverse shapes and orientations, as best exemplified today by the *Schollen* regime of Central Europe (Fig. 2.10; e.g., Lotze, 1937, 1971).

k5) Triple-Junction Rifts

Triple-junction rifts form at or near unstable triple junctions, at which plate evolution dictates the generation of "holes" owing to failure to create subduction zones along a plate boundary, commonly because one or more plates meeting at the triple junction consist of buoyant lithosphere. Şengör (1979a) pointed out that the FFT junction at the meeting point of the North and the East Anatolian faults at Karlıova in eastern Turkey had to generate the complex Karlıova extensional basin because the T (i.e., "trench") arm of the triple junction had to function as a broad zone of convergent high strain instead of a line of surface consumption. Şengör et al. (1980, 1985) presented another example at the Bitlis suture/East Anatolian fault/Dead Sea fault junction in southeastern Turkey (the former Kahramanmaraş triple junction), which led to the formation of the Hatay rift and the Adana basin in the later Miocene following the replacement of a trench with a suture zone along the Bitlis thrust belt. Ingersoll (1982a) interpreted the complex extensional tectonics of the western United States and Mexico, involving the Basin and Range province, Rio Grande rift, and the Gulf of California transtensional system, in terms of the non-rigid response of the North American plate to the instability of the Mendocino triple junction.

Dynamic (Genetic) Classification of Rifts (Fig. 2.5)

Rifts also may be classified according to the origin of forces that lead to rifting. Şengör and Burke (1978) proposed that stresses that cause rifting may be

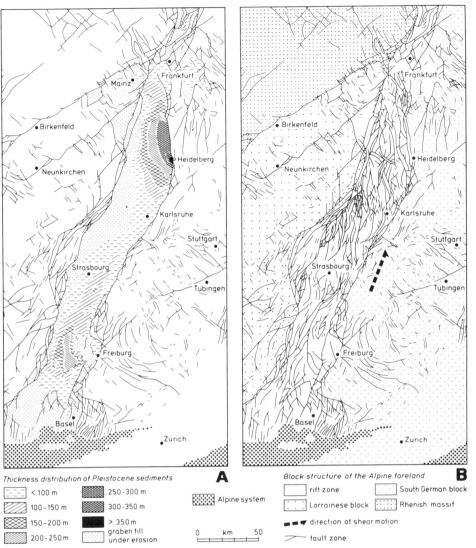

Figure 2.9 The type impactogen, the Upper Rhine rift.
A. Distribution of Pleistocene sediments outline modern Upper Rhine rift, and their thicknesses give an idea of activity of rift during Quaternary. (Reproduced with permission from Illies and Greiner, 1978, Fig. 7.)
B. Modern Rhine rift is functioning as a sinistral strike-slip zone with local shortening in its middle segment, and extension in south and in north. This map illustrates distribution of Quaternary faults. (Reproduced with permission from Illies and Greiner, 1978, Fig. 7.)

imposed on the lithosphere directly by the mantle beneath it or they may result from two-dimensional plate evolution. Accordingly, they termed these two modes of rifting *active* and *passive*. Although this subdivision has since been criticized by some (e.g., White and McKenzie, 1989), the criticism neglects the time dimension and thus is void. Moreover, this classification has proved very useful in understanding the underlying mechanisms of the multifarious styles and settings of rifting (e.g., see the contributions in Morgan and Baker, 1983a; also see Kazmin, 1987). In regional-geological (e.g., Ingersoll et al., 1990) and tectonic-modeling (e.g., Ingersoll, 1982a) studies, it is

also helpful. I shall accordingly use it here also. The proposal to call passive rifting "closed-system rifting" and active rifting "open-system rifting" by Gans (1987) is misleading because it is not necessarily obvious with respect to which parameter the system is considered open or closed (crustal addition, geochemical reservoir tapped, plate boundary network, or original sedimentary provenance). Consequently, it is avoided in the following paragraphs.

d1) Active rifting

"Active rifting" is rifting caused by mantle upwelling associated with hot spots in the mantle (Burke and

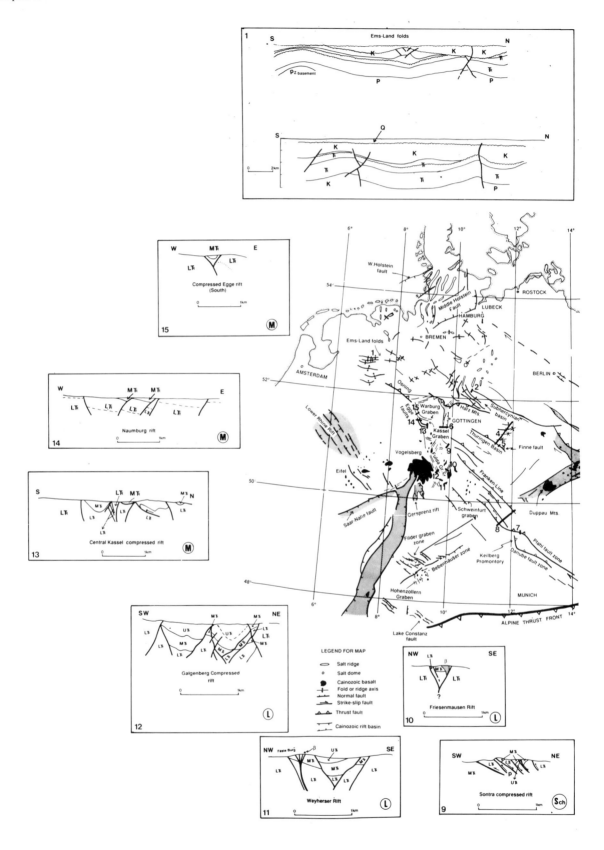

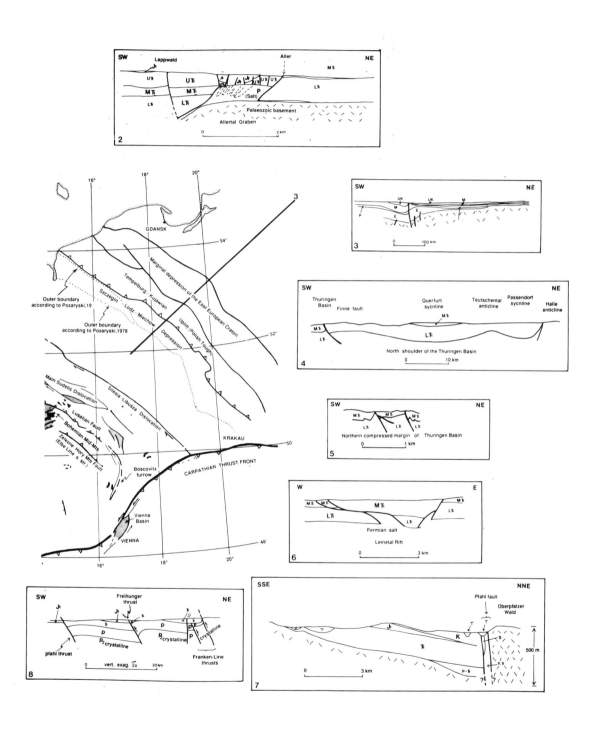

Figure 2.10 Cenozoic pack-ice-type rifts and related structures in central Europe. Stippled areas are Eocene to Recent rifts. All structures shown were active at some time during the Cenozoic, although their ancestry is variable. As a rule, most NW-SE-striking fault systems originated during the late Paleozoic, probably as right-lateral strike-slip faults (see Arthaud and Matte, 1977) and were reactivated in the Mesozoic, mainly as right-lateral strike-slip faults, until they all became transpressional structures in Late Cretaceous to the present, nucleating many pull-apart basins (e.g., those along the Main Sudetic dislocation: see Illies and Greiner, 1976). The NNE-trending basins mostly formed in Late Jurassic (in northern Germany) and in early Cenozoic (in south Germany and France) as extensional structures; and most remained extensional to the present (with exceptions, such as the compressed Egge rift).

Many features played more than one, not uncommonly contradictory role during their history, even within the Cenozoic, as exemplified most dramatically by Miocene and younger thrusting along the central part of the eastern master fault of the Upper Rhine rift. Alpine shortening has further accentuated the disjointed nature of the foreland and also led to alkalic mafic volcanicity.

The map was compiled from numerous sources, but notably from: Carlé (1950), Knetsch (1963), Svoboda (1966), Lotze (1971), Illies and Greiner (1976), Pozaryski (1977), and Meiburg (1982). Cross sections were taken from Schröder (Sch: 1925: Sontra rift); Lemcke (L: 1937: Friesenmausen, Galgenberg, and Weyherser rifts); Martini (M: 1937: Kassel, Naumburg, and Egge rifts); (Lotze, 1971: all the rest).

Dewey, 1973; Burke and Whiteman, 1973; Dewey and Burke, 1974; Thiessen et al, 1979; Morgan, 1981; White and McKenzie, 1989; Hill et al., 1992; also, see Sleep, 1987, and Davies and Richards, 1992). In such environments, rifting is thought to result from the tension created by the extrados stretching caused by doming (Cloos, 1939). In fact, Cloos (1939, p. 428–435) has shown that in a dome of 600 km diameter uplifting a brittle layer of about 12 km thickness to some 4 km, a stretching of some 4 km may be obtained. His experiments with wet clay cakes have further shown that, of this stretching, only 1/4 or even 1/5 was expressed in the faults that formed. His conclusion was that doming was enough to *initiate* stretching and graben building.

Studies since Cloos (1939) have shown that although doming is sufficient to *initiate normal faulting and graben building*, it is not sufficient to *maintain rifting* and to create anything like our present rift valleys in Africa or the Rhine Graben. Indeed, Illies (1967) found, during a palinspastic restoration of the Rhine Graben extension, that a minimum of 4.8 km of extension remained unaccounted for after he had eliminated all effects of the doming. This figure, Illies (1967) ascribed to actual lithospheric extension.

Two views have been advanced to explain the origin of the extension not related to extrados extension of domes rising above hot-spot jets. One ascribes the rifting to basal shear stresses induced by a spreading plume head beneath a dome (e.g., Sorochtin, 1974; Neugebauer, 1983). The other holds the potential energy of the rising dome responsible for driving the rifting (see Richter and McKenzie, 1978, section 2; White and McKenzie, 1989). All of these factors probably contribute to maintaining the active rifting

process at its habitually slow pace of less than 1 cm/y. Left to their own means, most active rifts quit eventually before leading to ocean generation (i.e., to thalassogeny). For a rift to grow into an ocean, extension along it must become compatible with, and required by, the evolving world-wide plate mosaic, whence extension rate grows above 1 cm/y. In the active mode of rifting, this generally happens when several hot spots are aligned along a line favorable to the generation of a future spreading center as Dewey and Burke (1974) and Crough (1983) have pointed out.

d2) Passive rifting

"Passive rifting" refers to a mode of rifting in which the mantle under the rifting area plays only a passive role. In the passive-rifting mode, extension is caused by two-dimensional motions of lithospheric plates. In this mode of rifting, there is no pre-rift doming related to a hot spot (Şengör and Burke, 1978). Kinematic mechanisms reviewed above under the headings k22, k31, k32, k33, k411, k412, k413, k421, k422, k423 and k5 all may form rifts in a "passive-rifting mode."

In the following section, I review the tectonic-sedimentation interplay in three types of fossil rifts only, using both fossil and modern examples of their particular classes. These are: 1) aulacogens, 2) collision rifts, and 3) intracontinental wrench basins. These three types of rifts constitute by far the commonest members of the terrestrial repertoire of fossil rifts, apart from the ones preserved along the Atlantic-type continental margins. This review uses the classification offered in this section as a framework because aulacogens, collision-related rifts, and intracontinental wrench-related extensional basins have numerous kinds and evolve through many

stages, illustrating the categories of rifts discussed above. Figure 2.5 can be used as a "flow chart" to follow the evolutionary histories of the various kinds of rift basins reviewed in the following sections.

Aulacogens

Brief History of the Term

The term *aulacogen* was introduced into the tectonic terminology in an originally unpublished lecture by Shatsky, delivered on May 21, 1960 (Shatsky, 1964).

In Shatsky's usage, the term *aulacogen* was intended to designate narrow, elongate, and fairly straight depressions striking into cratons, commonly from reentrants facing an adjoining larger basin or a mountain belt. Shatsky had recognized these structures as a distinct group of cratonic structures in the 1940s and called them "transverse basins and transverse fractures," considering them a subset of his "marginal transverse structures of old platforms" (Shatsky, 1946, p. 57; Schatski, 1961, p. 99–100). Shatsky noted that they appeared genetically related to the larger basin into which they opened and that they internally segmented cratons. Because Shatsky came to recognize the aulacogens in the course of his studies on the Riphean evolution of the East European platform (i.e., shortly after the craton was consolidated), he considered them among the most important structural elements of the *early stages of the development of the craton as craton*. He noted that they localized zones of maximum thickness of the Riphean and the lower Paleozoic sedimentary sections and volcanic rocks. Shatsky observed that the lower parts of the aulacogen fills corresponded with the weakly metamorphosed and "peculiarly folded" equivalents of the adjacent miogeosynclinal rocks!

Shatsky's recognition of aulacogens was not confined to the East European platform (cf., Shatsky, 1964, p. 544). He compared the Great Donbass basin with the Wichita basin in the United States (Shatsky, 1946; Schatski, 1961, p. 81–119) and considered aulacogens common structures on cratons.

Shatsky was a convinced fixist and interpreted the origin of aulacogens in the spirit of fixism. He believed that the cause of the subsidence that generated both the aulacogen and the larger basin into which it opened was a density increase in the lower crust and/or in the upper mantle. He held a pre-existing planetary network of fractures responsible for the geometry of the aulacogens. Owing to this predetermined location and orientation, aulacogens were subjected to repeated rejuvenations (what Shatsky called "posthumous movements"), which in some cases led to a weak folding and some metamorphism of the sedimentary fill of the aulacogen.

Following Shatsky's initial recognition, aulacogens were studied extensively and classified according to various criteria: Bogdanov (1962) and Bogdanov et al. (1963), for example, distinguished "early" and "late" aulacogens, the former corresponding with those that form early in the consolidation of a craton (e.g., the Pachelma aulacogen in the East European platform: Fig. 2.2; see Klubov and Klevtsova, 1981) and the latter with those that form substantially later than the formation of the craton (e.g., the Donetz basin in the East European platform: Fig. 2.2; see Tchirvinskaya, 1981). Bogdanov (1962) also classified aulacogens according to their positions in a craton: "intra-platformal, transverse, and longitudinal!" Later, Novikova (1964) emphasized that the aulacogens of the East European platform predated its syneclises.[3] Even some orogenic belts came to be interpreted as "aulacogens of an extreme case" (e.g., the Pyrenees and the Greater Caucasus: von Gaertner, 1969)! Despite the flurry of activity on aulacogens after Shatsky's death (for recent Soviet reviews of aulacogens, see Milanovsky, 1981, 1983, 1987; Khain and Michailov, 1985, esp. p. 150ff; p. 144–147 of the German translation), not much progress was made in understanding the origin of these structures until the rise of plate tectonics.

Origin of Aulacogens

Shatsky's explanation for the origin of the aulacogens was not only ad hoc, but also an unfalsifiable existential statement, which probably explains why not much progress occurred in understanding the origin of these structures after his death. By contrast, aulacogens have been interpreted in the framework of the

theory of plate tectonics in terms of testable hypotheses with considerable success (for general reviews, see Burke, 1976, 1977a, 1980; Milanovsky, 1981; Mohr, 1982; Şengör, 1987a).

No apology is needed in returning to Shatsky's original definition of aulacogens, not only because he invented the term, but also because his definition is so clear and useful. The following is a summary, compiled from several of his publications (esp., Shatsky, 1964): An aulacogen is a narrow, elongate and fairly straight depression trending into a craton commonly from a reentrant adjoining a major basin.[4] An aulacogen commonly contains a sedimentary section at least three times as thick (commonly 10–12 km: Khain and Michailov, 1985) as the section found on the surrounding craton (commonly between 1 and 5 km: Khain and Michailov, 1985) into which it trends and which it segments. This sedimentary section generally correlates with the even thicker sedimentary content of the larger basin toward which the aulacogen opens. Aulacogens may be simple or complex. Complex aulacogens contain several troughs separated by internal uplifts that may be mini-foldbelts.

Figure 2.11 illustrates and summarizes the most significant and the best-documented processes that are believed to lead to the formation of aulacogens. In the following discussion of these mechanisms, I illustrate the common sedimentary environments that form in aulacogens and their products by way of the specific examples I present.

1. Doming-rifting-drifting hypothesis (path d1–k2–k21–g3 in Figure 2.5: products commonly afrotype).

Figures 2.11A, C, D, and 2.12 illustrate the "doming-rifting-drifting" hypothesis of the origin of aulacogens. In the framework of the classification given above, this hypothesis (Burke and Wilson, 1972; Burke and Dewey, 1973; Burke and Whiteman, 1973) interprets the origin of aulacogens as rifts that form through the active participation of a mantle plume, creating a dome with common diameters ranging from hundreds of kilometers (Burke and Whiteman, 1973) to about 2000 km (White and McKenzie, 1989; for this size range, see also Kazmin, 1987). Thiessen et al. (1979) found that in the late Cenozoic history of Africa, hot-spot separations vary between 2 to 15

degrees (i.e., about 220 to 1650 km) with uplift diameters ranging between similar magnitudes. These observations agree well with the cell sizes obtained during convection experiments undertaken by Richter and Parsons (1975), assuming a depth of upper-mantle convection of about 700 km (for numerical modeling reaching similar conclusions, see White and McKenzie, 1989). (For the hypothesis of hot spots generated by whole-mantle convection, see Morgan, 1981 and Davies and Richards, 1992.) According to the doming-rifting-drifting model, taphrogeny and the consequent thalassogeny (or koilogeny) is preceded by the formation of several such domes rising to structural mean heights ranging from 1 to above 2 km (see Cloos, 1939; White and McKenzie, 1989), that may be expressed at the surface as topographic domes rising over 2 km, with peaks surpassing 4 km, as in Ethiopia.

It may be possible to document and date such uplifts in the geological record, long after their topographic expression has vanished, by means of the peripheral clastic wedges they shed. For example, the Miocene clastics of western Kenya and Uganda were produced by accelerated erosion during arching over the sites of the future western and eastern rifts and were later preserved beneath the younger volcanics (e.g., King, 1970). Figure 2.13 documents a similar situation from Patagonia, where Late Triassic to Early Cretaceous rifting of the Salado, Colorado, San Jorge, and other smaller rifts, which later led to opening of the South Atlantic ocean (cf., Burke, 1976), was preceded by uplift and erosion. Depositional products are today seen preserved in such fluviatile clastics as those of the Tralcan and those correlative with the upper part of the Puesto Piris porphyries, all of which are now blanketed by widespread Jurassic intermediate volcanic rocks in the region. Renewed doming was recorded by the Los Adobes, Las Heras, Cerro Barcino and La Rueda clastics partly on the Somoncura Massif, which are, in part, correlative with the rift fill (e.g., Colorado interior: Fig. 2.13C: see esp. Uliana and Biddle, 1987; Uliana et al., 1989). Such peripheral clastic wedges shed by plume-generated uplifts have been recognized even in cases where they were later involved in major orogeny! For instance, the pre-

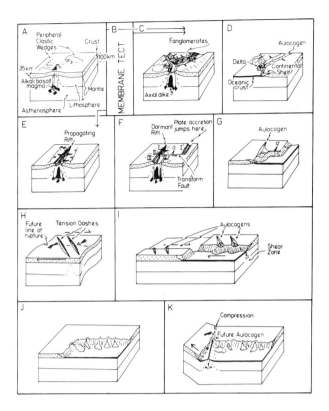

Figure 2.11 Schematic block diagrams illustrating most significant and best documented plate-tectonic processes that lead to formation of aulacogens (reproduced with permission from Şengör, 1987a, fig. 1). **A-C-D**: Doming-rifting-drifting hypothesis (see also fig. 2.13); **B-C-D**: Hypothesis of aulacogen formation through "membrane stresses" (see also fig. 2.16); **E-F-G**: Rift-tip-abandonment hypothesis; **H-I**: Strike-slip-related secondary-extension hypothesis; **J-K**: Continental-rotation hypothesis. See text for discussion and other hypotheses.

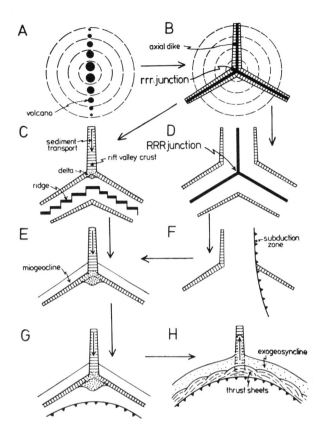

Figure 2.12 Schematic origin and evolution of mantle-plume-generated triple junctions, leading to doming-rifting-drifting sequence creating oceans and aulacogens (from Burke and Dewey, 1973, Fig. 2, reproduced with permission). **A.** Uplift develops over plume with crestal alkalic volcanoes; **B.** Three rift valleys develop at an rrr (rift-rift-rift) junction (e.g., Nakuru: see Fig. 2.6A); **C.** Two rift arms develop into a single plate margin (ridge) and continental separation ensues, leaving third rift arm as an aulacogen, down which a major river may flow and at the mouth of which a major delta may develop (e.g., Limpopo: see Fig. 2.2); **D.** Three rift arms develop into spreading centers meeting at an RRR (ridge-ridge-ridge) junction (e.g., Red Sea-Gulf of Aden-Wonji fault belt in Ethiopian rift system: Mohr, 1971; also see Burke and Dewey, 1973, Fig. 3A); **E.** Atlantic-type continental margin evolves with growth of delta at mouth of aulacogen and continental shelves (e.g., Mississippi: see Fig. 2.2); **F.** One arm of RRR system begins to close by subduction. If ocean is sufficiently wide (>1000 km), a magmatic arc will develop along its margin and any sediments in closing arm will be deformed (e.g., Lower Benue Trough); **G.** Atlantic-type continental margin with continental shelves and aulacogens approaches a subduction zone (e.g., present Sirte rift approaching the Hellenic subduction zone); **H.** Continental margin collides with subduction zone, collisional orogeny ensues, sediment transport in the aulacogen reverses polarity, and aulacogen may become tectonically rejuvenated (e.g., Athapuscow: Fig. 2.2)

Varanger shallow-marine sandstones and quartzites (late Riphean in age: 800–750 Ma) of the "Tosasf-jaellet basin" (now found in the Saerv Nappe of the Middle Allochthon of the Scandinavian Caledonides) were interpreted to have been shed from a plume-generated uplift, whose rifting later led to the birth of the Iapetus ocean (Kumpulainen and Nystuen, 1985, esp. Fig. 4).

Such plume-generated domes are, in places, sites of alkaline to peralkaline magmatism (see Thiessen et al., 1979, Fig. 3 for the distinction between high spots (i.e., those uplifts without magmatism) and hot spots (i.e., those with magmatism). The volcanic products of this magmatism preceding rifting are commonly preserved, as are the clastic sediments shed from the uplifts, usually in the form of volcaniclastics, more

rarely rhyolite flows or welded tuff (e.g., in Patagonia; see Uliana and Biddle, 1987; Fig. 2.13). In former domes, the volcanics crowning the dome are now

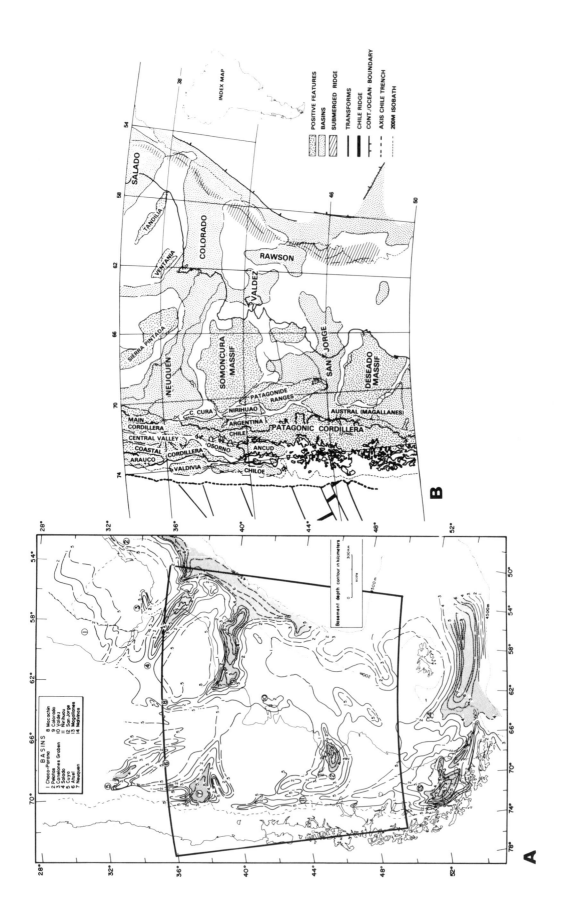

Figure 2.13 Mainly Early Cretaceous aulacogens of Argentine Atlantic continental margin (for a graphic history of their development see Uliana et al., 1989, Figs. 2,6,8,10, and 12).

A. "Technical" basement topography in Argentina, obtained by mapping all seismic velocities higher than 4.5 km/sec. Regions at depths greater than 4 km are stippled, and among them two prominent aulacogens, Salado and Colorado stand out (in Fig. 2.13B, they lie on both sides of the label "Argentina"). (Reproduced with permission from Urien and Zambrano, 1973, Fig. 4.)

B. Location map showing geological provinces, whose Mesozoic stratigraphy is summarized in fig. 2.13C. (Reproduced with permission from Uliana and Biddle, 1987, fig. 1.)

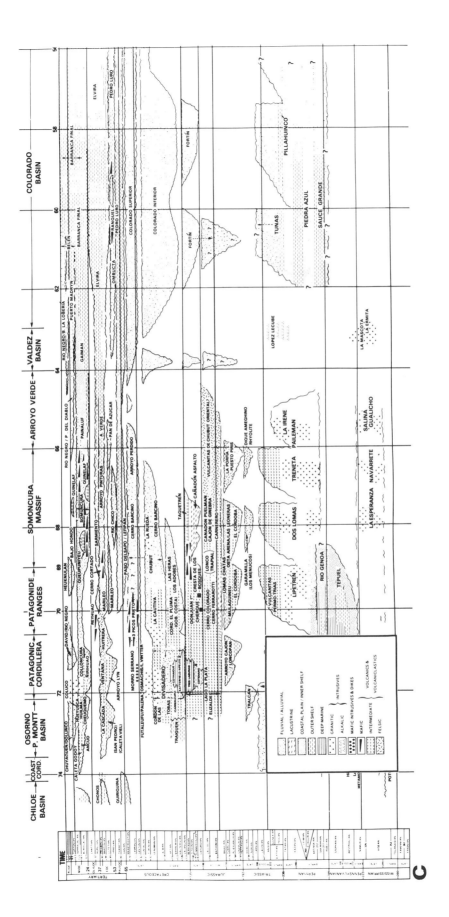

Figure 2.13 C. Chronostratigraphy of a strip on Fig. 2.13B, showing Late Triassic-Jurassic uplift at future site of Colorado rift (by the absence of corresponding deposition) and correlative erosional products represented by clastic rocks of Tralcan, those under the Puesto Piris porphyries, and those of the Osta Arena and Las Leoneras units to the south and west. (Reproduced with permission from Uliana and Biddle, 1987, plate 1.)

commonly eroded, but their former presence can be surmised from the preserved feeder dikes or other intrusions into the basement of the dome. One of the best places where such intrusions are today seen in the cores of former uplifts is Nigeria. There, the "younger granite magmatism" (cf., Jacobson et al., 1958; Whiteman, 1982) has produced igneous rocks ranging from alkalic granites and syenites to gabbros and basalts exposed as plutons and dikes (the ring complexes). In some cases, normal faulting accompanied granite intrusions, as a consequence of which, the rhyolitic carapaces are preserved (Jacobson et al., 1958). These intrusive rocks in Nigeria occur mostly on the northern shoulder of the Benue Trough (see Whiteman, 1982, Fig. 16; Benkhelil et al., 1988, Fig. 32–6).

The doming phase is succeeded by crestal rifting, and the initial dome-related normal faults commonly radiate away from the center of the dome in the form of a rift star (Figs. 2.11B, 2.12B; also compare Fig. 2.7A with Fig. 2.12; Burke and Whiteman, 1973, esp. Figs. 4 and 9). Cloos (1939), and following him, Burke and Dewey (1973) suggested that a triple-armed rift star with the three arms being ideally 120 degrees apart in an environment where the horizontal stress field is isotropic (similar to the ones shown in Figs. 2.6A, 2.11B, and 2.12B) may be a least-work configuration to release the dome-related stresses. This suggestion is corroborated by the many rift-rift-rift (rrr) junctions recognized on domes in Africa by Burke and Whiteman (1973).

Within these rifts, conglomerate, immature sandstone, and terrestrial mudrock, commonly representing floodplain deposits, and locally, evaporites in association with other playa facies, form. Young rifts do not possess well established drainage, and ongoing tectonism frequently creates and destroys lakes on the rift floor. Such lakes may deposit clastics, limestones, and diatomites in the Cretaceous and younger rifts, and even evaporites, with depositional sites ranging from deep lake (e.g., Lake Tanganyika, with 1.5 km water depth, the second deepest lake in the world today, is located in a rift valley along the East African rift chain: Burgess et al., 1988), to lake margin and swamp (e.g., the Middle Pliocene Awash lake in the southern Afar rift in Ethiopia: Williams et al.,

1986; for well described Mesozoic lakes in fossil rifts, see Lorenz, 1988, ch. 8; Olsen, 1988). Paludal environments in rifts are usually restricted and coal forms only under appropriate climatic conditions in marshy areas temporarily shielded from coarse clastic influx (e.g., Lorenz, 1988, ch. 7). All such sedimentary deposits together form what is commonly known as the "graben facies" (Bird and Dewey, 1970) or "rift valley facies" (Burggraf and Vondra, 1982). Ash-fall tuffs (e.g., de Heinzelin, 1982, esp. Fig. 2) and Milankovich rhythmites (called "Van Houten cycles": Olsen, 1984, 1986, 1991) provide the best stratigraphic markers for intra-rift correlations because the biostratigraphic record is usually poor in such terrestrial environments. In Cenozoic rifts, the fossils of rapidly evolving vertebrates have been used for detailed biostratigraphic calibration (e.g., suids in East Africa: White and Harris, 1977), but in older rifts, one is commonly faced only with a meager palynomorph record (e.g., Asmus and Ponte, 1973), or nothing at all (e.g., Uliana and Biddle, 1987). Some ancient rifts preserve a rich vertebrate fauna associated with lacustrine invertebrates and abundant plants (e.g., the Newark Supergroup in the Newark rift: Olsen, 1988). Moreover, Seyitoğlu and Scott (1991) have recently shown the great uncertainties that may plague the usage of palynomoprphs for stratigraphic calibration in rifts in the absence of isotopic age data even in the late Cenozoic. Magnetostratigraphy is of great help in calibrating terrestrial rift fills, but in old rifts, magmatism and burial may confuse the magnetic record. So far, the most reliable and most widely used method for dating old terrestrial sedimentary rift fills is isotopic dating of the associated igneous, especially volcanic, rocks. This becomes especially powerful when combined with biostratigraphic, magnetostratigraphic and paleoclimatologic data (e.g., Witte et al., 1991; Olsen et al., in press).

In rifts, marginal fanglomerates are particularly useful in stratigraphic analysis because they record the successive levels of exposure on the rift shoulders, thereby complementing the syngraben doming record of the clastic wedges outside the rift (for active and late Cenozoic examples, see Tiercelin, 1986; Williams et al, 1986; for a spectacular fossil example, see Steel and Gloppen, 1980; for further examples

from impactogens, see below). In this regard, the marginal clastics along main taphrogenic fault belts are analogous to the molasse deposits along main orogenic thrust belts, which record the uplift and unroofing history of the orogen (cf., Trümpy, 1973). Study of the marginal fanglomerates in many rifts has shown that rift-related morphogenesis and presumably related structure generation are locally episodic processes, but local episodes of relief formation (in most cases resulting from faulting) are not synchronous taphrogenwide, or even on both sides of the same segment of a single rift (e.g., Armijo et al., 1986).

Two of the arms of a rift star (which most commonly occurs in the form of an rrr junction: Burke and Dewey, 1973; Burke and Whiteman, 1973) may eventually propagate and link with similarly or conveniently orientated arms of other rift stars to form a single divergent plate boundary, along which an ocean may open (Figs. 2.11D, 2.12C and 2.13A; Burke and Dewey, 1973; Dewey and Burke, 1974; also see Siedlecka, 1975, Fig. 7). The third, "unspread" arm will be left as an aulacogen trending into a continent from an oceanic embayment (e.g., Gulf of Guinea and the Benue Trough; the North Caspian depression and the Donetz and Pachelma aulacogens). Global plate geometry determines the lines along which a new accreting plate boundary will develop, and therefore, which arms of a group of rift stars will not produce ocean.

2. Doming-rifting hypothesis (path d1–k1–g1 in Fig. 2.5: products commonly afrotype).

Through the doming-rifting mechanism, aulacogens need not form only as failed arms of triple-(or more-)spiked rift stars. They may form as solitary rifts at continental margins as a consequence of the fortuitous location of a hot spot at or near a continental margin (see Şengör et al., 1978, Fig. 1). White and McKenzie (1989, esp. Fig. 17) illustrated a well documented example from the former (and also partly, the present) western continental margin of India, where the Reunion hot spot responsible for Deccan trap magmatism also led to the rifting of the Seychelles fragment away from India (at magnetic chron 28 time; i.e., about 65 Ma). Van Houten (1983) suggested that the Cretaceous Sirte rift of north-central Libya (Fig. 2.14A) may be a fossil example of a solitary rift that formed above a hot spot at the previously formed North African continental margin, and which is now left as an aulacogen opening into the Eastern Mediterranean.

In such solitary rifts that formed at previously existing continental margins and were later abandoned as aulacogens, the rift fill commonly becomes marine much more quickly than in those rifts that go through the doming-rifting-drifting cycle. In the Sirte rift, for example, the rift fill is exclusively marine, with the transgressive Maragh clastic unit at the base (Fig. 2.14B).

3. Rift-tip-abandonment hypothesis (path d2–k2–k22–(g2 or g3) in Fig. 2.5: products commonly afro-, more rarely aegeotype).

Figures 2.11E–G illustrate another mechanism to form aulacogens which was put forward by Grant (1971). The tip of a propagating rift[5] (Figs. 2.11E, F, segment I) may become dormant as a result of the sideways jumping of the locus of extension, forming a new rift (Fig. 2.11F, segment I'). If ocean opening then takes place along the segments I'–II–III, then the original rift tip, I, may be left as an aulacogen trending into the continent from a reentrant formed as a result of transform fault/ridge (II–III) intersection (Fig. 2.11G). The best active example is the Gulf of Suez rift (Fairhead and Stuart, 1982).

The best known aulacogen of this type is probably the Connecticut Valley rift (Fig. 2.15), widely known for its spectacular dinosaur footprints.[6] It actually consists of two basins, namely those of Hartford (in the south) and Deerfield (in the north) (Fig. 2.15A; see Eardley, 1962, p.131ff. and Figs. 9.3 and 9.4; De Boer and Clifford, 1988; Olsen, 1988; Phillips, 1988). It trends southwards into yet another rift, called the New York Bight rift (Fig. 2.15C and D; Hutchinson and Klitgord, 1988), which opens into the New York Bight, the reentrant formed by the N 40 Kelvin Lineament, an early transform margin (see Hutchinson and Klitgord, 1988, Fig. 4–2). The sedimentary fill of the Connecticut rift valley spans the time interval from the late Norian to the Toarcian inclusive, and consists entirely of terrestrial sedimentary rocks and intercalated basaltic lavas (Eardley, 1962). At the bottom of the section are nearly 3 km of relatively coarse, fluvial gray and pink arkose, conglomerate,

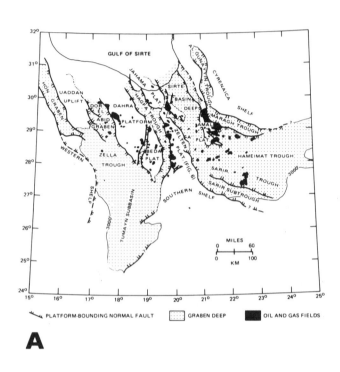

A

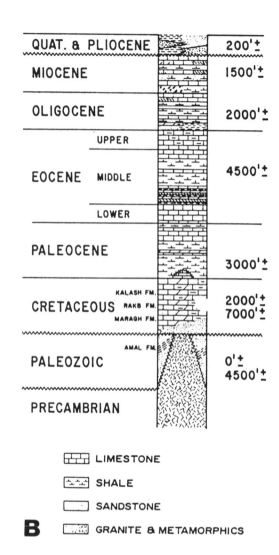

LIMESTONE

SHALE

SANDSTONE

B   GRANITE & METAMORPHICS

Figure 2.14 A. Sirte rift in Libya. Five major northwest-trending grabens have formed an unusually wide rift complex in northwest part of basin (reproduced with permission from Harding, 1983a, Fig. 5).

B. A generalized stratigraphy of the Sirte rift in Libya. Note dominance of marine section (brick pattern above Paleozoic-Cretaceous unconformity; reproduced with permission from Roberts, 1970, Fig. 7).

red feldspathic sandstone, and subordinate red siltstone and shale. These are overlain by nearly 1 km of thick, fine-grained lacustrine and paludal variegated and dark siltstone, shale, limestone, light feldspathic sandstone, subordinate coarse clastic rock, and three basaltic lava flows of Hettangian age (Ar-Ar dating: Sutter, 1988; older vintage K-Ar dates had indicated Hettangian and Sinemurian ages; see Manspeizer, 1988), which were erupted within an interval of only 6.7×10^5y, as determined by Milankovich-type lake cycles (Olsen et al. in press). In fact, Olsen et al. (in press) believe that the igneous rocks may be only

4.0×10^4y younger than the hitherto poorly constrained Triassic-Jurassic boundary.

The upper part of the section is a clastic deposit similar to the lower clastics, but is generally finer-grained. At all stratigraphic levels, the section is coarser toward the eastern master fault, along which are marginal fanglomerates called the Mount Toby conglomerate (Eardley, 1962). The total section thins from more than 5 km along the eastern margin of the rift to below 500 m in the west. As a whole, the Connecticut rift is a west-facing "half-rift", whose sedimentary fill consists of 10% conglomerate, 64%

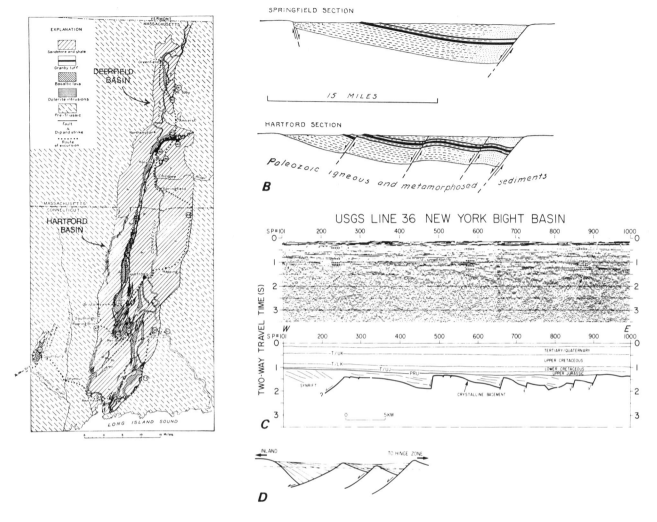

Figure 2.15 Largely onland example of an aulacogen formed by rift-tip-abandonment mechanism, Connecticut rift valley and New York Bight rift.

A. Simplified geological map of Connecticut rift valley, showing its two constituent rifts, namely Deerfield basin in the north and Hartford basin in the south. (Reproduced with permission from Eardley, 1962, Fig. 9.3.)

B. Two geological cross sections across Hartford basin, showing asymmetric nature of rift and distribution of clastic facies. See text for discussion. (Reproduced with permission from Eardley, 1962, Fig. 9.4.)

C. Submarine part of Connecticut rift valley system, New York Bight basin, likely separated from the former by a cross horst, forming Long Island (reproduced with permission from Hutchinson and Klitgord, 1988, Fig. 4-3,). Key to abbreviations: PRU: Post-rift unconformity; T/UK: top of Upper Cretaceous; T/LK: Top of Lower Cretaceous; T/UJ: Top of Upper Jurassic.

D. The asymmetric nature of New York Bight basin (compare with Connecticut rift valley; reproduced with permission from Hutchinson and Klitgord, 1988, Fig. 4-9).

sandstone, 25% shale and siltstone, and the rest limestone. In the New York Bight rift, the structure is similar to the Connecticut rift, but is more disrupted by numerous west-dipping faults, and the entire rift is unconformably overlain by an Upper Jurassic sedimentary blanket.

In older orogenic belts and/or in their forelands, such rift-tip propagation rifts are difficult to recognize. Thomas (1977, esp. Fig. 11) illustrated some possible examples related to former embayments that originated related to the ridge/transform fault intersections during the opening of the Iapetus ocean in the Appalachians.

4. Hypothesis of aulacogen formation through "membrane stresses" (path d2–k2–k22–(g2 or g3) in Fig. 2.5: products commonly afrotype).

Figure 2.11B suggests that the sequence of events shown in Figs. 2.11C–D and E–G may be initiated by

extensional membrane stresses set up in a continent as a result of the change of radius of curvature of a continental plate resulting from its longitudinal motion on a non-spherical earth (Fig. 2.16; Turcotte and Oxburgh, 1973; Turcotte, 1974). As the plate moves from the equator towards the pole, its radius of curvature must become larger (i.e., the plate must become "flatter") thus increasing the length of its perimeter. This induces tension around the periphery and compression in the center as shown in Fig. 2.16. Turcotte and Oxburgh (1973) thought that this may be one way of forming marginal aulacogens that may commonly appear as solitary rifts striking from the continental plate margins into the cratons, as, they contended, in the case of the Mississippi Embayment rift. They thought that it may have been associated with crack-tip propagation following a path along the St. Lawrence Valley, Missouri, and the New Madrid tectonic zone as the North American plate moved northward. They ascribed extension across the East African Rift chain to the same stresses generated by Africa's northerly motion (Oxburgh and Turcotte, 1974). Burke and Dewey (1974) disputed the role of membrane stresses in rifting processes, however, because it seems difficult to store the elastic stresses as long as required by the longitudinal motion of plates across distances sufficient to accumulate enough stress to initate rifting.

More recently, Solomon (1987) took up membrane stresses as a possible cause for rifting, but in the context of secular cooling of the earth. He contended that, as the earth cools, it shrinks and reduces its radius of curvature. This would have the equivalent effect on overlying plates as if they had been moving from the pole to the equator (cf., Fig. 2.16) (i.e., their interiors undergo extension). Solomon thought that this may cause the formation of rift stars, some of whose arms may link to disrupt a continent and lead to ocean and aulacogen generation.

In all cases of aulacogen formation in the framework of the membrane-stresses hypothesis, pre-rift doming is not a requirement, as $k22$ in their "formation path" expresses. But apart from this, it is difficult to test this hypothesis. In the interpretation of Turcotte and Oxburgh (1973), one obvious test is to check the apparent-polar-wander path of the continent in question to see whether it moved in a direc-

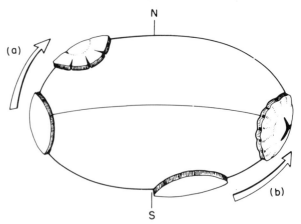

Figure 2.16 Schematic diagram illustrating mechanism of rifting through membrane stresses: a. Plate moving from equator to pole has center in compression and outer part in extension, creating radial aulacogens as solitary rifts (g1 in Fig. 2.5); b. Motion towards equator gives a central zone of extension and outer zones of compression. Central zone of extension gives rise to rifting, creating mainly rift stars (g2 in Fig. 2.5), which may propagate as plate continues to move to create rift chains (g3 in Fig. 2.5). Oxburgh and Turcotte (1974) maintained that this last configuration represents Neogene behavior of the African plate (but see Burke and Dewey, 1974).

tion expected from its rifting history. A corollary to the doming-rifting-drifting hypothesis is that the continent to rift is generally required to come to rest with respect to its underlying mantle convection pattern (cf., Burke and Wilson, 1972; Burke and Dewey, 1973). This requires that the plate move with respect to the magnetic pole during rifting only as fast as its underlying convection "pattern", which would be at most a few centimeters a year, judging from the relative motions of hot spots (Burke et al., 1973; Molnar and Franchetau, 1975; Molnar and Stock, 1987; but see Morgan, 1981). By contrast, the model of Turcotte and Oxburgh (1973) requires that the plate rift while moving fairly fast, so as not to dissipate the membrane stresses. As far as the present observations permit, the doming-rifting model for the origin of the East African Rift chain seems the more appropriate of the two, because the African plate seems to have been stationary with respect to the underlying convection pattern in the mantle since the early Miocene (Burke and Wilson, 1972).

Solomon's (1987) suggestion is almost impossible to test empirically. An incomplete test would be to ascertain that all major continental assemblies in the past began rifting in their middle regardless of the direction

of their motion on the surface of the earth and in the absence of any precedent doming around independent centers (hot- and high spots). This requirement is not met by the late Paleozoic Pangean disintegration (it began in the Carboniferous in the extreme east, in the present northeastern Australia).

From a sedimentological viewpoint, membrane-stress-driven rifting would ideally have no associated clastics anywhere shed from any pre-rift domes. Once the rifting commenced, the subsequent sedimentological evolution would be indistinguishable from the products of other mechanisms reviewed above.

5. Strike-slip-related secondary extension hypothesis (path d2–k3–k33–g1 in Fig. 2.5: products commonly aegeotype).

Transform-fault segments of ridge/transform intersections during the initial breakup of a continental mass may be substantial (e.g., the DeGeer line between the Barents shelf and Greenland and the Agulhas fracture zone between the Malvinas plateau and South Africa: Fig. 2.2). There will be an ever-shortening (along the strike of the transform fault) intracontinental strike-slip regime affecting the transform segments, while the separating continents are clearing each other along shear margins (cf., Emery, 1977, Fig. 2). Tension gashes and/or rotated and opened Riedel and anti-Riedel shears (cf., Wilson, 1960; Tchalenko, 1970; Harding et al., 1985, esp. Fig. 1; Sylvester, 1988) generate rifts at this stage at various angles to shear-zone strike (Fig. 2.11H). As the broad shear zone finally narrows to a throughgoing strike-slip fault zone, the previously formed rifts along the shear belt are cut by the new fault zone and left, on the resulting transform margin, as aulacogens opening into the ocean thus created (Fig. 2.11I). In this case too, pre-rift doming is not a requirement.

The Late Jurassic - Early Cretaceous rifts on- and offshore in South Africa (Figs. 2.17A,B: Rigassi and Dixon, 1972; Dingle, 1976; Tankard et al., 1982, ch. 12) and the post-Triassic rift of the Falkland Sound separating West Falkland from East Falkland (Halle, 1911; Ludwig et al., 1979) were likely formed as a result of the broad shearing along the future Agulhas fracture zone as South America began separating from South Africa.

In South Africa, the pre-rift stratigraphy is irregular because the mountains that formed during the Cape orogeny in the late Paleozoic - earliest Mesozoic stood high and their southern face sloped toward the present shoreline, which nearly coincided with the depositional limit of the post-Cape clastic Robberg Formation of probably Liassic to early Malm age (Fig. 2.17A: Rigassi and Dixon, 1972; Tankard et al., 1982, ch. 12). Wherever Robberg deposition coincided with later rifting, this formation is preserved within the rift trough (e.g., the southern part of the Plettenberg rift: Fig. 2.17A: Rigassi and Dixon, 1972, esp. Fig. 4), suggesting that there was no pre-rift doming, as expected from the postulated mode of rifting. The fact that in most South African rifts, the Robberg is not seen, is due to original nondeposition in those areas.

The rift sedimentary fill in South Africa consists of Uitenhage Group, consisting of three main units, namely, the Enon conglomerate, and the Kirkwood and Sundays River Formations (Figs. 2.17B and C; for review and literature, see Tankard et al., 1982, esp. p. 411ff). The Enon conglomerate is a coarse, red fluviatile deposit, with enormous boulders (up to 50 cubic meters!) at its base. It is interpreted to be associated with basement faulting (Fig. 2.17C: Tankard et al., 1982, p. 411) and is overlain by the shale, sandstone and local conglomerate of the Kirkwood Formation, which contains Lower Cretaceous plant remains (McLachlan and McMillan, 1976; corresponding with the Wood Beds which W. G. Atherstone recognized in the middle of the last century). Overlying these are the estuarine siltstone and shale with very subordinate limestone, overlain by the shallow-marine Sundays River Formation. The marine part of the Kirkwood and the Sundays River Formations have an age range of late Valanginian to Hauterivian, although some Kirkwood faunas have latest Jurassic affinities (McLachlan and McMillan, 1976). These formations also interfinger landward, indicating generally a rift-perforated piedmont environment grading into a shelf southward.

The rifts that form along large strike-slip segments may also be sphenochasms related to drag-related tearing of one of the future continental margins. The Cenozoic rifts of northwestern China around the

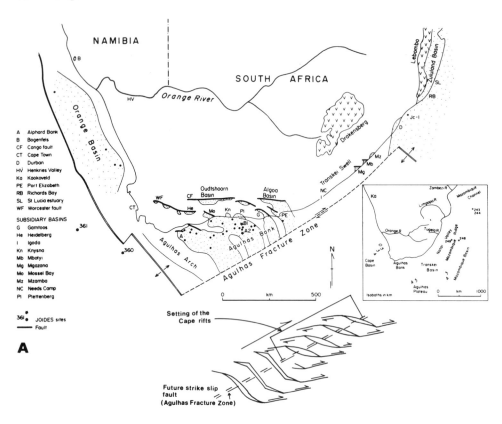

A   Alphard Bank
B   Bogenfels
CF   Congo fault
CT   Cape Town
D   Durban
HV   Henkries Valley
Ka   Kaokoveld
PE   Port Elizabeth
RB   Richards Bay
SL   St Lucia estuary
WF   Worcester fault

SUBSIDARY BASINS
G   Gamtoos
He   Heidelberg
I   Igoda
Kn   Knysna
Mb   Mbotyi
Mg   Mgazana
Mo   Mossel Bay
Mz   Mzamba
NC   Needs Camp
Pl   Plettenberg

361   JOIDES sites
——   Fault

Figure 2.17 Cape rifts as examples of aulacogens that formed through strike-slip-related secondary extension (path d2-k3-k33-g1 in Fig. 2.5). (Reproduced with permission from Tankard et al., 1982, Figs. 12-2, 12-4, and 12-5.) **A.** Map showing distribution of Mesozoic aulacogens in South Africa and their spatial relation to strike-slip margin formed by Agulhas fracture zone. Inset shows my interpretation of mechanism of formation of South African aulacogens as a combination of rotated tension gashes joined at their tips to Riedel shears (Modified from Wilson, 1960, Fig. 3B).

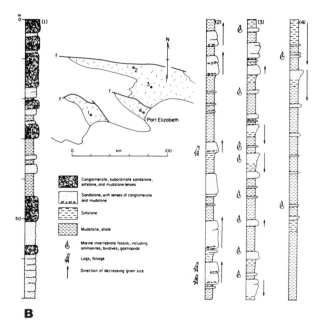

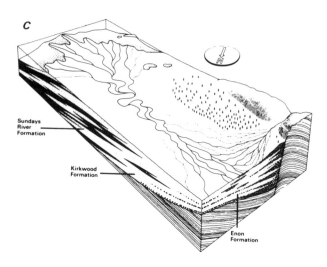

B. Sedimentary rock types deposited in Cape rifts, as exemplified by four sections observed in Algoa and Gamtoos basins (for location of the basins see Fig. 2.17A). (1) Enon Formation in Gamtoos basin; (2) and (3) Kirkwood Formation, and (4) Sundays River Formation in Algoa basin.

C. Schematic depositional model of Uitenhage Group in Algoa basin.

Ordos block may be a modern example of spheno-chasms forming along the Qin Ling strike-slip fault systems that do not extend beyond the Tan-Lu fault (Fig. 2.18; Ma et al., 1982; Burchfiel and Royden, 1991).

6. Continental-rotation hypothesis (path (d1 or d2)–k2–(k21 or k22)–(g2 or g3) in Figure 2.5: products commonly aegeotype).

Yet another way in which an aulacogen may form is shown schematically in Figs. 2.11J–K. If a continental piece were rifted and rotated for a limited amount from the main continent, perhaps similar to Iberia's rotating away fom the rest of Europe to open the Bay of Biscay (e.g., Ries, 1978), then the resulting rift scar (Fig. 2.11K) may be left as an aulacogen. For example, Burke and Dewey (1974) suggested that the Benue Trough and the associated central African rift system (cf., Burke, 1976; Fairhead, 1988) may have formed in response to an aborted attempt by northwestern Africa, in the latest Jurassic - Early Cretaceous, to rift and rotate away from the rest of the continent. This attempt (cf., also Pindell and Dewey, 1982) created a very complex rift net extending from the Benue in Nigeria to Chad in the north and to the Sudan in the east.

The sedimentological characteristics of the aulacogens resulting from this mechanism depend on whether the original rifting and rotation resulted from active (d1) or passive (d2) modes of rift formation and are similar to those of the corresponding aulacogen types.

7. The hypothesis of collision-related rifts that become aulacogens at the same time (path d2–k4–k42–(k421–k422–k423)–(g1–g3–g4) in Fig. 2.5: products may be of both afro- and aegeotype).

Şengör (1987a) suggested that a continent may break up upon collision with another continent, creating collision-related rifts. If any of these rifts extends as far as the uncollided margins of the continent in question, they will be "late aulacogens" with respect to that margin. One possible example that comes to mind is the Paleozoic Amazon rift that may have formed upon a Pan-African collision at its eastern end, along the São Louis craton suture in the latest Precambrian (with cooling ages well into the Ordovician: cf., Burke et al., 1978), and faced ocean at its western end (cf., Şengör and Burke, 1991). The Donets

aulacogen in the East European Platform (one of Bogdanov's, 1962, "late aulacogens") may have originated as an impactogen resulting from an early Paleozoic collision along the southern part of the Tornquist-Teisseyre lineament (cf., Blundell et al., 1991).

Naturally, two or more of the mechanisms reviewed above may operate at once in the formation of one or a group of aulacogens. For example, I have already mentioned under hypothesis 6 that the structural and sedimentological character of resulting aulacogens will be dependent upon whether the rotation commenced as a result of active or passive rifting. Of the suggested hypotheses, only the membrane-stress-related rifting may be inapplicable to the formation of aulacogens.

Post-Rifting Evolution of Aulacogens

Although somewhat dependent on their initial mode of formation, the subsequent geological evolution of all types of aulacogens ideally follows similar paths. Figure 2.19 is an idealized schematic block diagram showing the essential elements of a mature aulacogen. Subsequent to formation of the ocean, into which the aulacogen opens, the continental margin carrying the aulacogen subsides and shelf sediments cover the aulacogen, at least its more "distal" parts (e.g., the Reelfoot rift: Fig. 2.3A; the New York Bight rift: Figs. 2.15C and D). The aulacogen at this stage may localize continental drainage and nucleate large deltas at its mouth, as has happened to the Mississippi Embayment aulacogen (Fig. 2.2: Cenozoic delta progradation: McGookey, 1975; see also Moore et al., 1979) and the Benue Trough (Fig. 2.2: Niger delta: Burke, 1972) (see Figs. 2.11D, G, and I; for the localization of the world's major deltas by aulacogens, see also Audley-Charles et al., 1977). Such large, pre-orogenic deltas found in the stratigraphy of an orogenic belt against a rift striking at a high angle to it may help identify the rift as an aulacogen, as opposed to an impactogen (e.g., in the Vilyuy aulacogen: Parfenov, 1984, esp. p. 38ff and Fig. 13).

Burke (1975) suggested that, because ocean opening is a diachronous event, seawater may spill from the more advanced sectors into regions still at the rift stage and form very thick salt deposits (e.g.,

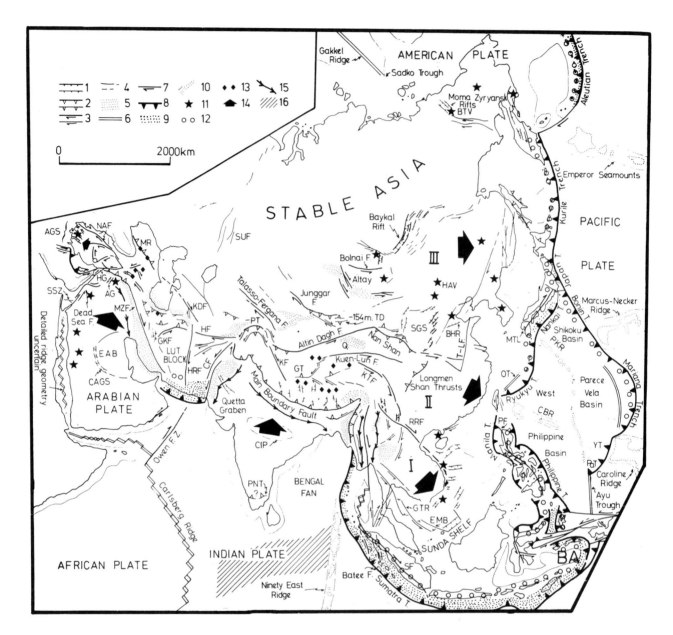

Figure 2.18 Neotectonic map of Asia, showing peri-Ordos spheno-chasms ("Shansi Graben System": SGS) that were opened by the locally inhibited movement of Nan Shan block with respect to northern Asia, Tibetan rift cluster, Baykal impactogen, and Moma Zoryansk rifts. (Reproduced from Şengör, 1987d, Fig. 10.)

Key to legend: 1, first- and second-order normal fault; 2, first- and second-order thrust fault; 3, first- and second-order strike-slip fault; 4, unspecified and/or suspected fault; 5, area of folding (commonly superficial); 6, sea-floor spreading center; 7, major transform fault; 8, active subduction zone; 9, active subduction-accretion complex; 10, aseismic submarine ridge; 11, intraplate volcanism; 12, subduction volcanism; 13, "Tibet-type" collisional volcanism; 14, approximate direction of movement relative to stable Asia of various continental fragments in Asia; 15, major course of present drainage.

Key to lettering: AG, Akcakale rift; AGS, Aegean rift system; BA, Banda Sea; BHR, Bo Hai rift; BTV, Balagan-Tas volcano; CAGS, Central Arabian graben system; CBR, Central Basin Ridge; CF, Chaman fault; CIP, Central Indian plateau; EAB, East Arabian Block; EMB, East Malaya basin; GKV, Great Kavir fault; GT, Gerze thrust; GTR, Gulf of Thailand rift; HAV, Hsing An fissure volcanoes; HF, Herat fault; HG, Hatay rift; HRF, Harirud fault; KDF, Kopet Dag fault; KF, Karakorum fault; KTF, Kang Ting fault; MR, Main Range of the Greater Caucasus; MTL, Median Tectonic Line; MZF, Main Zagros fault; NAF, North Anatolian fault; OT, Okinawa Trough; PaT, Palau trench; PF, Philippine fault; PKR, Palau-Kyushyu ridge; PNT, Palni-Nilgiri Hills thrust; PT, Pamir thrust; Q, Qaidam basin; RRF, Red River fault; SGS, Shansi graben system; SF, Semangko fault; SSZ, Sinai shear zone; SUF, South Ural seismic faults; TD, Turfan depression; T-LF, Tan-Lu fault; YT, Yap Trench.

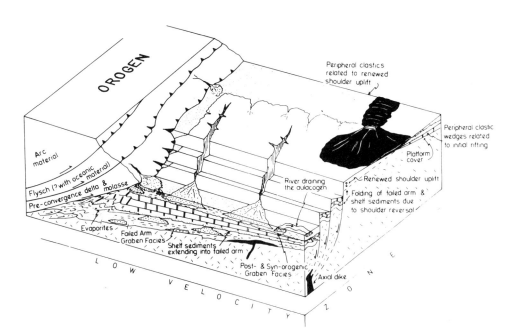

Figure 2.19 Idealized schematic diagram illustrating anatomy of an aulacogen. Sub-aerial erosion is ignored to emphasize structure and reactivation is assumed to revitalize original faults to keep diagram simple. See text for discussion. (Reproduced with permission from Şengör, 1987a, Fig. 4.)

in the "Mucuri Low" between the Cumuruxatiba and the São Mateus-Santa Barbara highs in the Espirito Santo basin in the Brazilian offshore, Fig. 2.2: Asmus and Ponte, 1973, Fig. 9). One of the most important roles of such evaporite horizons seems to be the trapping of oil and gas generated and preserved within the graben facies (e.g., in the Dnyepr-Donetz aulacogen, Fig. 2.2: Mirtsching, 1964, p. 127).

Whether or not doming predates rifting, rift formation is almost always followed by the uplift of rift shoulders owing to isostasy and elastic bending of the faulted plate and to extension-related pressure drop and consequent adiabatic partial melting and volumetric expansion in the underlying mantle. After the rift becomes inactive, uplift caused by the second mechanism commonly reverses and leads to subsidence, much like the young Atlantic-type rifted continental margins. This reversal may induce a very feeble axis-perpendicular compression of the rift contents and result in gentle axis-parallel folding (Figs. 2.19 and 2.20; see Borissjak, 1903; for many other examples see Milanovsky, 1981, esp. Fig. 24; Khain and Michailov, 1985, Fig. 52). However, a detailed consideration of the history of the Eurasian aulacogens has shown that axis-parallel folding of aulacogen strata is usually a result of collisional

orogeny near an aulacogen with an orogenic front parallel with the aulacogen axis (e.g., the Donetz folding phases in the late Paleozoic, middle Jurassic, and early Cenozoic correlate with Hercynian, Cimmeride and Alpide collisions to the south of the aulacogen; for this and other Eurasian examples, see Şengör, 1984; see also Siedlecka, 1975, for the Timan aulacogen/Ural orogen relationship).

When orogeny begins at a continental margin bearing an aulacogen, either owing to initiation of subduction, or the onset of continental collision, the aulacogen enters a new phase of its evolution. In Fig. 2.19 it is assumed, for graphic simplicity, that this reactivation is in the form of renewed rifting. In this particular case, this phase of the aulacogen is not dissimilar to the earlier phase of an impactogen (compare Fig. 2.21Ca with 2.21J: *path k4–k42–k422–g1* in Figure 2.5). Normal faults associated with this renewed rifting were drawn to be precisely the same as those that had initially created the aulacogen. Of course, this is an unlikely situation (but see the discussion in Etheridge, 1986), and is used here only to illustrate the principles involved.

The renewed-rifting phase has the same characteristics as the previous one, such as alkaline to peralkaline (or, if the extension is considerable, even tholeiitic) magmatism, new graben-facies development,

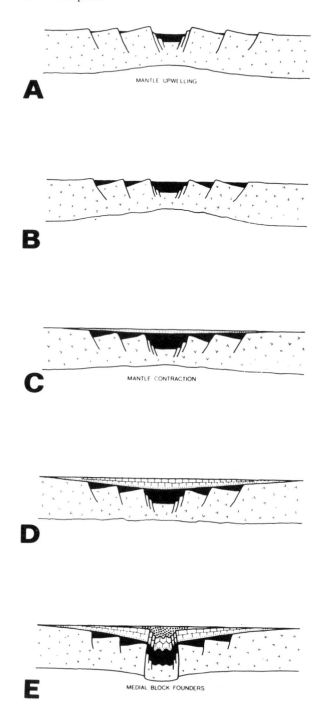

Figure 2.20 Cross-sectional evolution of Athapuscow aulacogen, showing shoulder subsidence-related shortening of rift contents. Compare this with Figs. 7 and 8 in Hoffman, 1973, showing the actual structural geometry in the aulacogen now. (Reproduced with permission from Hoffman et al., 1974, Fig. 10.)

and renewed shoulder uplift. An important difference between the sedimentary regimes of the initial rifting of the aulacogen and its "orogenic" extensional reactivation commonly is that the new mountain range provides an independent sediment source and induces a sediment transport direction along the axis of the aulacogen, opposite to the major axial sediment transport direction prevalent during the initial formation of the aulacogen. In the case of the Athapuscow aulacogen in Canada (Fig. 2.2; Hoffman, 1973) and the Southern Oklahoma aulacogen (Fig. 2.2; Ham and Wilson, 1967), coarse clastic sediments eroded from the rising collisional orogen were deposited in a trough larger than the original aulacogen. A similar situation exists in the Vilyuy aulacogen that was reactivated during the Verkhoyansk collision (Fig. 2.2; see Meyerhoff, 1982).

Reactivation of an aulacogen may happen by compressional deformation, if the aulacogen strikes at low to moderate angle to the direction of convergence of the associated orogen (Fig. 2.21Cc). The Timan mountains in northwestern Russia, a compressed aulacogen related to the terminal Ural collision, is an example of such compressional reactivation (Siedlecka, 1975). Strike-slip faulting parallel or subparallel with the axis of an aulacogen may also occur during reactivation of the structure under the influence of a nearby orogeny, again depending on the angle between the axis of the aulacogen and the direction of convergence (Fig. 2.21Cb). If the angle is fairly high, this strike-slip event may accompany some extension and be expressed as a transtensional reactivation. If the angle is moderate, it may be transpressional. In the Southern Oklahoma aulacogen, Wickham et al. (1975) recognized such large-scale strike-slip faulting, parallel with the aulacogen axis, related to Ouachita collisional orogenesis.

Finally, one of the more important characteristics of aulacogens that was first noted by Borissjak (1903) and Shatsky (1955; see Schatski, 1961, p. 175–196) is their *repeated reactivation* (see also Milanovsky, 1981, esp. Fig. 24). This characteristic has long been used as an argument for their "deep roots" reaching into the mantle and thus precluding continental drift, especially by certain schools of Soviet geologists (e.g.,

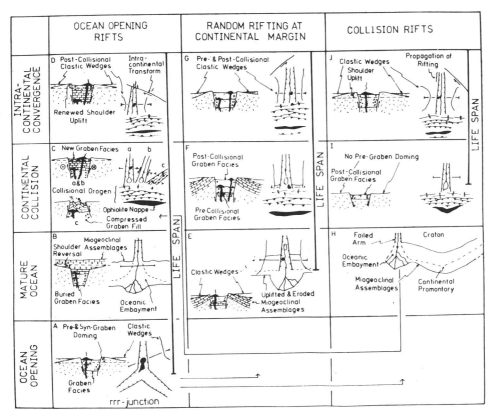

Figure 2.21 A schematic illustration of events during origin and evolution of an aulacogen (**A-D**), a random rift at high angle to a continental margin (**E-G**), and an impactogen. Cross sections show expected differences in stratigraphic evolution of three kinds of "high angle" rifts. Note that in **C**, three differently trending aulacogens, with respect to their related continental margin, are illustrated to show varied response they display to collision (ranging from re-rifting at **Ca** to compression across axis at **Cc**; reproduced with permission from Şengör et al., 1978, Fig. 1).

Beloussov, 1980, esp. p. 297ff). But repeated reactivations of aulacogens is most conveniently explained in terms of Wilson's (1968) model of repeated ocean closure, opening and reclosure along roughly the same lines (not necessarily along the same *locations,* an incorrect view that some "terrane analysts" have tried recently to ascribe to Wilson! See Şengör, 1990c, p. 21–22). Repeated ocean opening and closure (i.e., repetitive Wilson cycles) cause marginal structures such as aulacogens to nucleate new rifts. This situation also introduces the added complexity of confusing reactivated aulacogens with collision-induced rifts. I discuss field criteria to distinguish aulacogens from collision-related rifts below, after I review the geology of the latter.

Repeated compressional reactivation of aulacogens also frequently occurs if orogenic belts of different ages evolve parallel with one another and with the trend of the aulacogen, along the margin of the continent on which the aulacogen is located (e.g., Donetz aulacogen described above). Thus, repeated compressional reactivation of aulacogens is also caused by

Wilson-cycle-related events along other margins of the continents.

Aulacogens are important structures because they preserve the valuable pre- and syn-rifting stratigraphic record of now vanished oceans. Such early records are usually obliterated during orogenies that eventually consume the oceans. Aulacogens are also economically important, as they commonly contain rich economic reserves, such as hydrocarbons and rift-related minerals, as well as rich evaporite deposits.

Continental-Collision-Related Rifts

Brief History of the Concept

That orogeny leads to rifting at high angle to the orogenic trend is a twentieth-century idea (Fig. 2.22). Weber (1921, 1923, 1927) first pointed out that joint sets perpendicular both to axial planes of folds and to fold axes in orogenic belts are commonly seen to

propagate into forelands, where they bound exten- sional structures. He concluded from this that com- pression caused by orogeny also affects the cratons, forming forelands, and leads to their extension at high angle to the compression direction.

In the early days of plate tectonics, rifting was viewed almost entirely within the framework of con- tinental fragmentation and was associated with diver- gent plate boundaries, parallel with Wegener's initial interpretation of rift valleys in terms of continental

breakup. Only in the mid-seventies did it become fashionable *again* to consider rifting also as associated with orogeny, especially continental collision, and here it was once more the German school that took the lead (e.g., Illies, 1972, 1974a,b, 1975a,b; Illies and Greiner, 1978).

In 1975, Molnar and Tapponnier interpreted the Lake Baikal rift as a product of collision-related extension in terms of a plastic-rigid indentation model for Asia (Molnar and Tapponnier, 1975; also

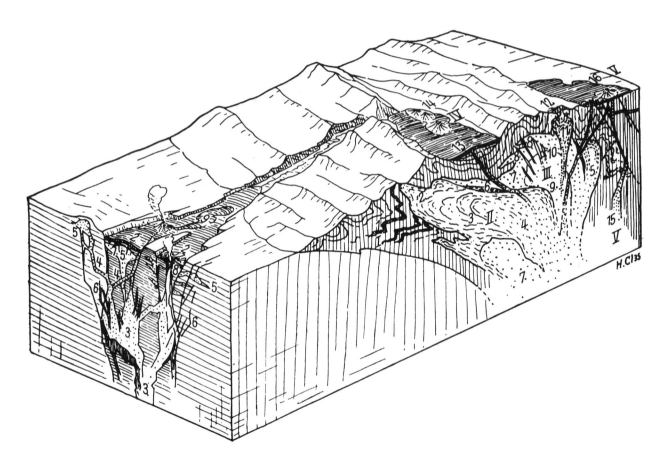

Figure 2.22 A collision-related rift, as illustrated by Cloos (1936, Fig. 120, reproduced with permission) to illustrate related magmatism (Cloos' interpretation of the origin of the rift was different, as explained in text, but the way he drew the orogen and the rift speaks for itself!). Key to numbers (all italicized annotations are mine): Left: Alkalic rocks in regions of extension: 1. Mafic parent at depth; 2. Mafic parent at surface; 3. Intermediate and felsic melts in branched plutons; 4, 5. Intermediate and felsic melts in dikes and subvolcanic plutons; 6,7. Late melts, mainly mafic, at depth and in shallow dikes; 8,9. Late melts, mainly mafic, in volcanoes and lava flows. Right: Alkalic rocks in folded mountains: I. 1. Mafic and ultramafic early melts at bottom of geosyncline (ophiolites, diabases, etc.) (*we now know that ophiolites are floor remnants of vanished*

oceans and they include diabase here separated by Cloos!); II. 5. Intermediate and felsic plutons of first, concordant, synorogenic main magmatism (*i.e., calc-alkalic arc-related magmatism in our present terminology*); III. 6. Mafic early magmas; 7,8,9. Intermediate and felsic plutons of second, discor- dant, late-orogenic main magmatism (*i.e., calc-alkalic late arc and collision- related magmatism in our present terminology*); 10. Late dikes; 11,12. Late- orogenic subvolcanic plutons and volcanoes (*post-collisional i.e., Tibetan, and/or pre-collisional i.e., Altiplano-type plateau magmatism*); IV. 13,14. Post- orogenic magmatism in intramontane basins (*e.g., in Eastern Turkey: see Pearce et al., 1990; Yılmaz, 1990*); V. 15, 16 Postorogenic magmatism in the hinterland, partly becoming alkalic (*"orogenic collapse" magmatism: cf., Lorenz and Nicholls, 1976*).

see Tapponnier and Molnar, 1986, esp. Fig. 4a). Şengör (1976a,b) proposed that continental collision commonly leads to rifting at high angles to the collision front and Şengör et al. (1978) termed such rifts *impactogens*, although not much attention originally was paid to the existence of various classes of collision-induced rifts with different genetic significance; thus, in many subsequent publications, all collision rifts were loosely referred to as "impactogens" (e.g., Burke, 1980, p. 45), until Şengör (1987b, p. 336) made a distinction between impactogens, which are "only those high angle rifts that result from extensional strain induced *directly* by continental collision without an intermediary process", and other collision rifts. This definition, thus took out of the impactogen concept those rifts that form in relation to other structures related to collision, such as strike-slip-related rift basins (e.g., the Akçakale Graben in SE Turkey: Tardu et al., 1987, esp. Figs. 1, 5a and 19).

In the eighties, the concept of collisional rifting grew in popularity and was not only applied to places where it originated, but also elsewhere (for reviews see Hancock and Bevan, 1987 and Şengör, 1987b), even including cases of ocean opening as a result of collisional rifting (e.g., Sawkins and Burke, 1980). In the following subsections, I review the tectonic and sedimentary relationships in the three main types of collision-induced rifts, namely impactogens (k421), rifts that form along intracontinental convergence belts (k422) and pack-ice-type rifts (k423).

Impactogens

Origin of Impactogens

Impactogens are thought to be produced by secondary tension set up at high angles to a zone of compression, similar to the situation in a "Brazilian" tensional strength test performed by loading a cylinder across a diameter between flat surfaces applying concentrated loads (Jaeger and Cook, 1971, p. 160).

Şengör (1976a,b) suggested that this "parting" mechanism (cf., Beloussov, 1962, p. 542, esp. Fig. 233) was responsible for the formation of impactogens. That the bounding faults of large impactogenal rifts are not vertical tension fissures, but normal faults, is a result of gravitational force that does not

significantly affect fractures forming in cylinders in experiments.

The model suggested by Molnar and Tapponnier (1975) and Tapponnier and Molnar (1976) is similar in principle, but assumes plastic, instead of elastic, behavior of the failing medium. If this medium is of finite extent (as continents such as Asia are) then a tensile state of stress will arise in the region adjacent to the boundary of the plastic medium opposite an indenting die (in our case, a colliding continent such as India).

Both Şengör's (1976a,b) and Tapponnier and Molnar's (1976) suggested mechanisms for impactogen formation may be valid, depending on the thermal state of the strained lithosphere and the rate of strain. Indeed, Şengör's (1976a,b) model was based largely on the Cenozoic tectonic history of the Alpine foreland in central Europe, where an initially cold lithosphere (last alpinotype orogeny was of late Paleozoic age) had been affected by a fairly rapid Alpine collision (see Trümpy, 1972, with the reservations expressed in Şengör 1991a). By contrast, Tapponnier and Molnar's (1976) model was developed on the basis of the Cenozoic tectonics of the Himalayan hinterland in central Asia, which consisted not only largely of Paleozoic/Mesozoic soft subduction/accretion material (cf., Şengör and Okuroğulları, 1991; Şengör, 1992), but also of blocks that had been affected by *late Mesozoic* alpinotype orogenies all the way from the southern tip of the Baikal rift to the Himalayan suture zone (see Şengör, 1987d; Şengör and Okuřogullari, 1991). This substrate is thus expected to behave more in a plastic fashion than the cold Alpine foreland, and perhaps, to induce a lower strain rate by making the straining area wide, despite the much higher convergence velocity of India with respect to Asia than was the case in the Alps.

Figure 2.23 summarizes in a schematic fashion ways in which impactogens are thought to originate in the framework of Şengör's model of elastic parting and some complications. Figure 2.23A displays two continents, *a* and *b*, that are converging at the expense of an intervening ocean that is being subducted along a trench located at the foot of the continental margin of continent *b*. In Fig. 2.23B, the promontory of continent *a* has collided with continent *b*. At this stage, the

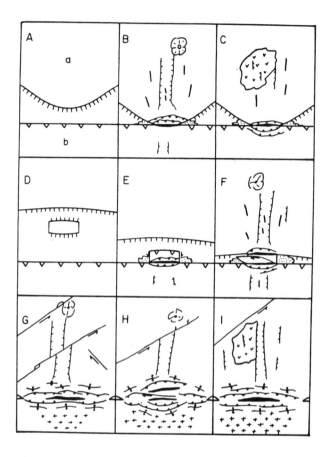

Figure 2.23 Sketch maps showing generation of Upper Rhine-type impactogens and some complications. The sequential maps are as follows: **A,B,G; A,C,I; D,E,F,H.**

Key to ornaments:

Lines with long hachures: Atlantic-type continental margin, hachures on the ocean side; lines with open triangles: Andean-type continental margins, triangles on overriding plates; lines with short hachures: normal faults with hachures on downthrown side; short, thick, black lines: mafic dikes; lines with black triangles: thrust faults with triangles on upper plate; dotted areas: flysch wedges; v's: flood basalts; lines with half-barbed arrows: strike-slip faults; black regions: ophiolitic sutures; lines with double-headed arrows across them: anticlines; plus signs: regions of Tibetan-type basement reactivation. (Reproduced with permission from Şengör, 1987b, Fig. 1)

collision is only effective at the *point* where the promontory has touched the continental margin of *b*. Such *point collisions* (in reality *line collisions* when one considers the thickness of the colliding continents) are analogous to the line loading of cylinders along their diameters in the Brazilian test discussed above, and similarly, produce tensional stresses perpendicular to the direction of convergence. If the collision is head-on, then rifting will occur perpendicular to the

collision front, and thus generate a high-angle rift (cf., Şengör et al., 1978). Figure 2.23B exhibits a situation in which neither of the collided margins has been thermally (by magmatism) or mechanically (by the growth of large accretionary complexes) weakened by long-lived subduction activity. This may be the case where the ocean that disappeared during continental collision had a small cross-convergence width (cf., Şengör, 1991b), as was likely the case in the Alps (<1000 km: Şengör et al., 1984; Şengör, 1991b). The collision of two such "strong" continental margins would result in rapid communication of the collision-induced stresses to the fore- and hinterlands and give rise to their deformation by convergence-perpendicular extension.

The structural expression of such deformation may be extension joints, mafic dikes and gravity faults, probably in that order of occurrence (Fig. 2.23B). In the case of the Upper Rhine rift, Bergerat (1983, 1987) and Villemin and Bergerat (1985, 1987) were able to document that the earliest structural event in the formation of the future Upper Rhine rift was north-south shortening, as expressed by NNE-SSW-striking conjugate strike-slip faults, NNE-SSW-striking tension gashes filled mainly with crystalline calcite, and E-W-striking horizontal stylolites. Although Bergerat and Villemin were unable to date this event with any precision, they did establish, by cross-cutting relationships, that these structures were the earliest in the history of the Upper Rhine rift. However, they noted that some of their tension gashes displayed a field continuum with some dikes (Villemin and Bergerat, 1985, p. 66). Illies (1974a, esp. Fig. 3) had earlier pointed out that the earliest harbinger of rifting along the present Upper Rhine structure was olivine-nephelinite magmatism that created dikes between 65 and 55Ma, whose broad locations and strikes are identical with those of the tension gashes mapped by Villemin and Bergerat. Data reported by Illies (1974a) date the earliest phase of the Rhine rift taphrogeny as Paleocene-Eocene. On the basis of regional considerations, Bergerat and Villemin also placed this phase in the Eocene.

Sedimentation that was coeval with this phase of tectonism was recorded by the 100m-thick fluviatile-limnic clastic rocks and rarer limestone and dolostone

(Doebl, 1970). The point emphasized by Bergerat (1983, 1987) and Villemin and Bergerat (1985, 1987), that pronounced normal faulting did not take place at this time, is indicated by the modest thickness of these early sediments and by the fact that they are distributed in an area larger than the future rift trough (Illies, 1977, p. 335). Illies (1977) pointed out, however, that even the earliest of the Rhine rift sediments show evidence for some normal dip-slip faulting.

The oldest sedimentary rocks of the Rhine rift, of Lutetian-Bartonian age on the basis of mollusc and vertebrate fossils, are dominantly fluviatile conglomerate at the base (clast composition being strongly controlled by the underlying rocks) and passing upward into claystone and sandstone with rarer marl; locally, there are limnic limestone and dolostone (Doebl, 1970). Freshwater limnic sediments of Lutetian age, distributed in an area not exactly coincident with the later rift trough, were laid down next, and contain freshwater limestone, marl, some dolomitic marl, rare sandy limestone, and lignite, indicating the presence of palludal environments (and thus limited topographic differentiation). The thickness of all these rocks nowhere exceeds 50 m and supports the inference that the rate of normal faulting was modest, in agreement with Bergerat and Villemin's structural observations.

By Priabonian time (late Eocene), rates of normal dip-slip faulting and related subsidence accelerated. The lower parts of these sequences consist of marl, marly limestone, anhydrite, and salt. In their upper parts, there is dolomitic marl (Sittler, 1965; Doebl, 1970). Toward the master faults of the rift trough, the rocks are conglomeratic. The thickness of the Priabonian sequence is up to about 900m to the southwest of Freiburg i.Br., about 500m near Karlsruhe, and close to zero near Mannheim (Fig. 2.24). Upper Eocene strata also indicate the first marine incursion into the rift trough that laid down the thick (about 300m) salt deposits.

Late Eocene subsidence correlates temporally with strong roughly east-west extension, indicated by numerous normal dip-slip faults and normal dip-slip movement on former strike-slip faults accompanied by continued opening of tension gashes (Bergerat, 1983, 1987; Villemin and Bergerat, 1985, 1987).

Villemin and Bergerat (1985, p. 68) pointed out that, among the normal dip-slip faults, easterly dips predominate. This observation is in excellent accord with the more pronounced subsidence at this time along the western margin of the rift trough, interpreted by Illies (1970, esp. Fig. 5) as asymmetric rifting by the formation of the (east-dipping) western master fault system first.

If the active continental margin in the collision system is a well developed Andean-type arc (e.g., Fig. 2.23C), then subduction-related magmatism is likely to have weakened it before the collision (cf., Armstrong, 1974; Burchfiel and Davis, 1975; Isacks, 1988; Şengör, 1990b; Dewey and Lamb, 1991, esp. Fig. 4). This leads, during the collision, to absorption of much of the convergent strain by the shortening of this "weak" margin. As a result, at least during the early stages of the continental collision, the foreland structures described in the preceding paragraphs may be only weakly developed. They may, for example, be confined only to a few gravity faults and dikes feeding flood basalts (Fig. 2.23C), as exemplified by the southeast Turkish foreland on the Arabian plate. Here, the soft Andean-type margin of Eurasia (cf., Yılmaz, 1985, 1990) took up much of the convergent strain by shortening and complementary thickening (cf., Şengör and Kidd, 1979; Dewey et al., 1986). The only foreland structures that developed in an impactogenal fashion are the N-S-striking fissures that fed the large Plio-Pleistocene basaltic rocks of the Karacadağ shield volcano (Şengör et al., 1978; Hancock and Bevan, 1987, Fig. 1b). Similarly, no major impactogens are known from the Indian subcontinent that collided with a rather "soft" Eurasian margin (Harrison et al., 1992). In the case of the latest Carboniferous - earliest Permian Oslo rift (Fig. 2.2), the earliest products of a much delayed impactogenal(?) rifting were the feeders of the alkalic basalt and rhomb-porphyry eruptions (Ramberg, 1976, p. 169).

Figures 2.23D–F show how a microcontinent between two larger converging continents in a collisional system may influence impactogen generation. Although in Fig. 2.23D there is no continental promontory to cause "point collisions", Figs. 2.23E and F show that the betwixt microcontinent functions as just such a promontory. The difference

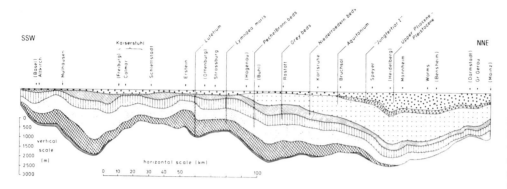

Figure 2.24 Thickness distribution of Tertiary sedimentary units in the main trough of the Upper Rhine rift displayed along a longitudinal section. As this figure illustrates, rift-related subsidence commenced in the south during Lutetian (medial Eocene). The center of maximum subsidence shifted northward during later evolution of the rift. Accelerated rift-floor subsidence is seen for the late Eocene - early Oligocene interval (Lymnea marls and Pechelbronn Beds), for Aquitanian (early Miocene), and late Pliocene-Pleistocene (for the last interval, see also Fig. 2.9B). (Reproduced with permission from Illies, 1975, Fig. 1.)

between the final picture that will result from the evolution shown in Fig. 2.23A and that in 2.23D is that in the former case the final picture contains a single suture (Fig. 2.23G), whereas in the latter, a double suture is seen (Fig. 2.23H). Such a case may have happened in the Alps, where the Briançonnais betwixt continents (cf., Trümpy, 1980, Fig. 15; Hsü and Briegel, 1991, Fig. 12.10) may have created a "promontory effect", where in reality, there was an embayment in the European continental margin (see esp. Hsü and Briegel, 1991, Fig. 12.2, frame 60 Ma).

Impactogenal rifting in the plastic-rigid indentation model of Tapponnier and Molnar (1976) also follows a similar evolutionary path as one that resulted from elastic parting, here described for the Upper Rhine rift. The Lake Baikal rift, for instance, seems to have opened in the early to medial Eocene (55 to 45 Ma), as suggested by the sudden deposition in its southern sub-basin of conglomerate, sandstone, and shale, whose cumulative thickness is 1100m (Grossheim and Khain, 1967). In the same places, Paleocene thicknesses were, at most, tens of meters. A shallow lake also formed around the present southwestern tip of the rift. In the late Eocene, similar conditions persisted, and in the Oligocene, the two distinct sub-depressions of the present rift basin, separated by the Akademichesky Ridge, became clearly recognizable. To the SE of the Akademichesky Ridge, 1000m of sandstone with some conglomerate

were deposited, indicating that the southern subbasin of the rift was the main venue of rifting in the later early Cenozoic (cf., Grossheim and Khain, 1967). Even if one followed Zonenshain and Savostin's (1981) slightly younger dating, rifting in the Lake Baikal area seems to have happened immediately after (if not synchronously with, if the later dating of the Himalayan collision at 45 Ma by Dewey et al., 1989a, is accepted) the closure of Neo-Tethys along the Indus-Yarlung suture.

It thus seems that impactogens form either synchronously with the collisions that give rise to them or that they follow after a short time, perhaps not longer than a maximum of 20 my (obtained by assuming that the Himalayan collision occurred at 55 Ma and that the Baikal rifting commenced during the Oligocene; i.e., at about 35 Ma).

The scheme of impactogen generation sketched above glosses over factors that may assist, complicate, or hinder impactogen generation. Şengör (1976b) summarized these as follows:

1. Generation of impactogens may be dependent on the rate of convergence of the two colliding continents and on the angle of collision. High rates of convergence and high angles of convergence with respect to convergent boundaries will assist impactogen generation.

2. The existence of zones of weakness in the fore- or hinterlands may nucleate new structures. Such

weak zones may prescribe, to a certain degree, their extent and orientation, and therefore, complicate the picture. Illies (1962), for example, pointed out that the orientation of the Upper Rhine rift had been influenced both by an Hercynian shear zone and some possible middle Jurassic rifting(?) (for the data see Illies, 1977, Fig. 6) only in the southern part of the future rift.

3. If crustal and/or lithospheric domes exist in the fore- or hinterlands, they may localize the formation of impactogens. These domes help impactogen formation in two independent ways. One is through the potential energy they store and thus predispose the rocks to rifting. The second is that they weaken the lithosphere either by crustal thickening or by lithospheric thinning (thus increasing the quartz+ feldspar/olivine ratio in any given cross section across them). These effects are thus cumulative. Since Cloos' (1939) classic study, all authors who have written about the Upper Rhine rift have pointed out that the Late Cretaceous swelling of the "Rhenic Shield" (Cloos, 1939) in one way or another must have helped the formation of the rift. Although I do not agree that the dome of the "Rhenic Shield" caused the rifting (as Burke and Dewey, 1973, and Illies, 1974a, following Cloos, 1939, thought), I certainly believe that its potential energy must have been a factor in localization of the rift.

4. The existence of a well developed continental-margin magmatic arc (i.e., "soft", convergent margin) on one or on both of the approaching margins may hinder the formation of impactogens by introducing a ductile zone between the two colliding continents, which may take up the impact of the collision (i.e., by reducing the effective rate of convergence of the brittle cratons). In other words, such a ductile zone reduces the "impact" of the collision on the brittle fore- and hinterlands.

Evolution of Impactogens

Once extension begins along the future site of an impactogen following a continental collision, rift nucleation succeeds the initial formation of individual extensional structures (Figs. 2.23B and F). Such rifts commonly form first near the collision site and prop-

agate later away from the collisional orogen. As a consequence, the amount of total extension and correlatively, the magnitude of the total subsidence, is commonly greatest near the suture and decreases away from it. In the above subsection, the thickness changes encountered in the Priabonian section of the Upper Rhine rift from about 900 m in the south to 0 in the north near Mannheim illustrate the decrease in subsidence northwards away from the Alpine front. This difference in rifting intensity between the north and the south in the Upper Rhine survived into the earliest Oligocene. In the case of Lake Baikal rift, rifting commenced in the south, nearer the collision front, and then propagated northwestwards away from the collisional orogen. Even today, the deepest part of the Baikal rift is located in the southern half of the rift trough.

Volcanism within the rift trough is not ubiquitously present in all impactogens, being totally absent in some (e.g., Lake Baikal: Logatchev et al; 1983), and overabundant in others (e.g., the Oslo rift: Ramberg, 1976). In all cases however, it is a post-rifting phenomenon, or at least it commences together with faulting and tension-gash formation. Within the Upper Rhine rift, major volcanicity commenced with the activity of the Kaiserstuhl volcanic center in the south about 18 Ma (Illies, 1974a,b, 1975b)(i.e., about 27 my after the Rhine rift had originated) and, in the south, subsided to more than 2.5 km. The subvolcanic breccias of the Kaiserstuhl and the close correspondence of volcanism and the rift suggest that faulting controlled volcanism (Baranyi, 1974). In the case of the Lake Baykal rift, volcanism occurred outside the rift trough and commenced with the eruption of trachybasaltic lavas in the Miocene (Florensov, 1965)(i.e., at least 10 my after the commencement of rifting, which may be as much as 35 my after rifting, if rifting began synchronously with the Himalayan collision and if 55 Ma is taken as the date of this collision; see above). By contrast, in the case of the Oslo rift (Ramberg, 1976) and in SE Turkey (Şengör et al., 1978), impactogen generation was coeval with the onset of volcanism.

Synchronous with rifting, shoulders of rift(s) begin to rise, likely as a result of a combination of

isostasy-driven elastic deformation on normal dip-slip faults (cf., Taber, 1927; Buck, 1988), and of partial melting and expansion of the underlying mantle as a consequence of pressure drop in the asthenosphere owing to rifting in the overlying lithospheric lid. This post-rifting rise of rift shoulders has been documented both from the Upper Rhine and the Lake Baikal rifts.

In the Upper Rhine rift, the Pechelbronn beds of dominantly evaporitic aspect were laid down in the early Oligocene (with the exception of the "Lower Pechelbronn Beds," whose age is now believed to be late Eocene: Doebl, 1970). At this time, the rifting reached as far north as the Hessen depression (Fig. 2.24; Lotze, 1971). The area of most rapid subsidence, however, was still in the south, as shown by the 1600 m thickness to the southwest of Freiburg and 900 m thickness near Karlsruhe of the Pechelbronn beds (Illies, 1974b). In Pechelbronn time, the so-called *Küstenkonglomerate* (i.e., "coastal conglomerates") began to be deposited along the master faults of the Upper Rhine rift. The important thing about these conglomerates is that they contain Jurassic material in the older pebbles, and older clasts with decreasing age of the conglomerates (Illies, 1967). In other words, the coastal conglomerates show an inverse relationship between the age of deposit and the age of its contained clasts. This indicates a progressive unroofing of the rift shoulders and the deposition of their erosion products at the foot of the normal dip-slip fault-bounded cliffs facing the rift trough. Both Illies (1970) and Şengör et al. (1978) interpreted this as late shoulder uplift.

The data I have at my disposal at this writing on the Baikal rift are not as detailed as they are for the Upper Rhine rift. Still, the equivalents of the "coastal conglomerates" do exist here too, especially along the master faults of the northern depression of the rift trough, and they are as young as medial Pliocene, indicating a very late rise of the rift shoulders here (Grossheim and Khain, 1967; Logatchev et al., 1978); the age of these deposits is now thought to be older, about medial Miocene (Dr. V. Kazmin, personal communication, 1991).

As the collision advances, the suture lengthens and the convergence becomes effective across a long front, as opposed to being concentrated at discrete nodes as shown in Figs. 2.23G, H, and I. One effect of ongoing convergence is to continue shortening the fore- or hinterlands, or both, in the direction of convergence and to extend them at high angles across the direction of convergence. This keeps the rifting active, but at a reduced pace than earlier, because the optimum stress conditions for rifting created by point collisions no longer apply. This is exemplified by the medial Oligocene evolution of the Upper Rhine rift. At this time, the graben subsidence levelled off, and a marine marl sequence was deposited all along the rift trough with more or less uniform thickness (Fig. 2.24; cf., Doebl, 1970). In the late Oligocene, the subsidence (and presumably the rate of extension) slowed down, and the deposits of this age are disconformably overlain, suggesting local episodes of erosion (Fig. 2.24). Villemin and Bergerat (1985) think that this slackening of rifting corresponded with a time of right-lateral strike-slip faulting along the rift trough, with little extension. Their structural studies indicate that the direction of maximum shortening at this time was NE-SW, slightly more easterly than the orientation of the main axis of the graben. This corresponds with the collision along the Alpine front having become effective all the way from the Pyrenees via the Provence chains into the Alps. In other words, the western part of the Mediterranean Alpides east of the Betic and the Riff cordilleras had sutured (cf., Cohen, 1980; Dercourt et al., 1986; Bethelsen and Şengör, 1993). This is also the time when the Bresse rift opened (Rat, 1975) and connected with the Upper Rhine rift via a transform fault system north of the Jura Mountains (Laubscher, 1970), and rifts in the Massif Central began opening under the influence of NW-SE-directed extension (e.g., Burg et al., 1982).

With the onset of the Miocene, two important changes occurred in the evolution of the Rhine rift: first, the center of subsidence shifted to the north, where today the thickest accumulations of Tertiary strata are found (Fig. 2.24; Sittler, 1969; Illies, 1974a,b); and second, major volcanism began in the rift trough in the south.

The Aquitanian section in the Upper Rhine rift consists of about 1.5 km of clastics and carbonates. However, after the Oligocene, the subsidence axis in

the rift shifted from a north-northeast trend to almost exactly northwest (cf., Illies, 1974a,b; Şengör et al., 1978), which is reflected by the distribution of the thickness of the rift fill. The coeval structures mapped by Villemin and Bergerat (1985) also show that the maximum shortening axis shifted to a NW-SE orientation and that the Upper Rhine rift began shearing left-laterally along its axis (i.e., in a sense opposite to what was dominant in the late Oligocene) (Fig. 2.10). This led, in its middle sector, to shortening associated with thrusting of the rift shoulders onto the rift floor (e.g., near Baden-Baden: Illies and Greiner, 1976). Today, the rift as a whole is a broad strike-slip zone with associated second-order extensional and compressional features (Ahorner, 1970; Ahorner and Schneider, 1974; see also Chorowicz et al., 1989).

As the history of the Upper Rhine rift shows, while the suture lengthened and compression became effective across a whole front, as opposed to at discrete nodes, the fore- and hinterlands began to fail by conjugate strike-slip faulting. At this stage, the impactogen may or may not remain in an orientation favorable to further extension. In that case, it may undergo shear (as happened to the Upper Rhine rift), some parts may go on extending (as did the northern part of the Upper Rhine rift), and others may become compressed (as did the middle part of the Upper Rhine rift). At this point in the development of the collisional system, the entire fore- or hinterland in which the impactogen is located is generally deformed by numerous faults of various orientations and kinds, disrupting the entire area into numerous blocks (*Schollen*: cf., Dewey and Şengör, 1979, p. 84, footnote 1). That is why it is appropriate to consider their further evolution under the headings of "Rifts that form along intracontinental convergence belts" (k422) and "Pack-ice-type rifts" (k423) below.

Rifts That Form Along Intracontinental Convergence Belts

Both continental collision and other kinds of orogeny create broad belts of intracontinental convergence (cf., Şengör, 1990a). Those associated with continental collision are particularly large as exemplified by the Tibetan plateau behind the gigantic Himalayan range

today (Dewey and Burke, 1973; Molnar and Tapponnier, 1975; Şengör and Kidd, 1979). Along collisional orogens, such belts of intracontinental convergence commonly form after the initial touch of the two continents, when the suture has achieved a certain length along the trend of the orogen. Along the Himalaya, for example, although the inital touch probably occurred sometime in the medial Eocene (about 55 Ma; but see Jaeger et al., 1989), the suture was not complete until sometime in the late Eocene, following at least 15° counterclockwise rotation of the Indian subcontinent (Klootwijk, 1981; see also Dewey et al., 1989).

Such intracontinental belts of convergence lengthen along their trend as they are shortened across it. In some orogens, this lengthening is simply expressed as joint sets or rare normal faults across the orogen (e.g., in the Alps: Weber, 1921, 1923, 1927; in the southern Appalachians, especially the "Gwinn-type" extensional lineaments: Gwinn, 1964; Kowalik and Gold, 1976; Krohn, 1976; Gold, 1980). In Tibet and in the Himalaya, along-trend lengthening of the orogen led to the formation of numerous rifts that trend at about 90° to the orogen (Figs. 2.2 and 2.18; Molnar and Tapponnier, 1978; Ni and York, 1978; Romanowicz, 1982; for the Himalaya, the classical example is the Thakkola-Mustang Graben in Nepal: Colchen et al., 1980; Fort, 1980, 1987, esp. Fig. 1; for the rest of Tibet, see Tapponnier et al., 1981, 1986; Han et al., 1984; Mercier et al., 1984; Rothery and Drury, 1984a,b; Armijo et al., 1986; Dewey et al., 1988; Kidd and Molnar, 1988).

The extensional basins in Tibet are probably both grabens and rifts. Many of them are asymmetric and there is no consistent direction of facing of the master faults of the basins. In fact, in many basins, facing of the master faults changes along trend. The basins are either tectonically isolated or they end on NW-SE- or NE-SW-striking strike-slip faults and form a kinematic whole with them in extending the plateau in an E-W direction (see esp. Han et al., 1984; Rothery and Drury, 1984a,b). Long, straight, continuous normal faults are rather the exception (e.g., along the Thakkola-Mustang basin: Fig. 2.2) than the rule. Most are bounded on one or both sides by discontinuous, commonly en-echelon faults with fault lengths on the order of 20 to 60 km (cf., Armijo et al., 1986). Most basins are short, rarely exceeding 100 km, but

the Yadong-Gulu rift system appears like a mini "rift chain", with a length exceeding 400 km (Fig. 2.2).

The Tibetan rifts and grabens are mostly filled with latest Pliocene-Quaternary lacustrine and fluvial silt, sand and gravel deposited in alluvial-fan/channel systems. Some contain coal seams, pointing to the presence of former palludal environments, possibly indicating a slackening of tectonic activity (Armijo et al., 1986). In these basins, thicknesses rarely exceed one kilometer and this agrees well with observed maximum throws of about 1600 m along normal faults bounding the basins (Armijo et al., 1986). Figure 2.25 exhibits a graphic synopsis of the evolution of the sedimentary environments in the south Tibetan extensional basins, which are dominated by an interplay between normal faulting, and fluvial and glacial erosion. Armijo et al. (1986) have demonstrated that the rate of normal faulting in most of these grabens commonly has exceeded the rate of glacial erosion, leading to the formation of hanging valleys or valleys with a "wine-glassprofile" and enabling the assessment of crude rates of faulting.

One characteristic of extensional basins in the Himalaya/Tibet border area in southern Tibet is that they originally were avenues of a north-directed drainage carrying debris from the rising young Himalayan range onto the endorheic basin of Tibet during the Plio-Pleistocene. This changed later, perhaps in the late Pleistocene-Holocene, by stream capture led by the evolving Himalayan drainage, when the basins began draining to the south into the Ganges system (Colchen et al., 1980, esp. Figs. 2–4; Armijo et al., 1986). This imparted onto the sedimentary record in these basins a two-phase fluvial history, which is the opposite of the drainage reversal seen in aulacogens later reactivated by collisions.

The average amount of extension of the 1100 km-long Himalayan front has been 1% per million years during the last 2 to 2.5 my This indicates an average extension per basin of about 1.5 mm/y, which is well below the average rate of extension in most rifts in the world (which is about 1 cm/y). As a consequence, none of these basins forms rift valleys comparable in size with those of the other environments, except some of the extensional basins in the pack-ice-type rift areas. Thus, it is highly unlikely that these rifts would eventually lead to sea-floor spreading as

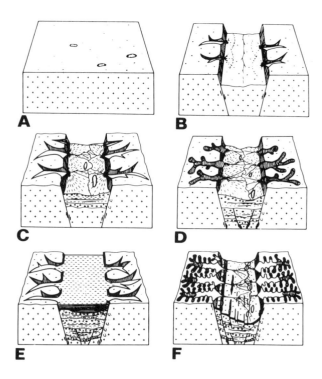

Figure 2.25 Schematic block diagrams illustrating geomorphological/ sedimentary evolution of Tibetan intracontinental convergence-related rifts.
A. In late Pliocene, flattish erosional surface of Tibetan plateau was formed at an altitude of about 3 km (cf., Dewey et al., 1988; only 1 km according to Han et al., 1984!). Bedrock is shown by plus signs.
B. In early Pleistocene (and/or latest Pliocene), several N-S-trending extensional basins formed. Erosional gullies formed across their shoulders and alluvial deposits dominated by conglomerates began forming at fault scarps.
C. Accelerated deformation in southern part of plateau during middle Pleistocene is believed to have caused unconformity between Q_1 and Q_2. This was also a time of major glaciation. During interglacials, gullies further developed in newly created fault scarps.
D. Further tectonization in late Pleistocene created unconformity between Q_2 and Q_3 and generated further faulting and folding. The sudden rise of plateau by nearly 2 km led to creation of both valley and piedmont glaciers on the plateau.
E. In middle Holocene, advance of the last "interglacial" led to formation of numerous lakes, that in places, occupied glacially modified rift basins.
F. In later Holocene, evolution of mature Himalayan drainage and diminution of precipitation led to disappearance of many lakes. Normal faulting continued and glacial level in many places retreated to above 5,800 m. (Reproduced with permission from Han et al., 1984, Fig. 5.)

assumed in some recent models of extensional orogenic collapse!

Rifts in intracontinental convergent belts get preserved only if the crust of the convergent belt is tectonically thinned *by an independent rifting event* immediately after the compression that created the

rifts, to protect them from erosional destruction. This happened, for example, in the case of western Turkey, where north-south shortening created roughly north-south-trending extensional basins between the late Oligocene and the ?early Miocene, which basins were then preserved by rapid north-south extension that began sometime in the early Miocene (Şengör, 1987c, but with the revised timing of Seyitoglu and Scott, 1991). Another possible example is the preservation of the Permo-Triassic Ecca basins containing Karoo sedimentary rocks by the independent extension that led to the opening of the Indian and the Atlantic Oceans in the Mesozoic (Tankard et al., 1982, p. 380ff and Fig. 11–1). Rift belts related to intracontinental convergence do not seem to accomplish sufficient extension to thin the continental crust below its average thickness in order to take its top below sea level (they did not accomplish this in any of the examples just cited, and in very few other cases, the north European Zechstein basins being one remarkable exception: Ziegler, 1990). This is another reason for my skepticism concerning the reality of extensional orogenic collapse driven by the potential energy difference between thickened continental crust and its unthickened surroundings.

Pack-Ice-Type Rift Basins

As the name implies, the pack-ice-type rifts form from the jostling of blocks, into which fore- and hinterlands of collisional orogens become divided following the formation of impactogens, other collisional rifts and collision-related conjugate strike-slip faults. The best example of a pack-ice-type rift area is the group of rifts that formed and/or reactivated following the Adria/Europe collision along the Alps in the Eocene (Fig. 2.10; see also Illies and Greiner, 1976, Fig. 8; Dewey and Windley, 1988, Fig. 1) and that now forms the classical *Schollentektonik* of Germany and surrounding regions (cf., Stille, 1925; Lemcke, 1937; Lotze, 1937, 1971, p. 204–357; Martini, 1937; for a recent re-evaluation of the *Schollentektonik*, see Illies and Greiner, 1976; Meiburg, 1982).

The *Schollentektonik* in Germany and surrounding areas (Fig. 2.10), reaching from the Polish Trough

(Pozaryski and Brochwicz-Lewinski, 1978) to eastern France (e.g., Burg et al., 1982) expresses a tectonic style characterized by the relative motion of rigid and semi-rigid blocks with respect to one another, along the margins of which compressive, distensive, and strike-slip deformation takes place (cf., Lotze, 1937, 1971, esp. p. 215). This tectonism is associated with dominantly alkaline magmatism (cf., Wimmenauer, 1974), and especially in the north, contains an important intracutaneous detachment horizon, namely the Zechstein salt (Lotze, 1953, 1957, 1971, esp. p. 216; Meiburg, 1982; Jaritz, 1987). In some places in northern Germany, salt-controlled detachment horizons also occur in the Triassic, and locally, the Zechstein and Triassic salt horizons may be mixed tectonically.

What created the multi-block structure shown in Fig. 2.10, and when, are still questions of dispute, but it seems that a general NW-SE-oriented broad right-lateral shear zone connecting the Carpathian area with the North Sea and reaching from the edge of the East European platform well into France and the Benelux countries is the most satisfactory explanation for the origin of these *Schollen* (e.g., Arthaud and Matte, 1977; see also W. H. Ziegler, 1975; for a different, but partly equivalent interpretation in terms of N-S compression, see Pratsch, 1982, esp. Fig. 12). The three major lineaments of the European basement, namely the Tempelburg-Kujawian uplift, the Fair Isle-Elbe Line fault zone, and the Danube-Pfahl-Frankian Line have functioned at least since the middle Jurassic as right-lateral shear zones[7] (perhaps since the Permian: see Arthaud and Matte, 1977; Pratsch, 1982; or even pre-Permian: see Ziegler, 1975, Fig. 6) and they were connected with each other through numerous cross elements oriented mainly NE-SW, compatible with tension-gash, conjugate-Riedel and rotated-tension-gash and conjugate-Riedel orientations (cf., Wilson, 1960; Tchalenko, 1970; see also Arthaud and Matte, 1977). Many of these cross elements have become extensional zones during the middle Jurassic to middle Cretaceous and accumulated thick strata. The main NW-SE-striking strike-slip systems also had, at this time, an extensional component. This extensional component along the main shear systems turned into compression in the Late

Cretaceous and remained so until the Eocene in many of them, extending well into the late Cenozoic in some. The NE-striking cross elements remained largely extensional through the Cenozoic, although a few of these also turned compressional (e.g., Martini, 1937; Meiburg, 1982). The cause of the Cenozoic (especially post-Eocene) events has been commonly ascribed to the Alpine collision in the south (Illies, 1974a,b; Illies and Greiner, 1976; Şengör, 1976b; Dewey and Windley, 1988). This collision reactivated many of the Mesozoic - earliest Cenozoic structures and created new ones within the larger blocks that had been already outlined by Mesozoic tectonics (e.g., the Filder Graben zone near Stuttgart or the Schweinfurt Graben north-northwest of Nürnberg: Fig. 2.10). The jostling of blocks created major grabens only in the Bohemian Massif (the Eger rift of late Oligocene age: Lotze, 1971, p. 190–193) and in the Rhenic Massif (the Lower Rhine Rift), plus the continuing rifting on the Upper Rhine impactogen. Smaller extensional structures, most likely only grabens, continued to extend along both sides of the Hessian depression (e.g., the Egge and the Leine grabens: cf., Meiburg, 1982) and farther north at the northwestern corner of the Subhercynian basin (Illies and Greiner, 1976).

The Rhenic and Bohemian massifs plus the South German block were internally disrupted elsewhere too, as indicated by alkaline basalts, phonolites, and other lavas. Small pull-apart basins also formed along the NW-SE-striking main shears in the Cenozoic, especially along the Fair Isle-Elbe line (Illies and Greiner, 1976).

The picture one obtains of the central European block mosaic in the Cenozoic is, then, one of basin formation in diverse orientations and through diverse mechanisms with variable magnitudes of extension across or indeed along them. Some basins pass along their trend into transpressional faults (e.g., pull-apart basins along the Fair Isle Elbe line), others pass into transtensional systems (e.g., grabens in the Hessian depression), and others have changed character during their evolution (e.g., Upper Rhine rift). In this mosaic, basins have formed at diverse times following the collision in the Alps from the Eocene to the Plio-Pleistocene! All of these observations can be most comfortably interpreted in terms of jostling in a

loose-block mosaic driven in front of the main compressional front of the Alps, indeed resembling drifting pack ice.

Differences Between Collision-Related Rifts and Aulacogens

Both aulacogens and most varieties of collision-induced rifts are kinds of "high-angle rifts" with respect to the associated orogenic belts and/or ocean margins (cf., Şengör et al., 1978). For this reason, many of them have been mistaken for the other in many parts of the world (e.g., Wilson, 1954 mistook the Ottawa-Bonnechere aulacogen [cf., Burke and Dewey, 1973] for an impactogen; Burke, 1977, misidentified the Upper Rhine impactogen as an aulacogen), in most cases simply because the authors did not appreciate the differences between the two fundamentally different classes of rifts. In the following paragraphs, I briefly outline their differences to give the reader rules of thumb to distinguish the two groups.

In sharp contrast to aulacogens, collision-induced rifts do not have any stratigraphic relation to the ocean from which the orogen adjacent to these rifts originated. The pre-rift basement of collision rifts is made up of the rocks of the orogenic fore- or hinterland, and their rift fill is clearly post-collisional. Post-rifting doming will give rise to clastic wedges that, along with the correlative fanglomerates of the rift fill, would reveal the age of doming, as in the case of the Rhine rift and the Baikal rift.

The decisive test of whether a high-angle rift is an aulacogen or an impactogen would be the detailed study of the stratigraphy of its sedimentary fill: sedimentary fill correlative with the continental-margin assemblages would indicate that the rift is an aulacogen, whereas one correlative only with the molasse sequences of the associated orogen would suggest an impactogen.

However, unless the rift is subsequently deformed, its pre-rifting basement and the sedimentary fill of its rift trough are commonly not accessible to field examination. Even if the sedimentary record is revealed, either by drilling or by geophysical methods, it is commonly nonmarine, and therefore, presents difficulties

for dating. Thus, identification in the stratigraphic record of the adjacent mountain belt opposite a high-angle rift of an exceptionally large delta and a deflection of the orogen toward the rift at this spot (e.g., the Vilyuy rift and the Verkhoyansk fold-and-thrust belt in eastern Siberia: Parfenov, 1984, Fig. 13; Gusev et al., 1985) may be used as another, possibly more practical, criterion for identifying the rift in question as an aulacogen.

Early, pre-rift doming and early syncollisional strike-slip movement parallel with the rift axis are additional, but risky criteria for the identification of aulacogens. Impactogens normally do not have associated pre-rift doming, and strike-slip deformation along the graben axis commonly begins late in their history, as shown by the history of the Upper Rhine rift.

Despite all these criteria, however, if ocean opening follows immediately a collisional orogeny and takes place parallel with the latter (e.g., as in the case of the Appalachians), then it may still be difficult to decide whether associated high-angle rifts belong to the collision phase and are thus impactogens, or to the ocean-opening phase, and are thus aulacogens. A further source of confusion would be if a rift originates as an impactogen related to one cratonic margin, and shortly thereafter is reactivated into an aulacogen by rifting along the opposite side. The geological history of the Dnyepr-Donetz aulacogen may be an example of such a "compound" case.

Intracontinental Wrench-Related Rift Basins

Brief History of Thought on Intracontinental Wrench-Related Rift Basins

The great Swiss geologist Arnold Escher von der Linth was possibly the first to have mapped a major strike slip fault in the Säntis Mountains south of the Wildkirchli in the Canton of Appenzell in the 1850's (Suess, 1883, p. 153–154). Suess (1883) was the first to note their widespread occurrence in mountain ranges. Kossmat (1926) and von Seidlitz (1931, esp. Chapters 40 and 41) were the first to imply that lateral motion of rigid or semi-rigid blocks may create extensional

basins of significant size. Becker (1934) and Lotze (1937) explicitly showed how strike-slip motion may generate such basins by extension parallel with strike slip and by transtension, respectively. These two types of rift basins constitute the two main types of intracontinental wrench-related rift basins that we recognize today. That is why I prematurely end my historical survey of the evolution of thought on these basins here and continue the narrative in the next section.

Two Fundamental Types of Intracontinental Wrench-Related Rift Basins

Rift basins that form along strike-slip faults in relation to wrench faulting may be divided into two fundamental classes: 1) Basins that form along wrench zones with a zero component of extension across the zone (Fig. 2.26A), and 2) basins that form along wrench zones with a greater-than-zero component of extension across the zone (transtensional zones in Harland's, 1971, p. 30 terminology: Fig. 2.26B). Of these two, class 1 is really the only purely "wrench-related rift basins". Class 2 may be best described as "wrench-modified rift basins". As Fig. 2.26 shows, these two classes correspond with the basin styles described by Becker (1934) and Lotze (1937), respectively.

Wrench-Related Rift Basins
Wrench-related rift basins result from imperfections encountered in the strike of the wrench fault or fault zone. These imperfections cause local divergences between the strike of the fault or the fault zone and the slip vector. If the imperfection is orientated in such a way as to lead to a local extension along the fault or the fault zone, then it is called a "releasing bend" (Crowell, 1974b) and is contrasted with a "restraining bend", where the imperfection is orientated in such a way as to lead to local shortening (Crowell, 1974b).

All releasing bends give rise to pull-apart basins (Burchfiel and Stewart, 1966), across which the amount of extension equals the cumulative slip along the strike-slip fault since the origination of the releasing bend. Depending on the mode of formation of the releasing bend, the geometry and character of the pull-apart basins that form across them change.

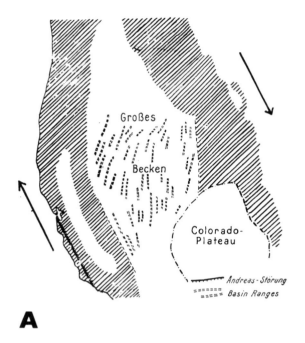

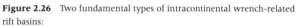

A

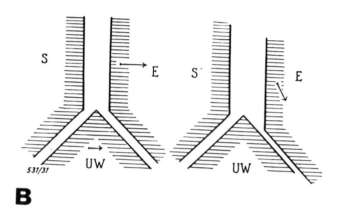

B

Figure 2.26 Two fundamental types of intracontinental wrench-related rift basins:
A. Basins that form along wrench zones with a zero component of net extension across wrench zone are illustrated here by Hans Becker's interpretation of San Andreas fault (*"Andreas-Störung"*)/Basin and Range region. As present Pacific/North America motion is essentially pure strike-slip (cf., Minster and Jordan, 1978, p. 5335), no net extension occurs *across* the shear zone depicted here, although enormous amounts

of extension oblique and parallel to the shear zone have been observed in the Basin and Range province. (Reproduced from Becker, 1934, Fig. 3.)
B. Basins that form along wrench faults, across which there is a net component of extension (i.e., transtensional wrench; for location, see Fig. 2.10). The triple junction depicted is the Eichenberg graben knot in central Germany and illustrates the kinematics of a rift star. Key to abbreviations: S: Solling block; E: Eichsfeld block; UW: Unterwerra block.

Accordingly, in the following paragraphs, I discuss the tectonic and sedimentary relationships in wrench-related extensional basins in the framework of the various modes of formation of the releasing bends.

Restraining bends may also give rise to wrench-related rift basins, by tearing at high angle to the slip vector of one of the fault walls inhibiting motion. In such basins, the amount of extension equals the strike-slip separation since the origin of the basin, minus the shortening across the restraining bend immediately along the fault and reduces to zero away from the fault. Such basins form Carey's (1958) sphenocasms and their formation leads to rotation around vertical axes of major portions of fault walls along large strike-slip faults.

Extensional basins caused by releasing bends along strike-slip faults
Along strike-slip faults, releasing bends may be either of *primary* or *secondary* origin. Primary releasing bends

are those that form along with a through-going strike-slip fault system. Holmquist (1932), Wilson (1960), Tchalenko (1970), Wilcox et al. (1973), Bartlett et al. (1981), and Naylor et al. (1986) discussed how a throughgoing strike-slip fault (or fault system) commonly develops (see also Sylvester, 1988). Figure 2.27 illustrates a general model that underlies the discussions cited above. A general tenet of this model is that before a throughgoing strike-slip fracture materializes, the sheared zone is deformed along a belt considerably broader than the ultimate strike-slip fault zone. This shear belt is deformed first by homogeneous straining in the area of the future shear belt (Tchalenko's, 1970, "pre-peak strength deformation"). If the sheared area has a vertically layered structure (as in any gently deformed or undeformed sedimentary basin), "en-echelon fold" trains may appear at this stage, with characteristic angles of 140° to 150° to the orientation of the shear belt (Fig. 2.27A; see Wilcox et al., 1973, Fig. 9, for photographs

of experimental folds created in a layered clay cake). Just before the peak shear strength is reached, a series of shear fractures with the same sense of shear as the main shear zone appears. These fractures make average angles of 12±1° with the main shear zone and are called *Riedel shears*. At peak strength, the Riedels are rotated to a maxi-

mum of 16°. During this stage also, another set of shears, with a sense of shear opposite to that of the main zone, forms. Its members are inclined to the main shear zone at angles of 78±1°, and they usually form either simultaneously with or just before the Riedels and in materials with a lower water content than those in which this second set does not develop

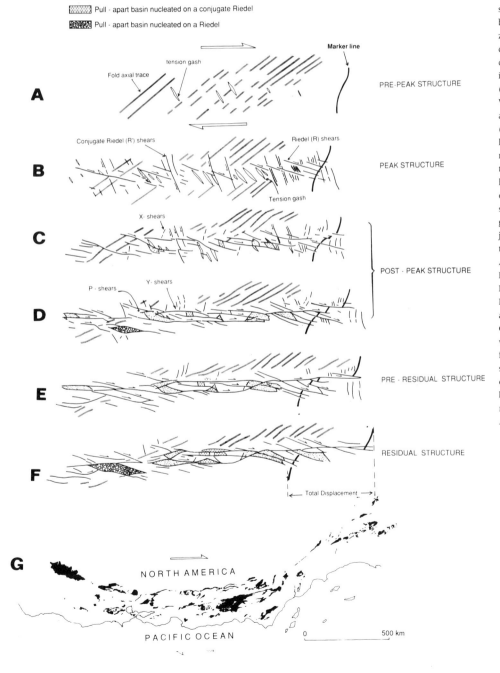

Figure 2.27 Sequence of structures, including pull-apart basins, that develop in a shear zone (**A** through **F**; see text for discussion). This sequence was developed on the basis of experiments and field work by Wilson (1960), Tchalenko (1970), Wilcox et al. (1973), Bartlett et al. (1981) and Naylor et al. (1986). Not all the structures had developed in any one experiment and the present figure illustrates "all possibilities". Notice how pull-apart basins form from openings of various shear-zone structures, such as tension gashes, Riedel shears, and conjugate Riedel shears, and grow to generate a string of basins. **A.** Pre-peak-structure stage; **B.** Peak-structure stage; **C.** and **D.** Post-peak-structure stages; **E.** Pre-residual-structure stage; and **F.** Residual structure stage. **G** shows a real geometry of wrench-related basin patterns from along the San Anrdeas system (however, showing both extensional and compressional basins! Reproduced with permission from the cover of Andersen and Rymer, 1983).

(Tchalenko, 1970). These are called the *conjugate Riedel shears,* which accomodate a maximum amount of shear between 0 and 50%. It is also at this stage that tension gashes may begin to open with strikes ideally orientated some 45° to the shear zone (Fig. 2.27B: Wilson, 1960) and with a direction of opening perpendicular to the strike of the tension gashes. When the peak strength is reached, both the Riedels and the conjugate Riedels are rotated to a maximum of about 16°, and also, some new shears form at angles of about 8°, along with the elongation of the Riedels with "flatter" tips (Fig. 2.27B). Tension gashes may continue to open and rotate into a sigmoidal form. The tips of such sigmoids may link up with Riedel shears and develop small pull-apart basins at this stage (Fig. 2.27B; see Wilson, 1960, esp. Fig. 3). The proportion of shear accommodated by these shears is about 75% (Tchalenko, 1970). At what Tchalenko (1970) termed "post-peak structure stage", an additional set of shears, called by him the *P-shears* form. They are usually formed at an angle of 170° (i.e., approximately symmetrically to the Riedels), and interconnect the latter, as seen in Fig. 2.27C. More than half of the shears are, at this time, inclined less than 4° to the shear zone and all the displacement takes place along the shears. The stage following this, what Tchalenko (1970) called the "pre-residual structure" stage, is characterized by formation of the first continuous horizontal principal displacement shears that isolate between them elongated lenses of essentially passive material (Fig. 2.27D).

It is clear from Fig. 2.27 that the principal displacement shears in a strike-slip zone come into being by interconnection of Riedels and P-shears, plus, in areas of incompatibility (i.e., where asperities form), by the additional generation of "flat" shears. A complex array of factors determines which particular Riedels and P-shears will link with one another to form the principal displacement shears (i.e., the throughgoing fault zones). These factors range from the anisotropies in the failing medium (both material-and structure-governed) through the displacement rate (Hempton, 1983) to the state of stress before wrenching begins (e.g., Naylor et al., 1986).

Figure 2.27D shows how both releasing and restraining bends may come into existence together with the main displacement shears, being nucleated on former Riedel shears or tension gashes and more rarely on the conjugate Riedels. Depending on the structure on which they nucleate, the resulting pull-apart basins will have various geometries and initial overlap distances of the master faults (*sensu* Rodgers, 1980, Fig. 3; Segall and Pollard, 1980, call the negative overlap *separation*, and they call *step*, what Rodgers, 1980, calls separation; it is Rodgers' terminology I follow here). In a numerical study, Rodgers (1980) has shown that the overlap and separation of the two master faults bounding a pull-apart basin are important parameters in determining vertical motions of the ground and the orientation of faults that will form around the ends of the principal faults. They therefore determine the geometry of the pull-apart basin and the topography of its surroundings.

One important aspect of Rodgers' models is division of the locus of extension into two nodes as the pull-apart basin lengthens parallel with the slip along the master faults. This is likely to happen in all pull-apart basins, as soon as the master-fault overlap equals the fault separation. Thus, in most well developed pull-apart basins, we expect to find two separate nodes of subsidence at the ends of the basin, separated by a "central horst", if the models of Rodgers (1980) apply.

Few pull-apart basins have been investigated in sufficient detail to see whether Rodgers' (1980) prediction is fulfilled. Rodgers (1980) was unable to cite any examples to corroborate his model with two loci of subsidence at the ends of a pull-apart basin. Both active and fossil examples of this situation, however, are known. The Dead Sea pull-apart basin now has two sub-basins, separated by the Lisan Peninsula swell (see Neev and Emery, 1967, Fig. 8, for a bathymetric map of the Dead Sea) and this geometry also existed during the Pleistocene, when the Dead Sea was occupied by the larger pluvial Lake Lisan (Neev and Emery, 1967). The similarly active Vienna pull-apart basin, a "thin-skinned" structure, also has two loci of maximum subsidence, located at both ends of the depression bounded by two major strike-slip faults (Royden, 1985, esp. Fig. 14). Finally, the Jurassic-Cretaceous fossil Soria basin in northern Spain has two loci of maximum sediment thickness at both its ends, although they are not exactly coeval.

The above considerations give us the first clues as to what sedimentation in a primary pull-apart basin ought to be like. For negative or null overlap, no high ground is created outside the pull-apart basin, unless it be a result of small fold development at the ends of the main fault strands (cf., Segall and Pollard, 1980, esp. Figs. 9d and 11d). The low ground that forms as a result of extensional faulting in the separation area has an "island" of no extension in the case of negative overlap (Segall and Pollard, 1980, esp. Figs. 9b and 11b). But as the main faults propagate to reach a condition of zero overlap, that "island" is eliminated and a zone of subsidence is created with a long axis connecting the tips of the propagating main shears (Rodgers, 1980, Fig. 4d). At this stage, still no high ground is necessarily created outside the pull-apart basin.

Figure 2.28 displays the predicted sedimentation patterns and the distribution of facies in a primary pull-apart basin. In the negative-overlap stage (Fig. 2.28A), normal faulting leads to clastic sedimentaton in the basin that is fed both from the basin edges and from the internal "island". Absence of highlands around the basin precludes prediction of preferred provenance sites for the clastics, which are functions of local geomorphic and climatic factors.

By the time the zero overlap stage is reached (Fig. 2.28B), the internal clastic source (the "island") is eliminated and the basin is fed only peripherally. At this time, the "ends" of the basin (i.e., the sides not delimited by the main shears) are the most likely avenues of sediment intake, because they lie lower than the "sides" of the basin (i.e., the edges of the basin bounded by the main shears) (Rodgers, 1980, Fig. 4d).

As overlap of the main shears reaches a value of 1/2 of the separation (Fig. 2.28C), two ridges trending almost perpendicularly away from the tips of the main shears begin rising, as the basin is elongated between the propagating main shears, still with a central depocenter. These ridges likely become sediment sources, and thus increase the clastic flux into the pull-apart basin and probably also the grain size of the incoming clastic sediments.

Further elongation of the pull-apart basin by increasing offset along the master shears results even-

tually in the overlap reaching a value equal to the separation (Fig. 2.28D). In this case, extension is localized at the "ends" of the pull-apart basin, creating two distinct nodes of subsidence that commonly divide the pull-apart basin into two "sub-basins". These sub-basins continue to be fed by the erosion of the highlands that trend away at high angles to the main shear zone from the tips of the master shears and may temporarily protect the center of the pull-apart basin from the axially fed sediment influx, in a way entirely reversing the situation at the "zero overlap" situation. A simple corollary to this is that, in an elongate pull-apart basin (Fig. 2.28E), a uniform total sediment thickness in the basin may not necessarily imply a layer-cake stratigraphy! As Rodgers (1980, Fig. 7) showed, further elongation of the primary pull-apart basin simply accentuates the situation attained by making separation equal to overlap.

Secondary releasing bends along a strike-slip fault form following a finite interval of time after a throughgoing fault strand is established. How this comes about has been investigated by Merzer and Freund (1975), J.F. Dewey (unpublished, 1979), Mann et al. (1983), and Şengör et al. (1985). The following is a brief synopsis of the main mechanisms for the formation of secondary releasing bends:

1) Strike-slip fault "buckles" during motion as a consequence of the compressional buckling of its bounding crustal boards (Merzer and Freund, 1975). This phenomenon commonly affects a family of parallel strike-slip faults, separated from one another by crustal boards of similar width (Merzer and Freund, 1975).

2) Strike-slip fault is intersected by a zone of high convergent and/or divergent strain. The segment affected by the intersection may be rotated away from its initial strike and thus may change its orientation with respect to the slip vector, causing either a releasing or a restraining bend (Mann et al., 1983, Fig. 3, mechanism "C"). Constriction of an escaping continental wedge from a zone of convergence (Mann et al., 1983, Fig. 3, mechanism "D"; e.g., the Anatolian Block or Scholle escaping from the Turkish-Iranian High Plateau convergent zone: Şengör and Kidd, 1979; Şengör et al., 1985) is a special case of such a situation. Another special case is

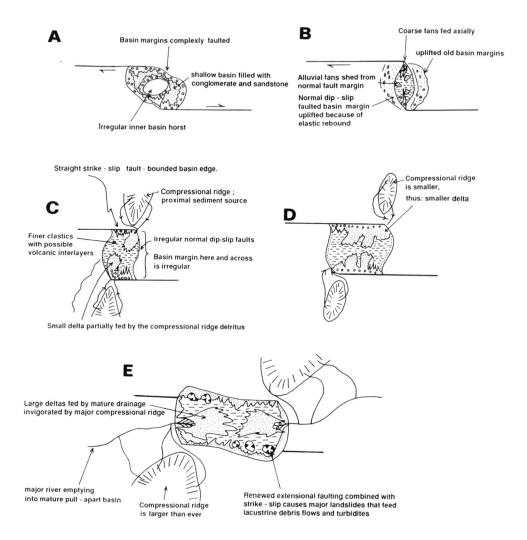

Figure 2.28 Evolution of sedimentary facies in a developing pull-apart basin, located in the temperate zone and in dominantly non-calcareous terrain.

A. Separation is 1/2 of negative overlap. Notice here that an "insular," unextended region functions as an inner-basin horst, providing intra-basinal sediment.

B. Zero overlap. This is equivalent to an "opening" conjugate Riedel shear.

C. Separation is twice the overlap. Note enlargement of highlands outside pull-apart basin and corresponding increase of influx of clastic sediment into the basin, commonly building deltas and subaqueous fans emanating from the two basin corners facing the highlands.

D. Separation equals overlap. The highlands are diminished, but they still supply material to commonly subaqueous fans in the basin, whose stratigraphy would exhibit an upward fining sequence with respect to the last frame.

E. Separation is twice the overlap. The relief between basin floor and highlands has increased dramatically compared with the last frame and normal faulting has begun invading territory outside bounding faults. Increased relief may increase clastic influx, and aided by juvenile normal faulting disrupting steep bounding master fault walls, may begin triggering massive landslides. *All this is most likely to be interpreted in sedimentary/structural record of pull-apart rift as a "phase of renewed rifting," wheraes in reality no increase in rate of strike-slip faulting is introduced.* This would thus give rise to an interpretation in terms of episodic tectonism, while in reality tectonism is perfectly uniform, but not its record!

the rotation of a strike-slip fault away from parallelism with the slip vector along its original course in a general area of quasi-homogeneous shortening as, for example, in Central Asia (Şengör, 1987d, Fig. 22) or in the Turkish-Iranian High Plateau (Şengör et al., 1985, Fig. 10). In such a situation, the original strike-slip fault may be broken into a series of pull-apart wrench-fault segments (mechanisms "A" and "B" in Fig. 3 of Mann et al., 1983).

3) Asperities along a strike-slip fault may cause local rotation of the strain field and create splay faults which begin to accommodate motion. These splay faults are at angles greater than zero degrees to the slip vector and may thus nucleate restraining and/or releasing bends (J.F. Dewey, personal communication, 1979).

The pull-apart basins that will form along secondary releasing bends are in principle, no different from those that form along primary releasing bends, except for the angle their long axes make with the slip vector. This angle is, in principle, more variable in the case of pull-apart basins forming along secondary releasing bends because these bends do not necessarily nucleate on structures that have fixed orientations with respect to the orientation of the slip vector, such as tension gashes or Riedel shears.

Extensional basins formed at restraining bends along strike-slip faults

Sphenochasms are the only major type of extensional basins that form at restraining bends along strike-slip faults. Sedimentation in such basins ranges from entirely continental to oceanic, depending on the amount of extension. Commonly, total thicknesses of sedimentary rocks decrease away from the major bounding fault, as seen, for example, in the case of the northwest Chinese graben complexes (Fig. 2.18: Ma et al., 1982; Burchfiel and Royden, 1991).

If no other factors interfere, then major relief in sphenocasms forms near the bounding strike-slip fault, both because of this faulting, and because it is near the bounding strike-slip fault that the amount of extension (and the presumably resultant shoulder uplift) is greatest (e.g., as in the Rio Grande sphenochasm: Baltz, 1978, esp. p. 211; Chapin, 1979, p. 3; see also Ingersoll

et al., 1990). This results in coarse clastics being deposited where the basin is widest and the grain size diminishes progressively away from the main bounding strike-slip fault (see for example the "Late Tertiary Palaeogeography of China" on p. 126 in Wang, 1985, for a map of the Northwest China Graben System showing a dimunition of grain size towards the north).

Wrench-Modified Rift Basins

Wrench-modified rift basins are those extensional basins in which extension occurs in an oblique orientation, imposing on the basin also a strike-slip component. Although many pull-apart basins and sphenochasms may be included in this definiton, I intend to exclude them by imposing a further restriction on the definition of wrench-modified extensional basins, viz. that the main tectonic grain in which they form must be parallel with their long axes. This excludes the presence of major strike-slip systems with strikes oblique to the basin, along which the basin may have opened as a pull-apart. In my sense, wrench-modified extensional basins are equivalent to Harland's (1971, p. 30, esp. Fig. 2) *transtensional* basins and also correspond with *leaky* strike-slip faults.

As I point out above, most rifts have transtensional components. However, when transtension becomes the characteristic regime of an extensional basin, it develops certain tectonic features peculiar to this environment that influence sedimentary processes and resultant sedimentary rock bodies. Figure 2.29A illustrates a simplified block diagram of a transtensional basin, in which normal-fault trains oblique to the main basin margins resemble tension gashes that form in strike-slip environments. Such oblique trends dominate many transtensional basins (e.g., Gulf of California: Kelts, 1981, 1982; Sea of Marmara: Şengör et al., 1985; Barka and Kadinsky-Cade, 1988) and the following discussion on the sedimentation and tectonics in such basins is mainly based on examples drawn from the now well exposed Central High Atlas basin in Morocco (Beauchamp, 1988; Laville, 1988).

Beauchamp (1988) and Beauchamp and Petit (1983) have shown, following earlier authors (e.g., Dewey et al., 1973; Van Houten, 1977; Manspeizer et

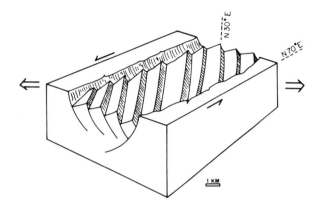

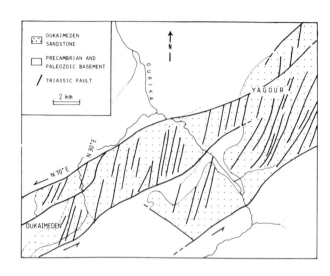

Figure 2.29 Geometry of faulting in transtensional basins:
A. A schematic block diagram illustrating the pattern of faulting in fossil intracontinental transtensional rift basin on the basis of data gathered from the Esçour basin in the High Atlas of Morocco. (Reproduced with permission from Beauchamp, 1988, Fig. 20-10.)

B. Pattern of Triassic faulting in a fossil transtensional intracontinental rift located in the Ourika region of the High Atlas of Morocco. (Reproduced with permission from Beauchamp, 1988, Fig. 20-11.)

al., 1978; Şengör et al., 1984), that the Triassic-Jurassic geological history of northwest Africa was dominated by transtension that detached the largely Hercynian deformed Moroccan and Oran mesetas from the Precambrian Saharan platform (Fig. 2.30). The present structure of the Central High Atlas mountains reflects the geometry following early Cretaceous through Cenozoic compressional deformation that moderately to gently overprinted Permo-Jurassic extensional structures. As Beauchamp (1988) and Laville (1988) have pointed out, following earlier authors, the structure of the Central High Atlas mountains is dominated by basins and ridges oriented obliquely to the main ENE grain of the mountain chain. Beauchamp (1988) has shown that the main oblique features correspond with NNE-trending ridges formed as normal-fault-bounded horst blocks or half horsts, as displayed in Fig. 2.30. Laville (1988) has shown that extensional structures form a more complicated network in the transtensional basin than the simple diagram in Fig. 2.29 would lead one to suspect. He distinguished three main kinds of oblique "ridges" that formed during transtension:

1) Extensional ridges. These ridges trend NNE to NE and consist of two distinct varieties:

1a) Extensional ridges without intrusions. These are commonly monoclinally tilted blocks or two-faced horsts bounded by normal faults. Some show a sigmoidal outline, ending at NNE-striking strike-slip faults (many now turned into steep thrusts) and others splay away from such faults.

1b) Extensional ridges with intrusions. These are anticlinal structures cored by early Mesozoic intrusions of gabbros, leucogabbros, and syenites in some places forming simple dikes and in others, ring complexes . In map view, they are sigmoidal to lozenge-shaped and appear to have filled in tension gashes opened as a result of transtension.

Einsele (1982, 1986) has developed a model to elucidate the evolution of such magmatic centers in transtensional systems and their relationship with basin sediments, which is illustrated by Fig. 2.31. Einsele's model shows that upwelling basaltic magma in pull-apart basins or in small spreading ocean basins in transtensional environments (e.g., the Guaymas basin in the Gulf of California: Curray et al., 1982) not only creates dike complexes or large intrusive bodies, but also sills in soft sediment accumulating in such basins. The thickness of the sill-sediment package is controlled chiefly by the rate of sedimentation. Extremely high sedimentation rate surpassing the spreading rate causes filling up of the basin and hampers the rise of magma to the surface (as in much of the High Atlas Mountains). Very low sedimentation rate (e.g., those prevailing at the major mid-oceanic

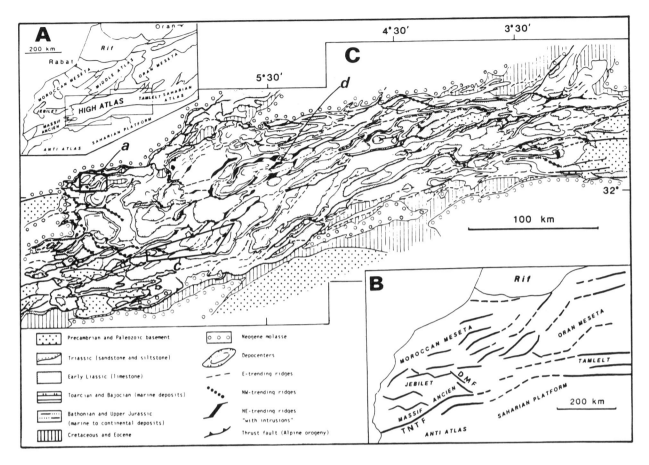

Figure 2.30 The present structural picture of the Central High Atlas Mountains, Morocco (reproduced with permission from Laville, 1988, Fig. 21-1,): **A.** Tectonic units of Morocco; **B.** Late Hercynian fault pattern in Morocco that has influenced nucleation of Alpide faults. (DMF = Demnat fault; TNTF = Tizi n'Test fault.) **C.** Simplified geological map of the Central High Atlas Mountains. Note on this map the spatial relationships between NE-trending "extensional" ridges with NW-trending "compressional" ridges, betraying a left-lateral strike-slip regime, even after intense N-S-directed Alpide shortening in Late Cretaceous and early Cenozoic.

spreading centers) hardly influence magma extrusion onto sea-floor. Intermediate rates that do not necessarily fill up the troughs may still generate sill-sediment thicknesses of up to a few-hundred meters (Einsele, 1986). In such basins, "sea-floor spreading" may not create conventional ocean floors, but instead generate what is commonly known under the various designations such as "slate-diabase complexes" (e.g., in the Greater Caucasus: Khain, 1975; Beridze, 1984) or "diabase-phyllitoid complexes" (e.g., in the Strandja Mountains in SE Bulgaria: Chatalov, 1991 and the literature cited therein) or indeed "schistes lustres" (Isler and Pantic, 1980) in the literature. The controversy created by the interpretation of Dewey et al. (1973) of the Atlas Moun-

tains in terms of a transtensional "ocean" although no "true ophiolites" have yet been found in them, likely has resulted from just such a situation. Many of the slate-diabase, or diabase-phyllitoid, or schistes lustres basins in the central and western Tethysides have been interpreted in terms of transtensional troughs (e.g., the Pennine realm in the Alps: Kelts, 1981; the Slate-Diabase Zone in the Greater Caucasus: Şengör, 1990b, 1991c); these "sill-sediment" packages appear to be a hallmark of transtensional basins in general.

2) Compressional ridges. These ridges trend NW and appear either as anticlinal folds or thrust-bounded uplifts (Fig. 2.30). They are commonly smaller and less common than extensional ridges. Like the latter,

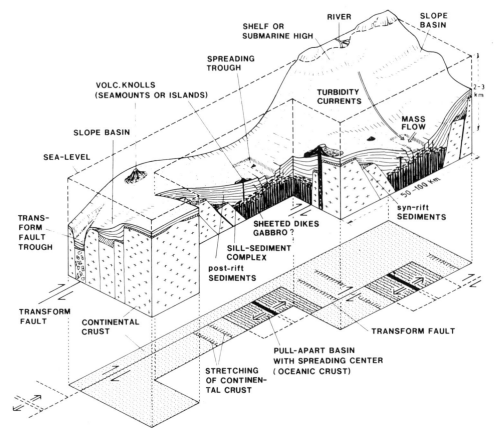

Figure 2.31 Einsele's (1986) model of pull-apart magmatism in transtensional systems. The top shows schematic block diagram, and the bottom shows the plan-view of Gulf of California-type basins (shown as simple pull-aparts for simplification; reproduced with permission from Einsele, 1986, Fig. 3).

they end on ENE-striking former strike-slip faults or splay away from them.

Numerous ENE-striking former strike-slip faults were active during Triassic-Jurassic transpression along the High Atlas Mountains and these acted as relay features connecting oblique elements with one another. These large ENE structures also had a normal dip-slip component (Beauchamp, 1988), reflecting the transtensional component of motion in the opening of the High Atlas basins, as do the considerably higher number and larger size of the extensional oblique ridges and basins than the compressional ones.

All of these structures were growing during Triassic-Jurassic extension, as shown by 1) thickness changes across them, 2) resedimentation by sliding material off their crests and flanks during the early and medial Jurassic (Lee and Burgess, 1978), 3) lateral facies changes between basins and ridges, indicating active uplift during deposition, and 4) local unconformities. Within the main trough of the Central High Atlas basin, all of these features were sediment sources

(Beauchamp, 1988), although internal dynamics of the depositing medium is likely to have controlled the dominant NE to SW dispersal pattern observed by Beauchamp (1988).

The main sedimentary rocks deposited during earlier phases of rifting in the Triassic in the High Atlas near Marrakesh show a progressively greater influence of marine environments, ranging from entirely terrestrial alluvial-fan/playa-lake deposits through shoreline, delta, and lagoonal deposits. During much of the Early and Medial Jurassic, sedimentation produced a thick sequence of limestone and marl (in places up to 7 km thick). In the later Early Jurassic and earlier Medial Jurassic (Aalenian to Bajocian), deposition changed gradually from deeper-water marl to shallower-water bioclastic and oolitic limestone, punctuated by coral-algal reef horizons (Laville, 1988); in the Bathonian, the basin entirely filled up, leading to fluvio-deltaic sediment deposition (DuDresnay, 1979; Jenny et al., 1981). Harmand and Laville (1983) have shown that greenschist metamorphism of the sedimentary rocks (cf., Studer, 1980)

shows concentric zonation away from a potassic center through an argillitic mantle to a prophyllitic periphery, and interpreted these as resulting from hydrothermal processes generated by a fossil geothermal field, similar to those seen in active and ancient fracture zones in the oceans (e.g., DeLong et al., 1979) or in the transtensional Gulf of California basin (e.g., Kelts, 1981; Einsele, 1986).

Discussion

The main purpose of this chapter is to show the tremendous variability in processes of rifting and to provide hints to help the reader to identify ancient rifts preserved in various states of completeness in the geological record. So far, emphasis in the literature has been placed mainly on individual rifts, rather than on systems of rifts. Even when large areas of rifting are covered by different authors, the general tendency has been to talk either about overall processes of stretching of the lithosphere, or about the structure of individual rifts. The main controversies concerning rifting lately have been concentrated either on whether rifting is an asymmetric process, creating a dominantly monoclinic structural fabric or a symmetric one with an orthorhombic fabric, or whether uniform or heterogeneous stretching best accounts for rifts. These polarizations of opinions have commonly come about between individuals or groups working on different rift systems or on different parts of the same rift system. Only later in the history of these controversies, the proponents of the two views have "invaded" a common ground and proposed radically different interpretations of the same tectonic objects.

To avoid falling into the trap of being blinded by peculiarities of any local structure, the first part of this chapter considers rifts as systems, as parts of larger structural entities called *taphrogens*. Such consideration has been common in the case of convergent structures, when they have been grouped into orogenic belts (or "folded belts"). That this has not occurred in studies involving regions of extension was mainly because taphrogens are commonly buried under their own fills, in contrast to orogens

that stand high and naked of sedimentary mantle, providing a much more attractive target for the hammer of the geologist. Indeed, when J.W. Evans took "regions of tension" as the topic of his presidential address to the Geological Society of London in 1925, he complained that "The results of compression have long been studied, but comparatively little attention has been given to the occurrence of tension in areas where it has left evidence of its existence in the form of joints, normal or slip faults (occasionally replaced by monoclinal folds), or of dykes and other characteristic igneous phenomena" (Evans, 1925, p. lxxx).

Once we recognized that rifts occurred in organized families, we had the need to distinguish those "thick-skinned" extensional structures from "thin-skinned" ones. We employed the historical evolution of the two common terms, namely *graben* and *rift*, to aid us in this and we found it convenient, and consistent with historical usage, to restrict the word graben to thin-skinned extensional structures and to use rift for thick-skinned ones. Taphrogens, like orogens, include both thin-skinned and thick-skinned structures, so they consist of both grabens and rifts.

Our next task in our pursuit of the tectonics of extensional regions was a classification of rifts. This we did from three, mutually not completely overlapping viewpoints. We first classified rifts in terms of the plan view of the taphrogens or parts of taphrogens in which they occurred, in other words from a purely *geometrical* viewpoint. Then, in order to elucidate their *kinematics*, we made use of the framework provided by plate tectonics and distinguished intraplate, plate boundary, and triple-junction-related rifts. Finally, from a *dynamic* viewpoint, we were able to distinguish only two classes, namely rifts formed through the active participation of the mantle through convection-driven hot jets, called mantle plumes underlying hot spots on the earth's surface, and those that form by stress systems generated by the two-dimensional evolution of the world plate mosaic, in which the mantle plays only a passive role.

The purpose of our classification is to make an overview of the tremendous variability in the

world's rift population easier, but this classification should not lead us to think that nature rigidly operates in its confines. As Stephen Jay Gould recently wrote "Classifications are *theories* about the basis of natural order, not dull catalogues compiled only to avoid chaos" (Gould, 1989, p. 98, italics mine). My classification also is a theory of rift origin, or, rather, a group of theories nested one within the other. It is proposed to allow us to recognize not only ancient individual rifts but also whole taphrogens even when only small parts of them or even parts of their constituent rifts and grabens are preserved. Figure 2.5 may be thought of as a flow chart within the framework of this identification process. In the text, I frequently stress the formation path of a rift or a taphrogen as a whole, when discussing its origin, for not *every* combination of items listed on this table seems allowed and this may aid us in filling in the missing parts from the records of ancient rifts.

I must stress the importance of recognizing the *continuous spectrum* that exists between the end members displayed on Fig. 2.5. Nature does not know about my classification scheme: the classes of this scheme *grade* into one another in nature and only rarely form well defined pigeon holes. Although pigeon holes are necessary for creating pedagogic aids and a structured history, rational thought demands continuity of process. It is within this consciousness of the continuity of process that the classes of this classification are to be understood only as aids to comprehension and retention.

Our scheme of classification guides us through a tour of the three most common types of ancient rifts encountered in the archive of historical geology, namely, aulacogens, collision rifts, and intracontinental strike-slip-related rifts. In each case, we find it useful to begin our discussion with a brief historical overview to enable us to see what sort of biases may exist in the handling of the specific problems we are to tackle. We then deal with the various representatives of the three types of ancient rifts, and in this we try, as much as possible, to keep a truly global perspective. The emphasis has been on what there is rather than what there ought to be, following implicit advice given to students of rifts by Warren Hamilton (1985a, p. 362).

Tectonic environments in which aulacogens, collision rifts, and intracontinental wrench-related rift basins form have been considered in greater detail than in most previous discussions of rift origin, and even among these three groups, *transitions* have been noted. In rift studies, especially in considering the historical geology of rifts, processes ought to be thoroughly considered and the record seen in rifts should be interpreted, as much as possible, in the light of processes of which we have first-hand knowledge (i.e., presently active processes). *The history of rifts does not consist only of the record preserved*, which forms only a fraction, and commonly a *small* fraction of the total history. To get the entire history, we must *interpolate* between record fragments and that can only be done soundly in the light of *actualistic* thinking.

The processes to be considered make up a spectrum. It ranges from purely structural ones, creating the rift valleys through igneous and sedimentary processes involved in filling the rifts, to weathering and erosion that destroy the rift. Such seemingly remote processes as the orbital controls on climate have recently opened up new and hitherto undreamed-of high precision in dating and correlating sedimentary contents of rifts. Similarly, hydrothermal-circulation studies along master faults have revealed thermal anomalies in the evolution of rifts not previously considered. All such truly multi-disciplinary studies on ancient rifts tell us that rift studies remain an exciting field that carries much promise for new and ground-breaking advances. This acquires added significance when we consider that Coppens' "east-side story" holds the formation of the east African rifts system directly responsible for generation of the hominids, and thus, of humans (Coppens, 1991)

Acknowledgments

Much of what I know about rifts has been learned from Kevin Burke during long years of collaboration, both while I was his student in Albany and later. His selfless devotion to geology and to his students, his inexhaustible knowledge about the geology of "our beloved planet" as he is fond of saying, and his contagious enthusiasm have been responsible for my inter-

est in rifts and for much of what I learned about them. I consequently take honor in dedicating this modest chapter to him. John F. Dewey also taught me an enormous amount about rifts. The late regretted J. Henning Illies was once my guide to European rifts and especially to the Upper Rhine rift. Warren Hamilton, Clark Burchfiel and Brian Wernicke have variously instructed me about the Basin and Range, Paul E. Olsen and Cahit Çoruh about the U.S. east coast rifts, and my old friends and mentors Professors İhsan Ketin and Sirri Erinç introduced me to the western Anatolian rifts. Nezihi Canitez, Naci Görür, James Jackson, Fuat Şaroğlu, and Yücel Yilmaz helped with actual collaboration in my studies on the Turkish taphrogens, both fossil and active. Dan McKenzie has been a constant aid and source of theoretical knowledge on extensional tectonics, although many a time in our discussions I had to stay behind, tangled in a thicket of equations! I am indebted also to William R. Dickinson, Kenneth J. Hsü, and Rudolph Trümpy for much instruction in general tectonics and sedimentation. Last but not least, I thank Ray Ingersoll, both for inviting this chapter and for bearing with me while its various drafts took shape.

Further Reading

Burke K, 1977a, *Aulacogens and continental breakup:* Annual Review of Earth and Planetary Sciences, v.5, p. 371–396.

Hamilton W, 1987, *Crustal extension in the Basin and Range Province, southwestern United States:* Geological Society of London Special Publication 28, p. 155–176.

Hancock PL, Bevan TG, 1987, *Brittle modes of foreland extension:* Geological Society of London Special Publication 28, p. 127–137.

Hempton MR, 1983, *The evolution of thought concerning sedimentation in pull-apart basins,* in Boardman SJ (ed), Revolution in the earth sciences: Kendal/Hunt, Dubuque, p. 167–180.

Hutchinson DR, Klitgord KD, 1988, *Evolution of rift basins on the continental margin off southern New England,* in Manspeizer W (ed), Triassic-Jurassic rifting, Part A: Elsevier, Amsterdam, p. 81–98.

Ingersoll RV, Cavazza W, Baldridge WS, Shafiqullah M, 1990, *Cenozoic sedimentation and paleotectonics of north-central New Mexico: implications for initiation and evolution of the Rio Grande rift:* Geological society of America Bulletin 102, p. 1280–1296.

Manspeizer W, 1985, *The Dead Sea rift: impact of climate and tectonism on Pleistocene and Holocene sedimentation:* Society of Economic Paleontologists and Mineralogists Special Publication 37, p. 143–158.

Şengör AMC, 1987c, *Cross-faults and differential stretching of hanging walls in regions of low-angle normal faulting: examples from western Turkey:* Geological Society of London Special Publication 28, p. 575–589.

Şengör AMC, Burke K, Dewey JF, 1978, *Rifts at high angles to orogenic belts: tests for their origin and the upper Rhine graben as an example:* American Journal of Science, v. 278, p. 24–40.

Wernicke BP, 1985, *Uniform-sense normal simple shear of the continental lithosphere:* Canadian Journal of Earth Sciences, v. 22, p. 108–125.

Endnotes

[1] I must emphasize that I am very skeptical about "extensional orogenic collapse" under the weight of uplands alone. Everywhere it has occurred, an additional process, such as tectonic escape, seems to have aided it. Moreover in many places where it is proposed to occur, the number of structures responsible for extension and the actual amount of stretching are far smaller than proposed (e.g. the Alps), and the direction of stretching is at variance with the solely gravity-driven extensional orogenic-collapse model (e.g. the Betic Cordillera).

[2] Jarrard's (1986a) seven or even five different types of arc behavior are far too detailed to be applicable to a general survey of the historical geology of arcs and are therefore not used here. For a discussion, see Şengör (1990a, p. 66).

[3] A synclise is essentially a commonly equant basin in the Soviet terminology: see Dennis (1967, p.147–148) for details; Shatsky even used it for what he later called aulacogens!

[4] Shatsky originally said "syneclise", but because most of his synclises since have turned out to be either oceans (e.g. the North Caspian "syneclise": see Burke [1977a]) or flexural foreland basins, there is no danger in exchanging "major basin" for the "syneclise" in the definition.

[5] For rift propagation in environments of continental separation, see esp. Courtillot (1982),Courtillot and Vink (1983), and McKenzie (1986); for a present-day example of a propagating rift, see the Sadko trough and its southern continuation in the Chersky Mountains where the Balagan Tas volcano and active normal faulting indicate that the Gakkel spreading center is propagating into Asia: Vogt and Avery (1974), p. 104ff and Fig. 14; Fujita and Cook, 1990, esp. Fig. 2

[6] Known since the beginning of the nineteenth century, when folklore referred to them as "the footprints of Noah's Raven" (see Colbert, 1968, p. 37ff).

[7] These major lineaments are not everywhere expressed as single fault lines. Especially in the north German/Polish basement, they are rather broad zones of complicated faulting and some folding, resulting from the distribution of shear in a broader zone than in the south, and thus do not show up on seismic-reflection lines. This should not, therefore, be interpreted as their being absent in the north!

Continental Rifts and Proto-Oceanic Rift Troughs 3

Michael R. Leeder

Introduction

Active extension or stretching of continental lithospheric plates causes surface deformation, volcanism and high heat flow due to the effects of normal faulting and changes in crustal and mantle thickness, structure and state. The elongate rift basins and marginal uplifts that result from these processes are prominent tectonic features of the continental surface. The major areas undergoing active extension and rifting are shown in Fig. 3.1. Evidence from these areas must be used to interpret the tectono-sedimentary evolution of formerly active rifts (see Chapter 2), such as those that bound Atlantic-type continental margins, and which are now overlain by thick sequences of younger marine sediments (see Chapter 4).

Classic linear rift valleys such as those of East Africa (eastern and western branches), Baikal, Rhine and Rio Grande, occur as narrow (50–100km), long (up to 1000km) features, much segmented in detail, atop major regional surface upwarps or domes covering many thousands of square kilometers. By way of contrast, wide (>1000km) rift provinces like the Basin and Range of the western United States include a plethora of individual narrow (15–30km) rift basins separated by range uplifts.

Rift basins and flank uplifts become depositional sinks and erosional sources, respectively, for sediment. The structural and sedimentary architecture of rift-basin infills is often referred to as the *syn-rift* phase of lithospheric extension. Once the process of active extension has ceased, the thermal anomaly due to the thinned lithosphere decays with time according to Fourier's law of heat conduction (McKenzie, 1978a). Regional thermal subsidence, broad on the scale of individual rift blocks, then results. This is called *post-rift* subsidence or, more evocatively, *sagging* (Leeder, 1982). The Tertiary North Sea basin of NW Europe is perhaps the best known example (Barton and Wood, 1983), although most continental margins go through this phase of subsidence. The creation of a new ocean necessarily proceeds by complete continental lithospheric separation. The ocean evolves via proto-oceanic rift troughs, such as the Red Sea or Gulf of California, active extension probably giving way to thermal subsidence and shelf sag once the first ocean crust has formed and extension in the continental/transitional region has ceased. The entire extensional spectrum, from rifting to oceanic crustal generation, may be seen in the Gulf of Suez/Red Sea/Ethiopian rift areas.

The process of lithospheric stretching that causes rifts may be due to several mechanisms; it takes place in several tectonic environments (see Chapter 2) and is characterized by different types of basin-bounding normal faults. The possible mechanisms are largely controlled by the thermal state of the lithosphere and asthenosphere, both locally and regionally, which also controls the modes of extension (Buck, 1991). The tectonic environment of stretching is controlled by regional plate motions. Extension may occur in areas of continental crust adjacent to young oceans, in backarc basins, in continental interiors and in thickened crustal orogens.

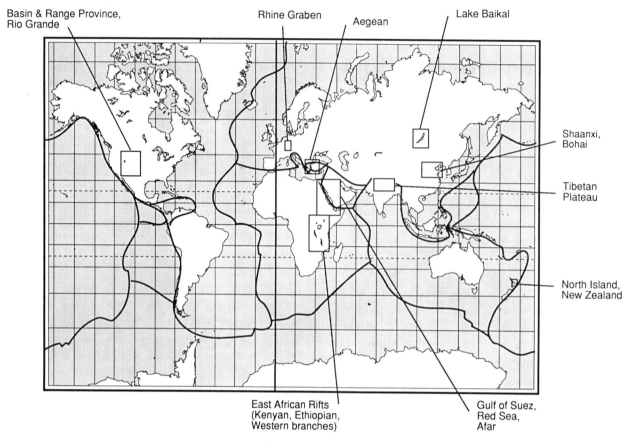

Basin & Range Province, Rio Grande

Rhine Graben

Aegean

Lake Baikal

Shaanxi, Bohai

Tibetan Plateau

North Island, New Zealand

East African Rifts (Kenyan, Ethiopian, Western branches)

Gulf of Suez, Red Sea, Afar

Figure. 3.1 World map (Mercator projection) with plate boundaries (solid lines), to show main areas of active continental extension. (Also see Fig. 2.2.)

General Aspects of Continental Extension

Rheology and Extension

Lithospheric stretching acts upon a compositionally and mechanically layered lithosphere (Fig. 3.2). Variations in material strength arise because of both compositional and thermal structure (Chen and Molnar, 1983; Lynch and Morgan, 1987). The most significant strength change relevant to the surface deformation explored in this chapter occurs in the crustal brittle-to-ductile transition. Above this zone, brittle deformation occurs by faulting. Below, extensional deformation is of pure-shear type, taken up by some form of slow, possibly continuous, plastic creep at a rate determined by imposed tectonic stresses and mineralogical composition of the lower crust. Olivine-dominated mantle rocks are initially more resistant to

deformation than the lower crust, but become progressively less so toward the base of the lithosphere. It should be noted that the depth of the brittle-to-ductile transition is extremely sensitive to thermal gradient; increased gradients cause the transition to migrate upward. It is, thus, possible that the maximum depth of normal faulting may be significantly different from the normal 10–15 km, depending on the regional thermal regime and crustal thickness. Much of the debate concerning the varying processes and products of extensional tectonics has arisen because workers have often applied their experiences in one thermal or thickness regime to those of another without due regard for changing boundary conditions. In particular, as noted above, the extension of continents can result in narrow or wide zones of rifting, dependent upon material parameters, thermal state and strain rate (Buck, 1991).

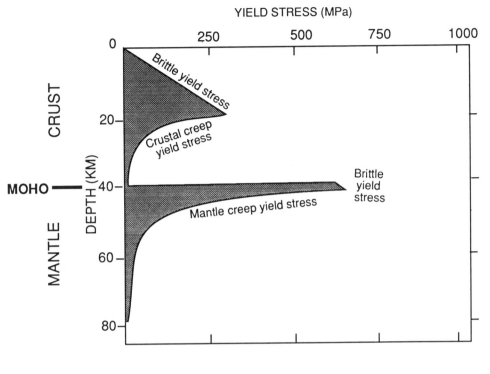

Figure. 3.2 Variation of strength (yield stress: heavy curve) with depth for lithosphere extending at 10^{-14} sec^{-1} with heat flow of 42 mW m^{-2} (after Lynch and Morgan, 1987). The integral of this profile over the full depth of the lithosphere is the total yield stress of the lithosphere. Creep rheologies are for 40 km-thick mafic-dominated crust and ultramafic mantle. Note two brittle-ductile transition zones, with brittle deformation at 0–19 and 40–42 km. Remaining areas flow by ductile-dislocation (power-law) creep. The lower crust in an extending regime behaves in ductile fashion and is assumed to be in a near-constant state of steady plane strain due to dislocation creep. Strain energy stored by brittle upper crust is then periodically released by fracture (causing stress drop) during earthquakes. Increase in the strain rate by a factor of 10 depresses creep curves and brittle-ductile transitions by about 4 km; vice versa for a decrease in strain rate. Brittle-ductile transition in the crust occurs at shallower depth for hotter lithosphere.

Extensional Mechanisms: Closed Systems

The simplest case (Fig. 3.3) is that of closed-system rifting, originally called passive rifting (Şengör and Burke, 1978), where local tensional stresses arise from plate-edge and lithostatic forces, and where the input of asthenospheric mass from outside the stretched lithospheric volume occurs passively *as a response to* lithospheric thinning (McKenzie, 1978a). The stretching process fractures the mid to upper crust, causing it to thin as fault blocks rotate. The lower crust and mantle lithosphere probably thin by plastic creep, causing warmer asthenospheric mantle to move closer to the surface. In closed systems, production of magma due to decompressive partial melting of the asthenosphere may occur (Fig. 3.4), initially in small amounts, but increasing as the amount of extension increases (Dixon et al.,1981; Foucher et

al., 1981; McKenzie and Bickle, 1988). Nonuniform extension by pure shear (e.g., Hellinger and Sclater, 1983; Fig. 3.3) describes the state where different rheological layers of the lithosphere have different responses to extension. Some more ductile layers may be stretched over wider areas, although net conservation of mass in the stretched volume must occur (White and McKenzie, 1988). Closed-system extension may also occur due to simple shear (Fig. 3.3), when a single low-angle fault cuts a significant part of the crust or whole lithosphere (Wernicke, 1985; see further discussion below). The overall effects of closed-system stretching are to cause rapid net subsidence in the rift basin, although local or regional flank uplifts may result from fault-block rotation, inherent asymmetry of nonuniform extension, flexure, lateral heat flow, or local magma production (see discussion below).

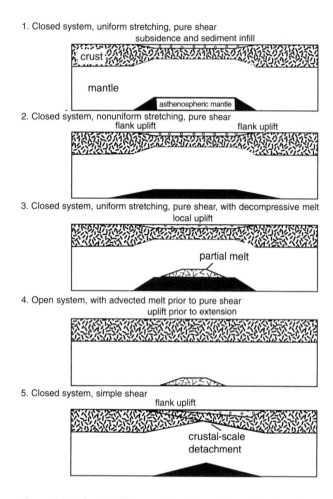

1. Closed system, uniform stretching, pure shear

2. Closed system, nonuniform stretching, pure shear

3. Closed system, uniform stretching, pure shear, with decompressive melt

4. Open system, with advected melt prior to pure shear

5. Closed system, simple shear

Figure. 3.3 Schematic diagrams to illustrate possible combinations of pure and simple shear, uniform or nonuniform stretching and magma production. Local (Airy) isostatic compensation assumed throughout (i.e., lithosphere has small elastic thickness). Surface and upper-crustal deformation by faulting not shown.

Extensional Mechanisms: Open Systems

Continental extension may also occur in an open system (Gans, 1987), originally referred to as active rifting (Şengör and Burke, 1978; Fig. 3.3), characterized by the eruption of voluminous volcanics. In contrast to the closed-system model, the asthenosphere can initially rise independently of the magnitude of lithospheric extension, although it may do so in areas already in a state of deviatoric stress. Production of melt does not occur as a result of the stretching process, but rather, melt is produced during asthenospheric upwelling or is advected into the system atop

rising diapiric plumes. As noted above, the chief effects in mid- to upper-crustal levels will be a regional elevation of the brittle-to-ductile transition. Minor extension occurs due to the updoming itself (Artyushkov, 1973); more important is the wedging effects of dike intrusion (Brown and Girdler, 1980; Baldridge et al., 1991; Davis, 1991). The uplift increases gravitational potential energy and induces deviatoric extensional stresses, and therefore, contributes to any pre-existing regional tension to cause surface normal faulting along the uplifted crustal dome (Neugebauer, 1978; Withjack, 1979).

Advected mantle melt may be added to the volume of stretched or normal lithosphere and asthenosphere by plumes or hotspots (White and McKenzie, 1989; Griffiths and Campbell, 1991; Hill, 1991). Thermal and dynamic effects of these plumes can cause regional uplift of up to 1 km at rates of up to 0.04 mm y^{-1} for typical plume heads of diameter 1000 km with a temperature contrast of 100°C (Hill, 1991). Uplift is postulated to begin long before the plume material intersects the base of the lithosphere. Regional magmatic underplating of pre-existing crust may result from open-system extension (McKenzie, 1984). This process may be responsible for characteristically strongly layered lower crust and very prominent Moho imaged by deep seismic-reflection studies in active or ancient extensional areas (Allmendinger et al., 1987; Warner, 1990).

Rift basins formed by open-system mechanisms are located on surface domal upwarps of regional extent or on continental plateaus kept high (1–2 km mean elevation) by buoyancy forces due to melting, dynamic flow in the asthenosphere, and basaltic underplating. The latter effect remains after the advected hot flux has dissipated by conduction. The late Tertiary development of the Basin and Range province of the western United States, the largest area of continental extension in the world (Fig. 3.5), is consistent with an origin due to open-system processes (McKenzie, 1984). However, it should be noted that the ultimate cause(s) of extension in the Basin and Range is fiercely debated (see reviews by Coney, 1987; Hamilton, 1987; Sonder et al., 1987; Ingersoll et al., 1990). Gans (1987), Gans et al., (1989), and Buck (1991) have provided unified con-

A.

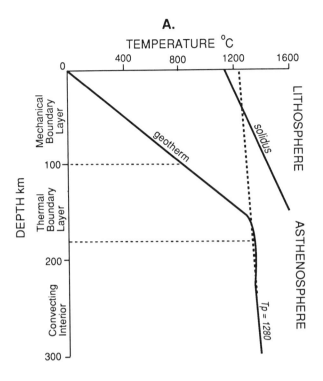

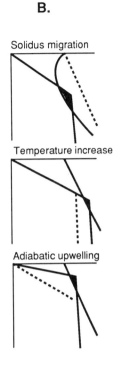

B.

Figure 3.4 Rifting and melting. A. Horizontally averaged thermal structure of lithosphere for potential temperature (Tp: the temperature on adiabatic gradient projected to surface pressure) of 1280°C, mechanical-boundary-layer (lithosphere) thickness of 100 km, and interior viscosity of 2.10^{17} m² s⁻¹ (after McKenzie and Bickle, 1988). B. Sketch graphs (after Latin et al., 1990) to summarize three possible mechanisms for producing melts during rifting from results of Fig. 3.4A. In B1, solidus migrates to left because volatiles like water are added to the system, as in island-arc environments. In B2, potential temperature is raised, causing geotherm to migrate to right, due to rising hot spot or plume (open-system melting). In B3, the lithosphere is thinned mechanically by closed-system stretching, with asthenosphere rising to be partially melted due to adiabatic decompression.

ceptual models for open-system, two-layer crustal stretching for the Basin and Range (Fig. 3.6; see also Fig. 3.18), which substantially explain: a) the dramatic lateral variation of measured extensional strains normal to regional strike in surface rocks, b) the spatial association of voluminous magmatism with areas that underwent large-magnitude extension, and c) the uniform present depth to the Moho (30–35 km) across the whole province. The pre-extensional crust is deduced to have been thick (c. 50 km) because of the effects of the Laramide orogeny. The upper crust deformed heterogeneously and was decoupled from a middle and lower crust that underwent uniform ductile deformation. A significant flux of subduction-related, mantle-derived magmas occurred into the lower crust under areas of highly extending upper crust, thus keeping crustal thickness more or less constant. More recently, it has been proposed that constant crustal thickness may be caused by ductile flow of lower-crustal rocks from horst-like areas of thick lower crust to areas of thinned crust (Block and Royden, 1990; Buck, 1991; Kruse et al., 1991).

The East African, Ethiopian and Rio Grande rifts also appear to have evolved, but not necessarily initiated, under open-system conditions. In contrast to the Basin and Range, the rift zones are relatively narrow, but they are still sited in the middle of broad regional upwarps (Ebinger, 1989a; Davis, 1991; Keller et al., 1991). These upwarps are due probably to plume dynamics since there is insufficient extension for the effects to be caused by passive asthenospheric upwelling.

Complete continental separation over hot spots caused by asthenospheric mantle plumes seems to have occurred in certain regions during the Jurassic to Tertiary breakup of the Pangea supercontinent (White and McKenzie, 1989). For example, the Late Cretaceous separation of India from southern Africa was accompanied by massive basaltic outpourings (Deccan Traps) and regional tilting of the Indian Plate to the east. Another commonly cited example is the Icelandic hot spot and plume, with its trailing aseismic ridges (White and McKenzie, 1989). It appears, however, that many continental break-up scenarios

Figure 3.5 Digital terrain image of western United States, illustrating the Basin and Range rift province and Rio Grande rift (RG), together with other major tectonic and physiographic provinces. Note small wavelength scale and complex longitudinal pattern of basins and ranges of Nevada, Utah, Idaho, and Montana. Note lack of evidence for extension in Colorado Plateau (CP), and granitoid of Idaho batholith (IB). Note active trace of Yellowstone hot spot (Y) at eastern end of Snake River Plain (SRP). Boxes show position of Central Nevada Seismic Belt (N), Idaho Seismic zone (IS), and Hebgen Lake neotectonic area (HL) (discussed in subsequent figures and text).

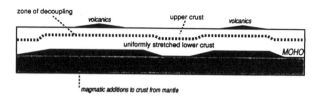

Figure 3.6 Summary sketch to illustrate mode of open-system, two-layer extension favored for Basin and Range province by Gans (1987). Discussion in text. See also Fig. 3.18 for supporting seismic-reflection evidence.

Closed versus Open Systems

The above discussion of closed- and open-system rifting has inevitably presented a rather black-and-white picture and it is probable that many rifts evolve under a combination or succession of processes. When trying to assess whether now-inactive rifts or the early stages of active rifts were closed or open systems, the chronology of magmatic, topographical, depositional and structural evolution is crucial (Şengör and Burke, 1978) and a satisfactory analysis can only come about by means of integrative studies. For example, interpretation of closed-system dynamics would be favored by observations of: a) early subsidence related to regional downwarping followed by the initiation of faulting, b) the presence of marine sediments in the earliest syn-rift basin infill if initial extension took place close to sea level, c) lack of geomorphological or thermal evidence for early regional upwarp and widespread erosion and d) volcanics in significant volumes only in the last stages of the syn-rift phase.

Studies in the Gulf of Suez (see below) clearly show that closed-system dynamics occurred early on, with geological evidence for regional uplift of the rift flanks post-dating initiation of rifting by several million years (Steckler, 1985; Garfunkel, 1988; Bohannon et al., 1989). In the East African rifts, there is good evidence for both early volcanism and flank uplift (Ebinger, 1989a), supporting open-system conditions (see Morgan, 1991). The open- or closed-system status of the regional rift basins comprising the Basin and Range is much more problematic to assess from sedimentary evidence, since the earliest basin infills were continental and have remained so to the present. It is very difficult to assess the elevation or elevation change of conti-

are not due directly to the effects of plumes since the increase of elevation caused by a plume is thought insufficient, by itself, to drive complete continental separation. Thus, many examples are known of plume-related uplifts and magmatic centers that did not give rise to separation (e.g., Siberian Traps, Columbia River volcanics, Karoo volcanics, and Cameroon line) (Fairhead, 1978; Fitton, 1983; Dixon et al., 1989).

As with closed-system rifting, the syn-rift evolution of open systems is accompanied by regional initial subsidence (Si in Fig. 3.10). This is of lesser magnitude than in closed systems with the same stretching factor due to magmatic additions to the crust of denser materials (White and McKenzie, 1988).

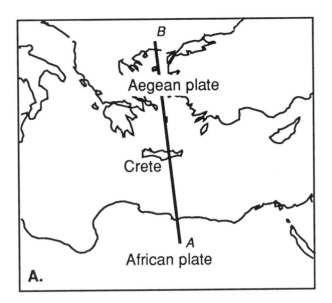

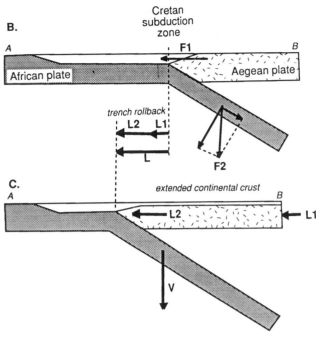

Figure. 3.7 Sketches to show mechanisms responsible for stretching of continental lithosphere in backarc environment (after LePichon and Angelier, 1981). Stretching is due to two force components, F1 and F2. F1 is the lateral component of gravitational force acting on overriding plate due to its hydrostatic head with respect to adjacent ocean crust. Part of this may be due to crustal thickening by accretion in the forearc or residual from previous crustal collisions. F2 is the negative buoyancy force arising from sinking lithospheric slab (slab-pull). The total displace-ment of overriding plate, L, relative to a fixed earth reference frame such as a hot spot, is given by the sum of the regional displacement of over-riding plate, L1, and displacement due to back-arc stretching, L2. Note that vertical motion, V, of the sinking slab does not include a rotational component in this sketch, though such motions are thought to occur during subduction. The figure is based on a model proposed for the Aegean Sea, but is clearly of more general applicability to backarc exten-sional regimes.

nental sediments from their faunas and floras. However, amongst other lines of evidence, abundant and voluminous early syn-rift volcanics in the Basin and Range attest to predominantly open-system dynamics in the early history of the region (Gans et al., 1989). Evidence from the Rio Grande rift suggests a two-stage process (Olsen et al., 1987): 1) an early phase (from about 30 to 20 Ma) of localized low-angle faulting (see below) and widespread open-system calc-alkaline magmatism and 2) a later phase (from 15 Ma to the present) of high-angle faulting and open-system bimodal volcanism. The later rift has evolved from early broad structural depressions to narrower fault-bounded half grabens with time, culminating in a phase of more recent regional uplift (Seager and Morgan, 1979; Morgan and Golombek, 1984; Morgan et al., 1986; Ingersoll et al., 1990).

Extensional Plate Motion and Dynamics in Backarcs and Orogens

The formation of rifts, either as open or closed systems, during the process of continental lithosphere separation to form new ocean basins has been noted above and is discussed further below. Other major areas adjacent to oceanic lithosphere where extensional continental rift basins are formed and may also be succeeded by oceanic crustal development, are backarc terranes (see Chapter 8). Backarc extension of an overriding plate at a subduction zone has been attributed to the phenomenon of trench rollback (Dewey, 1980; Le Pichon and Angelier, 1981; see explanation in Fig. 3.7).

It has been known for some time (Bott and Dean, 1972; Artyushkov, 1973) that regions of isostatically compensated thick crust that lie adjacent to areas of

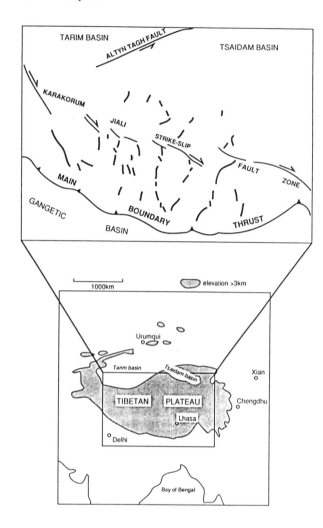

Fig. 3.8 Regional map showing location and extent of the Tibetan Plateau north of the Main Boundary Thrust of the Himalaya. Mean relief of the Plateau is around 5 km. The detailed map shows location of major normal and strike-slip faults (after Armijo et al., 1986).

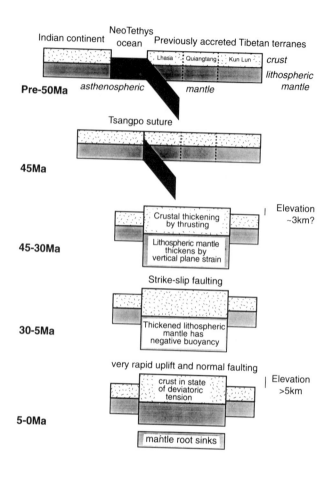

Fig. 3.9 Diagrams to illustrate evolution of thickened crust and modern extensional stress regime of the Tibetan Plateau (data taken from geological results of the 1985 Tibetan Geotraverse Dewey et al., 1988 and geophysical models of England and Houseman, 1989). Note that although the mechanism by which the thickened lithospheric root breaks off and sinks is unknown, the conversion of the root back to asthenosphere by thermal conduction is too slow to explain the (albeit scant) geological evidence for very rapid recent uplift (since 5 Ma) in Tibet.

compensated thin crust are in a state of extensional deviatoric stress. A major development in recent years has been the recognition that widespread gravitational collapse may occur due to this mechanism after periods of uplift due to crustal and lithospheric thickening following continental collision (see Bird, 1991, for an excellent account). The crustal-thickening model is most successfully applied to the 1300 km E-W-trending zone of the southern Tibetan Plateau north of the Himalayas (Fig. 3.8). Here, seven active extensional grabens cut the much-thickened crust across the structural grain created by Tertiary and earlier colli-

sions (Molnar and Tapponnier, 1975, 1978; Armijo et al., 1986; Dewey, 1988; Shackleton and Chengfa, 1988). The N-S-trending grabens cut early Tertiary geomorphic pediplanation surfaces and are thought to have formed during a phase of rapid surface uplift, possibly caused by detachment of an underlying slab of overthickened and dense lithospheric mantle into the asthenosphere (England and Houseman, 1989; Fig. 3.9). This mechanism is also thought to have contributed to the earliest extension in the Basin and Range, since the onset and development of Cenozoic extension occurred in areas that were previously

thickened during the Sevier and Laramide orogenies (Gans, 1987; Sonder et al., 1987).

Extensional Kinematics: Syn-Rift Subsidence and the Nature of Rift-Margin Faults

Active major basin-bounding faults from many areas of continental extension seem, from seismological evidence, to be steep (40°–60° dips) and essentially planar down to the brittle-to-ductile transition (Jackson, 1987). Curved, listric fault planes are also recorded along seismic-reflection lines across a few extensional basins (e.g., Gans et al., 1985). They are commoner within sedimentary fills as antithetic or synthetic faults flattening within the infill sequences (Jackson et al., 1988). Seen in crustal-scale 2-dimensional sections (Fig. 3.10), the characteristic pattern of normal faulting is one of tilted fault blocks

(so-called domino blocks). Extension causes rotation of these rigid crustal blocks about horizontal axes (Barr, 1987a,b; Jackson et al., 1988). Typically, the wavelength normal to strike of the individual blocks is the same order as the depth of the normal faults. Relative uplift of the immediate footwalls to the faults, around 10–15% of the total displacement (Jackson and McKenzie, 1983), occurs in addition to relative subsidence of the hangingwalls (Fig. 3.11).

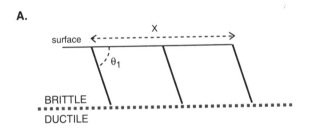

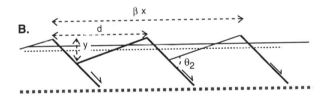

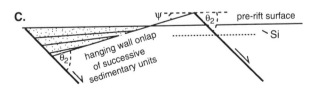

Figure 3.10 Sketches to illustrate horizontal-axis rotation of crustal domino blocks bounded by normal faults (after Jackson et al., 1988). The depth to brittle-ductile transition in the crust, and hence the maximum depth of major normal faulting, varies according to temperature gradient and strain rate (see Fig. 3.2), but in many areas of active extension lies between 10 and 15 km. See text for notation.

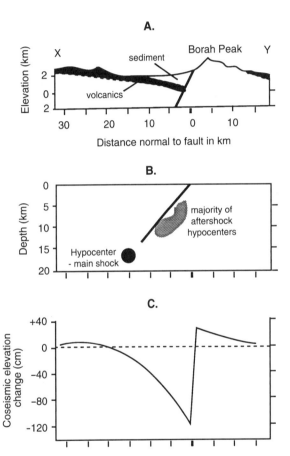

Figure 3.11 The Borah Peak earthquake of 28 October 1983 (after Stein and Barrientos, 1985). A) Sections to illustrate upper-crustal and surface structure and topography ; B) postulated fault-plane trace projected from surface break and location of main hypocenter and aftershock hypocenters; C) modelled curve for ground deformation. The observed ground deformation (from ground relevelling after the quake and not shown here) very closely follows the modeled fit. The most important result from this earthquake is confirmation that the elastic-dislocation theory can very closely model the co-seismic deformation experienced during an earthquake. For geologists and geomorphologists, the importance of the result is the apportionment of co-seismic displacement between footwall uplift and hangingwall subsidence, approximately at a ratio of 1:6. Section X–Y located in Fig. 3.21.

This co-seismic deformation occurs as an elastic response of the upper crust to faulting (Savage and Hastie, 1966). It is not known how this co-seismic deformation is modified during periods of seismic inactivity. Several authors have recently speculated upon the possible role of flexural (bending) deformation between seismic events or after active extension has ceased (King et al., 1988; Stein et al., 1988; Weissel and Karner, 1989; Kusznir et al., 1991). The models proposed by these authors are extremely difficult to test empirically since long-term post-seismic strain rates are small and impossible to detect.

Uniform extension (when crust and mantle extension are equal in magnitude) of typical continental lithosphere leads to an initial mean subsidence of $Si = 2.46(1 - 1/\beta)$, where β is the ratio of uniform extension, defined as the ratio of present crustal length to original length. The value of the constant depends upon the initial thermal structure, density and length scales of the crust, mantle and asthenosphere (McKenzie, 1978a). As shown in Fig. 3.10, this subsidence is superposed on a sawtooth topography of amplitude $y = d \sin \psi$ where ψ is the amount of rotation, given by the difference between the initial normal fault dip, θ_1, minus the present (rotated) fault dip, θ_2 (Barr, 1987a,b; Jackson et al., 1988). The lower surface space created by the keels of the tilt blocks in the middle crust is dissipated by plastic flow and by thermal re-equilibrium, the latter occurring on time scales rapid compared to observed periods of extension (Jackson

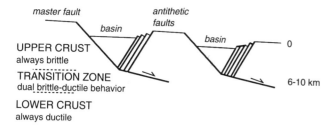

Figure 3.12 Sketches to illustrate change from steeply dipping faults in the brittle upper crust into shallow-dipping faults in the ductile lower crust and the generation of steep antithetic faults that progressively nucleate at the discordance (after Eyidogan and Jackson, 1985). The antithetic faults are large in the seismic sense that they generate big earthquakes, but do not appreciably affect the regional tilt of the basement and the synrift sediments. The resultant symmetrical form of the basin and horst blocks is seen in several examples in both the Aegean and the Basin and Range (see Leeder and Jackson, 1993).

and White, 1989). Important major antithetic faults often occur in active tilt blocks (Eyidogan and Jackson, 1985; Roberts and Jackson, 1991; Fig. 3.12), giving rise to more symmetrical cross sections.

Based on field observations, several authors (e.g., Morton and Black, 1975; Proffett, 1977) have postulated that as extension proceeds, rotated normal faults (whose initial dips are commonly around 60°) are abandoned and replaced by successive generations of more steeply dipping faults that rotate the older faults into lower angles of dip. The formerly active basin fills associated with the inactive faults are eroded by footwall uplift caused by the new fault systems (Jackson et al., 1982; Fig. 3.13). Forsyth (1992) has drawn attention to the paradox that although low-angle faults occur in many Tertiary and older extensional terranes, they are rare at present, giving no seismic records (Jackson, 1987). Many authors (e.g., Anderson, 1951; Sykes, 1978; Bruhn et al., 1982; Enfield and Coward, 1987; Huyghe and Mugnier, 1992) have pointed out that low-angle normal faults may nucleate from or form along weaknesses such as pre-existing thrusts. Forsyth (1992) proposed that, once formed, low-angle faults offer frictional and energetic advantages over steeper-dipping faults when extensional displacements are high.

In addition to the rotation of faulted blocks about horizontal axes, there is sound evidence from paleomagnetic (Kissel and Laj, 1988) and co-seismic-slip studies (McKenzie and Jackson, 1986) that fault-bounded blocks can also rotate about vertical axes (Fig. 3.14). Thus, the regional strike of a series of fault-bounded tilt blocks, seen in plan view, may change with time.

It has been postulated that regional extension of the crust and mantle may also occur as the result of simple shear (Wernicke and Burchfiel, 1982; Wernicke, 1985; Figs. 3.15–17). During such simple shear, a single low-angle shear plane cuts a significant part of the crust, the whole crust or even the whole lithosphere. Very large horizontal displacements then occur on the shear. In the shallow part of the crust, steeper seismogenic normal faults sole into this master fault (Fig. 3.15). A key difference between simple-shear deformation and pure-shear deformation is that, in the former, the locus of maximum thinning

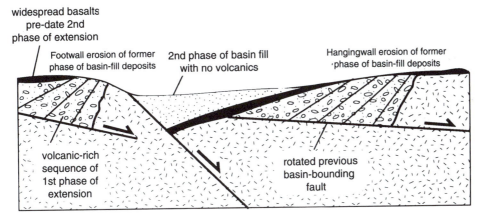

widespread basalts
pre-date 2nd
phase of extension

Footwall erosion of former
phase of basin-fill deposits

2nd phase of basin fill
with no volcanics

Hangingwall erosion of former
·phase of basin-fill deposits

volcanic-rich
sequence of
1st phase of
extension

rotated previous
basin-bounding
fault

Figure 3.13 Sketch to illustrate rotation and eventual abandonment of an initial set of normal faults, eruption of basalt and subsequent initiation and development of 2nd phase of high-angle normal faults (based upon sketches by Gans et al., 1989). Both first phase of basin infill and basalt cap are eroded and deposited in alluvial fans as second-cycle deposits in latest basin fill. This sequence of events is common in both the Basin and Range and the Aegean (although without volcanics in latter case), but the reader should note that there is also strong evidence for the existence of formerly active low-angle faults, such as those around metamorphic core complexes (see Fig. 3.15 and text).

of the crust is displaced laterally from that of the mantle lithosphere (Figs. 3.16, 3.17). This should cause markedly asymmetric and distinctive uplift, subsidence and heat-flow patterns (Buck et al., 1988) that up to now, have not been observed. Deep seismic-reflection and integrated geological/geochemical studies in the Basin and Range province have not revealed evidence for the significant variations in crustal thickness predicted by simple-shear models (Fig. 3.18; Gans, 1987). There is, however, support for the presence of low-angle detachments, which are the *vertically* uplifted parts of the cataclasite junction between crust that is deforming by brittle and ductile mechanisms (Gans et al., 1985; Gans, 1987). Above the detachments lie characteristically small tilt blocks. The low-angle detachments seem to have formed in a previous period when rapid open-system extension was accompanied by high heat flow and magmatic intrusion. Buck (1988), and Wernicke and Axen (1988) suggested that such uplifts occur in the footwalls of all low-angle listric shear zones, and that uplift and erosion of the tectonically denuded footwall causes exhumation of lower-crustal rocks (see Fig. 3.15). Such metamorphic core complexes, with Eocene to Miocene unroofing ages, are very common in the North American Cordillera and were originally defined by Coney (1980b) as "a group of generally domal or archlike, isolated uplifts of anomolously deformed metamorphic and plutonic rocks overlain by a tectonically detached and distended unmetamorphosed cover". The controlling shear zones are thus denudational faults that accommodate crustal extension by stripping away the upper plate to reveal mid-crustal rock masses (Figs. 3.15, 3.18). The progress of tectonic denudation is recorded in the changing assemblages of clast types in sedimentary strata derived by erosion of the evolving core complexes (Miller and John, 1988).

The closed-system simple-shear model (Wernicke, 1985) has also been applied as an alternative to pure shear for the evolution of the North Sea basin (e.g., Beach et al., 1987), but interpretation of deep crustal seismic sections and accurate estimates of spatial and temporal subsidence rates are inconsistent with this model (White, 1991). Similar tests incorporating heat flow and uplift observations in the Red Sea (e.g., Buck et al., 1988; Dixon et al., 1989) also do not favor the simple-shear model of Wernicke (1985).

As briefly noted above, it is evident that some form of flexural uplift may affect the footwall areas of major crustal normal faults (Weissel and Karner, 1989; Ebinger et al., 1991; Kusznir et al., 1991). Gravity studies in East Africa reveal substantial elastic thicknesses (Ebinger et al., 1989a) and thus, flexure

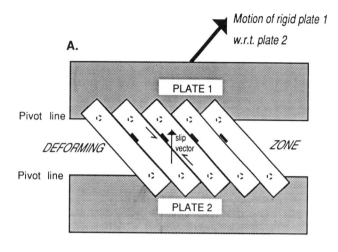

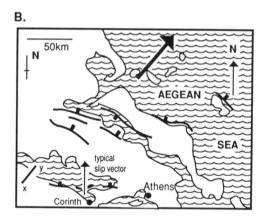

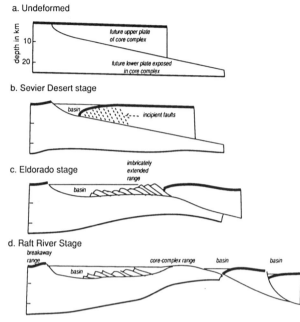

Figure 3.15 Evolution of a crustal section being extended under a regime of simple shear, according to the model of Wernicke (1985). Note that in the footwall of the sinuous low-angle normal fault, lower-crustal rocks are eventually exposed, forming core complexes now widely recognized in the western United States and elsewhere. For an alternative model of simple-shear deformation, see Gans et al., (1985).

Figure 3.14 A. Sketch to illustrate vertical-axis rotation of crustal blocks arranged between two rigid and undeformed plates (the rigid-slat model; after McKenzie and Jackson, 1986). The thin long arrow gives the mean slip vector measured on fault planes bounding the blocks and includes a strike-slip component arising from the difference between the vectors of rigid plate motion and block motion. Note that both vertical-axis and horizontal-axis rotations may occur at the same time, leading to tilt-block basins appearing as depressions adjacent to each active fault (see Fig. 3.10). Readers are encouraged to make their own models to illustrate the two possibilities - the author recommends that the vertical-axis model be attempted first with drawing pins and pieces of card. More recent models involve broken slats (Taymaz et al., 1991). B. Map of central Greek Aegean illustrates simple rigid-slat model in nature. X-Y marks section shown in Fig. 3.31.

may be a significant factor in the local elevation of rift flanks adjacent to rifts which are too narrow (<70 km) to generate small-scale convection under these areas (Ebinger et al., 1991). However, it is a mistake to assume that any rift flank uplift that persists after several tens of millions of years may be solely due to permanent flexure as modified by erosion (see Weissel and

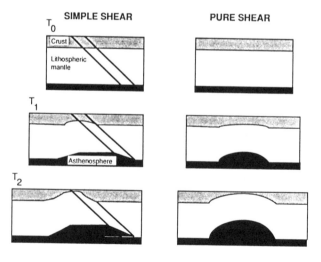

Figure 3.16 Sketches to illustrate extension of layered lithosphere by simple shear and pure shear (after Buck et al., 1988). The inclined lines in the simple-shear case represent a shear zone. Note the obvious asymmetry in the simple shear case and the symmetry in the pure-shear case.

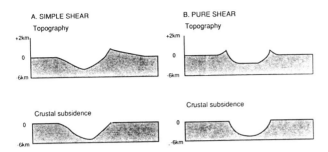

Figure 3.17 Sketches to illustrate models of topography and crustal subsidence due to simple- and pure-shear extension (see Fig. 3.16; after Buck et al., 1988). For modeling parameters, see original reference, but note that local isostatic compensation is assumed (i.e., zero flexural strength). Topography is thus supported by thermal effects.

Karner's 1989 hypothesis), since-small scale convection (e.g., Buck, 1986), magmatic underplating (e.g., Dixon et al., 1989) or non-uniform stretching (e.g., Bohannon et al., 1989) may cause similar effects. An alternative origin for the marginal upwarps adjacent to proto-oceanic rift troughs like the Red Sea/Gulf of Suez (Steckler, 1985) is in the thermal convective response of the underlying asthenosphere to extension and to the generation of significant quantities of mantle melt and subsequent basaltic underplating.

Studies in areas like the Basin and Range and the North Sea, where extension is more diffuse and where the surrounding nonrifted lithosphere is thinner,

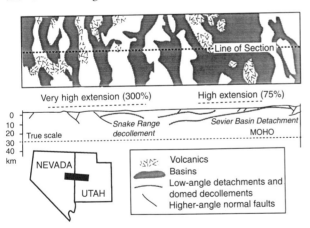

Figure 3.18 Map and crustal-scale section through the Basin and Range province of eastern-central Nevada and western-central Utah (after Gans, 1987). The section is based upon a deep seismic-reflection profile shot by COCORP and shows the apparently horizontal Moho at a depth of around 30 km across the whole area, regardless of amount of extension or abundance of volcanics. Such evidence has been used to support an open-system, two-layer model for Basin and Range extension (see Fig. 3.6).

reveal evidence for only small elastic thicknesses (<5km). In these areas, the role of flexure is controversial; some authors propose a minimal role for flexure (e.g., White and McKenzie, 1988), whereas others propose models incorporating significant flexural contributions to hangingwall subsidence and footwall uplift (e.g., Weissel and Karner, 1989; Kusznir et al., 1991).

Basin Structure and Subsidence Patterns

Surface Faulting

The surface length of individual historic normal-fault ruptures characteristically ranges from 10 to 15 km (Fig. 3.19), perhaps due to the approximate thickness of the brittle crustal layer (Jackson, 1987). Longer basin-bounding fault strands (up to 50 km) occur in parts of the East African rift (Ebinger, 1989a,b), perhaps reflecting the very thick old lithosphere in the area. It is not known, however, whether these long faults have ruptured along their total length or only in part during earthquake sequences. In active extensional areas, individual fault ruptures during major earthquakes show amplitudes of up to several meters (Figs. 3.11, 3.20). Faulting may trigger spectacular landslides, such as that which occurred during the 1959 Hebgen Lake earthquake in Montana (Myers and Hamilton, 1964). It seems likely that faults increase their length with time, since fault displacement and length are positively related (Watterson, 1986; Walsh and Watterson, 1988; Cowie and Schulz, 1992; Jackson and Leeder, 1994). A consequence of fault growth is that basins should also enlarge with time. Some of the stratigraphic effects of this enlargement are explored by Schlische (1991, 1992).

The space created between adjacent tilt blocks at the surface defines the characteristic half-graben form, with a major bounding fault separating the footwall upland from the hangingwall lowland where sediment progressively accumulates (Figs. 3.10, 3.12). The basin form and relief at any given time are results of the cumulative history of fault displacement since extension began, convolved with the regional isostatic response to extension, minus the effects of subaerial erosion and deposition.

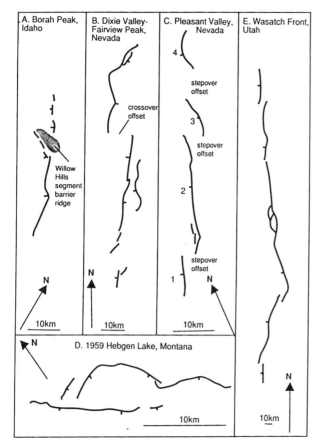

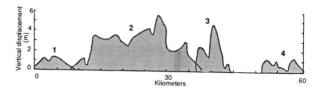

Figure 3.20 Graph showing variation of displacement with distance along the four 1915 Pleasant Valley fault segments (labeled 1-4 on Fig. 3.19; after Wallace, 1984).

Figure 3.19 A–D. Sketch maps to show arrays of normal faults produced during single historic earthquakes in the Basin and Range province. Map A is located on regional maps in Figs, 3.5 and 3.21. Maps B and C are located in Figs. 3.5 (Central Nevada Seismic Belt) and 3.22. Hebgen Lake area is outlined in Fig. 3.5. E. The normal-fault array at the front of the Wasatch Range, Utah. Note scale. No historic breaks have occurred here and it is thus not known exactly which fault segments were active at what time. Paleoseismic studies are the only way of deducing the pre-1840 rupture history. (All after Wheeler, 1989).

Evidence from active and young normal faults (Figs. 3.19, 3.21) reveals that individual ruptured fault strands are offset by stepover by up to several kilometers (en-echelon geometry), and that fault segments, comprising several offset strands, are themselves separated by more extensive transfer or accommodation zones (Crone and Haller, 1989; DePolo et al., 1989; Wheeler, 1989). Crossover offset or accommodation zones are areas where fault polarity reverses (Fig. 3.19B). The extension is taken up across numerous small normal faults, causing the offset zones to stand at higher elevations than the inter-

vening basins (Jackson and Leeder, 1994). Crossover accommodation zones are frequently sites of abundant volcanism in the East African rift system (Ebinger, 1989a,b; Ebinger et al., 1989b; Fig. 3.22). The ridges may cause hangingwall basins of individual tilt blocks to be topographically isolated from adjacent basins (Bosworth, 1985; Ebinger, 1989a). Although burial of the ridges by sedimentary onlap may subsequently occur, with axial drainage eventually connecting up adjacent basins (Figs. 3.23, 3.24), the ridges form subsurface highs recognizable from gravity or seismic data (Crone and Haller, 1989).

It should be noted that many geometrical relationships are possible between adjacent fault strands or segments if faults of differing ages are present. Offset geometries should only be described *if* fault activity on adjacent segments is synchronous or, at least, nearly so. This point has not been emphasized in the literature, and in my opinion, its neglect has encouraged highly detailed and overly complex classifications to be proposed (e.g., Rosendahl et al., 1986; Morley et al., 1990).

Multiple basin-margin faults are known from several areas (e.g., Dixie Valley, Nevada [Fig. 3.23; Bell and Katzer, 1990]; Yangbajain graben, Tibet [Armijo et al., 1986]; East African Rift [Fig. 3.22; Ebinger 1989b]). Examples are known where initially wide tiltblocks are fragmented into smaller blocks by new faults in the old hangingwall (Jackson et al., 1982; Roberts and Jackson, 1991). Cessation of activity on the old bounding fault is followed by footwall uplift, erosional dissection and transport of the old basin fill into the newly subsiding basin. As noted above, many hangingwalls are broken by subsidiary antithetic and synthetic normal faults (see Figs. 3.12, 3.23). The larger of these cause the formation of intra-graben horsts and subsidiary basins, whereas

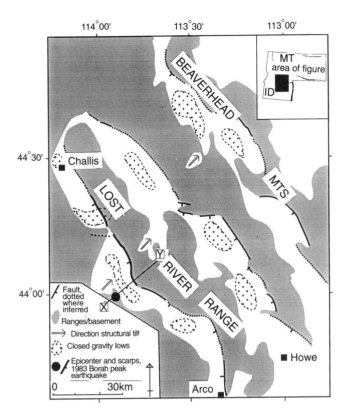

Figure 3.21 Sketch map of northern Idaho part of the Basin and Range (after Crone and Haller, 1989) to show straight to slightly sinuous fault segments and scoop-shaped basins (revealed by gravity lows) with their separating basement ridges and salients. Note the Borah Peak epicenter and line of section X-Y of Fig. 3.11.

the many smaller faults form as internal accommodation structures during extension.

Basin shape is partly controlled by the planform of the major bounding fault segments. Nearly all fault breaks recorded in historic earthquake sequences of the Basin and Range and in the Aegean Sea are more-or-less linear to slightly curved (Figs. 3.19, 3.21), albeit with some pronounced local curvature. One major exception is the Red Canyon strand of the 1959 Hebgen Lake earthquake sequence (Fig. 3.19D). It should be noted that linear-offset bounding faults do not give rise to linear basin forms, since basement ridges at crossover zones and the decay of fault throws toward fault-tip zones inevitably cause the sediment fill to thin laterally, giving the observed scoop-shaped basin fills in planform view. The strongly curvilinear bounding

normal-fault systems (e.g., Bosworth, 1985; Dunkelman et al., 1988) and their associated "scoop-shaped" sedimentary infills recognized in parts of the East African Rift seem unusual by comparison with Basin and Range and Aegean examples. Historical ruptures are rare in East Africa and mapping by remote sensing of rift-zone faults, transfer zones and lineaments may oversimplify a long and complex kinematic history. This leads to severe problems of interpretation of poorly dated multigeneration structures. For example, Ebinger (1989b, Fig. 12) presents evidence to suggest that adjacent faults in crossover zones may not be active at the same time. Such problems of kinematic reconstruction are also well illustrated in the Gulf of Suez, where highly curvilinear normal bounding faults mapped by early workers (e.g., Bosworth, 1985) are not confirmed by more recent detailed mapping (Jackson et al., 1988; Gawthorpe et al., 1990).

Structure of Rift-Basin Fills

The area of space created by tilt-block rotation about horizontal axes depends only upon the rate of rotation and the initial fault spacing (Barr, 1987a,b; Jackson et al., 1988). It is evident that the combined effects of deposition and continued rotation of older strata gradually cause hangingwall onlap and increased bed dips as seen in any planar vertical surface cut through that fill (Fig. 3.10). These features are evident on many seismic sections across tilt-block basins. Ignoring for the time being any changes in water balance or sediment supply, hangingwall onlap occurs as an inevitable response to basin infill by sediment derived from any source.

The hangingwall "rollover" geometry seen in many gravity-driven listric faults at delta fronts (e.g., Niger, Mississippi) and in intrabasinal listric faults (Fig. 3.25) that are asymptotic within the sedimentary basin fill is not an easy geometry to create by planar basin-bounding structures.

Subsidence following historic normal faulting (Fig. 3.11), although measured by releveling in only a handful of cases (e.g., Plovdiv, Romania [Richter, 1958]; Hebgen Lake, Montana [Myers and Hamilton,

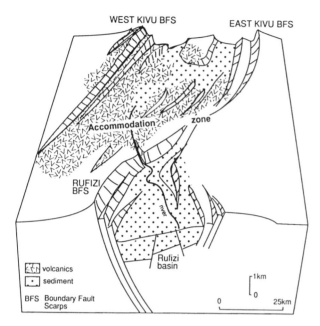

Figure 3.22 Block diagram of the Rufizi and Kivu grabens and their intervening accommodation zone in the Western Rift arm of East African Rift system (after Ebinger, 1989b). The upper Cenozoic volcanics are located in the region of the accommodation zone, either because that is close to their locus of eruption or because a higher preservation potential exists on the lower slopes present there. Note that a river has incised across the accommodation zone, connecting the two grabens. The listric form of the major bounding faults is intuitive; the exact shape awaits seismological investigation.

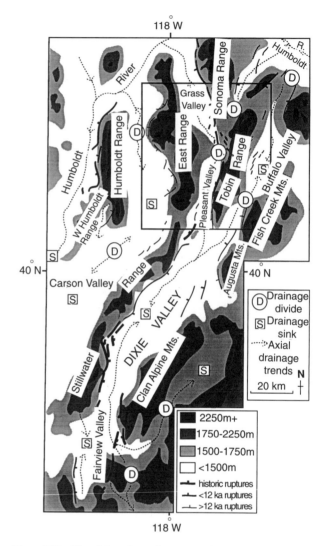

Figure 3.23 Map of Central Nevada Seismic Belt to show topography, faults, axial drainage trends, drainage divides between adjacent basins, and location of drainage sinks. Note that the originally asymmetric topography in these tiltblocks has been appreciably modified by the effects of erosion (Leeder and Jackson, 1993). Box outline is for Fig. 3.24.

1964] and Borah Peak, Idaho [Stein and Barrientos, 1985]), follows the trend expected from the elastic deformation of a domino block with finite horizontal extent (Savage and Hastie, 1966). Over time, the cumulative subsidence patterns of the graben as a whole follow the pattern of individual seismic events, leading to a crudely wedge-shaped accumulation in vertical section and a scoop-shaped section in plan view, with the rate of subsidence greatest along the axis drawn vertically up from the intersection of the bounding fault with the pre-rift basement/sediment contact. In general terms, the maximum subsidence reflects the degree of rotation of the fault blocks about horizontal axes. The pattern of subsidence also is strongly influenced by the development of intra-graben structures. The deposi-

tion rate in the basin reflects the erosion rate in the uplifting hinterlands, the axial sediment flux, and any sediment precipitated (chemically or biologically) within the basin. The basin relief at any instant (footwall crest to basin surface) is determined by the difference between rotation rate and deposition rate.

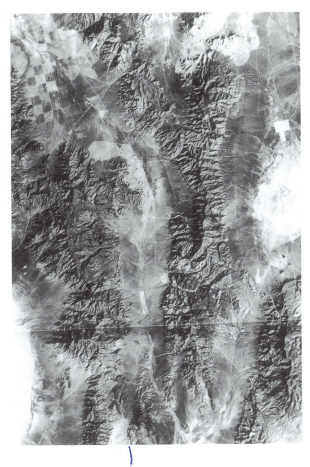

Figure 3.24 Satellite image of the northern part of Fig. 3.23 to show topography and drainage. Fine-grained playa sediments (pale areas) dominate the closed Buffalo and Buena Vista Valleys, whereas Pleasant Valley is an open system, whose axial drainage cuts through the Sou Hills offset zone to discharge into Dixie Valley to the south.

Figure 3.25 Prominent intrabasinal listric fault cutting fluviatile and megabreccia deposits, Megara basin, Greece. Author for scale. Exposure is c. 30 m high.

Sediment Derivation, Distribution and Facies Patterns

General

The characteristic structural asymmetry of many continental rift basins and their uplifted flanks exerts a fundamental control on the distribution of sedimentary environments and lithofacies (e.g., Surlyk, 1978; Frostick and Reid, 1987; Leeder and Gawthorpe, 1987; Morley, 1989; Mack and Seager, 1990; Schlische and Olsen, 1990; Lambiase, 1991; Leeder, 1991). This is particularly true along the basin margins, where transverse drainage systems evolve on the footwall and hangingwall uplands (Fig. 3.26), trans-

ferring clastic sediment toward the basin center, frequently via alluvial fans and fan deltas. The area of newly created tectonic uplands free to be drained by runoff is controlled by the length of the tectonic slope produced during extension (Leeder et al., 1991), via Hack's Law that relates drainage basin area to principal stream length (Fig. 3.27). Drainage-basin area is the primary control on the size of transverse-sourced alluvial fans (Fig. 3.26) that form along graben margins, although climate and bedrock lithology (Fig. 3.28) are important modifying influences (Leeder and Jackson, 1993).

A half graben may, thus, be envisaged as a sink for introduced clastic sediment and for biogenic and chemical sediments. The existence of isolated basins reflects the efficiency of crossover zones in preventing

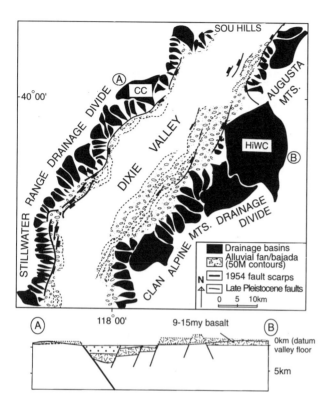

Figure 3.26 The drainage catchments, faults, and depositional topography of Dixie Valley, Nevada (after Leeder and Jackson, 1993). Structural data from Wallace (1978), Whitney (1980), Bell (1984) and Okaya and Thompson (1985). Inset cross section A-B from Okaya and Thompson (1985). HiWC: Hole-in-the-Wall catchment; CC: Cottonwood catchment. The footwall to Dixie Valley has a drained area of 393 km², with 58 main catchments, giving a mean catchment area of 6.8 km². The hangingwall has a drained area (excluding the Hole in the Wall catchment) of 347 km², with 20 main catchments, giving a mean drainage area of 17.4 km². If the Hole-in-the-Wall catchment is included then the total area is 624 km², and the mean is 29.7 km². The majority of footwall catchments are shorter and steeper than those in the hanging-wall. Gradients of the former are usually >1:6, whereas the latter are usually <1:8.

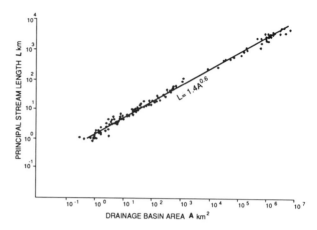

Figure 3.27 Relationship between drainage-basin area and drainage-basin length established by Hack (1957) and Leopold et al., (1964), with additional data points taken from measurements of drainage-basin char-acteristics in the tilt-block terrain of central Greece (Leeder et al., 1991). Since tectonics often controls the potential length of a drainage system, the Hack relationship is important in that it establishes a regularly spaced system of drainage basins, regardless of scale.

sediment transfer between adjacent depocenters (Fig. 3.23). Erosion and incision of crossover highs cause linkage of basins along structural strike (e.g., Figs. 3.22, 3.23), enabling discharge to become axial and transferring sediment out of individual basins (Leeder and Jackson 1993; Jackson and Leeder, 1994).

Basin-center environments are strongly controlled by climatic influences, with playa, semi-permanent and permanent lake systems forming according to the level of local freshwater influx relative to evapora-tion. Eolian sand complexes may form in some basins. Extension at or near sea level enables access to marine waters and the basin form becomes that of a marine gulf, fed by clastic systems from the sur-rounding uplands. Submarine fans are fed directly by transverse footwall-derived or hangingwall-derived fan deltas, or from axial sources.

In the discussion that follows, the sediment flux into rift basins is subdivided according to whether it is locally normal to the strike of the rift and its bounding faults (i.e., *transverse*, although not necessarily orthogo-nal) or parallel to the strike of the rift and its bounding faults (i.e., *axial*). It is important to stress that ulti-mately, the source of all axial flows, whether rivers, deltas or turbidity currents, is the sum of the transverse components gathered from up the regional paleoslope.

Transverse Sediment Flux

Footwall uplands adjacent to main basin-bounding normal faults (or multiple fault zones in some cases) increase their relief with cumulative slip. Short, nar-row drainage basins develop with steep gradients (Fig. 3.26). Individual faulting episodes cause knick-points to form in valley alluvium at the fault; they migrate upslope with time. Hanging valleys may form in resistant footwall rocks and the spurs sepa-rating drainage outlets become facetted as the

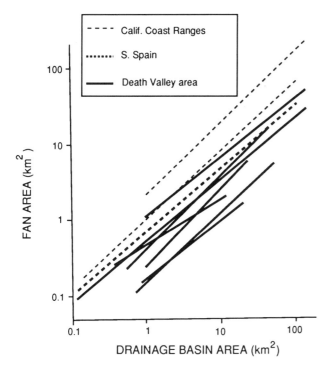

Figure 3.28 Relationship between alluvial-fan area and drainage-basin area for three locations (after Harvey, 1989). The overall positive relationship between the two variables is simply a reflection of the increasing magnitude of sediment load provided at the fanhead with increasing drainage basin area. The data from the three areas also indicate that more sediment is available for transport in the less arid and more easily erodeable coastal Californian drainage basins. In particular, the abundant fine-grained material in these Californian basins will be transported farther before deposition, thus increasing fan areas for given drainage basin areas (R. V. Ingersoll, personal communication, 1992).

drainage divides evolve (see Wallace, 1978). Generally, a larger drainage basin evolves at the locus of maximum throw along a fault segment, although local variations in footwall lithology also strongly influence drainage area. Fault offsets and transfer zones in the footwall may feature larger-than-average drainage basins (Crossley, 1984; Leeder and Gawthorpe, 1987) and hence, large alluvial-fan and fan-delta systems (Gawthorpe and Colella, 1990; Gawthorpe et al., 1990).

The decrease of gradient into a basin from the bounding fault usually causes rapid deposition and the construction of talus cones, alluvial fans, fan deltas and submarine fans. The area of the subaerial fans is partly controlled by the area of the drainage basin. The relatively steep depositional slopes of the

fans are decreased by some small amount at every increment of fault slip. This may soon be neutralized by renewed deposition on the fan surface. Knickpoint retreat occurs from the fault plane and the excess sediment deposited at the new gradient may form a new depositional lobe. More-or-less continuous bajadas may be formed by the coalescence of individual footwall-derived fans. Climatic and base-level changes may cause local progradational/retrogradational sequences to form. The development of multiple fan surfaces, with successive generations of paleosols, is characteristic of present bajada fronts, reflecting severe late Pleistocene/Holocene climatic changes. Such sequences are of great use in paleoseismic analysis, since trenching or natural exposures reveal the interplay between fault breaks and the various dateable layers of sediment. Multiple faulting may cross the bajada front, causing the higher fan segments to be cannibalized during footwall incision. Similar cannibalization and incision sequences result when a basin-center lake is progressively drawn down by evaporation.

The landslides that develop from steep footwall slopes during faulting episodes are introduced catastrophically into the basin as units of "megabreccia" (Longwell, 1951; Burchfiel, 1966), defined as "unsorted angular blocks and shattered masses of pervasively brecciated and transported bedrock" (see review in Dickinson, 1991, p. 52). These megabreccias are emplaced as dry debris avalanches or sturzstroms (Hsü, 1975; Krieger, 1977) and are important elements of basin architecture in many ancient rift basins (Yarnold and Lombard, 1989; Bentham et al., 1991).

By way of contrast to footwall uplands, hangingwall sourcelands are initially much broader, with gentler slopes normal to tectonic gradient. They become progressively smaller as basin infill proceeds, as witnessed by hangingwall onlap by younger sedimentary strata. The drainage basins are commonly larger than those that drain the footwall uplands, and they feed larger alluvial fans and fan deltas, whose gradients are normally gentler than footwall-sourced fans (Hooke, 1972; Leeder and Gawthorpe, 1987; Gawthorpe and Colella, 1990). Channel and

fan-surface gradients increase during each episode of fault slip, with local fan-surface incision. Fan incision may also result from purely climatic or runoff changes. Coalescence of hangingwall fans creates prominent low-angle bajadas, which also adjust to climatic or base-level changes.

Axial Sediment Fluxes

Closed basins with longitudinal tilt may develop short axial drainage channels fed by the transverse drainage

described above. More commonly, basins may become linked along strike to adjacent basins once crossover zones are breached (Figs 3.22, 3.23). Perhaps the most striking example is the Rio Grande rift of Colorado and New Mexico that, prior to a few million years ago, was a series of isolated half grabens (Kelley, 1979; Ingersoll et al., 1990). These were subsequently breached by the now through-flowing Rio Grande (Fig. 3.29).

The course of an axial river channel is controlled by the magnitude of alluvial fans issuing from transverse drainage basins. The axial channel, by fan toe-

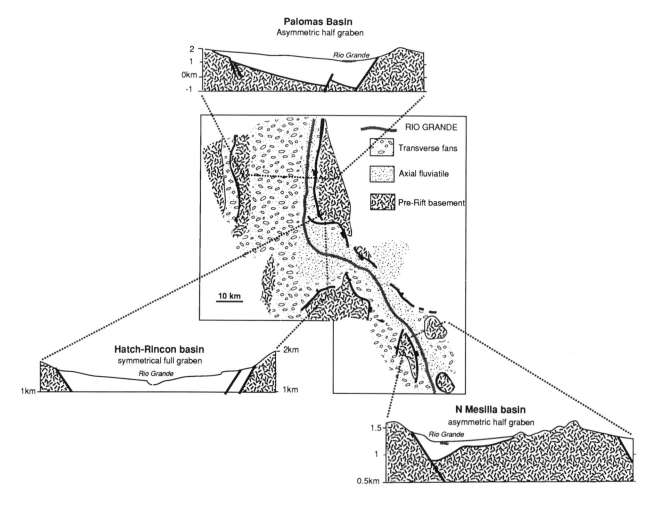

Figure 3.29 Southern Rio Grande rift north of Las Cruces, New Mexico: a field test for alluvial architectural models (simplified after figures in Mack and Seager, 1990). In the asymmetrical Palomas and North Mesilla basins, the axial fluvial facies is characterized by multistorey channel sand and is concentrated near the locus of maximum subsidence within a few kilometers of the footwall scarp. Fanglomerates derived from the footwall drainage basins extend only a few kilometers or less from the scarp, whereas alluvial-fan facies deposited on the hanging-wall dipslope occupy a much wider outcrop belt. In the symmetrical Hatch-Rincon basin, the axial-fluvial facies extends to within a few kilometers of both the northern and southern basin margins, indicating that the axial channels could migrate uncontrolled by tectonic bias. There is also a much higher percentage of fine-grained overbank deposits separating the channel facies, some of which contain calcareous paleosols.

trimming and alluviation, creates its own alluvial plain and facies architecture according to the nature of the channel (high/low sinuosity, etc.). Inevitably, the longitudinal, strike-parallel slope of the axial floodplain is perturbed during each period of slip along the basin-bounding faults. The axial channels adjust to this new slope component, either by slow migration or by channel-belt avulsion (Bridge and Leeder, 1979; Alexander and Leeder, 1987; Leeder and Alexander, 1987; Mack and Seager, 1990). Strongly asymmetric channel belts (Fig. 3.30) arise from both processes, the former producing oriented oxbow cutoffs, whose closed loops point up the tectonic slope. Major deltaic clastic depocenters result when axial channels reach lacustrine or marine environments. High-constructive deltas or fan deltas form when basin wave or tidal energy is relatively low, whereas coastal plains fronted by parallel beach ridges form when wave energy is high.

Basin Sinks: Lakes

Arid, closed continental basins commonly contain playa lakes, whose chemical deposits reflect the ionic composition of groundwaters and runoff from surrounding uplands. Gypsum, halite and other evaporite minerals result. Shallow permanent lakes in less evaporative climates, such as those in the Baringo-Bogoria part of the Gregory Rift, Kenya, are fringed by fan and axial deltas with a mixture of biogenic oozes (diatom-rich) and fine-grained clastic sediment (e.g., Tiercelin et al., 1987). Deeper lakes may develop permanent stratification, enabling good preservation of organic matter (and thus high hydrocarbon source-rock potential) and the development of seasonal varves. All lakes are sensitive to climatic changes controlling runoff and evapo-transpiration. The resulting rises and falls of lake level exert a fundamental control on subsurface facies mosaics, both in the marginal clastic fans, fan deltas and submarine fans, and in lake-bottom sediments (Scholz et al., 1990; Lambiase, 1991; Schlische, 1992). Sedimentary cycles resulting from climatic changes are prominent in the records of many Pleistocene to Holocene lake basins. Seismic-reflection studies in Lake Tanganyika and Lake Malawi have identified both highstand and lowstand coarse-grained facies, including prominent subaqueous channels formed by fluvial downcutting during

Figure 3.30 Oblique aerial view across the asymmetric meander belt of the South Fork Madison River in the Hebgen Lake area of Montana. The forested foreground comprises a late-Pleistocene outwash plain which lies about 10 m above the Holocene floodplain, separated from it by a prominent bluffline. The Holocene floodplain comprises a terraced expanse of abandoned meander loops and reaches which indicate that the South Fork Madison has gradually migrated toward the observer over the past 10^4 years (after Alexander et al., 1994).

lowstands (Scholz et al., 1990). These channels and their well developed levees are commonly parallel to the major border faults and their positions may be controlled by intra-rift synthetic faults. During highstands, coarse clastic deposition occurs in the deep lake basins by turbidity currents issuing from subaqueous channels and by downslope turbulent dilution of sediment gravity flows and slumps. Small highstand deltas develop, but tend to be prone to erosion during lowstands. Lowstand deltas prograding into the much-reduced lakes have the highest preservation potential and the greatest economic potential as reservoir sandbodies with efficient seals on all sides.

Particularly subtle Milankovitch cyclicity of lake bottom varves characterizes certain ancient lake basin infills, perhaps most spectacularly revealed by studies in the Triassic rifts of the eastern United States (Olsen, 1986; Olsen et al., 1989; Olsen and Kent, 1990).

Basin Sinks: Eolian

Wind reworking of alluvial and lake shoreline sands is prominent in many arid and semi-arid basins. The location of the resulting small ergs reflects the orientation of the basin relative to dominant prevailing winds. The eolian sands interfinger and are partially reworked by fluviatile, axial and transverse fluvial systems. A fine example occurs in the present San Luis basin of Colorado, where the Great Sand Dunes National Monument occupies the eastern end of the graben, close by a mountainous footwall.

Basin Sinks: Marine Siliciclastics

Marine gulfs occupying extensional rift basins occur in many areas and on various scales. The numerous gulfs surrounding the Aegean extensional province are, perhaps, the most impressive active examples. Here, seasonal runoff is sufficient to maintain active and locally large-scale axial and transverse-sourced alluvial fans, braid deltas, fan deltas and submarine fans. In the Gulf of Corinth, perhaps the most spectacular example (Ferentinos et al., 1988), many fan-delta sytems have developed over the past 2 million years. Along the southern margin, footwall-sourced fan deltas have formed. Gradual northward migration

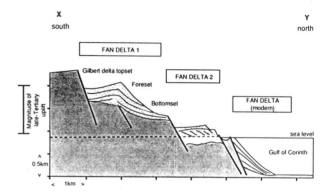

Figure 3.31 Section across the southern margin of the Gulf of Corinth, Greece (see Fig. 3.14 for location), showing the successively abandoned and uplifted coastal fan deltas caused by northward migration of active faulting (after Ori, 1989). The scale of the fan-delta deposits is awesome, with foresets over 500 m high.

of the area of uplift, and the concomitant abandonment and relocation of active normal faulting, have led to the partial preservation of fan-delta deposits, whose clinoform cross sets reach up to 700 m thick (Ori, 1989; Fig. 3.31). Modern fan deltas in the area are linked via submarine channels to submarine fans (Ferentinos et al., 1988).

By way of contrast, the presently arid Gulfs of Suez and Aqaba at the northern end of the Red Sea exhibit only relatively small-scale alluvial fans and fan deltas, whose largely inactive fringes are extensively colonized by patch reefs. These areas are considered further below.

The steep subaqueous slopes present in both marine and lacustrine basins are mantled by fine clastic and biogenic sediment that has slowly settled from suspension. They are cut by subaqueous channels that issue from both axial and transverse alluvial channels and fan deltas. These latter features end in base-of-slope submarine-fan bodies. The by-passed slopes and delta fronts are susceptible to mass failure during earthquakes. Sidescan and shallow seismic profiling reveals a plethora of slump scars, slumps and debris-flow lobes (e.g., Ferentinos et al., 1988). Some of these gravity flows are transformed to turbidity currents as they move downslope, the currents travelling over the basin floor to deposit parallel beds and to build up relatively smooth bottom topography (Brooks and Ferentinos, 1984). Opportunities for

oblique and normal turbidity-current reflection (Pantin and Leeder, 1987; Kneller et al., 1991) are plentiful in such basins. Direct access of fluviatile sediment in suspension as underflows is also possible. Basinal anoxic conditions lead to the development of organic-rich basin-floor mudrocks, with which the transverse and axial clastic systems interdigitate.

Surlyk (1989) presented useful conceptual models of clastic turbidite and other gravity-flow systems in syn-rift environments, stressing the important control that sea-level fluctuations have upon both the coarse clastic and fine-grained-organic fluxes found in periodically anoxic syn-rift basins.

Basin Sinks: Marine Carbonates and Evaporites

Climatic change or significant submergence causes uplands around marine-influenced half grabens located in low latitudes to cease shedding significant siliciclastic debris; chemical and biochemical sediments then accumulate. Since biogenic carbonate production and facies distributions, in particular, are strongly dependent on depth and slope, marked contrasts in facies occur across the half-graben system. Any regional or local restriction of marine circulation in arid climates should give rise to chemical evaporite deposition, both subaerial sabkha and subaqueous marine types. The distribution of facies in an idealized marine carbonate-dominated half graben is discussed further below.

Tectono-Sedimentary Facies Models for Rift Basins

The various sediment sources, fluxes and sinks described above may be combined with tectono-structural data into several architectural schemes for sedimentary-facies distribution and evolution (e.g., Leeder and Gawthorpe, 1987; Rosendahl, 1987). These schemes depend upon sometimes interrelated variables, including local climatic regime, hinterland geology, continentality, pre-rift elevation, topography, sediment flux magnitudes, extension magnitude and history, development of marine, lacustrine or eolian sinks, eustasy and volcanism. It is, thus, inevitable

that any particular rift basin, whether recent or ancient, will possess its own unique combination of factors that will control sedimentary architecture in the basin fill. Analysts of ancient syn-rift extensional basins should always bear this variability in mind. Despite this warning, some attempt is made below to summarize likely scenarios that may commonly arise in the geological record.

Continental Rift Basins With Interior Drainage and/or Lakes (Fig. 3.32)

These basins have drainage systems that are isolated from adjacent basins, sometimes because of bedrock ridges at crossover offset zones. A spectrum of examples exists, varying in the relative permanence and depth of the lakes. The prototype at the playa end of the spectrum is Death Valley and many other basins like it in the Basin and Range Province. During pluvial periods of the Pleistocene, widespread shallow lakes occupied many of the extensional basins in the area (e.g., Centennial Lake grabens of Idaho; Great Basin of Utah; Buffalo, Buena Vista and Dixie Valleys of Nevada; Figs. 3.23, 3.24). Their early Holocene shorelines may be easily mapped today, forming important yardsticks with which to measure post-highstand deformation by faulting and tilting (Wallace, 1978).

Lakes Malawi and Tanganyika in the East African rift are good examples of deeper lakes. As noted above, even these deep lakes (water depths up to 1500 m and 500 m, respectively) were not immune to climatically controlled fluctuations during the Pleistocene, with lake levels varying by several hundreds of meters and the resulting water dynamics fluctuating between shallow, saline during lowstand periods and deep, stratified during highstands.

Basin architecture is controlled by the interactions between basin-center environments, and encroaching alluvial fans and fan deltas sourced from footwall and hangingwall uplands. Periodic catastrophic denudation events are recorded by landslide megabreccias. In areas directly affected by Quaternary ice action, valley glaciers have left prominent terminal and side moraines that continue to strongly influence sedimentation. Climatically

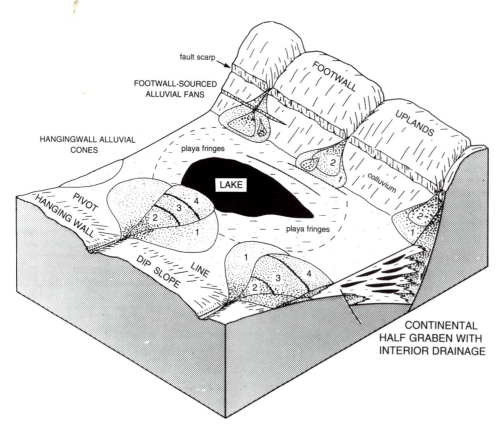

Figure 3.32 Block diagram to summarize the major clastic environments present in closed continental half grabens with interior drainage (after Leeder and Gawthorpe, 1987).

controlled cycles dominate the Pleistocene lake fills of many rift basins, with interesting analogs preserved in the Triassic grabens of the northeastern United States (see above).

Continental Rift Basins With Axial Through Drainage (Fig. 3.33)

Active examples are widespread in the Aegean (e.g., Vardar, Lamia, and Meander grabens), Rhine (Erft graben) and Basin and Range (e.g., Rio Grande, Lemhi, Beaverhead, and Madison). The position of the axial-stream facies reflects subtle controls exerted by periodic tilting, encroaching transverse fans, emplacement of landslide megabreccias, intrabasinal horst structures and mini-grabens. Perhaps the best example of these various models is the southern Rio Grande rift valley, where late-Pleistocene and Holocene incision has revealed the deposits and architecture of a previously aggrading phase (Mack and Seager, 1990; Fig. 3.29).

Coastal/Marine Siliciclastic Gulf (Fig. 3.34)

Fine examples of this model occur in the Aegean (Gulfs of Corinth, Euboea, Thermaios, etc.). Where axial fluviatile drainage is present, river- or wave-dominated delta lobes develop in this virtually tideless environment. Spectacular sections are documented as nearshore cyclical facies in a shallow-water, late Pleistocene/Holocene graben fill in the Corinth Canal, Greece (Collier, 1990). The very steep faulted footwall margins (Fig. 3.35) in other areas of this basin have led to the formation of talus cones, alluvial fans, gigantic fan deltas (Ori, 1989; Fig. 3.33) and submarine fans. Perhaps the best exposed ancient analogs are in the Mesozoic rifts of East Greenland (Surlyk, 1978, 1989).

Marine Carbonate Coastal Gulf (Fig. 3.36)

Syn-rift basins are usually dominated by siliciclastic deposition, but suitably old tiltblocks subsiding

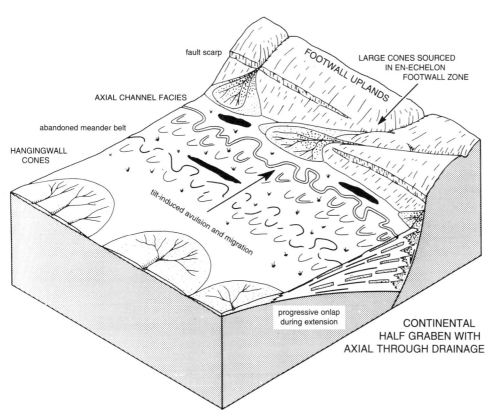

Figure 3.33 Block diagram summarizing the major clastic environments present in continental half grabens with through-flowing axial drainage (after Leeder and Gawthorpe, 1987).

thermally, or offshore tiltblocks in very arid climates may show variable development of biogenic and chemical sedimentation. Mixed siliciclastic/carbonate environments occur in some Holocene to late Pleistocene examples in central Greece, with tilt-block topography locally controlling the magnitude of tidal currents and causing rapid lateral facies variations (Collier and Thompson, 1991). Since carbonate production and facies distributions are strongly depth- and slope-dependent, marked lateral changes should occur within the structural template provided by the half-graben system (Gawthorpe, 1986; Leeder and Gawthorpe, 1987). The steep gradient caused by the main boundary fault ensures that the footwall slope transition is steep and abrupt, causing a structural bypass (e.g., McIlreath and James, 1978). The gentler hangingwall slope should initially develop into a ramp-type margin (e.g., Ahr, 1973), deepening toward the footwall scarp. Eventually, a rimmed-shelf and sedimentary-bypass margin is postulated to develop, following the evolutionary trend outlined by Read (1982, 1984). Axial basin environ-

ments may be starved of sediment and the facies developed are, thus, markedly condensed. Black shales will develop if the basin has a permanent anoxycline.

Proto-Oceanic Rift Troughs: The Red Sea

Introduction

The Red Sea/Gulf of Aden ocean basin (Fig. 3.37) is the chief natural laboratory where the early evolution of continental extension and its active transition to oceanic crustal generation may be studied. A similar scenario occurs in the Gulf of California, but here there is significant and rapid transform-fault motion, causing transtensional tectonics. The Red Sea area played a key role in early debates about the existence of continental drift (Girdler, 1958, 1962) and in establishing the basic concepts of the plate-tectonic Wilson Cycle in the 1970s (e.g., Gass, 1970; McKenzie et al., 1970).

The Red Sea is flanked by prominent marginal escarpments (Fig. 3.37, inset) that rise from narrow

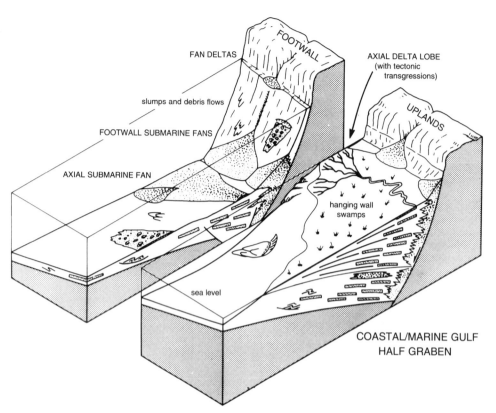

Figure 3.34 Block diagram summarizing the major clastic environments present in coastal or deep lacustrine half grabens (after Leeder and Gawthorpe, 1987).

Figure 3.35 View of prominent faulted footwall adjacent to the marine Corinth Basin near Skinos, Greece. The limestones of the footwall uplands source fan deltas and talus cones, which prograde into shallow lagoons fringed by barrier beaches.

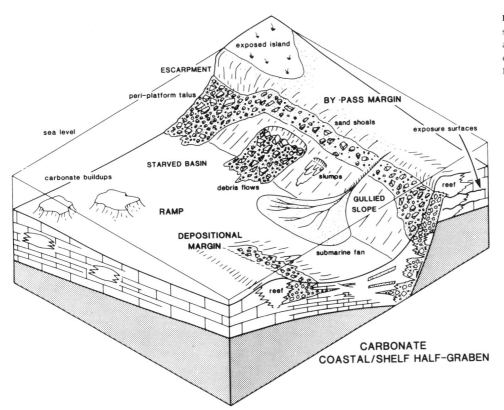

Figure 3.36 Block diagram summarizing the major carbonate environments present in coastal/shelf half grabens (after Leeder and Gawthorpe, 1987).

faulted coastal plains to over 2000 m elevation on both sides, although the Arabian flank has the greater mean elevation (up to 3200 m in the south). Gently sloping shelves, up to 150 km across, flank the 50 km-wide axial zone with its discontinuous rift valley and isolated axial deeps that descend to more than 3 km below sea level. The axial region is floored by continuous young (<5 my old) oceanic crust only in the southern third (south of 20°N) of the Red Sea. Farther north, isolated deeps are progressively less common. Geophysical studies (e.g., Egloff et al., 1991; Makris et al., 1991; Rihm et al., 1991) reveal that the shelves are underlain by stretched continental crust in central areas, but by an abrupt ocean/continental crust transition in the north (particularly on the western margin). It seems probable that stretched continental crust underlying the wide shelves has been modified (hybridized) by the intrusion of substantial amounts of mafic magma (Bonatti and Seyler, 1987; Bohannon et al., 1989; Baldridge et al., 1991; Bohannon and Eittreim, 1991), including dike swarms equivalent in age to those commonly found

intruded on land along both sides of the Red Sea (e.g., Gass, 1980; Coleman, 1984). To the south, the western tip of the slow-spreading (16 mm y⁻¹) Gulf of Aden rift comes onshore in Djibouti as the Asal rift (Stein et al., 1991). The Red Sea rift is also propagating westward into the Afar depression toward Asal.

Open or Closed Initial Rifting?

Prior to formation of the initial rift, the Red Sea area had low relief and was part of a northward-deepening epicontinental area from the Late Cretaceous to the early Oligocene. Concerning the sedimentary evidence for early uplift versus subsidence, Purser and Hotzel (1988) noted the disappearance of pre-rift marine Paleogene sediments about halfway along the rift on the Egyptian side and used the concomitant absence of facies changes to infer that a period of uplift and erosion occurred prior to the first stages of syn-rift sedimentation. This interpretation is at variance with the evidence of Bohannon et al., (1989), who recognized a southward facies change from

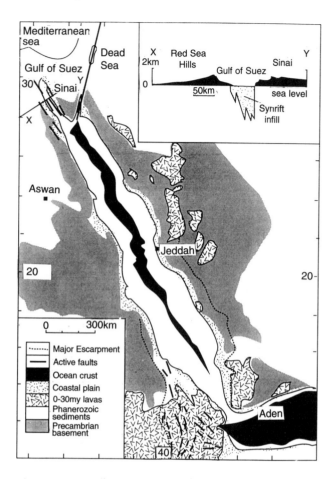

Figure 3.37 Map showing the Red Sea and its geological setting (modified from several sources). Detail shows a section across x-y in the Gulf of Suez (after Steckler, 1985).

of up to 2.5–5 km has occurred since the middle Miocene, about 15 my after the first extensional faulting and volcanism. The forces required to stretch the slowly NE-drifting Afro-Arabian plate were presumbly provided by slab-pull edge forces on the subducting plate descending beneath the Zagros Mountains of Iran and Iraq to the northeast. These forces had the effect of slightly accelerating the NE-drifting Arabian area of the plate, causing regional tensional stresses in the crust.

Sedimentation

Studies in the onshore and offshore Red Sea and Gulf of Suez (e.g., Angelier, 1985; Sellwood and Netherwood, 1986; Jackson et al., 1988; Purser and Hotzel, 1988; Gawthorpe et al., 1990) reveal, not surprisingly, that the earliest phase of syn-rift development (Oligocene to Miocene) was characterized by coarse alluvial-fan and fan-delta clastic deposition, with subordinate evaporites, in distinct rift basins. Detailed studies in areas close to major transfer zones along the faulted margins reveal the initial presence of successive and aggradational fan-delta-lobe deposits, presumably sourced from larger drainage basins located within transfer-zone uplands. Younger fan-delta lobes show evidence of periodic progradation under conditions of more stable long-term base level, although with plentiful evidence for periodic tectonic shocks and small-scale base-level changes. Marked onlap of tilt-block topography is recorded for the syn-rift phase. Away from areas of significant clastic input, spectacular reef and reef-talus development is recorded from several areas, with progradational clinoforms indicating water depths up to 100 m. Gross upward fining and onlap of the syn-rift fill was terminated by a regional phase of erosion and subsequent deposition of seaward dipping and thickening subaqueous sulphate-bearing evaporites, which in offshore areas, drape over the wedge-shaped syn-rift deposits and their basement topography. These Miocene evaporites may exceed 2 km in thickness in some areas (Hutchinson and Engels, 1972; Lowell and Genik, 1972).

Normal marine salinities and renewed extensional faulting occurred during the Pliocene and continued into the Holocene. The new faults fragmented the

marine to nonmarine sediments and ultimately to an area of thick continental laterite. Upwarp of the latter, recognized by workers for many years, occurred over large areas of SW Arabia and Sudan well after the first extensive volcanic eruptions that occurred between 30 and 25 Ma. The widespread presence of marine conditions at many localities along the Red Sea and Gulf of Suez margins prior to and shortly after the first evidence for extensional faulting (Purser and Hotzl, 1988; Bohannon et al., 1989; Gawthorpe et al., 1990) also implies that pre-rift doming was not important and that closed-system rifting conditions with decompressive mantle melting were operative. On the basis of fission-track dating (Garfunkel, 1988; Bohannon et al., 1989; Omar et al., 1989), flank uplift leading to basement exhumation

previous tilt blocks in some areas and caused differential uplift and subsidence of Miocene reefs. Both local tilt-block rotations and more regional uplift have led to the elevation, erosion and exhumation of much of the onshore extent of former syn- and post-rift sequences. Offshore these sequences are buried beneath marine deposits of the Red Sea.

Central Red Sea sedimentation in the Holocene has been dominated by pelagic foram/pteropod oozes with currently high sedimentation rates of up to 0.1 mm y^{-1} (Almogi-Labin et al., 1991). A highly stratified water column developed during late glacial and early deglaciation times (14–8 ka) due to poor connections with the Indian Ocean. Low productivity in the upper mixed layer, very high bottom salinities and accumulation of organic-rich sediments resulted. Subsequently, there has been a marked trend to increasingly well oxygenated conditions associated with higher productivities and sedimentation rates due to good mixing with the Indian Ocean today.

Tectono-Magmatic Models

Although the Red Sea rift itself is a more-or-less symmetrical structure, the voluminous volcanics erupted between 30 and 20 Ma are notably concentrated on the NE margin in Arabia (Fig. 3.37). They form a prominent zone 200–400 km at maximum away from the Red Sea axis, and as noted above, are associated with greater basement elevations than those observed on the opposite African margin. The volcanic zone swings southeast into Yemen and then across the southern Red Sea straits into the Afar part of the Ethiopian rift. It is reasonably well established that the latter area lies over a large mantle plume, reponsible for the massive amounts of mafic volcanics in the Ethiopian rift (the area originally inspired Cloos's (1939) classic experiment on rift domal uplift). The effects of the plume extend northward into Yemen, where eruption of early (31 to 26 Ma) flood basalts and silicic volcanics was followed by intense mafic dike intrusion, alkali granite plutonism and volcanism (24 to 20 Ma), and finally, by crustal extension along normal faults (Mohr, 1991). Here, it seems that magmatism was followed by stretching as the effects of the Red Sea extension

propagated southward into the area influenced by the Afar plume.

The simple-shear conceptual model has been invoked (Wernicke, 1985) to explain the asymmetry of the magmatic and topographic features of the Red Sea. Thus, a northeast-dipping lithospheric shear zone is postulated to have caused uplift and volcanism above the locus of fault intersection with the asthenosphere on the Arabian side of the Red Sea. However, seismological, heat-flow and sedimentary evidence (Buck et al., 1988; Dixon et al., 1989; Makris and Rihm, 1991) is inconsistent with this interpretation; in addition, the low-angle crustal detachment has not been discovered.

Recent models (e.g., Dixon et al., 1989) appeal to initial asthenospheric upwelling at around 30 Ma under the line of the volcanic zone, causing uplift due to crustal thickening associated with magmatic underplating and crustal intrusion. This was followed by northeastward lithospheric migration over the upwelled asthenosphere prior to establishment of the present rift along a zone of pre-existing crustal weakness. Although there is some evidence for ridge migration and asymmetric spreading in the axial zone of the Red Sea (the rate of speading on the western side of the axial zone is up to twice that in the east), the predicted timing and distribution of basement exhumation in this model do not seem consistent with the fission-track cooling ages obtained by Bohannon et al. (1989) and Omar et al. (1989) from both sides of the rift.

The first ocean crust was injected into the southern sector, at latitude 17°N, around 5 Ma, but more recently to the north and south. To the north, the axial rift is discontinuous. The regional pattern of isolated troughs and deeps strongly suggests a northward propogation of linear spreading segments (Courtillot, 1982; Cochran, 1983a), a trend attributed by Bonatti (1985) to a progressive south-to-north decay of "hot spot" emplacement of mantle diapirs serving as nuclei for axial propagation. The hot spots below the segmented deeps in the central region have an average 50 km spacing, and it was suggested by Bonatti that they might result from a Rayleigh-Taylor-type instability at the asthenosphere/lithosphere boundary. Such discontinuous

propagation of spreading centers is also apparent along the early margin of the east Atlantic Ridge and may be a general feature of the transition from continental rifts to oceanic crustal accretion at ridges.

Concluding Remarks

A persistent theme in this chapter has been the close interrelationship between extensional tectonics and sedimentation in continental rifts and basins. Normal faulting exerts an important control upon topography and drainage-basin development, and thus, helps determine the architecture of synrift sedimentary fill. Because of this control, it is stressed that facies and paleocurrent trends in ancient grabens may only be correctly interpreted when observations are made on a length scale of 10–20 km, comparable to that of the largest fault segments. Finally, it is also stressed that sediment flux from structurally controlled catchments and stratigraphic architecture are modulated in important ways by lithology and climate.

Acknowledgments

Numerous people have contributed to advancing my (still incomplete) knowledge of extensional tectonics in the past few years, in particular James Jackson, Nicky White and Cindy Ebinger. In addition, I grate-

fully thank the following persons for finding time to make numerous critical suggestions that have vastly improved successive drafts of this chapter—Cathy Busby, Cindy Ebinger, Ray Ingersoll, Greg Mack, Celal Şengör and Fyn Surlyk.

Further Reading

Cochran JR, 1983a, *A model for development of Red Sea:* American Association of Petroleum Geologists Bulletin, v. 67, p. 41–69.

Dickinson WR, 1991, *Tectonic setting of faulted Tertiary strata associated with the Catalina core complex in southern Arizona:* Geological Society of America Special Paper 264, 106p.

Ebinger CJ, 1989b, *Geometric and kinematic development of border faults and accommodation zones, Kivu-Rusizi rift, Africa:* Tectonics, v. 8, p. 117–133.

Gans PB, 1987, *An open-system, two-layer crustal stretching model for the eastern Great Basin:* Tectonics, v. 6, p. 1–12.

Leeder MR, Alexander J, 1987, *The origin and tectonic significance of asymmetrical meander belts:* Sedimentology, v. 34, p. 217–226.

Leeder MR, Gawthorpe RL, 1987, *Sedimentary models for extensional tilt-block/half-graben basins:* Geological Society of London Special Publication 28, p. 139–152.

Mack GH, Seager WR, 1990, *Tectonic control on facies distribution of the Camp Rice and Palomas Formations (Pliocene-Pleistocene) in the southern Rio Grande rift:* Geological Society of America Bulletin, v. 102, p. 45–53.

Ori GG, 1989, *Geologic history of the extensional basin of the Gulf of Corinth (?Miocene-Pleistocene), Greece:* Geology, v. 17, p. 918–921.

Rosendahl BR, 1987, *Architecture of continental rifts with special reference to East Africa:* Annual Review of Earth and Planetary Science, v. 15, p. 445–503.

Schlische RW, 1991, *Half-graben basin filling models: new constraints on continental extensional basin development:* Basin Research, v. 3, p. 123–141.

Continental Terraces and Rises

4

Gerard C. Bond, Michelle A. Kominz, and Robert E. Sheridan

Basin Position

Relation To Plate Boundaries and Tectonic Setting

Continental terraces and rises have a distinctive first-order structure that sets them apart from other basins. These enormous wedges of sediment are bounded on one side by a gently sloping to nearly flat free surface, the continental slope and rise, and on the other by a profound structural discontinuity (Fig. 4.1). The structural discontinuity coincides with the hinge zone, a lithospheric-scale boundary that separates "normal" continental crust from the deeply subsided and modified crust that lies beneath the terrace-rise system. In most cases, the free surface faces an active (or, in some examples, a once active) spreading ridge system.

It is now generally agreed that continental terraces and rises are a direct consequence of continental rifting and that they occur within passive margins initiated along divergent plate boundaries (see Chapters 2 and 3). In the early stages of the rifting process, rift valleys and proto-oceanic troughs evolve within or close to the plate boundary. Sedimentation of the continental terrace and rise begins after continental rifting is completed and as a new ocean basin has begun to form by sea-floor spreading. By this time, the plate boundary (the spreading ridge) is tens to hundreds of kilometers seaward of the basin, locking the basin into a relatively stable intraplate position at the edge of the rifted continent (Figs. 4.2, 4.3).

Continental Terraces and Rises Facing Orthogonally Divergent Ocean Basins

Probably the most commonly cited examples of continental terraces and rises are those facing orthogonally divergent ocean basins. The best known of these occur in the extensive passive continental margin off the east coast of the U.S.A. and Canada. Excellent summaries of the evolution of this margin can be found in Sheridan and Grow (1988) and Keen and Williams (1990). The passive margin formed during breakup of the supercontinent Pangea in early Mesozoic time and during subsequent opening of the modern Atlantic ocean basin. After the rifted continents separated, the ocean basin opened more or less at right angles to the trend of the passive margins so that most of the continental terraces and rises tend to face directly toward the plate boundary. The passive margins around eastern, western, and southern Africa, around western and southern Australia, in much of the Arctic Ocean, and around much of Antarctica formed in essentially the same way as a consequence of the breakup of Pangea (Fig. 4.3).

The eastern margin of North America was one of the first passive margins studied and is still one of the most thoroughly studied; it is not surprising that this passive margin and its system of continental terraces and rises has become one of the type examples. In fact, the classic stages leading to the formation of continental terraces and rises depicted in Fig. 4.2 are based largely on the evolution of the eastern margin of the U.S.A. There are many exceptions, however, to this relatively simple evolution of the basins and their relation to plate boundaries.

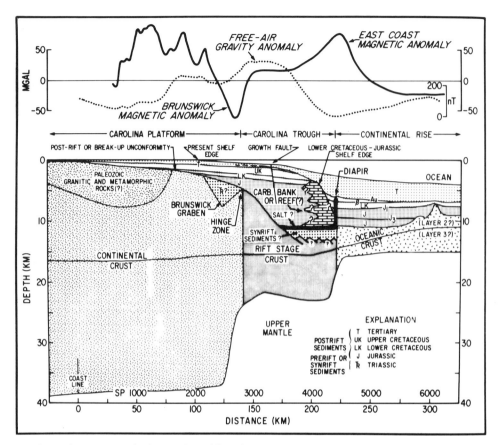

Fig. 4.1 Cross section of modern passive margin off coast of North Carolina (Hutchinson et al., 1982). Rift stage crust is equivalent to transitional crust of Dickinson (1976a). SP refers to shot point.

Continental Terraces and Rises Facing Obliquely Opening Ocean Basins

Many continental terraces and rises face obliquely opening ocean basins. Some of the best-known case histories are from along the rifted continental margins of the Gulf of California (e.g., Lonsdale, 1989; Fig. 4.4). Here, relative motion between the Pacific and North American plates is 10° to 20° oblique to the continental boundary, resulting in separation of Baja California from the continent, along with northward movement relative to North America. The plate boundary is exceedingly complex, consisting of long transform faults that parallel the relative plate motion and that are connected by small spreading ridges. The continental terraces and rises facing the ocean basin have complex structural relations to these transform faults and spreading ridges that set them apart from their much simpler counterparts in most orthogonally opening ocean basins (Fig. 4.4). Oblique opening of ocean basins may be relatively common. For example, oblique separation probably occurred early in the history of the passive margin along the eastern U.S.A.

(Swanson, 1982; Manspeizer and Cousminer, 1988) and southern Australian margins (Willcox and Stagg, 1990).

Continental Terraces and Rises Along Large Transform Faults

Other continental terraces and rises occur along large transform faults (e.g., Scrutton, 1979). This type is exemplified by the Grand Banks southwest of Newfoundland (Fig. 4.5; Todd et al., 1987; Keen et al., 1990), the margin of western Africa along the Ghana and Ivory Coast (Blarez and Mascle, 1988), and the Exmouth Plateau along the northwestern margin of Australia (Lorenzo et al., 1991). The continental terraces and rises in these places lie along steep-sided promontories of continental crust that jut into the ocean basin for tens to hundreds of kilometers. These margins probably are inherited from cross structures or accommodation zones that formed along horizontal offsets in continental rifts. As continental separation occurred, the accommodation zones evolved into transform faults that offset the spreading ridges.

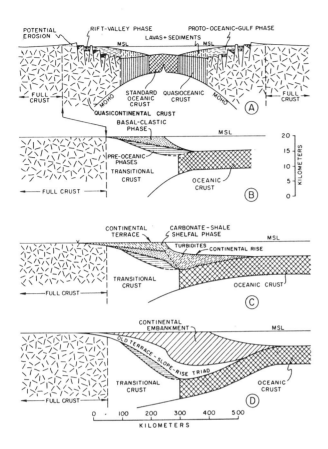

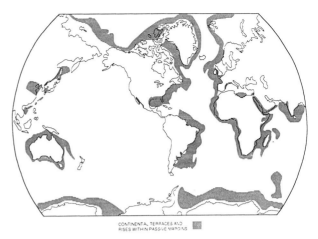

Fig. 4.3 Distribution of continental terraces and rises, showing mainly those that face orthogonally opening ocean basins and that formed during breakup of Pangea (from Bally, 1979; Jolivet et al., 1989; Channell et al., 1991). Large segments of Cenozoic-Mesozoic (CZ-MZ) active margins also contain continental terraces and rises, especially in the western Pacific and in the Mediterranean. A system of continental terraces and rises also occurs in the obliquely opening Gulf of California.

Fig. 4.2 Diagram showing evolution of a passive continental margin facing an orthogonally opening ocean basin (from Ingersoll, 1988b, modified from Dickinson, 1976a vertical exaggeration 10x). (Rift-valley stage not shown; see Chapters 1 and 3.) A) Proto-oceanic stage with syn-rift sediments overlying modified continental crust. During this stage, thermally controlled subsidence is beginning and new ocean floor has begun to form. B) Early post-rift stage when deposition of basal clastic wedge occurs as elevated rift shoulders are eroded. C) Formation of mature continental terrace and rise as thermally controlled subsidence continues and increased rigidity causes flexural bending of the unrifted continental crust and young oceanic crust. D) Continental embankment stage, reached only where major deltaic progradation occurs (see Chapter 1).

Along large transform margins, such as in western Africa, continental separation may have been oblique, producing structures comparable to those in the Gulf of California. Continental terraces and rises formed in this structural setting tend to be common in orthogonally opening ocean basins; however, they face away from the spreading ridges at high angles. In addition, because they formed along continental shear zones, these basins have rather unique thermal and mechanical evolutionary trends (Keen et al., 1990; Lorenzo and Vera, 1992), and they tend to be

influenced by motion along the transform boundary long after continental separation.

Continental Terraces And Rises Facing Back-arc Ocean Basins

Some of the most complex and poorly understood continental terraces and rises face small backarc ocean basins (also, see Chapter 8). Good examples of these are in the western and southwestern Pacific Ocean along the continental margins of Japan, South China and northwestern Australia. In contrast to the preceding examples, these continental terraces and rises lie close to convergent plate boundaries (e.g., Fig. 4.3). These basins were initiated by rifting within and behind active magmatic arcs. The rifting culminated in continental separation, and as the backarc basins opened, narrow fragments of continental crust beneath the arcs separated from the mainland (Fig. 4.6; Aubouin, 1990). These continental fragments and their volcanic cover may now form narrow ridges within backarc ocean basins. As the ocean basins widened, continental terraces and rises broadly similar to those in Atlantic-type ocean basins formed along the rifted edges of continental crust.

The exact nature of the divergent plate boundaries that initiated these continental terraces and rises is

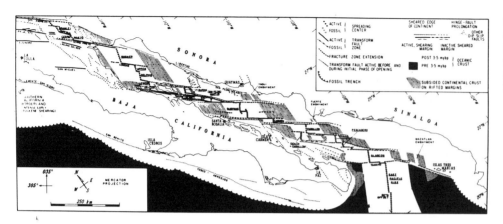

Fig. 4.4 Obliquely rifted margin in the Gulf of California (Lonsdale, 1989). Underlined names are spreading centers, and where inactive, are in brackets.

not well understood. Some view the boundaries as mainly a consequence of subduction. In this case, the rate of trench rollback may exceed the rate of trenchward motion of the overriding plate, mantle plumes may rise from the heated subducting slab initiating convection, or secondary convection may be induced by the downgoing slab (e.g., Tamaki and Honza,

1991; see Chapter 8). Others suggest that the boundaries have a much more complex origin broadly analogous to pull-apart structures in which local divergence is initiated by components of strike-slip deformation within an overall convergent plate boundary (e.g., Jolivet et al., 1989; Kano et al., 1990). Whatever their exact origin, these backarc ocean basins tend to be reactivated on time scales of a few tens of millions of years. Their thermal histories also tend to be complex, a consequence of recurring rifting and proximity to magmatic arcs and spreading ridges. The continental terraces and rises are, therefore, subjected to recurrent tectonism, producing complex subsidence histories and internal stratigraphy.

Continental Terraces and Rises in Zones of Continental Collision

The continental terraces and rises in the Mediterranean Ocean Basin are distinctive in that they lie

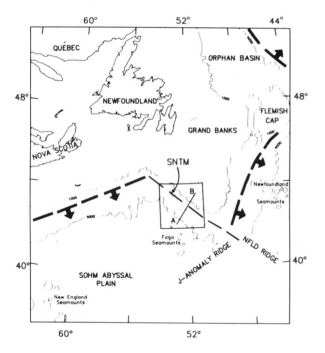

Fig. 4.5A Southwest Newfoundland Transform Margin (SNTM), and northwest continuation of the Newfoundland (NFLD) Ridge. This transform margin is a zone of shearing that offsets the stretched crust, the landward boundary of which is marked by the heavy dashed line. Arrows mark the side of the boundary containing stretched and faulted crust. AB is line of cross section shown diagramatically in Fig. 4.5B (from Reid, 1989).

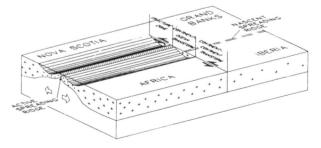

Fig. 4.5B Block diagram of the SNTM in Early Cretaceous time (from Todd et al., 1987). The active spreading ridge ends along the transform against the unrifted crust of the Grand Banks and Iberia. Lateral heat flow from the hot, oceanic crust causes thermally driven uplift of the unrifted crust along the transform.

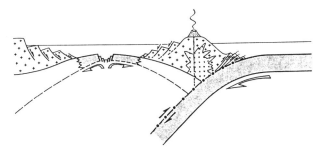

Fig. 4.6 Formation of rifted continental margins in backarc settings (from Aubouin, 1990). The volcanic arc originally lay along a continental margin. Extension within and behind the arc led to separation of the arc and its subjacent continental crust from the continent, culminating in the formation of a backarc ocean basin. Subsidence of the rifted crust occurs as the backarc ocean basin opens, resulting in the formation of continental terraces and rises along the edge of the continent and along the fragment of continental crust beneath the arc (see Chapter 8).

within a broad zone of convergence between two continents, Afro-Arabia and Eurasia (Fig. 4.3; Le Pichon et al., 1988; Channell et al., 1991). Changes in relative plate motions since Early Jurassic time have produced an exceedingly complex and poorly understood history of ocean-basin formation, entrapment, and destruction.

Along the southern margin of the Mediterranean Basin lie remnants of continental terraces and rises that formed in Jurassic and Early Cretaceous time, when the Tethys Ocean opened between Africa and Europe (Channell et al., 1979). These are exposed along the North African coast and occur in the subsurface of the North African continental margin. In Late Cretaceous time, change in plate motions initiated convergence between the two continents, and the spreading ridge in the Tethys Ocean was consumed. In the Eastern Mediterranean basin, the Tethys continental terraces and rises are now moving into an active subduction system (see Chapter 10).

Much younger continental terraces and rises, mostly mid to late Tertiary in age, face the small ocean basins along the complex northern edge of the Mediterranean Ocean. Here, after the destruction of the Tethys margin in early Tertiary time, extension behind a southeast facing arc system apparently caused rotation of Corsica and Sardinia away from Europe, opening the Balearic and Ligurian ocean basins (Smith and Woodcock, 1982). The Aegean and Tyrrhenian basins may have opened in a similar manner. Thus, although the continental terraces and rises

along the southern European margin of the Mediterranean Ocean have affinities with those in backarc basins in the Pacific, the Mediterranean examples have formed within an overall remnant ocean basin (see Chapter 10).

Crust Beneath Continental Terraces and Rises

Pre-Rift Crust

The nature of pre-rift crust beneath modern continental terraces and rises, to the extent that it is known, varies considerably along strike and from margin to margin, reflecting the diverse settings in which continental rifting has occurred. In places, such as along the Labrador Sea and Baffin Bay, the pre-rift crust is the continental shield, which, prior to rifting, had been stable for more than a billion years (Keen et al., 1990). More commonly, the pre-existing crust was severely deformed during collisional phases of the preceding Wilson Cycle. For example, along parts of eastern and western Africa, along parts of East Antarctica, and along northern and eastern South America, the pre-rift crust was severely deformed during the Pan African orogeny and assembly of Gondwanaland between about 800 and 600 Ma (Hoffman, 1991). In other places, such as the eastern U.S.A. and Canada, East Greenland and western Scandinavia, pre-rift crust was deformed and thickened during middle to late Paleozoic collisions that accompanied the assemblage of Pangea (Klitgord et al., 1988; Keen et al., 1990). Many of the fundamental structures produced by these collisional events have been traced into the pre-rift crust beneath continental terraces and rises, based on gravity and magnetic anomalies and on evidence from deep seismic profiles (Smythe, et al., 1982; Veevers, 1984; Klitgord et al., 1988; Keen at al., 1990). Some of these structures appear to have been reactivated as extensional faults during rifting (e.g., Fig. 4.7; Cheadle et al., 1987; Klitgord et al., 1988).

Syn-Rift Crust

The rifted crust beneath continental terraces and rises contains two fundamental boundaries, the hinge zone and the landward limit of oceanic crust. These boundaries divide the crust into three regimes,

normal continental crust, modified (or transitional) continental crust, and oceanic crust (Fig. 4.1).

Normal Continental Crust

The region of normal continental crust lies landward of the hinge zone beneath the relatively thin, inner edge of the continental shelf (Fig. 4.1). This thin wedge of sediment was formed when the weight of the sediment along the margin became great enough to cause flexural bending of the edge of the craton. "Normal" means that the crust is similar in thickness and seismic velocity to the crust that can be traced into the continent. There are examples, however, where at least the upper part of this crust was modified by normal faulting during the rifting stage of the margin. One of the best documented of these is the rift complex preserved in the Triassic to Middle Jurassic Newark Supergroup along the east coast of the U.S.A. (Fig. 4.8; Manspeizer and Cousminer, 1988). The rift basins are up to 10 km deep, and were formed by rotation of crustal blocks above major listric normal faults. A similar rift complex lies inboard of the hinge zone along parts of the conjugate margin in west and northwest Africa (Jansa and Wiedmann, 1982). This wide system of rift basins, extending in places 300 to 400 km landward of the hinge zone, is thought to have been initiated by clockwise rotation of Gondwanaland with respect to Laurentia, causing shearing and oblique extension along pre-existing basement structures in the Variscan-Alleghanian orogen. Data from a few seismic profiles and from gravity anomalies along the eastern coast of the U.S.A. have led some to suggest that these rift basins formed above a system of midcrustal detachments that propagated landward from the hinge zone (e.g., Bell et al., 1988; Klitgord et al., 1988; McBride, 1991).

Hinge Zone

The normal crust beneath the continental terrace terminates at the hinge zone, a region where basement deepens rapidly and the overlying sediment abruptly thickens (Fig. 4.1, 4.2). Based on data from deep seismic profiling and refraction experiments, the hinge zone marks a zone of rapid crustal thinning and abrupt decrease in the depth to the Moho (Fig. 4.1). This boundary, then, is the landward limit of substan-

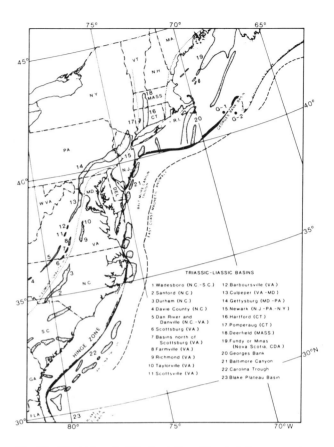

Fig. 4.7 Triassic-Jurassic rift basins, indicated by stippled pattern, at the surface and in the subsurface in eastern North America. Piedmont gravity high, East Coast magnetic anomaly (ECMA), hinge zone of the margin and locations of COST wells are shown (from Manspeizer and Cousminer, 1988). Note syn-rift basins over 300 kilometers landward of the hinge zone.

tial crustal thinning and rapid subsidence of the margin. Beneath nontransform margins, the hinge zone typically is 10 to 30 km wide and contains faulted blocks of syn-rift sediment and pre-rift crust. Along the east coast of the U.S.A. and Canada, the hinge zone coincides in some places with half-graben structures (tilted blocks) bordered by seaward-dipping faults and in other places with tilted blocks bounded by landward-dipping faults. Similar structures have been documented along hinge zones in the Irish margin (deGraciansky et al., 1985), in the North Biscay margin (Montradert et al., 1979), and along the western Ligurian Tethys margin (Lemoine et al., 1986).

As might be expected, the hinge zone is quite different along transform margins and obliquely opening basins. For example, along the southern edge of the

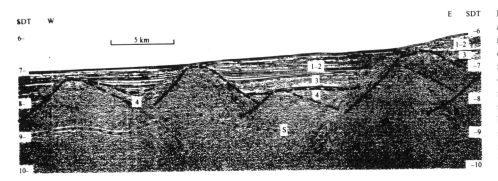

Fig. 4.8 Rotated crustal blocks above listric normal faults in modified continental crust along continental terraces and rises. Seismic profile from just west of Galicia Bank (Spain), showing rotated basement blocks below sediment layer 3. Listric faults bounding these blocks appear to sole into a major crustal detachment fault inferred to lie along the reflector S. Vertical scale is in seconds and numbers 1 to 4 refer to stratigraphic intervals (from de Charpal et al., 1978).

Grand Banks (Keen et al., 1990) and along the southern Exmouth Plateau (Lorenzo et al., 1991), the zone of crustal thinning is abrupt, and a narrow, highly fractured zone separates normal continental crust and oceanic crust. Several magmatic intrusions occur within this zone in the Exmouth Plateau. The narrow boundary along these transform margins is interpreted to be a zone of shearing, which has thoroughly disrupted the fabric of the pre-rift crust. Commonly, the crust landward of the hinge zone is strongly uplifted, forming a prominent basement high or elevated rift shoulder. This basement high has recently been interpreted as a consequence of thermal expansion of lithosphere as heat flows laterally across the sharp discontinuity separating thinned continental crust from unrifted crust (e.g., Fig. 4.5b). Along transform faults in the obliquely opening Gulf of California, sharp drops occur in topography and in the gravity field, suggesting abrupt transitions from old continental to rifted continental crust. Along the rifted segments facing the offset spreading ridges (Fig. 4.4), the hinge zone is deeply buried by sediment but appears to be broader, based on more gradual gradients in depth, gravity and magnetics.

Modified or Transitional Continental Crust

Thinned and compositionally modified continental crust lies beneath much of the thick wedge of sediment in continental terraces, extending from about 100 to over 400 km seaward of the hinge zone. Typical thicknesses of the modified crust range from about 25 km near the hinge zone to 10 km or less at the transition to oceanic crust. The nature of this crust is commonly obscured by thick overlying sedi-

ments. The best data are from the upper few kilometers of the crust, where seismic-reflection profiles reveal evidence of brittle deformation. Much of this deformation is characterized by rotation of blocks of basement and syn-rift sediment above systems of listric normal faults (Fig. 4.8). In some margins, normal faults that step down to the ocean are observed; in others, the structure is dominated by half grabens. In parts of some margins, deep seismic profiles suggest that the listric faults sole into a system of mid- to upper-crustal detachments (e.g., "extensional allochthons" Fig. 4.9; Klitgord, et al., 1988; Etheridge et al., 1989; Mutter et al., 1989; Lister et al., 1991), or into a highly reflective lower crust that is interpreted to be a thick zone of simple shear (Reston, 1990). The structural style of the modified crust closely resembles that observed in many of the world's active rift basins (see Chapters 2 and 3).

In parts of some margins, normal faults appear to be only locally developed within the thinned crust beneath continental terraces. Examples are the margin off Nova Scotia and the northern part of the Baltimore Canyon trough. Here, deep seismic profiling indicates that the continental crust is thinned by more than a factor of 3 (Keen and Williams, 1990). Keen et al. (1990) suggested that the absence of pervasive normal faulting may be evidence that the basins formed above a large fault block underlain by a single, large fault that flattens seaward into a decollement. This idea is compatible with recent detachment models of passive margins. These models are based on simple shear along decollement or low-angle detachment faults. Separation of continental blocks along the detachment faults leads to the

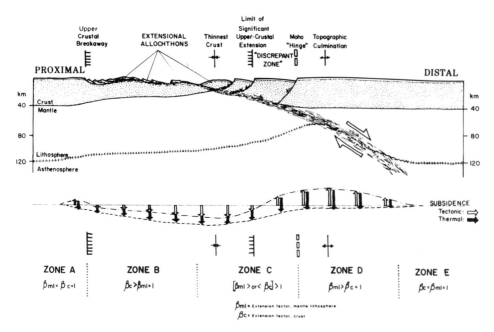

Fig. 4.9 Detachment model for extension of lithosphere (from Wernicke, 1985). Vertical arrows give hypothetical subsidence and uplift. As applied to passive margins, Zone A to the middle of Zone C constitute a lower-plate margin and middle of Zone C to Zone E constitute an upper-plate margin.

development of upper-plate (distal) and lower-plate (proximal) margins with different uplift and subsidence histories (Wernicke, 1985; Lister et. al., 1986). Differences in amounts of extension and in thicknesses of syn-rift and post-rift sediment from the Georges Bank basin southeast of New England to the Baltimore Canyon trough off the New Jersey coast are thought to reflect the change from a lower- to an upper-plate margin (Klitgord et al., 1988).

Much less is known about modified crust in other types of passive margins. Seismic profiling along the Grand Banks and the Exmouth transform margins suggests that the zone of thinned and rifted crust is very narrow, if present at all. This is compatible with the extensional model for passive margins in that separation along transform boundaries would be dominated by shearing rather than by stretching.

Recent technical advances in deep seismic-reflection profiling have made it possible to probe deeper levels of modified continental crust in a few margins. One of the most intriguing discoveries is the presence of a lower-crustal layer with an unusually high velocity, ranging from 7.1 to 7.5 km/s (Fig. 4.10; LASE Study Group, 1986; Tréhu et al., 1989). This layer has been observed in the margins of Norway and East Greenland (Zendar et al., 1990), beneath the transform margin along the Exmouth Plateau off northwestern Australia (Lorenzo et al., 1991) and

along parts of the east coast of Canada and the U.S.A. (Klitgord et al., 1988; Keen et al., 1990). The layer is lensoid, being very thin or absent landward of the hinge zone, reaching a maximum thickness of over 10 km beneath the continental terrace, and thinning seaward to a few kilometers and merging with oceanic crust. The presence of this unusual layer makes this type of margin different from "normal" margins, such as the Bay of Biscay (Figs. 4.10B, 4.11). There is a growing consensus that the high velocities of the lower-crustal layer are evidence of large amounts of mafic igneous material, which was underplated beneath, or intruded into, thinned continental crust during the final stages of rifting, producing what have been termed "volcanic margins" (e.g., Beaumont et al., 1982a; Mutter et al., 1982; Keen, 1985; LASE Study Group, 1986; White et al., 1987). The mechanism responsible for this massive igneous activity in some margins but not in others is not well understood. White and McKenzie (1989) suggested that volcanic margins formed where rifting occurred close to hot spots, such as beneath Iceland. On the other hand, Mutter et al. (1988) and Zehnder et al. (1990) have argued that if extension occurs rapidly enough, then strong temperature gradients initiate small-scale mantle convection, leading to massive igneous intrusion beneath the ruptured margins. Tréhu et al. (1989) point out, however, that neither

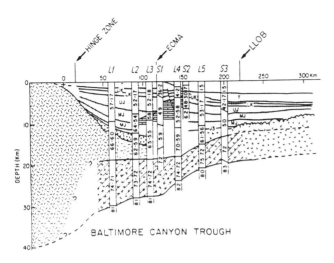

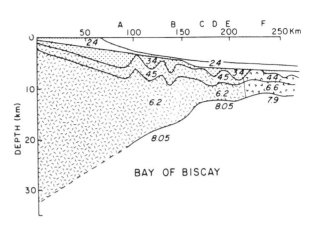

Fig. 4.10 Example of a volcanic margin A) off the coast of New Jersey and a nonvolcanic margin B) from the Bay of Biscay in the western Mediterranean (diagrams from Tréhu et al. (1989). ECMA is the East Coast magnetic anomaly and LLOB is the landward limit of oceanic basement. L1-L5 are LASE ESP profiles and S1-S3 are Sonobuoy refraction lines. Seismic velocities in km/second are given in the vertical columns in A and in italics in B. Note thick layer of high-velocity (7.1 to 7.5 km/s) crust beneath nearly the entire volcanic margin in A, a layer that is not present in the nonvolcanic margin in B. The high-velocity lower crust is interpreted to be a layer of basaltic material that was underplated during rifting.

model completely accounts for the nature of the lower high-velocity layer in parts of the margin off the east coast of the U.S.A.

Oceanic Crust

Along much of the eastern margins of Canada and the U.S.A., the boundary between modified continental crust and oceanic crust is taken to be coincident with a prominent magnetic anomaly at the continental rise, known as the East Coast Magnetic Anomaly (ECMA, Fig. 4.7). This prominent anomaly may result from a combination of relief on magnetic basement and contrasting magnetization between continental and oceanic crust (Klitgord et al., 1988; Keen et al., 1990); it may also represent the magnetic signature of an Alleghenian suture or sutures, along which later rifting occurred (Nelson et al., 1985 a,b). The continent-ocean boundary also is marked by a steep gradient between the large, positive and negative free-air gravity anomalies that lie along the edge of the continental shelf and above the continental rise (Fig. 4.1). This gravity gradient has been interpreted as a consequence of the isostatic adjustment of the lithosphere to the large sediment load in the margins (Beaumont et al., 1982b;

Karner and Watts, 1982). In contrast, the large transform segments of the Canadian-U.S.A. margin, such as along the Grand Banks, lack prominent magnetic and gravity anomalies at the continent-ocean transition.

In parts of margins lacking thick, underplated mafic material, thicknesses and structure appear to be essentially the same as those of seaward oceanic crust (e.g., Fig. 4.10). Where the high-velocity underplated layer is present beneath the modified continental crust, however, the landward part of the oceanic crust is different and referred to as marginal oceanic crust (Fig. 4.10). Here, it is more than 10 km thick and contains a 7.1 to 7.5 km/s layer. Deep seismic profiles suggest that this crust is continuous with the underplated layer beneath the modified continental crust. One of the most distinctive features of marginal oceanic crust is the presence of prominent seaward-dipping reflectors in its upper part. These have been interpreted to be subaerial lava flows extruded onto thinned continental crust (Hinz, 1981) or, more likely, structures within subaerial oceanic crust that formed during the earliest phases of sea-floor spreading along thermally elevated volcanic margins (see Mutter et al., 1982).

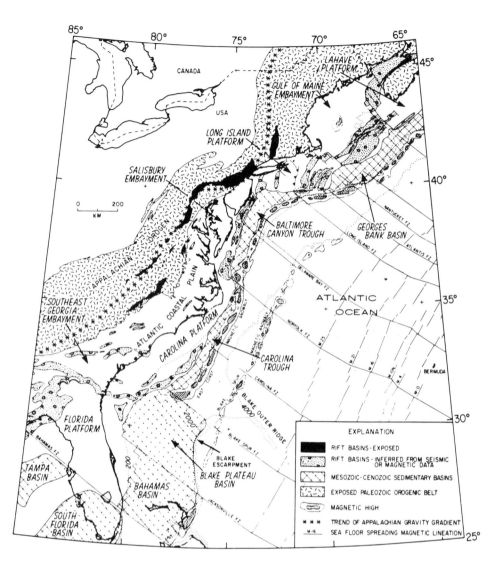

Fig. 4.11 Regional structure of the eastern United States continental margin (from Sheridan and Grow, 1988). This margin, as well as many others, consists of deep, sediment-filled basins, plateaus, platforms, and embayments that are complexly faulted and overlain by only a thin veneer of post-rift sediment. The platforms lie landward of the hinge zone and in places project across the margin, forming transverse boundaries to the basins.

Basin Architecture

Dimensions

The dimensions of continental rises and terraces reflect, to first order, their formation along the rifted edges of continents. Their lengths, determined by the lengths of the rifts that formed when the continents split apart, are enormous, ranging from hundreds to thousands of kilometers. Their widths and thicknesses depend on the widths of the rift belts, and on the amounts and mechanisms of crustal thinning, ranging from minimum values along transform margins to maximum values along parts of margins facing orthogonally opening ocean

basins. Commonly, they are 100 to 200 kilometers wide and contain several kilometers of postrift sedimentary strata. Much larger basins are not uncommon, however; for example, the basin along the Scotian margin in Canada is over 300 kilometers wide and the Blake Plateau basin off the coast of Florida is over 400 kilometers wide. In the Baltimore Canyon basin, off the coast of New Jersey, postrift sediments are thought to be over 13 kilometers thick, perhaps reflecting the large amount of sediment supplied to the basin from the young Appalachian Mountain system (Sheridan and Grow, 1988). According to one estimate, present continental terraces and rises constitute over 18% of the total area of basins worldwide (Klemme, 1980).

The dimensions of continental terraces and rises are also strongly influenced by processes that occur following rifting, the most important ones being the effects of sediment loading, the effects of flexure due to cooling of the lithosphere, and the landward onlap and seaward progradation of sediment. The weight of sedimentary strata significantly adds to the subsidence caused by extension and crustal thinning, commonly enhancing it by a factor of about 3 (see section on models below). As the rifted lithosphere cools, its strength increases, causing flexural bending and subsidence over a broader region with time, both landward and seaward of the rift belt. This, together with the landward and seaward deposition of sediment as the basin subsides, increases the width of the basin well beyond that produced solely by crustal extension.

Rift-Drift Transition

The lower stratigraphic boundary of the continental terrace-rise deposit is taken as the level coinciding with the transition from rifting to drifting. This corresponds to a time of major change in tectonics of the basin from block faulting and formation of grabens during extension to regional subsidence controlled by lithospheric cooling and sediment loading at the onset of sea-floor spreading. The identification of this level in the basin stratigraphy, however, is not everywhere straightforward. It is commonly regarded as coinciding with a regional unconformity, the breakup unconformity, that can be identified on many seismic profiles (Falvey, 1974). This unconformity is the deepest surface that is mappable across the margin. It is best defined in block-faulted areas at the edges of platforms, where faulted syn-rift sedimentary rocks appear as landward or seaward dipping reflectors truncated by postrift near-horizontal reflectors. In parts of the U.S. margin, the surface has been traced from an erosional unconformity near the hinge zone to a more or less conformable surface at the outer part of the basin (Klitgord et al., 1988). In these places, there is relief on the surface, in the form of small faults and changes in dip, that are the result of differential subsidence. These structures can be confused with syn-rift faults, making the distinction between syn-rift and postrift strata difficult. In addition, some margins, such as the one off Nova Scotia,

appear to lack a prominent unconformity at the rift-to-drift transition (e.g., Keen and Williams, 1990). Although eustatic sea-level changes may contribute to the formation of these surfaces, this is not regarded as their fundamental cause. Their ages are broadly correlative with the onset of sea-floor spreading in the adjacent ocean basin rather than to eustatic sea-level changes. An important consequence of the identification of breakup unconformities was the development of nonuniform passive-margin models (e.g., Falvey, 1974; Royden and Keen, 1980), which, in contrast to the uniform-stretching model of McKenzie (McKenzie, 1978a), could simulate uplift at the onset of sea-floor spreading (see section on modeling below).

Structure

Because continental terraces and rises form after rifting as the lithosphere cools and subsides, their structure mainly is inherited from the larger structures that formed during continental extension. One of the most distinctive structural features in margins facing orthogonally opening oceans is a system of basins, plateaus, and embayments (Fig. 4.11). The basins are bounded on the landward side by the hinge zone; along the oceanward side, basin strata merge with strata on oceanic crust. These basins overlie rifted and thinned continental crust, and typically contain 8 to 13 kilometers of postrift sediments.

Platforms and embayments lie landward of the hinge zone and have a less obvious relation to structures formed during rifting. The platforms project partly across the margins, forming sheared along-margin boundaries of the basins. Embayments are recessed into the margins and form only the landward edges of the basins. The platforms and embayments typically are covered by less than 5 km of postrift strata. Where the platforms project across the margin, they appear to be separated from basins by steep-sided faults; in many places, platforms can be traced seaward into oceanic fracture zones (Fig. 4.11). The northern edge of the Long Island platform, which lies between Baltimore Canyon and Georges Bank may coincide with a major shear zone along the New England coast. Where detailed seismic studies have been completed across the platforms (e.g., Long

Island platform), the structure consists of a complex system of deep, crustal-penetrating faults with rift basins near the upper ends of several of them. Some of these faults have been active long after rifting ended and appear to be the loci of some seismicity (e.g., Hutchinson and Grow, 1985). The embayments, on the other hand, are rimmed by crustal-scale faults and rift structures, but appear to be underlain in their central parts by largely unfaulted continental crust.

Although most of the basins merge seaward with strata of the oceanic floor, there are exceptions. Off the east coast of Newfoundland, the Orphan Knoll and Flemish Cap are deeply submerged (1 to 2 km) basement highs situated between the basins and oceanic crust. These appear to be fragments of continental crust that were thinned less than the crust beneath the basins, and hence, did not subside as much as the sub-basin crust. Possible outer basement highs along other margins have been discussed by Schuepbach and Vail (1980).

Subsidence History

The presence of shallow-water and nonmarine sedimentary rocks at depths of more than 10 km in continental terraces and rises attests to the large magnitude of subsidence in these settings. The subsidence can be attributed mainly to three factors: tectonic subsidence, sediment loading, and sediment compaction (Fig. 4.12). Tectonic subsidence is the subsidence of the pre-rift crust in the absence of sediment and is controlled by driving or tectonic forces associated with the formation of the basin. In the case of passive margins, the tectonic subsidence is usually described in terms of two phases. The initial or syn-rift phase occurs during rifting and is essentially an isostatic response to extension and thinning of the pre-rift crust. During this phase, rift basins form (see Chapters 2 and 3). The second phase is postrift subsidence, which begins at the onset of sea-floor spreading, and hence, is the component of subsidence that occurs during the growth of continental terraces and rises. "Sediment loading" refers to the effect of the weight of sediment as it is deposited along the margin. As sediments are shed into the basin, their weight amplifies the tectonic subsidence, creating

new space for additional subsidence. Finally, as sediments lithify, they create additional new space for sediment, a process that continues throughout the growth of the basin. Where sediments are lithified by the addition of cement, the weight of cement is a component of sediment loading.

The forms and magnitudes of the different components of subsidence can be estimated from well logs or outcrop sections using a procedure called "backstripping" (Steckler and Watts, 1978a; Bond and Kominz, 1984; Allen and Allen, 1990). This procedure involves removing from each layer the effects of sediment lithification and sediment loading. This reduces the cumulative-thickness curve (observed stratigraphic section) to a curve that measures only tectonic subsidence. An example of backstripping from the COST B2 well in the margin off the coast of New Jersey demonstrates typical forms and magnitudes of the components of subsidence in a continental terrace (Fig. 4.12). The effect of sediment loading, given by the difference between the delithified curve and the tectonic-subsidence curve (with the water-depth correction removed), is quite large; it commonly amplifies tectonic subsidence by a factor of 2 or 3. When the section is not fully lithified, two of the delithified sediment thicknesses are well constrained, that at the base of the section (no sediment means no porosity) and that at the top of the section (the present porosity is known). Thus, the difference between the observed sediment thickness and the delithified sediment thickness increases and then decreases with depth of burial (e.g., Fig. 4.12). For a fully lithified sediment column, the difference between the observed and delithified sediment thickness increases with depth of burial (Bond and Kominz, 1984).

One of the significant contributions from backstripping analyses of continental terraces and rises is that, in nearly all of the examples studied, the postrift tectonic subsidence decays with time in an exponential fashion (Figs. 4.12, 4.13). These results are strong support for the idea originally suggested by Sleep (1971) that postrift subsidence is controlled by thermal contraction of lithosphere that was heated during rifting (also, see the following sections on models). Sleep (1971) also clearly recognized that heating fol-

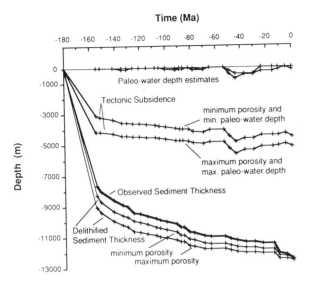

Time (Ma)

Fig. 4.12 Main components of subsidence in a continental terrace, revealed by backstripping the COST B2 well in the Baltimore Canyon trough off the coast of New Jersey. Observed sediment thickness is cumulative stratigraphic thickness measured in the well (data from Steckler and Watts, 1978a). One subsidence component is sediment compaction, which is the difference between observed and delithified curves. Delithified thickness curves are produced after removing effects of compaction. Values are from maximum and minimum estimates of porosity as a function of depth and lithology (Bond and Kominz, 1984). Another subsidence component is sediment loading, which is depression of the crust by weight of sediment. Magnitude of this component is given by difference between delithified curves and tectonic subsidence curves (without paleo-water-depth corrections). Third component is the tectonic or driving force, which is approximated by tectonic-subsidence curve. For passive margins, this component is produced by cooling and increase in density of lithosphere that begins at onset of sea-floor spreading. Other factors, paleo-water depths and eustatic sea level, must also be estimated to accurately isolate tectonic subsidence (see equation below).

lowed by cooling could not, by itself, cause the observed subsidence because the crust would simply return to its initial elevation when cooling ended. These two pieces of evidence have led to the current view that the rate and magnitude of postrift tectonic subsidence in passive margins are, to first order, functions of the amount of crustal thinning and that the form of postrift subsidence is due to thermal contraction of the lithosphere as the thermal anomaly generated during rifting decays, restoring the equilibrium thermal gradient (e.g., Figs. 4.13 ,4.14).

Although backstripping is a useful technique for quantifying subsidence history, it calls for specific information that is not always well constrained in continental terraces and rises. The method requires knowledge of the lithostratigraphy and chronostratigraphy, water depths of deposition, lithology-dependent porosity-depth relationships, magnitude of sea-level changes, and method of sediment-load compensation. One approach to the uncertainty in porosity-depth relations has been to specify a range in porosity-depth curves for different rock types, and to derive maximum and minimum values for the components of subsidence (Bond and Kominz, 1984). The correct model for compensation of loads (e.g., airy versus flexural) is difficult to determine for a given margin. Possible errors introduced by assuming incorrect mechanisms of compensation are discussed below.

Subsidence Models

Active versus Passive Rifting

Although this chapter does not focus on the syn-rift or rifting phase of passive-margin generation, it is worth noting that two basic types of models have been proposed to account for them (see Chapters 2 and 3). In one case, active rifting supposes the presence of active heating from below. This may be due to a thermal plume and may have considerable impact on the extension process. Models of mantle diapirs have been proposed to account for thinning prior to rifting along passive margins (e.g., Neugebauer, 1983; Spohn and Schubert, 1983; Moretti and Froidevaux, 1986). These models predict subcrustal lithospheric heating in excess of that of the crust, which may be thinned little or not at all. Alternatively, passive-rifting models assume that the driving force for extension is located at some distance from the locus of extension. In this case, heating of the lithosphere and

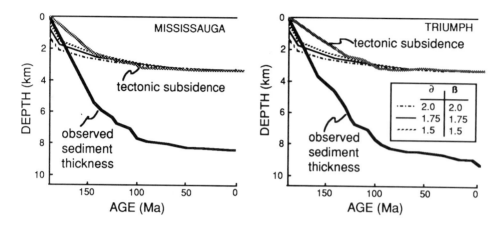

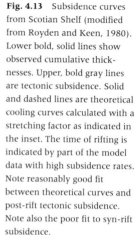

Fig. 4.13 Subsidence curves from Scotian Shelf (modified from Royden and Keen, 1980). Lower bold, solid lines show observed cumulative thicknesses. Upper, bold gray lines are tectonic subsidence. Solid and dashed lines are theoretical cooling curves calculated with a stretching factor as indicated in the inset. The time of rifting is indicated by part of the model data with high subsidence rates. Note reasonably good fit between theoretical curves and post-rift tectonic subsidence. Note also the poor fit to syn-rift subsidence.

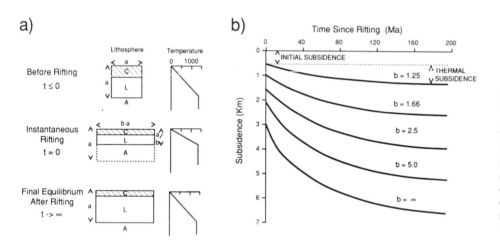

Fig. 4.14 a) Uniform-stretching model (modified from McKenzie, 1978a). A is asthenosphere, L is lithosphere and C is crust. Lengths are unit dimensions and ß is amount of stretching. As unit area is stretched, isotherms are compressed, changing thermal structure as shown on right. b) Subsidence curves calculated from stretching model (modified from Watts, 1981). Subsidence during rifting is initial subsidence. Form of post-rift subsidence is exponential after first 16 my of cooling (McKenzie, 1978a).

thinning of the crust are a passive process. The postrift subsidence for either passive or active mechanisms can be simulated with the uniform- and nonuniform-extension models described below.

Uniform-Stretching Models

One of the earliest models developed to account for the thermal form of subsidence in passive margins is the uniform-stretching model of McKenzie (1978a). This model is still widely used and has served as a basis for many subsequent modifications. The model begins with a continental lithosphere of thickness **a**, including continental crust of thickness **c** and upper mantle of thickness **a-c** in isostatic equilibrium at sea-level (Fig. 4.14a). The thermal gradient is linear, 0°C at the surface and 1300°C at depth **a**. For a crustal den-

sity of 2.8 g/cm³, mantle density of 3.3 g/cm³(at 0° C), a thermal decay constant of 3.3x10⁻⁵ per degree Kelvin, and an equilibrium lithosphere of 125 km, an unstretched continental crustal thickness of 27.6 km is required if its surface is to be at sea level. Formation of a passive margin is modeled as being the result of instantaneous stretching of the continental crust and upper mantle by a factor of ß, so that the stretched lithosphere thickness is **a/ß** and the stretched crust thickness is **c/ß** (Fig. 4.14a). The amount of tectonic subsidence resulting from the stretching (termed syn-rift subsidence) is readily calculated by assuming isostatic conditions (Fig. 4.14b). Densification of the lithosphere resulting from stretching of the crust (replacement of low-density crust with high-density mantle) is offset somewhat by a lightening of the section due to raising the tempera-

ture of the entire lithosphere. Subsequent subsidence (postrift or thermal subsidence), resulting from cooling as the thermal regime returns to equilibrium temperature conditions, is calculated assuming Airy isostasy (Fig. 4.14b; McKenzie, 1978a). The advantage of this model is its simplicity. Given a specified amount of stretching (ß) the magnitude of subsidence is easily calculated (Fig. 4.14b), or knowing the postrift subsidence, one can estimate the amount of stretching.

Subsequent models for the formation of passive margins modify or change the distribution of heat due to rifting (see following sections). Some models focus on the details of syn-rift extension, ranging from simple kinetic models, which include finite rift times, to elaborate finite-element modeling. Models have also been developed that focus on two-dimensional aspects of rifting, invoking lateral heat flow, flexural response to sediment loads and small-scale convection. These models make specific predictions for syn-rift subsidence and uplift. Although all of these models predict that postrift subsidence is exponential (thermal) in form, the amount of postrift subsidence differs markedly from model to model for the same amount of stretching of crust or lithosphere.

Nonuniform-Stretching Models

Nonuniform-stretching models were developed to explain the observation along many passive margins that the ratio of syn-rift to postrift strata is incompatible with predictions of the uniform-stretching model. In general, there has been less subsidence than should have occurred during the rifting stage. To account for this discrepancy, Royden and Keen (1980) and Royden et al. (1980) suggested modifications of the McKenzie (1978a) model, in which stretching of the crust was not equal to stretching of the upper mantle. In their model, the magnitude of initial subsidence (or uplift) is a function of both increased density due to crustal thinning and decreased density due to lithospheric thinning. As such, it is dependent upon the amount of crustal and mantle stretching (Fig. 4.15a). The difference in density between pre-rift lithosphere and final lithosphere after cooling is proportional to the thickness of the crust alone; thus, the total subsidence is dependent only on the amount of crustal-stretching (Fig. 4.15b).

It follows, then, that the combination of total subsidence and initial subsidence uniquely defines appropriate stretching factors for crust and subcrustal lithosphere.

Critics of this model suggest that it is not conservative of space, since extension values in the crust differ from those in the upper mantle. However, this problem has been addressed in several ways. First, the rheological difference between crust and upper mantle (brittle vs ductile) suggests that stretching should be quite different in the two regions. That is, stretching in the crust might be more localized than in the mantle (e.g., Dunbar and Sawyer, 1989). Alternatively, active heating from below may cause isotherms to rise in the mantle faster than in the crust. Postrift subsidence for either of these mechanisms can be simulated with the nonuniform-extension model.

Melt-Segregation Models

An alternative uniform-stretching model (McKenzie and Bickle, 1988; White and McKenzie, 1989) considers the thermal regime at the time of rifting. This model accounts for evidence of syn-rift sedimentary deposits that are thin or absent relative to that predicted by McKenzie's (1978a) uniform-stretching model. In this model, the volume of crustal- or intermediate-density magma generated in the asthenosphere during extension is a function of the amount of crustal stretching and temperature of the asthenosphere (Fig. 4.16). For example, normal 1280°C asthenosphere and infinite stretching produce an appropriate volume of melt to reproduce oceanic crustal thickness, about 6 km. For lower stretching factors (ß), much less magma is generated and the uniform-stretching model of McKenzie (1978a) applies (Fig. 4.16). However, if the extension occurs above a hot spot or mantle plume, then elevated temperatures result in increased magma generation. Because this material is lighter than lithospheric mantle, it underplates and rethickens the stretched crust. This model is compelling because it is internally consistent with testable predictions. As in the nonuniform-stretching model, the relative thicknesses of syn-rift and postrift sediments can be used to estimate the crustal thickness and the initial

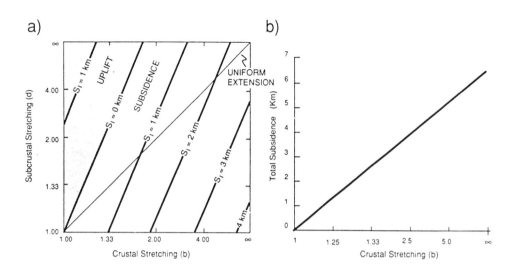

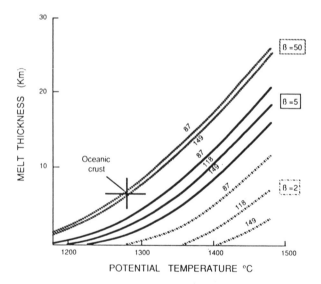

Fig. 4.15 a) Amount of initial subsidence, S^i, (or uplift) for the nonuniform-extension model as a function of ß (stretching factor for crust) and d (stretching factor for subcrustal lithosphere). Initial thickness of lithosphere = 125 km; initial thickness of crust = 35 km; density of crust = 2.9 g/cm³, density of mantle = 3.3 g/cm³; density of water = 1.03 g/cm³ (diagram modified from Keen and Beaumont, 1990). b) Total subsidence (syn-plus post-rift) for the same parameters. Note that total subsidence is entirely dependent on the crustal stretching, ß.

heating of a margin. The melt-segregation model makes additional predictions both for the proximity of the margin to a mantle plume at the time of rifting and for the expected thickness of high-velocity lower crust (7.1 to 7.5 km/s), which may be observed using deep seismic-imaging techniques (e.g., the Baltimore Canyon trough) (Fig. 4.10a).

Finite-Rifting Models

The one-dimensional models described above do not simulate syn-rift (rift-basin) processes; instead, they specify that rifting is instantaneous. Models that consider temporal duration in the rifting process are termed "finite-rifting models." The duration and rate of rifting vary considerably. The primary effect of finite rifting is additional cooling of the margin during stretching. This effect was calculated in a one-dimensional model by Jarvis and McKenzie (1980), who showed that if rifting occurs in less than 15 my, then postrift subsidence is essentially equivalent to that obtained in the uniform-stretching model. Longer periods of extension significantly reduce the magnitude of postrift subsidence but do not change the form.

Simple-Shear Models

The above models can all be classified as pure-shear models, in which shear is uniform throughout the lithosphere. In contrast, Wernicke (1985) suggested

that extension does not occur uniformly, but rather along a single zone of detachment, or by simple shear, leading to the formation of upper- and lower-plate margins. Asymmetry is predicted for the two margins in their geometry, syn-rift structures, thermal histories, and syn-rift and postrift subsidence (Fig. 4.9). Quantitative models of extension by simple shear have been developed that focus on thermal,

Fig. 4.16 Adiabatic decompression of asthenospheric mantle undergoing extension generates melt thicknesses that are a function of uniform-stretching factors (b), potential temperature and initial thickness of the thermal lithosphere (given for each curve in kilometers). Cross shows limits of normal asthenospheric temperatures calculated from range of observed oceanic igneous crustal thickness at spreading centers (modified from White and McKenzie, 1989).

topographic, and sedimentological effects (e.g., Bell et al., 1988; Buck et al., 1988; Issler et al., 1989). In contrast to pure-shear models, simple-shear models predict strong asymmetric patterns of subsidence and uplift (Fig. 4.17). Subsidence in the pure-shear model is similar to that predicted by the uniform-extension model. Thermal subsidence in the simple-shear model, however, is more like that predicted for nonuniform extension because the magnitude of extension in the crust and in the subcrustal lithosphere are different for any vertical profile (Figs. 4.9; 4.17). In addition, the pattern of syn-rift sedimentation is unique in the simple-shear basin and is another predictive feature of this model (Fig. 4.18; Issler et al., 1989).

Kinetic-Rifting Models

Finite-element models for rifting have been developed that quantitatively address the effects of lithosphere rheology, inherited zones of weakness, and uneven crustal thickness during extension. Melosh (1990) has developed a mechanical model to determine the consequences of extending crust composed of granite, which becomes ductile at lower temperatures and pressures than the olivine-rich subcrustal lithosphere. When crust with this rheology is extended, a detachment is predicted at mid-crustal depths (Melosh, 1990).

In an elaborate rheological model, Dunbar and Sawyer (1989) examined the effects of pre-existing weaknesses in crust and lithosphere on the locus of syn-rift extension (Fig. 4.19). A weak mantle was produced by increasing crustal thickness, which increases the temperature and pressure of olivine in the mantle (Fig. 4.19). A weakness in the crust was modeled by considering the crust to be composed not of quartz-diorite alone, but of granite overlying quartz-diorite. Their results show that although crustal extension begins at the zone of crustal weakness, most subcrustal extension is centered beneath the zone of mantle weakness (that is, beneath the thicker crust). Thus, the locus of crustal extension shifts to the zone of mantle weakness and it is here that final rifting is most likely to occur (Fig. 4.19). Postrift subsidence of a passive margin formed in this way could be simulated with the nonuniform-extension model.

Models of the Thermal Effects of Sediment Deposition

All of these models assume that the lithosphere cools beneath water, which maintains the upper-surface temperature of the lithosphere at about 0°C. The deposited sediment is of uniform density with no effect on the thermal structure of the lithosphere. This assumption was modified somewhat by Watts

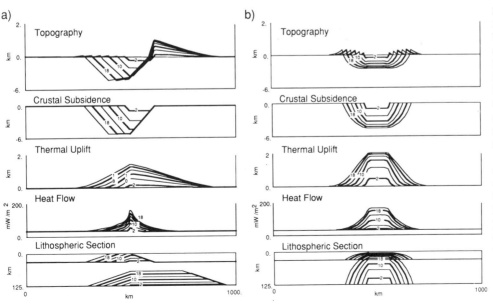

Fig. 4.17 Model results at 2, 6, 10 14, 18, and 22 my after beginning of rifting for a) simple-shear model and b) pure-shear model (modified from Buck et al., 1988). For both models, total plate separation rate is 1 cm/y. Top panels are Airy-compensated topography, which is the sum of crustal subsidence and thermal uplift (below). Distances are in km and heat flow is in mWm⁻². In a) detachment dips 15°. In b) initial rift width is 100 km. Note strong asymmetry in all parameters for simple-shear model in contrast to symmetry for pure-shear model.

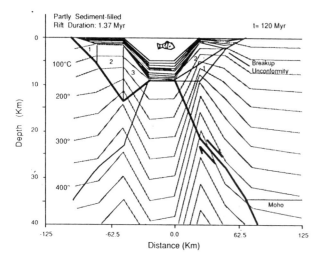

Fig. 4.18 Evolution of stratigraphy, and thermal structure in a basin produced by simple-shear extension (modified after Issler et al., 1989). Cross section of stratigraphy 120 my after start of rifting (rifting phase lasts 1.37 my). Grey lines are isotherms at 120 my in 50°C contours. Shaded region is crust. Detachment fault is bold line marked by arrows indicating relative fault motion. Airy isostasy results in current shape of detachment. Thin solid lines are isochronous boundaries in sedimentary fill. Syn-rift sediment packages are numbered in order of age, oldest to youngest. These are vertical above lower plate because basin is sediment-filled as soon as it is exposed. Simple-shear extension produces thick syn-rift and thin post-rift sediments. For the upper plate, rapid extension thins the subcrustal lithosphere more than the crust. Thus, a breakup unconformity in the upper plate is a direct consequence of simple-shear mechanism. Thermal structure of the isotherms is due to thermal blanketing of sedimentary fill and lack of horizontal heat flow in model and is not indicative of large thermal anomaly at 120 my after rifting.

and Thorne (1984), who considered variation of sediment density as a function of porosity, and therefore, of sediment thickness. The thermal history of sediments in this model was calculated by assuming equilibrium gradients and the heat flow from the nonuniform-extension model (Watts and Thorne, 1984).

Beaumont et al. (1982b) incorporated sediments into their two-dimensional finite-difference model, treating the entire sediment package as a single finite-difference block. Their model suggests that sediment acts as an insulator, and hence, retards the cooling of the margin. Their model compares uniform, nonuniform, and melt-segregation models, including tests for the effects of ranges of lithospheric flexure, and the presence of radioactivity within the sediments. An improved thermomechanical model was developed by Keen and Beaumont (1990), who discussed sensitivity tests of parameters and models.

The effects of thermal blanketing are considered in a one-dimensional model by Zhang (1993). Sediments are added to a nonuniform-extension model at temperatures in equilibrium with lithospheric heat flow. Syn-rift subsidence is reduced, as compared to a nonuniform-extension model, due to the inclusion of radioactivity, adding heat to the stretched lithosphere. The total subsidence is further reduced due to the inclusion of thermal blanketing as well as the contributions of radiogenic heat in both the sediments and the crust (Fig. 4.20).

Models of the Effects of Flexure and Lateral Heat Flow

On passive margins, flexural rigidity is a second-order characteristic, which modifies the subsidence history

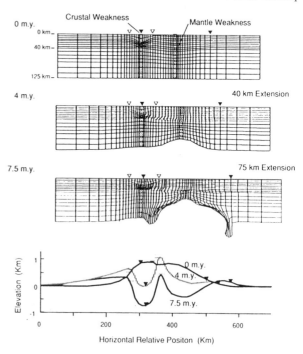

Fig. 4.19 Rheological model of extension with horizontally offset weakness in crust and mantle (modified from Dunbar and Sawyer, 1989). Crustal weakness is between open triangles and is produced by a granitic body tapered to 25 km thick embedded within a quartz-diorite crust. Mantle weakness is between solid triangles and is produced by a gradual increase in crustal thickness from 40 to 50 km. A 30% reduction in strength of lithosphere is associated with each weakness. Beneath is shown the topography of model at each time step. Extension rate is 1 cm/y. Note that horizontal detachment forms between the two zones of weakness as most subcrustal extension occurs beneath the zone of mantle weakness. Ocean lithosphere will form at zone of mantle weakness if rifting continues.

predicted by one-dimensional models of cooling passive margins. In two-dimensional models, the flexural rigidity of the postrift continental lithosphere is generally assumed to increase with time as a result of cooling of the lithosphere, in much the same way as oceanic lithosphere (Watts et al., 1980; Karner and Watts, 1982). This has several consequences for postrift subsidence. The margin subsides tectonically as a result of stretching oceanward of the hinge zone, and thus, serves as a locus of sediment loading. Because the lithosphere behaves as a semi-rigid flexural beam, sediment loading serves to bow down the unstretched lithosphere both landward of the hinge zone, generating the coastal-plain wedge (Fig. 4.21), and seaward of the depocenter on adjoining oceanic lithosphere. As flexural rigidity increases, the width of the zone bowed down also increases, causing onlap of the coastal plain wedge (Watts, 1982).

Lateral heat flow within the lithosphere also modifies postrift subsidence. One-dimensional models only consider conductive heating in a vertical profile; however, horizontal gradients also result in conductive heat flow. Near the hinge zone, heat flows laterally from stretched lithosphere into unstretched lithosphere, causing uplift landward of the hinge zone, and at the same time, increasing the subsidence rate of the stretched lithosphere. Because thermal gradients are greatest just after rifting, the effects of lateral heat flow are most pronounced at this time.

The combined effects of lateral heat flow and flexure substantially change postrift subsidence predicted by one-dimensional models (Fig. 4.21b). Landward of the hinge zone, subsidence with an apparent stretching factor of about as much as 1.5 is seen, although no tectonic subsidence occurred. Within the extended margin, the rigidity of the lithosphere causes the basin to be held up by its sides, so that tectonic subsidence is reduced relative to that predicted by a one-dimensional model (Fig. 4.21; Steckler, 1982). It is interesting to note, however, that in a more recent study by Watts (1988), detailed gravity modeling of the Baltimore Canyon area suggests that the stretched crust beneath the margin has not behaved flexurally, nor is there any indication of increasing rigidity with time.

Erosion is predicted landward of the coastal-plain wedge when rigidity of the margin is low and lateral heat flow is high (Fig. 4.21). However, quite rapidly, flexural effects overwhelm the effects of lateral heat flow in this region. Seaward, the more rapid subsidence due to lateral heat flow is counteracted by the increasing flexural strength of the lithosphere, which reduces subsidence rates. Again, lateral heat flow is most significant in the early history of the passive margin. These effects are particularly pronounced along a transform margin, where the zone of attenuated continental lithosphere is very narrow, so that a spreading ridge is juxtaposed with unstretched continental lithosphere (e.g., Lorenzo and Vera, 1992).

A consequence of lateral thermal gradients within the lithosphere is the establishment of small-scale convection in the asthenosphere beneath the margin. Models by Steckler (1985) and Buck (1986) show that induced convective cells at the margin enhance uplift at the hinge zone. They also show that this is a transient phenomenon, decreasing in intensity as the extended margin and the ocean crust cool. More recently, Mutter et al. (1988) have called on convection beneath extended margins to account both for uplift and for elevated mantle temperatures, and thus, increased partial melting at volcanic margins.

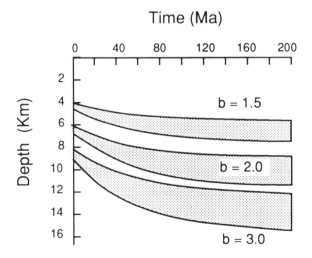

Fig. 4.20 Effect of thermal blanketing on syn- and post-rift subsidence with sediment load included (modified from Zhang, 1993). Three different uniform-extension results are shown. The upper curve is model subsidence with sediments and radioactivity included in thermal model. The lower curve is subsidence of McKenzie's (1978a) uniform-extension model with same stretching factor, b. Differences are highlighted by shading.

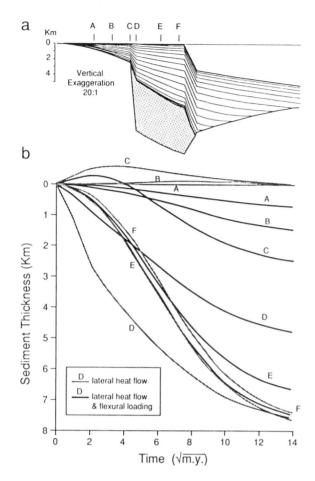

Fig. 4.21 Two-dimensional model of a passive margin showing effects of flexure and lateral heat flow (modified from Steckler, 1982). a) Model based on uniform-extension model of McKenzie (1978). For locations A through C crust is unstretched, ß= 1.0. For locations D through F crust is stretched, ß= 3.0. Seaward of F oceanic crust is formed. Stippled pattern represents syn-rift sediments. Contours in post-rift sediments show sediment thickness at 4, 9, 16, 25, 36, 49, 64, 81, 100, 121, 144, 169 and 196 my (2^2, 3^2, ..., 14^2 my) after the rifting event. Loading is flexural. b) Post-rift subsidence, with sediments (S*, in Fig 13), of margin at points A through F. Results are given for the case of Airy loading with lateral heat flow (grey lines) and for the case of both flexural loading and lateral heat flow (solid lines). Elastic thickness is modeled in terms of T_e, (depth to 500° isotherm). Increase in rigidity as the basin cools causes unrifted edges of basin to bend downward and sediments to progressively onlap continental margin and ocean plate. Lateral flow of heat causes thermal expansion of the lithosphere adjacent to margin, leading to uplift of rift shoulder (hinge zone) and nondeposition of sediment in early part of post-rift subsidence (with flexure) or erosion throughout thermal cooling phase (lateral heat flow alone).

Sediment Distribution and Facies Patterns

Rift Basins

Rift basins exposed around the margins of the North Atlantic Ocean, such as the Triassic/Jurassic Newark basin of New York, New Jersey, and Pennsylvania (Fig. 4.22; Schlische, 1992), provide many common tectonic elements (see Chapter 3): 1) asymmetry along low-angle and listric border faults, resulting in half-graben form, with large cumulative throw on one side; 2) fault movement contemporaneous with sediment infill, resulting in differential thickening and facies coarsening at border fault; 3) relatively rapid subsidence and rapid infill at rates of 500 m/my or more; 4) relative uplift of adjacent rift shoulders of basement rocks that eroded rapidly and served as local sources of half-graben sediments; and 5) basaltic lava flows, dikes, and sills.

In many cases, such as the Newark Series basins (Fig. 4.22), the rift deposits record nonmarine intermontane deposition. In the early stages of individual basins, sedimentation is controlled by strain rates, throw rates on border faults, lengths of border faults, local climate, sediment influx rates, and basin outlet levels. If all these factors remain constant, fault length and basin width increase with time (Schlische, 1992). With constant sediment influx, the narrower early rift basin is filled to base level and fluvial deposition occurs. As the basin widens, sediment influx is spread over a wider area and basin topography may increase, leading to deposition of lacustrine laminated black shale. As strain rate decreases, or deeper lakes drain or evaporate, fluvial deposition may recur. This pattern of deposition in terrestrial rift basins leads to a tripartite sequence of fluvial/lacustrine/fluvial deposits (Schlische, 1992). Rift sediments consist predominantly of quartz, feldspar, and clay minerals, reflecting continental basement source rocks (e.g., Dickinson, 1985).

Organic-rich lacustrine deposits are common in rift basins where the climate was humid. Burial and compaction of these deposits to depths of 2 to 4 km and temperatures of 100–200°C lead to coalification. Coal deposits are common in the Triassic Richmond basin (Fig. 4.22).

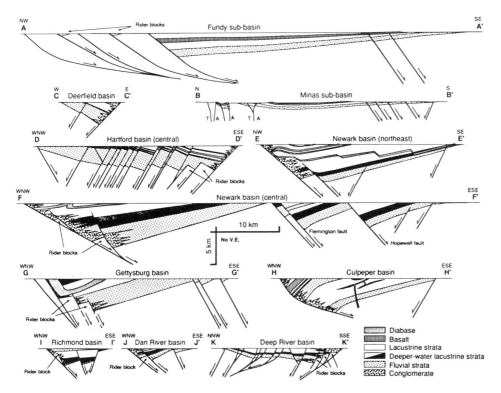

Fig. 4.22 Sealed cross sections of the rift basins of eastern North America, showing the tri-partite stratigraphy of fluvial/lacustrine/fluvial successions. Also shown are basalt flows, and diabase dikes and sills (from Schlische, 1992).

In the late stages of rifting, passive-margin rift basins may subside to form proto-oceanic troughs (see Chapter 3) or epi-continental seaways. In these narrow seaways, if in the right latitude, evaporation can lead to the deposition of thick halite deposits. The thick salt deposits at the bases of drift-stage strata along the North American Atlantic (Sheridan, 1989) and the Gulf of Mexico (Worrall and Snelson, 1989) continental margins probably originated in this way.

Volcanics in the rift basins of passive margins commonly occur as 300-400m-thick basaltic lava flows or sills (Puffer et al., 1981), with subordinate rhyolite.

Volcanic Wedges

Seismic-reflection studies of passive continental margins have documented the common occurence of seaward dipping reflections, referred to as "volcanic wedges" generally over magmatically underplated crust at the continent/ocean boundary. An example of such a wedge is found on the U.S. Atlantic margin, where subaerial basaltic flows intercalated with clastic sediments are probably

Jurassic (Sheridan et al., 1993). Apparently, this sudden voluminous volcanism occurred at breakup and the beginning of sea-floor spreading.

In many cases, the volcanic wedges, which reach >20 km thick locally, thin seaward and merge oceanic basement with hummocky seismic character, interpreted as pillow lava. This implies that marine pillow basalts and subaerial volcanic wedges were coeval lateral equivalents for a short time as the continental margins migrated away from the spreading centers. The volcanic wedges appear to have been erupted rapidly, based on stratigraphic controls on the U.S. Atlantic margin. The wedges overlap the rifted basins, which rifted until the Aalenian (~175 Ma), and the wedges are overlapped by earliest drift-stage strata of Bathonian (~165 Ma) age. Rates of extrusion of 10^6 km^3/my are probable and approach rates large basalt outpourings such as the Cretaceous Deccan Traps.

Early Drift Stage

In this phase of deposition along passive margins, there generally remains a continental hinterland, providing a source for detrital sediments. Topography

is subdued, so that sediments are compositionally mature (e.g.,, Dickinson, 1985). These sediments are dominated by quartz, feldspar, clay minerals, and other siliceous or phosphatic authigenic minerals, such as glauconite. Subdivisions of these siliciclastic facies are controlled by morphology of the coastal land, position of the shoreline relative to the shelf-slope break, climate and dynamic oceanographic conditions partly related to water depth. Complex interplay of these factors results in distinct environments of deposition. Broadly, these environments can be identified as coastal plain, shelf, slope, and basin (rise) (Fig. 4.23; Brown and Fisher, 1977).

Nonmarine Coastal-Plain Facies

Because passive margins enjoy a long history of gentle subsidence, the hinterland is often eroded down to a subtle piedmont (Fig. 4.23). In addition, drift-stage subsidence involves flexure, which tilts the hinterland; as a result, the coastal-plain wedge onlaps.

Thus, the emerged coastal plain can be broad (~100–200 km). During low stands of sea-level, the coastal plain may be emergent all the way to the shelf edge; during high stands of sea-level, the coastal plain may be completely submerged. Dominant sediments of the nonmarine coastal plain are fluvial sand and gravel, intertongued with lacustrine and fluvial mud, and upland-swamp mud and peat.

Marginal-Marine Coastal Facies

At the seaward edge of the coastal plain, distinct sediment types are deposited from the high-tide limit to below the shoreface. Intertidal-marsh deposits of mud and peat grade seaward into clay-rich lagoonal deposits, which grade seaward into the sand deposits of barriers and beaches. Where there are inlets through the barriers, coarse sand and shelly gravel interfinger with crossbedded sand of tidal deltas. On the shoreface, wave erosion cuts into barrier sand to 10-20 m depths and creates erosional ravinement

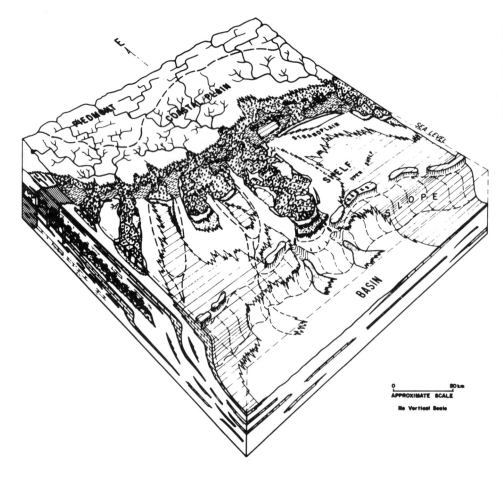

Fig. 4.23 Block diagram showing typical depositional environments on passive margins, including nonmarine coastal plain, marine coastal, continental shelf, slope, and deep basin (from Brown and Fisher, 1977).

surfaces. Nearshore marine sand may be deposited below fair-weather wave-base on the ravinement surface; this sand eventually buries and preserves transgressed lagoonal and estuarine mud and marsh peat (Sheridan et al., 1974).

Shelf Facies
On the inner and outer shelf (water depths of 10 to 200 m), the environment is dynamic, being forced by storm, wave motions, and tidal and longshore currents. The dominant sediment is well sorted and crossbedded sand. Only in a few places on modern shelves does mud accumulate in local sinks controlled by dominant shelf current circulations. Shell fragments are common in shelf sand, and bivalve and crustacean burrowing is common. Large ridge and shoal structures (kilometers across and 10 m in amplitude) consist of crossbedded sand.

Coastal and shelf facies interfinger across the shelf (Fig. 4.23), and are interlayered under the shelf as a result of the migration of environments across the shelf. Consequently, coring or drilling through the shelf recovers vertical stacks of all the common coastal and shelf facies (Fig. 4.23; Ashley et al., 1991).

During times of sediment starvation, there are several authigenic facies common to shelves. These include glauconitic deposits, phosphorite, diatomite, and chert. The presence of nutrients and proper oxidation conditions favor precipitation of these minerals; the lack of dilution by terrigenous clastic sediments favors their concentration.

Slope Facies
Hemipelagic terrigenous mud dominates this facies. Off-shelf transport of mud by nepheloid layers results in a pelagic rain of mud, which drapes the slope. Formation of floccules or fecal pellets delivers the fine-grained sediments more rapidly to the bottom. These draping muds produce the foreset clinoforms of large-scale sigmoidal prograding wedges.

Slump deposits and associated chaotic debris flows may accumulate along the base of the slope. While most gravity-flow deposits bypass the slope via canyons, local slumps and locally derived disturbed sediments may be found anywhere on the slope.

Basin (Rise) Facies
The ultimate depository for off-shelf transported sediment is the continental rise and the abyssal basin. Clay of pelagic biogenic oozes and is interbedded with gravity deposits of turbidites and debris flows (Sheridan et al., 1978). Turbidites are characteristically graded beds, with gravel, sand and silt underlying pelagic clay, whereas debris-flow deposits are characterized by matrix-supported extra- and intra-basinal clasts in chaotic pebbly mudstone.

Geostrophic boundary currents may winnow rise and basin deposits, and transport mud great distances along slope. At the margins of these currents, decreasing velocities result in deposition of great volumes of mud as contourites.

Later Drift Stage

During the later part of the drift-stage waning of siliciclastic sources, sea-level rise and trapping of siliciclastics in estuaries leave shelf environments prone to chemical deposition. In latitudes below 30° and with the proper oceanographic conditions, carbonate facies commonly dominate. Also associated with these carbonates are evaporites, such as anhydrite and halite.

Nonmarine Coastal or Island Facies
In humid climates, karst topography may form, with lakes, ponds, swamps and associated peat. Eolian deposits may be common in less humid settings. In evaporitic environments, anhydrite and salt pans may form (Kendall and Schlager, 1981).

Platform-Interior Facies
Several distinct sediment types occur on typical carbonate platforms, such as the modern Great Bahama Bank (Schlager and Ginsburg, 1981). Peloidal sand, made of particles cemented in "grapestone" and fecal pellets, may be winnowed by wave action over the bank interior. In the lee of islands and in atoll lagoons, carbonate mud, which is detritus from algae sheath and coral spicules, may accumulate. In areas of strong currents, oolite sand shoals may form. Where hard substrate is available, patch reefs can form. On the lee side of islands and in protected bays and lagoons, the carbonate mud facies grades landward into tidal mudflat, where mangrove swamps

trap and baffle the sediments. Some of the mud forms stromboid, upward convex banks. In protected environments, such as lagoons, algae mats cover large areas, leading to aggradation of the carbonate platform interior.

Platform-Edge Facies

Organic reefs of frame-building plants and animals border the edges of many carbonate platforms. Local debris from these reefs produces skeletal sand of coral, mollusk, and foraminiferal fragments. In high-energy environments of the platform edge, skeletal sand is dispersed over marginal zones associated with the reefs. Forereef talus blocks and debris are deposited on the upper marginal escarpments of carbonate platforms in water depths of 80–200 m (Schlager and Ginsburg, 1981).

Platform-Slope Facies

Gravity transport dominates slope deposition; chaotic slump deposits, debris-flow deposits, and turbidites are interbedded with periplatform ooze (Cook and Mullins, 1983). Periplatform ooze is a mixture of biogenic pelagic ooze and the off-bank sediment of aragonite mud. Debris-flow deposits and slump masses collect on the lower slope as resedimented periplatform ooze, whereas the upper slope only contains predominantly ooze.

On some slopes, slope-parallel currents winnow periplatform ooze into nodular limestone, and the winnowed carbonate sand eventually may be deposited in contourite mounds or noses on the slope. Deep-water biothermal reefs may form in water depths of 600–1000 m (Cook and Mullins, 1983).

Carbonate Basin Facies

Pelagic carbonate ooze and siliceous basinal claystone and chert are commonly interbedded with distal carbonate turbidites and debris-flow deposits. Thin-bedded, fine-grained carbonate turbidites dominate the facies. Some thin-bedded, rippled, crossbedded, lenticular bedded, well laminated carbonates are deposited as contourites on the rise as well as on the slope. Displaced oolites and shallow water fossils and peloids are common in the carbonate turbidites (Sheridan et al., 1978).

Sediment Facies Architecture

Rift-Basin Asymmetry

Differential subsidence of half-graben rift basins leads to initial dips of the beds toward border faults. Divergence of stratal boundaries toward border faults also results, with onlap toward the hinge or ramp margin of the basin (Schlische, 1992) (see Chapter 3). Differential uplift of the footwall of the border fault causes erosion of the footwall block and shedding of coarse debris into the steeper rim of the basin. In terrestrial settings, the steep rim of the basin is the locus of alluvial fans, on which debris flows and mud flows spread basinward. Resultant fanglomerates dominate border-fault margins (Fig. 4.22), and these coarse facies grade basinward into the central basin dominated by finer-grained fluvial and lacustrine facies (Manspizer, 1980). But, as Leeder and Gawthorpe (1987) demonstrate, in some cases, the finest material is close to border faults, and the coarse material is derived from the hangingwall block, where it might have a much greater drainage network (see Chapter 3).

Drift-Stage Sedimentary Prism

Differential subsidence across passive continental margins results in more rapid and greater amounts of subsidence at the continent-ocean boundary (Fig. 4.1). Assuming that sediment influx is adequate to fill the tectonically produced accommodation space, then more and thicker sediment accumulates here, and less subsidence occurs at the hinge zone. Consequently, drift-stage shelf strata form seaward thickening prisms (Figs. 4.1, 4.21; Watts, 1981). In addition, flexural subsidence landward of the hinge zone causes progressive onlap of the feather edge of the prism, the coastal plain. In addition, the seaward edge of the sediment prism thins seaward, where it merges with open-ocean sediments of the abyssal plain.

Eustatic Sea-Level Changes and Depositional Sequences

An important contributor to the architecture of drift-stage prisms is eustatic sea-level change. During relative highstands, influx of sediment results in sigmoidal clinoform progradational deposition in the slope areas while coastal onlap occurs in the coastal plain (Fig. 4.29; Vail et al., 1977). During lowstands,

the shelf is exposed and sedimentation occurs in deltas at the shelf break. Lowstand delta wedges cover the slope, with marine onlap occurring on a marine-cut submarine unconformity (Fig. 4.24). Highstand tracts (HST) and lowstand tracts (LST) created by each change in sea-level cycles form unconformity-bounded sequences (Vail et al., 1977).

A more detailed sequence-stratigraphy model has been developed (e.g., Haq et al. 1987). Subaerial hiati that result from sea-level falls are type-one sequence boundaries (SB1). Type-one (SB1) sequence boundaries form the bases of sea-level cycle sequence. SB1 surfaces form while low stand fans are forming on the rise and deep basin (Fig. 4.25). During maximum rate of sea-level rise, the transgressing shoreline crosses the SB1 and deposition of the transgressive system tract (TST) occurs. In some places, fluvial valleys are filled with lowstand sediments and overlain by transgressive-system-tract sediments. During the highest rate of sea-level rise and transgression, the shelf may become starved and a submarine hiatus forms as a nondepositional surface or condensed horizon. As sea-level rise slows, approaching maximum highstand, the influx of sediment may cause regression and progradation, producing downlapping progradational clinoforms (Fig. 4.25). Type-two sequence boundaries (SB2) may form during subsequent fluctuations where the shelf is not totally exposed and the unconformity extends only partially across the shelf.

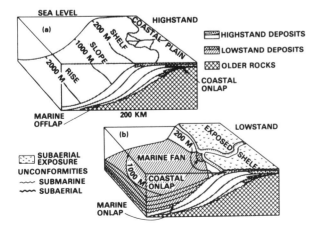

Fig. 4.24 Progradational depositional pattern during highstands of sea level (a) and coastal onlap patterns during lowstand of sea level (b) (from Vail el al., 1977).

Deltas

Large rivers deliver terrigenous sediments to a few points along passive continental margins. More than in most other tectonic settings, deltaic deposition can act for long time intervals and dominate passive margins. Prograding across shelves, deltas deliver sediments to the slope and create deep-sea fans at the base of the slope and in the basin (rise). In extreme cases (primarily at ancient failed rifts, such as the Benue Trough of Africa and the Mississippi Embayment), deltaic progradation may lead to development of continental embankments, with shorelines regressing beyond continent-ocean boundaries (Fig. 4.2D) (Burke, 1972; Dickinson, 1976a; Ingersoll, 1988).

The process of progradation of fluvial environments across shelves leads to the accretion of laterally shifting channels of meandering distributary channels (Fig. 4.26; Winker and Edwards, 1983). Coarse, sandy fluvial deposits dominate topset beds of the delta (Fig. 4.26). At the slope, finer sediment is deposited as progradational foreset beds, forming the thickest part of the prism. The foreset beds of the prodelta thin seaward into basinal bottomset beds, including deep-sea fans. Detailed side-scan studies of passive-margin deep-sea fans reveal that they form by the lateral accretion of narrow, serpentine channel deposits that extend as meandering features for hundreds of kilometers across the continental rise (Damuth et al., 1983). The deep-sea fan channels allow confined flow, such that coarse gravel is found hundreds of kilometers from the delta mouth. Prodelta mud is so water-saturated and rapidly deposited as to be unstable locally, leading to extensive slumping of the delta slope (Coleman et al., 1983). In fact, slumping is so prevalent that slump scars cover the entire slope, thus creating submarine canyons (Fig. 4.26).

Carbonate-Platform Architecture

Carbonate mud is deposited on broad platforms along the edges of continental shelves; these areas experience rapid tectonic subsidence, combined with great distance from terrigenous sources. Thus, tectonics is one of the basic controls on carbonate-platform development. The shelf edge is the most sediment-starved part of the margin, and given suitable climatic

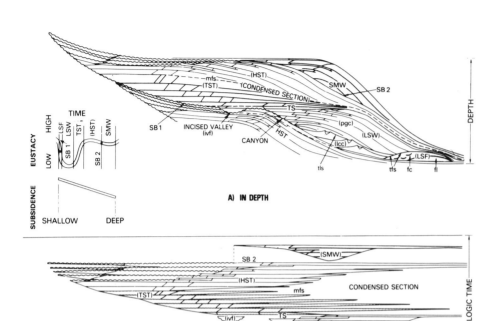

Fig. 4.25 Sequence-stratigraphy model, showing architecture of facies tracts controlled by changes of sea level on a gently subsiding passive margin (from Haq et al., 1987).

A) IN DEPTH

B) IN GEOLOGIC TIME

LEGEND

SURFACES

(SB) SEQUENCE BOUNDARIES
  (SB 1) = TYPE 1
  (SB 2) = TYPE 2
(DLS) DOWNLAP SURFACES
  (mfs) = maximum flooding surface
  (tfs) = top fan surface
  (tls) = top leveed channel surface
(TS) TRANSGRESSIVE SURFACE
  (First flooding surface above maximum regression)

SYSTEMS TRACTS

HST = HIGHSTAND SYSTEMS TRACT
TST = TRANSGRESSIVE SYSTEMS TRACT
LSW = LOWSTAND WEDGE SYSTEMS TRACT
  ivf = incised valley fill
  pgc = prograding complex
  lcc = leveed channel complex
LSF = LOWSTAND FAN SYSTEMS TRACT
  fc = fan channels
  fl = fan lobes
SMW = SHELF MARGIN WEDGE SYSTEMS TRACT

conditions (low latitudes), carbonate deposition can occur at rates equal to subsidence, thus maintaining and building the platform (Fig. 4.27; Kendall and Schlager, 1981).

A common evolutionary pattern for carbonate platforms on passive margins is the development first of a carbonate ramp, and later, an escarpment-bounded carbonate platform. The ramp is a gently seaward sloping carbonate shelf with peloidal sand facies, oolite shoals and some patch reefs. The marginal-barrier shelf facies is not well developed. Modern examples are the Campeche and West Florida shelves. The carbonate platform is a flat, reef-bounded and escarpment-bounded megabank with restricted bank-interior peloidal mud facies in the leeward shadow of barrier reefs or islands. Great Bahama Bank is a modern example (Fig. 4.28; Schlager and Ginsburg, 1981). Once the reef-bounded platform edge exists, orientation of the prevailing wind to the edge controls the wind-shadow effect and the locus of deposition of fine-grained carbonate mud.

The continued subsidence of passive margins allows accommodation space. The reef-building organisms deposit carbonate at a sufficient rate to keep up

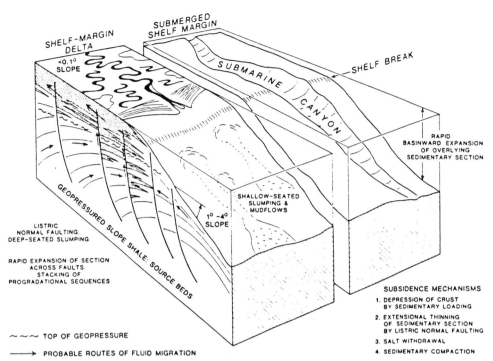

Fig. 4.26 Block diagram model of a growth-faulted deltaic shelf margin during high and low stands of sea level (from Winker and Edwards, 1983).

with relative sea-level rise. Meanwhile, the deep basin beyond the platform is starved of terrigenous clastics, and with subsidence, it deepens. The relief between basin and platform edge increases, reaching 2-4 km (Fig. 4.27). Steep carbonate escarpments evolve, oversteepening leads to slumping and biochemical erosion of the escarpment hastens the mass wasting. The modern Blake-Bahama escarpment is an example.

Once the carbonate escarpment is built, geostrophic contour currents commonly develop along these slopes (Cook and Mullins, 1983). Large amounts of off-bank carbonate mud are carried by these currents and eventually deposited.

Sediment Ridges (Outer Ridges)
The presence of continental slopes on passive margins leads to the development of boundary currents. Geostrophic forces acting on water masses of different densities, due to salinity or temperature differences, generate slope-parallel currents, commonly called "contour currents" because of their coincidence with

GROWTH POTENTIAL MATCHES OR EXCEEDS RELATIVE SEA LEVEL RISE.

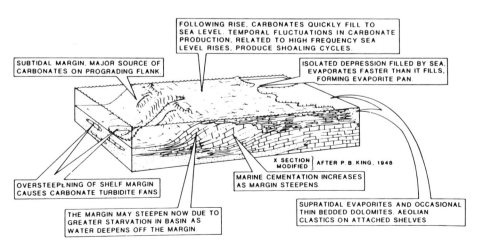

Fig. 4.27 Block diagram illustrating depositional environments of carbonate platform/platform margin, and off-bank basin (from Kendall and Schlager, 1981).

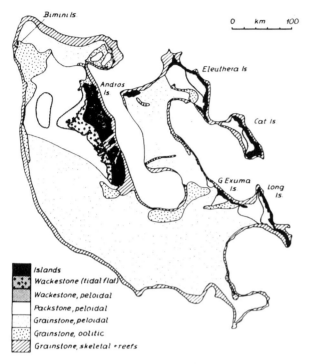

Fig. 4.28 Sediment distribution on Great Bahama Bank (from Schlager and Ginsburg, 1981).

bathymetric contours. In modern oceans, contour currents transport dense muddy water, or a nepholoid layer, from which is deposited an enormous amount of muddy sediments. These sediments appear in seismic images as mounded hummocky reflectors that downlap the planar continuous reflectors of flat-lying sea floor (Fig. 4.29; Sheridan, 1981).

Intrastratal Structures

Major intrastratial structures caused by gravity tectonics control the sedimentary architecture of many passive margins (Fig. 4.29; Sheridan, 1981). Low-angle, rotational slumps involve lenses of water-saturated mud and mudstone of the continental slope. The slumps are associated with slump scars and chaotic toe-of-slump deformation (Figs. 4.26, 4.29). Downslope from the slumps are debris-flow deposits. Water-saturated, rapidly deposited mud is susceptible to gravity failure on low slopes (<1°).

Deeply penetrating growth faults are common along passive margins, especially near shelf edges (Figs. 4.26, 4.29; Sheridan, 1981; Winker and Edwards, 1983). Growth faults are listric normal faults characterized by differential thickening of

sequences on the down-thrown blocks. As fault motion continues sediment is deposited on the shelf; faults offset the shelf and contribute to local topography, thus influencing depositional patterns. Such faults may be active for long intervals (e.g., Cretaceous to Holocene along the Gulf Coast); these faults penetrate more than 10 km below the surface. The growth faults sole out in units that accommodate the fault displacement by ductile flowage, commonly water-saturated mud of prodelta slope deposits (Fig. 4.26).

Salt or shale may flow into diapirs, forming anastomozing linear highs and lows along the continental slope. Gravity loading on the over-pressured shales forces the diapirs toward the free surface (Fig. 4.30). Not only does the flow of the diapirs facilitate growth faulting under the shelf edge, which controls local sand deposition, but diapiric ridges on the slope locally trap sand in the irregular bathymetry.

Economic Deposits of Passive Margins

The sediments of passive margins contain various economic resources. The siliciclastic facies include sand and gravel that are valuable borrow sites

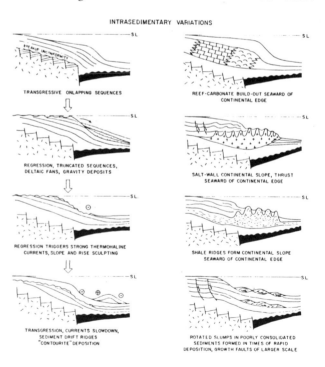

Fig. 4.29 Some intrastratal features common to many passive margins (from Sheridan, 1981).

(Duane and Stubblefield, 1988). In the Bahamas, the oolite shoals are mined for ooids that are a pure form of carbonate used for agricultural lime and cement. On the Blake Plateau, manganese nodules occur near land and at sufficiently shallow depth to, perhaps, warrant mining in the future (Riggs and Manheim, 1988).

In some areas off South Africa, incised valleys across the continental shelf contain diamond placers; off Alaska, there are gold placers on the continental shelf. In general, where there are rich placers on land, placers are also likely on neighboring shelves. Titanium-bearing minerals are in significant concentrations in Tertiary beach and nearshore sand of New Jersey to be economic (Riggs and Manheim, 1988).

Sulfur in cap rock of salt domes in the Gulf of Mexico has been recovered for economic use. Dissolution of anhydrite ($CaSO_4$) associated with evaporite deposits yields sulfur that is eventually concentrated in the cap rocks.

Under the North Carolina shelf are Neogene deposits of phosphorite similar to those now mined on the Coastal Plain (Riggs and Belkamp, 1988). These offshore deposits have concentrations and environments of deposition similar to those on land.

By far the most economically important resource of passive margins is petroleum. Billions of barrels of oil and trillions of cubic feet of gas have been extracted from passive margins such as the Gulf of Mexico. Although still relatively unexplored, the Baltimore Canyon exhibits many of the petroleum traps and situations that have been tested and exploited elsewhere (Fig. 4.30; Benson and Doyle, 1984).

Domal structures above salt diapirs are commonly productive on passive margins. Normal faults radiating from the centers of domes produce local traps (Fig. 4.30). On the flanks of salt diapirs, petroleum in upturned permeable sandstone is trapped at the salt contact. Igneous intrusions may dome up sedimentary strata, and petroleum can be trapped in the domes. Both igneous domes and salt domes have local thermal aureoles related to the magmatic heat and higher salt thermal conductivity, respectively. These thermal aureoles contribute to the local maturation of kerogen in source rocks. Growth faults generally have rollover anticlines on down-thrown blocks, and these anticlines form traps for petroleum. Deltaic sand associated with shelf-edge growth faults (Fig. 4.26) offer good reservoir rocks, and fluid migration from pro-delta mud is a good source of petroleum. Carbonate-platform edges form sealed, buried escarpments. The cavernous porosity in platform edges can trap petroleum, and the platform-interior facies dominated by algae offer good source rocks for petroleum. Beneath postrift unconformities (Fig. 4.1), buried rift basins offer fault-tilted

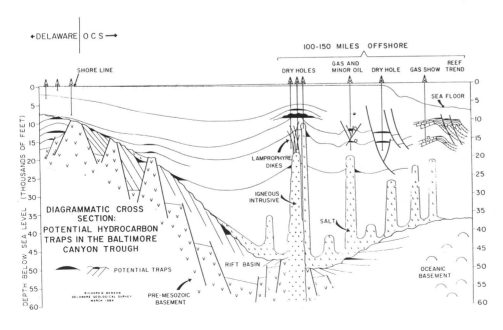

Fig. 4.30 Schematic cross section of Baltimore Canyon trough, showing potential petroleum traps. Wells summarize exploration history (from Benson and Doyle, 1984).

structures in permeable fluvial sandstones, and deep lacustrine black shales are good potential source rocks.

An unconventional source of petroleum in the form of methane gas occurs in the gas hydrates of some continental rises (Mountain and Tucholke, 1984). Muddy sediment to depths as great as 600 m beneath the sea floor has pores partially filled with ice formed of methane hydrates. In ice form, methane is about one fiftieth the volume it would be in gas form at the surface. Drilling and seismic studies on the Blake Outer Ridge indicate that gas hydrate occupies about 30% of the pore space (Sheridan et al., 1983). Given the volume of hydrate on the United States Atlantic margin, and conversion of hydrate to gas volumes at the surface, trillions of cubic feet of gas reserves exist in this form.

Further Reading

Keen CE, Beaumont C, 1990, Geodynamics of rifted continental margins, in Keen MJ, Williams GL (eds), *Geology of the continental margin of eastern Canada* [The Geology of North America, v. I-1]: Geological Society of America, Boulder, p. 391–472.

Klitgord KD, Hutchinson DR, Schouten H, 1988, U.S. Atlantic continental margin: structural and tectonic framework, in Sheridan RE, Grow JA (eds), *The Atlantic continental margin U.S.* [The Geology of North America, v. 1–2]: Geological Society of America, Boulder, p. 19–55.

LASE Study Group, 1986, Deep structure of the U.S. East Coast passive margin from large aperture seismic experiments (LASE): *Marine and Petroleum Geology*, v. 3, p. 234–242.

Lister GS, Ethridge MA, Symonds PA, 1986, Detachment faulting and the evolution of passive continental margins: *Geology*, v. 14, p. 246–250. (See Comment and Reply, v. 14, p. 890–892.

Lonsdale P, 1989, Geology and tectonic history of the Gulf of California, in Winterer EL, Hussong DM, Decker RW (eds), *The eastern Pacific Ocean and Hawaii* [The Geology of North America, v. N]: Geological Society of America, Boulder, p. 499–521.

McKenzie DP, 1978a, Some remarks on the development of sedimentary basins: *Earth and Planetary Science Letters*, v. 40, p. 25–32.

Sleep NH, 1971, Thermal effects on the formation of Atlantic continental margins by continental breakup: *Geophysical Journal of the Royal Astronomical Society*, v. 24, p. 325–350.

Worrall DM, Snelson S, 1989, Evolution of the northern Gulf of Mexico, with emphasis on Cenozoic growth faulting and the role of salt, in Bally AW, Palmer AR (eds), *The Geology of North America—an overview* [The Geology of North America, v. A]: Geological Society of America, Boulder, p. 91–138.

Trenches and Trench-Slope Basins

5

Michael B. Underwood and Gregory F. Moore

Introduction

The primary goal of this chapter is to describe the dynamic interplay between deformation and sedimentation within subduction zones. Physical processes of sedimentation in subduction zones are no different than those operating in other deep-marine environments (Pickering et al., 1989). The dominant sedimentary processes include steady vertical settling of suspended detritus (e.g., Gorsline, 1984), sluggish movement of bedload and suspended load by thermohaline bottom currents (e.g., Stow and Lovell, 1979), and punctuated movement by sediment gravity flows, including submarine slides, debris flows, grain flows, and turbidity currents (e.g., Middleton and Hampton, 1976). Although links between modern and ancient data sets are poor for many depositional environments in the marine realm, the study of subduction zones is unusually problematic, for the following reasons. First, direct sampling of submarine stratigraphy (through drilling and coring) has been restricted to roughly the top kilometer of the crust. This information permits reasonable assessment of sedimentary facies relations, but the tectonic processes that occur at greater depths within subduction margins are poorly understood. Consequently, many of the geodynamic models and hypotheses for evolution of inferred ancient analogs remain speculative and virtually untested by direct imaging or sampling of modern examples (e.g., Cloos, 1982; Pavlis an Bruhn, 1983; Stockmal, 1983; Shreve and Cloos, 1986; Cloos and Shreve, 1988a,b). Second, most strata successions preserved within ancient subduction zones are highly disrupted; in most cases, polyphase structural overprints inhibit or prevent recognition of primary sedimentary-facies relations and early-phase structural architecture. Finally, technological improvements in geophysical methods have allowed oceanographers to define the gross structure of subduction zones with much improved three-dimensional clarity; even with these improvements, however, resolution remains at dimensions comparable only to regional-scale field studies. We note that a similar scale problem applies to comparisons of virtually all types of modern and ancient turbidite systems (Mutti and Normark, 1987), but the dilemma is compounded in subduction zones because one must also analyze deformation fabrics and structural geometries.

Deep-sea trenches of the world (Fig. 5.1) are surface expressions of subduction boundaries. Each example is located where a plate of oceanic lithosphere descends into the mantle along the Benioff-Wadati zone. As defined by Uyeda (1982), subduction may occur beneath either oceanic lithosphere (Mariana type) or continental lithosphere (Andean type). In either case, *the outer trench slope* (or seaward slope) forms in response to downward flexure of the subducting slab. A wide range of sedimentary environments can develop outboard of the trench as the oceanic crust is rafted toward its subduction boundary. In most cases, the oceanic-plate sediments are dominated by pelagic ooze and hemipelagic mud of the abyssal plain, although turbidites and submarine-fan deposits are also possible. Normally, the abyssal deposits are buried subsequently by considerable thicknesses of terrigenous trench-wedge turbidites as the downgoing plate enters the subduction zone (Piper et al., 1973). Although most of the oceanic

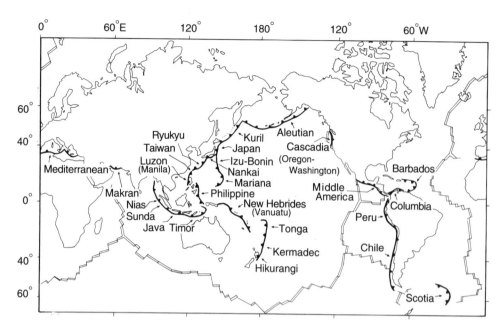

Fig. 5.1 Global distribution of active oceanic subduction zones.

lithosphere is subducted into the mantle, large components of the sedimentary cover and fragments of basement highs in the downgoing plate are retained in the crust by mechanical transfer into the overriding plate (Fig. 5.2).

In accretionary subduction zones (Fig. 5.3), trench-floor and oceanic-plate deposits are added to the toe of the *landward trench slope* (or inner slope) by imbricate thrusting (Karig, 1974; Seely et al., 1974; Karig and Sharman, 1975). A detachment surface, or *décollement* (Fig. 5.2), separates the upper part of the accreted section (i.e., the zone of offscraping) from material that is underthrust beyond the base of the slope (Moore, 1989). Above the décollement, off-scraped sediment is transferred to the so-called accretionary prism (or accretionary wedge), and this prism displays a rugged and irregular seafloor morphology governed by numerous tectonic ridges that form by

folding and fault dislocation (Fig. 5.4). Each successive thrust fault normally propagates seaward from the décollement through the wedge of trench sediments. At any given time, a deformation front separates the undisturbed trench sedimentary environment from the progressively disrupted accretionary prism (Fig. 5.2), but this structural boundary migrates seaward through time as the prism grows. Recognition of the deformation front is not always straightforward, particularly if incipient thrust faults have not broken through to the seafloor. With high-quality seismic-reflection data, the deformation front can be identified on the basis of broken and/or offset reflectors within the wedge of trench deposits; in other cases, the operational boundary of the accretionary prism is placed at the base of the landward trench slope.

Toward the magmatic arc, a structural high known as the *trench-slope break* serves as the second opera-

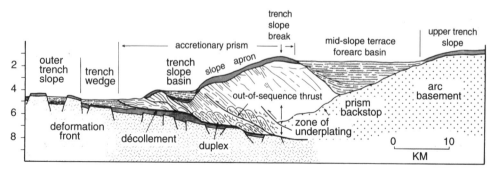

Fig. 5.2 Schematic illustration of major bathymetric domains, regional-scale structural features, and sites of deposition within subduction zones.

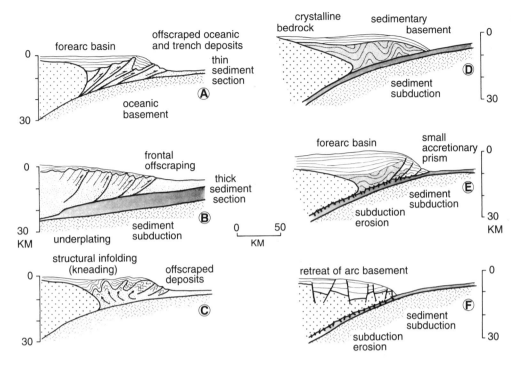

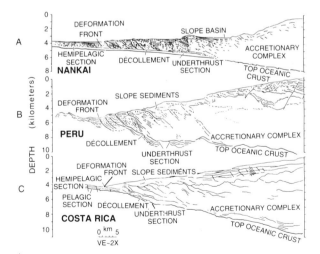

Fig. 5.3 Conceptual models of accretion by offscrapping, sediment subduction, and subduction erosion (after Scholl et al., 1980). (A) Formation of accretionary prism by stacking of thrust slices derived from a relatively thin sequence of oceanic deposits. (B) Frontal offscrapping by imbricate thrusting within a thick sequence of oceanic and trench-wedge deposits; partial subduction and underplating are likely below the decollement. (C) Formation of structurally chaotic accretionary mass, together with structural infolding of trench-slope deposits (tectonic kneading). (D) Tectonic consumption of oceanic sediments beneath bedrock of a subduction margin; this model could be incorporated into models A or B, with structural partitioning along the decollement. (E) Subduction of oceanic deposits, mechanical erosion of the bedrock of the margin, and temporary outgrowth of a small accretionary wedge. (F) Advanced stages of subduction erosion, leading to retreat and exposure of igneous and metamorphic basement to volcanic arc or continental massif

tional boundary of the accretionary prism (Fig. 5.2). This bathymetric feature is also called the outer ridge or outer-arc high. Regions seaward of the trench-slope break are usually referred to as the *lower trench slope*. We further subdivide the rugged lower slope into two types of bathymetric domains: zones with relatively steep inclination (*slope apron*), and small, flat-floored benches (*trench-slope basins*). The trench-slope break also migrates with time because of seaward prism expansion, coupled with subsidence of the accreted material along its landward margin (Karig, 1977). In response to several tectonic and lithologic factors, the trench-slope break may exhibit one of many common bathymetric and structural configurations (Seely, 1979). For example, parts of the basement complex landward of the trench-slope break can consist of accreted sedimentary material, trapped slices of oceanic crust, continental crust, or seaward extensions of the arc massif.

Fig. 5.4 Comparison of trench wedges and accretionary prisms from: (A) Nankai Trough (after Moore et al., 1990); (B) Peru Trench (modified from von Huene et al., 1985a); and (C) Middle America Trench offshore Costa Rica (from Shipley et al., 1990). These profiles are line drawings of migrated depth sections, based on digital processing and interpretation of seismic-reflection data.

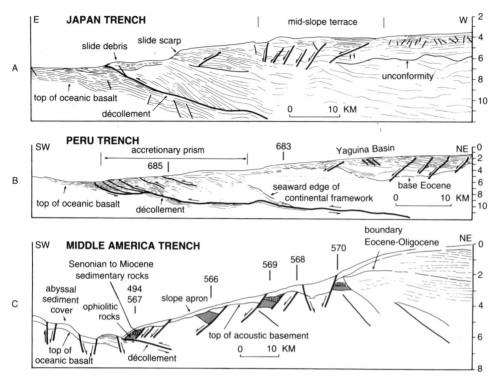

Fig. 5.5 Line drawings of migrated depth sections showing trench wedges and landward trench slopes for nonaccretionary subduction margins. (A) Japan Trench in the vicinity of DSDP transects (von Huene and Culotta, 1989); normal faults occur mostly within the mid-slope and upper-slope environments. (B) Peru Trench, in a region of forearc subsidence and inferred tectonic erosion (von Huene et al., 1985a); Yaquina Basin (forearc) is characterized by normal faults, whereas the lower slope is cut by thrusts. (C) Guatemala segment of the Middle America Trench, where DSDP boreholes show the slope apron to be underlain by Senonian to Miocene sedimentary and ophiolitic rocks possibly associated with retreating oceanic basement of the island arc (Aubouin et al., 1985a); normal faults occur throughout this subduction zone.

Virtually all convergent margins contain basins immediately landward of the trench-slope break, at either mid-slope or shelfal water depths (Dickinson and Seely, 1979). These flat-floored basins, many of which contain huge volumes of sediment, are referred to herein as *forearc basins* (see Chapter 6). Forearc basins at mid-slope water depths are separated from the continental shelf or arc platform by a steep *upper trench slope* with a comparatively thin sediment cover (Fig. 5.2). We stress here that regional-scale facies relations and patterns of sediment dispersal for the lower slope and trench cannot be divorced from those of the upper slope and forearc basins. For purely organizational reasons, this chapter concentrates on forearc segments located seaward of the trench-slope break, but when appropriate, ties are also made to upper parts of the forearc.

Some modern subduction margins appear to be zones of nonaccretion, at least at the base of the trench slope (Fig. 5.3). These examples fail to conform to the general description of accretionary prisms outlined above, and most are associated with intra-oceanic subduction zones (Fig. 5.5). Because rates of terrigenous sediment influx are unusually low along

these plate boundaries, all of the trench-floor and oceanic-plate deposits may be subducted (Scholl et al., 1980; Hilde, 1983; Scholl and Vallier, 1983; von Huene, 1984). In some cases, subduction of oceanic lithosphere apparently is accompanied by landward retreat of the trench, extensional faulting, and subsidence of the entire landward slope (Aubouin et al., 1985a,b; Aubouin, 1989; von Huene and Lallemand, 1990). In addition, local rates of tectonic erosion can accelerate where large bathymetric features, such as seamounts and aseismic ridges, collide with a subduction zone (e.g., Ballance et al., 1989; von Huene and Lallemand, 1990; Fisher et al., 1991). Although these intriguing examples are important in the global mass-balance equation, subsidence of a lower trench slope and/or persistent retreat of a continental margin or island-arc platform will not leave much behind for geologists to study unless the erosional phase is followed by accretion and uplift. It certainly is possible to preserve thick sedimentary successions and volcanic rocks in extensional forearc basins, but the subaerial rock record cannot retain a record of rocks lost from the crust by net erosion of the lower trench slope and subduction of that mate-

rial into the mantle. Consequently, erosional margins are of trivial importance to the central theme of this anthology.

Trench-slope sediments accumulate on all types of subduction margins, even those that experience nonaccretion or subduction-erosion (Figs. 5.2, 5.3). In this chapter, we subdivide the overall trench-slope setting into two distinct components: (1) the inclined slope apron, which is dominated by relatively thin deposits of hemipelagic mud, and (2) flat-floored intraslope basins located seaward of the trench-slope break. This segregation of depositional domains is fundamental because seafloor gradients exert considerable influence on the mechanical behavior and competence of turbidity currents (Komar, 1970), which in turn, dictate facies patterns. Intraslope basins are not unique to subduction margins; they can also form, for example, behind salt diapirs along mature rifted margins (e.g., Bouma et al., 1978). To clarify, therefore, we favor the more precise term *trench-slope basin*. In most instances, slope-basin detritus becomes ponded behind large tectonic ridges that form as bathymetric by-products of thrust faulting and diapirism within the underlying accretionary prism (G.F. Moore and Karig, 1976). Stratal successions within trench-slope basins, therefore, tend to be considerably thicker than those of the steeply inclined slope apron; this is because gentle seafloor gradients promote decay of sediment kinetic energy within gravity flows, thereby leading to deposition, whereas steep slopes promote acceleration and sediment bypassing.

Ultimately, the abyssal-plain, trench-floor, and lower-trench-slope deposits overlap one another in terms of both sedimentologic and structural character. The three depositional environments are separated by time-transgressive angular unconformities, but the geometric discordance can be subtle to begin with and later modified by a broad spectrum of tectonic processes. Large-scale models of prism evolution, based on Coulomb failure theory, assume that deformation occurs by brittle failure; the wedge angle or taper depends dynamically on the strength of the wedge material and the amount of shear stress concentrated at the base of the wedge (Davis et al., 1983; Dahlen et al., 1984). The shape and internal structure of the prism are also influenced somewhat by the geometry of its rigid *backstop*, which is formed by the

edge of a rifted continent, the arc platform, or an older accretionary complex (Byrne et al., 1993). In most cases, seismic-reflection records show backstops dipping toward the trench (Silver et al., 1985; Westbrook et al., 1988).

Within the interior of an accretionary prism, progressive deformation results in significant modification of the internal three-dimensional facies relations and facies boundaries. In theory, each thrust fault should become older toward the volcanic arc, and rates of displacement should increase toward the deformation front (Karig and Sharman, 1975). Several deviations from this pattern are possible, however. For example, new faults can propagate through the prism toe arcward of the deformation front; these structures are referred to as *out-of-sequence thrusts*. The *vergence* of thrust faults near the prism toe (i.e., direction of tectonic transport) can be in either a landward or a seaward sense, and vergence can change both through time and along the strike of a single subduction zone (Seely, 1979). Other complexities include formation of ramps (i.e., steep fault segments cutting upsection) and flats along the décollement, development of *duplex structures* (defined by Boyer and Elliot [1982] as an imbricate family of subsidiary contraction faults asymptotically curving downward to a sole or floor thrust and upward to a roof thrust), backthrusting, seamount subduction, and development of wide shear zones (Silver et al., 1985; Moore and Silver, 1987; Cloos and Shreve, 1988a; Silver and Reed, 1988; Moore, 1989; Lundberg and Reed, 1991).

The term *underplating* (or subcretion) is generally used to describe the addition of material to the bottom of a prism, at an intermediate position between the magmatic arc and the trench; underplating leads to thickening and uplift of the accretionary wedge without additional shortening (J.C. Moore et al., 1982b; Cloos and Shreve, 1988a). In essence, any material added arcward of the frontal zone of offscraping is classified as underplated. One likely mechanism of underplating is through down-stepping of the décollement and duplex formation (Silver et al., 1985). A second possible process is viscous flow (Cloos, 1982, 1984; Pavlis and Bruhn, 1983; Shreve and Cloos, 1986). Deformation within the zone of

underplating, however, is not necessarily intense. Powerful seismic-reflection systems, for example, have been able to image coherent packets of reflectors thought to represent underplated sediment at lower crustal depths (Moore et al., 1991).

Prism uplift and oversteepening of the seafloor can lead to instability of near-surface sedimentary strata, and material recycling often occurs due to gravitational failure (Moore et al., 1976; Jacobi, 1984). Slope failures are particularly frequent along the frontal ridge of the accretionary prism (i.e., the lowest ridge on the lower slope), and associated debris lobes commonly extend across the trench floor (e.g., Davis and Hyndman, 1989). These mass-wasting events are capable of remobilizing both the slope cover and previously accreted rocks.

Modern subduction zones are diverse and complicated. Ancient accretionary complexes present additional challenges. Most subaerial analogs, for example, have been exhumed from beneath several kilometers of tectonic overburden, and the seafloor drilling record does not even come close to these depths. The initial identification of fossil subduction zones, in fact, has been based largely on occurrences of extreme structural dismemberment and/or high-pressure, low-temperature (blueschist facies) metamorphism (e.g., Hamilton, 1969; Ernst, 1970; Uyeda and Miyashiro, 1974). Some tectonostratigraphic units within inferred subduction complexes are dominated by relatively coherent slates, together with map-scale duplex structures; these examples are thought to represent underplated material (e.g., Fisher and Byrne, 1987; Sample and Moore, 1987). More troublesome aspects of interpreting the rock record involve *mélange zones*, loosely defined here as mappable units of fragmented strata in which blocks are imbedded in a fine-grained matrix. The study of mélanges has been so riddled with controversy that scientists have been unable to agree on either an operational definition of the term or a universal classification system (Raymond, 1984). Cowan (1985) defined four end members within a spectrum of mélange types: Type I = stratified sequences of sandstone and mudstone progressively disrupted by layer-parallel extension; Type II = progressively disrupted sequences of mudstone, tuff, chert, and sandstone;

Type III = block-in-matrix mudstone chaos; and Type IV = mudstone-dominated brittle fault zones (Fig. 5.6). Unfortunately, most of the criteria used to decipher mélange genesis are ambiguous. Interpretations of mélange belts generally fall into one of four categories: (1) tectonic deformation in shear imbricate fault systems; (2) mixing within deep-seated channels of ductile flow; (3) diapiric movement within the accretionary prism; and (4) olistostromes formed by gravitational failure and downslope movement of the slope apron and upper accretionary prism (for representative examples, see Underwood [1984]; Larue and Huddleston [1987]; Lash [1987a, b]; Moore and Byrne [1987]; Cloos and Shreve [1988a]; Brandon [1989]; Brown [1990]; Orange [1990]; and Okamura [1991]).

One important conclusion is that ancient subduction complexes are characterized at all structural levels and at all scales by a broad spectrum of deformation styles, which in turn, are produced by consequent variations in deformation mechanisms (Karig, 1985; Moore, 1989). This diversity makes the formulation of a single actualistic model describing accretionary-prism stratigraphy awkward, at best. Even under the best of circumstances (i.e., relatively

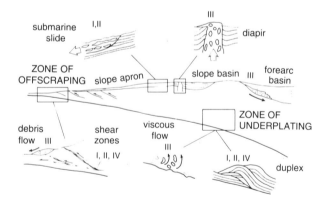

Fig. 5.6 Schematic cross section through an idealized accretionary wedge at an active subduction margin showing inferred processes responsible for mélange genesis. Roman numerals refer to the four end-member types of mélange, as classified by Cowan (1985). Type I = stratified sequences of sandstone and mudstone progressively disrupted by layer-parallel extension; Type II = progressively disrupted sequences of mudstone, tuff, chert, and sandstone; Type III = block-in-matrix chaos; and Type IV = mudstone-dominated brittle fault zones. Modified from Cowan (1985).

mild mesoscopic deformation and shallow depths of maximum burial), virtually all ancient subduction complexes have been subjected to complicated histories of polyphase deformation, such that recognition and unequivocal interpretation of the primary sedimentary facies relations become all but impossible. Presented with this dilemma, we have chosen to place greater emphasis on Quaternary examples and marine data sets, rather than the more controversial examples in the rock record. Philosophically, we believe that one should begin with the simplest and most tightly constrained case studies before applying those results to investigations of their dismembered (inferred) analogs.

Structural and Tectonic Controls

All sedimentary environments within subduction zones are governed by a complex array of interdependent structural and tectonic variables. First, relief on the subducting oceanic basement is inherited from the constructive phase of volcanism and the extensional stress regime of mid-ocean ridges and/or intraplate hot spots, then modified by extensional rupture as the lithosphere bends through the outer swell of the trench (e.g., Schweller and Kulm, 1978a; Hilde, 1983; Aubouin, 1989). The amount of relief on basement structures as they approach a subduction front will be tempered by the rate and duration of sediment accumulation on the abyssal plain. Water depth and bathymetric geometry of the trench floor are affected by the rate of subduction, the dip angle of the subducting slab, and the age of the subducting lithosphere (Jarrard, 1986a). Seismic zones are flatter beneath wide accretionary prisms; the size of the prism evidently affects the dip-angle of the shallow part of the descending lithosphere (Karig et al., 1976). Finally, the near-surface character of the landward trench slope is generated mostly by folding and faulting within the underlying accretionary prism. Because there are so many variables to consider, it is not always possible to separate cause and effect.

Deep-sea trenches vary greatly in their depths, widths, and types of sedimentary successions (Uyeda, 1982). Although the depth and width parameters are controlled, in part, by the age of subducting litho-

sphere and subduction rate (Hilde and Uyeda, 1983; Jarrard, 1986a), perhaps the most important factors that dictate the overall geometry of a subduction margin are the amount of sedimentary fill in the trench and whether the forearc is accretionary or erosional. Erosional trenches typically are isolated from large amounts of terrigenous sediment; intraoceanic examples such as the Mariana and Tonga trenches (Fig. 5.1) are deep (>9,000 m) and narrow (<5 km wide). At the other end of the spectrum, accretionary trenches typically are choked with terrigenous sediment; examples such as the western Sunda, eastern Aleutian, Nankai, Makran, and southern Barbados trenches (Fig. 5.1) are shallow (3,000 to 5,000 m) and wide (>10 km). Accretionary trenches with smaller amounts of terrigenous sediment, such as the northern Chile Trench, the Peru Trench, and the northern Middle America Trench off Mexico (Fig. 5.1), are intermediate in depth (5,000 to 7,000 m) and width (5 to 10 km).

In general, a trench with thick accumulations of sediment, such as the Nankai Trough (Fig. 5.4) or the eastern Aleutian Trench, displays a nearly flat surface topography with gentle axial gradients along strike. Trenches with little sediment fill, on the other hand, exhibit pronounced bathymetric irregularities along strike. For instance, the Middle America Trench off Mexico is segmented into numerous discrete basins, each approximately 40 to 60 km long; this segmentation has been caused by the intersection of bathymetric highs on the subducting plate with the trench fill (Underwood and Bachman, 1982; Shipley and Moore, 1985; Moore and Shipley, 1988). Smaller segmented basins in the Middle America Trench off Guatemala (2 to 15 km long) formed because of fault-bounded ridges that intersect the trench obliquely (Aubouin et al., 1982a; Moore et al., 1986). The Java and Japan trenches (Fig. 5.1) are similarly segmented by subducting topography, including large seamounts (Moore et al., 1980; Cadet et al., 1987). The most extreme examples of trench segmentation are associated with zones of ridge-trench interaction, involving either active spreading ridges, such as the Chile Rise (Cande and Leslie, 1986), or extinct aseismic ridges, such as the Nazca Ridge (Kulm et al., 1974; Schweller et al., 1981).

Outer Trench Slope

The outer trench slope, by definition, forms the sea-ward wall of every trench (Fig. 5.2), but the relief and structural character of the outer slope can be extremely variable. The ideal stratigraphy of a convergent margin includes the pelagic and hemipelagic sediment carried into the trench on the subducting plate, in addition to the overlying wedge of trench deposits (Piper et al., 1973; von Huene, 1974; Schweller and Kulm, 1978b). The subducting plate usually is composed of oceanic lithosphere, although in the case of a peripheral foreland basin (Ingersoll, 1988b), continental crust may be pulled into the subduction zone during incipient crustal collision (see Chapters 10 and 11). Examples of this type include the Timor Trough (Hamilton, 1979; Jacobson et al., 1979; von der Borch, 1979; Karig et al., 1987) and the Taiwan collision zone (Suppe, 1987; Lundberg and Dorsey, 1988; Huang et al., 1992). Abyssal, plain sediments range in thickness from a few tens of meters in intraoceanic trenches such as Tonga and Mariana (Uyeda, 1982; Hilde, 1983), to more than 5 km adjacent to continental collision zones such as the northern Sunda Trench (Moore et al., 1982) and the Makran margin of the Gulf of Oman (White and Klitgord, 1976; White and Louden, 1982). Rates of sediment accumulation on the outer trench slope are generally less than 100 m/my (Table 5.1). Many thicker abyssal-plain successions are dominated by terrigenous turbidites, large channel-levee complexes, and either active or abandoned submarine fans (Ness and Kulm, 1973; Curray and Moore, 1974; Stevenson et al., 1983; Emmel and Curray, 1985; Kolla and Coumes, 1985; Stevenson and Embley, 1987). Along arc-continent collision zones, such as the Timor Trough, the underthrusting continental crust can include great thicknesses of passive-continental-margin strata (Jacobson et al., 1979; Karig et al., 1987). The stratigraphy of the oceanic plate is important because some of those rocks and sediments eventually are incorporated into the accretionary prism, thereby forming part of the "basement" beneath the slope apron and trench-slope basins.

Unless buried by thick abyssal-plain sediments, the outer slope of a trench should be dominated by normal faults that are related both to pre-existing plate fabrics created at mid-ocean spreading ridges and to crustal rupture during flexure as the subducting plate passes through the outer swell or bulge (Schweller and Kulm, 1978a; Aubouin et al., 1982a; Hilde, 1983; Warsi et al., 1983). The apex of the peripheral bulge is located 120 to 150 km seaward of the trench axis, and the maximum uplift with respect to the flat abyssal plain is generally 300 to 500 m (e.g., Bodine and Watts, 1979; Forsyth, 1980). The amount of vertical movement may be sufficient to produce prominent unconformities, and judging from both modern and ancient examples (Dubois et al., 1975; Jacobi, 1981), atolls and reefs on aseismic ridges can be uplifted sufficiently to expose rocks to near-shore and subaerial erosive processes. In addition, the occurrence of strong bottom currents in deeper water may produce a substantial depositional hiatus (e.g., Hussong and Uyeda, 1981). On a smaller scale, a horst-and-graben morphology will be produced by extension of the upper crust, and these structures exert considerable control on the sedimentary facies relations of the seaward slope by concentrating sediment accumulations in downfaulted grabens and half grabens.

Trench Floor

Trench-floor basins are wedge-shaped in transverse section, and most seismic-reflection profiles display clear seaward onlap of the trench sediments onto the underlying oceanic-plate succession (Fig. 5.2). These sediment wedges are highly variable in three-dimensional geometry, and they range in maximum thickness from less than 100 m in intraoceanic systems (e.g., Costa Rica, Tonga, and Mariana) to between 500 and 2,000 m in continental-margin systems (e.g., Nankai, eastern Aleutians, Oregon, Washington, western Sunda, and southern Barbados). The width of the trench floor maintains dynamic equilibrium (steady-state) only when subduction rate, dip angle, and trench-sedimentation rate are held constant. For example, if the subduction rate increases or the sedimentation rate decreases, then the trench wedge shrinks; either a lower subduction rate or a higher sedimentation rate results in expanding trench wedges. Subduction rates vary from about 1 cm/y (Oregon and Nankai) to 8–10 cm/y (Japan, Middle

Table 5.1. DSDP and ODP Drilling Sites in Subduction Zones, with Rates of Quaternary Sediment Accumulation, Uncorrected for Compaction

Drill Site	Depth (meters)	Location	Depositional Environment	Accum. Rate (m/my)	Reference
178	4218	Alaska	abyssal plain	122	Kulm et al., 1973
297	4458	Shikoku	abyssal plain	101	Karig et al., 1975
436	5240	Japan	outer slope	50 to 56	Shipboard Party, 1980
442	4639	Shikoku	abyssal plain	31	Klein et al., 1980
443	4372	Shikoku	abyssal plain	75 to 76	Klein et al., 1980
444	4843	Shikoku	abyssal plain	24 to 48	Klein et al., 1980
487	4764	Mexico	outer slope	126	Watkins et al., 1982
495	4140	Guatemala	outer slope	37	Aubouin et al., 1982b
672	4975	Barbados	abyssal floor	20	Mascle et al., 1988
828	3087	Vanuatu	abyssal ridge	60	Collot et al., 1992
831	1066	Vanuatu	abyssal guyot	7	Collot et al., 1992
174	2815	Cascadia	trench fan	370 to 940	Kulm et al., 1973
180	4923	Aleutian	trench floor	200 to 2,845	Kulm et al., 1973
298	4628	Nankai	trench floor*	733	Karig et al., 1975
486	5142	Mexico	trench floor	unknown	Watkins et al., 1982
499	6105	Guatemala	trench floor	300	Aubouin et al., 1982b
500	6094	Guatemala	trench floor	70	Aubouin et al., 1982b
542	5016	Barbados	trench floor*	55	Biju-Duval et al., 1984
543	5633	Barbados	trench floor	24	Biju-Duval et al., 1984
582	4879	Nankai	trench floor	340 to 1,340	Kagami et al., 1985
583	4634	Nankai	trench floor*	225 to 315	Kagami et al., 1985
676	5052	Barbados	trench floor*	40	Mascle et al., 1988
808	4676	Nankai	trench floor*	787 to 1,381	Taira et al., 1991
175	1999	Cascadia	slope basin	125 to 238	Kulm et al., 1973
181	3086	Aleutian	slope basin	422 to 636	Kulm et al., 1973
182	1411	Aleutian	slope apron	128	Kulm et al., 1973
434	5986	Japan	slope apron	4	Shipboard Party, 1980
435	3401	Japan	slope apron	52	Shipboard Party, 1980
438	1552	Japan	slope apron	29	Shipboard Party, 1980
440	4509	Japan	slope basin	105 to 270	Shipboard Party, 1980
441	5655	Japan	slope apron	4	Shipboard Party, 1980
460	6452	Mariana	slope basin	18 to 37	Hussong et al., 1981
461	7034	Mariana	slope bench	30	Hussong et al., 1981
488	4254	Mexico	slope bench	364	Watkins et al., 1982
490	1761	Mexico	slope apron	89	Watkins et al., 1982
491	2883	Mexico	slope apron	18	Watkins et al., 1982
492	1935	Mexico	slope apron	36	Watkins et al., 1982
494	5472	Guatemala	slope apron	100	Aubouin et al., 1982b
496	2049	Guatemala	slope apron	141	Aubouin et al., 1982b
497	2347	Guatemala	slope apron	99	Aubouin et al., 1982b
498	5478	Guatemala	slope apron	132	Aubouin et al., 1982b
541	4940	Barbados	slope apron	56	Biju-Duval et al., 1984
565	3099	Costa Rica	slope apron	165	von Huene et al., 1985b
566	3745	Guatemala	canyon floor	75	von Huene et al., 1985b
567	5500	Guatemala	slope apron	40	von Huene et al., 1985b
568	2010	Guatemala	slope apron	117	von Huene et al., 1985b
569	2744	Guatemala	slope apron	33	von Huene et al., 1985b
570	1698	Guatemala	slope bench	130	von Huene et al., 1985b
671	4914	Barbados	slope apron	16	Mascle et al., 1988
673	4661	Barbados	slope apron	17	Mascle et al., 1988
674	4550	Barbados	slope apron	28	Mascle et al., 1988
682	3788	Peru	slope apron	26	Suess et al., 1988a
685	5070	Peru	slope apron	100	Suess et al., 1988a

188   Chapter 5

Table 5.1. Continued.

688	3820	Peru	slope basin	98 to 350	Suess et al., 1988a
783	4648	Izu-Bonin	seamount	22	Fryer et al., 1990a
784	4900	Izu-Bonin	slope basin	31	Fryer et al., 1990a
827	2803	Vanuatu	slope basin	65 to 344	Collot et al., 1992
830	1018	Vanuatu	guyot collision	<115	Collot et al., 1992

*Trench-wedge facies recovered by drilling into toe of accretionary prism.

America, and Peru-Chile Trenches). Rates of accumulation on the trench floor (Table 5.1) are strongly controlled by sedimentary processes (including the number and spacing of sediment-input points), as well as sediment composition (terrigenous influx versus pelagic settling).

As mentioned above, horst blocks produced by normal faulting in the subducting plate may act as structural boundaries of segmented trench basins. Normal faults are less conspicuous in trenches where young oceanic crust is being subducted (e.g., northern Middle America Trench); this is because the outer swell is more subdued due to a temperature-dependent buoyancy and lower dip-angle of subduction. In addition, the thickness of trench sediment may be great enough to bury the basement relief completely, as shown by the Cascadia subduction zone (Kulm and Fowler, 1974a; Carlson and Nelson, 1987; Davis and Hyndman, 1989; Hyndman et al., 1990) and the northern Sunda Trench (Moore et al., 1982). Even with high rates of sediment accumulation, however, high-angle faults sometimes propagate through the sediment cover and influence patterns of sediment dispersal on the trench floor (e.g., Thornburg et al., 1990).

The sedimentary facies boundary between the trench wedge and the underlying oceanic-plate stratigraphy is abrupt if the trench floor is narrow and abyssal deposits are dominated by lithified pelagic oozes. Under these circumstances, one should expect a sharp contrast in bulk density and shear strength between pelagic rocks (e.g., chert, radiolarian mudstone) and overlying poorly consolidated clastic sediments (Moore, 1975; Karig, 1985, 1986). Conversely, with a wide, terrigenous-dominated trench, such as the Nankai Trough, a gradual facies change occurs across the outer trench slope, with thin silty turbidites interfingering oceanward with hemipelagic deposits of the abyssal plain (e.g., Piper

et al., 1973; Taira et al., 1992). Variations in physical properties have been discussed in detail for the Nankai Trough example by Bray and Karig (1988). Among other responses, burial-induced consolidation beneath the overburden of the trench wedge causes arcward thinning within the hemipelagic unit as it moves into the subduction zone. Drilling results show that depth gradients in critical physical parameters (i.e., porosity, water content, and bulk density) are not necessarily deflected at depths corresponding to the trench-to-basin facies transition (see also, Bray and Karig, 1985; Karig, 1986; Taira et al., 1992). On the other hand, sharp changes in these physical properties typically occur across the décollement (Moore, 1989), even in cases where the position of the décollement does not correspond to a lithofacies boundary (Taira et al., 1992).

Sediment Apportioning at the Décollement

The décollement marks the structural boundary between sediment that is offscraped and added to the toe of the accretionary prism and material that is carried toward the arc and underthrust beyond the toe of the prism (Fig. 5.2). The depth of the décollement is of paramount importance to the facies character of the accretionary prism because it determines the amount and type of sediment that is accreted, as opposed to the amount and type that is underthrust or subducted to greater depths. In trenches without net accretion, the décollement intersects the sea floor at or near the deformation front, such that virtually all of the incoming sediment is subducted. In some accretionary systems, the décollement is located near the base of the trench wedge. Under these circumstances, the trench-floor deposits are accreted but the underlying pelagic and hemipelagic sediments should be detached and selectively subducted (Moore, 1975),

as exemplified by the Middle America Trench off Mexico (Moore and Shipley, 1988). In other trenches, the décollement propagates through the pelagic/hemipelagic deposits of the oceanic plate, such that the upper part of the abyssal facies is accreted along with the trench wedge (Moore, 1989). Prominent examples of this type include northern Barbados (Westbrook and Smith, 1983; J.C. Moore et al., 1982b; Moore et al., 1988) and the Nankai Trough (Moore et al., 1990; Taira et al., 1992). In all of these cases, some of the subducted material, including pieces of igneous basement, may become incorporated into the rear portion of the accretionary prism by underplating mechanisms (i.e., via duplexing, flow mélange, etc.).

Landward Trench Slope

Siliciclastic, volcaniclastic, and/or biogenous sediment accumulates on the landward slope of any trench, whether it is adjacent to a continental margin or an intraoceanic volcanic arc. In some arc/trench systems (e.g., Costa Rica [Fig. 5.4], Mexico, Guatemala, and northern Barbados), distinct basins do not develop on the lower trench slope. This is probably because vertical offsets along individual faults are insufficient to form prominent ridges. Instead, a blanket of relatively homogeneous hemipelagic sediment simply drapes the imbricated accretionary prism (Fig. 5.2). This slope apron can reach several-hundred meters in thickness, and the sediments are prone to submarine sliding and downslope creep (Baltuck et al., 1985). In addition, the sediment carapace may be deformed by folding and thrusting of the underlying accretionary complex (Shipley et al., 1990).

Most intraoceanic arc-trench systems, such as the Mariana Trench (Karig and Rankin, 1983), the Izu-Bonin Trench (Honza and Tamaki, 1985; Horine et al., 1990), the Tonga Trench (Herzer and Exon, 1985; Ballance, 1991), and the southern Middle America Trench (von Huene et al., 1980; Shipley and Moore, 1985, 1986) are characterized by thin trench sediments and relatively small accretionary prisms. However, significant volumes of slope sediments commonly accumulate behind irregular ridges of igneous basement; some of these ridges form by diapiric

emplacement of serpentinite and/or normal faulting. For example, ODP Leg 125 recovered up to 300 m of glass-rich mud and vitric siltstone behind a diapiric serpentinite seamount on the lower slope of the Bonin Trench (Fryer et al., 1990a). Similarly, accumulations of volcaniclastic sediment up to several-hundred meters in thickness occur in basins on the Mariana lower slope (Hussong et al., 1981). Seismic reflectors beneath the basin floor display low angles of tilt toward the arc, plus downward increasing dips (Karig and Rankin, 1983); these reflector geometries are consistent with offset along seaward-dipping normal faults.

In most accretionary systems (e.g., Nankai, Cascadia, Aleutians, Sunda, Makran, and New Zealand), hanging-wall anticlines develop as individual thrust sheets slide over ramps. Three-dimensional morphology of an accretionary prism (e.g., initial surface slope, prism width, spacing between thrust faults, fault dip, fold style, etc.) clearly changes, however, in response to variations in sediment thickness outboard of the deformation front (Breen, 1989; MacKay and Moore, 1990). In addition, thrust faults can verge in either a landward or seaward sense, depending on the stress regime and the shear strength of the basal layer of the wedge (Seely, 1977). Some tectonic ridges have antithetic normal or high-angle reverse faults. The anticlinal structures generally form linear bathymetric highs that define the structural grain of the accretionary prism (e.g., Lewis, 1980; Lewis et al., 1988; Davis and Hyndman, 1989). Compared to the background of hemipelagic deposition which forms the slope apron, much greater thicknesses of sediment accumulate behind the tectonic ridges in trench-slope basins (Fig. 5.7). This is largely because the flatter seafloor gradients promote momentum decay in turbidity currents as they pass across a basin floor. In addition, if seaward bounding ridges are high enough, they can block trenchward flow paths and initiate flow reflection. Trench-slope basins generally increase in size upslope because of sedimentary progradation over extinct thrusts, and there is a concomitant upslope increase in sediment thickness, reaching 3,000 m or more (e.g., Davey et al., 1986). Seismic-reflection profiles also show that slope-basin sediments are usually back-tilted, and the landward-dipping seismic

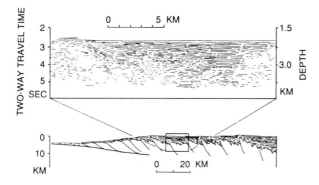

Fig. 5.7 Schematic diagram (below) showing the structural geometries responsible for formation of sedimentary basins on the lower trench slope. This general model is a composite based on seismic-reflection profiles, deep-sea drilling, and field work on Nias Island (modified from G.F. Moore and Karig, 1976). Expanded-scale line drawing (above) is based on a migrated multichannel seismic-reflection profile across the Hikurangi margin of New Zealand (from Davey et al., 1986). This actual trench-slope basin displays evidence for several episodes of thrusting and back-tilting of sedimentary sequences, as well as a blurring of the basal contact between accreted deposits and slope-basin deposits.

reflectors are steeper with increasing sub-bottom depth because of continued vertical movement of the marginal tectonic ridges (Fig. 5.7).

According to some evolutionary models of accretionary systems, underplating mechanisms lead to uplift of deep-seated rocks and trigger near-surface extension near the rear of the wedge (e.g., Pavlis and Bruhn, 1983; Platt, 1986). Although some lower-trench-slope basins could develop as grabens or half grabens under these circumstances, most of the extensional faults predicted by the Platt (1986) model are confined to the upper part of the slope where underplating dominates (Fig. 5.8). This model helps explain patterns of uplift within many orogenic belts, but the application to submarine accretionary prisms remains less certain. Seismic-reflection profiles from the Peru subduction margin, for example, display expected landward-dipping thrusts on the accretionary lower slope and high-angle normal faults on the upper trench slope (Fig. 5.5B; von Huene et al., 1985a). Similar interpretations of seismic profiles have been made for the Middle America Trench offshore Costa Rica (Lundberg, 1983; Shipley et al., 1990). Thus, at first glance, the model and observations seem to agree. On the other hand, the continental margin of Peru has an enigmatic history of

widespread subsidence and apparent tectonic erosion (von Huene et al., 1988; von Huene and Lallemand, 1990), rather than underplating and forearc uplift. Seismic interpretations for the Guatemala segment of the Middle America Trench, as a third example, show seaward-dipping normal faults throughout the lower-slope environment, as well as grabens and half grabens filled with slope sediment (Fig. 5.5C). This

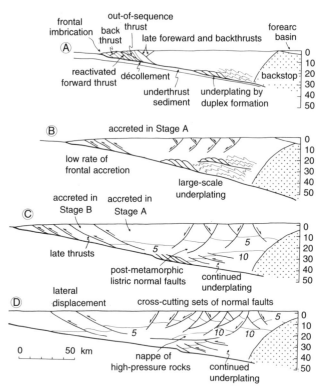

Fig. 5.8 Evolution model of an accretionary wedge (modified from Platt, 1986). (A) Early stage during which frontal accretion is dominant; specific structural features within the zone of offscraping include imbricate thrusts, backthrusts, out-of-sequence thrusts, and reactivated forward thrusts. Underplating mechanisms include duplexing, folding, thrusting, and (possibly) ductile flow. (B) Intermediate stage with low rate of frontal accretion and acceleration of the rate of underplating. Rear portion of the wedge extends by normal faulting and possibly by ductile flow at depth. Deeper parts of the wedge undergo high-pressure, low-temperature metamorphism (blueschist-facies). (C) Continued underplating and extension cause movement of high-pressure rocks toward the surface, lateral displacement from the rear of the wedge, and late thrusting toward the prism front. Note the uplifted isobars corresponding to approximately 5 and 10 kbar paleo-pressure. (D) Mature prism in which underplating and extension have brought high-pressure rocks to within 15 km of the surface, accessible to future erosion and exposure. Numbers refer to 5-kbar and 10-kbar isobars. Uplifted rocks will have several generations of accretionary structures overprinted by multistage normal faults.

intraoceanic margin also has been regarded as a site of tectonic erosion (von Huene, 1984; Aubouin et al., 1985a; Aubouin, 1989) rather than accretion.

In conclusion, based on the Peru and Guatemala examples, it seems plausible that normal faults can control slope-basin evolution for at least two reasons: (1) gravitational collapse triggered by forearc uplift; and (2) upper-plate extension due to tectonic erosion. In either scenario, progressive rotation of beds in a normal sense along seaward-dipping listric faults should tilt seismic reflectors in an arcward direction, but it is important to note that this same direction of rotation will occur in thrust-controlled basins with seaward vergence. Consequently, unless the faults themselves can be imaged, one must look for additional types of evidence to constrain the offset history of slope-basin deposits. Grabens preserve lithofacies and the paleontologic record of subsidence, whereas a progressive uplift history should characterize thrust basins, and this distinction might be tested through deep-sea drilling and recovery of benthic fossils.

Post-Sedimentation Structural Modification

Along accretionary subduction margins, strata are transferred from the oceanic plate and trench wedge into the toe of the accretionary prism, but deformation does not end there. The accreted sediments are subjected to progressive structural modification, with deformation rate diminishing with distance from the deformation front. Thrusting and folding continue upslope, at least to the position of the trench-slope break, and this tectonic activity causes considerable disruption of the slope cover, particularly near the basal unconformity of the slope apron and/or slope basins (Moore and Karig, 1976). Scholl et al. (1980) have referred to structural blurring of this important facies boundary as *tectonic kneading* (Fig. 5.3C). Contact relations also can be modified by diapiric intrusions of mélange into the slope-apron and slope-basin sediments (Becker and Cloos, 1985; Brown, 1990). In addition, out-of-sequence thrusts can propagate through the prism and cause late-stage segmentation of larger slope basins, by emplacing older slope-basin strata above younger sections of the slope apron. Strike-slip and oblique-

slip faults can likewise cut across the lower slope at transverse angles to the dominant grain of the accretionary prism (Karig, 1980; Ryan and Scholl, 1989; Goldfinger et al., 1992; MacKay et al., 1992). As discussed above, normal faulting locally causes segmentation of the trench-slope stratigraphy late in its evolutionary history, as uplift in the rear of the prism triggers extension of the upper slope. Finally, post-accretion deformation caused by subduction of seamounts can create new basins, modify existing basins, or trigger large-scale mass-wasting events (Yamazaki and Okamura, 1989; Okamura, 1991). The net effect of late-stage structural modification is to obscure or destroy the original three-dimensional architecture of both sedimentary lithofacies and structural domains.

Sediment Distribution and Facies Patterns

Sediment Provenance in Subduction Zones

Interpretations of sediment provenance within both modern and ancient subduction zones are complicated by the likelihood of long-distance movement of sediment gravity flows parallel to the strike of the margin, as well as the potential for post-depositional offset both prior to and following accretion. Within the context of this chapter, turbidite sands tend to be partitioned into two fundamental types of depositional environments, as dictated by their relatively flat seafloor gradients: (1) small basins perched on the lower trench slope; and (2) sediment wedges located outboard of the deformation front, including both the synorogenic trench deposits and older turbidite bodies carried on the subducting oceanic plate (Underwood and Bachman, 1982). The spatial linkage between the sources of this sediment and their sites of deposition can be unpredictable, especially for the accreted stratigraphic successions. Petrofacies comparisons between inferred slope-basin deposits and sandstones from associated mélange units can show similarities, if both were derived from arc-related sources transverse to the trench (e.g., Underwood and Bachman, 1986); conversely, sandstone suites can display marked differences if the trench

wedge is supplied by long-distance axial flow from an orogenic or continental source area unrelated to the active arc (e.g., Moore, 1979).

The coeval magmatic arc is the most likely source for transverse influx of both trench-floor and trench-slope sediment (Dickinson, 1982; Valloni, 1985; Marsaglia and Ingersoll, 1992), but long-distance transport parallel to the strike of the subduction zone is also commonplace (Lash, 1985; Underwood, 1986a). Hundreds of kilometers of margin-parallel transport via turbidity currents can occur either within the bathymetric confines of a trench (e.g., Schweller et al., 1981) or outboard of the trench on the abyssal plain (e.g., Curray and Moore, 1974). Sedimentary debris can be recycled from older accreted terranes within a nearby continental margin (Moore, 1979), and even compositionally mature, quartz-rich sands can be transported into subduction zones (Lash, 1985). For example, Eocene turbidite sands both within and seaward of the Barbados accretionary prism, which is an intraoceanic subduction zone, were transported northward from continental sources located on the South American craton (Velbel, 1985; Baldwin et al., 1986; Kasper and Larue, 1986; Dolan et al., 1990). Abyssal-plain sediments, moreover, can be translated thousands of kilometers, at any angle to the strike of the trench, after they are deposited (e.g., Stevenson et al., 1983). Thus, *a priori* assumptions regarding sediment provenance and dispersal pathways within subduction zones are seldom valid.

Trench-Slope Deposits

Sedimentation rates for depositional sites located between the trench-slope break and the trench floor are highly variable in both time and space, owing to differences in sediment dispersal mechanisms, sediment supply, and seafloor morphology (Table 5.1). For example, sediment accumulation rates are approximately 25 m/my on the inclined slope apron of the Peru Trench (Site 682); rates increase by more than tenfold in a nearby flat-floored slope basin (Site 688). Both the slope-apron environment and the slope-basin environment are exposed to a steady rain of fine-grained suspended sediment. The local sedimentologic responses to tectonism, however, are profound; this is because changing seafloor gradients

exert a powerful influence on the mechanical behavior of all types of sediment-gravity flows (Komar, 1970; Middleton and Hampton, 1976). The collective seafloor manifestation of thrust faults, anticlinal folds, normal faults, and mélange/mud diapirs is a complex morphologic system of ridges and intervening troughs. Steeper seafloor gradients promote transport and/or erosion by high-density turbidity currents, rather than deposition. Because of this, steeply inclined segments of lower trench slopes typically are mantled by veneers of hemipelagic to pelagic mud that accumulate slowly from suspension fallout (Underwood and Bachman, 1982). Hemipelagic muds are not the only products possible, however. In many cases, the muds are interspersed with layers of wind-blown volcanic ash derived from the adjacent magmatic arc (e.g., Lewis, 1980); rare interbeds of thin sand also can originate from thick turbidity currents that spill over the banks of large channels and submarine canyons. Sand deposition is particularly favored where gradients flatten or pathways for turbidity flows are blocked or deflected by bathymetric obstructions. Flat-floored basins that develop behind tectonic ridges, therefore, serve as the most favorable sites for the selective concentration of coarse-grained mass-flow deposits (Moore and Karig, 1976; Underwood and Bachman, 1982). Finally, if vigorous bottom-water circulation prevails (e.g., where bottom currents funnel through a bathymetric sill in the trench-slope break), the slope sediments can be reworked into drifts of contourites (Reed et al., 1987).

Examples of Modern Trench-Slope Basins and Slope Aprons

Sunda Fore-arc

Perhaps the most detailed geophysical survey of modern trench-slope basins was completed by Stevens and Moore (1985) in the western Sunda forearc, near Nias Island (Fig. 5.1). We consider their results to be representative of slope-basin evolution in general, even though the sampled record of lithofacies relations is poor. The total sediment thickness of the basin fill reaches values in excess of 1,000 m (Fig. 5.9). Juvenile basins near the toe of the Sunda accretionary prism are narrow and contain less sediment than do

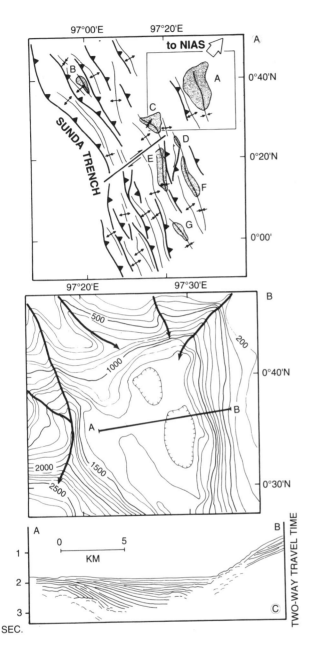

Fig. 5.9 (A) Regional structural map of the lower trench slope seaward of Nias Island, Sunda Trench, northeastern Indian Ocean. Ridges are identified by diverging arrows, troughs by converging arrows, thrust faults by heavy barbed lines, and strike-slip cross faults by opposing arrows. Letters are used to identify specific trench-slope basins (highlighted by stippled pattern). (B) Bathymetric map (100–meter contour interval) showing the details of basin A on the lower trench slope. Location of track-line A–B for seismic-reflection profile is shown by heavy line. Submarine canyons are highlighted by arrows. (C) Line drawing of single-channel seismic-reflection profile from basin A on the lower trench slope. Hummocky zone near the landward edge of the basin probably is due to submarine sliding of the slope apron. From Stevens and Moore (1985).

older and broader basins located higher on the slope (Karig et al., 1980a,b). This is largely because the structural depressions fill, and the bounding faults and anticlines become less active with increasing distance from the deformation front. The effects of landward tilting and out-of-sequence faulting also are more pronounced toward the magmatic arc.

The Sunda convergent margin contains a partly subaerial outer-arc high (which includes Nias Island), and there is a large forearc basin between the magmatic arc and the trench-slope break (Karig et al., 1979; Beaudry and Moore, 1985). Slope basins in close proximity to the trench-slope break are connected directly to sources of terrigenous turbidites through a network of small submarine canyons that head in shallow water (Fig. 5.9B). In one such basin, a small fan-like sediment body (without distributary channels) emanates from the mouth of a feeder canyon. Local slumping of hemipelagic debris off the adjacent ridges provides a secondary component of the basinal sediment budget (Fig. 5.9C).

Cascadia

The southern Cascadia continental slope (offshore Washington and Oregon) displays the same type of classic ridge-and-trough architecture as the Nias region of Indonesia (Fig. 5.10). Most of the linear ridges are anticlinal and/or cored by thrust faults, and these structural features formed as a direct consequence of sediment offscraping at the subduction front (Kulm and Fowler, 1974b; Carson, 1977; MacKay et al., 1992). Ten major submarine canyons, plus countless smaller channels and gullies, are superposed upon this rugged accretionary bathymetry (Fig. 5.10), and there is a comparable system of canyons and channels off Vancouver Island (Davis and Hyndman, 1989). Only Quinault Canyon, however, provides unequivocal evidence of extensive late postglacial activity; supporting evidence comes from numerous layers of overbank sand in piston cores recovered close to the canyon rim (Barnard, 1978), as well as comprehensive studies of suspensions within the water column and sediment accumulation on the canyon floor (Carson et al., 1986, and related papers). Other canyons, in contrast, appear to be filling up with lutite turbidites and hemipelagic mud. This inactivity is a result of the present highstand of sea

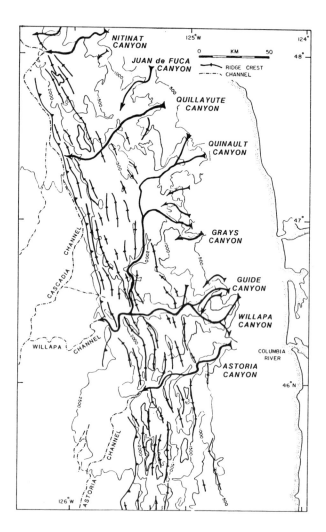

Fig. 5.10 Simplified map of bathymetry (in meters) and structure for the landward trench slope and Cascadia Basin offshore Washington and northern Oregon. Submarine canyons are highlighted by arrows. Stippled pattern corresponds to well defined slope basins. Based on interpreted GLORIA side-scan sonar data (EEZ SCAN 84 Scientific Staff, 1986).

level and the positions of the canyon heads; most of the canyons terminate at mid-shelf positions rather than near the present shoreline, thereby separating them from littoral supplies of sand.

Most of the near-surface sediments on the lower slope of the Cascadia margin are dominated by hemipelagic mud (Barnard, 1978; Kulm and Scheidegger, 1979; Krissek, 1984); likewise, hemipelagic muds (together with rare thin sand/silt turbidites, pebbly lutites, and pebbly sands) typify trench-slope

basins. DSDP Hole 175 (offshore Oregon) completely penetrated one such basin at a position just landward of the frontal ridge (Kulm et al., 1973), but this site appears to be isolated from direct sediment influx through submarine canyons. Late Quaternary sedimentation rates for the upper part of the mud-dominated basin fill have been no higher than 240 m/my, which is close to the average accumulation rate for fine-grained deposits all along the Oregon-Washington slope (Krissek, 1984). The position of the unconformable contact between the basin fill and uplifted trench deposits was not clearly defined by drilling, but it probably occurs at a sub-bottom depth of about 100 to 120 m, based on analyses of microfossils and mineral assemblages (von Huene and Kulm, 1973; Kulm and Fowler, 1974b). Piston cores from basins higher on the slope contain abundant near-surface turbidite sands (Kulm and Scheidegger, 1979); presumably, the long-term influx of coarse sediment increases upslope.

Eastern Aleutian Trench
Another trench-slope basin was drilled during DSDP Leg 18 on the lower slope of the eastern Aleutian Trench (Site 181). This basin, which formed behind the frontal ridge of the accretionary prism, is connected directly to the shelf edge by one small submarine canyon (Underwood and Norville, 1986). Many similar basins occur within the region (von Huene, 1979; Lewis et al., 1988). The average rate of late Quaternary sedimentation at Site 181 has been at least 420 m/my (von Huene and Kulm, 1973), which is considerably higher than the rate of background sedimentation on the inclined slope (Table 5.1). In addition, the basin fill contains a much larger component of turbidite sand than does the Aleutian slope apron or the Site 175 basin offshore Oregon. Based on changes in physical properties and an obvious increase in the degree of mesoscopic deformation, the basal unconformity of the slope basin evidently is located at a subbottom depth of about 170 m (Kulm et al., 1973). Petrographic analyses of the silt and sand fractions of the turbidite deposits show considerable variations in detrital modes (Slatt and Piper, 1974; Underwood and Norville, 1986), so it is doubtful that the single feeder canyon provided a

point source for the basin fill. Instead, unconfined turbidity currents triggered in different source areas contributed to the local sediment budget (Underwood and Norville, 1986).

Middle America Trench

As mentioned above, some accretionary margins do not contain discrete basins on the lower slope. As an example of this, the lower slope of the Middle America Trench (offshore Mexico, Guatemala [Fig. 5.5C], and Costa Rica [Fig. 5.4C]) displays a fairly uniform seaward gradient with only a few subdued benches or breaks in slope (von Huene et al., 1980; Aubouin et al., 1982a; J.C. Moore et al., 1982a; Shipley et al., 1982, 1990). This gentle relief is probably a consequence of relatively small vertical offsets along individual faults. Gravity and piston cores from the Mexican and Guatemalan slope aprons consist almost exclusively of laminated hemipelagic mud (Ross, 1971; McMillen et al., 1982). Thin beds and/or laminations of sand are common only in close proximity to large canyons (e.g., Ometepec Canyon), and these sandy horizons undoubtedly originated from canyon spill-over events (Fig. 5.11). Layers of volcanic ash also are common as interbeds throughout the slope apron (Aubouin et al., 1982b; Watkins et al., 1982; von Huene et al., 1985b). Variations in the abundance of ash probably are related to fluctuations in the explosive activity of volcanoes in the arcs of Mexico and Central America.

Drill cores recovered during DSDP Legs 66 and 67 demonstrate that the muddy slope apron is actually quite thick (up to several-hundred meters), particularly in the vicinity of Ometepec Canyon, offshore Mexico (J.C. Moore et al., 1982a,c) and near San Jose Canyon, offshore Guatemala (von Huene et al., 1980). Quaternary sedimentation rates on steeper slope segments are consistently less than 100 m/my in the northern Middle America Trench, whereas rates increase to 360 m/my on flatter benches such as at Site 488 (Table 5.1). Rates of Quaternary sedimentation on the Guatemala slope are 30 to 140 m/my (Table 5.1). Appreciable quantities of coarse-grained slope deposits were recovered only at Site 570, which is located at the seaward margin of a small upper-slope basin off Guatemala (Baltuck et al., 1985). This site confirms

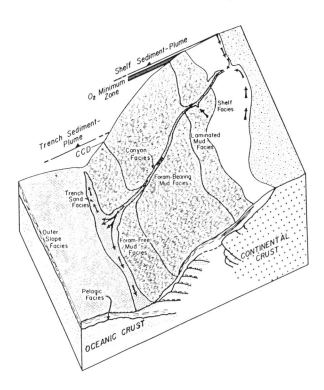

Fig. 5.11 Sedimentary facies model for the northern Middle America Trench, offshore Mexico, showing the influence of Ometepec Canyon on forearc bypassing, the formation of a trench fan at the canyon mouth, and the development of a plume of suspended sediment rising several hundred meters above the trench floor. Foram-free mud facies corresponds to water depths below the calcite compensation depth (CCD). Laminated mud facies (minimal bioturbation) corresponds to water depths of the oxygen-minimum zone. (Modified from J.C. Moore et al., 1982a.)

the idea that local structural depressions trap sandy detritus virtually anywhere on the trench slope.

Because of an extensive sampling program, it is possible to subdivide the Mexican slope apron into three depth-dependent sub-facies on the basis of foraminifera and color laminations (Fig. 5.11). The foram-free mud facies of the lowest slope contains abundant thin interbeds of graded silt and graded mud turbidites; these sediments were deposited from trench-floor plumes that routinely rise several-hundred meters above the seafloor, as turbidity currents enter and traverse the trench floor (J.C. Moore et al., 1982c). The depth of the CCD (calcite compensation depth) controls the upper limit of the foram-free mud facies. A third factor to influence lithologic character is the degree of bioturbation, which is greatly inhibited within the oxygen-minimum zone due to a

reduction in benthic faunal activity. The mud facies with preserved laminae is restricted to water depths within the oxygen-minimum zone (Fig. 5.11). Between these two limits imposed by oceanographic conditions, the slope muds contain abundant forams and are homogenized by vigorous bioturbation.

Structurally below the lower and middle segments of the slope apron, there is a sandy turbidite facies (Fig. 5.11). The age of these deposits increases from Quaternary to middle-late Miocene as a function of distance from the present trench axis. This is the expected stratal-age pattern for accreted trench sands within an imbricate-thrust system. The tectonic-accretion interpretation is supported by three additional lines of evidence: (1) there is a spatial coincidence between the sandy strata and a zone of landward dipping seismic reflectors; (2) benthic fossils show that the abyssal turbidite deposits and abyssal-to-bathyal slope muds have been uplifted progressively, such that water depths inferred for deposition at each site decrease up-section; and (3) there are close lithofacies and petrofacies matches between the older subsurface sands and the modern trench wedge (J.C. Moore et al., 1982a).

Barbados Ridge
The Barbados Ridge accretionary prism is mantled by a relatively thin apron of pelagic and hemipelagic deposits. The southern edge of the subduction margin is characterized by high rates of sediment influx from South America and a thick abyssal-plain section (Westbrook, 1982; Brown and Westbrook, 1987). Frontal accretion in this region has led to the formation of several trench-slope basins containing sediment up to 650–700 m thick (Mascle et al., 1990). Farther to the north, where rates of sediment influx are much lower, the slope apron displays subdued relief; drilling results (J.C. Moore et al., 1982b, 1988) and conventional piston/gravity coring (Cleary et al., 1984; Wright, 1984) both show that the slope sediments are dominated by bioturbated marly calcareous ooze. Superposed on the background of pelagic mud are turbidites composed of arc-derived calcareous and volcanic sand (particularly common in small bathymetric depressions), lutite turbidites remobilized from higher positions on the landward slope, and volcanic

ash transported from explosive centers along the Lesser Antilles Arc (Sigurdsson et al., 1980). Although eolian transport and air fall are the most likely mechanisms responsible for ash emplacement, submarine observations along the western flank of the arc platform suggest that some of the ash deposits may have originated as subaerial pyroclastic flows which entered the ocean and continued to move downslope as turbidity currents (Sparks et al.,1980a,b).

Japan Trench
In comparison to the examples cited above, the lower slope of the Japan Trench is starved of siliciclastic sediment (Table 5.1). This is largely because the terrigenous influx from northern Honshu is low, especially if compared to a major fluvial delivery system such as the Columbia River, or the glaciated margin of southern Alaska. In addition, large volumes of sediment are either transported into the Sea of Japan (backarc basin) or trapped in shallow-water forearc basins (Arthur et al., 1980; Boggs, 1984). DSDP boreholes and conventional piston/gravity cores show that typical rates of Quaternary deposition on the landward slope have been between about 5 and 50 m/my (Arthur et al., 1980; Boggs, 1984); these rates are comparable to rates for the steepest depositional surfaces of the Middle America Trench slope (Table 5.1). The upper slope of the Japan Trench contains several channels, but there are no deeply incised submarine canyons (Arthur et al., 1980). The largest erosional feature, Hidaka Channel, appears to be the sole connection between the shelf edge and the trench.

Most of the trench slope is dominated by diatom-rich hemipelagic muds. Higher rates of accumulation, plus sporadic sand and gravel deposits, are common only on the broad upper-slope terrace (i.e., the forearc basin), as well as some of the smaller mid-slope terraces (e.g., where DSDP Site 440 is located). Rare patches of muddy sand and gravel elsewhere on the slope may be related to debris flows (Boggs,1984) and/or ice-rafting (von Huene and Arthur, 1982). DSDP cores contain several biostratigraphic inversions, and seismic-reflection data (Fig. 5.5) show that the near-surface muds and underlying consolidated mudrocks are particularly prone to gravitational failure and resedimentation by large-scale mass movements

(Arthur et al., 1980; von Huene et al., 1982; Cadet et al., 1987; von Huene and Culotta, 1989). In addition, benthic foraminiferal stratigraphy shows that the Japan forearc probably has been a site of subsidence and tectonic erosion since late Oligocene time, and the arcward retreat of the trench has been punctuated by the effects of subducting numerous seamounts (Lallemand and Le Pichon, 1987; von Huene and Culotta, 1989; von Huene and Lallemand, 1990).

Summary of Trench-Slope Facies

Figure 5.12 summarizes the spectrum of possible facies relations within the lower trench-slope environment; this model is most appropriate for continental-margin systems with relatively high rates of sediment influx (Underwood and Bachman, 1982). Background sedimentation on the slope and within slope basins is dominated by slow and steady hemipelagic fallout, together with sporadic high-energy events involving remobilization of the muds by submarine slides, debris flows, and downslope creep. Deposition of wind-blown volcanic ash on the lower slope is likewise possible. In theory, if a given slope basin is linked directly to suitable sources of focused terrigenous influx (i.e., "mature" basins), it could serve as a growth site for a small submarine fan, complete with distributary channels and channel-mouth lobes. We note, however, that true submarine fans, with all of the expected morphologic components described by Mutti and Normark (1987) and Pickering et al. (1989), have not been documented within these environments. Instead, seismic character suggests that most slope basins contain sheet-like turbidite ramps. This pattern of deposition probably is influenced by either or both of two factors: (1) transport may be dominated by unrestricted (rather than channelized) turbidity currents; and (2) basin geometries are severely restricted by structural architecture, thereby preventing the development of all fan-facies components.

Sediment gravity flows tend to be deflected around or reflected off bathymetric highs, and because of this, some isolated troughs lack a significant influx of terrigenous sands, particularly on the lowermost slope. These immature basins fill largely with hemipelagic muds, lutite turbidites, and muddy submarine-slide deposits derived from the flanks of adjacent ridges. With an immature example, it may be quite difficult to discriminate between the slope-basin

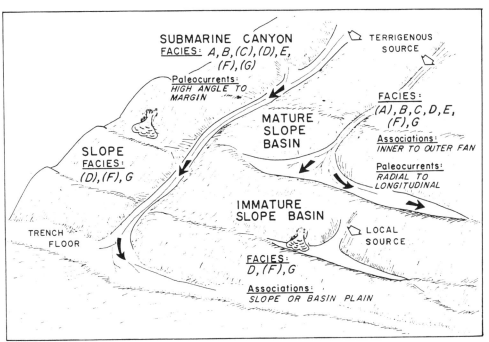

Fig. 5.12 General model of sedimentary facies relations within trench-slope environments. See Table 5.2 for brief descriptions of turbidite lithofacies (letters), as defined by Mutti and Ricci Lucchi (1978); lithofacies associations for typical subenvironments of a submarine fan were likewise described by Mutti and Ricci Lucchi (1978). "Immature" slope basin refers to a depositional site that is not linked to terrigenous sources via submarine canyons. Siliciclastic influx is higher in "mature" slope basins with channel connections to the shoreline or arc platform. From Underwood and Bachman (1982).

deposits and the background hemipelagic facies of the slope apron using either lithofacies criteria or sedimentation rates. Conversely, the sedimentologic connection between a basin and a terrigenous sand source becomes more efficient if submarine canyons and smaller slope gullies establish and maintain active conduits to bypass upslope obstructions. Rates of sedimentation increase by a factor of ten or more when a consistent connection is maintained between a slope basin and the edge of the continental shelf or arc platform, and this condition becomes more likely as the basin is uplifted. Although canyons are important in this overall scheme of slope sedimentation, the possibility also exists for transport of sandy sediment into basins via unconfined turbidity currents.

Inferred Analogs in the Rock Record

Slope-apron and slope-basin deposits have been identified in a wide variety of fossil subduction zones. Prominent examples for which facies analysis has been attempted include: Nias Island in Indonesia (Moore et al., 1980); the Makara basin, North Island of New Zealand (van der Lingen and Pettinga, 1980); parts of the Sitkalidak Formation, Kodiak Islands of Alaska (Nilsen et al., 1979; Moore and Allwardt, 1980); the Whatarangi Formation, Torlesse terrane of New Zealand (George, 1992); the Cambria slab and Point San Luis slab, Franciscan Complex of central California (Smith et al., 1979); the Yager terrane and eastern unit of the Franciscan Coastal Belt, northern California (Bachman, 1982; Underwood, 1985); the Shijujiyama Formation, which is structurally associated with the Tertiary Shimanto Belt of southwest Japan (Hibbard and Karig, 1990; Hibbard et al., 1992); and several Neogene examples from the Japanese island of Hokkaido (Klein et al., 1979).

One common observation that has been used to discriminate between the deposits cited above and associated rocks of the inferred coeval accretionary prism is the style of deformation. The slope-basin deposits typically display tight asymmetric folds, mesoscale thrust faults, and increasing amounts of shear toward contacts with adjacent tectonostratigraphic units. Structural disruption is less severe in the slope-apron and slope-basin strata as compared to

strongly deformed units of mélange (e.g., Moore and Allwardt, 1980; Moore and Karig, 1980; Bachman, 1982). This criterion, however, is by no means unique to slope basins, as strain partitioning can occur within deeper levels of a subduction complex. For example, if deeply buried and strongly lithified abyssal-plain turbidites enter a subduction zone and become underplated by duplexing, then the degree of internal deformation between the roof and floor thrusts may be minimal, and the facies characteristics might be indistinguishable from those of a slope-basin succession. Another criterion is thermal maturity, as measured, for example, by vitrinite reflectance or metamorphic mineral assemblages. Ideally, the slope-basin deposits should maintain ranks or grades of thermal alteration lower than those of the underlying accretionary basement, and in some cases, this type of pattern has been, at least partially, preserved (e.g., Underwood and Howell, 1987; Underwood et al., 1988). However, in many instances, there is no clear difference in thermal maturity between highly deformed and mildly deformed subunits of a subduction complex (e.g., Moore and Allwardt, 1980). One possible reason for this is the progressive uplift of tectonic ridges, combined with deep burial of the basal slope-basin deposits. In addition, subsequent structural modification of contact relations and/or late-stage increases in regional geothermal gradients can destroy the primary thermal structure of the subduction zone (Underwood et al., 1992). Heat flow can increase dramatically through time because of progressive decreases in the age of subducting lithosphere (Underwood, 1989), or in extreme cases, interaction between an active (or recently extinct) spreading ridge and the accretionary prism (Cande et al., 1987). Perhaps the most reliable evidence for slope-basin deposition is a fossil assemblage indicative of intermediate water depths, progressive uplift, and shoaling (Moore et al., 1980a,b).

The lithofacies patterns displayed by ancient slope-basin analogs are quite variable, and we have been unable to recognize any features common to all. Many ancient sequences (e.g., the Cambria slab) contain impressive deposits of massive and thick-bedded sandstone, with structural thicknesses over 4,000 m (Smith et al., 1979). It is important to note, however,

that this type of facies association has never been recovered from a modern trench-slope basin because of technological problems associated with the recovery of thick, unconsolidated sand. Virtually all of the subaerial sections can be described within the organizational framework of turbidite lithofacies classifications (Table 5.2) and fan-related vertical cycles and facies associations (e.g., Mutti and Ricci Lucchi, 1978; Mutti and Normark, 1987; Pickering et al., 1989). Individual measured sections of turbidites can be explained in terms of lateral migration of distributary channels and levees, for example, if they display upward thinning and fining cycles, whereas progradation of depositional lobes at the channel mouths are thought to produce upward thickening and coarsening cycles. Many vertical sections in turbidites,

however, are noncyclic or ambiguous (interchannel, lobe-fringe, etc.), and for these stratigraphic intervals to be interpreted properly, they must be placed within the context of coherent three-dimensional regional-scale facies changes. Because of the polyphase folding and faulting that characterize subduction complexes, typically coupled with poor three-dimensional exposure, it is seldom feasible to correlate among individual isolated rock exposures with any degree of confidence. Nevertheless, mapping the spatial distribution of dominant lithofacies or facies associations may document crude temporal changes caused by movement of regional-scale depositional subenvironments, such as channel, interchannel, and channel-mouth lobe (Moore et al., 1980a,b; Underwood, 1985; George, 1992).

Table 5.2. Summary of Lithofacies Characteristics for Turbidites and Related Deep-Marine Deposits, as Defined by Mutti and Ricci Lucchi (1978)

Facies A	*Description:*	coarse sandstone and conglomerate
		irregular cut-and-fill structures
		size grading and shale rip-up clasts
	Interpretation:	grain flows and high-concentration turbidity currents
Facies B	*Description:*	thick sandstone, coarse to medium-fine
		lenticular beds
		parallel or broadly undulating current laminae
		dewatering features
	Interpretation:	grain flows and high-concentration turbidity currents
Facies C	*Description:*	thick-bedded, medium to fine sandstone; minor shale
		good lateral continuity, low-relief channels
		sandstone: shale ratio > 1
		rip-up clasts, complete Bouma sequences
	Interpretation:	classical turbidity currents
Facies D	*Description:*	thin-bedded, fine-grained sandstone; shale interbeds
		marked lateral continuity
		sandstone: shale ratio < 1
		base-missing Bouma sequences
	Interpretation:	dilute, low-density turbidity currents
Facies E	*Description:*	thin sandstone and shale interbeds
		discontinuous beds with wedging and lensing
		ripple marks, cross-laminae, sharp tops and bases
		sandstone: shale ratio > 1
	Interpretation:	overbank deposition on levees, interchannel
Facies F	*Description:*	chaotic deformation, soft-sediment folds
		blocks in fine-grained matrix
	Interpretation:	submarine slides, mudflows, debris flows
Facies G	*Description:*	mudstone, shale, marl
		indistinct to even, parallel bedding
	Interpretation:	hemipelagic and pelagic deposition

Influence of Submarine Canyons in Forearc Bypassing

Canyon Activity

We have alluded to the fact that submarine canyons are important agents of sediment bypassing in the depositional environments of the landward trench slope (Fig. 5.13). Several transport mechanisms must be considered. The evolution of any particular canyon depends on several variables (Shepard, 1981; Pickering et al., 1989), but the most important factor is probably the spatial separation between the canyon head and the continental shoreline, or in the case of an intraoceanic system, the platform of arc volcanoes (see Chapters 6 and 7). Today, many submarine canyons head in water depths corresponding to the middle or outer parts of continental shelves (Shepard et al., 1979); these canyons have been stranded with respect to their primary sources of fluvial and littoral sediment because of Holocene rise in sea level. Other canyons, however, particularly those along active continental margins with high coastal relief and rapid rates of uplift, begin very close to the modern shoreline. As just one such example, Rio Balsas Canyon in the northern Middle America forearc remains active even under highstand sea-level conditions because the canyon head is positioned right at the shoreline

to intercept rip currents and cells of longshore drift (Reimnitz, 1971); in addition, slumps, grain-flows, and self-accelerating turbidity currents can be triggered in shallow water on a fairly regular basis by storm waves (Inman et al., 1976; Marshall, 1978; Fukushima et al., 1985).

As background to the occasional high-energy downslope surges (mostly via turbidity current), there seems to be almost continuous activity in canyons involving low-velocity currents that oscillate up and down the canyon axes (Shepard et al., 1979). These currents typically exceed threshold velocities required for transport of fine sand on a flat substrate (i.e., greater than 20 cm/sec), and they undoubtedly result in a net transfer of sand and mud down the canyons. Shoaling internal waves (i.e., gravity waves propagating along the pycnocline) are also capable of generating bedforms and suspensions of sediment in canyons (Karl et al., 1986), thereby providing another triggering mechanism for downslope mass movement. Finally, it is well known that turbidity currents can be triggered by earthquakes (e.g., Morgenstern, 1967; Piper et al., 1985), and seismic events are frequent along subduction margins. In summary, sediment transport down submarine canyons can be initiated by several common processes, particularly in shallow water.

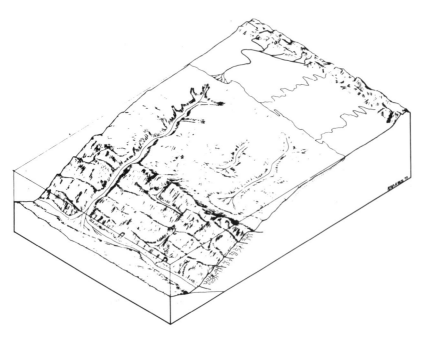

Fig. 5.13 Perspective diagram showing large submarine canyons as agents of sedimentary bypassing in accretionary forearc terranes. From Underwood and Karig (1980).

Canyon Evolution

Structural controls are very important in the evolution of forearc canyons. Major faults within the accretionary prism or the arc basement commonly provide the nucleation points for large canyons (Shepard, 1981), and structural features on the subducting plate appear to influence the canyon debouchment points as they enter trench-floor basins (Thornburg et al., 1990). Most major river-fed submarine canyons (e.g., Columbia River linked to Astoria Canyon) head on the continental shelves and erode their way downslope with successive high-energy turbidity-current events. However, many canyons nucleate in deeper water and extend themselves in both directions. Canyon growth is accomplished through the combined effects of headward erosion by mass failure and down-cutting by previously unconfined turbidity flows once they are captured and focused through evolving depressions (Farre et al., 1983; Klaus and Taylor, 1991). Recent investigations of the Cascadia margin suggest that dewatering of the accretionary prism plays a critical role in the birth of new canyons near the deformation front (Orange and Breen, 1992). According to these workers, seepage forces act on the solid matrix of a saturated sediment in the direction of fluid flow, and the magnitude of this force is proportional to the hydraulic gradient. Seepage forces increase along zones of fluid venting, thereby causing local reductions in sediment shear strength; lowered shear strength triggers slope failures and headless submarine canyons are born. Once an initial excavation has taken place via submarine sliding, the erosive power of subsequent turbidity currents will be deflected toward and funneled into the axis of a juvenile depression. The over-steepened head region, moreover, will continue to be a favored site for additional submarine slides. Headward erosion, therefore, provides a mechanism for breaching bathymetric obstructions that lie between the arc and the trench.

Facies Partitioning

Along most Andean-type subduction margins, there is a dramatic textural contrast between the hemipelagic muds that typify all steeply inclined trench slopes (i.e., slope aprons) and sandy trench-floor deposits (Underwood and Karig, 1980). This sedimentologic paradigm can be explained by looking at the spatial distribution of and activity within submarine canyons. Most Quaternary trench-wedge sands have been funneled across their respective forearcs as high-density turbidity currents confined within large, through-going submarine canyons (Fig. 5.13). As outlined above, comparatively minor amounts of Quaternary sand have been recovered from some trench-slope environments, but nearly all of those deposits either were trapped within slope basins (i.e., where sea-floor gradients flatten and/or smaller canyons and slope gullies are blocked by ridges and diapirs) or they resulted from dilute overbank flows that escaped the confines of nearby canyons. The overall pattern of sand dispersal along subduction margins includes two distinct components. First, efficient bypassing of forearc basins and the lower trench slope occurs via large, through-going submarine canyons. Second, local upslope trapping also takes place, especially behind the trench-slope break (i.e., in the forearc basins; see Chapter 6), but also within the smaller basins perched on the lower slope (Fig. 5.13).

Modern Trench-Floor Deposits

Once coarse-grained siliciclastic debris reaches a trench floor, a wide variety of depositional geometries can develop (Schweller and Kulm, 1978b; Underwood and Bachman, 1982). Figure 5.14 illustrates four members within a generalized spectrum of possible facies relations: (1) trench fans; (2) axial channel-levee complexes; (3) sheet-flow (non-channelized) turbidites; and (4) starved trenches. All of these facies associations can be described within the framework of turbidite lithofacies classifications (Table 5.2) and models of submarine-fan deposition (e.g., Mutti and Ricci Lucchi, 1978; Mutti and Normark, 1987; Pickering et al., 1989). Unlike a typical submarine fan, however, the three-dimensional continuity of a trench wedge depends on the interplay that develops among rates of sediment accumulation, distances of axial transport, rates of plate convergence, and the inherited basement relief of the subducting oceanic plate. All of these factors can change considerably

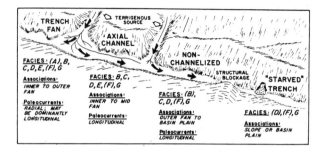

Fig. 5.14 General model of sedimentary facies relations within trench-floor environments, with turbidite lithofacies and submarine-fan facies associations as defined by Mutti and Ricci Lucchi (1972, 1978). From Underwood and Bachman (1982).

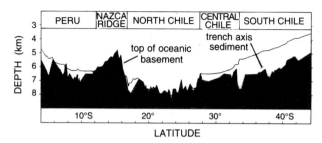

Fig. 5.15 Axial profile of the Peru-Chile Trench, eastern Pacific Ocean. Trench–axis sediment is highlighted by stippled pattern. Note the bathymetric blockage formed at intersection of trench and Nazca Ridge, as well as the dramatic decrease in wedge thickness toward the north away from the glaciated coastline of Chile. Modified from Schweller et al. (1981).

along strike; they can also vary through time within any given corridor of a subduction margin. Rates of terrigenous influx are particularly sensitive to longitudinal zonation in climate and spacing of major fluvial drainage systems.

Modern Trench-Wedge Examples

Chile Trench

The plan-view depositional morphology of the Chile Trench is actually quite intricate and variable. Many segments of the Peru-Chile trench north of about 31°S latitude are nearly devoid of sediment because of extremely arid climatic conditions in the adjacent continent, and the thin ponds of trench sediment are separated from one another by structural highs inherited from the subducting oceanic basement (Fig. 5.15). As an extreme case in point, axial transport in the Peru-Chile Trench is blocked completely by the Nazca Ridge. In contrast, the trench wedge is a thick and continuous feature south of 31°S latitude, where cool-humid conditions and extensive glaciation have provided abundant sediment throughout the Quaternary (Schweller et al., 1981). The southernmost segment shows no evidence of channelized flow, but individual turbidite sheets extend for at least 30 km across the width of the trench (Thornburg and Kulm, 1987a). A prominent axial channel begins near latitude 41°S (Fig. 5.16); this impressive channel trends down the axial gradient with apparent continuity to beyond 33°S, a dis-

tance of roughly 950 km (Schweller and Kulm, 1978b; Schweller et al., 1981; Thornburg and Kulm, 1987a; Thornburg et al., 1990). The landward trench slope between latitudes 41°S and 32°S is incised by several major submarine canyons, and small lobate fans have been constructed at the mouths of these canyons (Figure 5.16). Petrographic data demonstrate that each trench fan is linked directly to a specific fluvial drainage basin through its kindred submarine canyon (Thornburg and Kulm, 1987b).

Facies relations change abruptly where the canyons debouch onto the floor of the trench, for several reasons. First, the surficial morphology of the trench floor responds to the physical interaction between the through-going axial channel and the smaller distributary channels belonging to trench fans. A second factor is the difference in behavior between up-gradient and down-gradient flows once turbidity currents escape from the canyon mouths (Thornburg and Kulm, 1987a). As a flow moves up gradient, it tends to generate constructional morphology during deceleration, whereas down-gradient currents maintain velocities high enough to remobilize sediment and erode gullies. Basement relief and syndepositional faults also influence patterns of sediment dispersal and depositional morphology (Thornburg et al., 1990).

Northern Middle America Trench

The landward slope of the northern Middle America Trench likewise is incised by numerous submarine

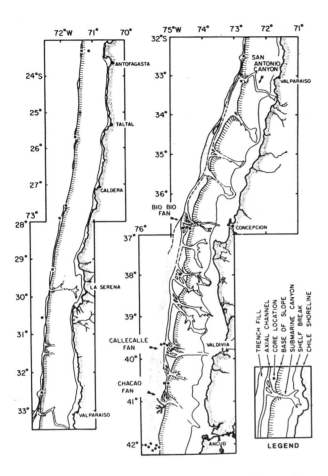

Fig. 5.16 Map showing major submarine canyons, trench fans, and axial channel of the Chile Trench. From Thornburg et al. (1990).

canyons (Underwood and Karig, 1980). There is a tendency for smaller canyons and slope gullies to coalesce downslope and debouch into mid-slope (fore-arc) basins (Karig et al., 1978). Only the largest canyons (Rio Ameca, Manzanillo, Rio Balsas, and Ometepec) have eroded headward to the shoreline and extend downslope to the base of the accretionary prism. Direct measurements from current meters placed in Rio Balsas Canyon prove that turbidity-current activity has been maintained during the Holocene highstand of sea level (Reimnitz, 1971; Shepard et al., 1979). There is no evidence for a major axial channel in the Middle America Trench. Seismic-reflection records show that the thickness of the trench wedge increases at the canyon mouths, where lobate fans with distributary channels and levees have developed (Underwood and Karig, 1980;

Moore and Shipley, 1988). For example, a radiating system of distributary channels clearly curves away from the Ometepec Canyon mouth in both directions as the fan progrades into the segmented trench-floor basin that was sampled during DSDP Leg 66 (Fig. 5.17). Piston cores and drill cores from the Ometepec trench fan contain thick layers of sand (McMillen et al., 1982; J.C. Moore et al., 1982c). On a regional scale, however, the sandy facies appears to be compartmentalized within several discrete trench-floor basins; each compartment has formed where axial flow paths are blocked by bathymetric highs in the subducting oceanic basement (Underwood and Bachman, 1982). Although the coring record is spotty, sand-layer thickness appears to decrease abruptly away from canyon sources. For example, localities more than about 25 km from Ometepec Canyon contain mostly hemipelagic and/or turbiditic mud (McMillen et al., 1982). Thus, facies continuity within the Middle America Trench is rather limited.

Oregon-Washington Margin
As mentioned above, the continental margin of Oregon and Washington has several impressive submarine canyons (Fig. 5.10). Barnard (1978) calculated the late Pleistocene sediment volume for Cascadia Basin (seaward of the deformation front) and compared that value with the estimated volume for the Washington slope; he concluded that roughly

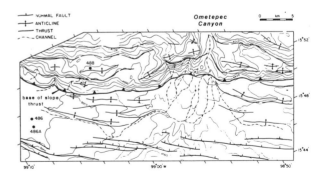

Fig. 5.17 Detailed map of bathymetry (in meters) and structure for the lower slope and trench in the vicinity of Ometepec Canyon, northern Middle America Trench. Numbers refer to DSDP drill sites. Note the coalescing distributary channels that collectively define the trench fan; most of these channels curve away from the canyon mouth in the down–gradient direction. Simplified from Moore and Shipley (1988).

two-thirds of the sediment delivered to the shelf edge bypasses the slope completely through the dominant submarine canyons. Rates of turbidite accumulation within Cascadia Basin have been exceptionally high, particularly during glacial low-stands (Table 5.1). The most important features are the Nitinat and Astoria fans, and the Vancouver and Cascadia channels (Griggs and Kulm, 1970; Carlson and Nelson, 1987; Karl et al., 1989). Smaller lobes of sediment have accumulated at the mouths of several other canyons, particularly off Vancouver Island (Davis and Hyndman, 1989). The basin floor is unusually shallow (2,400 to 3,000 m) because the subducting oceanic crust (generated by sea-floor spreading along the nearby Gorda/Juan de Fuca Ridge system) is young, warm, and buoyant; the rate of plate convergence is also relatively slow. As a consequence, the trench is completely obscured as a bathymetric feature.

Astoria Canyon, which serves as the largest sediment conduit along the Cascadia margin, begins in a mid-shelf position at a water depth of about 100 m (Fig. 5.18). At present, there are no surface manifestations of a connection between the canyon and the

Columbia River mouth, but buried channels are evident on some seismic-reflection profiles (Carlson and Nelson, 1969). The canyon cuts directly through at least half-a-dozen sizeable accretionary ridges (Fig. 5.18). The dimensions of Astoria Canyon are imposing: maximum relief of the walls is 915 m, and the average relief is 550 m; maximum width from rim to rim is 13.3 km, and the maximum floor width is 10.2 km (Carlson and Nelson, 1969). Cores of Holocene sediment recovered from the canyon floor consist mostly of hemipelagic and turbidite mud, together with a few thin interbeds of sandy turbidites (Nelson et al., 1970; Kulm and Scheidegger, 1979). Post-glacial sedimentary activity within the canyon, therefore, appears to be subdued due to separation of the canyon head from the shoreline and the Columbia River estuary.

Cascadia Channel clearly has altered its course by migrating seaward in response to progradation and uplift of the adjacent accretionary prism (Karl et al., 1989). Presumably, the same overall pattern of facies migration has occurred on the leveed fan valleys that radiate from the mouth of Astoria Canyon. Two prominent U-shaped channels dominate the

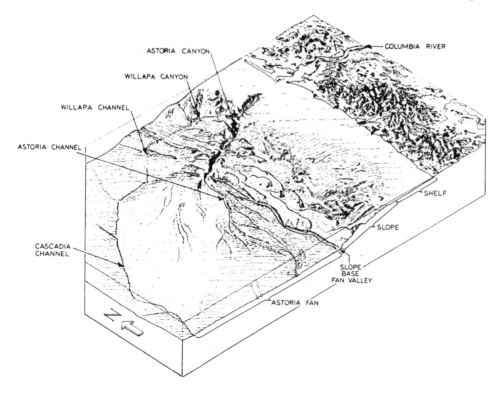

Fig. 5.18 Physiographic diagram of the Astoria submarine canyon and fan complex, offshore Oregon. From Nelson (1976).

upward concave inner-fan surface: the base-of-slope fan valley and Astoria Channel (Fig. 5.18). Other channels appear to be remnants of older, abandoned fan valley systems. The elongate fan is more than 1,000 m thick at its apex. With a surface area of 20,000 km, Astoria represents the second largest submarine fan of any active subduction margin (Barnes and Normark, 1985), exceeded in size only by Magdalena Fan in the Caribbean Sea (Lu and McMillen, 1982; Kolla and Buffler, 1985; Breen, 1989). The two main fan valleys break into distributaries within the middle fan sub-environment, which is characterized by an upward convex morphology and subdued surface gradients. Terrigenous sand and gravel are especially common within the upper- and middle-fan channels, but most of this coarse near-surface debris was deposited during the last glacial lowstand (Nelson, 1976). Petrographic analyses of drill cores from the distal edge of the fan (DSDP Site 174) show that the Columbia River drainage basin has provided the detritus, at least since early Pleistocene time (Kulm and Fowler, 1974b; see also Gergen and Ingersoll, 1986). Astoria Fan has served as one of the classic case studies of deep-marine depositional morphology and lithofacies; Late Quaternary growth patterns seem to have been controlled more by sedimentary parameters (number, volume, and grain size of sediment sources feeding the fan) than tectonic factors associated with the subduction zone (Nelson, 1985).

Japan Trench
The Japan Trench, with its low rate of sedimentation, is at the other end of the spectrum of facies relations (Fig. 5.5). Very few samples have been recovered from the floor of the Japan Trench because its depth (over 7,000 m) is beyond the limits of DSDP/ODP drilling; most seismic profiles, however, show a virtual absence of trench-wedge sediment (Arthur et al., 1980; Matsuzawa et al., 1980; von Huene et al., 1982; Cadet et al., 1987; Lallemand et al., 1989). Piston cores from the trench floor contain mostly pelagic clay, hemipelagic mud, and mud turbidites, plus traces of fine sand (von Huene and Arthur, 1982; Boggs, 1984). A large, irregular sediment mass at the northern end of the trench displaces the trench axis seaward (Fig. 5.5). Von Huene and Arthur (1982)

suggested that much of this sediment body probably moved down Hidaka Channel. However, recent seismic-reflection profiles show truncated reflectors and steep slide scars immediately upslope of the sediment mass, so this feature evidently is a large slump deposit (Cadet et al.,1987; von Huene and Culotta, 1989). Thus, the volumetric flux through Hidaka Channel probably is surpassed by remobilization from the lower slope by mass wasting and hemipelagic fallout. Because the trench wedge is so thin, there is very little sedimentary material available to accrete. The subduction of seamounts, therefore, plays a dominant role in the structural evolution of the forearc by promoting tectonic erosion via direct abrasion, mass wasting from the lower slope, and subduction of submarine-slide deposits (Lallemand et al., 1989; von Huene and Culotta, 1989; Yamazaki and Okamura, 1989; Okamura, 1991; von Huene and Lallemand, 1990).

Barbados Ridge
The Barbados subduction margin actually contains very little in the way of Quaternary trench sediments. Sediments on the abyssal floor of the Atlantic (east of the deformation front) are as old as Cretaceous but reach a cumulative thickness of less than 1,000 m at the latitude of the DSDP/ODP transects (Fig. 5.19). Higher rates of sediment accumulation result in thickening of the total sediment package to approximately 7,000 m toward the coastline of South America (Westbrook, 1982); most of the terrigenous sediment has been transported from the Orinoco and Amazon drainage systems (Embley and Langseth, 1977). Drilling has demonstrated that the Pliocene to lower Pleistocene section, which forms the uppermost seismic unit of the subducting plate, consists of calcareous clay and mud,

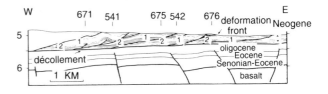

Fig. 5.19 Simplified structural cross section through the toe of the Barbados Ridge accretionary prism. Small numbers (1 and 2) refer to the lower boundaries of seismic units. Large numbers correspond to DSDP and ODP drill sites. Modified from Moore et al. (1988).

marl, and layers of volcanic ash (Moore et al., 1988). Near-surface (Holocene) sediments similarly range from pelagic clay (distal to South America) to pelagic-hemipelagic calcareous mud with increasing proximity to the continent (Wright, 1984). The near-surface calcareous mud overlies a Pleistocene hemipelagic unit that contains interbeds of terrigenous silt and sand, which were deposited as both turbidites and contourites (Damuth, 1977). The activity of turbidity currents emanating from South America evidently has been greater during periods of sea-level lowstand (Wright, 1984).

In spite of the absence of a true Quaternary trench wedge, the Barbados Ridge is a site of active frontal accretion via imbricate thrusting (Biju-Duval et al., 1984; Brown and Westbrook, 1987; Moore et al., 1988; Westbrook et al., 1988). The décollement zone, which was cored successfully at three drilling sites, initiates at a depth of approximately 200 m near the deformation front and deepens progressively to the west beneath the thickening accretionary prism. The décollement propagates through a unit of upper Oligocene to lower Miocene radiolarian-bearing claystone, and this stratigraphic horizon is situated well above the contact with underlying igneous crust (Fig. 5.19). Geochemical, heat-flow, physical-property, and structural data collectively show that enhanced fracture permeability at the base of the prism leads to development of a principal pathway for episodic lateral migration of fluids derived from deep within the accretionary prism (Moore et al., 1988; Fisher and Hounslow, 1990; Foucher et al., 1990; Gieskes et al., 1990).

Nankai Trough

The Japanese margin changes dramatically southwest of the Izu-Bonin Arc, where youthful oceanic crust of the Shikoku Basin is being subducted beneath the Nankai Trough (Fig. 5.20). The depth of this trench floor is less than 5,000 m, and the gradient dips gently to the southwest. Total flux of sediment is not as large as that of the Cascadia Basin, but local accumulation rates are up to 1380 m/my (Table 5.1). The high rate of sediment influx is due to rapid uplift and erosion within the collision zone of the Honshu Arc and the Izu-Bonin Arc (Taira and Niitsuma, 1986; Soh et al., 1991). Tectonic accretion has

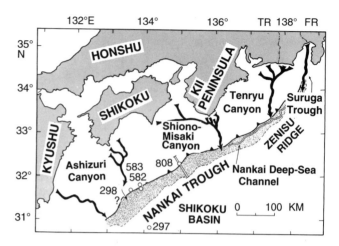

Fig. 5.20 Simplified map of the Nankai Trough and vicinity, modified from Shimamura (1989). Submarine canyons are highlighted by arrows. Important fluvial drainage systems within the source region for trench sediments include the Tenryu River (TR) and the Fuji River (FR). Numbers refer to DSDP/ODP drill sites. See Fig. 5.22 for seismic-reflection profile through Site 808.

generated a familiar ridge-and-trough bathymetry on the landward slope (Le Pichon et al., 1987a,b; Moore et al., 1990). Several prominent forearc basins at mid-slope water depths intercept flows from small canyons and slope gullies (Fig. 5.20). The largest through-going sediment conduit is Suruga Trough, which is a fault-controlled canyon that extends from the shoreline of the collision zone on the west side of Izu Peninsula, along the northern flank of Zenisu Ridge, and into the northeast end of the Nankai Trough (Le Pichon et al., 1987b; Nakamura et al., 1987). A second noteworthy feature is the Nankai deep-sea channel, which has been traced in exceptional detail by Shimamura (1989) from its inception point in Suruga Trough down the axis of Nankai Trough to its termination point offshore the Kii Peninsula (Fig. 5.20). Small fan-like sediment mounds occur at the mouths of Tenryu (Fig. 5.21) and Shiono-Misaki Canyons; the axial channel is displaced seaward around the buildup of Tenryu Fan (Fig. 5.21). It is uncertain whether or not Ashizuri Canyon reaches the base of slope (Fig. 5.20), but a fifth (unnamed) canyon enters the western end of the trench from Kyushu (Fig. 5.20). Seismic-reflection records show that the canyon-supplied trench wedge is typically 450 to 700 m thick (Fig. 5.22); the turbidites overlie 600 to 750 m of hemipelagic strata,

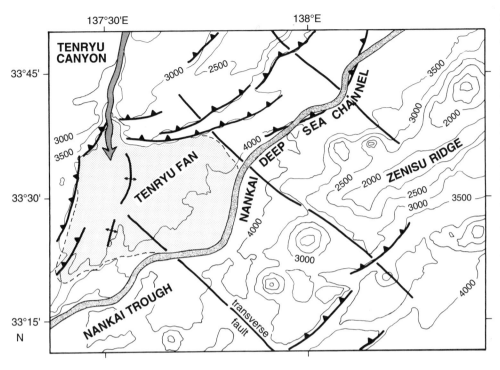

Fig. 5.21 Bathymetric and structural map of the Nankai Trough near its northeast limit (modified from Soh et al.,1991). Zenisu Ridge formed as a consequence of rupture and uplift of the Shikoku Basin oceanic basement due to intraplate compression near the zone of collision between the Honshu Arc and the Izu-Bonin Arc. Important sedimentary features on the floor of Nankai Trough include a prominent axial channel (Nankai deep-sea channel), which emanates from the Suruga Trough (submarine canyon, see Fig. 5.20), and Tenryu Fan, which is supplied by transverse flows through Tenryu Canyon. Note the seaward deflection of the axial channel around the fan. Note also the structural complexity of the collision zone, with thrust faults on both seaward and landward trench slopes, strike-slip transverse faults, and anticlines within the trench.

volcanic ash, and (at Site 297) thin siliciclastic turbidites that were deposited during the Pliocene in the Shikoku Basin (Fig. 5.23).

The stratigraphy of the Nankai Trench wedge (which is preserved in the toe of the accretionary prism) has been documented in great detail by drilling (Fig. 5.23). The stratal succession at DSDP Site 298, for example, is an upward coarsening sequence of accreted trench turbidites and slump deposits, underlain by the thinner, finer-grained,

abyssal-plain deposits inherited from the Shikoku Basin (J.C. Moore and Karig, 1976). The accreted sections at DSDP Site 583 and ODP Site 808 also contain thick successions of turbidites, and core recoveries from Site 582 confirm that the *in situ* axial fill of Nankai Trough comprises more than 500 m of graded sand interbedded with hemipelagic mud (Coulbourn, 1986). Magnetic-fabric analyses show that most of the trench sand at Site 582 was transported by southwest-directed axial flows, and petrographic analyses

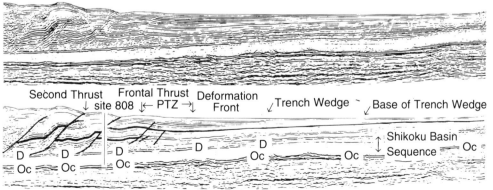

Fig. 5.22 Interpreted multichannel seismic-reflection profile (migrated depth section) through ODP Site 808, central Nankai Trough. See Fig. 5.20 for location. Modified from Moore et al. (1990).

PTZ = Protothrust zone

D = Décollement

Oc = Oceanic crust

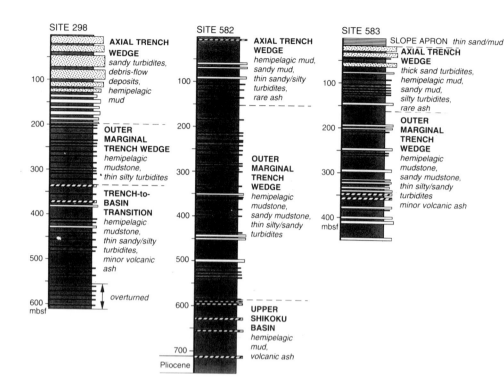

Fig. 5.23 Simplified stratigraphic columns for DSDP and ODP boreholes near Nankai Trough (from Pickering et al., 1993). Note the complete upward coarsening cycle and the position of the decollement within the abyssal-plain (Shikoku Basin) section at Site 808. Compare this borehole stratigraphy with the seismic-reflection profile in Fig. 5.22.

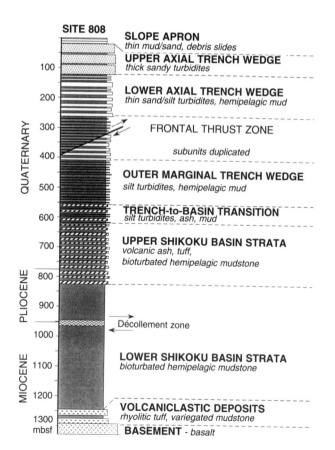

point to a provenance in the Tokai drainage basin, which includes the Tenryu and Fuji Rivers (Fig. 5.20). This principal source area is located in the collision zone north of the Izu Peninsula (Taira and Niitsuma, 1985; Marsaglia et al., 1992). Petrographic data from Site 808 confirm this interpretation (Taira et al., 1991). Thus, most Quaternary trench-floor sediment was funneled through Suruga Trough before streaming down the trench axis for hundreds of kilometers in the low-relief axial channel and as unconfined sheet flows.

Forearc Bypassing via Unconfined Mass Flows

Most of the studies cited above provide convincing evidence in support of the canyon-funneling hypothesis for forearc bypassing (Underwood and Karig, 1980; Underwood, 1991). Not all trenches, however, receive sediment in this manner. The Aleutian Trench, for example, contains a very thick wedge of sediment, yet through-going canyons are present only along its easternmost edge. Glaciation at high latitudes clearly influences this pattern of sediment dispersal. In addition to increasing rates of trench sedimentation (von Huene and Kulm, 1973), glacial

transport to the continental shelf edge provides a more uniform line source rather than a typical sporadic pattern of fluvial point sources. Consequently, there is less likelihood of focused sediment discharge from major submarine canyons at widely spaced sites, and unconfined turbidity currents probably are more common. A similar linear delivery system exists along the glaciated margin of southernmost Chile, where canyon/fan systems are likewise absent on the trench floor (Schweller et al., 1981; Thornburg and Kulm, 1987a). Although supporting evidence is most convincing for examples at high latitude (see below), we suspect that the physical mechanism of unconfined downslope movement is important in the total sediment budget of virtually all arc-trench systems.

Bypassing of the Aleutian Forearc

The absence of large, through-going canyons does not inhibit large volumes of sediment from entering the Aleutian Trench. Parts of the eastern segment of the trench wedge contain an axial channel (Piper et al., 1973; von Huene, 1974), although it is unclear exactly where this channel begins or ends. Drilling at Site 180 failed to recover much coarse-grained sediment (Kulm et al., 1973); this could be due to the site location on the broad seaward levee of the axial channel, but it may be that the entire trench wedge is muddy. The Shumagin segment bordering the Alaska Peninsula displays a steeper axial gradient than the eastern segment, and this tilting of the depositional surface evidently promotes west-directed transport rather than sedimentation (Scholl, 1974). Seismic profiles show that the central trench segment is dominated by sheet-flow turbidites (Fig. 5.24; Scholl et al., 1982a; McCarthy and Scholl, 1985). The maximum thickness of this turbidite wedge (at the intersection with the Amlia fracture zone) is at least 3,500 m (Scholl et al., 1982b), but most segments are roughly 1,500 m thick.

Most workers have concluded that the glaciated margin of mainland Alaska has served as the principal late Quaternary sediment source for the central trench segment, which would require distances of axial transport up to 2,000 km (Scholl, 1974; Scholl et al., 1982a,b). Petrographic analyses of near-surface sediments, however, show that there are fundamental regional variations in Aleutian detrital modes, and the same basic longitudinal variations exist for the landward slope, the trench wedge, and the seaward trench slope (Underwood, 1991). Some interdigitation is evident on the trench floor between polymictic axial-flow deposits and arc-derived volcaniclastic sands, but the trench-sand compositions generally mimic the landward-slope petrofacies within common longitudinal corridors. This is the pattern expected for transverse delivery systems, so the

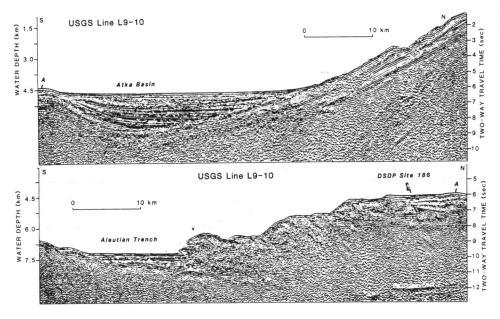

Fig. 5.24 Unmigrated multi-channel seismic profile through Atka Basin and the central Aleutian Trench. See Fig. 5.25 for location of trackline. The two segments of this profile join at point A. Note the position of DSDP Site 186. See Scholl et al. (1987b) for additional interpretation.

record of Holocene deposition fails to support the axial-flow hypothesis. We simply do not know whether or not the Holocene sediments are representative of the entire trench wedge.

The Aleutian arc platform and upper slope are heavily dissected by submarine canyons (Fig. 5.25). All of these erosional features, however, terminate along the Aleutian Terrace, which contains several major forearc basins at mid-slope water depths (Figs. 5.24, 5.25). Transverse turbidity currents obviously begin their trenchward journey in a confined state, but given the absence of through-going submarine canyons, transport from the mid-slope position to the trench floor must occur in an unconfined state (Underwood, 1986b). To reach the trench floor, the transverse flows must move over or around the bathymetric highs associated with the trench-slope break and smaller accretionary ridges of the lower slope.

Seismic-reflection records (e.g., Fig. 5.24) and DSDP drilling at Site 186 show that the Aleutian forearc basins are underlain by a tremendous volume of arc-derived sediment, yet there is no evidence for erosion of channels or aggradation of constructional levees (Marlow et al., 1973; Scholl et al., 1982a, 1987b). In effect, the arc platform and upper slope have served as a line source for the forearc-basin sediment, and the basins are occupied by prograding turbidite ramps rather than discrete submarine fans. South of Atka Basin, the structural high that forms

the trench-slope break has been buried nearly to its crest (Fig. 5.24), so there is very little seafloor relief remaining to impede additional south-directed movement of sediment down the lower slope. Ridge heights in the Shumagin corridor (offshore the Alaska Peninsula) are likewise subdued (Lewis et al., 1988). In other forearc basins, the seafloor relief of the outer structural high is more impressive (e.g., Marlow et al., 1973). Even maximum relief of 500 to 1,000 m, however, appears to be insufficient to block arc-derived turbidity currents completely.

Mechanics of Ridge Bypassing

Bathymetric obstructions can be bypassed by turbidity currents in three ways. First, unconfined turbidity currents can move around a ridge via deflection and escape through the lowest gap in the sill. Second, bypassing can occur because of the momentum retained from downslope acceleration; this kinetic energy will lead to a vertical shift in the flow's center of gravity as the mass encounters a reversal in seafloor gradient. We refer to this phenomenon as *upslope flow* (Fig. 5.26). Finally, the flow thickness simply may be greater than the ridge height. Under these circumstances, *flow stripping*, as defined by Piper and Normark (1983) for overbank flows in channel-levee complexes, allows some of the sediment suspended in the entrained layer to pass over the crest of the barrier (Fig. 5.26). With certain bathymetric

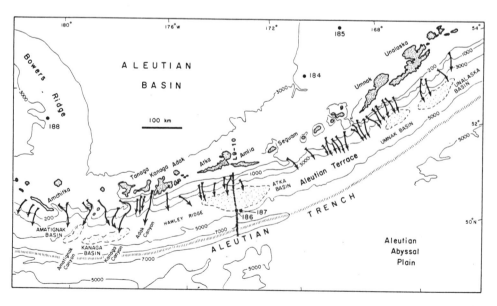

Fig. 5.25 Simplified bathymetric map (in meters) of the central Aleutian forearc region. Note that the upper slope is incised by numerous submarine canyons (arrows). Numbers refer to DSDP drill sites, and major forearc basins are highlighted by stippled pattern. Note the trackline position for seismic profile L910 (Fig. 5.24). Modified from Scholl et al. (1987b).

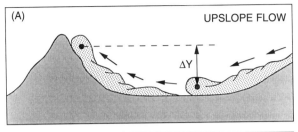

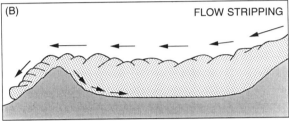

Fig. 5.26 Schematic illustrations of ridge-bypassing mechanisms within a forearc. (A) During upslope flow, the center of gravity of an arc-derived turbidity current shifts up (△Y) when intersected by an opposing slope due to momentum retained from the preceding phase of trenchward downslope acceleration. (B) Flow stripping occurs when obstruction height is less than the thickness of the turbidity current; the entrained layer detaches, continues to move across the ridge crest, and accelerates down the opposite flank. In both cases, flow reflection (back-sloshing) is likely within the basal turbidity current.

configurations, Coriolis and centrifugal effects also can enhance the vertical run-up of a turbidity current (on one side of a basin or large channel) by creating a cross-flow surface slope (Komar, 1969).

Several physical parameters control the magnitude of upslope movement and flow behavior during reflection of density currents (Komar, 1977; Allen, 1985; Pantin and Leeder, 1987; Muck and Underwood, 1990; Kneller et al., 1991). Dilute, muddy suspensions are best suited for upslope transfer, but a finite amount of upslope flow occurs whenever a moving turbidity current encounters a reversal in gradient, as kinetic energy is both transformed into potential energy and lost due to frictional heating. Muck and Underwood (1990) concluded that a reasonable upper limit for the vertical run-up of a natural turbidity current moving orthogonal to an obstruction is equal to 1.54 times the flow thickness. If this limit is exceeded by the height of an obstruction, then most of the mass will slosh back into the adjacent basin, as illustrated by Pickering and Hiscott (1985) and Marjanac (1990) using examples from the

rock record. In many cases, the angle of approach will be oblique to the strike of a ridge, and deflected flows should move back toward the adjacent basin along a line of maximum dip once the upslope phase of movement has been arrested (Kneller et al., 1991). Direct evidence for this trajectory of oblique reflection was obtained by Pickering et al. (1992), through measurements of paleocurrent indicators within the outer trench-wedge facies at ODP Site 808. The dip azimuths of ripple cross-laminae, corrected for core rotation using paleomagnetic data, showed that most of the silty turbidity currents moved back toward the axis of the Nankai Trough during the final stages of flow collapse. In cases where ridge height is lower than the top of the rising turbidity current, the entrained layer moves over the crest, strips itself from the remaining mass, and accelerates down the opposite flank. This mechanism allows sandy and silty turbidity currents to reach lower-slope basins and the trench floor, even in the absence of through-going canyons.

Empirical Evidence for Upslope Deposition

The magnitude of upslope transfer can be constrained further by examining cases in which turbidites have been deposited on bathymetric highs. Empirical evidence for this phenomonon comes from several types of tectonic settings, such as the Mid-Atlantic Ridge (van Andel and Komar, 1969), Ceara Rise, located seaward of the Amazon Fan (Damuth and Embley, 1979), and the Hawaiian Ridge (Moore et al., 1989; Garcia and Hall, 1994). More relevant to this chapter are sites where sands have been carried to elevated positions on the seaward side of a trench. A recent study by Dolan et al. (1989), for example, provides outstanding documentation of epiclastic turbidite sands (derived from South America) that moved 1,500 m upslope onto the Tiburon Rise during the Eocene; Tiburon Rise is now being thrust beneath the northern Barbados trench (Moore et al., 1988). These pulses of upslope sedimentation are remarkable because they followed more than 1,000 km of transport across the Atlantic abyssal floor. Hemipelagic sediment with interlayers of silty terrigenous turbidites have been recovered from the floors of several grabens on the seaward slope of the Peru

Trench as high as 1,000 m above the trench axis (Prince et al., 1974). There are similar sand beds on the outer slope of the Middle America Trench across from Ometepec Canyon, up to 700 m above the trench floor, which have been attributed to a trench sediment plume that routinely rises hundreds of meters above the trench floo (McMillen et al., 1982 ; Moore et al., 1982c). Sediment waves on the outer slope of the Manila Trench are likely due to axial turbidity currents that reached thicknesses of 900 m or more (Damuth, 1979). The Aleutian Abyssal Plain contains rare layers of arc-derived sand at coring sites as high as 2,000 m above and as far as 120 km south of the trench axis (Underwood, 1991).

The examples cited above are important to the central theme of this chapter for two reasons. First, because upslope deposition at elevations of 1000 m or more seems fairly commonplace, we conclude that thick unconfined turbidity currents in the forearc (both upper and lower trench slopes) should be able to bypass small bathymetric obstructions (500 m or less) with relative ease and frequency. Submarine canyons, therefore, are not a prerequisite for bypassing of the forearc. Second, because turbidity currents have the ability to move up the seaward slope of a trench, the facies character of the oceanic plate will be affected. The trench-to-basin transition may encompass 50 to 100 m of the stratigraphic succession entering a subduction zone (Fig. 5.23). Particularly in localities such as the Nankai Trough, where a youthful subducting slab and high rates of sedimentation reduce the relief of the seaward slope, large trench-floor turbidity currents will move tens of kilometers onto the abyssal floor, thereby blurring the lithofacies boundary between the trench wedge and the oceanic plate.

Seaward Trench Slope and Abyssal Plain

Although abyssal-floor deposits of the incoming oceanic plate are generally dominated by pelagic clays and biogenic oozes, other types of deposits occur. As explained above, turbidites typically pond in the grabens which form as a consequence of normal faulting as the downgoing lithosphere flexes through the outer bulge (e.g., Prince et al., 1974). Even where

sands are absent, an increase in the suspended-sediment fallout from a large trench-floor sediment plume (Fig. 5.11) raises rates of hemipelagic sedimentation on the seaward slope. For example, at DSDP Site 487 (located roughly 380 m above and seaward of the axis of the Middle America Trench), Quaternary muds have been deposited at a rate of 126 m/my (Watkins et al., 1982). With increasing distance from the trench, sediment grain size and rates of sedimentation should decrease gradually.

The expected style of abyssal sedimentation will be perturbed strongly if submarine fans and channels prograde in a direction other than transverse to the adjacent arc-trench couplet. One noteworthy example of this fan/channel geometry is the Surveyor-Baranoff fan system, which is located south of the Aleutian Trench but maintains sediment sources located in southeastern Alaska and British Columbia (Ness and Kulm, 1973; Stevenson and Embley, 1987). An even more impressive example is the Bengal-Nicobar fan system, which is located west of the Sunda Trench but derived from erosion of the Himalayas to the north (Curray and Moore, 1974; G.F. Moore et al., 1982; Emmel and Curray, 1985; see Chapter 10). In addition, older abandoned turbidite successions can be buried within the stratigraphy of the subducting oceanic plate, as exemplified by the Eocene Zodiac fan which is now being subducted along the central Aleutian Trench (Stewart, 1976; Stevenson et al., 1983). Once abyssal-plain turbidites are added to the accretionary prism by offscrapping and/or underplating (e.g., Moore et al., 1991), they may be impossible to discriminate from the trench-wedge sediments using lithofacies criteria alone. Under ideal circumstances, a distinction in sediment provenance may exist between the two turbidite bodies, but because of the complications imparted by long-distance axial flows and post-depositional tectonic translation, provenance interpretations may be equivocal as well. Finally, one might expect to find a significant contrast in physical properties at the deformation front between deeply buried and strongly compacted abyssal-plain turbidites and overlying trench turbidites. This could trigger a change in the respective mechanical responses to subduction-accretion, thereby leading to a contrast in intensity of

deformation or development of specific structural fabrics. Specifically, underplated abyssal-plain turbidites probably are more likely to form coherent slate belts, rather than chaotic mélanges (Sample and Fisher, 1986; Fisher and Byrne, 1987; Sample and Moore, 1987).

Hemipelagic-Mud Dispersal versus Turbidite-Sand Dispersal

Few comprehensive studies have compared regional patterns of hemipelagic mud dispersal along subduction margins with those of interbedded turbidite sands. In most cases, such as the Nankai Trough, the sand and mud components of the total sediment budget seem to share a common source (e.g., Underwood et al., 1993). Data from the Aleutian forearc and trench, however, show that this scenario is not always applicable. One reason for this departure from the norm is that, in addition to transport by bottom-seeking density flows, dispersal patterns for hemipelagic muds can be influenced by strong currents in the surface waters, thermo-haline bottom (contour) currents, and repeated episodes of particle flux in and out of the bottom nepheloid layer (Gorsline, 1984). In the case of the Aleutians, sluggish bottom currents move sediment toward the west on the forearc slope and toward the east above the trench floor and seaward slope (Fig. 5.27). Because of this contour-parallel drift of the nepheloid layer, clay minerals eroded from continental sources in mainland Alaska (illite- and chlorite-rich) have been transported as far west as Atka Basin, where they

have been homogenized with smectite-rich muds from the intraoceanic volcanic arc (Underwood and Hathon, 1989; Hathon and Underwood, 1991). Thus, the clay mineralogy of the central forearc is consistent with a mixed tectonic provenance, whereas the interlayers of turbidite sand display the pristine provenance signature of the intraoceanic volcanic arc (Stewart, 1978; Underwood, 1991). On the trench floor, the pattern is complicated further by the effects of both axial and transverse turbidity currents, combined with the eastward movement of bottom water (Fig. 5.27). In conclusion, a common source for interbedded turbidite sands and hemipelagic muds should be demonstrated, rather than assumed, for each forearc-trench system.

Examples from the Accretionary Rock Record

A summary of findings from structural analyses of fossil subduction complexes is beyond the scope of this chapter. However, the following list provides some of the more prominent case studies in which sedimentary facies analysis has been attempted: the Franciscan Complex of California (Aalto, 1976, 1982, 1989; Bachman, 1982; Underwood, 1984); the Kodiak/Chugach system of southeast Alaska (Nilsen et al., 1979; Moore and Allwardt, 1980; Nilsen and Zuffa, 1982; Nilsen, 1985); the Shimanto Belt of southwest Japan (Taira et al., 1982, 1988; Hibbard et al., 1992); the Rangitata, North Island, and Torlesse systems of New Zealand (Carter et al., 1978; van der Lingen, 1982; MacKinnon and Howell, 1985); the

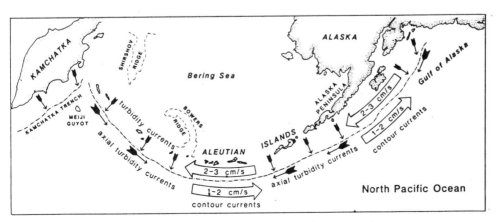

Fig. 5.27 Diagram showing sources of clay minerals, velocities of bottom-water circulation, approximate boundaries of bottom-water masses, and inferred dispersal networks of hemipelagic muds in Aleutian forearc-trench region. From Underwood and Hathon (1989).

New England Fold Belt of Australia (Fergusson, 1985); the Eocene accretionary complex of Barbados (Pudsey and Reading, 1982; Larue and Speed, 1983; Larue, 1985); the Southern Uplands of Scotland (Leggett, 1980; Leggett et al., 1982); and the central Appalachians of Pennsylvania (Lash, 1986a,b).

Given the fact that modern trenches display such a wide range of lithofacies associations (Fig. 5.14), and that regional-scale patterns of sediment dispersal can be axial, radial, transverse, reflected, deflected, or some combination thereof, sedimentologists who examine ancient subduction complexes should expect to encounter almost anything in the field. Intact sections measuring several tens of meters certainly can be interpreted within the conceptual framework of fan-facies associations (Mutti and Normark, 1987; Pickering et al., 1989). In all cases, even in highly deformed mélange terranes, it should be possible to recognize most or all of the basic rock types of the oceanic-plate, trench-floor, and trench-slope environments (i.e., basalt, radiolarian chert, variegated mudstone, shale, and turbidite sandstone), and the relative proportions of these components should allow recognition of the basic style of regional sedimentation. On the other hand, regional-scale three-dimensional reconstructions of the primary facies distributions require increasing amounts of imagination as the degree and complexity of rock deformation intensify. In particular, paleocurrent data will have little bearing on reality unless the entire history of rock-fabric rotation can be identified; a simple rotation to paleohorizontal about the strike of bedding is seldom correct. Thus, facies interpretations within subduction complexes should be tempered with a healthy dose of skepticism.

Economic Deposits

Metals in Subduction Complexes

In general, the potential for economic exploitation of accretionary prisms and trench-slope basins is relatively poor. Most deposits of base and precious metals, for example, are small, and most were produced by hydrothermal or magmatic overprints long after accretion, deformation, and regional metamorphism (e.g., McLaughlin et al., 1985). Massive-sulfide deposits are particularly unusual within turbidite-rich mélanges, but one example of this type has been discovered in the Franciscan Complex of California (Koski et al., 1993). Massive-sulfide deposits typically occur within basalts that have been subjected to hydrothermal alteration along mid-ocean ridges (Sawkins, 1976), although precipitation can be concurrent with siliciclastic sedimentation if an active ridge is close to a continental margin; the Gorda - Juan de Fuca ridge system seaward of the Cascadia subduction zone is a modern example of a sedimented ridge (Davis et al., 1987; Morton et al., 1987; Koski et al., 1988). Incorporation of sediment-hosted massive-sulfide deposits into a subduction complex seemingly requires unusual circumstances. One possibility is for the mineralized rocks to be delaminated from the oceanic plate and underplated to a mélange zone within the interior of a subduction zone; alternatively, in situ mineralization could occur at the intersection between a trench and an active spreading ridge (e.g., Cande et al., 1987).

Several significant deposits of sediment-hosted gold are associated with fossil subduction zones and/or allochthonous accreted terranes (Kerrich and Wyman, 1990). Late-stage gold-bearing quartz veins are widespread, for example, in structurally deformed turbidites of the Valdez Group, which is part of the Chugach subduction complex of south-central Alaska (Goldfarb et al., 1986). There are similar mélange-hosted and turbidite-hosted gold deposits in the Franciscan Complex of central California (Hart, 1966), the Hodgkinson field of Queensland, Australia (Peters et al., 1990), and within greenschist to amphibolite facies accreted terranes in the Juneau gold belt of southeast Alaska (Goldfarb et al., 1988). Mesothermal deposits of this type precipitate from low-salinity waters at temperatures of about 250°C to 340°C. Two mechanisms have been proposed to explain their origin: (1) deep circulation of meteoric water in the vicinity of major fault zones (Nesbitt et al., 1986); and (2) upward infiltration of metamorphic fluids generated by rapid dewatering within the rear of an accretionary wedge (Goldfarb et al., 1988, 1991).

Important occurrences of mercury are fairly widespread within the Franciscan Complex of the Califor-

nia Coast Ranges (e.g., Eckel and Meyers, 1946; Everhart, 1950; Bailey and Everhart, 1964). These hydrothermal deposits, which occur in both fractured clastic rocks and serpentinite, evidently formed as late Cenozoic centers of volcanism and geothermal activity migrated northward in the wake of the Mendocino triple junction (Fox et al., 1985; Peabody and Einaudi, 1992). Franciscan sedimentary rocks are the probable source of the mercury (as well as associated petroleum), but the process of mineralization, in this and similar cases around the world, is unrelated to the phase of subduction that formed the accretionary complex (Peabody and Einaudi, 1992).

Hydrocarbons in Subduction Complexes

Except for small reservoirs of oil on Barbados (Larue et al., 1985; Speed et al., 1991), we know of no proven petroleum production coming directly from an accretionary prism. One reason for this limited production is poor source-bed potential, particularly where rates of siliciclastic influx are high, such as Cascadia, the Aleutians, and Nankai Trough. Total organic carbon (TOC) for both trench-floor sediments and trench-slope sediments in these regions is consistently below 1 wt-% (Dow, 1979; Kvenvolden and von Huene, 1985; Mukhopadhyay et al., 1986), and these values are marginal, at best, for petroleum source rocks. Analogous rock units within ancient subduction zones, such as the Franciscan Complex of California, typically contain 0.5 to 1.5 wt-% TOC (Larue, 1991). Higher contents of organic carbon (1 to 4 wt-%) have been documented, however, for uplifted rocks exposed on Barbados (Larue et al., 1985) and in the slope aprons of the Middle America Trench offshore both Mexico (Summerhayes and Gilbert, 1982a) and Guatemala (Summerhayes and Gilbert, 1982b; Gilbert and Cunningham, 1985). TOC values are highest (1 to 8 wt-%) for lower-slope sediments offshore Peru (Suess et al., 1988a), which is located in a region of intense coastal upwelling (Suess et al., 1988b). Thus, under favorable oceanographic conditions and biogenic productivity, good petroleum source beds can accumulate in subduction zones.

A second important consideration in any evaluation of petroleum potential is the thermal maturation of prospective source beds. Because crustal heat flow in ocean basins decays as a function of lithospheric age (Lister, 1977; Parsons and Sclater, 1977), isotherms within an accretionary prism should be depressed in zones where old oceanic lithosphere is subducted (Watanabe et al., 1977; Wang and Shi, 1984; Yamano and Uyeda, 1988). Reduced heat flow (<50 mW/m^2) and abnormally low geothermal gradients (5 to 15°C/km) mean that temperature windows for peak oil and gas generation will shift to deeper levels of the accretionary stratigraphy (perhaps 5 to 6 km or more), where stratal disruption tends to be more extreme. On the other hand, in localities such as Cascadia and the Nankai Trough, the subducting oceanic lithosphere is relatively young. Geothermal gradients will rise substantially under these circumstances (to greater than 40°C/km); heat flow reaches calculated values as high as 130 mW/m^2 near the prism toe and decreases toward the upper forearc to less than 70 mW/m^2 (Yamano et al., 1982, 1984; Shi et al., 1988). Heat is also transferred through the accretionary wedge via advective movement of pore waters, which are derived from mechanical consolidation and dewatering of the accreted sediment, as well as from dehydration reactions involving clay minerals (Reck, 1987; Davis et al., 1990; Fisher and Hounslow, 1990). Thus, it is possible to find geothermal conditions favorable for petroleum generation at relatively shallow depths of an accretionary prism (i.e., less than 2 km).

A final requirement for large petroleum accumulations is the formation of an adequate reservoir and trap. As it turns out, the prospects for trapping significant quantities of oil and gas within subduction zones are far worse than the prospects for petroleum generation. One dilemma involves the inverse relationship between source-bed potential and reservoir potential. For example, a fine-grained slope apron, even if it contains higher contents of TOC, is ill-suited as a hydrocarbon trap because the compacted mudrocks maintain low primary porosities and low permeabilities. In addition, the reservoir quality of arc-derived sandstones within trenches and slope basins tends to be poor because pseudomatrix is formed by chemical and mechanical alteration of labile rock fragments (Dickinson, 1970b, 1982; Galloway, 1974).

Even if good stratigraphic traps are inherited from a trench wedge (e.g., well sorted quartzose channel sands), those sedimentary geometries ultimately are modified and/or destroyed by faulting and folding as progressive deformation occurs landward of the deformation front. Many seismic-reflection profiles from lower trench slopes display anticlinal geometries, particularly in association with thrust ramps. Some authors have pointed out that these structures seem fitting, at least superficially, as targets for oil or thermogenic gas (e.g., Thompson, 1976; McCarthy et al., 1984; Kvenvolden and von Huene, 1985). However, the development of high secondary porosities due to intense fracturing of the accreted sediment means that the cap rocks for these structural traps are probably leaky. Deformation also crosses the basal unconformity of the slope apron, thereby reducing the quality of that potential seal. Most of the hydrocarbons that are generated in accretionary wedges, therefore, probably percolate slowly through the prism and vent into the ocean water (Larue, 1991).

The most favorable circumstances for long-term hydrocarbon storage in a subduction zone might be found where thick, sand-rich slope-basin accumulations are in a position to intercept oil and gas as they migrate toward the seafloor. Similar traps will form where forearc-basin turbidite sequences prograde seaward and lap onto subsiding strata of the accretionary prism near the trench-slope break (see Chapter 6). The West Sumatra forearc basin (Beaudry and Moore, 1985) and the Eel River basin of northern California (Underwood, 1985; Clarke, 1987, 1992) represent viable examples of this type. Given favorable thermal conditions and proper timing of thermal maturation, hydrocarbon migration, and sediment accumulation above the prism unconformity, facies-controlled reservoirs should be able to trap the hydrocarbons.

Interpretive Stratigraphic Summary

Accretionary Prisms

The internal stratigraphy of accreted trench-wedge and oceanic-plate turbidite deposits is at least as variable and complicated as that of submarine fans, and all of

the same caveats must be applied to comparisons of modern and ancient examples (Mutti and Normark, 1987). In particular, one must appreciate the differences in physical scale and the great dissimilarities in the types of data used to describe modern and ancient deposits (i.e., multi-channel seismic-reflection versus piston-core or outcrop). We see no reason to refute the basic model of trench-wedge stratigraphy, which was first established two decades ago by Piper et al. (1973). This model involves a gradual coarsening and thickening upward as the oceanic igneous basement and overlying mud-rich abyssal-plain section collectively approach the subduction zone, to be buried by sandy turbidite deposits of the trench wedge (Fig. 5.28). At the same time, we have stressed in this chapter the plan-view variability in trench-floor facies relations. The uppermost part of the turbidite section can consist of one or more of the following: (1) sheet-flow deposits; (2) submarine fans complete with depositional lobes and radial arrays of distributary channels; (3) large channel-levee complexes oriented parallel to the trench axis; and (4) hemipelagic muds and debris flows (Fig. 5.14). Spatial and temporal changes in facies character can be extreme. Comparable diversity exists for the abyssal/oceanic sections, which range from a thin cover of biogenic pelagic ooze above oceanic basalt to a thick section of channel/levee or submarine-fan deposits. In addition, we have shown that even mud-rich abyssal-plain deposits can be punctuated by sporadic influx of sand or silt turbidites when mass flows move upslope onto the seaward margin of a trench.

The position of the décollement at the base of an accretionary prism plays a critical role in determining which parts of trench-wedge and oceanic-plate strata are transferred into the evolving subduction margin (Moore, 1989). The ideal upward coarsening succession is seldom, if ever, preserved intact. Drilling results from Leg 131 of the Ocean Drilling Program provide a representative and complete example of accreted strata for a setting with high rates of sedimentation and a relatively thick abyssal-plain section (Fig. 5.22). The décollement near the toe of the Nankai accretionary prism is located in the middle of a thick hemipelagic mud-and-ash section that was deposited outboard of the trench in the Shikoku

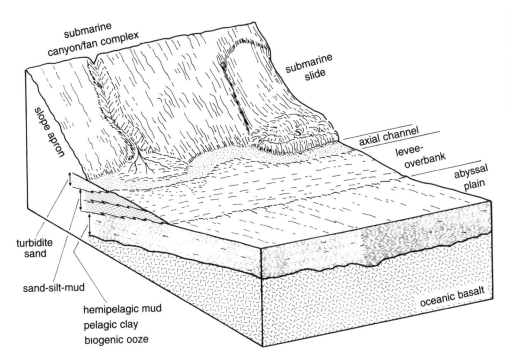

Fig. 5.28 Paradigm upward coarsening stratigraphy outboard of the deformation front of a subduction margin, based largely on DSDP drilling data from the eastern Aleutian Trench. Modified from Piper et al. (1973).

Basin. Consequently, only the upper part of the Shikoku Basin section is being accreted, together with the trench-wedge deposits. Presumably, the remainder of the abyssal-plain section is underplated within a vaguely defined domain located landward of the frontal thrust, where the accretionary wedge thickens (Moore et al., 1990).

In theory, particularly in intraoceanic subduction zones where low rates of trench sedimentation prevail, thrust packets within an accretionary prism could include the entire abyssal-pelagic section (e.g., Lash, 1985) and even delaminated fragments of basaltic basement (Fig. 5.29). However, we stress here that these hypothetical variations on the basic theme of accretionary stratigraphy have never been documented by ocean drilling. For example, even in the case of the Barbados Ridge accretionary prism, which maintains low rates of sediment accumulation, the décollement at the prism toe is located well above the top of subducting oceanic basalt, in the middle of a pelagic-clay succession (Fig. 5.19). The likelihood of adding large slabs of igneous basement to the accretionary wedge probably increases where seamounts and other bathymetric highs collide with the trench (for an ancient example, see MacPherson, 1983). On the other hand, some workers have invoked seamount collisions as a way to increase the local rate of subduction erosion and arcward retreat of the trench (e.g., von Huene and Lallemand, 1990), rather than ophiolite accretion.

In conclusion, a continuous and intact stratigraphic succession, from basalt to pelagic ooze to hemipelagic mudstone to turbidite sandstone, should *not* be expected within ancient subduction complexes. Paradoxically, tectonic mixing among mesoscopic blocks of these components seems to be commonplace, judging from the numerous examples of polymictic mélange in the rock record (e.g., Raymond, 1984; Cowan, 1985). In spite of the best efforts of the research community, the processes responsible for accretion of igneous basement and the formation of mélange at deep levels of the accretionary wedge remain poorly understood.

Trench-Slope Basins and Slope Aprons

A general model for the stratigraphy of trench-slope basins is shown in Figure 5.30. Initial formation of the basin (due to thrust faulting, folding, or diapirism within the underlying, more highly

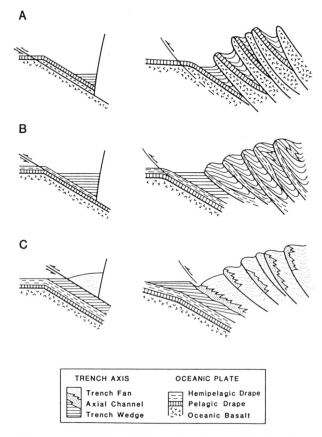

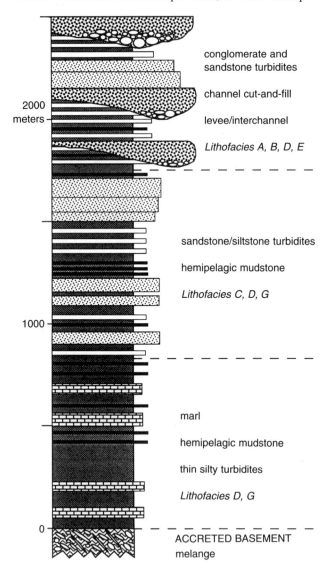

Fig. 5.29 Hypothetical model of accretion and preservation of trench-wedge and abyssal-plain deposits along a subduction margin. The position of the decollement and the stratigraphic assemblage that is preferentially accreted are both influenced by the lithofacies succession outboard of the deformation front. From Thornburg and Kulm (1987a).

coarsen and thicken upward as the basin undergoes uplift, the actual pattern may be quite complicated or noncyclic.

To the best of our knowledge, a well defined upward coarsening cycle has never been documented within a modern trench-slope basin, so this concep-

deformed accretionary basement) is followed by the accumulation of hemipelagic mud, together with remobilized slope deposits. The products of this early phase of basin evolution are difficult to distinguish from those of the slope-apron facies. Individual basins experience progressive uplift as the accretionary prism continues to grow, and the combined effects of headward erosion of lower-slope canyons and downslope erosion of river-fed canyons should increase the basinal supply of terrigenous turbidites through time. Changes in sea level influence rates and spatial patterns of sediment delivery, with canyon activity decreasing during highstands. Turbidite deposition in slope basins, moreover, can occur even if the basins are not connected to the continental shelf or island-arc platform via submarine canyons. Thus, even though one might expect basinal strata to

Fig. 5.30 Hypothetical model for stratigraphic evolution of a trench-slope basin, based largely on the observations of Moore et al. (1980b) on Nias Island, Indonesia. See Table 5.2 for explanation of lithofacies designations. The upward coarsening and thickening megacycle is the result of basin uplift and temporal increases in terrigenous influx, as submarine canyons enhance the efficiency of transport from the continental shelf or arc platform. Smaller-scale cycles are due to progradation or lateral migration of depositional lobes (upward thickening) and migration/abandonment of channels (upward fining and thinning).

tual model represents a collage of several modern and ancient examples (particularly Nias Island). Recognition of the slope-basin facies in the rock record is hampered if fossil evidence is insufficient to place limits on paleodepth or changes in depth through time. Moreover, progressive deformation (folding, faulting, shearing, and diapiric intrusion) blurs the structural distinction between basin fill and the underlying or laterally adjacent accreted basement. As a final caveat, structural geologists do not know enough yet about strain partitioning within accretionary prisms to draw an unequivocal connection between the style of deformation and the spatial domain of deformation. Thus, just because a particular stratal succession within an ancient subduction complex displays mild internal disruption relative to neighboring units, one should not conclude, categorically, that the less-deformed rocks represent a slope basin.

Further Reading

Bouma AH, Normark WR, Barnes NE (eds), 1985, *Submarine fans and related turbidite systems*: Springer–Verlag, New York, 351p.

Burk CA, Drake CL (eds), 1974, *The geology of continental margins*: Springer–Verlag, New York, 1009p.

Dott RH, Jr, Shaver RH (eds), 1974, *Modern and ancient geosynclinal sedimentation*: Society of Economic Paleontologists and Mineralogists Special Publication 19, 380p.

Leggett JK (ed), 1982, *Trench–forearc geology: sedimentation and tectonics on modern and ancient active plate margins*: Geological Society of London Special Publication 10, 576p.

Pickering KT, Hiscott RN, Hein FJ, 1989, *Deep–marine environments: clastic sedimentation and tectonics*: Unwin Hyman, London, 416p.

Stanley DJ, Kelling G (eds), 1978, *Sedimentation in submarine canyons, fans, and trenches*: Dowden, Hutchinson and Ross, Stroudsburg, 395p.

Talwani M, Pitman WC, III (eds), 1977, *Island arcs, deep–sea trenches, and back–arc basins* [Maurice Ewing Series 1]: American Geophysical Union, Washington, 470p.

Watkins JS, Drake CL (eds), 1982, *Studies in continental margin geology*: American Association of Petroleum Geologists Memoir 34, 801p.

Watkins JS, Montadert L, Dickerson PW (eds), 1979, *Geological and geophysical investigations of continental margins*: American Association of Petroleum Geologists Memoir 29, 472p.

Forearc Basins

6

William R. Dickinson

Introduction

Forearc basins lie between the axes of trenches, which mark subduction zones, and the parallel magmatic arcs, where igneous activity is induced by inclined descent of oceanic lithosphere into the underlying mantle. The term "forearc" reflects the convention that arc-trench systems, with inherent polarity defined by the asymmetry of subduction, are said to "face" from arc toward trench (Fig. 6.1). Geodynamic controls on forearc subsidence may vary, however, even for different parts of individual forearc basins.

Recognition of forearc basins as a generic category of sedimentary basin came only with the advent of plate tectonics. Previously, forearc basins were either (a) widely ignored as offshore sediment prisms unfamiliar to most geologists, (b) alternatively viewed as "epieugeosynclines" or "miogeosynclines" by different workers, or (c) termed "interdeeps" from their occurrence between volcanic "inner arcs" and non-volcanic "outer arcs" (see Dickinson and Seely, 1979). Because forearc basins occur midway along the irregular topographic and bathymetric gradient between the crest of the volcanic arc and the deep keel of the trench, some refer to them as "midslope basins" (see Chapter 5). However, major forearc basins form deep troughs 50 to 100 km wide or underlie comparably broad shelves or deeper terraces. After Hamilton (1979), forearc basins are also termed "outer-arc basins" by some workers, but Hamilton (1989) has subsequently adopted the term "forearc basin."

Forearc basins occur along the flanks of intra-oceanic island arcs that fringe marginal seas and also along active continental margins, where the associated volcanic chain may stand either within the continental landmass or along an offshore island chain near the edge of the continental block. The longitudinal continuity of forearc basins along the trends of arc-trench systems is highly variable. At one extreme are elongate troughs or submarine benches that

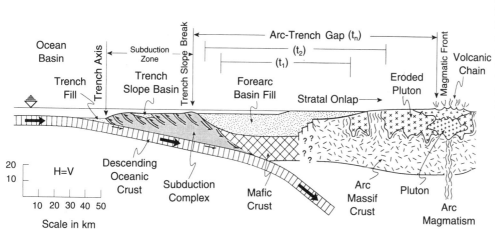

Fig. 6.1 Geotectonic features of forearc basins in transverse profile. See Fig. 6.5 for internal structures of subduction complex and Fig. 6.6 for origins of forearc ophiolites (mafic crustal substratum). Stipples indicate undeformed sediment accumulations. Diagram shows ideal case for which sedimentary infilling of forearc basin keeps pace with accretionary growth of subduction complex, and omits potential faults (thrust, normal, strike-slip) that may cut forearc sediment prism. Cases lacking subduction complex not depicted (see text for discussion).

extend for hundreds or even thousands of kilometers along strike within the "arc-trench gap" (Dickinson, 1971a). In other cases, forearc basins of more limited individual extent form an array of discrete depocenters arranged along the belt between arc and trench like beads on a string.

The inherent variability of forearc basins is difficult to treat systematically because both structural relations and sediment fill are so diverse. In this chapter, general concepts are developed in early passages discussing basin architecture and the nature of basin fill. Later passages discuss specific forearc regions selected to illustrate key variants of forearc evolution.

Tectonic Setting

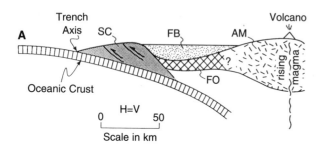

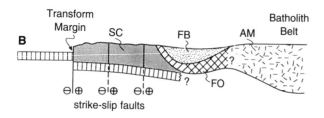

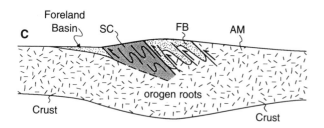

Fig. 6.2 Tectonic settings of forearc basins (schematic): A, origin within active arc-trench system (after Fig. 6.1); B, preservation of basin remnant after conversion of subduction to transform slip (e.g., Great Valley of California); C, deformation of basin fill by incorporation into suture belt of collisional orogenic system (e.g., Midland Valley of Scotland). Symbols: AM, arc massif; FB, forearc basin; FO, forearc ophiolite; SC, subduction complex (but forearc substratum not everywhere ophiolitic; see text for discussion).

Forearc basins typically lie between deep trenches and active magmatic arcs. Forearc sedimentation is thus typically coeval with volcanism and plutonism in the arc massif, and with deformation and metamorphism of the subduction complex (Figs. 6.1, 6.2A). Basin evolution may, however, include phases during which the active subduction zone is overfilled, thereby obscuring trench morphology (see Chapter 5), or during which magmatic activity along the arc trend is dormant even though subduction continues. Forearc sediment prisms may also persist as prominent sedimentary basins after arc-trench tectonism has ceased or shifted in position (Fig. 6.2B,C). The location of the associated magmatic arc is then marked by a linear trend of granitoid batholiths and their wallrocks, exposed as erosion removes surficial volcanic cover (Dickinson, 1970a). The position of the parallel subduction zone is indicated by exposures of exhumed tectonites including blueschists, mélanges, and dismembered ophiolites as especially diagnostic structural elements (Ernst, 1970). The thermal regime of forearc basins derives from their proximity to subduction zones, and basinal geotherms generally are low, analogous to subduction complexes rather than to active magmatic arcs behind the volcanic front (Dumitru, 1988).

Where subduction zones along continental margins or the flanks of island arcs are supplanted by transform systems of mainly lateral slip between crustal blocks and adjacent seafloor, forearc basins may be preserved with only minor structural modification (Fig. 6.2B). Similar degrees of preservation occur for other special cases of basin evolution not illustrated: (a) where subduction ceases at the associated trench without development of a transform plate boundary, (b) where forearc basins are uplifted and deformed as subduction migrates farther offshore during progressive widening of the subduction complex, and (c) where reversal of island-arc polarity transfers a forearc basin to a backarc position. Where subduction of adjacent oceanic lithosphere results in the suturing of island arcs or active continental margins to other crustal blocks, however, the sedimentary assemblages of forearc basins are incorporated, with varying tectonic overprints of deformation and metamorphism, into resulting suture belts (Fig. 6.2C). These derivative basins are here termed "sutural"

forearc basins. Although forearc basins form as integral elements of arc orogens, modified basin fill of forearc origin may thus survive to preserve a record of arc evolution within a collisional orogen, the development of which terminated arc activity.

Basin Architecture

Forearc basins are typically bounded on their arc flanks by the volcano-plutonic assemblages and associated metamorphic rocks of arc massifs, and on their trench flanks by uplifted subduction complexes composed of varying proportions of deformed and partly metamorphosed oceanic crust, seafloor sediments, trench fill, and trench-slope deposits (Fig. 6.1). Subduction complexes typically include structurally juxtaposed belts or interleaved sheets composed variously of stratally disrupted mélanges and coherently stratified sequences (Cowan, 1974; Moore and Allwardt, 1980; Byrne, 1982). The "trench-slope break" (Dickinson, 1973), defined as the transition between the steep inner slope of the trench and gentler slopes within the forearc belt, marks the arcward limit of deformation related directly to subduction (G.F. Moore and Karig, 1976).

Basin fill commonly laps unconformably upon the eroded flank of the arc massif along the arcward side of a forearc basin, but normal faults may step the floor of the basin downward from the elevated tract of the arc in cases where extension occurs across the forearc belt. Along the trenchward edge of forearc basins, more complicated stratal relationships develop between basin fill and the evolving structural high atop the subduction complex. Progressive stratigraphic overlap of forearc deposits across deformed strata of the subduction complex is characteristic, but faulted contacts between basin fill and uplifted parts of the subduction complex are also common because deformation of the latter is still in progress during forearc sedimentation. Prograde accretion of subduction complexes and retrograde migration of arc activity tend to widen forearc basins with time as subduction continues (Dickinson, 1973; Ingersoll, 1988b).

In some arc-trench systems, especially those involving intraoceanic island arcs (Hawkins et al., 1984), wholesale subduction of oceanic sediment at the trench may prevent growth of an accretionary

subduction complex. This behavior discourages the formation of forearc basins for lack of a confining tectonic dam, and the forearc sediment prism may be no more than an inclined slope apron draped across the forearc substratum. If, however, tectonic erosion of forearc crust accompanies sediment subduction at the trench ("ablative subduction" of Tao and O'Connell, 1992), then severe extensional deformation akin to incipient seafloor spreading may affect parts of the forearc belt (Fig. 6.3). Where extensional rifting controls the evolution of forearc basins, conglomeratic facies confined to multiple grabens or half grabens may form a prominent component of the forearc sediment assemblage (Morris et al., 1989; Smith and Busby, 1993).

Beyond the trench-slope break, small basins perched on the trench slope above a substratum of deformed subduction complex are transient forearc basins that lose an identity separate from the subduction complex as deformation of the trench slope continues. Because these intraslope basins have close affinities with trench fill and its deformation, they are treated in Chapter 5. Trench-slope basins begin to form at the trench floor, where subduction detaches packets of strata from the descending plate, and they continue to evolve as processes of accretion uplift them toward the trench-slope break. Where uplift is strong, trench-slope basins may be exposed on land while subduction continues offshore (van der Lingen and Pettinga, 1980).

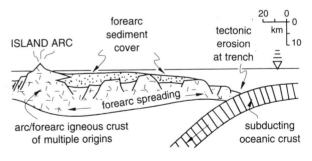

Fig. 6.3 Schematic transverse profile (vertical exaggeration 2:1) of extensional forearc, derived from concepts of Hussong and Uyeda (1981), Hawkins et al. (1984), Fryer (1990), and Tao and O'Connell (1992) for Izu-Mariana and Tonga-Kermadec intraoceanic arc-trench systems; preservation of gently deformed forearc sediment cover dating from near the inception of subduction (Scholl and Herzer, 1992) implies that principal episodes of forearc extension occurred early in basin evolution (soon after initiation of new subduction zone).

Basin Morphology

The aggregate length of modern arc-trench systems is approximately 40,000 km, roughly half along continental margins and half along intraoceanic island arcs (Dickinson, 1975). The global spectrum of modern forearc basins encompasses a wide range of basin morphology, but most can be grouped into a limited number of basic variants controlled by two key factors (Fig. 6.4): (a) the elevation of the trench-slope break relative to sea level, and (b) the sedimentation rate in the forearc basin relative to the evolving structural relief of the trench-slope break.

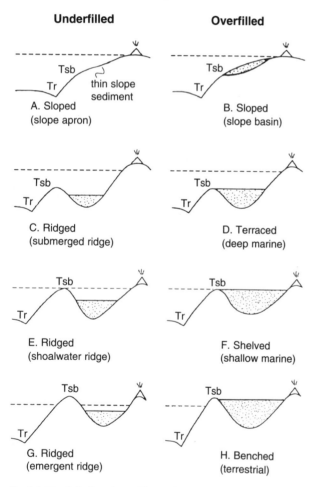

Underfilled

A. Sloped (slope apron)

thin slope sediment

Tsb
Tr

C. Ridged (submerged ridge)

Tsb
Tr

E. Ridged (shoalwater ridge)

Tsb
Tr

G. Ridged (emergent ridge)

Tsb
Tr

Overfilled

B. Sloped (slope basin)

Tsb
Tr

D. Terraced (deep marine)

Tsb
Tr

F. Shelved (shallow marine)

Tsb
Tr

H. Benched (terrestrial)

Tsb
Tr

Fig. 6.4 Morphologic variants of forearc basins, with basin fill stippled and vertical exaggeration for clarity: left, "underfilled" (most common for island arcs); right, "overfilled" (most common for continental margins). Symbols: dashed lines = sea level, Tr = trench, Tsb = trench-slope break (coincides with shelf-slope break for shelved forearcs). Modified after Seely (1979) and Dickinson and Seely (1979). Distribution of morphologic types in modern arc-trench systems indicated by Table 6.1.

Where tectonic accretion occurs along the trench slope, the trench-slope break lies at the crest of a tectonic dam formed by the subduction complex, and serves as the trenchward sill of the forearc basin. If the basin is filled to sill level, the surface of basin sediment forms a slope apron, deep-marine terrace, marine shelf, or terrestrial plain, depending upon the elevation of the trench-slope break relative to sea level. If the structural basin between arc massif and trench-slope break is underfilled with sediment, then the forearc basin is a trough flanked on one side by highlands or insular volcanoes of the magmatic arc and on the other side by a submerged bathymetric barrier or an emergent topographic ridge, again depending upon the elevation of the trench-slope break relative to sea level. Where tectonic erosion occurs along the trench slope, the trench-slope break lacks the same tectonic significance, or structural relief above the forearc basin, but still delimits the trenchward margin of undeformed slope sediment cover within the forearc belt.

In transverse profile, forearc regions can be described as sloped, terraced, shelved, benched, or ridged (Fig. 6.4), as relief of the trench-slope break and thickness of sediment fill in the forearc basin vary. Each morphologic type of forearc basin must be understood as potentially just one phase in the geologic evolution of an individual basin. Where rates of forearc sedimentation and uplift of the trench-slope break vary with time, a single basin may evolve through multiple morphologic stages.

Longitudinally along forearc belts, transverse structural features commonly subdivide tracts of forearc sedimentation into separate depocenters that are aligned as discrete basins or sub-basins along the trends of arc-trench systems (Table 6.1). This segmentation of forearc regions into multiple depocenters evidently reflects some combination of differential subsidence and nonuniform deformation along strike. Individual basins or sub-basins, linked along strike within a forearc belt, may display varying morphology, dependent upon local structural evolution and on position with respect to sediment supply.

The tectonic evolution of some arc-trench systems has given rise to multiple forearc basins, here termed "compound" forearc basins, which lie parallel to one

another in transects across forearc regions (Table 6.1). Their transverse spacing is typically 125 to 150 km, as measured between the centroids of parallel depocenters. The multiple parallel depocenters become younger in age of inception from the arc toward the trench. They are separated by bathymetric or topographic barriers that reflect successive positions of the trench-slope break through time. Sedimentation in compound forearc systems may be partly coeval in the separate basins, or almost entirely sequential from basin to basin.

Basin Dimensions

The sizes and configurations of both modern and ancient forearc basins are highly variable (Tables 6.1, 6.2), but their dimensions define a spectrum of variability that is comparable the world over, at least for Phanerozoic time. Most specific figures in Tables 6.1 and 6.2 should be treated with caution, for information about almost all forearc basins is incomplete. Few modern forearc basins have been drilled extensively, and the interpretive tectonic reconstructions necessary to define ancient forearc basins introduce an element of uncertainty into all conclusions about them. It is clear, however, that typical forearc basins are narrow and elongate, with thick sediment prisms confined to deep structural keels.

Most forearc basins are only 25 to 125 km wide, in the direction transverse to an arc-trench system; two-thirds of the modern examples (Table 6.1), and half of their ancient counterparts (Table 6.2) are 50 to 100 km wide. The lengths of basinal depocenters defined by closed sediment isopachs range widely, from less than 50 km to more than 500 km. Half of the modern basins or sub-basins are 100 to 250 km long, whereas half of their ancient counterparts are 250 to 500 km long; the contrast probably stems from the greater difficulty of distinguishing individual depocenters within the deformed sediment prisms of inactive forearc belts. Strings of linked forearc depocenters commonly extend for 2000 to 4000 km along modern arc-trench systems; the centroids of separate depocenters are spaced 100 to 750 km apart along tectonic strike, with a median spacing of about 250 km.

The maximum thickness of sediment fill in modern forearc basins ranges from the order of 1000 m to more than 10,000 m (Table 6.1). Bear in mind, however, that the tabulation omits most forearc regions where only thin aprons of slope sediment are draped across the forearc substratum (Fig. 6.4A). Half of the forearc basins tabulated from continental-margin arc-trench systems contain 3000 to 6000 m of basin fill along their structural keels, whereas half of the forearc basins of intraoceanic island arcs contain only 1000 to 3000 m. The contrast presumably reflects generally more voluminous sediment sources along continental margins. Similarly, half of the forearc basins tabulated from continental-margin arc-trench systems are shelved basins (Fig. 6.4F), where sediment cover has built up nearly to sea level, whereas the majority of intraoceanic forearc basins are ridged or terraced basins with the sediment-water interface in deep water (Fig. 6.4C,D). Few modern forearc basins contain 7500 to 12,500 m of sediment fill, but nearly half of the ancient forearc basins tabulated fall into that category (Table 6.2). The contrast is interpreted to reflect both a tendency for preservation of fully mature forearc basins, for which the passage of time has allowed all available accommodation space to be filled with sediment, and also a bias toward the recognition and description of thick forearc sediment prisms in the geologic record.

Net accumulation rates along the structural keels of forearc basins range widely, from perhaps 25 m/my to more than 250 m/my (Tables 6.1, 6.2). Rates for approximately half of both modern and ancient forearc basins fall in the range 75 to 175 m/my, but rates for half of modern intraoceanic forearc basins are less than 125 m/my. The volume of sediment fill within individual basins or sub-basins also varies widely, but nearly all values lie between 2.5×10^3 km^3 and 2.5×10^5 km^3. Half of the modern forearc basins tabulated contain only $1-5 \times 10^4$ km^3 of sediment fill (Table 6.1), whereas a majority of fully mature ancient counterparts contain $1-5 \times 10^5$ km^3 (Table 6.2). Rates of growth of sediment volume in forearc basins lie generally in the range of 2.5 to 7.5 km^3 per longitudinal kilometer of arc-trench system per million years.

Arc Massif

The geological nature of the adjacent arc massif varies markedly from intraoceanic island arcs to

Table 6.1 Morphology and Approximate Dimensions of Fifty Selected Modern Forearc Basins in Fifteen Active Arc-Trench Systems Categories (I-VII) of Arc-Trench Systems after Dickinson (1975): CM, Continental Margin; IA, Intra-Oceanic Letters (a-h) denote multiple depocenters linked along strike within individual arc-trench systems

Arc-Trench System (CF=compound forearc)[1]	Basin[2] or Sub-Basin[3]	Morphology[4] (Fig. 6.4)	Age of Basal Strata	Thickness of Axial Fill[5] (m)	Net Accumulation Rate (m/m.y.)	Basin Width in Km (transverse)	Basin Length in Km (longitudinal)	Volume[6] of Basin Fill (Km3)	Basin Spacing[7] in Km	References
I. CM with mainland volcanoes within continental landmass										
Alaska Range (CF)	Cook-Shelikof	G (to E)	mid-Mesozoic	7500–12,500	25–75	75–125	750–1000	5×10^5	100–125 (t)	Fisher et al., 1987
	outer shelf belt (inc. a-c)	F	mid-Miocene	1000–6000	65–400	50–75	550–600	7.5×10^4	100–125 (t)	von Huene et al., 1987
	(a) Tugidak	F	mid-Miocene	~5000	300–350	50–60	~75	1×10^4	~175 (l)	von Huene et al., 1987
	(b) Albatross	F	mid-Miocene	3000–4000	200–250	~50	~150	2×10^4	175–225 (l)	von Huene et al., 1987
	(c) Stevenson	F	mid-Miocene	4000–6000	250–400	50–75	~150	3×10^4	~225 (l)	von Huene et al., 1987
Cascades, USA-Canada (CF)	Williamette-Puget	H	Eocene (?)	~4000	~100	25–50	~450	4×10^4	125–175 (t)	Keach et al., 1989
	shelf basins (inc. a-e)	F (to H)	mid-Tertiary	1000–4500	25–175	25–50	850–875	7.5×10^4	125–175 (t)	Snavely, 1987
	(a) Tofino	F	Oligocene	~3500	~100	~50	~275	3×10^4	225–250 (l)	Shouldice, 1973
	(b) Willapa	F	mid-Miocene (?)	~2000	~200	30–40	100–125	8×10^3	150–250 (l)	Snavely, 1987
	(c) Astoria	F	mid-Miocene (?)	~2000	100–150	20–30	100–125	4×10^3	100–150 (l)	Snavely, 1987
	(d) Newport	F	mid-Miocene (?)	~2500	150–175	~25	~50	2×10^3	100–125 (l)	Snavely, 1987
	(e) Coos Bay	F	mid-Miocene (?)	2000 (?)	100–150	30–40	~100	5×10^3	125–250 (l)	Snavely, 1987
	(f) Eel River	F & H	mid-Miocene	3000–4500	200–300	40–60	200–225	2.5×10^4	250–275 (l)	Clarke, 1992
Central Andes[8]	aggregate (inc. a-h)	D, E, H	Paleogene	1000–3000	25–100	75–100	~2850	2.5×10^5	—	see (a) to (h)
	(a) Tumaco	H (to F)	Eocene (?)	~2500	50–75	35–55	125–150	1.5×10^4	~300 (l)	Lonsdale, 1978
	(b) Manabi	H	Eocene	~3000	~75	65–115	~250	5×10^4	300–500 (l)	Lonsdale, 1978
	(c) Sechura	H & F	Oligocene	~3000	~100	65–90	~275	5×10^4	425–500 (l)	Thornburg & Kulm, 1981
	(d) Salaverry	F	Oligocene (?)	~2000	~75	55–80	450–475	4×10^4	425–675 (l)	Thornburg & Kulm, 1981
	(e) Pisco	F	Eocene	~2500	~50	35–55	~275	2.5×10^4	500–675 (l)	Thornburg & Kulm, 1981
	(f) Arequipa	D	mid-Tertiary (?)	~1750	~50	35–55	~325	2×10^4	225–500 (l)	Coulburn & Moberly, 1977
	(g) Arica	D	mid-Tertiary (?)	~1500	25–50	25–45	125–150	6×10^3	150–225 (l)	Coulburn & Moberly, 1977
	(h) Iquique	D	mid-Tertiary (?)	~1250	25–50	25–35	~125	4×10^3	~150 (l)	Coulburn & Moberly, 1977

Table 6.1 Continued.

Region	Subregion	Code	Age							Reference
II. CM with mainland volcanoes along peninsula or isthmus										
Alaska Peninsula	aggregate (inc. a-b)	F	mid-Miocene	1000–6000	65–400	50–75	~450	5×10^4	—	Bruns et al., 1987
	(a) Sanak	F	mid-Miocene	5000–6000	300–400	25–50	~200	2.5×10^4	~250 (l)	Bruns et al., 1987
	(b) Shumagin	F	mid-Miocene	~2500	~150	~75	~125	1.5×10^4	~250 (l)	Bruns et al., 1987
Central America	Middle America	F	Cretaceous (?)	~10,000	~100	80–90	450–550	4×10^5	—	Seely et al., 1974
III. CM with insular volcanoes on offshore island arc										
Sunda (Indonesia)	aggregate (inc. a-h)	C, E, F, G	Oligocene	2500–6500	100–250	75–150	~4225	12.5×10^5	—	Hamilton, 1979
	(a) Simeulue	C & E	Oligocene (?)	3000–6000	100–250	75–100	500–525	1.5×10^5	~425 (l)	Hamilton, 1979
	(b) Nias	F & G	Oligocene (?)	3000–5000	100–200	75–110	~225	0.5×10^5	400–425 (l)	Hamilton, 1979
	(c) Siberut	E & G	Oligocene (?)	3500–5500	125–225	~100	~425	1×10^5	400–600 (l)	Hamilton, 1979
	(d) Enggano	C & E	Oligocene (?)	3500–5500	125–225	~100	~475	1.5×10^5	600–650 (l)	Hamilton, 1979
	(e) SW Java	C	Oligocene (?)	4000–6500	150–250	100–150	~525	2×10^5	400–650 (l)	Hamilton, 1979
	(f) SE Java	C	Oligocene (?)	4500–5500	150–225	~115	300–325	1×10^5	450–650 (l)	Hamilton, 1979
	(g) Lombok	C	Oligocene (?)	~3500	~125	~150	425–575	1.5×10^5	650–750 (l)	Hamilton, 1979
	(h) Savu	E & G	Oligocene (?)	3000–6500	100–250	75–150	400–425	1.5×10^5	~750 (l)	Hamilton, 1979
IV. Stationary IA without backarc spreading										
Aleutian	Aleutian Terrace (inc. a-c)	D (to C)	basal Oligocene (?)	1000–6000	25–75	25–75	~1300	1×10^5	—	Scholl et al., 1987b
	(a) Kanaga	C	basal Oligocene (?)	2500 (?)	~75	~25	125–175	7.5×10^3	~325 (l)	Scholl et al., 1987b
	(b) Atka	D	basal Oligocene (?)	5000–6000	~150	50–75	150–175	3.5×10^4	325–475 (l)	Scholl et al., 1987b
	(c) Unalaska	D	basal Oligocene (?)	4000 (?)	100–125	~50	~200	2.5×10^4	~475 (l)	Scholl et al., 1987b
V. Migratory IA with continental basement beneath arc										
Ryukyu[9] (Japan)	Shimanjari-Miyazaki	A & B	mid-Miocene	2000–4000	125–250	25–75	~1000	7.5×10^4	?	Letouzey & Kimura, 1986
Southwest Japan	aggregate (inc. a-d)	C & D	Miocene (?)	750–1750	25–75	50–125	~700	6×10^4	—	Taira et al., 1988; Okuda & Honza, 1988
	(a) Hyuga	D	Miocene (?)	~1500	50–75	85–125	~150	1.5×10^4	150–175 (l)	see a-d
	(b) Tosa	D	Miocene (?)	1250–1750	50–75	75–115	~120	1.5×10^4	125–175 (l)	see a-d
	(c) Muroto	C & D	Miocene (?)	~1500	50–75	~50	~135	1×10^4	125–150 (l)	see a-d
	(d) Kumano	C	Miocene (?)	1500–1750	50–75	50–65	~115	1×10^4	~150 (l)	see a-d
Northeast Japan	Tohoku Forearc	B & D	Cretaceous	~7500	50–100	75–125	425–450	2.5×10^5	—	Shiki & Misawa, 1982

Table 6.1 Continued.

Arc-Trench System (CF=compound forearc)[1]	Basin[2] or Sub-Basin[3]	Morphology[4] (Fig. 6.4)	Age of Basal Strata	Thickness of Axial Fill[5] (m)	Net Accumulation Rate (m/m.y.)	Basin Width in Km (transverse)	Basin Length in Km (longitudinal)	Volume[6] of Basin Fill (Km3)	Basin Spacing[7] in Km	References
VI. Migratory IA without continental basement in arc										
Banda (Indonesia)	Weber Deep	E & G	mid-Tertiary	~2500	~100	65–130	425–450	5×10^4	—	Bowin et al., 1980
Kermadec	Raukumara	D	Oligocene (?)	~1000	25–50	~150	~225	2.5×10^4	—	Karig, 1970
Tonga	Tonga Ridge	F (to D)[10]	Upper Eocene	3000–5000	75–125	60–75	~500	1×10^5	—	Scholl et al., 1985; Scholl & Herzer, 1992
Mariana	Mariana Ridge	A & B	mid-Tertiary	500–2500	25–50	75–125	~725	1×10^5	—	Mrozowski & Hayes, 1980; Lundberg, 1983
Lesser Antilles	Tobago Trough	C & E	mid-Eocene	5000–10,000	100–200	~100	~200	1.5×10^5	—	Speed et al., 1989
VII. Migratory IA with history of polarity reversal										
Vanuatu[11]	Cumberland-Big Bay	B & C & G	mid-Miocene	1250–2250	75–150	~25	~100	4×10^3	—	Greene & Wong, 1988
Vanuatu[11]	Vanicolo	D (to E)	mid-Miocene (?)	4000–6000	~250 (?)	50–75	~200	5×10^4	—	Greene & Wong, 1988
Solomons	Queen Emma[12]	C	mid-Oligocene	5000–7500	175–250	35–65	300–325	8×10^4	—	Vedder et al., 1989
Luzon (CF)	Luzon Central Valley	H (to G)	Oligocene	12,500–15,000	400–600	40–70	175–250	8×10^4	~125 (t)	Bachman et al., 1983
Luzon (CF)	West Luzon Trough	D	Late Miocene (?)	2000–5000	150–400	30–40	~250	2×10^4	~125 (t)	Lewis & Hayes, 1984
Luzon	North Luzon Trough	C	intra-Miocene (?)	2500 (?)	~150 (?)	20–30	450–475	1.5×10^4	~500 (l)	Lewis & Hayes, 1984

[1]Compound forearc basins (see text) are defined as multiple parallel forearc basins lying side by side within forearc belts

[2]Aggregate data are summations for entire integrated basins (including lettered sub-basins and the sill areas between them)

[3]Sub-basins of longitudinally integrated forearc basins are denoted by letters

[4]A. sloped (slope apron); B. sloped (slope basin); C. ridged (submerged ridge); D. terraced (deep marine); E. ridged (shoalwater ridge); F. shelved (shallow marine); G. ridged (emergent ridge); H. benched (terrestrial)

[5]Figures cited are maximum values pertaining to keels of elongate depocenters

[6]Figures cited are valid only as comparative order-of-magnitude estimates (incomplete subsurface data preclude full accuracy)

[7]Distances from centroid of basin fill to centroids of basin sediment prisms in adjacent depocenters along tectonic strike (l) and/or transverse (t) to tectonic strike across compound fore-arc belts (i.e., basins 100 km long or wide spaced 200 km apart would be separated by 100 km of forearc lacking basinal depocenters)

[8]Belt of prominent trench-slope basins lying farther offshore not tabulated

[9]Multiple forearc depocenters not well defined along strike

[10]Shallow carbonate platform along trenchward edge gently backtilted to ~1000 m depth toward arc

[11]Other small basins may occur but are poorly studied (the two tabulated are en echelon in relative position)

[12]Interpreted as composite forearc basin with sedimentation both before and after reversal of arc polarity (cf. Table 6.2C)

Table 6.2. Approximate Dimensions of Twenty Selected Ancient Forearc Basins

Basin	Age Range of Basin Fill	Thickness of Axial Fill (m)	Net Accumulation Rate[1] (m/m.y.)	Basin Width in Km (transverse)	Basin Length in Km (longitudinal)	Estimated Volume[2] of Basin Fill (Km^3)	Coeval Magmatic Arc	Subduction Complex	References
A. Stranded within inactive forearc belt by cessation of subduction at associated trench									
Tamworth Trough, Australia	Devonian to Carboniferous	~12,500	~100	~50	~400	2×10^5	buried volcanic chain	Tablelands subduction complex	Leitch, 1975; Korsch, 1977
Sarawak[3] (offshore & onshore)	mid-Oligocene to mid-Pliocene	~7500	~250	~175	575–625	5×10^5	Sabah-Sulu paleoarc	northwest Borneo paleo-trench	Hamilton, 1979
Southeast Palawan, Philippines	Paleogene	3000–4000	100–125	125–150	450–550	2×10^5	Cagayan Ridge paleoarc	Palawan Ridge complex	Hamilton, 1979
Iloilo, Panay, Philippines	mid-Oligocene to mid-Pliocene	~3000	~100	30–40	200–225	2×10^4	Negros volcanic chain	Negros paleo-trench	Hamilton, 1979
B. Deformed and uplifted during continued or renewed subduction offshore									
Central Kalimantan, Indonesia	Paleogene	2000–3000	75–100	125–150	450–500	1.5×10^5	southern Borneo igneous belt	Danau-Crocker-Rajang terrane	Hamilton, 1979
Nemuro, Hokkaido (Japan)	Late Cretaceous to Paleogene	1250–2750	25–75	25–50	~250	2×10^4	ancestral Kuril arc	Nemuro Peninsula uplift	Kimura & Tamaki, 1985; Okada, 1974
Oregon Coast Range (USA)	Paleogene	5000–7500	250–300	50–100	475–525	1.5×10^5	Challis-Clarno-Cascades arcs	offshore subduction complex	Heller & Ryberg, 1983; Heller, 1983
Ochoco, central Oregon (USA)[4]	Late Triassic to Late Jurassic	12,500–15,000	150–200	50–60	~225	1×10^5	Olds Ferry arc terrane	Baker melange terrane	Dickinson, 1979
Hornbrook, California-Oregon (USA)	Late Cretaceous	~1250	~50	25 (?)	~85	2.5×10^3	buried granitic batholiths	Klamath accretionary system	Nilsen, 1984, 1985a
C. Transferred with minor deformation to backarc setting by reversal of arc polarity									
New Ireland[5], Papua New Guinea	Oligocene (?) to Holocene	3500–5500 [6] (3000?–4000?)	100–150 [6] (150?–200?)	35–85	~500	1.2×10^5 (1×10^5)	New Ireland paleoarc	Emirau-Feni Ridge (Manus-Kilinailau Trench wedge)	Marlow et al., 1988; Exon & Marlow, 1988

Table 6.2 Continued.

D. Modified by conversion from subduction to transform margin

Hokonui, New Zealand	Permian to Jurassic	~12,500	~100	25–125	~1150	5×10^5	western basement province	Torlesse subduction complex	Dickinson, 1971a,b
Great Valley[7], California (USA)	Late Jurassic to Paleogene	12,000–15,000	125–150	100–125	~675	5×10^5	Sierra Nevada batholith	Franciscan subduction complex	multiple (see text)
Vizcaino, Baja California	Late Jurassic to Late Cretaceous	~12,500	~175	60–110	300 (?)	3×10^5	Alisitos volcanic belt and Peninsular Range batholith	Vizcaino-Cedros melange belt	Boles, 1986 Busby-Spera and Boles, 1986

E. Incorporated into suture belt by arc or continent collision[8]

Coastal Range, Taiwan	Miocene to Pliocene	~5000[9] (~1000)	~1000[9] (~100)	? (collapsed)	~150	$1.5 (?) \times 10^5$	Taiwan Coastal Range arc	ancestral Taiwan Central Range	Lundberg & Dorsey, 1988
Yezo, Hokkaido (Japan)	Upper Cretaceous	5000–6500	100–200	~25	~400	5×10^5	Oshima-Rebun igneous belt	Kamuikotan subduction complex	Okada, 1974, 1982, 1983
Magog, Quebec	Ordovician	5000 (?)	~100	10–20	~500	$2.5 (?) \times 10^4$	Ascot-Weedon volcanic arc	St. Daniel melange	St. Julien & Hubert, 1975
Midland Valley, Great Britain	Ordovician to Devonian	~2500	~50	50–75	~500	8×10^4	Northern Highlands basement	Southern Uplands complex	Dewey, 1971
Sanandaj, western Iran	Cretaceous	~2500	25–50	~75	225–250	2×10^5	Sanandaj-Sirjan Belt	Zagros Range melanges	Cherven, 1986
Indus Trans-himalayan, India	mid-Cretaceous to mid-Tertiary	~3000[9] (~1750)	~50[9] (30–35)	~25	~500 (?)	$2.5 (?) \times 10^5$	Ladakh paleoarc	Dras-Nindam accretionary prism	Garzanti & van Haver, 1988
Burmese Lowland, Burma[10]	Cenozoic	5000–10,000	75–175	75–125	~825	6×10^5	Shan Plateau & Burma arc	Indoburman Ranges uplift	Mukhopadhyay & Dasgupta, 1988

[1] Figures cited are maximal values pertaining to keels of elongate depocenters

[2] Figures cited are valid only as comparative order-of-magnitude estimates (incomplete preservation precludes full accuracy)

[3] Forearc sediment buried offshore by 1000–2500 m of Neogene shelf and deltaic deposits

[4] Unconformably overlying Cretaceous strata not included in tabulation

[5] Includes three linked depocenters spaced 200–225 km apart along tectonic strike

[6] Basin fill includes both forearc and backarc phases of sedimentation; figures in parentheses are estimates for forearc phase only

[7] Partitioned into Sacramento and San Joaquin sub-basins during latest Cretaceous and Paleogene time

[8] Termed sutural forearc basins (see text)

[9] Basin fill includes both pre-collisional (forearc) and syncollisional to post-collisional phases of sedimentation; figures in parentheses are estimates for forearc phase only

[10] Collision suture of Indoburman Ranges with Indian subcontinent incomplete beneath deltaic foreland basin of Bangla Desh

.ive continental margins. Crustal thicknesses vary from only 15 to 25 km for young intraoceanic arcs to values most commonly in the range of 30 to 50 km for arc massifs along continental margins; exceptional thicknesses of 60 to 75 km are known for the central Andes (Dickinson, 1975). Common to all active magmatic arcs is a surficial cover of volcanic and volcaniclastic rocks, averaging perhaps 5 to 10 km in cumulative thickness, capping a crustal root of cogenetic plutons that intrude part of the basal volcanic cover (Hamilton, 1981). The topographic or bathymetric feature of positive relief formed by the volcanogenic edifice of superposed or overlapping volcanoes, which cap the arc massif but may themselves be partly eroded, is termed the arc platform (see Chapter 7).

The volcanic components of the igneous suite are commonly andesitic and dacitic rocks of stratovolcano chains. Basaltic lavas are also abundant, particularly for intraoceanic arcs, and more silicic ignimbrites erupted from collapse calderas are characteristic associates in continental-margin arcs. The plutonic roots of arcs are granitoid intrusions that commonly form elongate composite batholiths along the arc trend. Batholithic wallrocks may include varied metasedimentary strata and metamorphosed basement of pre-arc origin, as well as metavolcanic rocks that grade upward to unmodified volcanic cover. The cumulative effect of magmatic heat advection associated with the emplacement of plutons into the crustal roots of an arc produces a metamorphic belt, coincident with the arc trend, composed of low-pressure facies series (Barton and Hanson, 1989).

Along the arcward flank of forearc basins, volcaniclastic strata of the arc assemblage may interfinger with sediments of the forearc basin (Kuenzi et al., 1979). This gradational relationship implies that distinction between the sediment fill of forearc and intra-arc basins can be problematical in the geologic record. Restriction of the term "intra-arc basin" to depocenters confined within the arc platform keeps the distinction clear, at least in concept. Isolated depocenters lying within the arc massif trenchward of the coeval volcanic front can be regarded as intramassif forearc basins, but they are probably indistinguishable in morphology and sedimentology from similar intra-arc basins (see Chapter 7).

Accretionary Sill

The geodynamic nature of the accretionary sill at the trench-slope break, forming the structural crest of a subduction complex typically 10 to 15 km thick, is a controversial topic despite intensive past research. Initial interpretations of the trench slope envisioned a simple stack of imbricate thrust panels offscraped in succession from the downgoing slab of oceanic lithosphere (e.g., Seely et al., 1974). The locus of detachment of offscraped sedimentary materials from the descending slab is the foot of the trench slope at the deepest keel of the asymmetric trench, where oceanic crust enters the subduction zone (Fig. 6.5A). Listric thrusts that reach the seafloor at the trench axis curve downward to merge with a master decollement at the upper surface of subducting oceanic crust at some distance beneath the toe of the subduction complex (Hyndman et al., 1990). Arcward tilted slabs of ophiolitic rocks are buried under thin sediment cover beneath some trench slopes (Aubouin et al., 1985b); these are interpreted as slices of oceanic crust peeled successively off descending oceanic lithosphere by listric thrusts that disrupted the master decollement surface at irregular intervals during evolution of the subduction complex (Bernstein-Taylor et al., 1992a).

The concept of progressive thrust stacking within a subduction complex is commensurate with mechanical models of thrust tectonics (e.g., Davis et al., 1983). The tectonically adjustable trenchward inclination of the trench slope conforms to the critical taper required to support slip on the decollement along which oceanic lithosphere passes beneath the subduction complex. By the tenets of theory, the relief of the trench-slope break above the trench floor grows progressively in the manner required to maintain the critical taper as additional thrust panels are accreted to the imbricate stack beneath a gradually widening subduction complex. The requisite internal thickening of the accretionary wedge can be achieved by some combination of thrusting out of sequence and antithetic backthrusting within the subduction complex. Once the maximum relief that can be sustained is achieved, the trench-slope break is expected to migrate gradually trenchward as some arcward part of the subduction complex becomes tectonically

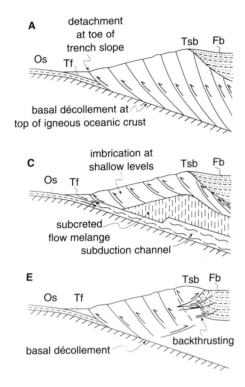

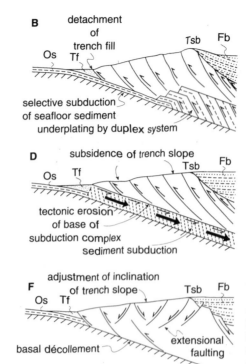

Fig. 6.5 Postulated mechanism for structural evolution of accretionary sills flanking forearc basins, with vertical exaggeration for clarity (Os = oceanic seafloor sediment, Tf = turbidite trench fill, Tsb = trench-slope break, Fb = flank of forearc basin fill): A, accretionary off-scraping of successive imbricate thrust panels (Seely et al., 1974); B, selective subduction and underplating of thrust duplexes (Sample and Fisher, 1986); C, subcretion of flow mélange from subduction channel (Shreve and Cloos, 1986; Cloos and Shreve, 1988a,b); D, subduction erosion along decollement beneath subduction complex (von Huene et al., 1982); E, backthrusting of subduction complex toward forearc basin (Silver and Reed, 1988); F, denudational normal faulting accompanied by tectonic exhumation (Platt, 1986).

inactive. As this model of frontal accretion by imbricate thrusting predicts, calculated sediment budgets for some well studied subduction zones indicate that the cumulative volume of the subduction complex closely matches the incremental volume of sediment entering the trench through time (Davis and Hyndman, 1989).

Further analysis has shown, however, that many or most subduction zones are considerably more complicated than implied by this simple model (Moore and Silver, 1987; Lundberg and Reed, 1991). Several complex processes influence patterns of uplift or subsidence at the trench-slope break and help control the structural configurations of forearc basins (Fig. 6.5B-F):

(a) Lower horizons of the sediment cover on the downgoing oceanic plate may not be detached from igneous oceanic basement at the trench, but instead can be selectively subducted beneath at least some trenchward part of the subduction complex (Moore, 1975). Because of its inherent physical properties, pelagic sea-floor sediment is more prone to undeformed passage beneath the basal decol-

lement of the subduction complex than is turbidite sediment. Consequently, the decollement beneath the toe of the subduction complex may develop along the sedimentary interface between rafted sea-floor sediment and clastic trench fill, rather than along the contact between sedimentary strata and igneous oceanic crust (Hayes and Lewis, 1984; Lewis and Hayes, 1984). The undeformed sediment that is drawn beneath the subduction complex by selective subduction may be underplated to the base of the subduction complex by structural duplexing at depth beneath the trench slope (Fig. 6.5B), rather than being accreted to its frontal edge (Moore et al., 1985; Silver et al., 1985). As the distance of underthrusting prior to underplating can be at least 75 km in some instances (Westbrook et al., 1982), uplift of the trench-slope break may be controlled as much by progressive underplating as by successive accretion of imbricate thrust panels. Forearc uplift produced by underplating has the potential to reduce the width of a forearc

basin as its trenchward flank is backtilted along the margin of the rising subduction complex (Seely, 1979). The potentially immense scale of underplating and its scope for influencing uplift are revealed by recent reflection profiles that have imaged crustal layers interpreted as underplated masses beneath modern subduction complexes at depths of 10 to 35 km and distances of 100 to 250 km from the trench (Moore et al., 1991).

(b) Subduction complexes may enlarge characteristically by subterranean addition, here termed subcretion (Fig. 6.5C), of underplated flow mélange (Cloos, 1982, 1984; Shreve and Cloos, 1986; Cloos and Shreve, 1988a,b), as well as by accretionary offscraping of imbricate thrust panels and underplating of structural duplexes. This insight stems from the postulate that stratally disrupted rock masses termed "mélange" and "broken formation," both of which are common within and largely restricted to subduction complexes, are generated by subterranean flowage of subducted materials. The deformation that produces mélange is inferred to occur within a planar or wedgelike zone, termed the subduction channel, which forms an inclined layer analogous to a lubricant between the descending oceanic plate and the overriding crustal block of a subduction zone. Bearing theory implies that the direction of flowage of material within the subduction channel can be both downward and upward, depending upon buoyancy relations and the ratio of subducted materials to the volumetric capacity of the subduction channel. Theory further implies that masses of flow melange may be either offscraped (accreted) or underplated (subcreted) to the subduction complex. Resultant uplift of the trench-slope break may then bear no simple relation to gradual enlargement of the subduction complex.

(c) Tectonic erosion of the edge and undersurface of the overriding crustal block may occur through interaction with the upper surface of the descending oceanic plate at the decollement beneath the subduction complex (Scholl et al., 1980). Seismic-reflection profiles of selected modern forearc regions have been interpreted to support this suggestion of subduction erosion (e.g., von Huene and Lallemand, 1990), which may be accomplished by subduction of flow melanges. Wherever it does occur, subduction erosion would be expected to trigger an episode of subsidence of the trench-slope break and concomitant lowering of the sill level for the forearc basin (Fig. 6.5D). If tectonic erosion were dominant throughout the evolution of a forearc region, then it might suppress the growth of a subduction complex entirely, and thus discourage the formation of prominent forearc basins for lack of a confining tectonic dam. Moreover, wholesale sediment subduction beneath the flanks of arc-trench systems may well alter the geochemical balance of the mantle (von Huene and Scholl, 1991).

(d) Where structural relief develops between the trench-slope break and the sediment surface within the forearc basin (Fig. 6.5E), backthrusting (with vergence toward the arc) may carry part of the subduction complex structurally over the flank of the forearc basin (Silver and Reed, 1988). For structures of this character to develop, an arcward slope sufficient to maintain a critical thrust taper (e.g., Davis et al., 1983) must presumably exist above the backthrust wedge of subduction complex. Consequently, major backthrusts along the trenchward flanks of forearc basins are expected mainly in cases where forearc basins are underfilled with sediment. If backthrusting does develop, however, it should enhance the tendency for the basin sill at the trench-slope break to remain at a higher elevation than the sediment fill of the forearc basin. Strata along the trenchward flank of the forearc basin may be folded and partly incorporated into the backthrust belt (Unruh et al., 1991).

(e) The global distribution of blueschist metamorphic belts composed of high-pressure facies series implies their generation within deep levels of subduction complexes (Ernst, 1971, 1975). Explanations for eventual exposure of blueschists at the surface include the suggestion that extensional deformation within evolving subduction complexes involves tectonic denudation that exhumes blueschist horizons from beneath higher structural levels (Platt, 1986). Lateral extension, allowing denudational normal faults to develop, might be driven by adjustment of the inclination of the trench slope to a shallower angle in order to recover a critical taper in the aftermath of massive underplating (or subcretion of flow melange). This process could maintain the relief of the trench-slope break at a quasi-stable level, and would cause normal faulting to overprint thrust faulting within the accretionary prism (Fig. 6.5F).

Basin Substratum

The nature of the substratum beneath the buried keels of modern forearc basins is controversial because basin floors are hidden beneath undisturbed basin fill. Subsurface geophysical methods cannot distinguish with confidence among thinned continental crust, overthickened oceanic crust, and metamorphosed subduction complex within the forearc region (Grow, 1973). In some cases, extensions of the arc massif and the subduction complex may be in direct contact in the subsurface at the floor of basin fill (Karig, 1977, 1982). Well exposed stratigraphic bases of the deformed sedimentary fill of many ancient forearc basins indicate, however, that forearc basement is commonly composed, in part, of ophiolitic sequences resembling oceanic crust (e.g., Bailey et al., 1970; Coombs et al., 1976). Ophiolitic basement also lies beneath thin sediment cover in forearc regions of modern intraoceanic arc-trench systems (e.g., Lundberg, 1983), and forearc ophiolites are regarded as a characteristic tectonic element of many collisional orogenic systems (e.g., Gealey, 1980).

Suggested origins of forearc ophiolites are varied (Fig. 6.6). Oceanic lithosphere trapped beneath arc-trench gaps by initiation of subduction at positions offshore from continental margins or island arcs may include several variants (Fig. 6.6A-C): (a) ophiolite

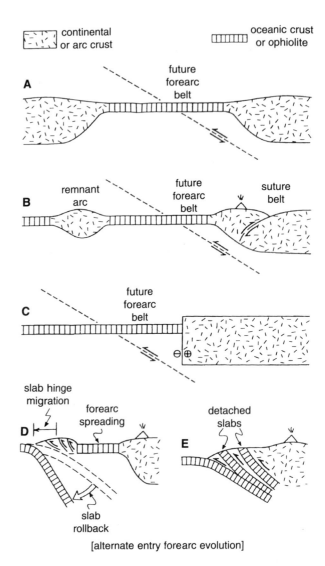

Fig. 6.6 Potential origins of forearc ophiolites (dashed lines for A,B,C denote subsequent locus of inclined subduction to incorporate ophiolite into forearc belt): A, rift ophiolite formed adjacent to a continental margin by seafloor spreading during continental separation that formed a passive continental margin, later activated to form an arc-trench system; B, inter-arc ophiolite from arc rifting prior to arc-continent collision and subsequent reversal of subduction polarity; C, midocean ophiolite from transform slip prior to initiation of subduction; D, nascent ophiolite from forearc spreading; E, tilted ophiolite slabs representing oceanic lithosphere imbricated by subduction accretion. See text for discussion and references.

formed during the rift phase along a passive continental margin that was later converted to an active continental margin (Dickinson and Seely, 1979; Seely, 1979), (b) interarc ophiolite formed by backarc spreading prior to polarity reversal that converted part of a backarc region to a forearc belt (Schweickert and Cowan, 1975; Ingersoll and Schweickert, 1986), or (c) midocean ophiolite emplaced adjacent to an arc massif by transform strike slip prior to initiation of subduction (Karig, 1983a). Moreover, incipient spreading (Fig. 6.5D) caused by extensional tectonism within the forearc belt may generate a special variety of ophiolitic forearc lithosphere (Natland and Tarney, 1981), formed in place and termed "supra-subduction-zone ophiolite" (Pearce et al., 1984). Finally, subduction accretion (Fig. 6.5E) of tilted slabs of oceanic lithosphere can emplace ophiolitic masses within a forearc belt prior to the onset of forearc sedimentation (Karig and Ranken, 1983; Bernstein-Taylor et al., 1992b).

Ophiolitic masses of diverse origins may occur together beneath some complex forearc regions. Within the Mariana forearc, for example, incipient forearc spreading evidently emplaced parallel belts of supra-subduction zone ophiolite between residual belts of oceanic lithosphere that was present prior to the initiation of subduction (Hawkins et al., 1984). Recent work suggests that deformed ophiolitic bodies which represent tectonically accreted oceanic lithosphere also occur within the Mariana forearc belt (Johnson et al., 1991). In addition, the Mariana forearc is dotted with nonvolcanic seamounts that formed in response to serpentinite diapirism, which brought disrupted ophiolitic materials upward to the sea floor (Fryer et al., 1985).

Igneous generation or tectonic emplacement of forearc ophiolites evidently precedes the principal phases of forearc sedimentation because forearc-basin fills are typically lacking in volcanic materials, except for windblown pyroclastic layers and volcaniclastic detritus derived from the adjacent magmatic arc. Moreover, petrogenetic studies suggest that supra-subduction-zone ophiolites of forearc regions are generated during early stages of intraoceanic subduction prior to the development of well defined magmatic arcs (Pearce et al., 1984). Transient exceptions to the amagmatic character of forearc regions arise where subduction of an oceanic spreading ridge alters the thermal state of the forearc region (Marshak and Karig, 1977; Forsythe et al., 1986). During episodes of ridge subduction, granitic plutonism and intermediate volcanism may affect the subduction complex and forearc basin near the trench-slope break (Forsythe and Nelson, 1985; Barker et al., 1992). Ridge subduction may also give rise to forearc ophiolite successions that resemble midocean-ridge assemblages but were formed in place within the forearc belt as the oceanic spreading system impinged upon it (Kaeding et al., 1990).

Along the trenchward flanks of forearc basins, sediment fill commonly rests concordantly on forearc ophiolite sequences along contacts that are gradational between pillow lavas and the overlying sediment column (Dickinson and Seely, 1979). This relationship seemingly favors tectonic models in which forearc ophiolites exist undeformed within the arc-trench gap at the time of initiation of forearc sedimentation. In other cases, however, such as coastal Central America (Lundberg, 1982), ophiolitic masses near the trench-slope break are deformed slabs of oceanic crust broken from the descending plate near the trench axis and carried upward during the succeeding tectonic evolution of the trench slope. Onlap of forearc-basin fill over the trench-slope break may then involve a stratigraphic hiatus at the depositional contact between forearc-basin fill and ophiolitic rocks. Contact relations may not be clear-cut, however, because hemipelagic slope deposits laid down during the transit of the ophiolitic rocks up the trench slope may intervene between the ophiolite sequence and the base of onlapping forearc-basin fill. Close scrutiny of the varied structural and stratigraphic relationships between forearc ophiolites and forearc-basin fill is required to resolve the issue of ophiolite origin in each specific case.

Subsidence Mechanisms

Quantitative studies of forearc subsidence are rare because reliable constraints for modeling are difficult to establish. Although backstripping of sediment fill is straightforward, appraisal of net subsidence through

time is complicated by the difficulty of establishing rigorous paleobathymetric controls for inferred basin depth during phases of turbidite sedimentation in deep water (Dickinson et al., 1987). Moreover, gauging the isostatic balance of crustal masses and mantle bodies beneath and adjacent to forearc basins presents a severe challenge. Particularly obscure is the nature of flexural coupling among the descending oceanic lithosphere, the subduction complex, the forearc substratum, and the arc massif. Even more fundamental isostatic parameters are also commonly uncertain because the surroundings of forearc basins are not static masses of crust or mantle. In devising basin models, allowance must be made for (a) the changing thickness and expanse of the subduction complex, (b) the changing crustal profile and thermal regime of the arc massif (and perhaps changes in the nature of the mantle beneath it as well), and (c) the varying buoyancy of the slab of oceanic lithosphere descending beneath the forearc region as the age of the seafloor being subducted changes through time.

Subsidence histories of forearc basins are thus less amenable to geodynamic analysis, and less predictable from first principles, than those of rifted continental margins and foreland basins, where single or closely linked mechanisms of subsidence predominate during basin evolution. Moreover, different parts of individual forearc basins may respond primarily to different controls on subsidence. In the Great Valley forearc basin of California, for example, patterns of subsidence differed substantially across the forearc belt during Cretaceous and Paleogene time (Moxon and Graham, 1987). The arcward flank of the basin underwent gradual thermotectonic subsidence as arc lithosphere cooled during onlap by basin fill. By contrast, the trenchward flank of the basin experienced variable rates of both tectonic subsidence and uplift as the angle of descent and the buoyancy of the subducted slab of oceanic lithosphere passing beneath the Franciscan subduction complex varied through time. Isostatic subsidence of the basin floor in response to the growing sediment load deposited along the deep keel of the forearc trough was also a major factor in the overall subsidence of the Great Valley forearc basin as it filled with sediment. Little or no net tectonic subsidence is required to explain the

accumulation of much of the central part of the sediment prism as it evolved from deep-water turbidite facies at lower horizons to shallow-water shelf and deltaic facies at upper horizons (Dickinson et al., 1987). In other cases, however, net rates of tectonic subsidence in forearc basins along active continental margins influenced by processes related to subduction can clearly exceed rates of thermotectonic subsidence along passive continental margins (Heller, 1983).

In summary, the subsidence of forearc basins can be ascribed to some combination of the following influences (Fig. 6.7):

(a) Bulk subsidence of the whole forearc region as older and less buoyant oceanic lithosphere is drawn into the subduction zone and descends beneath the arc-trench system; this factor is inherently transient and potentially reversible to induce rebound of the basin floor.

(b) Flexural subsidence of forearc substratum under the growing tectonic load of the subduction complex; this factor would influence subsidence mainly below some trenchward part of the forearc basin, but could be especially significant where backthrusting of subduction complex occurs near the trench-slope break.

(c) Passive isostatic subsidence of initially thin forearc crust under the growing sediment load of the forearc basin itself; flexural coupling of forearc sediment fill to the adjacent arc massif and subduction complex may also

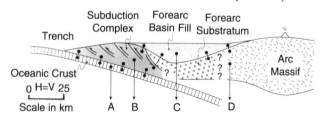

Fig. 6.7 Key factors (diagrammatic) at crustal levels influencing subsidence of forearc basins (arrows): A, negative buoyancy of slab of descending oceanic lithosphere (only capping oceanic crust shown); B, isostatic tectonic load of subduction complex; C, isostatic sedimentary load of forearc-basin fill; D, thermotectonic subsidence of flank of magmatic arc massif (arrows A,B,D potentially reversible to induce uplift of basin flanks and rebound of basin floor). Barbells denote flexural couplings of oceanic lithosphere and subduction complex, subduction complex and forearc-basin fill, and forearc-basin fill and arc massif.

warp the flanks of the forearc basin downward as the sediment prism thickens along its keel.

(d) Thermotectonic subsidence of the flank of the arc massif as retrograde migration of arc magmatism carries the principal locus of heat advection farther away from the keel of the forearc basin; this factor could directly influence subsidence only below some arcward part of the forearc basin.

Internal Structures

Deformation during sedimentation within forearc basins, and along their flanks, is varied and not yet well understood. Subsurface belts of recurrent normal or reverse faulting may occur along the intracrustal join between the forearc substratum and basement of the arc massif (Dickinson and Seely, 1979), but are not present for all forearc basins. For intraoceanic forearcs, this locus of potential deformation is termed the "upper-slope discontinuity" (Karig and Sharman, 1975). Differential uplift of the subduction complex along the trenchward flank of a forearc basin may backtilt the floor of the basin toward this structural discontinuity along its arcward flank (Scholl and Herzer, 1992). Along the trenchward flank, basin fill may be underthrust by the subduction complex (Dickinson and Seely, 1979), or backthrusts may foster the opposite structural relationship (Silver and Reed, 1988).

Within the interior of forearc basins, either contractional or extensional deformation may occur during forearc sedimentation. Syndepositional folding about an axis parallel to tectonic strike is perhaps best shown by seismic reflectors within the Tofino basin off Vancouver Island (Hyndman et al., 1990). In other forearc belts, normal faults that trend parallel to tectonic strike are widespread (Lundberg, 1983). Conglomeratic facies of Cretaceous forearc strata in Baja California have been interpreted as the sedimentary record of distributed normal faulting that produced half-graben sub-basins across nearly the full width of a major forearc basin (Morris et al., 1989). Little is known in detail about the relationship of intrabasinal structures to relevant subduction parameters, but such factors as the plate convergence rate at the trench, the dip of the subducted slab of oceanic lithosphere beneath the forearc belt, and the motion of the arc massif relative to the rollback of the subducted slab into the asthenosphere may all influence tectonism within forearc basins.

Strike slip is also significant on faults that trend longitudinally along some forearc belts, or cut obliquely across them. Although plate convergence is roughly normal to most trenches (Scotese and Rowley, 1985), arc-parallel strike-slip fault systems occur in half of all modern arc-trench systems in response to some obliquity of subduction (Jarrard, 1986b). In many cases, the principal slip zone lies along the thermally softened trend of the arc itself, and serves as a transform to separate a forearc sliver plate from the bulk of the overriding plate; the obliquity of subduction is apparently resolved or partitioned into nearly normal convergence at the trench and lateral slip along the arc-parallel transform (Fitch, 1972; DeMets, 1992). Ideally, the forearc basin rides along undeformed as part of the forearc sliver plate. Where the locus of arc-parallel strike slip is not wholly confined to the arc massif, however, splay faults or subsidiary faults may slice through the forearc region to influence depositional systems of forearc basins.

Where the obliquity of subduction is variable along the trend of an arc-trench system, longitudinal extension may be induced within the forearc belt as linked segments of the arc-trench gap slip laterally at different rates (McCaffrey, 1992). Longitudinal buckling may also logically occur within forearc regions, where the variation in obliquity of subduction is the reverse of that required to induce extension. It is tempting to infer that much of the segmentation of elongate forearc basins into multiple depocenters may stem from such influences related to variable obliquity of subduction along strike. In response to longitudinal extension or contraction within the forearc belt, transverse zones of diffuse crustal stretching might allow differential subsidence of discrete depocenters, or broad transverse arches might serve to separate discrete depocenters. At present, however, clear evidence for such behavior is generally lacking.

Basin Fill

As the subduction complex grows beneath the basin sill and forearc deposition proceeds, the net tendency of basin evolution is typically toward depositional systems that shoal through time. Accordingly, the most characteristic stratigraphic column for major forearc basins is an upward shoaling megasequence, with turbidites overlain by shelf and nonmarine strata. Because the crests of arc massifs commonly form regional drainage divides, adjacent magmatic arcs are the main provenance for most forearc basins. Exceptions may occur where major river systems cut through the arc trend from more interior parts of continents or where complex tectonic history fosters atypical patterns of sediment dispersal. Where subduction complexes are strongly uplifted, supplemental sources along the trench-slope break or in uplifts between compound forearc basins may also contribute recycled sediment and ophiolitic debris in varying proportions to adjacent forearc basins. Deep burial of immature detritus promotes widespread burial metamorphism or advanced diagenesis within forearc-basin fill. Consequently, porosity is commonly reduced at a relatively rapid rate during progressive burial (Galloway, 1974).

Lithofacies Patterns

Because highlands of the arc massif typically provide a nearby source of abundant detritus, clastic sediments are generally predominant in forearc basins. Water depths and locations outside the tropical belt of reef growth also inhibit carbonate sedimentation in many forearc basins. In the tropical intraoceanic setting of Tonga, however, where only a few volcanic islands project above sea level, a forearc carbonate platform 250m-thick caps an underlying sequence (~2500 m) of pelagic and volcaniclastic sediment deposited along the flank of the arc massif (Scholl et al., 1985). The proportion of pyroclastic material and the abundance of ashfall tuff layers within forearc basins is largely dependent upon the orientation of the magmatic arc with respect to prevailing wind directions (Sigurdsson et al., 1980).

The proportions of conglomeratic, sandy, and shaly sediment in forearc basins are highly variable

and probably dependent largely upon weathering regimes within the arc massif, although eustatic fluctuations in sea level and episodicity of arc tectonism may also be significant influences. Most forearc-basin fill is composed dominantly of interbedded sandstone and shale, with conglomeratic intervals largely confined to proximal sites near basin margins and to transport paths of high energy along submarine or fluviodeltaic channels. Many forearc basins function as efficient sediment traps that pond large volumes of arc-derived sediment. There is no question, however, that varying proportions of forearc sediment can also be bypassed into the trench, either by unconfined turbidity currents or down trench-slope submarine canyons that head in the forearc region (Underwood, 1991; also see Chapter 5).

As in other sedimentary basins, the ratio of turbidite sediment to shallow-marine or terrestrial deposits is governed by the interplay among initial water depth, subsidence rate, and sedimentation rate. Forearc basins with sills that remain in deep water, or associated with intraoceanic arc-trench systems where the supply of terrestrial sediment is minor, may accumulate only turbidites and hemipelagites throughout their depositional histories. Where the subduction complex grows through time to produce ridged forearcs, however, progressive sedimentation can produce shelved forearc basins, where shallow-marine and deltaic depositional systems are dominant. Where accretion and aggradation are pronounced, forearc basins may even become wholly terrestrial depocenters (e.g., Farhoudi and Karig, 1977).

Transverse to basin trend, slopes of the sediment-water interface are typically asymmetric, with the deepest water at or near the bounding sill formed by the subduction complex (Fig. 6.8). Where uplift of the trench-slope break is sufficient, ponding of sediment during underfilled phases (Fig. 6.4) of basin evolution produces basin-plain turbidite environments of limited areal extent along the keel of a deep forearc trough. Lateral progradation of depositional systems causes shelf-slope assemblages and related turbidite fans to encroach upon the basin plain from the arcward flank of the basin (e.g., Ingersoll, 1979; Beaudry and Moore, 1985). Submarine-canyon fills

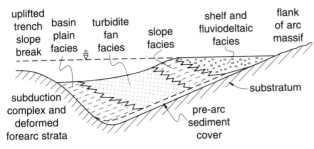

Fig. 6.8 Facies framework of ideal forearc basin shown in transverse profile (not to scale); character of depositional systems and widths of facies tracts highly variable (see text); pre-arc sediment cover may range from oceanic pelagite and hemipelagite to foundered shelf deposits; influence of intrabasinal structures not depicted.

composed of detritus derived from the arc massif are also prominent along the arcward flanks of some forearc basins (Morris and Busby-Spera, 1988). In tropical settings, shelf assemblages may include prominent carbonate buildups with associated slope aprons of carbonate turbidite deposits, and basinal successions in deeper water include calcareous pela-gites as well as clastic hemipelagites (Beaudry and Moore, 1981, 1985). Much of the sediment within intraoceanic forearc basins of the tropical belt may consist of chalk and marl interstratified with volcani-clastic intervals (Exon and Marlow, 1988; Marlow et al., 1988).

Contrasts in the subsidence history of the basin axis and in the uplift history of the trench-slope break may cause successive forearc depocenters and associated facies belts to migrate either arcward or trenchward during forearc sedimentation, and no single evolutionary model is applicable to all forearc basins (Beaudry and Moore, 1985). Consequently, shallow-marine and turbidite depositional systems may evolve in complex patterns to produce varied facies frameworks in both time and space. As mature forearc basins evolve toward overfilled phases (Fig. 6.4) of sedimentation, fluviodeltaic complexes and their derivative turbidite fans or foredelta ramps may prograde either laterally or longitudinally across the basin floor (Cherven, 1983; Heller and Dickinson, 1985). Moreover, where extensional deformation is severe during basin evolution, half-graben sub-basins may be filled by locally derived fan-delta and tur-bidite-fan complexes shed from intrabasinal fault blocks oriented either transverse or longitudinal to

the trend of the arc-trench system (Busby-Spera and Boles, 1986; Morris et al., 1989; Smith and Busby, 1992).

Petrofacies Relations

Sandstones derived from arc massifs form volcano-plutonic suites with subquartzose frameworks rang-ing from feldspatholithic to lithofeldspathic (Dickin-son, 1982). Aphanitic lithic fragments are derived principally from volcanic and metavolcanic rocks, although contributions from sedimentary and meta-sedimentary wallrocks in the roots of the arcs are also characteristic. Feldspar sand grains are derived both from plutonic rocks and from phenocrysts in lavas or tuffs, but abundant K-feldspar commonly reflects plutonic derivation. Volcanic quartz, typi-cally bipyramidal with straight extinction, occurs in minor volume; most quartz grains in more quartzose variants of forearc sandstone suites are derived from granitoid intrusive rocks. Some recycled or meta-morphic quartz may also be present. Clasts in con-glomerates include not only volcanic, metavolcanic, and plutonic rocks of the igneous arc assemblage, but also metasedimentary rocks of varying grade from the wallrock assemblage in the roots of the magmatic arc.

The degree of dissection of the arc massif through time is the principal factor controlling petrofacies variation (Fig. 6.9) from feldspatholithic frameworks of dominantly volcanic derivation to lithofeldspathic (and even quartzofeldspathic) frameworks of domi-nantly plutonic and metamorphic derivation. Undis-sected arc massifs are volcanic chains (continental, peninsular, or insular) that deliver mainly quartz-poor volcaniclastic detritus to adjacent forearc basins. As erosion carves into uplifted arc massifs, subjacent plutons begin to supply arkosic detritus to the forearc region. The transition from volcaniclastic to plutoni-clastic sands is achieved more easily for magmatic arcs along continental margins, where crust is thick, than for island arcs where the arc massif may remain partly submerged for long intervals. Rupture of the arc massif by arc-parallel transforms or related faults may enhance exposure of plutons and their wallrocks (Marsaglia and Ingersoll, 1992).

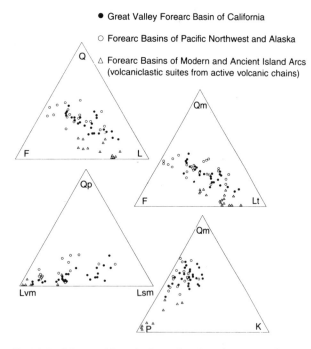

Fig. 6.9 Modal compositions of volcanoplutonic sandstone suites from forearc basins, where Q (total quartzose grains) = Qm (monocrystalline quartz grains) + Qp (polycrystalline quartzose lithic fragments), L (total unstable lithic fragments) = Lvm (volcanic and metavolcanic lithic fragments) + Lsm (sedimentary and metasedimentary lithic fragments), and F (total feldspar grains) = P (plagioclase grains) + K (K-feldspar grains). Points plotted are reported or calculated means for selected forearc sandstone suites from Dickinson (1982), Dickinson et al. (1982), Ward and Stanley (1982), Cawood (1983), Heller and Ryberg (1983), Ingersoll (1983), Johnson (1984), Korsch (1984), Short and Ingersoll (1990), and Lundberg (1991).

The interplay of eruption and erosion in mature arc massifs leads to mixed volcanic-plutonic provenance, whose evolving character cannot be predicted in detail but can be specified by study of forearc-basin fill. For ancient forearc basins, the petrofacies at different horizons form the only reliable record of structural levels eroded from the arc massif (Dickinson and Rich, 1972). Recent isotopic evidence from the Great Valley Group of California showing that the bulk of its arc detritus was derived from near the volcanic or magmatic front, where igneous activity is at its peak at any given time, suggests that the changing petrologic character of arc magmas through time can be monitored by study of forearc-basin fill (Linn et al., 1991, 1992).

Within arc-trench systems that terminate at one end against transforms bounding a continental block, or pass along strike into an evolving collisional orogen, forearc basins may derive sediment from rivers draining orogenic highlands or cratonic lowlands apart from the arc massif. No examples are well documented at present, but these provenances are analogous to those that feed remnant ocean basins (see Chapter 10). Extra-arc sources would also be significant where major river systems terminate in forearc basins.

Burial Metamorphism

Recognition of the zeolite mineral facies, which bridges the gap between diagenesis and metamorphism (Boles and Coombs, 1977), was based upon studies of volcaniclastic strata in the Hokonui forearc basin of New Zealand (Coombs, 1954); the concept of "burial metamorphism" (Coombs, 1961) stemmed from the same body of work. "Burial metamorphism" refers to pervasive mineralogical reconstitution of strata on a regional scale at low temperatures attained by simple burial without the influence of either igneous intrusion or penetrative deformation. Much forearc-basin fill is especially susceptible to widespread burial metamorphism because geochemically immature, and therefore reactive, clastic materials are carried to great depths along the keels of major forearc basins without undergoing strong deformation or being affected by advective magmatic heat. As quite low geothermal gradients are characteristic of many forearc basins, secondary mineral assemblages including zeolites, prehnite, pumpellyite, and albite are common at depth within basin fill. Most burial metamorphism does not interfere with the general provenance analysis of sandstones because primary detrital textures are clearly visible in thin section through a diagenetic screen of pseudomorphous replacement (Dickinson, 1970b). Even the internal textures of volcanic rock fragments, for example, are typically well preserved. All plagioclase, however, whether occurring as separate detrital grains or within igneous lithic fragments, is commonly converted to zeolite or to albite, with or without microscopic inclusions of hydrous lime silicates. Provenance criteria that depend upon plagioclase composition cannot then be applied. Mafic accessory minerals may also be converted wholly to chlorite,

with additional loss of otherwise useful information. Locally, widespread replacement of sand frameworks by calcite may render provenance analysis impossible for selected rock samples.

Hydrocarbon Resources

Only three of the more than 500 giant oil and gas fields known in the world are located in forearc basins (Carmalt and St. John, 1986), and forearc coal reserves are similarly restricted in comparison to those of the major coal-bearing sequences of the world. Known forearc petroleum and natural-gas reserves are most voluminous in three continental-margin forearc basins widely spaced along the eastern flank of the Pacific Ocean:

(a) The Cook Inlet part of the Cook Inlet-Shelikof Strait forearc basin in coastal Alaska, where nonmarine clastic reservoirs of mid-Tertiary age occur along an inner belt of forearc depocenters within a compound forearc system developed by subduction that was initiated in mid-Mesozoic time (Fisher and Magoon, 1978; Fisher et al., 1987).

(b) The Sacramento sub-basin of the Great Valley forearc basin in California, where uppermost Cretaceous to Paleogene clastic reservoirs of shelf and deltaic facies occur within a relict forearc basin preserved after conversion of the adjacent continental margin to a transform plate boundary during Neogene time (Morrison et al., 1971; also see subsequent passages in text).

(c) The Talara forearc basin of coastal Peru, where intricately deformed clastic reservoirs of Tertiary age closely associated structurally with the subduction complex occur along an outer belt of forearc depocenters within a compound forearc system flanking the Mesozoic-Cenozoic Andean orogen (Stabler, 1990).

Although exploration is still too incomplete in many other forearc basins to rule them out as potential hydrocarbon provinces, several unfavorable factors combine to limit attractive prospects in typical cases: (a) deposition either under marine conditions or on narrow coastal plains, where environments for coal deposition are either absent or not areally extensive; (b) widespread deposition in open marine environments, where attractive facies rich in preserved organic matter are confined to narrow belts of slope facies deposited within the oxygen-minimum zone; (c) the characteristic dominance of immature clastic sediment and consequent diagenctic loss of porosity during burial of potential sandstone reservoirs for fluid hydrocarbons; and (d) low geothermal gradients, which suppress or delay maturation of buried organic matter past the times during which adequate reservoir porosity might be preserved within basin fill. Preservation of porosity may be the most critical factor for successful exploration, as the reservoirs in all three known forearc oil and gas provinces occupy comparatively shallow horizons of basin fill; deeper horizons are thus far unproductive in each case.

The best chances for additional major hydrocarbon discoveries probably lie in forearc basins where (a) gas-prone deltaic depositional systems are extensive; (b) arkosic sandstones, derived from denuded granitoid rocks of the arc massif and able to preserve adequate porosity despite deep burial, are present in addition to more lithic varieties that lose porosity rapidly with burial; or (c) carbonate buildups that might represent attractive reservoir facies occur as part of basin-flank shelf assemblages.

Modern Forearc Basins

The structural framework and varied morphology of forearc basins can be understood best with reference to modern settings, for which regional tectonic relations are unambiguous. Because drilling is not extensive, however, in most modern forearc regions, the internal stratigraphy of forearc-basin fill is known mostly from reflection profiling. Detailed sedimentological relationships within forearc basins are consequently best known for ancient analogs, where exposures are available for study on land.

The global spectrum of modern forearc basins (Table 6.1) is illustrated by the following selected examples: (1) forearc basins off Sumatra and Java in Indonesia, where oceanic lithosphere passes beneath the partly submerged continental crust of the Sunda Shelf projection of Eurasia, (2) forearc basins of varied

origin along the South American continental margin adjacent to the Andean orogen, (3) the Tobago Trough of the intraoceanic Lesser Antilles arc-trench system, and (4) forearc basins that entirely overlie previously accreted subduction complexes along the flank of the migratory Japanese island arc.

Sunda Forearc Basins

The Sunda arc-trench system (Fig. 6.10), more than 5000 km long, reflects subduction of the Indian or Australian plate beneath the southeast Asian portion of the Eurasian plate (Moore et al., 1982). The subduction system extends along strike to the northwest into the continental collision belt of the Indoburman Ranges and to the southeast into the incipient collision zone between the Banda arc and Australia. Parallel to the insular volcanic chain of Java and Sumatra, which border the submerged Sunda Shelf, a ridged forearc basin approximately 100 to 125 km wide is continuous for at least 2500 km along strike, although it is broken into multiple sub-basins (Table

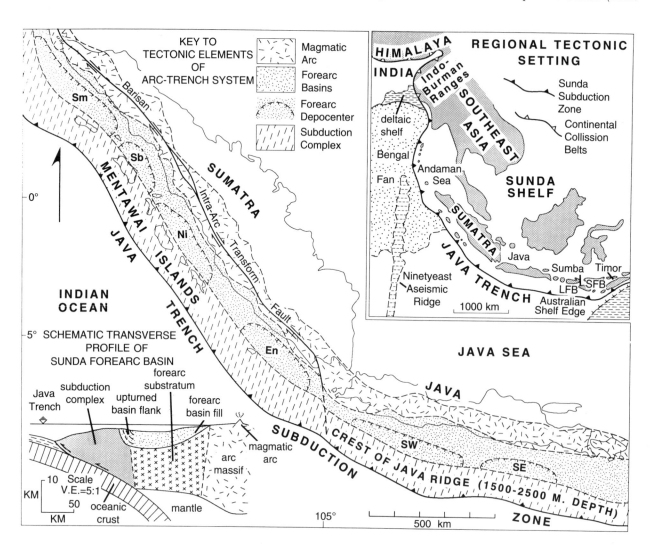

Fig. 6.10 Configuration of elongate forearc basin, with multiple depocenters, in Sunda arc-trench system of Sumatra and Java (Hamilton, 1979; Karig et al., 1979, 1980a; Kieckhefer et al., 1981). Inset shows relation of Sunda subduction zone (note position of Sumatra and Java) to Himalayan suture belt and incipient arc-continent collision at Timor (LFB, Lombok forearc basin; SFB, Savu forearc basin). Forearc depocenters off Sumatra-Java (Table 6.1, III): Sm, Simeulue; Ni, Nias; Sb, Siberut; En, Enggano; SW, southwest Java; SE, southeast Java.

6.1, III), each 250 to 500 km long, by transverse structural highs (Hamilton, 1979). All three tectonic elements of the forearc region display a strike-parallel regional slope to the southeast, with a net gradient of nearly a meter per kilometer (Hamilton, 1979; Moore et al., 1982): (a) the trench floor deepens from 4000 m near northern Sumatra to 6000 m near eastern Java, (b) the trench-slope break at the crest of the subduction complex on the trenchward flank of the forearc trough stands near sea level in the Mentawai Islands off Sumatra, but typically lies at depths of 1500 to 2500 m off Java, and (c) the forearc basin underlies a structural trough that ranges in water depth from 1000 to 2000 m in sub-basins off Sumatra to 3000 to 4000 m in sub-basins off Java. The longitudinal change in forearc morphology is controlled by the increasing thickness of seafloor sediment entering the trench from the northwest (Moore et al., 1980a). This greater sediment thickness reflects southward progradation of immense submarine fans down the floor of the Bay of Bengal from sources in the Himalayan collision belt (see Chapter 10), and has led to both a shallower trench and construction of a bulkier subduction complex toward the northwest. The southeastward slope of the forearc trough reflects more voluminous sediment sources in Sumatra than in Java.

Although the age of the basal horizons of forearc-basin fill is uncertain, sedimentation probably began during mid-Tertiary time (Beaudry and Moore, 1985). The forearc morphology was initially terraced (locally shelved), but evolved into the present ridged morphology as the subduction complex grew in bulk

during Neogene time (Matson and Moore, 1992). The total thickness of sediment fill reaches 5000 to 6000 m beneath prominent sub-basins, but is only 2500 to 3500 m thick over the structural highs that separate them (Hamilton, 1979; Moore et al., 1982). Seismic reflectors are sharply backtilted where they overlie the flank of the subduction complex (see transverse profile of Fig. 6.10), but are mostly flat-lying beneath the keel of the forearc trough. Strata onlap the flank of the arc massif along a gently inclined unconformity that buries a narrow insular shelf bordered by a paleoshelf break (Karig et al., 1980a).

The provenance for most terrigenous detritus in the forearc basin is inferred to be the arc massif and active volcanoes of Sumatra and Java. The character of reflectors on seismic records implies that turbidite sedimentation has been dominant along the trough floor throughout basin evolution, but shelf deposits flank the arc massif, and both deltaic and reefal systems form progradational intervals at multiple horizons within the basin fill (Beaudry and Moore, 1981). Prominent buried carbonate platforms developed locally above the paleoshelf break. Figure 6.11 depicts an interpretive facies framework that appears to be characteristic for much of the Sunda forearc basin, but local variations are clearly common along strike. Areal and temporal variations in subsidence rate were evidently controlled largely by the changing configuration of the flexural load represented by the subduction complex (Matson and Moore, 1992).

Strike-slip faults splay into the forearc region from the intra-arc transform of the Barisan fault system in

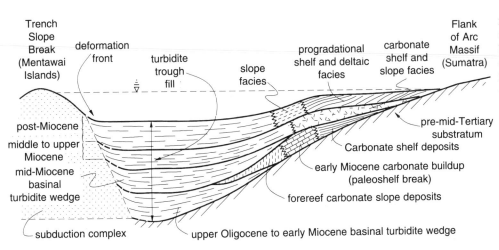

Fig. 6.11 Schematic facies framework of Sunda forearc basin off Sumatra (Fig. 6.10); adapted from seismic-reflection profiles interpreted by Karig et al. (1980a), Beaudry and Moore (1981, 1985), and Matson and Moore (1992).

Sumatra, and slice obliquely through the forearc sliver plate (Moore et al., 1980). These faults may be responsible for segmentation of the forearc basin into discrete sub-basins (Karig et al., 1980a; Matson and Moore, 1992). Recent evidence for longitudinal strike slip along a buried fault, located on the arcward flank of the crest of the subduction complex in the Mentawai Islands (Fig. 6.10), suggests that the fore-arc sliver plate may also be sliced into multiple parallel domains by distributed strike slip (Diament et al., 1992). Transpressional and transtensional effects related to slip on forearc strike-slip faults may have influenced the tectonic evolution of individual sub-basins, but such relations are largely undocumented at present. Extensional deformation that apparently controls the location of the topographic and structural break at the submerged strait between Sumatra and Java has been ascribed, however, to variation in the rate of arc-parallel strike slip in response to a change in the obliquity of subduction along the trend of the arc-trench system (McCaffrey, 1991).

The nature of the substratum beneath the keel of the Sunda forearc basin is still uncertain (Kieckhefer et al., 1981), but has been interpreted variously as (a) anomalously thick oceanic crust trapped within the forearc region (Curray et al., 1977), (b) thinned continental crust along the flank of the Sunda Shelf continental block (Kieckhefer et al., 1980), and (c) a metamorphosed accretionary prism that has subsided between the arc massif and the uplifted modern subduction complex (Karig et al., 1980a). Along the coast of Sumatra, deformed strata of a buried Paleogene forearc basin may underlie the shelf flank of the Neogene forearc basin (Karig et al., 1979), and their occurrence offers tentative support for the notion that a buried Paleogene accretionary prism underlies the keel of the modern forearc basin.

Near the point where the Sunda and Banda arc-trench systems join along strike east of Java, incipient arc-continent collision has drawn the edge of the Australian continent northward into the subduction zone (Hamilton, 1979; von der Borch, 1979; Johnston and Bowin, 1981). As a result, the accretionary prism has been uplifted to form the large island of Timor (Jacobson et al., 1979; McBride and Karig, 1987), and prominent backthrusting has carried the arcward edge of the subduction complex over the flank of the Savu forearc basin to the north (Reed et al., 1986). The Timor subduction complex includes not only frontally accreted imbricate thrust panels representing a pre-collisional oceanic forearc, but also underplated thrust sheets composed of strata peeled at depth off the underthrust Australian continental margin (Charlton et al., 1991). Deformation within the forearc region has also enhanced forearc segmentation and isolated the Savu forearc basin north of Timor from the Lombok forearc basin to the west by uplift of the island of Sumba (van Weering et al., 1989). These tectonic relations near the junction of the Sunda and Banda arcs represent an early stage of transition from an active forearc basin within an arc orogen to a sutural forearc basin within a collisional orogen (see also Chapters 10 and 11).

Andean Continental Margin

The Peru-Chile Trench extends for more than 5000 km along the western margin of South America, where the oceanic Nazca plate descends beneath the continent, and is associated with a shelved and terraced forearc (Fig. 6.12). Geophysical data, supported by calculations comparing the cumulative volume of seafloor and trench sediments available for offscraping with the volume of the accretionary prism offshore, suggest that sediment subduction and tectonic erosion have been important processes along the trench axis (Hussong and Wipperman, 1981). A wedgelike fringe of continental basement underlies a belt of forearc basins beneath the narrow continental shelf, and beneath structurally analogous offshore terraces at depths of 1200 to 1500 m, where their sediment fill is ponded behind the trench-slope break (Kulm et al., 1981b, 1982; Thornburg and Kulm, 1981). Parallel but smaller trench-slope basins farther offshore rest upon the subduction complex. On average, the trench slope between the trench axis and the trench-slope break is 50 to 75 km wide, and the fore-arc shelf or terrace forms an adjacent belt 75 to 100 km wide (Coulbourn, 1981; Kulm et al., 1981b). Individual depocenters of the forearc shelf-terrace belt are 150 to 450 km long (Table 6.1, I). Extensional forearc tectonism is recorded by multiple nor-

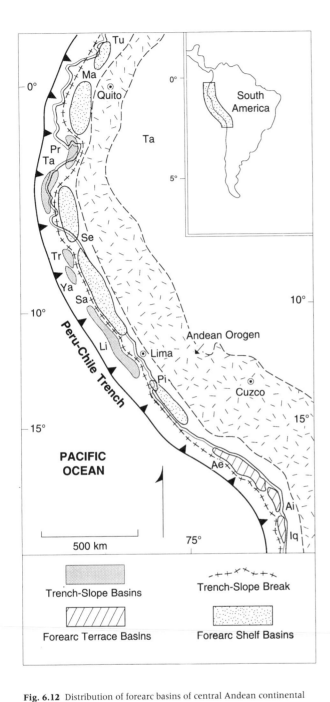

Fig. 6.12 Distribution of forearc basins of central Andean continental margin (Lonsdale, 1978; Coulbourn, 1981; Thornburg and Kulm, 1981). Double line indicates coastline. Inset shows location off South America. Basin names (Table 6.1, I): Ae, Arequipa; Ai, Arica; Iq, Iquique; Li, Lima; Ma, Manabi; Pi, Pisco; Pr, Progresso; Sa, Salaverry; Se, Sechura; Ta, Talara; Tr, Trujillo; Tu, Tumaco; Ya, Yaquina (onshore shelf basins include terrestrial deposystems). See Figure 6.4 for comparative morphology of shelf and terrace basins.

mal faults that repeatedly offset the stratigraphic succession in several of the forearc basins (Moore and Taylor, 1988; von Huene and Miller, 1988).

Reflection profiles suggest that forearc shelf basins along much of the coast are symmetric sags, but forearc-basin fill beneath deeper marine terraces is backtilted toward the arc massif, and subterrace depocenters migrated away from the trench through time (Coulbourn and Moberly, 1977; Moberly et al., 1982). The evolutionary trend is attributed to progressive uplift of the trench-slope break as accretion and underplating gradually enlarged the subduction complex. Total sediment thickness in typical Andean forearc basins ranges from 1500 to 2500 m (Thornburg and Kulm, 1981; Moberly et al., 1982). Little is known in detail of the depositional systems within the basins, but their settings suggest that shelf and slope deposits are abundant, although turbidites underlie the floors of the terrace basins. Limited drilling shows that terrigenous and carbonate contourite mudstones are also widespread in the forearc terrace basins (Ballesteros et al., 1988; Kulm et al., 1988). On the north, where the trench-slope break lies above sea level along the coast (Fig. 6.12), thicker basin fill beneath lowland plains farther inland is composed, in large part, of intercalated deltaic and turbidite strata (Lonsdale, 1978). A staircase of normal faults separates these terrestrial forearc basins from the arc massif, and normal faults may partly define the arcward flank of the backtilted terrace basins farther south.

Antillean Tobago Trough

The subduction complex in the intraoceanic arc-trench system of the Lesser Antilles (Fig. 6.13), where Atlantic seafloor descends beneath the Caribbean plate, is one of the largest in the world. The Barbados Ridge along the trench-slope break is underlain by approximately 20 km of accretionary crustal material and the width of the trench slope from the toe of the subduction zone to the crest of the ridge is fully 200 km along a transect crossing the island of Barbados (Westbrook, 1975; Boynton et al., 1979). The great size of the subduction complex is ascribed to offscraping and underplating (Westbrook et al., 1988) of voluminous seafloor

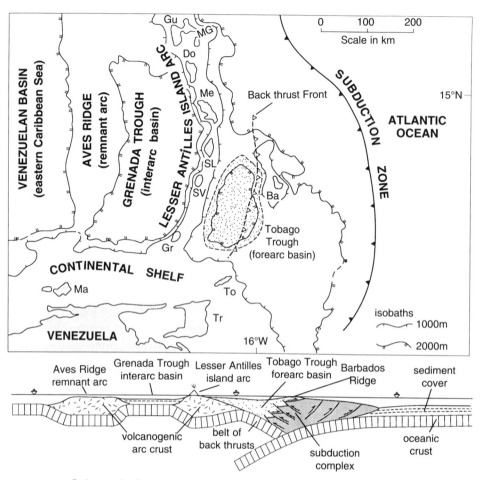

Fig. 6.13 Tectonic setting (map) and inferred structural relations (section) of Tobago Trough fore-arc basin (Table 6.1, VI) within Lesser Antilles arc-trench system (Westbrook, 1975, 1982; Speed et al., 1989; Torrini and Speed, 1989); arcward vergent folds deform forearc-basin fill at shallow levels above zone of backthrusting at depth. Islands shown: Ba, Barbados; Do, Dominica; Gr, Grenada; Gu, Guadeloupe; Ma, Margarita; Me, Martinique; MG, Marie Galante; SL, St. Lucia; SV, St. Vincent; To, Tobago; Tr, Trinidad.

Schematic Crustal Profile Across Tobago Trough (V.E. = 2:1)

turbidites delivered to the Atlantic seafloor from orogenic and cratonic sources within the South American continent, which lies to the south (Velbel, 1985; Kasper and Larue, 1986). The width and structural relief of the subduction complex decrease systematically from south to north along the trend of the arc-trench system, which in effect, has subducted a huge sediment wedge sideways with respect to its depositional isopachs. The prominently ridged Neogene forearc probably evolved from a terraced Paleogene forearc as growth of the subduction complex proceeded (Westbrook, 1982). The oldest exposed strata of the accretionary prism on Barbados are Eocene (Speed and Larue, 1982).

The Tobago Trough (Table 6.1, VI) is a forearc basin of undeformed and gently deformed sediment reaching thicknesses in excess of 5000 m in the southern

Lesser Antilles forearc region, where the subduction complex is largest (Westbrook, 1982). The basin fill is ponded between the Barbados Ridge and the volcanic chain of the Lesser Antilles, where volcanogenic crust is 30 to 35 km thick (Boynton et al., 1979). The main depocenter is not markedly elongate, being 100 km wide and only 200 km long. The floor of the forearc trough lies at a depth of almost 2500 m, whereas the trench-slope break stands at depths of less than 1000 m. The insular ridge along the volcanic island arc lies within a few-hundred meters of sea level. The provenance of the forearc-basin fill is not well established, but may include the continental margin along tectonic strike to the south in addition to nearby volcaniclastic sources within the adjacent volcanic chain. The substratum beneath the keel of the Tobago Trough has seismic velocity characteristic of mafic

crust, and its upper surface slopes consistently trenchward across the basin, as the thickness of crustal basement decreases from 25 to 30 km along the flank of the arc massif to only 8 to 12 km in the subsurface near the trench-slope break. Although the origin of the forearc basement is uncertain (Speed et al., 1989), it may represent a remnant of pre-arc oceanic crust (Boynton et al., 1979) partly covered by a wedgelike flank of the arc massif.

Combined reflection profiling and seismic-refraction studies suggest that the Tobago Trough depocenter contains approximately 10,000 m of flat-lying basin fill, including mid-Eocene strata near its base (Speed et al., 1989). Gently dipping basin fill laps depositionally onto the flank of the arc massif, but the trenchward flank of basin fill is abruptly upturned against the structural flank of the subduction complex. Moreover, much of the thickness of the basin fill within a belt 50 km wide adjacent to the trench-slope break is involved in arcward-vergent folds that are inferred to reflect incipient backthrusting of the subduction complex toward the arc massif (Westbrook, 1975, 1982; Speed and Larue, 1982; Speed et al., 1989).

Backthrusting is not, however, a full explanation of structural relations along the trench-slope break. Reflection profiling at sea and field observations on Barbados indicate that trenchward-vergent thrusting has placed nappe-like thrust sheets and duplexes composed of deformed forearc-basin strata above structurally underlying subduction complex (Speed and Larue, 1982; Torrini et al., 1985; Torrini and Speed, 1989). This structural relationship, which reflects underthrusting of strata in the forearc basin by the subduction complex, dominates the shallow subsurface beneath the trench-slope break. Balanced geometric models imply that folds in deeper horizons of the forearc basin along the flank of the uplifted subduction complex are related to arcward-vergent backthrusts rooted at deeper levels of the subduction complex (Torrini and Speed, 1989). The deeper thrusts are inferred to place structural wedges of subduction complex over the substratum and lower horizons of forearc-basin fill. Forearc downflexure in response to tectonic loading by these backthrust wedges may be responsible for the persistent trenchward tilt of the forearc substratum (Fig. 6.13). All interpretations of

the deep structure remain ambiguous, however, because neither strongly deformed forearc-basin strata nor rock masses within the accretionary prism of the subduction complex display coherent reflectors on seismic records, and the two are thus indistinguishable on reflection records of the subsurface.

Southwest Japan Forearc

Along the trenchward margin of southwest Japan, a string of nearly equant to slightly elongate forearc basins underlies a segmented structural terrace lying between the island arc and the actively accreting subduction zone of the Nankai Trough (Fig. 6.14). Water depths in these terrace basins are generally 1000 to 2000 m, and each contains approximately 1500 m of Neogene sediment (Table 6.1, V) that rests unconformably on deformed Cretaceous and Paleogene strata. The basins are separated by transverse structural sills that project across the forearc terrace from peninsular promontories of the adjacent islands.

The substratum of the forearc terrace basins is composed entirely of deformed subduction complex in the Shimanto belt, and the trenchward sills of the basins are formed by the trench-slope break at the crest of the active subduction zone (Taira, 1985; Taira et al., 1988). The Shimanto belt was formed by progressive accretion of trench and trench-slope deposits to the flank of southwest Japan from Cretaceous to

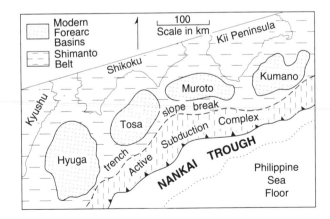

Fig. 6.14 Sketch map of forearc region off southwest Japan (Okuda and Honza, 1988; Taira et al., 1988) showing distribution of Neogene forearc-terrace basins resting unconformably on Cretaceous-Paleogene subduction complex of Shimanto belt; barbed line denotes base of trench inner slope.

early Miocene time (Taira et al., 1982, 1988). The belt consists of multiple imbricate structural panels that are successively younger trenchward and are bounded by thrust zones that dip arcward (Nishi, 1988). Seafloor lavas detached from underthrusting oceanic lithosphere are incorporated into some of the structural panels, and melange structure is widespread (Kimura and Mukai, 1991). The most common lithologic assemblage in the Shimanto belt is turbidite sandstone and shale intercalated with hemipelagites, but tectonic slivers of pillow basalt, radiolarian chert, and nannofossil limestone are also abundant (Taira et al., 1988).

Early Miocene and younger forearc-basin deposits rest unconformably across truncated structures and steeply dipping strata of the Shimanto belt (Taira et al., 1982, 1988; Taira, 1985). Mid-Miocene to mid-Pliocene extensional deformation within the forearc region apparently affected subsidence patterns (Okuda, 1985; Okuda and Honza, 1988). The tilted flank of a forearc terrace basin is locally exposed in the lower to middle Miocene Kumano Group along the eastern side of the Kii Peninsula (Fig. 6.14); outcrop studies suggest that basin fill is composed mostly of muddy turbidites and contourites (Chijiwa, 1988; Hisatomi, 1988), although strandline deposits occur along the eroded edge of the Kumano basin (Fig. 6.15).

Compound Forearc Basins

Some arc-trench systems contain multiple forearc depocenters, which occupy subparallel belts that are initiated sequentially from the arc toward the trench

LITHOLOGY	FACIES	ENVIRONMENT
massive mudstone with coal seams	backshore	coastal swamp and lagoon
laminated and cross-laminated sandstone	foreshore/shoreface	shoreline
interbedded sandstone and siltstone	nearshore bar	inner shelf
gray siltstone	offshore shelf	outer shelf
interbedded sandstone and mudstone	turbidite	upper slope
massive mudstone and siltstone	contourite	lower slope

Fig. 6.15 Depositional facies along northern flank of Kumano forearc basin off southwest Japan; adapted from Chijiwa (1988) and Hisatomi (1988).

as progressive accretion of crustal material at the subduction zone widens the arc-trench gap. These compound forearc systems provide a conceptual bridge between modern and ancient forearc basins because arcward belts of forearc basins in compound forearc systems are bounded on their trenchward flanks by uplifted but inactive parts of associated subduction complexes. Compound forearc basins are prominent in the Alaskan forearc, where massive accretionary processes have widened the arc-trench gap markedly compared to its counterpart along strike in the Aleutian forearc. This process has also affected onshore and offshore forearc basins of western Luzon.

Alaskan-Aleutian Forearc

The arcuate Aleutian Trench extends for 4000 km from the Kamchatka Peninsula on the west to the Gulf of Alaska on the east (Fig. 6.16). The parallel magmatic arc is an intraoceanic arc along the Aleutian Islands, but lies along the continental margin on the Alaska Peninsula. The Aleutian segment of the arc-trench system was formed in Paleogene time (Marlow et al., 1973), when oceanic crust of the Bering Sea was trapped in the backarc region by initiation of subduction offshore (Cooper et al., 1987); continuous or episodic subduction has been underway along the Alaskan segment of the arc-trench system since mid-Mesozoic time (Dickinson, 1982). The Aleutian Ridge is flanked by a simple terraced forearc (Table 6.1, IV), but incremental growth of a broad composite accretionary prism farther east has formed a compound ridged forearc off Alaska (Table 6.1, I). This difference in forearc morphology may partly reflect the greater age of the arc-trench system on the east, but also reflects the delivery of more voluminous sediment to the eastern end of the trench from sources within the adjacent continental block. Sediment delivery to the eastern end of the trench occurs by a combination of sedimentary and tectonic processes: (a) longitudinal dispersal of turbidites down the axis of the trench, and (b) lateral tectonic transport of the sediment prism which has been shed westward onto the seafloor south of the trench from the transform continental margin that trends southward from the eastern extremity of the trench

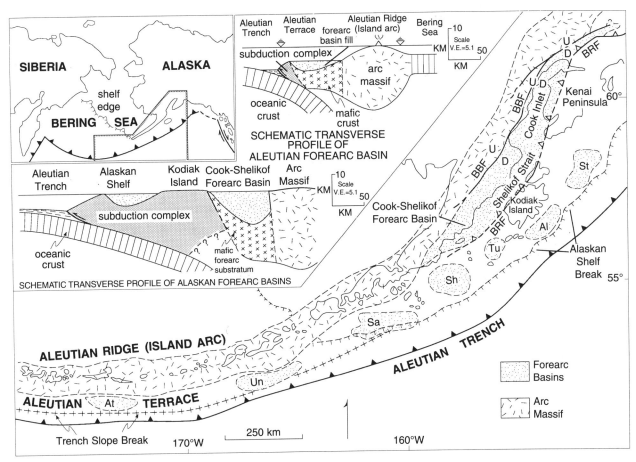

Fig. 6.16 Relations of multiple Alaskan and Aleutian forearc basins (Bruns et al., 1987; Fisher et al., 1987; Scholl et al., 1987b; von Huene et al., 1987): Aleutian terrace basins (At, Atka; Un, Unalaska), Alaskan shelf basins (Sa, Sanak; Sh, Shumagin; Tu, Tugidak; Al, Albatross; St, Stevenson), and Cook-Shelikof basin (Table 6.1, I, II, IV; Kanaga basin is west of edge of map); inset map shows relations to continental blocks and linked transform system along continental margin. Mafic crust beneath Aleutian forearc basins can be regarded as buried flank of arc massif (Geist et al., 1988). BBF, Bruin Bay fault; BRF, Border Ranges fault.

(Dickinson, 1982). Farther west, in the Aleutian forearc region, transverse infilling of the trench with less voluminous volcaniclastic debris derived from sources along the Aleutian island arc is dominant (Underwood, 1986b).

Uplifted older components of the subduction complex on Kodiak Island and the Kenai Peninsula separate inner and outer belts of forearc basins off Alaska (von Huene, 1979). The inner forearc basin of Cook Inlet and Shelikof Strait includes both Mesozoic and Cenozoic strata, and masks the subsurface juncture between the arc massif and the subduction complex. The outer forearc basins that underlie the continental shelf contain only Cenozoic strata, which rest depositionally on accreted materials of the subduction complex lying arcward from the trench-slope break. These offshore forearc basins occur as discrete depocenters of nearly equant to slightly elongate shape in a belt aligned along strike with the belt of forearc basins of similar shape and size beneath the Aleutian terrace farther west. Forearc basins are absent or subdued along the westernmost 1000 km of the Aleutian Ridge, where subduction is highly oblique to the subduction zone. Farther east, where terrace basins are prominent in the Aleutian forearc region, the obliquity of subduction has induced longitudinal strike-slip faulting, accompanied by related transpressional and transtensional deformation, along or near the trench-slope break (Ryan and Scholl, 1989).

The shelf and terrace basins are 50 to 75 km wide, 100 to 200 km long, and filled by 2500 to 5000 m of undeformed to mildly deformed sediment (Bruns et al., 1987; Scholl et al., 1987b; von Huene et al., 1987). The sediment-water interface is at shelf depths of 250 to 500 m for the Alaskan shelf basins (Table 6.1, I and II), but lies at a mean depth of 4000 m for the Aleutian terrace basins (Table 6.1, IV). Basin sills at the trench-slope break lie in water depths of 3500 to 4500 m at the edge of the Aleutian terrace, but form the continental shelf break off the Alaska Peninsula. The floor of the Aleutian Trench also deepens from 4000 km at its eastern end to 7000 m in the region of the Aleutian terrace. As in the Sunda fore-arc of Indonesia, longitudinal slopes of the forearc-basin floors, the trench-slope break, and the trench axis display a common regional gradient, averaging one meter per kilometer, inclined downward from the end of the trench where tectonic accretion has been greatest (i.e., the east).

Along the arcward margins of the Aleutian terrace basins, strata onlap the arc massif of the Aleutian Ridge, but the Alaskan shelf basins are bounded on their arcward flanks by emergent or submerged structural highs of uplifted subduction complex in the ridged Alaskan forearc. The Alaskan shelf basins are underlain entirely by subsided subduction complex. The substratum beneath the terrace basins, however, is mafic crust that apparently grades laterally into the basement of the Aleutian Ridge, and the basal sediment layers above this arc-flank basement are evidently slope deposits laid down prior to development of a basinal configuration by growth of the subduction complex (Scholl et al., 1982b). The sediment fill of the terrace basins is dominantly volcaniclastic (Stewart, 1978; Scholl et al., 1982b), but dispersal systems for the shelf basins tap varied continental sources. Combined underthrusting and uplift of the subduction complex beneath the trench-slope break has deformed trenchward flanks of some shelf-terrace basins, but aggradation and lateral progradation of basin fill has draped sediment cover across the trench-slope break along the margins of others (Bruns et al., 1987; Scholl et al., 1987b; von Huene et al., 1987).

The inner Cook-Shelikof forearc basin (Table 6.1, I) is 100 km wide and extends for at least 750 km from the coastal fringe of the Alaska Peninsula on the southwest to Matanuska Valley beyond Cook Inlet on the northeast (Fisher et al., 1987). Its elongate depocenter contains 10,000 m of marine Mesozoic and nonmarine Cenozoic strata (Fisher and Magoon, 1978; Fisher et al., 1987) of volcanoplutonic provenance (Dickinson, 1982). The arcward margin of the basin is defined by reverse faults, flanking the arc massif, along which the Alaska Range batholith, composed of plutons emplaced between mid-Jurassic and mid-Tertiary time, has been uplifted by up to 2500 m with respect to basin fill. The trenchward margin of the basin is defined by the Border Ranges fault, along which the subduction complex of the broad Alaskan accretionary prism was underthrust beneath the basin fill during Mesozoic time. The bounding fault zones were dormant while continuing Cenozoic subduction accreted a widening expanse of subduction complex to the continental margin farther south, but Tertiary underplating of the subduction complex contributed to tectonic uplift of Kodiak Island and the Kenai Peninsula along the flank of Cook-Shelikof basin (Byrne, 1986; Fisher and Byrne, 1987). In transverse profile, the basin forms a broad, internally faulted megasyncline with gentle limbs, but the nature of the forearc substratum beneath the keel of the basin is essentially unknown. The thickness of the sediment prism within the basin suggests thin mafic crust (Fig. 6.16), similar to the arc-flank basement beneath Aleutian terrace basins.

Western Luzon Forearc

West of Luzon, the Manila Trench extends for almost 1000 km from Mindoro northward toward Taiwan (Hayes and Lewis, 1984). The trench reaches its maximum depth in excess of 5000 m off westernmost Luzon south of Lingayen Gulf (Fig. 6.17). Parallel to this segment of the trench lie two deep forearc basins (Table 6.1, VII), separated by the uplifted block of the coastal Zambales Range. The offshore West Luzon Trough, 250 km long and 30 to 40 km wide, underlies a forearc terrace, with its surface at depths of 2000 to 3000 m; it contains nearly 5000 m of turbidite fill (Lewis and Hayes, 1984). The onshore forearc basin, 50 to 75 km wide and at least 175 km long, underlies the terrestrial lowland of the Luzon Central

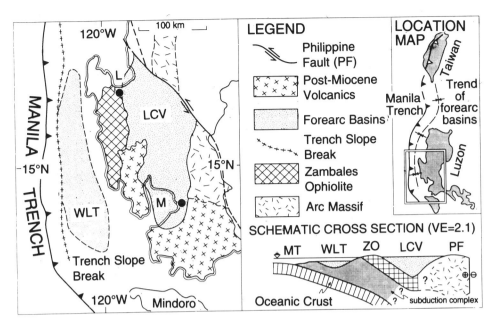

Fig. 6.17 Forearc basins associated with Manila Trench (MT) west of Luzon (Table 6.1, VII) in Philippine Islands (Bowin et al., 1978; Bachman et al., 1983; Hayes and Lewis, 1984; Lewis and Hayes, 1984): West Luzon Trough (WLT) and Luzon Central Valley (LCV) separated by Zambales ophiolite (ZO) of coastal Zambales Range; L, Lingayen; M, Manila; untextured areas underlain by varied rocks of diverse ages. Double line indicates coastline.

Valley and contains 12,000 to 15,000 m of varied sediment fill along its keel (Bachman et al., 1983). The coastal ridge separating the two parallel forearc basins is underlain by the Zambales ophiolite sequence, tilted eastward at an angle of 15 to 25 degrees.

The western Luzon arc-trench system was established in mid-Tertiary time, when westward subduction under the eastern side of Luzon was supplanted by eastward subduction under its western side (Bowin et al., 1978). Associated arc magmatism gave rise to plutons emplaced along the Luzon Central Cordillera east of the Luzon Central Valley from Oligocene through Pliocene time (Bachman et al., 1983), and to Pliocene and younger volcanoes erupted along the eastern flank of the Zambales Range (deBoer et al., 1980). The volcanoes along the post-Miocene volcanic front overlie the western (trenchward) margin of the Luzon Central Valley forearc basin. Trenchward migration of the magmatic arc has thus superposed the youngest elements of the arc platform above the basin fill. The eastern (arcward) margin of the Luzon Central Valley is sliced obliquely by the sinistral Philippine fault system, which has offset pre-Pliocene increments of the magmatic arc by about 100 km.

The western flank of the Luzon Central Valley forearc basin rests concordantly and gradationally upon pillow basalt of the Zambales ophiolite (Schweller and Karig, 1982; Bachman et al., 1983). Along the

eastern flank of the Zambales Range, basal strata are upper Eocene to lower Oligocene pelagic limestone (50 m) overlain by lower to middle Oligocene volcaniclastic strata (75 m). These thin strata were evidently deposited in a sediment-starved marginal sea or interarc basin prior to initiation of subduction at the ancestral Manila Trench by polarity reversal of the Luzon arc (Schweller et al., 1983, 1984). Gabbro, diabase, and basalt of the ophiolite, which has a pseudostratigraphic thickness of 7000 m, represent the oceanic crust of that basin. The Zambales ophiolite may have been emplaced adjacent to the arc massif by strike slip along a shear zone now buried in the subsurface beneath the Luzon Central Valley (Karig, 1983a), or it may have floored a backarc basin behind the east-facing Paleogene arc of the Sierra Madre in eastern Luzon. The present eastward tilt of the ophiolite slab beneath the flank of the Luzon Central Valley forearc basin is attributed to uplift above underthrust materials subducted at the Manila Trench.

Along the eastern flank of the Luzon Central Valley forearc basin, Miocene shelf deposits (2500 m) lie unconformably upon the arc massif, whereas Miocene turbidites (2250 m) along the western flank of the basin rest concordantly on the sediment cover of the Zambales ophiolite (Bachman et al., 1983). Miocene strata thin toward both the arc massif and the Zambales Range as lenticular wedges of strata pinch out between multiple internal surfaces of stratal

onlap and overlap. Marine strata along both basin margins are capped by Pliocene and younger nonmarine deposits (250-500 m). Within the roughly symmetrical synclinal keel of the basin, Miocene turbidites are overlain gradationally by Pliocene shelf and younger nonmarine deposits. The location and nature of the subsurface contact between the Zambales ophiolite and the arc massif are uncertain, but tilt of the ophiolite was synchronous with deposition of the overlying forearc succession, beginning late in Oligocene time and continuing into Miocene time (Schweller et al., 1983).

Although the ages of strata offshore are not well constrained, the West Luzon Trough probably did not form until late in the depositional history of the Luzon Central Valley depocenter. Comparison of the volume of the subduction complex offshore with the volume of sediment present along the trench floor suggests that construction of the offshore accretionary prism did not begin until late in Miocene time (Hayes and Lewis, 1984). If so, development of the tectonic sill at the trench-slope break and ponding of turbidites within the forearc terrace basin could not have begun earlier. Backtilting of basin fill, with greater dips at lower horizons, indicates that forearc sedimentation in the West Luzon Trough accompanied structural uplift of the trench-slope break (Lewis and Hayes, 1984). Multiple wedges of trough turbidites thin by onlap and overlap of the trench-slope break, but strata along the eastern margin of the trough onlap the submerged flank of the Zambales Range along a buttress unconformity with several kilometers of depositional relief. Initiation of sedimentation within the West Luzon Trough apparently was roughly contemporaneous with the transition from marine to nonmarine deposition within the Luzon Central Valley, and with prograde migration of the magmatic arc into the region of forearc sedimentation. Prior to development of the West Luzon Trough, the trench-slope break of an ancestral Manila Trench evidently lay at the crest of the Zambales Range adjacent to the Luzon Central Valley.

Along strike to the north, the west Luzon arc-trench system trends into the collisional belt of eastern Taiwan (Bowin et al., 1978), where the continuation of the Luzon island arc was sutured to the continental margin of China by orogeny that began early in Pliocene time (also see Chapter 10). As crustal suturing developed, the continuation of the North Luzon Trough, a forearc basin along strike from the West Luzon Trough, was converted into the "coastal range collisional basin," which now underlies the longitudinal valley of Taiwan (Lundberg and Dorsey, 1988). Inward-facing thrust or reverse faults mark both flanks of the lowland belt where deformed basin fill is exposed. Miocene volcaniclastic deposits of the ancestral forearc basin were buried under thick synorogenic strata, at least 5000 m thick, derived from mainly sedimentary and metasedimentary sources in the adjacent subduction complex, which was uplifted above the continental margin during crustal collision (Dorsey, 1988). Little sediment fill was inherited from the forearc phase of evolution, even though the position of the elongate basin that existed during the collisional phase of evolution was controlled closely by the previous forearc morphology.

Ancient Forearc Basins

Forearc basins of modern arc-trench systems serve as actualistic analogs for ancient forearc basins related to subduction systems now dormant or extinct. Although the tectonic settings of ancient forearc basins are interpretive, erosional exposures of basin fill allow the study of forearc depositional systems in more detail. As the California Great Valley forearc basin of late Mesozoic and early Cenozoic age is the best known in terms of lithofacies and petrofacies, its depositional history provides a conceptual template for the evolution of a forearc basin through time. Structural and stratigraphic relations of the uppermost Paleozoic to Mesozoic Hokonui assemblage in New Zealand and the Paleozoic Tamworth trough in Australia document analogous development of older forearc basins.

California Great Valley

The Great Valley of California extends subparallel to the continental margin for 675 km, from the Klamath Mountains on the north to the Transverse Ranges on the south (Fig. 6.18). In late Mesozoic and early Cenozoic time, the Great Valley forearc basin (Table

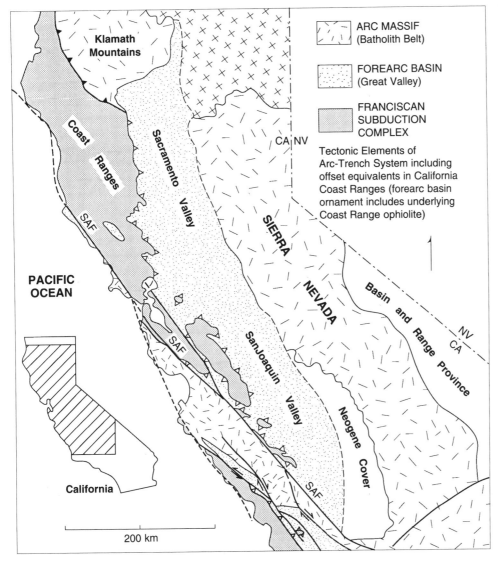

Fig. 6.18 Geologic setting of Late Jurassic to Paleogene Great Valley forearc basin (Table 6.2D) in California (Dickinson and Seely, 1979; Dickinson et al., 1979; Ingersoll, 1982b, 1988a). San Andreas (SAF), San Gregorio-Hosgri (SGF), and related Neogene strike-slip faults in California Coast Ranges obliquely disrupt older forearc trends. South Fork Mountain thrust between arc massif of Klamath Mountains and Franciscan subduction complex indicated by solid triangles, and Coast Range fault between Great Valley forearc-basin fill (including ophiolitic substratum) and Franciscan subduction complex indicated by open triangles.

6.2D) lay between the Sierra Nevada batholith and magmatic arc to the east and the Franciscan subduction complex to the west (Dickinson, 1970a). Other forearc basins of similar age farther north in the Pacific Northwest (Nilsen, 1984; Heller and Dickinson, 1985; Heller et al., 1987) and farther south in Peninsular California (Nilsen and Abbott, 1981; Busby-Spera and Boles, 1986) reflect comparable forearc evolution related to subduction of Pacific sea floor along at least 4000 km of the continental margin (Hamilton, 1969; Dickinson, 1976b). Well established paleontologic and isotopic ages indicate that Great Valley forearc sedimentation was coeval with arc magmatism in the Sierra Nevada and with deposi-

tion and metamorphism of the Franciscan subduction complex (Dickinson, 1981). The forearc basin evolved from a deep marine trough to a shelf basin scored by submarine canyons, as its morphology was modified by sedimentary and tectonic processes (Ingersoll, 1978a, 1982b; Dickinson and Seely, 1979; Dickinson et al., 1979a).

Provenance and Petrofacies
Clastic sedimentation of detritus derived from the Sierra Nevada arc terrane was dominant throughout basin evolution, and petrofacies of sandstones within the basin fill record the magmatic and erosional history of the adjacent arc (Dickinson and Rich, 1972;

Mansfield, 1979; Ingersoll, 1983; Linn et al., 1991, 1992). Volcanic and plutonic sources in the arc massif were the principal provenance, contributing both quartz and feldspar grains and volcanic lithic fragments in varying proportions throughout the depositional history (Fig. 6.9). Recycled quartzose detritus from metamorphic wallrocks of the plutons and chert-argillite grains from uplifted pre-Franciscan accretionary prisms of the Sierra Nevada foothills and Klamath Mountains are also present at all horizons, and are major components in some areas and at selected horizons (Dickinson et al., 1982; Bertucci, 1983; Ingersoll, 1983; Short and Ingersoll, 1990). The net trend of provenance variation during basin evolution, from largely volcaniclastic detritus at lower horizons to largely plutoniclastic detritus at higher horizons (Dickinson and Rich, 1972; Ingersoll, 1978c, 1983, 1988a), reflects enhanced dissection of the arc massif through time. An upward decrease in the ratio of plagioclase to K-feldspar is a key facet of this time-dependent trend. Areally within the forearc basin, petrofacies of the San Joaquin Valley on the south contrast systematically with petrofacies of the Sacramento Valley on the north by containing higher ratios of silicic to andesitic volcanic lithic fragments and of metamorphic to volcanic lithic fragments, and higher contents of detrital mica flakes of sand size (Ingersoll, 1978c, 1983, 1988a). These differences reflect a southward increase in the continentality of the Sierra Nevada province.

Post-subduction History
Beginning in mid-Tertiary time, the geodynamic setting of the Great Valley has been altered by development of the San Andreas transform system along the edge of the continental block (Graham et al., 1984). Strata that represent the preceding forearc phase of basin evolution now form an asymmetric megasyncline less than 100 km wide, but the width of the forearc basin prior to late Cenozoic deformation was more nearly 150 km. Forearc strata forming the gently inclined eastern limb of the Great Valley megasyncline onlap the deeply eroded Sierra Nevada block, whereas forearc strata forming the steeply upturned western limb are faulted against intensely deformed and partly metamorphosed strata of the Franciscan assemblage in the uplifted core of the California Coast Ranges. The tectonic contact between the two coeval assemblages may be a time-transgressive thrust system (Ingersoll and Dickinson, 1981) that tectonically emplaced progressively younger components of the Franciscan subduction complex beneath the aggrading trenchward flank of the Great Valley forearc basin.

Isolated remnants of imbricated forearc-basin fill exposed within the Coast Ranges structurally overlie Franciscan rocks along strands of the Coast Range fault system, which is now folded and structurally dismembered (Bailey et al., 1970). This composite structural feature first accommodated underthrusting of the Franciscan subduction complex, but structural telescoping was later overprinted by denudational normal faulting as the subduction system evolved (Jayko et al., 1987; Harms et al., 1992). Still later dismemberment by strike-slip and reverse faulting accompanied transpressional folding during evolution of the San Andreas transform system.

The trend of the Great Valley megasyncline, controlled by late Cenozoic development of the Coast Ranges, is oblique to the trend of the pre-existing forearc basin. The latter was oriented more nearly north-south (in present coordinates), at an angle of approximately 20 degrees to the San Andreas fault and the associated Coast Ranges uplift (Ingersoll, 1978a, 1988a). Consequently, transects across the northern part of the preserved forearc basin (Sacramento Valley and northern Coast Ranges) include more trenchward portions of the sedimentary fill than transects across its southern part (San Joaquin Valley and southern Coast Ranges).

Basin Evolution
Stratigraphic studies indicate that the Great Valley forearc basin widened markedly during sedimentation (Fig. 6.19); the morphology changed in response to the combined effects of (a) oceanward migration of the trench-slope break in response to continued accretion of the Franciscan subduction complex and (b) concomitant continentward migration of the locus of arc magmatism (Ingersoll, 1978a, 1982b). The aggregate rate of basin expansion averaged a kilometer per million years (Ingersoll and Dickinson, 1981). As one result, only Upper Cretaceous and Paleogene strata are present within a belt 50 to 60 km wide above the

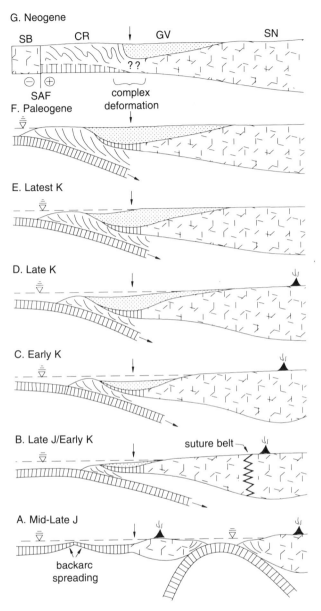

Fig. 6.19 Schematic diagrams (no vertical exaggeration) to illustrate sequential tectonic evolution of Great Valley forearc basin and related geologic features in California (Dickinson and Seely, 1979; Ingersoll, 1982b); successive stages of basin evolution (A–F) keyed by letter to superposed phases of sedimentation (Fig. 6.20). Legend: SB, Salinian block; SAF, San Andreas fault; CR, California Coast Ranges underlain by deformed Franciscan subduction complex; GV, Great Valley; SN, Sierra Nevada. Symbols: jackstraw pattern denotes arc massif, stipples denote forearc-basin fill, vertical lines ornament oceanic crust (ophiolite). Vertical arrows indicate position of homoclinal outcrop belt along flank of Great Valley. Triangles show position of arc magmatism, but Laramide magmatism (E–F) lay farther east (right) than edge of diagrams. Note that Great Valley forearc basin evolved from slope basin (B) through ridged morphologies (C–E) to shelved basin (F) as trench-slope break (Fig. 6.1) shoaled and migrated westward (left).

onlap unconformity with Sierran basement beneath Neogene deposits of the Great Valley. Deformed Upper Jurassic and Lower Cretaceous strata of the Great Valley megasyncline are present only in the steep homocline that flanks the sedimented floor of the Great Valley and delineates the eastern flank of the Coast Ranges. The structural keel of the Great Valley megasyncline thus developed near the locus of mid-Cretaceous overstepping of older parts of basin fill by younger horizons, but the depositional keel of the forearc basin lay farther west and has been disrupted by Coast Range deformation. Consequently, the thickness of basin fill resting depositionally on Sierran basement beneath the gentle eastern limb of the megasyncline is less than 5000 m (Bartow and Nilsen, 1990), but the composite thickness of the more complete succession in the homocline of the western limb flanking the Coast Ranges reaches at least 12,000 to 15,000 m (Moxon, 1988). Subsidence beneath the eastern side of the forearc basin was controlled by thermotectonic subsidence of the flank of the adjacent magmatic arc as the locus of magmatism retreated eastward, but subsidence beneath the western side of the basin reflected more complex flexural effects related to the changing configuration of the subducted slab as it responded to the load of the subduction complex, and to varying buoyancy of descending oceanic lithosphere through time (Moxon and Graham, 1987).

Ophiolitic Substratum

West of the onlap of the Sierran block by mid-Cretaceous and younger strata, the substratum beneath forearc-basin fill is the Upper Jurassic Coast Range ophiolite (Hopson et al., 1981), which is structurally attenuated in most exposures by internal disruption and by basal truncation along the Coast Range fault system. Where the Coast Range ophiolite is least disrupted, its pseudostratigraphic thickness is approximately 5000 m (Moxon, 1988), comparable to oceanic crustal profiles. The nature of this oceanic crust beneath the western flank of the Great Valley forearc basin is controversial (Robertson, 1989), but it has the geochemical character of supra-subduction-zone ophiolites formed either by forearc extension or as the floors of marginal seas or interarc basins formed by the rifting of magmatic

arcs (Evarts, 1977; Shervais and Kimbrough, 1985; LaGabrielle et al., 1986; McLaughlin et al., 1988; Shervais, 1990).

The ophiolite may have formed by forearc spreading (Stern and Bloomer, 1992), but most probably formed by backarc spreading (Fig. 6.19A) behind an east-facing intraoceanic island arc involved in arc-continent collision to produce the Late Jurassic Nevadan orogeny of the Sierra Nevada foothills (Schweickert and Cowan, 1975; Ingersoll and Schweickert, 1986). By inference, accretion of the arc structure to the continental margin caused subduction to step sharply oceanward from the Sierra Nevada foothills and initiated Franciscan subduction. Part of the accreted arc is exposed in the Sierra Nevada foothills, but much of its width is apparently represented by mafic crust now buried beneath the eastern limb of the Great Valley megasyncline. The crustal thickness (20 km) of this hidden substratum is intermediate between oceanic and continental profiles (Holbrook and Mooney, 1987). Its components are inferred here to be some combination of ophiolitic and island-arc basement intruded by granitic plutons emplaced during early phases of Franciscan subduction.

Although erosional scour or stratal discordance occurs locally in many places along the contact between the Coast Range ophiolite and overlying sedimentary strata, the transition is locally gradational between pillow basalt and chert or argillite. These strata at basal horizons of the Great Valley forearc basin probably accumulated within the interarc basin that existed prior to initiation of Franciscan subduction (Ingersoll, 1982b). Ophiolitic breccias and olistostromes that occur locally at or near the base of the sedimentary succession (Phipps, 1984; Robertson, 1990) may record extensional deformation during either closing phases of backarc spreading or initial phases of Franciscan subduction.

Facies Succession

The forearc-basin morphology changed steadily during deposition in response to evolving basin architecture, and different types of depositional systems occupied the basin at different stages of its evolution (Fig. 6.20).

The oldest clastic strata of the Great Valley forearc basin are uppermost Jurassic (Tithonian) and lower-

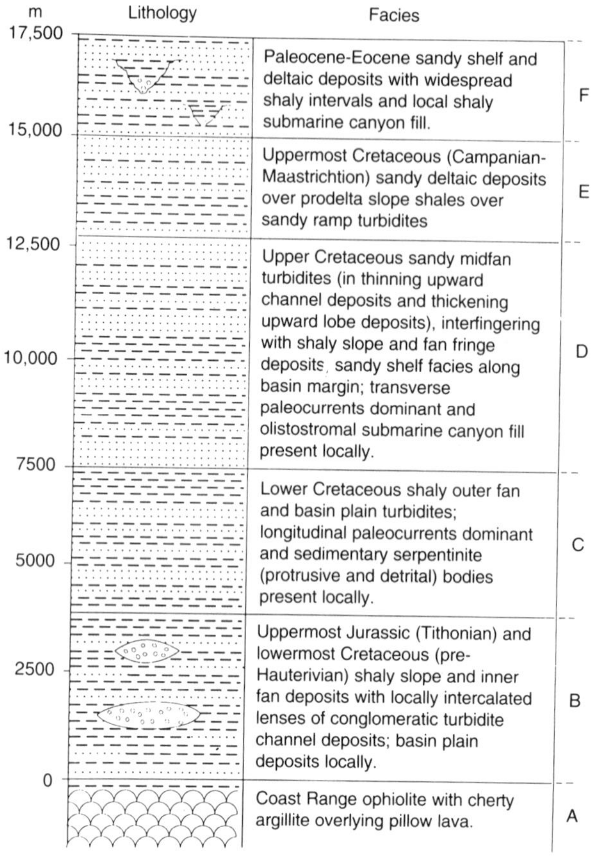

Fig. 6.20 Typical stratigraphic column along keel of Great Valley forearc basin (Ingersoll, 1978a,b, 1979, 1982b; Ingersoll and Dickinson, 1981; Cherven, 1983; Moxon, 1988, 1990); most sandy intervals are also conglomeratic locally. Letters on right (A-F) key phases of sedimentation to stages of basin evolution (Fig. 6.19).

most Cretaceous (pre-Hauterivian) hemipelagites and turbidites. These strata were deposited above the Coast Range ophiolite during the same time span that blueschist metamorphism affected older components of the Franciscan subduction complex (Fig. 6.19B). Local variations in thickness, which ranges from 2000 to 6000 m (Moxon, 1988, 1990), seemingly reflect syndepositional faulting that may have been associated with continued extensional deformation as the tectonic setting of the basin changed from a backarc to a forearc regime. The facies exposed in the homocline along the eastern flank of the Coast Ranges are chiefly slope deposits (Ingersoll, 1978a) composed of lenses of conglomeratic turbidites encased in finer-

grained strata, although basin-plain facies associations are also present (Suchecki, 1984). The sedimentology of these strata suggests accumulation across a sloped forearc belt, which led downward to a forearc trough with its keel adjacent to a subdued trench-slope break located near the subduction zone (Ingersoll, 1982b).

Continued forearc sedimentation produced Lower Cretaceous (post-Valanginian) strata, as thick as 3500 m (Moxon, 1988), composed of widespread basin-plain deposits that interfinger with turbidite associations typical of outer submarine fans (Ingersoll, 1978a, 1982b). The sedimentology of these strata suggests that the subduction complex had achieved enough bulk and structural relief to pond arc-derived sediment within a prominent forearc trough in a ridged forearc with a deeply submerged trench-slope break (Fig. 6.19C). This inference is supported by the dominance of longitudinal paleocurrent indicators, suggesting that turbidite sedimentation was influenced by the presence of a bathymetric barrier lying to the west (Ojakangas, 1968; Ingersoll, 1978a, 1982b, 1988a). Initially, the Early Cretaceous forearc trough was less than 50 km wide but probably 5000 m deep (Ingersoll, 1978a). Locally, massive bodies of protrusive serpentinite enclosed within Lower Cretaceous strata reflect the diapiric movement of ophiolitic serpentinite upward to the floor of the forearc trough along the flank of the subduction complex behind the trench-slope break (Carlson, 1984). By mid-Cretaceous time, the Great Valley forearc basin was 100 km wide (Ingersoll, 1978a) and the aggregate thickness of basin fill was approximately 7500 m (Ingersoll and Dickinson, 1981; Moxon, 1990) beneath a ridged or terraced forearc.

By Late Cretaceous time (Fig. 6.19D), progradation of depositional systems along the arcward flank of the forearc basin resulted in deposition of sandy midfan assemblages now exposed along the homoclinal outcrop belt forming the upturned western limb of the Great Valley megasyncline. Both progradational and retrogradational depositional systems of submarine fans intertongue in complex fashion with slope and fan-fringe deposits (Ingersoll, 1978a,b, 1979). Westerly to southwesterly paleocurrent indicators oriented transverse to the trend of the arc are characteristic in

these strata (Ingersoll, 1979; Mansfield, 1979). By Campanian time, the forearc basin was approximately 150 km wide (Ingersoll, 1978a), but its keel still lay below the CCD at depths estimated to have been 4000 m or more (Ingersoll, 1979). Depositional systems along the arcward side of the basin included deltaic complexes and shelf deposits resting unconformably on the eroded flank of the Sierra Nevada block, and the eastern slope of the basin evidently had an inclination in the range of 5 to 10 degrees (Ingersoll, 1979). Uppermost Cretaceous (Campanian to Maastrichtian) deposits (Fig. 19E), known mostly from subsurface data, include delta-platform associations that prograded longitudinally as well as transversely across advancing muddy slopes and sandy submarine fans of the residual forearc trough (Cherven, 1983; Nilsen, 1990). Upper Cretaceous strata total approximately 7500 m, of which 5250 m are pre-Campanian turbidites and 2250 m represent the progradational Campanian-Maastrichtian delta-related stratal assemblages.

By Paleogene time (Fig. 6.19F), the crest of the trench-slope break had been exposed by uplift to allow minor recycling of Franciscan detritus into a forearc shelf basin traversed at intervals by prominent submarine canyons (Allmgren, 1978; Dickinson et al., 1979). Perhaps 3000 m of marine Paleogene strata are present locally. The submarine canyons that dissected the forearc shelf were 75 to 250 km long, 5 to 25 km wide at their rims, and 750 to 1000 m deep at their mouths in the Delta depocenter (located near the junction between the present Sacramento and San Joaquin Valleys). Extensive trench-slope basins developed farther west along the oceanward flank of the present Coast Ranges in a belt beyond the trench-slope break (Underwood and Bachman, 1986). Neogene cover above the Great Valley forearc basin is entirely nonmarine in the Sacramento Valley, but transform-related Neogene basins beneath the San Joaquin Valley include thick marine deposits above a substratum composed of forearc-basin fill (Ingersoll, 1982b, 1988a; Graham, 1987). Recent seismic-reflection profiling suggests thrusting of the Franciscan subduction complex along the steep limb of the Great Valley megasyncline during Neogene deformation of the Coast Ranges (Unruh et al., 1991).

New Zealand Hokonui

Permian to Lower Cretaceous strata of the Hokonui and Torlesse assemblages (Fig. 6.21), analogous respectively to the Great Valley and Franciscan assemblages of California, form parallel curvilinear belts, offset by the Alpine fault, that extend north-south for 1500 km along the length of New Zealand (Dickinson, 1971a,b; Landis and Bishop, 1972). Close counterparts occur along tectonic strike in New Caledonia, which lies 1500 to 2000 km farther north. Although the tectonic analogy with California is not exact (Blake et al., 1974), the Hokonui-Torlesse pair represents the stratigraphic record of an arc-trench system that lay along the margin of the Australia-Antarctica segment of Gondwana prior to rifting that formed the Tasman Sea in Late Cretaceous time (Sporli, 1978; MacKinnon, 1983).

Remnants of the deformed and dismembered Hokonui forearc basin (Table 6.2D) are commonly less than 50 km wide, but the cumulative thickness of preserved basin fill, which consists mainly of volcaniclastic graywacke and argillite, exceeds 12,500 m. Although turbidites are present locally, much of the sequence was deposited in shelf and slope environments, and nonmarine strata are also present. Along the trenchward flank of the forearc basin, the base of the sedimentary fill locally rests depositionally on volcanic rocks forming the uppermost member of the structurally concordant Dun Mountain ophiolite belt (Coombs et al., 1976), which is in tectonic contact with the Torlesse subduction complex. Thick Permian volcanogenic strata exposed along the arcward flank of the forearc basin were once thought to represent part of the arc massif, from which the volcaniclastic strata of the basin were derived, but recent isotopic studies indicate that provenance relations were more complicated and are not yet well understood (Frost and Coombs, 1989).

Australian Tamworth Trough

The Tamworth Trough (Fig. 6.22) of northeastern New South Wales is an elongate forearc basin (Table 6.2A), 50 km wide and 400 km long, adjacent to the Tablelands subduction complex within the New England foldbelt (Leitch, 1975; Korsch, 1977; Lindsay,

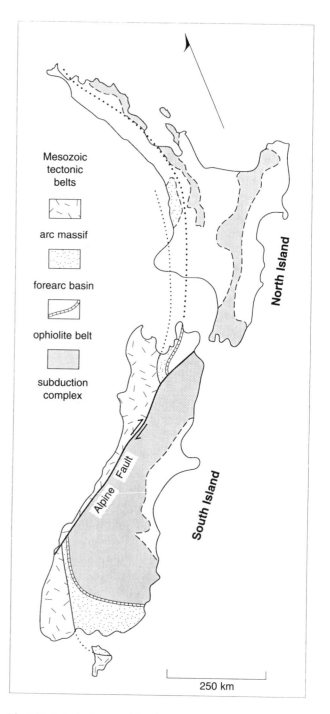

Mesozoic tectonic belts

arc massif

forearc basin

ophiolite belt

subduction complex

North Island

South Island

Alpine Fault

250 km

Fig. 6.21 Tectonic elements of Permian to Cretaceous arc-trench system, offset by Cenozoic Alpine fault, in New Zealand (Dickinson, 1971a,b; Landis and Bishop, 1972; Blake et al., 1974; Sporli, 1978): (a) flank of arc massif is western basement province, (b) forearc-basin belt (Table 6.2D) is Hokonui assemblage and other strata of Murihiku Supergroup, (c) ophiolitic assemblage is Dun Mountain ophiolite and associated strata, (d) subduction complex is Torlesse assemblage and related metamorphic rocks.

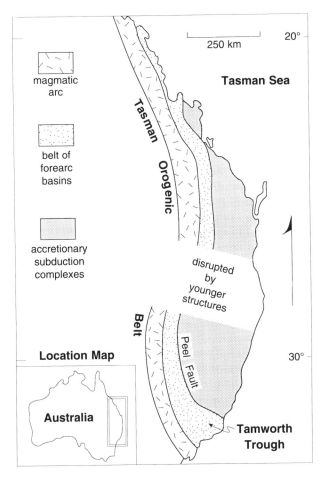

Fig. 6.22 Location of Tamworth Trough forearc basin (Table 6.2A) in mid-Paleozoic orogen of eastern Australia (Leitch, 1975; Murray et al., 1987; Lindsay, 1990); ophiolitic forearc substratum in structural contact with Tablelands subduction complex along Peel fault.

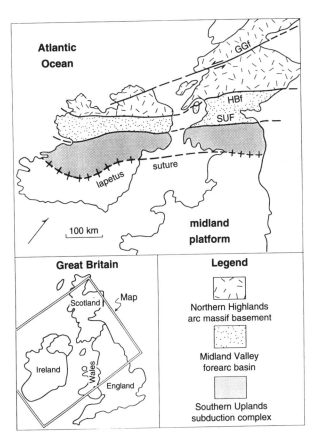

Fig. 6.23 Tectonic relations of remnant lower Paleozoic Midland Valley forearc basin (Table 6.2E) parallel to collisional suture marking closure of Iapetus Ocean within Caledonian orogen of British Isles (Dewey, 1971; Leggett et al., 1982; McKerrow and Soper, 1989); GGF, sinistral Great Glen fault; HBF, highland boundary fault; SUF, southern uplands fault zone.

1990). The contact between the two assemblages is the Peel fault system. The forearc-basin fill, principally of Devonian and Carboniferous age, reaches a net thickness of 12,500 m. Turbidites low in the forearc succession are overlain by shelf and nonmarine deposits. Basal strata of the forearc basin overlie an ophiolite sequence, which is interpreted as oceanic substratum in fault contact with the accretionary subduction complex lying to the east. Along the western flank of the basin, Devonian strata include lavas and volcanic breccias representing volcanogenic strata erupted along the edge of an inferred magmatic arc lying to the west. The Tamworth Trough occupies the locally preserved segment of a structurally dismembered Paleozoic arc-trench system that extended for

at least 1500 km along the eastern margin of Australia (Murray et al., 1987).

Sutural Forearc Basins

Survival of forearc basins within complexly deformed suture belts is illustrated by the Midland Valley (Table 6.2E) and its tectonic continuations along strike in Great Britain (Fig. 6.23), where forearc-basin fill forms part of the mid-Paleozoic Caledonian orogen (Dewey, 1971; Leggett et al., 1982). The Midland Valley sedimentary assemblage of Scotland occupies a belt 50 to 75 km wide and extends westward along strike for 500 km across Ireland. The Paleozoic forearc basin was bounded on the south by the uplifted subduction complex of the Southern Uplands and on the north by the arc massif of the Northern Highlands,

but little is known directly of its substratum. The basin fill is poorly exposed, but at least 2500 m of Silurian strata are preserved (Leggett, 1980), and both Ordovician and Devonian beds are present locally. Turbidite facies in lower horizons of the forearc sequence pass upward to shelf and nonmarine strata, deposited in part after suturing had begun. The provenance was initially the arc massif, but eventual uplift of the trench-slope break exposed sediment sources in the subduction complex as well.

Little note has been taken to date of pre-collision forearc basins within the intercontinental mountain systems of southern Eurasia, but sedimentary assemblages caught within the Zagros and Himalayan suture belts form a record of arc evolution prior to continental collision. In Iran, the Cretaceous Sanandaj forearc basin (Table 6.2E) adjacent to the Zagros suture contains 2500 m of little-studied arc-derived turbidites (Cherven, 1986). In the Transhimalayan forearc basin (Table 6.2E), the Indus Group includes pre-collision (mid-Cretaceous to mid-Paleogene) turbidites and fluviodeltaic deposits (1750 m) overlain by post-collision (middle Eocene and younger) fluvial and lacustrine strata (1250 m), which were deposited along an intramontane lowland that evolved from the forearc basin during closure of the collisional suture (Garzanti and van Haver, 1988). On the north, the Indus Group unconformably onlaps the Ladakh granitoid batholith of the arc massif, but is in tectonic contact to the south with backthrust elements of the subduction complex. Most clastic detritus was derived from the arc massif, with volcanic sources dominant for the pre-collision succession and plutonic sources dominant for the post-collision succession. Closer study of these and other sutural forearc sequences within ancient orogenic belts offers the means to document early phases of tectonic evolution that have been overprinted by collisional deformation. Recent work, for example, has begun the recognition of sutural forearc basins within Precambrian collision belts (de Kock, 1992; Hefferan et al., 1992).

Interpretive Summary

The diagnostic signals of forearc basins in the geologic record are (a) an elongate trend of thick sedimentary accumulations, expressed either as a continuous megasyncline or as a string of related depocenters, flanked on one side by (b) the volcano-plutonic association, mainly calc-alkalic, of a coeval magmatic arc, and on the other side by (c) the deformed strata, including melanges and blueschists, of a coeval subduction complex; although other trends in facies occur in specific instances, forearc-basin fill typically (d) shoals upward from turbidite associations to deltaic or shelf sequences and (e) includes a succession of petrofacies that record unroofing and dissection of the adjacent arc massif. Hydrocarbon resources in forearc basins are modest on a global scale, but coal, petroleum, and natural-gas reserves are all locally important on a regional scale.

Where forearc basins are incorporated, together with associated magmatic arcs and subduction complexes, into complex collisional orogens, their original features are overprinted and obscured to varying degrees by folding, thrusting, metamorphism, and erosion. Recognition of deformed forearc basins can improve regional tectonic interpretations of complex orogenic belts because the basin fills form significant elements of local crustal profiles (Fig. 6.24), and their positions are guides to overall orogenic architecture.

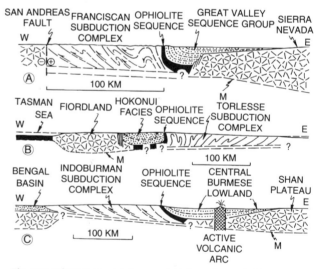

Fig. 6.24 Schematic crustal diagrams (M is base of crust) indicating scale of forearc-basin fill (stippled with dashes to show bedding) in relation to arc massifs (star-dash pattern) and subduction complexes (thrusts shown schematically): A, coastal California; B, New Zealand (South Island); C, Bangla Desh and Burma (Bengal Basin is peripheral foreland basin formed by collision of arc-trench system with underthrust Indian subcontinent); no vertical exaggeration.

Acknowledgments

Concepts developed in this chapter stem from joint research on forearc basins with V.B. Cherven, S.A. Graham, P.L. Heller, R.V. Ingersoll, C.F. Mansfield, R.W. Ojakangas, E.I. Rich, P.T. Ryberg, R.J. Stewart, Win Swe and E.V. Tamesis. Reviews by C.J. Busby, R.V. Ingersoll, G.A. Smith and M.B. Underwood improved the manuscript.

Further Reading

Dickinson WR, 1973, *Widths of modern arc-trench gaps proportional to past duration of igneous activity in associated magmatic arcs*: Journal of Geophysical Research, v. 78, p. 3376–3389.

Dickinson WR, Seely DR, 1979, *Structure and stratigraphy of forearc regions*: American Association of Petroleum Geologists Bulletin, v. 63, p. 2–31.

Hamilton W, 1979, *Tectonics of the Indonesian region*: United States Geological Survey Professional Paper 1078, 345p.

Ingersoll RV, 1983, *Petrofacies and provenance of late Mesozoic forearc basin, northern and central California*: American Association of Petroleum Geologists Bulletin, v. 67, p. 1125–1142.

Ingersoll RV, Nilsen TH (eds), 1990, *Sacramento Valley symposium and guidebook* [Book 65]: Pacific Section, Society of Economic Paleontologists and Mineralogists, Los Angeles, 215p.

Kulm LD, Dymond J, Dasch EJ, Hussong DM (eds), 1981a, *Nazca plate: crustal formation and Andean convergence*: Geological Society of America Memoir 154, 824p.

Leggett JK (ed), 1982, *Trench-forearc geology: sedimentation and tectonics on modern and ancient active plate margins*: Geological Society of London Special Publication 10, 576p.

Scholl DW, Grantz A, Vedder JG (eds), 1987a, *Geology and resource potential of the continental margin of western North America and adjacent ocean basins: Beaufort Sea to Baja California:* Circum-Pacific Council for Energy and Mineral Resources Earth Science Series, v. 6, 799p.

Intra-Arc Basins

7

Gary A. Smith and Charles Landis

Introduction

Intra-Arc Basin Definition and Position

Most overviews of convergent plate margins recognize volcanic arcs as sites of characteristic magmatism, but provide little consideration to the structure of arcs and, particularly, the common presence of basins in which arc volcanics and derivative sediments are preserved. Sufficient knowledge of the structure of circum-Pacific arcs has been gained in the past decade to provide an actualistic basis for the interpretation of modern and ancient intra-arc basins (Table 7.1).

Volcanic arcs are generally arcuate or linear constructs, typically exceeding 1000 km in length and ranging from 50 to 250 km in width, which parallel subduction-zone trenches. The trenchward side of the arc is characterized by a sharply defined *volcanic front* (Fig. 7.1), which typically corresponds to the axis of the largest and most active volcanoes, above where the subducted slab is 100 to 125 km deep (Gill, 1981). A diffuse zone of volcanism may extend more than 200 km behind this axis, with most volcanism concentrated within 100 km of the volcanic front, such that the boundary from intra-arc to backarc settings may be arbitrary. The asymmetry in distribution of volcanoes is probably related to processes of magma generation above the subducted slab (Tatsumi et al., 1986). The *arc massif* (Dickinson, 1974a,b; Fig. 7.1) refers to that region underlain by crust generated by arc magmatic processes. Some arcs are spatially stable for more than 150 my (e.g., Patagonian batholith; Bruce et al., 1991). Other arcs migrate, either toward or away from the trench (Ingersoll, 1988b; Priest, 1990; Avila-Salinas, 1992; Walker et al., 1992) in response to changing subduction angles,

plate rollback, and other processes that control the position of the volcanic front and width of the arc (Kay et al., 1992). In long-lived arcs, therefore, the arc crust may underlie a broader area than can be associated with volcanism for any narrow time interval. Arc crust typically underlies part or all of forearc basins (see Chapter 6) and may extend right to the trench wall (Hussong and Uyeda, 1981; Schweller et al., 1981). The topographic or bathymetric feature, typically of positive relief, formed by the volcanogenic edifices of superposed or overlapping volcanoes, which cap part or all of the arc massif, is termed the *arc platform* (Fig. 7.1). The arc massif, therefore, is composed of volcanogenic and plutonic rocks formed along and below arc platforms, whose position may change through time.

Although intra-arc or summit basins have been described in the literature for many years, they have neither been subjected to the same rigorous treatment or review as forearc and backarc basins, nor explicitly defined. Some depictions of forearcs and backarcs show each extending continuously, but in opposite directions, from a hypothetical narrow linear arc, leaving no room for an intra-arc environment; a different approach is taken here.

Intra-arc basins are herein defined as basins located within or including the arc platform at sites where thick accumulations of volcanic and sedimentary material, the latter largely volcaniclastic, are found. The diverse origins, structures, and subsidence mechanisms for intra-arc basins are considered in detail below. Intra-arc basins are associated with both intraoceanic and continental-margin arcs. These arcs are constructed on variable crustal types, including normal oceanic crust, backarc oceanic crust, continental crust, and accreted crustal fragments of varied origins (Hussong and Uyeda, 1981; Larue et al., 1991;

Table 7.1 Characteristics of Selected Intra-Arc Basins

Basin	Age	Thickness of Fill	Sed./Volc. Accum. Rate	Max. Subsid Rate	Basin Dimensions	References
Volcano-bounded Basins						
Kallinago Depression, Lesser Antilles	m. Mio.- Holo.	1 km	100 m/my		20–50 km wide 250 km long 1.5–3.0 km deep	Bouysse, 1984
Izu-Bonin-Mariana Arc	l. Mio. Holo.	?	?	?	600–1000 km long 50–75 km wide 1–2 km deep	Karig, 1971a; Taylor, 1992.
Fault-bounded Basins						
Semangko fault basins, Sumatra, Indonesia	Mio.- Holo.	?	?	?	15–25 km wide 75–140 km long	Williams, 1941 Westerveld, 1953, 1963; Posavec et al., 1973; Katili, 1974.
Aleutian Summit Basins, Alaska, USA	l. Mio.- Holo.	0.3–1.0 km	≤200 m/my	0.2–0.7 km/my	15–40 km wide 50–75 km long 0.5–2.0 km deep	Geist et al., 1988; Geist and Scholl, 1992; Scholl et al., 1975; Marlow et al., 1970.
Miembro Medio basin, Chile	L. Penn- E. Perm.	600 m	?	?	>100 km wide ~300 km long	Bahlburg & Breitkreuz, 1991; Brietkreuz, 1991.
Guatemalan arc-transverse basins, Central America	Plio.- Holo.	?	?	?	20–35 km wide 45–60 km long	Williams, 1960; Burkhart & Self, 1985.
Median Trough/Nicaraguan Depression, Central America	Plio.- Holo.	1 km	?	?	40–50 km wide ~900 km long >2 km deep	McBirney & Williams, 1965; Burkhart & Self, 1985; Weinberg, 1992.
High Cascades grabens, Oregon, USA	l. Mio.- Pleist.(?)	1–2 km	>400 m/my	>1 km/my	25–40 km wide 75–110 km long 1–3 km deep	G. A. Smith et al., 1987; Priest, 1990; Taylor, 1990.
Kamchatka grabens, USSR	Mio., Quat.	>1 km	?	?	50–100 km wide ≤350 km long 1.5–2.0 km deep	Erlich, 1968.
Taupo Volcanic Zone, New Zealand	Pleist.- Holo.	>3 km	>2 km/my	>2 km/my	60–90 km wide 200 km long 1–3 km deep	Grindley, 1965; Wilson et al., 1984; Cole, 1986.
Northeast Japan basins (evolved into back-arc basin)	m. Mio.	0.2 km	<1 km/my?	0.2–3.0 km/my	5–10 km wide 30–45 km long <3 km deep zone 150 km wide x 400 km long	Yamaji, 1990.
Aso-Beppu depression, Southwest Japan	Pleist.- Holo.	1.0–1.5 km	>1 km/my	1 km/my	30 km wide ~200 km long 2 km deep	Yamazaki & Hayashi, 1976.
Sierran arc depression, Western Cordillera, USA	L. Tri.- E. Jur.	≤11 km	80–250 m/my	200–1000 m/my	~100 km wide ~1000 km long 2–12 km deep	Busby-Spera, 1988b; Riggs & Busby-Spera, 1990; Tosdal et al., 1989.
Altiplano-Puna basin, central Andes	m. Mio.- Holo.	≤10 km	250–600 m/my	≤600 m/my	75–200 km wide ~1500 km long	Zeil, 1979; Sebrier et al., 1985; Alonso et al., 1991; Sebrier and Soler, 1992.
Marinduque basin Philippines	Mio.- Plio.	>1 km	?	?	38 km wide 74 km long	Sarewitz and Lewis, 1991.
Hybrid basins:						
Central Solomons Trough, Solomon Islands	l. Mio.- Holo.	4–8 km	≤500 m/my avg. <100 m/my	>1 km/my (?)	<120 km wide ~600 km long 6 km deep	Bruns et al., 1986; Cooper et al., 1986; Ryan and Coleman, 1992.
Vanuatu Central Basins New Hebrides Islands	l. Mio.- Holo.	1–6 km	≤1 km/my	≤1 km/my	max: 35 km wide 60 km long 8 km deep	Johnson & Greene, 1988; Greene & Johnson, 1988.

Table 7.1 Continued.

Ancient Intra-arc Basin Sequences (specific classification unknown)

Troodos Complex, Cyprus	L. Cret.	0.75 km	?	?	?	Robertson, 1977.
New Hebrides Is.	l. Olig.-e. Milo.	>7 km	~500 m/my	?	?	Mitchell, 1970; MacFarlane et al., 1988.
Port Sandwich basins, Vanuatu, New Hebrides	m. Mio.	2.5 km	800 m/my	?	>75 km wide >400 km long	MacFarlane et al., 1988; Greene et al., 1988a.
South Shetland Is.	L. Jur.-E. Cret.	~2 km	~100 m/my	?	?	Smellie et al., 1980.
Fiji	Plio.	~0.5 km	<500 m/my	?	?	Dickinson, 1968.
E. Klamath Basins, California, USA	Dev.-Perm.	~2.5 km	~1 km/my	>1 km/my?	?	LaPierre et al., 1985; Watkins and Flory, 1986; Miller, 1989.
Brook Street terrane,	Permian	18 km	~1 km/my	~1–1.5 km/my	?	Houghton and Landis, 1989.

Rubin and Saleeby, 1991; Taylor, 1992). The presence of vent-proximal volcanic rocks and/or related intrusions is critical to the recognition of intra-arc basins in the geologic record. Volcaniclastic sediments derived from the arc may be spread for great distances into forearc, backarc, and foreland basins, but in these locations, they will not be found with coeval arc-platform volcanic rocks.

Forearc and backarc basins are located adjacent to the arc platform; basins immediately in front of or behind the active volcanic front, but *located on the arc platform*, are intra-arc basins. At the time of formation, intra-arc basins are spatially distinct from forearc and backarc basins. Many backarc basins originate with rifting within the arc platform (see Chapter 8), and are therefore, intra-arc basins during early stages of development. This relationship of intra-arc and backarc basins is discussed below. Note that *intra*-arc basins should be distinguished from *inter*-arc basins. The latter term, introduced by Karig (1970a), has been widely superseded by *backarc*. The distinction of forearc and intra-arc basins is discussed in Chapter 6.

Because of arc migration, a single site may change among intra-arc, backarc, and forearc settings. The Luzon Central Valley, in the Philippines, illustrates the complexity that can arise during the evolution of convergent margins and the difficulty of classifying basins. Changes in subduction polarity and other processes affecting the position of the arc axis over the past 40 my have placed the Central Valley successively in backarc, forearc, and intra-arc positions (Fig. 7.2; Bachman et al., 1983; see Chapter 6). Strike-slip deformation, as a consequence of oblique convergence, has influenced basin structure and deposition during most of this time. Similar difficulties in classifying convergent-margin basins have been discussed by Jordan and Alonso (1987) and Flint et al. (1993), and highlight the ambiguities that can arise when defining or recognizing forearc, intra-arc, and backarc basins and their sedimentary and volcanogenic fill.

Many ancient volcanic-arc successions are deformed by intrusions or collision-accretion processes along plate margins. Because of the severity of this deformation and subsequent uplift and erosion, the spatial relationships of volcanogenic materials to the arc axis and

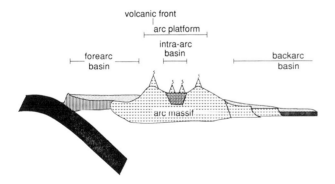

Fig. 7.1 Diagramatic cross section through a convergent plate margin, showing location of arc platform relative to forearc and backarc basins. Areas underlain by arc crust, termed "arc massif" by Dickinson (1974a,b), include basement to parts of forearc and backarc basins. Intra-arc basins form within active and typically high-standing arc platforms. Arc volcanoes are typically dispersed over a wide zone perpendicular to plate boundary, but most active volcanoes are aligned along a distinct volcanic front.

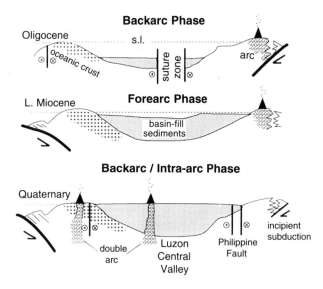

Fig. 7.2 Schematic cross sections illustrating tectonic evolution of Luzon Central Valley (Philippines). Reversal of subduction polarity and shifting arc positions have produced sequential backarc, forearc, and intra-arc stages in evolution of basin, which is also strongly influenced by strike-slip motion along Philippine fault. (Modified with permission from Bachman et al., 1983). Such compound basins are common along convergent plate margins and make simple distinction of basin type difficult.

the distinction of forearc, intra-arc, and backarc positions is not always clear (e.g., the Ordovician of Wales; Kokelaar, 1992, p. 1454). Basins containing volcanogenic material may, more generally, be referred to as *arc-related basins*, especially where spatial relationship to the arc platform is not clear from structural and stratigraphic data. Some arc-related basin fills have been interpreted as backarc-basin fills despite the abundance of proximal extrusive and intrusive rocks of arc compositional affinity that may indicate an intra-arc position (e.g., Cas and Jones, 1979; Cole, 1984; LaPierre et al., 1985).

Because of the complexities outlined above, this chapter emphasizes the structural and lithologic aspect of late Cenozoic intra-arc basins, whose tectonic setting is relatively unambiguous (Table 7.1). Inferences are made regarding the origins of some older arc-related assemblages as probable intra-arc basin fills. In contrast, aspects of volcanic stratigraphy and sedimentation in volcanic arcs is necessarily biased toward older, uplifted arc sequences, whose relationship to intra-arc basins cannot always be ascertained.

We stress that *intra-arc basin* is a descriptive term carrying no genetic connotation other than association with a volcanic arc and convergent plate margin. Thus, many different, possibly unrelated, mechanisms may be responsible for the origin of basins described in this chapter.

The recognition of intra-arc basins is important for reconstruction of the tectonic evolution of convergent margins. Many ancient magmatic arcs are marked by uplifted and deeply eroded plutonic belts that afford little or no proximal volcanic record. Where intra-arc basins form, direct (volcanic edifices) and indirect (volcaniclastic aprons and basin fills) records of the composition, eruptive style, and history of volcanic activity are preserved. Because intra-arc basins contain the most proximal volcanic record, their stratigraphy may be more complete and less ambiguous to interpret than distal sections preserved in adjacent forearc and backarc basins. Evidence for initiation of magmatism may be interpreted from intra-arc basin stratigraphy (Larue et al., 1991) and may be used to estimate the timing of onset of subduction at a convergent margin, which should precede initial arc magmatism by no more than 5 my (Gill, 1981). The nature and history of basin-bounding structures can be used to infer the time-varying state of stress in the arc, itself related to changing subduction parameters that reflect relative plate motions (Dewey, 1980; Jarrard, 1986a), asthenospheric circulation (Apperson, 1991), and collisions of buoyant crustal features with arc-trench systems (Fisher, 1986; Cloos, 1993).

Types of Intra-Arc Basins

The nature of topographic and bathymetric boundaries of intra-arc basins suggests two end member types, *volcano-bounded* and *fault-bounded* (Fig. 7.3).

Some volcano-bounded basins are coalesced aprons of volcanic and sedimentary debris that accumulate in the low areas among volcanoes on the arc platform. These basins have only ill-defined physiographic margins of constructional volcanic origin. Significant thicknesses of basin fill seldom form on relatively high-standing continental-margin arc platforms; however, island arcs in intraoceanic settings commonly accumulate thick volcanogenic fills in volcano-bounded basins (e.g., Larue et al., 1991).

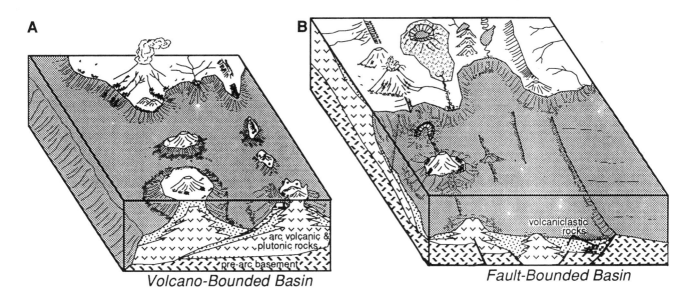

Volcano-Bounded Basin

Fault-Bounded Basin

Fig. 7.3 Schematic block diagrams of morphology and structure of: (A) volcano-bounded and (B) fault-bounded intra-arc basins. In each case, volcanic and sedimentary rocks accumulate in troughs along arc platform. Margins of trough are defined by constructional volcanic features in case of volcano-bounded basins and by faults for fault-bounded basins. Hybrid basins, not illustrated, are bordered by both constructional and structural ridges. In (A), more recently active arc axis is located on left. Volcanics and coarse-grained sediments from this active line of volcanoes onlap volcaniclastic apron related to slightly older line of volcanoes on right and interfinger with finer-grained sediments, partly derived from fringing coral reefs that flank the inactive volcanoes.

Other volcano-bounded basins are linear arc-parallel troughs between volcanic ridges on the arc platform. Multiple linear chains of volcanoes form in arcs for various reasons, including reversals in subduction polarity, migration of the volcanic front in response to changes in the angle of subduction of the underriding plate, and processes controlling melting in the asthenosphere. Discrete shifts in the arc axis because of changes in subduction parameters controlling arc position or development of coeval, parallel arc axes will form a trough between closely spaced arc ridges. Because these basins form between volcanic ridges, and may be in front of or behind the active volcanic front, it might be argued that they be termed inter-arc, forearc, and backarc basins. These troughs are on the arc platform, however, and as described in greater detail below, are spatially separate from forearc and backarc basins; they are not associated with arc rifting and formation of new oceanic crust, as is the usual case for inter-arc basins (Dickinson, 1974a; see Chapter 8).

In contrast, fault-bounded basins (Fig. 7.3B) are rapidly subsiding, fault-bounded, arc-parallel, or arc-transverse basins caused by tectonically induced subsi-

dence of segments of the arc platform. Tectonic structures, not constructional volcanic features, account for most of the topographic/bathymetric relief along the basin margins. Normal-fault-bounded intra-arc basins are found in many modern and ancient volcanic arcs (e.g., Erlich, 1968; Dengo et al., 1970; Burkhart and Self, 1985; Smith et al., 1987; Busby-Spera, 1988b; Geist et al., 1988; Yamaji, 1990). The tectonic origin, structural characteristics, and stratigraphy of several of these basins are considered below.

Hybrid basins, with elements of both volcano-bounded and fault-bounded types, also form. These are best represented by troughs with normal, reverse, or strike-slip faults along part or all of their margins, but where most of the topographic/bathymetric relief is defined by constructional volcanic ridges.

Arc platforms undergo complex histories of alternating uplift and subsidence related to angle, obliquity, and rate of subduction, in turn partly related to age, thickness, and crustal type of subducting lithosphere. These parameters typically change on time scales of 5 to 20 my and are the first-order controls on the structural history of the arc platform. The structural histories of intra-arc basins, therefore, vary through time

and may include evolutionary paths through any or all of the three basin types defined above.

Basin Architecture and Subsidence Patterns

Introduction

Table 7.1 summarizes many important characteristics of numerous intra-arc basins. Subsidence and accumulation rates have been estimated where possible from primary references. Geohistory analysis, however, is precluded by a lack of high-resolution chronostratigraphic control, strong dependence on seismic profiles rather than outcrop or bore-hole data, and general lack of paleobathymetric data for marine strata. Accumulation rates are typically on the order of 0.2 to 1.0 km/my, with comparable subsidence rates. Fault-bounded intra-arc basins are typically 1 to 6 km deep. Some intra-arc basin sedimentary sequences (e.g., Mesozoic Cordilleran arc graben-depression, Permian Brook Street terrane, Neogene Altiplano-Puna basin) are more than 10 km thick and accumulated over time intervals in excess of 10 my.

Subsidence Mechanisms

Uplift in arcs may be associated with crustal thickening, and thermal and physical effects of rising magma (e.g., Tobisch et al., 1986). Mechanisms for generating subsidence within the constructionally high arc platform are poorly understood, largely hypothetical, and more complex than can be explained by thermal-contraction and flexural-loading models typically applied to other basin types. The proposed mechanisms can be divided into six categories, which are discussed below. These mechanisms, acting singly or in combination, may account for both regional and localized subsidence of arc platforms, of either broad synclinal, or graben-faulted character.

Plate-Boundary Forces

Long-wavelength, long-duration and relatively slow deformation of the arc-forearc region can be accounted for by the dynamic interaction of plates at subduction zones. Several hypotheses have been presented as a result of geophysical modeling or to explain long-term alternations of uplift and subsidence indicated by geological data. Mechanisms associated with plate-boundary forces are probably the first-order regulators of subsidence of intra-arc basins.

Davies (1981) argued that the surface of the earth is pulled downward by the excess weight of the subducted slab in the asthenosphere. The required depression is partly accounted for by the formation of the trench, but about 50% of the mass of the plate is compensated for by stresses transmitted through the viscous mantle wedge, causing a broad, shallow depression in the overriding plate. The maximum amplitude of the modeled depression is about 1 km at a distance of 100 km from the trench, for a slab descending at a 45° angle, and has a deflection of about 0.5 km at a distance of 500 km.

A more rigorous finite-element analysis of stresses and plate-boundary forces associated with convergent plate boundaries was undertaken by Whittaker et al. (1992) and yielded models for horizontal deviatoric stresses for conditions of a locked or free-sliding subduction faults. Compression and downflexure in the arc-trench region occur when the fault is locked. In contrast, tension occurs if the fault is unlocked and trench suction causes extension, and possible crustal thinning and subsidence. Under either condition, therefore, subsidence is predicted for the arc platform.

Scholl et al. (1987b) and Geist et al. (1988) proposed a different model to explain wholesale uplift, followed by subsidence of the Aleutian Ridge arc platform (Fig. 7.4). Uplift of the previously submerged arc platform to near or above sea level during early to middle Tertiary is attributed to the subduction of relatively young, buoyant Kula plate lithosphere. Following subduction of the Kula Ridge at about 10 Ma, subduction of progressively older oceanic crust, presumably at a steeper angle than previously, resulted in broad subsidence, erosion of a wave-cut surface across the platform, and development of 100 to 200 m water depths, at which marine volcanogenic sediment has subsequently accumulated. Although this hypothesis of uparching of the arc platform is at odds with the downbending predicted by the finite-element model of Whittaker et al. (1992), it may

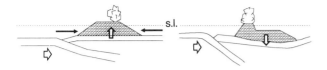

Fig. 7.4 Model for alternating uplift and subsidence of an arc platform related to changing plate-boundary forces, based on late Cenozoic evolution of Aleutian Ridge. (Modified with permission from Geist et al., 1988.) Isostatic subsidence can be inhibited by horizontal forces associated with subduction (e.g., rapid low-angle subduction of relatively young, bouyant lithosphere). Steepening subduction angle and/or decrease in convergence rate leads to rapid subsidence to achieve isostatic equilibrium. Note inference of wave planation of arc platform during subsidence.

explain the broad geanticlinal doming along arcs that is described in Indonesia (e.g., van Bemmelen, 1970; Aldiss and Ghazali, 1984) and Nicaragua (McBirney and Williams, 1965).

Relative Plate Motions

Deformation within volcanic arcs may partly reflect the rate and direction of relative convergence. Oblique convergence is typically resolved into a strongly orthogonal compressional component at the trench and a strike-slip component expressed by structures in the forearc and arc itself (Fitch, 1972). The largest and most continuous arc-parallel strike-slip faults are present in the arc, because of the thermally weakened state of the crust there, and are better developed in continental-margin arcs because continental crust is weaker and better coupled to the subducting slab than is thin oceanic-arc crust (Dewey, 1980; Jarrard, 1986b; Ryan and Coleman, 1992). Many intra-arc basins are related, in one fashion or another, to arc-parallel, intra-arc strike-slip zones (e.g., Semangko fault zone, Sumatra; Nicaraguan Depression; Marindique basin, Philippines; Central Solomons Trough), block rotations related to oblique subduction (Aleutian summit basins, Sunda Strait), or trench-transform boundaries (Guatemala). Examples of such basins are discussed below.

The low-lying arc of El Salvador and Nicaragua rests in a fault-bounded depression that is at least partly transtentional in origin, based on seismic focal mechanisms and mapped structures (Carr, 1976; Weinberg, 1992), but may also be related to other complex plate motions in the Caribbean region

besides the oblique subduction of the Cocos plate (Fig. 7.5). Neogene sinistral displacement on the order of 450 km along the Cayman Trough transform requires a large component of eastward drift of the Caribbean plate, away from the Central American Trench. Burkart and Self (1985) hypothesize that these relative plate motions lead to extension perpendicular to the arc and the formation of a subsiding, intra-arc graben about 700 km long.

To account for alternating phases of compression and extension in the Andean arc that lack correlation to changes in Nazca plate convergence rate, Sebrier and Soler (1991) appeal to instabilities in the dynamic equilibrium between the westward motion of the South American plate, shortening in the sub-Andean retroarc thrust belt, and rollback of Nazca lithosphere (Fig. 7.6; see Chapter 9). In their model, the rate of rollback of the subducted slab controls Andean tectonics. When the rollback rate is low relative to the rate of westward motion of the South American plate, intraplate compression causes extreme shortening in the retroarc thrust belt and compressional deformation and uplift in the arc. When the

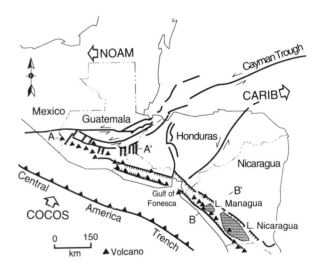

Fig. 7.5 Map showing relationship of arc-transverse basins, in Guatemala, and arc-parallel Medial Trough-Nicaraguan Depression to relative plate motions in Central America. (After Burkhart and Self, 1985; Dengo et al., 1970.) Caribbean (CARIB)-North American (NOAM) relative motion and spreading in Cayman Trough has produced pull-apart basins along western part of arc. Because Caribbean "absolute" motion is away from Cocos plate and Central American Trench, arc-parallel extension has occurred along arc platform in Nicaragua and El Salvador. Sections A-A' and B-B' are illustrated in Fig. 7.14.

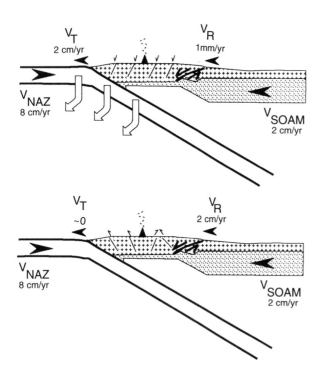

Fig. 7.6 Model for alternating extension and compression in Andean arc (after Sebrier and Soler, 1991) related to dynamic equilibrium between fixed rates of relative motion of Nazca (V_{NAZ}) and South American (V_{SOAM}) plates, and variable rates of trench rollback (V_T) and retroarc foreland thrusting (V_R). When V_T is nearly equivalent to V_{SOAM} (upper diagram), V_R is low and extension occurs within arc massif. When V_T is low, NAZ–SOAM relative motion is absorbed by increased rates of foreland thrusting and reverse-fault uplift of the arc massif.

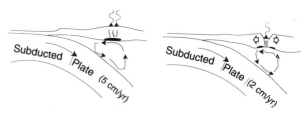

Fig. 7.7 Model for intra-arc extension related to asthenospheric corner flow (Apperson, 1991) induced by subduction (modified from Taylor, 1990). When relative convergence rate is high, circulation cells are elliptical in cross section (left diagram), but are hypothesized to be more circular and to impinge upon base of lithosphere when convergence rates decrease (right diagram). Taylor (1990) hypothesized that this process caused late Miocene extension in High Cascade Range.

rollback rate and the westward velocity of the South American plate are similar, the rate of shortening decreases in the thrust belt but thrusting continues. As a result, the rollback is faster than the net westward movement of South American lithosphere, including retroarc shortening, and the arc-forearc region extends and subsides while spreading westward. This mechanism may be most effective for strongly coupled subduction faults, for which plate rollback induces a larger trench-suction force in the arc-forearc region (Forsyth and Uyeda, 1975; Dewey, 1980).

Asthenospheric Flow

Apperson (1991) summarized seismic-strain-field data from 17 volcanic-arc regions and showed that most convergent margins are characterized by subhorizontal extension perpendicular to the arc regardless of the obliquity of convergence. She interpreted her data to imply that plate-boundary stresses generated by coupling between overriding and subducting plates may not be transmitted to the volcanic arc. Instead, the strain field of the arc is thought to be influenced by asthenospheric flow in the wedge above the subducted slab. This flow is induced by the downward and lateral motion of the subducting slab. This explanation for arc extension is similar to the model proposed by Toksöz and Bird (1977) for the initiation of backarc spreading.

Taylor (1990) appealed to asthenospheric corner flow to explain many features of magma generation and tectonism in the Cascade Range throughout the Cenozoic. He hypothesized cylindrical flow cells that develop elliptical cross sections when subduction rates are high, and more nearly circular cross sections when rates are low (Fig. 7.7). Wide elliptical cells induce melting over a wider region of the asthenosphere wedge and favor development of a wider arc, as also discussed by Tatsumi et al. (1986). Conversely, smaller circular cells focus magmatism along a narrow arc; these cells may directly impinge on the base of the overriding lithosphere and cause extension and graben formation in the arc platform.

Regional Isostasy

Isostatic generation of accommodation space for intra-arc sediment is a result not only of local compensation for the accumulating sediment pile, but also the flexural adjustment for the large, denser volcanic edifices, if they are not buoyantly compensated by rising magma or partly supported by horizontal forces. Smith et al. (1989) investigated this effect (Fig.

7.8) using an analogy to the flexural trough and fore-bulge that have formed adjacent to the Hawaiian-Emperor chain (Watts, 1978). For a rigid, unbroken plate loaded with a 2-km high, 20-km wide arc and flanking volcaniclastic aprons, a downward deflection of about 1 km is expected under the arc axis, diminishing to about 400 m at a distance of 40 km on either flank of the arc (Smith et al., 1989). Such a modest load could be generated in most arcs in 1 to 10 my and provide a self-sustaining mechanism for accumulating thick volcanic and volcaniclastic sequences on arc platforms.

Smith et al. (1989) also considered the consequences of a broken plate. If tensional strain of any origin induces faulting in a previously rigid plate, then the load of volcanic arc is compensated only locally within a fault block, rather than flexurally over a broad region (Fig. 7.8). Subsidence on the order of several kilometers can be expected at rates on the order of 1 km/my without any crustal thinning and at strain rates of only a few percent.

Hildebrand and Bowring (1984) noted that most continental-margin volcanic arcs differ from the classic Andean arc by remaining at low elevations, usually with basal elevations within 1000 m of sea level, throughout their history. Uplift that would be expected as a consequence of crustal underplating by magmatic processes is not prevalent, requiring maintenance of isostatic equilibrium throughout arc history. Hildebrand and Bowring (1984) propose that this equilibrium is accomplished by a balance between the addition of dense mafic granulite in the lower crust, and eruption of a large mass of

silicic pyroclastic material into the atmosphere, where it is lost from the arc system. The result is net long-term subsidence and the development of non-extensional synclinal arcs and intra-arc basins (Fig. 7.9).

Both hypotheses of Hildebrand and Bowring (1984) and Smith et al. (1989) are consistent with the generally synclinal form of some arcs (Hammond, 1979; Levi and Aguirre, 1981), but are not consistent with the geantclinal form of others (McBirney and Williams, 1965; van Bemmelen, 1970). The Hildebrand and Bowring model does not account for the prevalence of normal faulting within many arcs or intra-arc depressions forming not only on continental crust, but also in intraoceanic arcs where large volumes of silicic pyroclastic material are not erupted and lost from the arc environment. In some cases, the basins have subsided very rapidly over short time periods (Table 7.1) and have not been long-lived, slowly subsiding features. Most arcs appear, instead, to be characterized by alternating periods of uplift and subsidence throughout their history (e.g., Merier, 1981; Scholl et al., 1987b; Greene et al., 1988a; Priest, 1990), suggesting complex interactions between isostatic effects and those associated with lithospheric and asthenospheric processes at plate margins.

Magmatic Withdrawal

Eruption of large volumes (10 to 10^3 km^3) of magma, typically as intermediate to silicic pyroclastics, leads to collapse and formation of *calderas*. Most calderas are only a few kilometers across and form at the top of

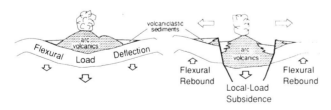

Fig. 7.8 Models of isostatic-load depression in volcanic arcs (modified from Smith et al., 1989). Load of volcanic edifices may be flexurally compensated to produce depression for accommodating volcanic sequences and flanking volcaniclastic aprons. If extension causes faulting of arc platform, then rapid subsidence of central fault block is driven by change from flexural isostasy to Airy-type, local-load compensation.

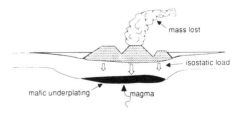

Fig. 7.9 Flexural isostatic subsidence below and adjacent to arc axis may, in some cases, be facilitated by magmatic withdrawal and loss of magmatic components by distant dispersal of fallout tephra while mafic magmas add dense layer to lower crust (after Hildebrand and Bowring, 1984).

composite volcanoes; these features are of little consequence for net subsidence of the arc platform. Other large calderas in arcs are not associated with single volcanic centers, have diameters of 10 to 100 km, and have subsided from 1 to 4 km. Most calderas are filled with ignimbrite and collapse breccia derived from the caldera-forming eruption, but these are commonly capped by post-collapse sedimentary and volcanic units (Lipman, 1984). Calderas, therefore, are an important form of localized subsidence on arc platforms. Although typically viewed as downfaulted cylinders, Walker (1984) points out that calderas are known, or may be inferred, to be expressed by a wide variety of structural styles, including broad downsags with little or no structural relief formed along ring faults.

In some cases, depressions related to eruption of voluminous ignimbrites appear to be partly or wholly bounded by regional tectonic faults. Large-scale linear depressions controlled by both tectonic and volcanic processes were named *volcano-tectonic depressions* by van Bemmelen (1933) (also see Williams, 1941). Often cited examples of such depressions are the Toba caldera in Indonesia, which is 100 km long but only 18.5 km wide and elongated parallel to regional faults (van Bemmelen, 1933; Aldiss and Ghazali, 1984), and the Taupo volcanic zone in New Zealand, where many overlapping and nested calderas, combined with graben faulting, have resulted in 1.5 to 4.0 km of subsidence over a 150-by-60 km area over the last 1 to 2 my (Wilson et al., 1984). Regional normal faults are believed to have been involved in the collapsed of some Taupo calderas (Walker, 1984; Wilson et al., 1984).

The difficulty of distinguishing between subsidence related to regional faulting and subsidence due to magmatic withdrawal is further exacerbated by the common correlation of large-volume caldera-forming eruption with extension and normal faulting in volcanic arcs (e.g., Westerveld, 1963; McBirney and Williams, 1965; Oide, 1968; Pichler and Weyl, 1973; Burkart and Self, 1985; Smith et al., 1987; Busby-Spera, 1988b, Crawford et al., 1988). In some situations, single calderas or complexes of nested or overlapping calderas are recognized within intra-arc grabens (Aramaki, 1984; Wilson et al., 1984; Rose et al., 1987) and careful stratigraphic study is required

to distinguish the cause of subsidence (e.g., Riggs and Busby-Spera, 1991). The relation between extension and caldera formation is discussed further below and is probably the result of larger-volume magma production and development of larger, higher-level crustal magma chambers under extensional conditions. In extending arcs, subsidence in response to magma withdrawal is probably an important component of the total subsidence of the arc platform.

Gravitational Collapse

Extensional collapse of orogenic belts is widely recognized as a consequence of tectonic overthickening within orogens (Molnar and Tapponnier, 1978; Dewey, 1988), and may be focused in or near the magmatic arc, where crust is thermally weakened and the potential for deep ductile deformation is enhanced. Most continental-margin volcanic arcs remain topographically low during most of their history (Hildebrand and Bowring, 1984), with the notable exception of the modern Andes, with a mean elevation of 4 km. North-south extension of a 400 km-long belt in the central Peruvian Andes is demonstrated by focal mechanisms and kinematic features along Quaternary fault scarps, indicating 1 to 8% of extension perpendicular to the arc over the last 2 my (Mercier, 1981; Sebrier et al., 1985; Cabrera et al., 1987). Sebrier et al. (1985) interpreted the extension to have resulted from gravitational collapse of the high-standing Andes and lateral spreading to generate foreland thrust sheets to the east. The total magnitude of subsidence that is produced by this process is not clear, but up to 10 km of Cenozoic sedimentary and volcanic material have accumulated in the Andes (James, 1971).

Volcano-Bounded Basins

First-order estimations of ancient basin dimensions and shapes in most tectonic settings are determined from facies patterns and paleocurrent data that permit designation of basin-bounding positive features of tectonic origin. Within volcanic arcs, basins are commonly bounded by volcanoes; constructional rather than structural features, thus strongly influence facies patterns and sediment dispersal. Topographic or bathymetric lows between constructional volcanic features,

therefore, may define basin boundaries within arc plat-forms (Fig. 7.3), whose overall subsidence is governed by processes discussed above. In this section, emphasis is placed on the morphological expression of volcano-bounded lows within modern oceanic arcs, where relief is entirely a consequence of volcano growth. Basins whose boundaries are defined by both volcanoes and faulted uplifts are discussed below.

Dimensions and outlines of volcano-bounded basins are determined by the spacing and size of volcanic cen-ters. The larger edifices in most volcanic arcs are spaced at distances of 30 to 50 km, although the range is 10 to 80 km (Williams and McBirney, 1979; Gill, 1981). Volcaniclastic aprons derived from flanking vol-canic centers partly fill these intervening areas. The thickest volcano-bounded basin fills are found in intra-oceanic arcs developed in deep water. The bathy-metric/topographic profiles in Fig. 7.10 illustrate the variation in volcano spacing, volcano size, and water depth that control the sedimentological characteristics of volcano-bounded basins in relatively narrow oceanic arcs.

Another variety of volcano-bounded basin is rep-resented by linear to arcuate, arc-parallel troughs located between volcanic ridges on the arc platform. Multiple linear chains of volcanoes form in arcs for various reasons, including reversals in subduction polarity and migration of the volcanic front in response to changes in the angle of subduction of the underriding plate. Coeval parallel arcs, separated by tens of kilometers, may result from melt generation at two specific depths above the subducted slab, as Defant et al. (1988) proposed for the parallel volcanic zones in the Bataan and Mindoro arcs in the Philip-pines (Fig. 7.17A). A similar mechanism may account for the ponding of sediment between coeval arcs in the Bolivian Altiplano during the early to middle Miocene (Jordan and Alonso, 1987).

If discrete shifts in the arc axis within the arc plat-form occur because of major changes in subduction parameters controlling arc position, then a depression is formed between the older and newer arc axes. Two late Cenozoic examples of such volcano-bounded basins are described below.

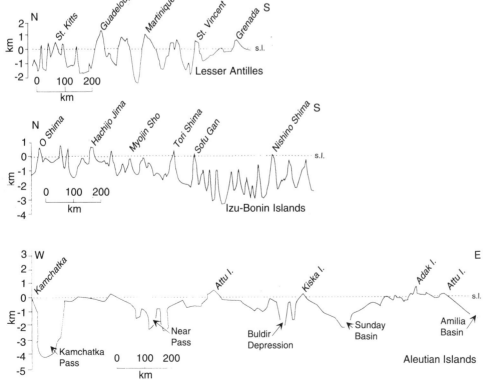

Fig. 7.10 Along-arc bathy-metric/topographic profiles of Lesser Antilles (after Wadge, 1986), Izu-Bonin Islands (after Taylor, 1992), and Aleutian Islands (after Geist et al., 1988). Construction of volcanic centers produces intervening basinal areas that are as deep as 3 km. The depth of volcano-bounded basins in Aleutian Islands is generally less than 1 km, because of late Cenozoic uplift and wave planation of arc plat-form (Scholl et al., 1987b). Deeper Aleutian basins, marked by arrows, are fault-bounded extensional structures produced by strike-slip-related block rota-tions along arc axis (Geist et al., 1988; see also Fig. 7.13). Gener-ally deeper water along Izu-Bonin island arc between Sofu Gan and Nishino Shima is related to grabens that appear genetically related to backarc rift basins to north and south (Tay-lor, 1992).

Lesser Antilles

The Kallinago Depression in the northern Lesser Antilles has formed in the last 10 my between an outer (trenchward) Eocene to early Miocene volcanic arc and an inner late Miocene to Holocene arc (Fig. 7.11). The westward shift in the arc was probably related to arrival of an aseismic ridge at the trench, with consequent changes in the rate of subduction and the angle of plate descent (Bouysse, 1984). This complication in subduction dynamics affected only the northern part of the Lesser Antilles; the Kallinago Depression terminates southward, where the two island chains merge into a single arc axis. Seismic-reflection profiles (Bouysse, 1984) show a broadly synformal basin fill without obvious structural margins (Fig. 7.11). Locally jumbled reflections probably represent gravity-driven deformation along the very steep margins.

Izu-Bonin-Marianas Arc

Paired arcuate ridges with an intervening trough also characterize segments of the Izu-Bonin-Mariana arc, western Pacific Ocean. Cenozoic arc rifting has dispersed arc fragments over an area as wide as 450 km from the trench wall to the backarc region (Karig, 1971a; Hussong and Uyeda, 1981; Taylor, 1992). The arc platform (frontal arc of Karig, 1971a) consists mostly of a single chain of Quaternary volcanoes. In some arc segments, however, the active, largely submarine volcanoes parallel nearby uplifted Eocene - lower Miocene arc volcanoes capped by Pliocene-Quaternary limestones (Karig, 1971a; Fig. 7.12). The morphologically defined fore-arc basin lies trenchward (east) of these older arc rocks (Karig, 1971a; Taylor, 1992); the trough between the active volcanoes and the older Tertiary volcanoes is best described as a volcano-bounded intra-arc basin.

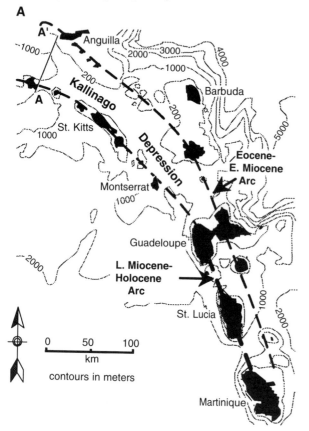

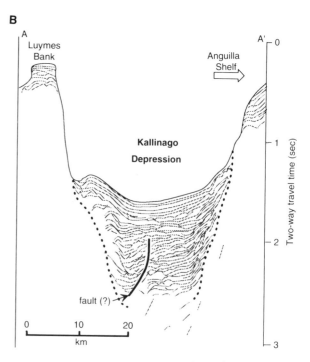

Fig. 7.11 (A) Map of northern Lesser Antilles and Kallinago Depression, which is a volcano-bounded intra-arc basin formed between inactive Ecocene to early Miocene arc and younger late Miocene to Holocene arc of the Lesser Antilles. Volcanic axes converge southward to a single arc. (B) Interpreted seismic-reflection profile (location in A), showing slightly deformed sedimentary fill within Kallinago Depression (from profile in Bouysse, 1984). Paucity of reflectors adjacent to depression is probably result of poor penetration of seismic waves though volcanic rocks comprising adjacent arc axes.

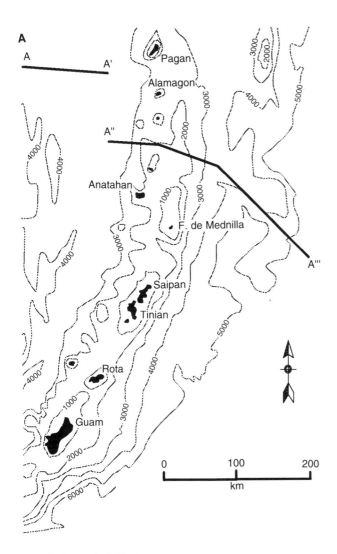

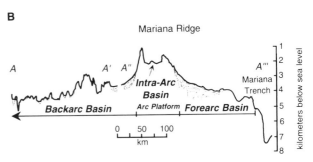

Fig. 7.12 (A) Map and (B) interpreted seismic profile across largely submarine Marianas Ridge and adjacent backarc and forearc basins (after Karig, 1971a). Platform locally consists of both active arc (west) and uplifted older arc volcanoes (east) with intervening volcano-bounded intra-arc basin.

Fault-Bounded Basins

As observed by Dickinson (1974a), most intra-arc basins are bounded by normal faults. Consideration of several late Cenozoic fault-bounded intra-arc basins (Table 7.1) indicates a wide variety of structural characteristics and probable origins. Recognition of fault-bounded intra-arc basins is difficult in subsequently deformed and uplifted arc terranes. Volcanic and volcaniclastic sequences that are only a few kilometers thick and accumulated over periods of several millions or tens of millions of years (e.g., Robertson, 1977; Smellie et al., 1980; Table 7.1) may be accommodated by broad subsidence of the arc platform without faulting. Thicker sequences, such as those in the early Mesozoic Cordilleran arc (11

km; Busby-Spera, 1988b), and Brook Street terrane in New Zealand (16 km; Houghton and Landis, 1989) suggest accumulation within rapidly subsiding basins. In the case of the Cordilleran arc graben depression, stratigraphic evidence exists for the bounding faults (Busby-Spera, 1988b), whereas in other cases, this evidence is lacking. Although inferences might be drawn from the deposition of thick arc successions within fault-bounded basins in ancient orogenic belts, the discussion in this section draws primarily on a few late Cenozoic examples. The reader is referred to Table 7.1, and references cited therein for more examples of fault-bounded intra-arc basins.

Aleutian Ridge

Eight basins are present along the summit of the central Aleutian Ridge arc platform (Table 7.1, Figs. 7.10, 7.13; Perry and Nichols, 1966; Marlow et al., 1970; Scholl et al., 1975, 1983). Geist et al. (1988) and Geist and Scholl (1992) proposed that these basins resulted from oblique subduction, causing discrete block rotations along the eastern and central arc platform, with a through-going strike-slip fault zone restricted to the western platform, where the convergence vector is more highly oblique (Fig. 7.13). Each block is about 100 km wide and blocks range in length parallel to the trench from 100 to 400 km. As each block has rotated clockwise, compression has occurred near the leading corner, "tear away" zones marked by submarine canyons in the forearc have formed at block boundaries, and triangle-shaped

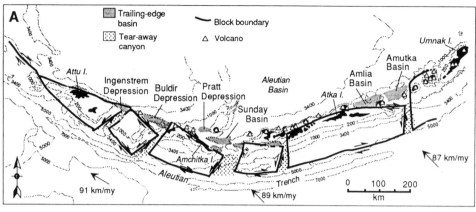

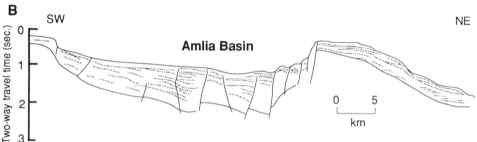

Fig. 7.13 (A) Map of intra-arc basins along Aleutian Ridge (after Geist et al., 1988). Oblique subduction, shown by arrows near bottom of map, causes clockwise rotation of discrete arc-forearc blocks, producing triangular, asymmetric, normal-faulted trailing-edge basins along arc axis and tear-away canyons in forearc. Note absence of discrete blocks eastward, where convergence is nearly normal to orientation of margin, and to west, where volcanic arc is replaced by through-going strike-slip fault. (B) Interpreted seismic profile, showing typical asymmetry of basins formed along arc platform by hypothesized block rotations (after Geist et al., 1988).

basins have formed along the trailing edges of blocks within the arc platform. The intra-arc basins vary in size (Table 7.1) and are locally divided into multiple depocenters by intra-basinal volcanic highs (Scholl et al., 1983). The basins are asymmetric and deepest to the north (Scholl et al., 1987b; Geist et al., 1988) against a major fault (Fig. 7.13B). Faulting presumably has progressed trenchward with time, synchronous with basin deepening as the forearc block has detached and rotated southward.

Central America Arc

The Central America arc contains both arc-parallel and arc-transverse intra-arc basins (Fig. 7.5). The arc-transverse basins are restricted to the Guatemala portion of the arc, which is landward of a trench-trench-transform triple junction. Sinistral movement along a series of southward convex faults produces counterclockwise block rotation and development of north-south-trending grabens within the nearly east-west-trending volcanic arc (Fig. 7.14A; Williams, 1960; McBirney and Williams, 1965; Burkart and Self, 1985). Bimodal volcanism along north-south vent alignments

within the grabens (McBirney and Williams, 1965), major caldera systems, and a very high level of shallow crustal seismicity are indicative of intra-arc extension (Burkart and Self, 1985).

In El Salvador and Nicaragua, the arc is largely confined within a nearly continuous, 1000km-long arc-parallel graben system known as the Medial Trough and Nicaraguan Depression (Figs. 7.5, 7.14B). This basin developed in the late Miocene or early Pliocene atop a broad geanticline, forming the forearc and arc platform (Dengo et al., 1970; Weyl, 1980). Quaternary volcanic centers, which are dominantly basaltic (McBirney and Williams, 1965; Weyl, 1980), are present mostly along the southwestern margin and in the center of the basin, with older volcanic rocks found along the northeastern side (McBirney and Williams, 1965). The Medial Trough-Nicaraguan Depression is possibly related to the complexities of Cocos-Caribbean plate interactions (Fig. 7.5; Burkart and Self, 1985). Focal mechanisms and field studies document dextral movement along the bounding structures, with sinistral movement along northeast-trending transverse faults (Carr, 1976; Burkart and Self, 1985). Weinberg (1992) interpreted strike-slip deformation and forma-

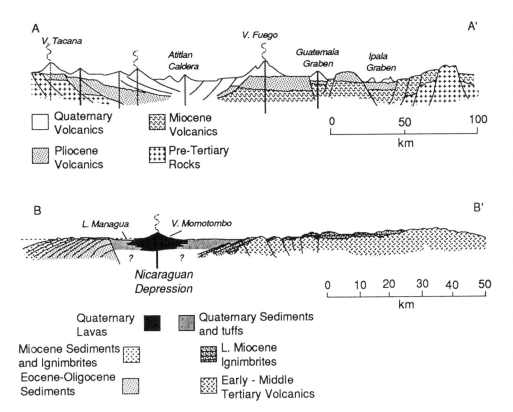

Fig. 7.14 Schematic cross sections of intra-arc grabens in Central America; locations of sections shown in Fig. 7.5. (A) Volcanic centers in Guatemala formed within or between arc-transverse pull-apart basins related to strike-slip motion along Caribbean-North America plate boundary (after Burkart and Self, 1985). (B) Cross section through Nicaraguan Depression (after McBirney and Williams, 1965), an arc-parallel graben related to oblique-convergence strike-slip faulting (Carr, 1976; Weinberg, 1992) or eastward motion of Caribbean plate away from Central American Trench (Fig.7.5, Burkart and Self, 1985). Subsidence followed a late Miocene period of explosive eruptions near position of modern volcanic axis that formed an ignimbrite plateau to east and also deposited pyroclastic rocks within now-uplifted forearc sediments to west of arc.

tion of the Lake Managua pull-apart graben within the Nicaragua Depression as a younger phase of deformation, possibly related to oblique subduction (Burkart and Self, 1985; Jarrard, 1986b) or southeastward increase in the convergence rate (Carr, 1976).

Oregon High Cascade Range
Development of the late Miocene central Oregon High Cascade graben (Fig. 7.15) has been well documented by geological (GA Smith et al., 1987; Priest, 1990; Taylor, 1990) and geophysical (Couch et al., 1982; Keach et al., 1989; Livelybrooks et al., 1989; Stanley et al., 1990) studies. This basin, which formed within the late Miocene arc and contains the present arc, rapidly subsided at least 2 km in no more than 1 my (Table 7.1; GA Smith et al., 1987; Taylor, 1990). Subsidence was probably contemporaneous with uplift of older arc volcanics in the western Cascades (Fig. 7.15; Priest, 1990). Other depressions of similar or younger age occur both north and south of the central Oregon High Cascade graben (Taylor, 1990). Many schematic cross sections of the central Oregon High Cascade graben depict a symmetrical

depression (e.g., GA Smith et al., 1987; Priest, 1990; Taylor, 1990), but seismic-reflection (Keach et al., 1989) and subsurface data (Sherrod and Conrey, 1988; Oregon Department of Geology and Mineral Industries, unpublished data) favor a series of asymmetric grabens or half grabens (Fig. 7.15B). Accommodation zones between tilt domains are buried by younger volcanics within the intra-arc graben, but appear to correspond to long-lived, regional northwest-trending dextral shear zones (Fig. 7.15; Smith, 1989). The northwest-trending faults may have passively defined tilt domains during extension, or the intra-arc grabens may be pull-apart basins generated by movement on the dextral faults.

North Island, New Zealand
Volcanoes in central North Island, New Zealand, referred to as the Taupo Volcanic Zone, are the southward onshore extension of the Kermadec-Tonga intraoceanic arc (Fig. 7.16). Volcanism began in the Taupo zone less than 2 my ago (Cole, 1979; Wilson et al., 1984), forming a 250km-long subaerial arc largely bounded by normal faults. Geophysi-

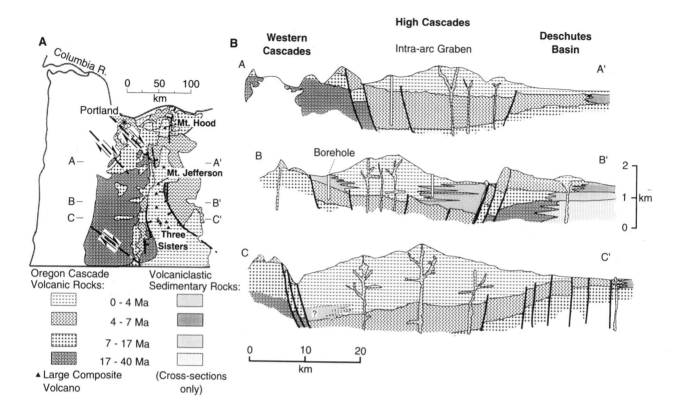

Fig. 7.15 (A) Map and (B) cross sections through central Oregon High Cascades graben. Sections A-A′ and B-B′ are partly constrained by borehole-core data; C-C′ is partly constrained by seismic-reflection profile of Keach et al. (1989). Extensional basins formed between 5.4 and 4.0 Ma. Different tilt domains within intra-arc basin (B) may be bounded by northwest-trending dextral shear zones (A). Another half graben is present farther north in vicinity of Mount Hood (A).

cal data suggest rates of extension of about 7 mm/y and crustal thinning to 15 km, as opposed to a thickness of 35 km in adjacent areas (Cole, 1984). Subsidence of 1 to 4 km in the Taupo Volcanic Zone is indicated by geophysical data (Cole and Lewis, 1981). Maximum subsidence has occurred in the central part of the zone, where rhyolitic caldera-dome centers dominate (Wilson et al., 1984; Fig. 7.16). Active andesitic and dacitic volcanoes are most common at the northern and southern ends of the zone, and are also known from borehole and geophysical data to underlie the rhyolites in the central part (Stern, 1986; Browne et al., 1992). Rapid rates of extension and subsidence in the Taupo Volcanic Zone may be a transtentsional response to oblique subduction (Cole and Lewis, 1981) or be related to backarc rifting farther north (Cole, 1984).

Central Philippines

Convergent-margin tectonics in the Philippines is extraordinarily complicated, as a result of oppositely facing subduction zones, 500 km apart, and the arc-parallel Philippine fault, a strike-slip zone that weaves along or between the two volcanic arcs (Fig. 7.17A). The two volcanic arcs are as close as 140 km, making the present forearc, backarc, and intra-arc settings difficult to define (Sarewitz and Lewis, 1991, p. 597); the complicated evolution of the arcs through time further complicates the distinction (Bachman et al., 1983). Late Cenozoic pull-apart basins have formed along and near the Philippine fault. Sarewitz and Lewis (1991) referred to all of these as intra-arc basins, although some are positioned between the arcs, do not contain volcanoes, and probably are not located on older arc massifs; the classification of these basins is problematic.

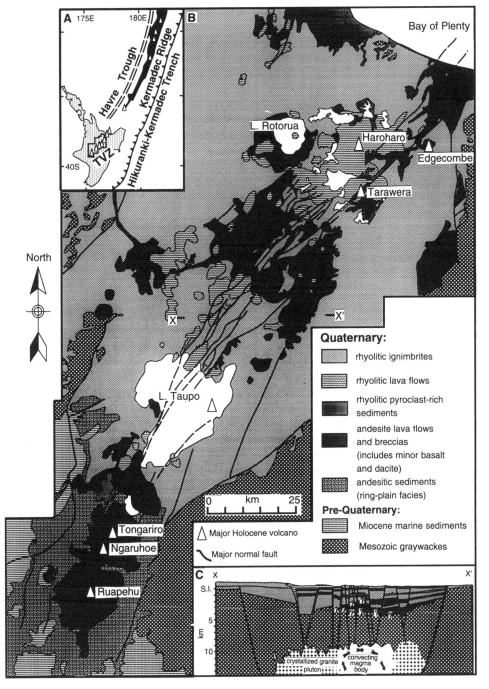

Fig. 7.16 (A) Index map of North Island, New Zealand, illustrating colinearity of Taupo Volcanic Zone (TVZ) with Kermadec Ridge arc, not with Havre Trough backarc basin. (B) Geologic map (simplified from Grindley, 1960, and Healey et al., 1974) of Taupo Volcanic Zone (area outlined in A). Rhyolitic lavas and lakes mark positions of major caldera centers that are sources for plateau-forming ignimbrites that flank central volcanic zone. Holocene andesitic volcanism has mostly occurred in southern part of TVZ, but has also occurred at scattered centers farther north. TVZ is bounded by normal faults that juxtapose Quaternary volcanics and derivative sediments against Mesozoic graywackes. (C) Cross section of central TVZ (based on Grindley, 1960, 1965, and Browne et al., 1992), illustrating low-relief of the volcanic arc resulting from high rates of subsidence, caused both by extensional tectonism and caldera collapse. Thickness and extent of andesitic lavas within depression are not known with certainty (see Brown et al., 1992). High heat flow and geophysical data support presence of magma body at shallow depth, thus accounting for widely dispersed geothermal features between Lake Taupo and Haroharo.

The Marinduque basin is the most clearly defined intra-arc basin in the central Philippines. Based on seismic-reflection and side-scan-radar data, Sarewitz and Lewis (1991) interpreted the Marinduque basin as a rhombic pull-apart depression, approximately 75 km long and 38 km wide, bounded on the east by the Philippine fault zone. Structural relief exceeds 2 km and at least 1 km of fill is seismically imaged in the basin. The basin lies along the trend of the east Mindoro arc (Defant et al., 1988) and includes a submarine volcano in its center (Sarewitz and Lewis, 1991). A symmetrical pattern of magnetic anomalies across

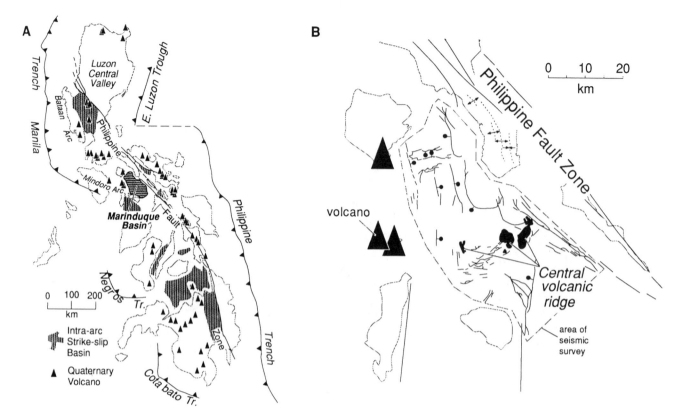

Fig. 7.17 (A) Tectonic map of Philippines (modified from Sarewitz and Lewis, 1991), showing complexity of volcanic arcs associated with opposing subduction zones to the east and west, and arc-parallel strike-slip deformation along Philippine fault, which has produced several intra-arc basins. Parallel volcanic alignments in Bataan and Mindoro arcs may have formed by melt generation at two specific depths above sub-ducted lithosphere (Defant et al., 1988). (B) Structural map of Marinduque intra-arc basin, based on seismic-profile data of Sarewitz and Lewis (1991), illustrating normal faults produced by Neogene (?) transtension and subsequent transpressional anticlines, along which northeastern basin margin has been uplifted. Alternation of extensional and compressional deformation is typical of tectonism along Philippine strike-slip fault. Volcanics of central volcanic ridge are hypothesized by Sarewitz and Lewis (1991) to record arc-transverse sea-floor spreading during transtensional stage.

this central volcanic ridge suggests that spreading occurred along an axis perpendicular to the trend of the arc. This period of intra-arc rifting coincided, with transtension and was followed, probably beginning in the Pliocene, by transpression along the east side of the basin. The shape of the Marinduque basin, altering phases of compression and extension, and magmatism are all similar to pull-apart basins in other tectonic settings (see Chapter 12), this represents one type of fault-bounded intra-arc basin.

Mesozoic Cordilleran Arc

An early Mesozoic arc graben depression is inferred to have formed within the Sierra Nevada and Mojave-Sonoran Deserts of eastern California, western Nevada, and southern Arizona (Busby-Spera, 1988b). The depression was filled with up to 11 km of continental and marine volcanic and sedimentary material while the arc was maintained at low relief and elevation (Fig. 7.18; Busby-Spera, 1988b; Tosdal et al., 1989; Riggs and Busby-Spera, 1990). Syndepositional normal faults and interpreted fault-scarp talus breccias are exposed discontinuously along the 1000-km trend of the magmatic arc, although spatial continuity and temporal equivalence of extensional structures are not demonstrated throughout the region (Busby-Spera, 1988b). This deformation was part of a broad zone of supra-subduction-zone extension postulated for fore-arc to backarc settings throughout the Cordillera at this time (Saleeby and Busby-Spera, 1992).

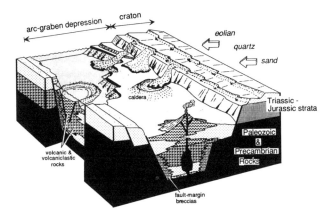

Fig. 7.18 Schematic diagram (after Busby-Spera, 1988b), illustrating early Mesozoic arc graben depression that formed at western margin of North American craton. This intra-arc basin was filled with up to 11 km of continental and marine volcanic and volcaniclastic strata, including composite volcanoes and caldera complexes. Jurassic eolian quartz sand derived from craton was blown into depression and intercalated with volcanic deposits.

Jurassic and Early Cretaceous intra-arc basins in eastern California were ruptured by middle Cretaceous regional tumescence above rising Sierra Nevada batholith magmas. The older basin-fill sequences were tilted to high angles along listric normal faults above the rising plutons, and thinned by as much as 50% by strain induced by the high-level inflation of magma chambers (Tobisch et al., 1986).

Saleeby (1981) suggested that Triassic - Early Cretaceous extension within the arc was a consequence of wrench tectonics associated with oblique subduc-

tion. Intra-arc strike-slip faulting, however, has only been demonstrated during the Late Cretaceous (Busby-Spera and Saleeby, 1990). Busby-Spera et al. (1990) speculated that arc rifting reflected initiation of subduction of old, cold oceanic lithosphere and trench rollback.

Hybrid Basins

Some intra-arc depressions exhibit features characteristic of both volcano-bounded and fault-bounded basins. These hybrid basins are partly or wholly fault-bounded, but are also bounded by parallel volcanic chains.

Central Solomon Islands

The central Solomon Trough (Fig. 7.19), in the southwest Pacific Ocean, originated following subduction polarity change resulting from the failure to subduct the Ontong Java Plateau into the North Solomons Trench (Bruns et al., 1986). This arrested subduction caused strong uplift of the northern, now inactive arc. Part of the younger, southern arc has also been uplifted as a consequence of subducting young, buoyant Woodlark basin crust. In addition, the trough between these arcs is partly bounded by seismically imaged steep, reverse-faults (Cooper et al., 1986) and is divided by transverse fault zones into four depocenters: Queen Elizabeth, Shortland, Russell, and Indispensable basins, that contain 4 to 8 km of basin fill (Fig. 7.19; Bruns et al., 1986).

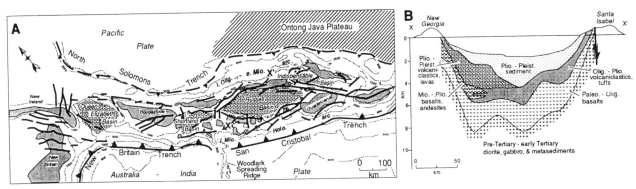

Fig. 7.19 (A) Structural map, showing major strike-slip-related faults (modified from Bruns et al., 1986 and Ryan and Coleman, 1992) and (B) interpretive cross section (Cooper et al., 1986) of Solomon Islands and Central Solomons Trough. This hybrid basin formed as a consequence of: (1) arc-polarity reversal following arrested subduction of Ontong Java oceanic plateau on north and resultant formation of a new

subduction zone and arc on south; and (2) along-arc strike-slip faulting that produced Queen Elizabeth, Shortland, Russell, and Indispensable pull-apart basins (Ryan and Coleman, 1992). Seismic stratigraphic units shown in (B) are asymmetrically distributed in Russell Basin because of diachroneity of volcanism in adjacent arcs.

The nature of faulting in and marginal to the Central Solomons Trough remains poorly known. Greene and Wong (1989) interpreted the transverse structures as upper-plate responses to subduction of spreading-ridge axes and an aseismic ridge. Ryan and Coleman (1992) interpreted the four basins as pull-apart structures and interpret the faults in the Solomon Islands as east-west-trending, sinistral strike-slip features with associated normal and reverse faults, all related to oblique convergence into the New Britain-San Cristobal trench system. Although there is little field evidence for strike-slip faulting, the Ryan and Coleman (1992) hypothesis is supported by focal-mechanism solutions along the east-west faults, the rhomboid shape of the Russell Basin, counterclockwise vertical-axis rotation of eastern Guadalacanal, and rapid rates (3–4 km/my) of vertical deformation. Relief along the Central Solomons Trough, therefore, is a combination of tectonic effects and the near juxtaposition of two arc axes.

New Hebrides Islands

The Central Basin of the Vanuatu segment of the New Hebrides island arc (Fig. 7.20) was initiated by late Miocene westward shift in the arc axis following a volcanic hiatus in the now inactive eastern arc (Greene et al., 1988a; Macfarlane et al., 1988). The cause of this shift and the hiatus in volcanism is not clear. Controversy exists over whether or not a change in subduction polarity occurred (Greene et al., 1988b), but this scenario is favored by recent studies (Greene et al., 1988a). Late Cenozoic volcanoes formed along the eastern margin of, and within, the basin. The Central Basin was initially an eastward sloping shelf that subsequently broke into deep, asymmetric, east-tilted grabens (Fig. 7.20B); Greene and Johnson, 1988). The intra-arc trough is segmented into discrete depocenters by islands of the central volcanic chain, arc-parallel graben faults, transverse faults associated with fracture zones on the overriding plate, and transverse structures developed subsequent to middle Pliocene impingement of

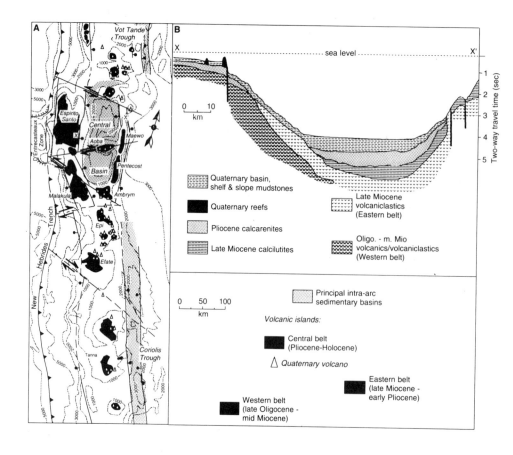

Fig. 7.20 (A) Map and (B) interpretive cross section of intra-arc basins along Vanuatu segment of New Hebrides arc (after Greene and Johnson, 1988). Active volcanism occurs within hybrid Central Basin, which is bounded both by high-angle faults and islands composed of older arc volcanics. Vot Tande and Coriolis Troughs are fault-bounded basins on the arc platform.

the D'Entrecasteaux seamount chain into the New Hebrides Trench (Fisher, 1986; Greene and Wong, 1989). Alternating periods of subsidence and uplift of both the eastern and western arcs are recorded by unconformities in onshore sections that can be traced part way into the Central Basin on seismic sections.

Reconstruction of onshore stratigraphic sections in the New Hebrides indicates that such a hybrid intra-arc basin has probably existed through the Neogene (Fig. 7.21; Macfarlane et al., 1988). During the growth of multiple arcs, changes in subduction polarity, rifting to generate the North Fiji Basin backarc basin, and other episodes of uplift and subsidence, a central trough has been maintained for more than 15 my (Fig. 7.21). Thorough stratigraphic study of the basin fill is lacking, but Mitchell (1970) described more than 5 km of lower to middle Miocene, largely volcaniclastic sedimentary rocks that were deposited in this basin and subsequently uplifted in the western arc.

Relationship Between Intra-Arc and Backarc Basins

Many, though not all, backarc basins originate by rifting of the arc platform, followed by the generation of new oceanic crust (Karig, 1971a; Taylor and Karner, 1983). The origin of backarc-basin rifting and spreading is not clearly understood (Taylor and Karner, 1983), and the possible relationship between intra-arc extension and backarc-basin development has not been fully explored. Distinction between intra-arc and backarc basins may not always be possible in ancient records, although basins containing active arc volcanoes are best termed intra-arc basins. Evolution of intra-arc to backarc basins is an expected consequence of intra-arc rifting. Fault-bounded intra-arc basins may form contemporaneously with spatially distinct backarc rifts (e.g., New Hebrides, Fig. 7.20). In other cases, such as the Izu-Bonin arc (Figs. 7.10; Taylor, 1992; Chapter 8, Fig. 8.4C), backarc rifts may trend into segments of the arc.

Middle Miocene opening of the Sea of Japan backarc basin was heralded by a 3my period of rapid (up to 3km/my) subsidence and development of

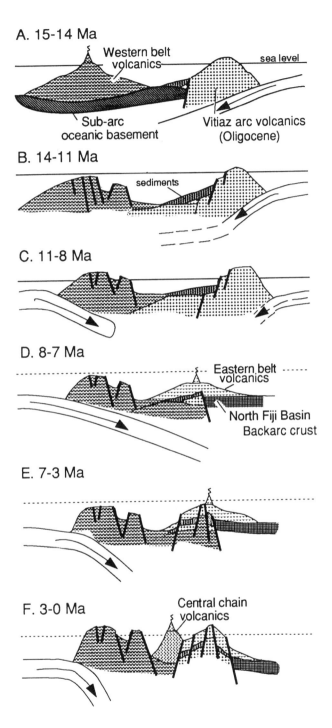

Fig. 7.21 Schematic diagram (after Greene et al., 1988b) of evolution of New Hebrides arc in vicinity of Vanuatu Central Basin. Changes in subduction polarity and other factors controlling arc position have caused development of multiple parallel arc ridges. Development of these parallel arcs, and also faulting of arc platform, have generated a series of hybrid intra-arc basins. Notice that arc massif is built on oceanic and backarc crust of different ages.

narrow (10 km) grabens over a broad region exceeding 1000 km in width and including the arc (Yamaji, 1990). This brief interval of rapid subsidence was followed by 3 to 4 my of backarc-basin spreading. Although the structural style of the initial intra-arc grabens resembles that of intra-arc basins described above, the broad region of deformation distinguishes the northeast Japan basins and other immature backarc basins (e.g., Gust et al., 1985) from intra-arc basins that have not evolved into mature backarc basins. In contrast, later Quaternary extension has formed the Aso-Beppu intra-arc graben in southwestern Japan (Table 7.1; Yamazaki and Hayashi, 1976), which shows no obvious connection to backarc-basin spreading (Taira et al., 1989).

The Taupo Volcanic Zone (Fig. 7.16) has been interpreted as a southward extension of the Havre Trough backarc basin by Karig (1970a), Cole and Lewis (1981), Cole (1984, 1986), and Stern (1985). This interpretation is at odds, however, with the colinearity of the Taupo zone with the Tonga-Kermadec arc, not the backarc basin (Fig. 7.16; Ballance, 1976; Sporli and Ballance, 1989), and the petrogenetic similarity of Taupo and Tonga basaltic-andesite, and andesite lavas (Ewart et al., 1977). Cole (1986) attributes the andesites to arc volcanism in front of a rhyolite-dominated backarc basin. The andesitic and rhyolitic centers, however, are colinear and rest in the same fault-bounded depression (Fig. 7.16). The Taupo depression encloses the arc and is best termed an intra-arc basin.

To the extent that many backarc basins are initiated by extension within the arc, some modern extensional/transtentsional intra-arc basins (e.g., Taupo Volcanic Zone, Aso-Beppu depression) may be incipient backarc basins. The other extensional basins described above are however, either 1) short-lived without evolution to a backarc spreading stage (e.g., Aleutian summit basins, central Oregon High Cascade graben, Nicaraguan Depression); 2) include possible spreading centers that were short-lived and failed to generate a basin wider than the active arc (e.g., Marinduque basin); or 3) have been too long-lived (e.g., Cordilleran arc graben-depression; Central Solomons Trough), by comparison to intra-arc rifting that initiated opening of the Sea of Japan (Yamaji, 1990). Some of these basins might have

evolved into backarc basins if the extensional strain had been larger or if extensional deformation had continued longer. Extensional intra-arc basins may, therefore, represent incipient or arrested backarc basins. The designation of *intra-arc basin* should be retained if these basins are relatively narrow and restricted to the arc platform.

Sedimentary and Volcanic Facies Patterns and Distribution

Composition and Nature of Volcanic Rocks

A diverse array of volcanic rocks characterizes most convergent-margin arcs. Although volcanic arcs are generally thought of as dominated by calc-alkaline andesites, some arcs contain mostly basalts and basaltic andesites, and others are largely rhyolitic; oceanic and some extensional continental-margin arcs contain substantial volumes of tholeiitic volcanics. The variations are due to heterogeneity of mantle source areas, degree of melt or volatile contribution from subducted lithosphere, thickness, composition and extent of assimilation of overlying lithosphere, and depth of fractionation. Detailed consideration of the petrogenesis of arc magmas is beyond the scope of this treatment of intra-arc basins; the interested reader is referred to Gill (1981) and Thorpe (1982) for more information.

The presence of volcanic and related intrusive rocks is critical to the recognition of intra-arc basins in the geologic record. Volcaniclastic sediments of arc derivation may be spread for great distances into trench, fore-arc, backarc, and foreland basins, but in those locations, they will not be associated with characteristic arc-platform volcanic rocks. Volcanism is virtually unknown in foreland basins and deep-sea trenches, and is uncommon in forearc basins. Backarc-basin volcanism is, of course, also voluminous and may be difficult to distinguish from arc volcanic products in ancient sequences. Mature backarc-basin basaltic crust is similar to mid-ocean-ridge basalt (MORB) in many aspects, but may be transitional in character to arc basalts (Pearce et al., 1981; Saunders and Tarney, 1984). The principal geochemical differences between MORB and arc magmas are relative enrichment in large-ion-lithophile and rare-

earth elements and ^{87}Sr in arc magmas (e.g., Gill, 1981). These components are thought to be derived from the subducted lithosphere and are depleted relative to MORB source areas by previous melting events. Backarc-basin basalts closely resemble MORB, but contain varying degrees of arc-magmatic influence, especially at initiation of spreading, depending on proximity of the backarc spreading axis to the arc and to the mantle wedge that has been metasomatized by fluids from the subducted slab (Saunders and Tarney, 1984). Nonetheless, in many cases, careful geochemical studies permit the distinction of arc and backarc volcanic rocks.

Many convergent margins and arc platforms contain fault slivers of volcaniclastic rock, commonly along with disrupted ophiolite sequences. Straightforward identification of depositional setting by reference to geochemical data bearing on the origin of the associated ophiolite crust is commonly impossible however, particularly in view of widespread hydrothermal and weathering processes. Ophiolites within volcanic arc settings may be of mid-oceanic, arc-root, or backarc-basin origin, even where the ophiolite represents the foundation for arc construction (Hussong and Uyeda, 1981; Taylor and Karner, 1983; Pearce et al., 1984). Dike and sill complexes in these ophiolites, often interpreted as evidence for spreading-center activity, can also be composed of arc magmas that were injected during arc extension that may or may not have culminated in formation of a backarc basin (Pearce et al., 1981; Sarewitz and Lewis, 1991).

Several classical ophiolite sequences (e.g., Oman, Josephine, Dun Mountain, Coast Range, and Troodos) can be shown to be intimately related to island-arc volcanism and to have formed in arc-related basins. Although original spatial relations between many ophiolite basins and associated arcs tend to be ambiguous, examples cited by Beets et al. (1984), Pearce et al. (1984), and Harper (1989), specifically raise the case for ophiolite generation in intra-arc basins. Recent study of the Marinduque Basin, Philippines (Sarewitz and Lewis, 1991), provides a compelling case for generation of ophiolite in a modern intra-arc basin. Thus, ophiolites representing a brief period of origin and flooring of basins of small regional extent that also contain contemporaneous arc magmas must be seriously considered as intra-arc basins.

Extension in arcs may have a profound effect on the composition of erupted lavas. In the Aleutian arc, tholeiitic lavas are uniquely associated with the extensional summit basins, whereas calc-alkaline lavas prevail elsewhere (Kay et al., 1982). High-level, virtually closed-system, fractionation produced compositionally distinctive magmas in areas of extended crust (Singer and Myers, 1990). High-level fractionation during extension is also responsible for the large volume of tholeiitic basalts and andesites, the latter unusually iron- and titanium-rich for arc settings, that were erupted during and shortly following formation of the central Oregon High Cascade graben (GA Smith et al., 1987). Some of the basalts erupted at this time closely resemble MORB, perhaps because diminished convergence rates had led to depletion of components derived from the subducted slab (Hughes, 1990). Other anomalous mafic lavas may represent very small-degree, sub-arc-mantle partial melts that would not have successfully risen to the surface if extensional pathways had not formed (Conrey, 1990). Basalt and basaltic-andesite lavas dominate the stratigraphic record of most late Cenozoic extended continental-margin arcs and oceanic arcs (Erlich, 1968; Weyl, 1980; GA Smith et al., 1987). Dike propagation, rather than diapirism, is required for large volumes of dense melt to rise to the surface (Shaw, 1980; Marsh, 1982); therefore, the abundance of mafic lava is consistent with the extensional tectonic setting.

The intrusion of melts into high crustal levels of extensional arcs also facilitates large degrees of crustal melting to produce silicic magmas, which may be tapped at high levels to form voluminous ignimbrites and related calderas. The rapidly extending Taupo Volcanic Zone is dominated by ignimbrite sheets and overlapping caldera complexes, partly bounded by tectonic faults (Wilson et al, 1984). Geochemical studies suggest that 16,000 km^3 of Quaternary rhyolite has formed from high-level partial melting of Mesozoic graywackes (Ewart et al., 1977). Similarly, silicic caldera volcanism characterized igneous activity contemporaneous with extension and preceding penultimate graben formation in the High Cascades (Smith et al, 1987) and the Nicaraguan Depression (McBirney and Williams, 1965; Pichler and Weyl, 1973). In many other cases, caldera systems within volcanic arcs are

associated with extensional tectonic conditions (West-erveld, 1963; Oide, 1968; Burkart and Self, 1985; Busby-Spera, 1988b; Crawford et al., 1988).

The nature of volcanic eruptions and resulting products is not only a function of magma composition (i.e., viscosity, volatile content), but is also affected by the extent of interaction between magma and external water. This influence is important in the construction of oceanic arcs and in the evolution of some continental arcs where low-lying lake basins may occupy segments of the arc platform. Water-lava interaction forms characteristic pillow lavas, pillow breccias, and associated *hyaloclastites* generated by spalling of glassy pillow rinds (e.g., Carlisle, 1963; Hanson, 1991; Yamagishi, 1991).

Water-magma interaction also causes explosive fragmentation of the melt to generate considerable volumes of morphologically distinctive *hydroclastic* fragments (Fisher and Schmincke, 1984). Magmatic vesiculation and generation of steam from water-magma interaction are inhibited, however, by the extreme hydrostatic pressure of deep ocean settings (McBirney, 1963; Fisher, 1984). The construction of an oceanic arc, therefore, may be lithologically complex, even if the magma composition changes little with time. Early, deep-water stages are characterized by poorly vesicular pillow and sheet lavas overlain by increasingly voluminous explosive hydroclastic debris, formed in shallower water, and eventually by pyroclastic products that reflect little or no magma-water interaction erupted from sub-aerial-island vents (Fisher and Schmincke, 1984; Stix, 1991).

The volume and composition of arc volcanic rocks vary spatially in response to crustal differences, and temporally in response to changes in rate and angle of subduction and the state of stress in the volcanic arc (Dewey, 1980). It is common for the longterm stratigraphic record of an arc to be divided into *volcanic episodes* that are characterized by differing compositions of erupted materials and/or the rates at which magma was extruded (Pitcher and Weyl, 1973; Kennett et al., 1977; Zeil, 1979; Scholl et al., 1983; Priest, 1990). Variations in these characteristics should be common in intra-arc basin fills. Periods of little or no volcanic activity will be represented stratigraphically by a paucity of volcanic rocks, and in marine settings, by thick accumulations of pelagic and hemipelagic sediments or disconformities on the arc platform. These hiatuses are often related to low-angle subduction and/or accretion of buoyant crustal elements (Nur and Ben-Avraham, 1981) and may also correspond to periods of backarc spreading (Klein, 1985a).

Composition and Origin of Sediment

Most sediment in intra-arc basins is, not surprisingly, volcaniclastic. Volcaniclastic fragments originate by variable processes (Fig. 7.22; Fisher, 1961; Schmid, 1981; Fisher and Schmincke, 1984). Epiclastic fragments, formed by the weathering and erosion of pre-existing rocks, are often subordinate to those generated by volcanic processes. This latter group includes pyroclasts generated by magmatic vesiculation, hydroclasts produced by magma-water interaction,

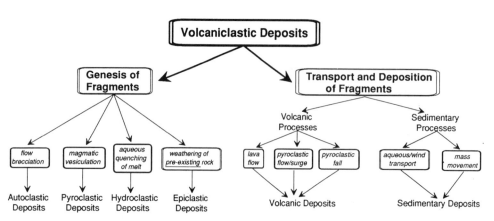

Fig. 7.22 Nomenclature of types of volcaniclastic deposits. Volcaniclastic deposits are all those fragmental materials of volcanic composition regardless of origin or process of deposition (Fisher, 1961; Fisher and Schmincke, 1984). Nomenclature is based on genesis of fragments or processes of fragment transport and deposition (based on classifications of Fisher, 1961; Schmid, 1981; Fisher and Schmincke, 1984).

and autoclasts related to fragmentation of flowing lava (Fisher, 1961; Schmid, 1981). Pyroclastic, hydroclastic, and autoclastic debris is initially emplaced by volcanic processes, but because of its unconsolidated condition, is typically reworked and redeposited by surficial processes that are generally regarded as sedimentary. In the classifications of Fisher (1961), Schmid (1981), and Pettijohn et al. (1987), all these deposits are referred to as pyroclastic, hydroclastic, and autoclastic, provided mixing with epiclastic sediment is minimal. This reflects the fact that the nomenclature is based on the origin of the fragments, not the process of final deposition (Fig. 7.22). An alternative view, not adopted here, redefines epiclasts to include not only particles released by weathering of older rocks, but also those redeposited by normal surface sedimentary processes irrespective of mechanism of fragmentation (Cas and Wright, 1987). Regardless of the nomenclatural approach that is chosen, the fundamental characteristic of sedimentation in volcanic environments, compared to other clastic settings, is the production of sediment independent of weathering processes. In addition, sediment volumes and dispersal distances are typically much larger than noted in other clastic depositional systems (e.g., Vessel and Davies, 1981; GA Smith, 1987; Houghton and Landis, 1989). Petrographic distinction of pyroclastic and hydroclastic fragments from epiclastic grains may permit distinction of sediments that accumulated synchronously with volcanism from those derived from dissection of inactive and possibly quite ancient volcanic massifs (Fisher, 1961; Stix, 1991).

Nonvolcaniclastic sediments may also be locally significant in intra-arc basins. These sediments include quartzofeldspathic or lithofeldspathic material derived from exposed plutons within the arc (e.g., Scholl et al., 1975b), or from high-standing plutonometamorphic terranes located peripheral to the arc (e.g., James, 1971; LaPierre et al., 1985). Busby-Spera (1988b) and Riggs and Busby-Spera (1990) described the unusual situation of accumulation of locally thick craton-derived, eolian quartz arenites within a continental-margin arc depression (Fig. 7.18). Low-latitude oceanic arcs often contain fringing reefs or atolls, which are not only represented by shallow-water carbonate facies, but are also the source for resedimented calc-arenites and limestone breccias within deeper parts of intra-arc basins (e.g., Jones, 1967; Mitchell, 1970; Cooper et al., 1986; Watkins and Flory, 1986; Miller, 1989; Fig. 7.22). Rapid subsidence permits the aggradation and propagation of carbonate-platform sequences that may be 0.5 to 1.0 km thick but only a few tens of kilometers wide (Watkins, 1993). Because of positions remote from continental margins, island-arc limestones typically contain biogeographically endemic shallow-water benthic faunas (Watkins, 1993).

Biogenic and nonbiogenic pelagic and hemipelagic sediment, as well as chemical sediment, accumulates on parts of oceanic arc platforms and in deep intra-arc basins between widely spaced volcanoes. Depending on water depth, these facies are commonly dominated by calcareous or siliceous microfossils intercalated with fine grained volcanic ash (Wright, 1984; Pollock, 1987; Miller, 1989). Submarine alteration of ash and hydrothermal processes may also contribute to chert formation (Pollock, 1987) and deposition of bedded sulfates (Watkins and Flory, 1986). Hydrothermal alteration also contributes to the formation of evaporite sediments in arid, continental intra-arc basins (Alonso et al., 1991). Manganese crusts on volcaniclastic sediment have been described from the flanks of a small basin on the flanks of the Tonga-Kermadec Ridge (Cronin et al., 1984).

Principal Facies Associations

In a perceptive early study of arc volcaniclastic sedimentation, Jones (1967, p. 1286) described the marine volcanic arc and its environs as a "self-sufficient, centrifugal sedimentary system within which the central volcanic edifice serves as the clastic source and provides environmental conditions which favor the movement of clastic debris under the influence of gravity to an outer zone of sedimentation." Thus, sedimentary facies associated with arc volcanoes can be expected to define a relatively straightforward pattern. In this review, we arrange sedimentary and volcanic deposits associated with arc volcanism into three general facies associations: central, apron, and distal (Table 7.2, modified from Dickinson, 1974a; Figs. 7.23–7.25).

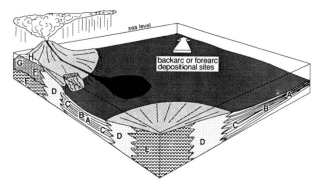

Fig. 7.23 Schematic diagram illustrating facies distribution around both subaerial and subaqueous island-arc volcanoes. Facies are described in Table 7.3. Not that as a consequence of radial dispersal of sediment, relatively distal facies (e.g., A and B) can be present between volcanic centers on arc platform or in forearc and backarc basins adjacent to the arc (modified from Houghton and Landis, 1989).

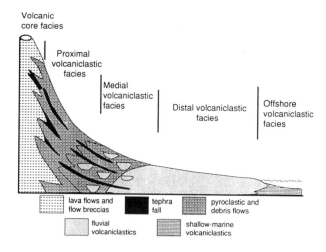

Fig.7.24 Generalized cross section of composite volcano and flanking volcaniclastic apron, illustrating distribution of major facies. Volcaniclastic materials are present in both proximal cone-forming and medial-to-distal sediment-dominated aprons. Distribution of tephra-fall deposits is dependent on prevailing wind directions (modified from Vessel and Davies, 1981).

The *central facies* comprises the volcanic edifices and consists primarily of lava flows, autoclastic and pyroclastic breccias, and high-level intrusions (Figs. 7.24, 7.25, Table 7.3). Locally thick hydroclastic tuffs are characteristic in oceanic arcs and some low-lying continental arcs. Caldera-fill ignimbrites, collapse breccias, and epiclastic sediments may also occur in this assemblage (see Lipman, 1984 for a review). Relatively steep initial dips characterize the flanks of volcanoes (e.g., Hackett and Houghton, 1989) composed of lava flows, autobreccias, and possibly the deposits of pyroclastic flows and volcanic debris flows (Fig. 7.24). *Peperites* may be present in island-arc sequences or in association with accumulations of tuff or fine-grained lacustrine sediment (e.g., caldera fills) in continental arcs. Peperites are formed by the mixing of magma and wet sediment by burrowing of lavas or intrusions into less dense sediment (Duffield et al., 1986), often facilitated by fluid expulsion and explosive fragmentation of the magma in contact with interstitial water (Busby-Spera and White, 1987). Richly calcareous sediments and reefs may also develop on the flanks of partially submerged islands (Jones, 1967) and extensive shallow marine clastics can be expected where wave plantation has been widespread (e.g., Scholl et al., 1987b).

The *apron facies* includes rapidly deposited volcanic and sedimentary materials that encircle individual volcanoes or form prism-shaped accumulations of debris flanking the arc and contributed from many vents (Figs. 7.23, 7.25). The extent of these aprons is variable and depends on the size and steepness of the volcanoes and the volume of easily eroded volcaniclastic debris generated by eruptive activity. Apron-facies sediment is predominantly volcaniclastic, with varying proportions of biogenic debris in marine facies. Alternations in the proportion of pyroclastic and epiclastic debris reflect the rapidly changing sediment supply at the source area (Houghton and Landis, 1989; GA Smith, 1991). Most deposits are texturally very immature, and pyroclastic and hydroclastic fragments may retain characteristic grain morphologies because most sediment is rapidly deposited and experiences little or no subsequent reworking. Stratigraphic units, although laterally persistent for 100 km or more, tend to exhibit striking facies and thickness variations along strike (Miller, 1989; Fig. 7.26).

Sedimentary strata in the apron facies are composed mostly of event deposits related directly or indirectly to the stripping of loose pyroclastic, hydroclastic, and autoclastic debris from the steep volcanic slopes during or shortly following eruptive episodes (Vessel and Davies, 1981; GA Smith, 1987; Palmer and Walton, 1990). The resulting poorly sorted facies include the products of debris avalanches (Palmer et al., 1991), debris flows (Rodolfo, 1989), and hyper-concentrated floods (Smith, 1986; Walton and

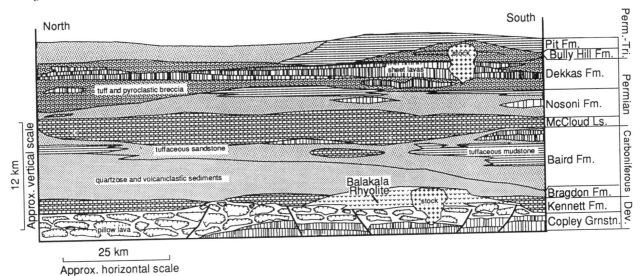

Composite Volcano

cinder cone

lava flow

autobreccias, pyroclastic breccias, debris-flow deposits

proximal apron of debris-flow deposits

ignimbrites

Shield Volcano

cinder cone

Eroded Composite Volcano

medial to distal apron of volcaniclastic sediments (finer grained away from volcanic edifice)

post-volcanism epiclastic volcaniclastic sediments

Fig. 7.25 Schematic illustration of facies relationships in a subaerial volcanic arc (modified from JG Smith, 1987). Sequential overlapping of volcanic edifices and their flanking aprons generates a complex vertical succession of facies.

Palmer, 1988; Smith and Lowe, 1991), and coarse-grained sediment-gravity-flow deposits in very thick to massive beds in subaqueous settings (Fisher, 1984; Ballance and Gregory, 1991). The dominance of debris-flow and flood facies in continental volcaniclastic aprons, or *ring plains* as they are named in New Zealand (Hackett and Houghton, 1989; Fig. 7.16), results in deposition of sediment that may resemble nonvolcanogenic alluvial fans. Alluvial fans, however, rarely extend more than 10 to 15 km from the bases of volcanoes, and most subaerial volcaniclastic aprons are characterized by nonfan alluvial plains (GA Smith, 1987, 1991). The large volume and discharge of volcaniclastic debris account for the wider

dispersal of their deposits than is customarily seen in nonvolcanic regions.

In marine or lacustrine settings, volcanic islands have a bipartite apron composed of subaerial and subaqueous parts (Fig. 7.23; Sigurdsson et al., 1980; Houghton and Landis, 1989). Rapid and extensive delta progradation may occur when sediment loads are extreme during and following eruptions, although preservation of these deposits is limited along wave-dominated coasts (Kuenzi et al., 1979). Submarine volcanic slopes are generally steep and very little sediment accumulates at inner-shelf to shoreface depths except during intervals of rapid wave erosion, particularly when volcanism is dormant (e.g., Smellie et al., 1980;

Fig. 7.26 Schematic cross section through mid-to-late Paleozoic volcanic and volcaniclastic sedimentary assemblages of the Klamath Mountains, illustrating continuity of defineable rock units, albeit with considerable variations in thickness and lithofacies, in an intra-oceanic-arc terrane (after Miller, 1989).

Scholl et al., 1987b; Cas et al., 1989). The resulting deposits show at least moderate rounding and sorting, and are commonly fossiliferous and bioturbated. Extensive shallow-marine volcaniclastic sediments are most likely to be associated with shallowly submerged volcanic complexes that periodically build hydroclastic cones near or above sea level, with subsequent wave redistribution of most of the products (Ross, 1986). Depositional slopes on the volcano-flanking aprons may be sufficient to cause mass movement or down-slope soft-sediment deformation, especially in submarine sequences (Fiske and Tobisch, 1978; Fisher, 1984). Considerable deformation may also accompany the emplacement of thick sediment-gravity-flow deposits (e.g., Ballance and Gregory, 1991).

Primary volcanic components in the apron facies include relatively low-viscosity lava flows, relatively thin ignimbrite outflow sheets (as compared to thick intracaldera ignimbrites), and fine- to coarse-grained fallout tephras (Fig. 7.25). Although subaqueous welding has been demonstrated (Kokelaar and Busby, 1992), there is considerable debate concerning mechanisms of subaqueous eruption of pyroclastic flows or the possibility of subaerially erupted pyroclastic flows remaining hot following their entry into the marine environment (see Fisher, 1984; Cas and Wright, 1987, 1991, for discussion and reviews). Massive, pumiceous ignimbrite-like deposits are intercalated with turbidite facies in many arc-related sedimentary sequences, but evidence for high-temperature emplacement is equivocal for most cases where the depositional setting is unambiguously marine, leaving open the possibility that high-temperature, gas-fluidized pyroclastic flows were transformed into cooler water-saturated debris flows and high-concentration turbidites upon passage from land to sea.

Distal facies represent terminal deposition at sites that receive little or no direct volcanic or indirect sedimentary impact from eruptive events. In subaerial settings, primary volcanic products are generally restricted to relatively fine-grained fall deposits, with far-traveled basaltic lavas or ignimbrites comprising rare contributions to the stratigraphy. Volcanic debris-flow and flood deposits may travel 100 km or more from source (GA Smith, 1986, 1991), but distal subaerial facies are composed primarily of normal fluvial deposits, and may contain considerable admixed non-volcanic sediment (Smith, 1988). In contrast, distal marine facies contain relatively thin, low-concentration turbidity-current deposits (including tuffaceous turbidites), fine-grained fallout tephras (mainly of silicic composition), and pelagic and hemipelagic sediments (Fig. 7.23; Table 7.3; Houghton and Landis, 1989; Larue et al., 1991).

These three general facies associations are not unique to intra-arc basins. Distal facies commonly extend into other basinal settings adjacent to the arc and are only likely to characterize those parts of intra-arc basins that lie along the arc-platform between widely spaced volcanic centers (Fig. 7.23) (e.g., Lesser Antilles, Wright, 1984; Aleutians, Scholl et al., 1983, 1987b). Where the arc platform is narrow or where the arc axis is near the edge of the platform, the apron facies may extend into forearc (e.g., Kuenzi et al., 1979; Vessel and Davies, 1981) or backarc extensional (Karig and Moore, 1975; Farquharson et al., 1984; Busby-Spera, 1988a) and nonextensional (GA Smith, 1987, 1988) basins. Recognition of the central facies containing arc volcanic lithofacies is essential for the recognition of intra-arc sequences.

Terrestrial facies patterns in emergent arcs

The facies architecture of volcaniclastic aprons is typically complex, reflecting alternating periods of high-volume sedimentation in response to eruptive activity, and low-volume sedimentation during inter-eruption

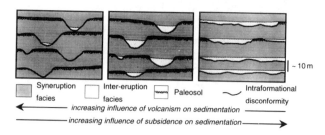

Fig. 7.27 Facies relations in continental volcaniclastic sequences. Syneruption facies are dominated by tuffaceous sandy flood and debris-flow deposits; inter-eruption facies are dominated by conglomeratic, epiclastic alluvial deposits. Syneruption deposits aggrade rapidly in response to eruptions, followed by incision of channels and soil development on abandoned surfaces when sediment supply diminishes. Relief on erosion surfaces decreases and relative abundance of inter-eruption facies increases as subsidence rates increase and/or eruption frequency decreases (after GA Smith, 1991).

Table 7.2 Facies Association in Volcanic Terranes

Facies Association	Continental Arcs	Oceanic Arcs	
		Largely Emmergent Volcanoes	Largely Submerged volcanoes
Central Facies 0–~10 km	Lava flows and domes; proximal pyroclastic and autoclastic breccias; locally thick accumulations of hydrovolcanic tuffs and breccias; thick caldera-fill ignimbrites and and sediments; high-level intrusions; locally steep (>15°) primary dips		Pillow and sheet lavas; hyaloclastite; proximal pyroclastic and hydroclastic breccias; caldera fills, high-level intrusions; locally steep primary dips, peperites; wave worked volcaniclastics; possibly reefs in shallow submerged cases.
Apron Facies ~5–~35 km	Mafic lava flows; ignimbrite outflow sheets; coarse-grained sediments dominated by products of debris avalanches, debris flows, and floods, fine- to coarse-grained fall deposits.	Subaqueous ignimbrites or coarse-grained pumiceoussediment-gravity-flow deposits; high-density turbidites and debris-flow deposits; possibly subaqueous debris-avalanche deposits; fine- to coarse-grained fall deposits; calcarenite calcarenite turbidites where shallow-water carbonates are present.	
Distal Facies ~20–~70 km	Fluvial sediments of mixed volcaniclastic-nonvolcaniclastic composition; fine-grained fall deposits; may include lacustrine deposits in fault-bounded basins.	Fine-grained volcaniclastic turbidites; pelagic/hemipelagic muds and oozes, fine-grained fall deposits.	

Table 7.3 Island-Arc Lithofacies Associations (modified from Houghton and Landis, 1989)

Association	Principal lithofacies	Subordinate lithofacies	Inferred environment
A	mudstone, siltstone, tuff, thin-bedded sandstone	basaltic hyaloclastite, lava flows	extremely distal margins of arc-flanking apron or basin
B	thin-bedded sandstone, mudstone, tuff	thick-bedded sandstone, conglomerate	distal volcaniclastic apron or fan
C	thick-bedded and thin-bedded sandstone and conglomerate	tuff, lava flow, tuff breccia, agglomerate	proximal volcaniclastic apron
D	tuff breccia, graded sandstone	sandstone, shale, lignite, agglomerate, tuff, lava flow	shallow-marine to nonmarine environment immediately adjacent to subaerial volcanic center
E	sheet lava, pillow lava, pillow breccia, hyaloclastite	tuff, graded sandstone	deep-water submarine vent
F	sheet and pillow lava, scoriaceous pillow breccia	hyaloclastite, sandstone	shallow-water vent
G	aa and block lava flow, autobrecciated lava, ballistic-fall ejecta	welded fall tephra, pyroclastic-flow deposit, debris-flow breccia	subaerial vent (basaltic to andesitic magma)
H	lava dome and flow with associated breccias, ignimbrites, ballistic-fall ejecta	Plinian- and Vulcanian-type fall tephra, pyroclastic-surge deposit	subaerial vent facies (dacitic to rhyolitic magma)

intervals (Fig. 7.27). These latter are sometimes erosional rather than depositional (Vessel and Davies, 1981; GA Smith, 1987, 1991; Palmer and Neall, 1991). Debris-flow and flood facies, therefore, alternate with more typical fluvial sediments. Short-term aggradation in response to volcanism typically exceeds the accommodation space provided by long-term subsidence, so that inter-eruption deposits disconformably overlie erosion surfaces, with relief of as much as 50 m, developed on the syneruption sediments. The relative abundance of syneruption and inter-eruption facies and the extent and scale of erosion surfaces between them depend on the frequency of eruptions, the scale of volcanism-driven sedimentation, and the subsidence rate in the basin (Palmer and Neall, 1991; Smith, 1991). Where subsidence rates are low, the stratigraphic record may consist almost entirely of syneruption deposits in discrete units separated by erosion surfaces and paleosols; inter-eruption conditions are recorded by regarding of stream profiles and re-establishment of landscape stability in the apron without a mechanism for causing aggradation in the absence of extreme eruption-related sediment loads (Fig. 7.27).

Lacustrine deposits, commonly characterized by tuff and diatomite, may occupy various depositional sites in intra-arc sequences. Shallow, usually rapidly infilled lakes of limited lateral extent form on aprons due to drainage disruption by lava flows, ignimbrites and landslides, or rapid valley aggradation that impounds tributaries (e.g., Kuenzi et al., 1979). Thick, laterally restricted lacustrine deposits within the central facies may record crater or caldera lakes. Lacustrine sequences that are laterally extensive and often hundreds of meters thick are characteristic

of many nonmarine intra-arc-basin fills. These sequences are composed of fine-grained lacustrine clastics and diatomites that are finely laminated and intercalated with lava flows, hydroclastic tuffs, and debris-flow deposits derived both from the arc axis and from fan deltas along the basin-margin fault scarps (e.g., Brietkreuz, 1991; Fig. 7.28).

In the Central Oregon High Cascade graben, basinal lavas and pyroclastic deposits are predominantly basaltic. Here, longitudinal fluvial systems also contributed to the early filling of the lake basins but were later captured by headward erosion of rivers that dissected the arc platform.

Similar fluvio-lacustrine sequences, with a greater proportion of rhyolitic, rather than basaltic, tuff characterize much of the fill of the intra-arc graben along the Taupo Volcanic Zone (Grindley, 1965). Although high-standing andesitic volcanoes and associated ring plains occur along the southern part of the Taupo zone (Fig. 7.16), the central and northern part has very low relief, generally 300 to 500 m above sea level. High rates of tectonic subsidence and frequent caldera collapse have caused topographic inversion of this part of the arc, so that drainage is largely centripetal to lake basins in volcano-tectonic depressions.

The high arid Altiplano-Puna basin is hydrologically closed, and large playa and playa-like systems have formed. Middle Miocene to Recent intra-arc sedimentary sequences are up to 5 km thick, and although dominated by fluvial clastics, contain evaporite sections locally exceeding 1 km in thickness (Alonso et al., 1991).

Intrabasinal shifts from continental to marine facies may reflect transitions from emergent to submergent

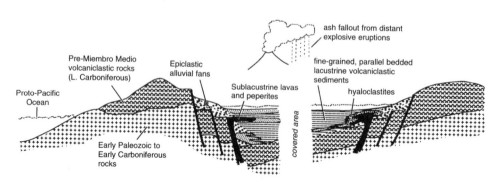

Fig. 7.28 Schematic cross section through Carboniferous intra-arc basin in Chile that contains lacustrine sediments and volcanics of the Miembro Medio (after Brietkreuz, 1991).

parts of island arcs or partial flooding of continental intra-arc depressions. Partial submergence of parts of continental-margin arcs can be observed today at the Gulf of Fonseca in Central America (Fig. 7.5), and in the Aso-Beppu depression in southwestern Japan (Yamazaki and Hayashi, 1976). The Sumatra-Java arc, built partly on oceanic crust, is entirely emergent except where extensional subsidence has caused collapse of the arc platform at the Sunda Strait (Huchon and LePichon, 1984). Parts of the early to middle Mesozoic Sierra Nevada arc that were seated on continental crust were also flooded during periods of intra-arc graben subsidence (Busby-Spera, 1988b; Fig. 7.18).

Marine Facies in Intra-Oceanic Arcs and Flooded Continental Arcs

An outstanding problem in stratigraphic and sedimentologic studies of intra-arc basins concerns the limited knowledge of modern arc roots. Drilling and seismic-reflection investigations that have yielded significant data on other basin types have produced only limited and shallow results from arc settings. Most data bearing directly on the underpinnings of modern arcs are obtained from seismic-refraction and gravity studies. In terms of their contribution to details of lithology and structure, these studies are not specific; many uncertainties remain regarding the deeper parts of intra-arc basins, particularly those in intra-oceanic settings. Therefore, our understanding of intra-oceanic-arc underpinnings is based largely on exposures of deformed, uplifted, and eroded ancient arcs (Jones, 1967; Mitchell, 1970; Houghton and Landis, 1989; Miller, 1989; Larue et al., 1991; Rubin and Saleeby, 1991). Isostatic response to the load of the volcano produces downwarping of the crust, with the result that the volcano, its apron and surrounding sea floor define a basin (Fig. 7.8) that grows upward as well as down. In cross section, such basins appear as doubly convex lenses (Fig. 7.29).

At the same time, arc magmas crystallizing within volcanic edifices form extensive dikes, sills, and larger plutons. Although these intrusive rocks are commonly regarded as forming the bulk of arc roots, Larue et al. (1991) argue compellingly that exhumed island-arc volcanoes in the Caribbean consist mainly

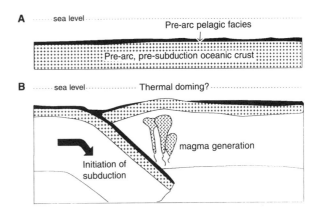

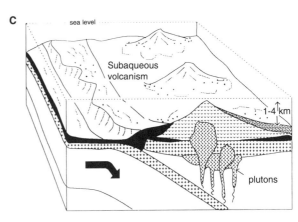

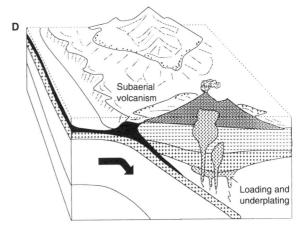

Fig. 7.29 Generalized sequence of intra-oceanic-arc construction (after Larue et al., 1991). Thermal doming and sub-sea erosion of pelagic sediments (B) as a consequence of asthenospheric upwelling following initiation of subduction are hypothetical. Early submarine arc growth (C) features lava flows and hyaloclastites with no products of explosive volcanism generated until volcanoes shoal sufficiently to permit magma vesiculation and steam bubbles to form. Combination of crustal thickening by magmatic underplating and extrusion of volcanics leads to cross section of doubly convex lens for mature arcs (D).

of sedimentary materials, with true igneous rocks forming relatively minor constituents. Similar relations also prevail in the Permian Brook Street terrane of New Zealand (Houghton and Landis, 1989).

Thick volcaniclastic sediment-gravity-flow aprons interlayered with lava flows dominate the marine record of proximal arc volcanism. These sequences are known best from deformed ancient volcanic arcs (e.g., Houghton and Landis, 1989; Larue et al., 1991; Rubin and Saleeby, 1991; Marsden and Thorkelson, 1992) because of technical problems in coring poorly consolidated coarse-grained sediment in proximal areas of modern arcs. Because of the abundance of pyroclastic sand, apron turbidite facies are characterized by high sandstone: shale ratios (e.g., Busby-Spera, 1985); facies patterns characteristic of subenvironments on epiclastic submarine fans are blurred by the direct input of fallout tephra, and the sudden voluminous influx of eroded materials from numerous point sources or a line source following eruptions (Houghton and Landis, 1989; Miller, 1989). Unusually thick and coarse-grained sediment-gravity-flow deposits may represent the subaqueous equivalents of subaerial debris avalanches caused by the collapse of inherently unstable volcanic edifices (Ballance and Gregory, 1991). Bedded and typically graded tuffs of subaqueous fallout origin are also common close to volcanic centers (e.g., Fiske and Tobisch, 1978; Busby-Spera, 1986; Houghton and Landis, 1990). These facies, however, are similar to thick hydroclastic tuffs that may accumulate in low-lying continental arcs and it is necessary to establish the association of these tuff facies with unambiguous marine deposits in order to be certain of the marine setting. For example, well bedded tuffs in the Oligocene Ohanapecosh Formation in Washington were originally interpreted as being deposited subaqueously (Fiske, 1963), but subsequent work has shown that these deposits are more likely subaerial (Vance et al., 1987).

In general, intra-arc-basin fills tend to coarsen upward and record the evolution of nearby vents (Figs. 7.29–7.31). Volcaniclastic turbidite aprons typically prograde over distal pelagic mudstones or biogenic sediments (Fig. 7.30; e.g., Mitchell, 1970; Robertson, 1977; Smellie et al., 1980; Houghton and Landis, 1989; Larue et al., 1991). Early stages of sub-

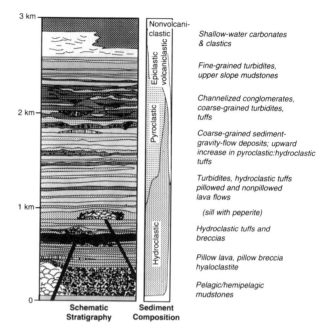

Fig. 7.30 Schematic depiction of stratigraphic record adjacent to growing intra-oceanic-arc volcanic center. Except for restricted accumulations of coarse breccias associated with small, local volcanic centers, section generally coarsens and thickens upward. These trends reflect growth of steep volcanic edifice nearby. Sedimentary composition also track development of volcano. Pyroclastic debris is introduced after evolution to a subaerial vent stage. As the island grows, increasing volume of epiclastic sediment is generated by subaerial weathering. Nonvolcaniclastic and epiclastic volcaniclastic sediments dominate after cessation of eruptive activity.

marine volcanism are marked by accumulation of peperites, pillow lavas and breccias, and intrusions, followed by development of voluminous turbidite aprons as vent complexes grow, steepen, and generate greater volumes of fragmental debris by eruptions in shallow water or subaerially (Fig. 7.30; Mitchell, 1970; Fisher, 1984; Larue et al., 1991). In the lower Miocene of the New Hebrides, upper intervals of the sequence are marked by terrestrial plant debris, clearly epiclastic fragments, and well rounded clasts with a history of fluvial or littoral abrasion (Mitchell, 1970), indicating the construction of a subaerial island-vent system. A Mesozoic sequence in the South Shetland Islands shoals uniformly form basal radiolarian mudstone, to apron-facies turbidites, to shallow-marine fossiliferous clastics, and finally to a section of terrestrial lava and sediment (Smellie et al., 1980). Smaller-scale deepening and shallowing cycles are typical, however, and are less likely caused

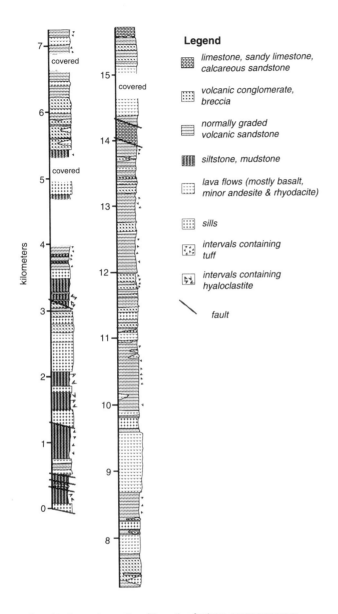

Legend

limestone, sandy limestone, calcareous sandstone

volcanic conglomerate, breccia

normally graded volcanic sandstone

siltstone, mudstone

lava flows (mostly basalt, minor andesite & rhyodacite)

sills

intervals containing tuff

intervals containing hyaloclastite

fault

Fig. 7.31 Composite stratigraphic section for intra-arc sequence preserved in Permian Brook Street terrane of New Zealand (after Houghton and Landis, 1989). Abundance of lavas and sills indicates accumulation within arc platform, although there are no cone-forming sequences to suggest close proximity to volcanic vents. Note that section coarsens upward, hyaloclastites are restricted to lower part of section, and pyroclastic tuffs are more abundant in upper part of section, as also illustrated schematically in Fig. 7.30. Sandstone and limestone near top of section were apparently deposited following cessation of volcanism.

by eustasy than by abrupt alterations in subsidence and uplift, at varying time scales, that are characteristic of arcs (Dobson et al., 1991; Sloan and Williams, 1991).

Extensional basin-fill sequences are generally thicker and record higher subsidence rates than other basin types (Table 7.1). Facies patterns indicate alternating long-term shoaling and deepening patterns that reflect the tectonic instability of the basin and, to some degree, eustasy (Figs. 7.31, 7.32; Busby-Spera, 1984b; Houghton and Landis, 1989). Individual submarine composite volcanoes and calderas have been well preserved in the thick Sierran intra-arc depression (Fiske and Tobisch, 1978; Busby-Spera, 1984a, 1986). Some intra-arc basin margins are separated from the volcanic centers by broad, deep shelves. The apron facies is largely confined to these shelves, so that the deep basins slowly accumulate fine-grained turbidites, slump deposits from the steep structural margins, and pelagic clays and oozes (Scholl et al., 1983, 1987b; Greene and Johnson, 1986).

Apron facies that coarsen and shallow upward are also typical of the near-arc margins of backarc basins (Karig and Moore, 1975; Klein, 1985; Busby-Spera, 1988a). These upward coarsening trends, however, are in some cases, believed to follow rapid upward fining and deepening trends associated with initial rifting of the arc (Robertson, 1977; Klein, 1985). During the Miocene opening of the Sea of Japan, early intra-arc rifting permitted accumulation of andesitic volcanics and coarse rift-basin alluvium in relatively small, fault-bounded basins; this was followed by rapid subsidence to bathyal depths, where backarc basalt and black mudstone were deposited (Yamaji, 1990). This transition from terrestrial rift basins to a more continuous marine basin resembles the rift-to drift stratigraphic transitions seen along passive margins (see Chapter 3 and 4).

Economic Geology

Geothermal Energy

Many of the world's most productive geothermal energy resources are associated with fault-bounded intra-arc basins in New Zealand, Japan, and Central America (Grindley, 1965; Yamazaki and Hayashi, 1976). Favorable conditions for geothermal resources that are afforded by intra-arc basins

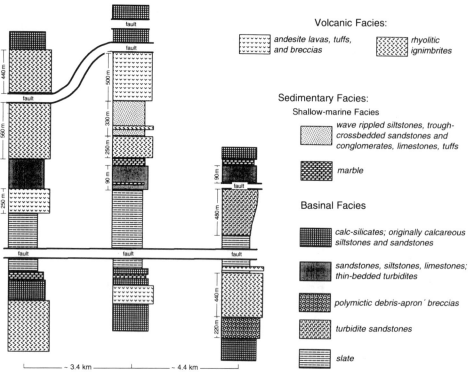

Volcanic Facies:

andesite lavas, tuffs, and breccias

rhyolitic ignimbrites

Sedimentary Facies:

Shallow-marine Facies

wave rippled siltstones, trough-crossbedded sandstones and conglomerates, limestones, tuffs

marble

Basinal Facies

calc-silicates; originally calcareous siltstones and sandstones

sandstones, siltstones, limestones; thin-bedded turbidites

polymictic debris-apron´ breccias

turbidite sandstones

slate

~ 3.4 km   ~ 4.4 km

Fig. 7.32 Generalized stratigraphy of metamorphosed Lower (?) Triassic to Lower Jurassic volcanic and volcaniclastic rocks formed in submerged continental-margin arc and now preserved as Mineral King roof pendant within Sierra Nevada batholith of California. Faults of uncertain displacement cut steeply dipping (75°-89°) section. Present-day, tectonically flattened thickness of selected stratigraphic units are shown. Despite metamorphism and structural complexity, general stratigraphic relations are coherent and illustrate dramatic lateral facies and thickness changes. Some thick ignimbrites were deposited in calderas. Vertical variations in shallow-marine and basinal sediments attest to active tectonism and possibly eustasy (modified from Busby-Spera, 1984b).

include: 1) high regional heat flow; 2) high-level intrusions, often associated with calderas, that provide shallow heat sources; 3) stratigraphically complex basin fills that include relatively permeable sediments and non-welded tuffs, alternating with welded tuffs and thick lava flows that serve as aquicludes; and 4) normal-fault zones that facilitate convective fluid flow and rapid ascent of hot waters from great depth, and may juxtapose aquifers and aquicludes, leading to development of pressurized reservoirs. High-level heat sources, generally associated with silicic volcanism, are critical. Most Quaternary volcanism in the Oregon High Cascade intra-arc depression, in contrast, has featured the ascent of mafic magmas from great depth, and although the volume of extrusions has been large, there are few high-level intrusions and geothermal reserves are consequently insignificant. A recent summary of geothermal energy and its relation to physical volcanology is presented by Wohletz and Heiken (1992).

Economic Minerals

Circulation of hot fluids, expressed by geothermal production in modern arcs, is also responsible for metallic-mineral deposits in uplifted and dissected ancient arcs. Hydrothermal fluids are enriched in relatively incompatible metals derived by aqueous or volatile transfer from crystallizing plutons, and by leaching of country rock. Most sizable gold, silver, and copper deposits are associated with convergent-margin arc magmatism, including porphyry copper deposits, Cordilleran vein-type deposits, and epithermal silver-gold deposits (Guilbert and Park, 1986). Important resources of copper, zinc, lead, silver, and gold occur in massive-sulfide deposits generally associated with submarine silicic volcanism in extensional terranes. These volcanogenic massive-sulfide deposits are best known as Kuroko-type mineralization in Japan. They are hosted in Miocene marine silicic pyroclastic rocks in association with calderas and faults related to arrested intra-arc rifting that followed the major period of opening of the Sea of Japan (Ohmoto and Skinner, 1983). Similar submarine rhyolite-hosted sulfide ores are associated with Devonian intra-arc grabens in the eastern Klamath Mountains of northern California (Albers and Bain, 1985).

Hydrocarbon Resources

The hydrocarbon potential of intra-arc basins is largely untested. Structural and stratigraphic traps are likely numerous in faulted basins and in sequences of intercalated volcanic and sedimentary strata. Volcaniclastic materials are generally viewed as poor hydrocarbon reservoirs because of the high reactivity of volcanic minerals and glass, and resulting pore occlusion by authigenic phases (e.g., Galloway, 1979; Surdam and Boles, 1979). Easily altered or dissolved minerals are variably removed by weathering in most sediment source areas and are consequently depleted in epiclastic sediment relative to source-area abundances in most settings. Because most volcaniclastic sediments are not generated by weathering processes, there is virtually no selective destruction of unstable phases prior to deposition.

Mathisen and McPherson (1991) emphasize, however, that conventional wisdom dismissing the reservoir capabilities of volcaniclastic sediments is not always correct. Dissolution of volcanic framework grains can produce considerable (at least 30%) secondary porosity. Early grain-boundary cementation by clays may also form a rigid framework that holds open pores formed by later grain dissolution. The most favorable diagenetic conditions are associated with meteoric waters. Nonmarine volcaniclastic sedimentary rocks typically lack extensive early diagenetic carbonate cementation and replacement that are common in marine volcaniclastic rocks. The lower ionic strengths of fresh waters also contribute to framework-grain dissolution. Locally important carbonate facies, epiclastic volcanic and plutonic sediments, or even craton-derived sediments may provide better reservoir potential in some intra-arc basins. The presence of low-seismic-velocity rocks in the Aleutian arc substrate indicate that alteration processes have not fully sealed potential reservoirs (Scholl et al., 1987b).

Of greater uncertainty is the availability of adequate hydrocarbon source rocks. Rapid sediment accumulation and high heat flow favor preservation and maturation of organic matter. Upwelling along some intrta-oceanic arcs may lead to high biological productivity, but organic carbon contents are considerably diluted by volcaniclastic sediment in most intra-arc basins (Scholl et al., 1987b; Fisher 1988).

Interpretive Summary

Basins along arc platforms are varied in their dimensions, subsidence histories, mechanisms of formation, and stratigraphy. There are no common denominators in the tectonic conditions that lead to their development and evolution. Fault-bounded basins contain the thickest intra-arc basin fills and provide the best preserved record of volcanic edifices in ancient arc sequences. The origin of these basins is related, in most cases, to complex relative plate motions, including strike-slip deformation along oblique convergence zones. Extensional arcs may represent initial or arrested efforts at inter-arc rifting to create backarc basins. Few of the Tertiary examples considered in this chapter evolved into backarc basins, however, stressing the need to carefully distinguish between backarc and intra-arc basins in the geologic record. Intra-arc deformation is generally confined to a narrow arc platform, rarely more than 50 to 100 km wide; initial backarc-basin rifting in Japan (Yamaji, 1990) and South America (Gust et al., 1985) affected much wider areas.

Of the basin types considered by most students and workers involved in basin analysis, intra-arc basins remain the poorest known. Very few volcanic arcs are well mapped and few sedimentological studies have been undertaken in what is generally considered volcanic terrane. In active arcs, young volcanic rocks may largely obscure older stratigraphic units and structures. Although seismic studies have delineated the structural characteristics and seismic stratigraphy of several modern circum-Pacific intra-arc basins, there is very little direct knowledge of the nature of the sedimentary and volcanic fills. Ancient volcanic arc sequences are commonly highly deformed and or metamophosed by later tectonic dismemberment or plutonism. These complexities impede the ability to determine which structures were active during the development of the arc platform. Despite these obstacles to the understanding of intra-arc basins, recent work, synthesized here, shows that volcanic arcs are of greater use in regional tectonic and basin analyses than as simply the zone of magmatism or as convenient delineators of forearc and backarc regimes. As the body of knowledge increases, it is likely that

more intra-arc-basin sequences will be recognized in the geologic record.

Acknowledgments

G. Smith expresses his appreciation to C. Busby, G. Ross, R. Hildebrand, D. Scholl and B. Houghton, who, through discussions over the past 10 years, sustained his faith that intra-arc basins existed despite the notable paucity of references to them in the literature. Conversations with Bill Dickinson were very beneficial for developing a classification of intra-arc basins and defining them separately from forearc and backarc basins. The manuscript was improved by the reviews of the editors, Cathy Busby and Ray Ingersoll, and also by David Scholl and an anonymous reviewer.

Further Reading

Cas RAF, Wright JV, 1987, *Volcanic successions modern and ancient:* Allen and Unwin, Boston, 528p.

Dickinson WR, 1974a, *Sedimentation within and beside ancient and modern magmatic arcs:* Society of Economic Paleontologists and Mineralogists Special Publication 19, p. 230-239.

Fisher RV, Smith GA (eds), 1991, *Sedimentation in volcanic settings:* SEPM (Society for Sedimentary Geology) Special Publication 45, 257p.

Larue DK, Smith AL, Schellekens JH, 1991, *Oceanic island arc stratigraphy in the Caribbean region: don't take it for granite:* Sedimentary Geology, v. 74, p. 289-308.

Interarc and Backarc Basins

8

Kathleen M. Marsaglia

Introduction

Backarc basins occur behind volcanic island arcs and are common along continental margins, as well as along convergent plate margins; hence, these basins are often termed "marginal" basins (Karig, 1971a). Over 75% of the world's backarc basins are found in the circum-Pacific region, and most are concentrated on the western side of the Pacific Ocean (Fig. 8.1; Tamaki and Honza, 1991). Four major types of backarc basins are known (Karig, 1983b; Ingersoll, 1988b). 1) Extensional backarc basins, which form by rifting and sea-floor spreading (Tamaki and Honza, 1991). This type of basin may initiate by rifting within or behind continental-margin (e.g., the Japan Sea and Okinawa Trough) or intraoceanic (e.g., Mari-

ana and Lau basins) arcs. In intraoceanic settings, the term "interarc" basin is used because rifting produces a basin flanked by a remnant arc and an active arc (Fig. 8.2). There are two types of nonextensional backarc basins. 2) The first type consists of old ocean basins trapped during plate reorganization, which causes a sudden shift of the subduction zone (e.g., the Bering Sea, and the West Philippine Basin; Ben-Avraham and Uyeda, 1983; Tamaki and Honza, 1991). 3) The other type develops on continental crust (e.g., Sunda Shelf and Java Sea; Hamilton, 1979), and is transitional into retroarc foreland basins (see Chapter 9). 4) "Boundary" basins (Taylor and Karner, 1983) may be produced by extension along plate boundaries with translational components. This type may or

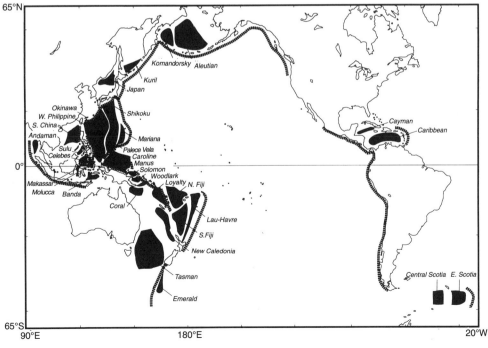

Fig. 8.1 Distribution of trenches (blocked lines) and backarc basins (solid areas) in the circum-Pacific region. After Tamaki and Honza (1991).

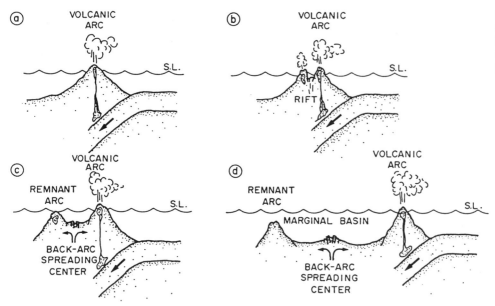

Fig. 8.2 Generalized tectonic evolution of an intraoceanic backarc basin. From Carey and Sigurdsson (1984). After Karig (1971a,b).

may not be associated with contemporaneous arc magmatism (Karig, 1983b). Examples of this type of backarc basin are the Andaman Sea, where oblique rifting has resulted from aseismic-ridge subduction associated with a nearby continental collision (Eguchi et al., 1980), and the Gulf of California, which is associated with nearby triple-junction interactions (Taylor and Karner, 1983). Basins in the last three groups are treated briefly herein, whereas small ocean basins flanked by Atlantic-type passive margins (e.g., South China, Coral and Tasman seas: Taylor and Hayes, 1983; Celebes Sea: Silver and Rangin, 1991) are not included in this discussion.

Of the various basin types included in this volume, backarc basins are among the least understood because modern counterparts are entirely submarine, and subaerial exposures of ancient examples are difficult to recognize. Our knowledge of these basins has been obtained primarily from marine geophysical data (bathymetry, seismic surveys, sidescan sonar, heat flow, magnetics, and gravity), with lesser sedimentologic and petrologic data. The latter are provided by dredge samples, piston cores, deep-sea drilling, and rare instances of direct sampling by submersibles. Consequently, models of intraoceanic backarc-basin evolution have only emerged during

the last 20 years, based largely on results of the Deep Sea Drilling Project (DSDP) and the Ocean Drilling Program (ODP). A recent series of ODP legs in the western Pacific Ocean (Legs 124 to 135, 1988–1991) have greatly expanded our knowledge and permit refinement of previous models.

Extensional Backarc Basins

Origin

Karig (1971a) provided a summary of pre-plate-tectonic theories of backarc-basin formation, and the post-1960 testing of these theories during the formulation of plate-tectonic theory. Karig (1971a) and Packham and Falvey (1971) proposed that backarc basins initiate by crustal extension, producing first rifts and then new ocean crust by sea-floor spreading (Fig. 8.2); this theory is now widely accepted (Tamaki and Honza, 1991). The extensional origin of these basins is indicated by the presence of normal faults, high heat flow (active basins), and magnetic lineations (Karig, 1971a; Packham and Falvey, 1971; Taylor and Karner, 1983). Backarc spreading may occur sporadically (or not at all) during the subduc-

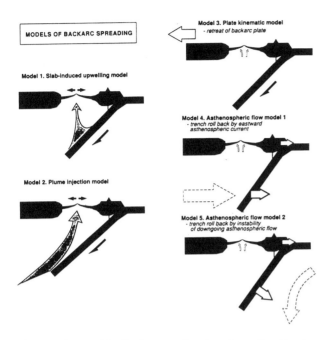

Fig. 8.3 Various models of backarc spreading. According to Tamaki and Honza (1991), Models 1 and 2 are active-opening models, Model 3 is a passive-opening model, and Models 4 and 5 may be classified as either passive, in terms of plate kinematics, or active, in that they depend on dynamic asthenospheric flow. Model 1 proposes mantle upwelling induced by slab subduction (Karig, 1971a; Sleep and Toksöz, 1971). In Model 2, backarc spreading is induced by injection of a mantle plume (Miyashiro, 1986). Model 3 is a kinematic model, where backarc spreading results from the retreat of the overriding plate from the subduction zone (Dewey, 1980). Backarc spreading in Model 4 is induced by eastward asthenospheric flow, possibly associated with the earth's rotation (Uyeda and Kanamori, 1979), and in Model 5, it is induced by global downward asthenospheric flow (slab pull)(Glatzmaier et al., 1990). From Tamaki and Honza (1991).

tion history of an arc segment (Taylor and Karner, 1983). Backarc extension/spreading appears to be associated with rapid subduction of older (>80 my old) oceanic lithosphere (Molnar and Atwater, 1978; Gill, 1981; Furlong et al., 1982), and steeply dipping subduction zones (Cross and Pilger, 1982).

The mechanisms of backarc-basin formation are still debated (Taylor and Karner, 1983; Tamaki and Honza, 1991). Various active and passive models have been proposed (summarized in Fig. 8.3). No one theory adequately explains the formation of all backarc basins. Models for backarc spreading include extensive magma intrusion (Hasebe et al., 1970), mantle convection or mantle-wedge flow induced by the

subducting slab (McKenzie, 1969; Sleep and Toksöz, 1971; Toksöz and Bird, 1977; Toksöz and Hsui, 1980; Hsui and Toksöz, 1981), and thermal upwelling of a mantle diapir (Karig, 1971a,b; Oxburgh and Turcotte, 1971). To accommodate the formation of a backarc basin, the associated trench must migrate oceanward through time (Chase, 1978). This migration may be a function of trench suction due to sinking of the subducted plate (Forsyth and Uyeda, 1975; Molnar and Atwater, 1978), or in part, due to ridge push, where spreading on the overriding plate pushes the trench oceanward (Hsui and Toksöz, 1981).

Both modern and ancient backarc basins are nonuniformly distributed along subduction zones (Ben-Avraham and Uyeda, 1983). These variations may be a function of the degree of coupling between the subducting and overriding plates (Uyeda and Kanamori, 1979; Brooks et al., 1984). When plates are weakly coupled, backarc extension and spreading occur (Uyeda and Kanamori, 1979). In turn, the degree of coupling may vary through time for any given arc segment. Uncoupled and coupled subduction may be different evolutionary stages of a single process (Kanamori, 1971; Uyeda and Kanamori, 1979). Coupling may be a function of: 1) age of the subducting plate (coupled when young and hot, and decoupled when old and cold; Molnar and Atwater, 1978), or 2) motion of the landward plate (coupled when the landward plate moves toward the trench and decoupled when the landward plate moves away from the trench; Chase, 1978; Uyeda and Kanamori, 1979). The former has been termed the "retreating trench" model (trench line retreats due to gravitational pull exerted on subducting slab), and the latter has been termed the "anchored slab" model (subducting slab and trench line remain fixed, and overriding plate retreats landward) (Seno and Maruyama, 1984). Seno and Maruyama (1984) favor the retreating-trench model for the opening of the Shikoku-Parece Vela and West Philippine basins.

Alternate models have been proposed to explain the formation of backarc basins in the tectonically complex Mediterranean region (e.g., Horváth et al., 1981; Malinverno and Ryan, 1986; Boccaletti et al., 1990b). Malinverno and Ryan (1986) proposed an arc-migration model to explain the formation of the

Tyrrhenian basin, which could be applied to other non-Mediterranean backarc basins. Drilling results from Ocean Drilling Program Leg 107 in the Tyrrhenian backarc basin support a model of either asymmetrical rifting associated with a low-angle detachment fault, or migration of the zone of maximum extension due to rollback of the hinge zone of the subducting slab (Kastens et al., 1988). Spreading may also have been caused by an adjustment of the descending-slab profile resulting from a change in subduction rate (Hynes and Mott, 1985).

Miyashiro (1986) proposed a link between large-scale asthenospheric upwelling and backarc-basin formation in the western Pacific. He correlated spatial and temporal shifts in backarc-basin spreading to lateral migration of a large "hot region." According to Miyashiro (1986), this migration accounts for the short duration of spreading within any given backarc basin. The main question with this type of model is what is cause and what is effect (active [mantle-driven] versus passive [plate-driven] models).

Crustal Structure

Rift Phase

Backarc basins may develop on oceanic, continental, or transitional crust. Extensional backarc basins are predominantly associated with intraoceanic arcs. These backarc basins often exhibit episodic spreading, and as a result, spreading centers may initiate on backarc crust formed during a prior phase of spreading (Saunders and Tarney, 1984). For example, the Cenozoic history of the Philippine Sea has been characterized by backarc-basin formation. Rifting and sea-floor spreading behind the Izu-Bonin-Mariana arc system produced the Shikoku and Parece Vela backarc basins during the Oligocene to early Miocene, and the Mariana Trough along the southern part of the arc system during the Miocene to Recent (Fig. 8.4; Mrozowski and Hayes, 1979; Seno and Maruyama, 1984; Chamot-Rooke et al., 1987); this second phase of extension has recently begun in the northern segment of this arc system (e.g., Volcano Arc, southern Iwo-Jima Ridge: Stern et al., 1984; Izu-Bonin arc, northern Iwo-Jima Ridge: Taylor et al., 1992; Fig. 8.4).

Other backarc basins (e.g., Japan Sea and the Okinawa Trough, Fig. 8.1) initiated behind continental-margin arcs, and in the early rift phase, were floored by extended continental crust (see discussion of Okinawa Trough below). Additional complex basement types have been suggested; for example, Nichols and Hall (1991) proposed that a backarc basin formed along the Neogene Halmahera arc, eastern Indonesia, by subsidence of thickened crust composed of imbricated Mesozoic-Paleogene arc and ophiolitic rocks.

Taylor et al. (1991) and Taylor (1992) hypothesized that initial rifting occurs adjacent to the main arc axis, where thicker crust and higher heat flow results in a zone of weaker lithosphere. As a result, rifting is not limited to the backarc, but may also occur in the forearc and within the arc (Taylor and Karner, 1983; Hawkins, 1984; Taylor et al., 1991; Taylor, 1992). The implications of backarc versus forearc rifting are discussed below.

Seafloor-Spreading Phase

When a backarc basin passes from the initial-rifting phase to a period of sea-floor spreading, the attenuated pre-spreading basement gives way to new ocean crust. In general, backarc-basin crust is thin, ranging from 5 to 15 km (Brooks et al., 1984), and according to Cas and Wright (1987), should show an ophiolitic profile of pelagic sediments, pillow lavas, volcanic breccias and flows, and sheeted dikes, overlying upper-mantle basement.

Although arc rifting, from initiation to spreading, may take only a few million years, rift propagation requires more than 5 my (Hussong and Uyeda, 1981; Stern et al., 1984; Taylor, 1992). Active, northward rift propagation is occurring in the Mariana backarc basin (Stern et al., 1984). Rift propagation rates of 10 to 30 cm/y have been proposed for the Japan Basin (Tamaki et al., 1992).

Magnetic-anomaly patterns suggest that sea-floor spreading within backarc basins occurs along linear

Fig. 8.4A Deep Sea Drilling Project site locations and physiography of the Philippine Sea, from Carey and Sigurdsson (1984), after Hussong et al. (1981). Land areas are indicated by horizontal lines, trenches are hachured, and submarine ridges and rifts are dotted. Various symbols show DSDP site locations. Star is approximate location of Sumisu Rift and ODP Leg 126 sites shown in C.

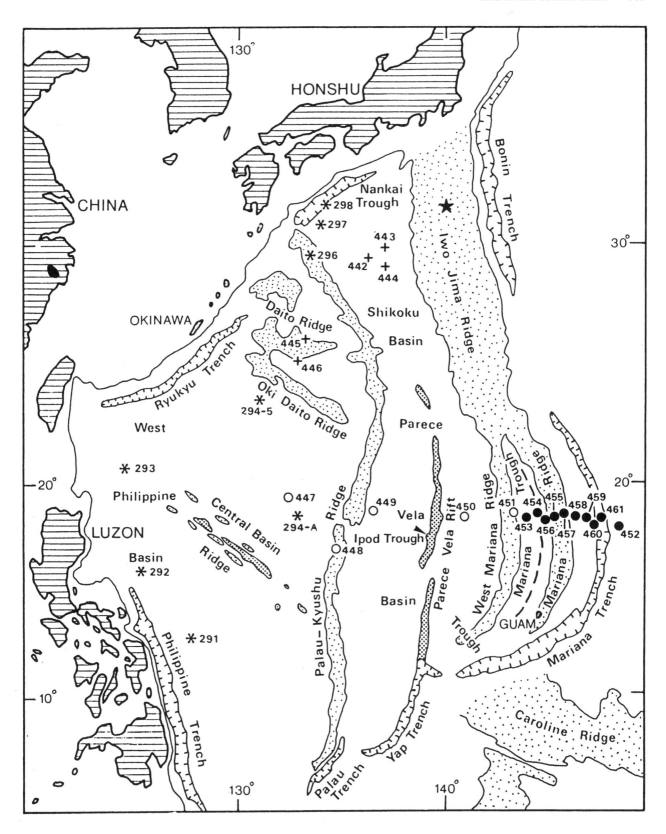

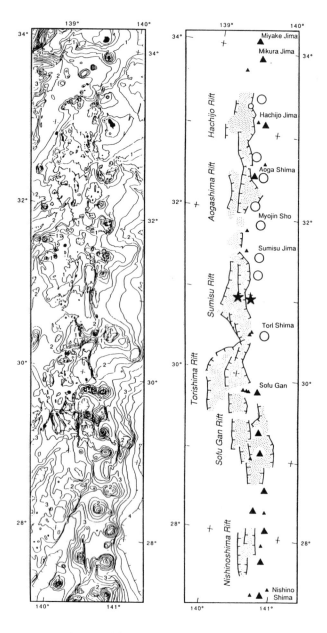

Fig. 8.4B Bathymetry of the Izu-Bonin arc (northern Iwo Jima Ridge). Contour interval is 250 m (from Taylor, 1992); **Fig. 8.4C** Tectonic interpretation of B with arc volcanoes (solid triangles), arc calderas (open circles), major normal faults (ticked lines), and rift basins (stippled regions) as indicated (from Taylor, 1992). ODP Leg 126 sites in the Sumisu Rift are also shown: Sites 790/791 star to the left within the rift basin, and Site 788 star to the right on the rift flank.

ridges, which can be offset along transform segments, or alternatively, by overlapping spreading centers, similar to midocean ridges (Yamazaki et al., 1991). These spreading ridges are arranged in en-echelon

segments with characteristic rift valleys. Spreading may be symmetric or asymmetric, and the spreading axis may shift, complicating magnetic-anomaly patterns. For example, magnetic lineations in the Shikoku Basin were interpreted in terms of symmetrical spreading (e.g., Klein and Kobayashi, 1981), and more recently, in terms of an eastward shift in ridge position (Taylor, 1992). In other instances, where magnetic anomalies are symmetric, the spreading axis may have remained relatively stable through the history of the basin. Spreading half rates may range up to 6.5 cm/y (Taylor, 1979), but may be highly variable within a given backarc region (Leitch, 1984).

Backarc and major ocean basins show similar crustal structure and magnetic-lineation patterns, indicating similar modes of crustal generation in each setting (Taylor and Karner, 1983). Magnetic lineations in backarc basins, however, tend to be less well developed than those in major ocean basins in terms of intensity and fidelity, suggesting different modes of spreading. According to Taylor and Karner (1983), disorganized magnetic-lineation patterns in backarc basins may be a function of smaller-scale spreading processes easily affected by changes in stress, resulting in frequent spreading-center jumps. Alternatively, these disorganized patterns may result from diffuse spreading, overlapping spreading centers, low geomagnetic latitude of equatorial basins coupled with north-south strike of basin axes, irregular basement topography, or thick sedimentary cover (Saunders and Tarney, 1984; Yamazaki et al., 1991). Some backarc basins, such as the basin behind the South Sandwich arc and the Parece Vela Basin, exhibit distinct spreading axes and well developed magnetic lineations (Barker, 1972; Scott and Kroenke, 1980; Barker and Hill, 1981).

Poor magnetic-anomaly record in some continental backarc basins with higher sedimentation rates is attributed by Tarney et al. (1981) to irregular spreading processes, manifested in a higher incidence of sills. Poor magnetic-anomaly patterns in parts of the Japan Sea have been attributed to the presence of transitional or thickened crust (Katao, 1988; Tamaki, 1988). Sill-flow-sediment complexes have been drilled in the Yamato backarc basin in the southern Japan Sea, where they represent acoustic basement (Tamaki et al., 1990). True oceanic crust in the Japan

Sea is limited to the Japan Basin, in the northern Japan Sea (Fig. 8.1).

Basin Tectonics

Most backarc-basin studies have focused on the sea-floor-spreading stage of backarc-basin formation. Recent investigations of rifting within the Izu-Bonin arc system and within the Okinawa Trough provide insights into the early stages of intraoceanic and continental-margin backarc-basin formation, respectively. As a result, a more detailed picture of the evolution of backarc-basin architecture has emerged.

Rift Phase of Intraoceanic Backarc Basins

The early rift phase of backarc-basin formation is best illustrated within the Izu-Bonin arc system, which extends along the Iwo-Jima Ridge, south of central Honshu, Japan (Fig. 8.4). The Izu-Bonin arc is characterized along its 700-km length by a series of rifts thought to represent incipient backarc basins (Fig. 8.4; Taylor et al., 1990b, 1991). These rifts are asymmetric structural grabens ranging from 25 to 110 km long and 25 to 40 km wide, separated along strike by chains of volcanoes and structural highs (Taylor et al., 1990b, 1991; Klaus et al., 1992b). Oblique transfer zones link opposing master faults and/or rift-flank uplifts (Taylor et al., 1990b). A switch from backarc to arc-axis rifting occurs at 29° (Fig. 8.4) along the southern Izu-Bonin arc (Taylor, 1992). Fault-bounded rifts are better developed between arc volcanoes rather than adjacent to them in the Izu-Bonin system, suggesting that extensional strain may be accommodated by intrusions associated with arc volcanoes (Taylor et al., 1991; Klaus et al., 1992).

The best-documented active intraoceanic backarc rift segment is the Sumisu Rift, located near the center of the Izu-Bonin arc system (Fig. 8.4C). This rift has been investigated extensively using SeaMARC II sidescan and bathymetry swathmapping, seismic-reflection, magnetic and gravity profiles, direct observation and sampling using the submersible ALVIN, dredging, gravity coring and drilling by the Ocean Drilling Program. These data are summarized by Brown and Taylor (1988), Murakami (1988), Nishimura and Murakami (1988), Nishimura et al.

(1988), Yamazaki (1988), Ikeda and Yuasa (1989), Fryer et al. (1990b), Hochstaedter et al. (1990a), Nakao et al. (1990), Smith et al. (1990), Taylor et al. (1990a,b, 1991), Urabe and Kusakabe (1990), Klaus et al. (1992b), and Taylor (1992).

The following description of the Sumisu Rift is summarized from Taylor et al. (1990b, 1991), and Klaus et al. (1992). The 100km-long and 40km-wide Sumisu Rift lies a few kilometers to the west of the active arc axis (Fig. 8.5). It is a young rift (approximately 2 my old) that has accommodated up to 5 km of extension in its southern extremes. High- to moderate-angle (25–75°) normal faults, some of which offset surficial sediments (Fig. 8.6), characterize the Sumisu Rift; the major normal faults that bound the basin exhibit bathymetric relief of approximately 1000 m, with up to 2500 m throw. It is not clear whether the major bounding faults offset the entire crust and lithosphere, or are detached at mid- to upper-crustal levels. Normal faults at all scales within the Sumisu Rift exhibit a zig-zag pattern, ranging from concave to convex curvilinear shapes; these faults are similar to those in continental rift systems. Where observed directly by submersible, fault scarps are flanked by sandy sediment aprons. Fault systems are thought to have been developed in two phases: first, during the half-graben phase, a series of normal faults developed after an earlier sag phase; the second, full-graben stage, was characterized by development of antithetic faults in the hanging wall that concentrated subsidence in an inner rift (Fig. 8.6). The direction of rift extension is orthogonal to the arc axis and the regional trend of the rift system and basin widening is achieved by footwall collapse. Pre-rift structures in the Sumisu Rift region do not appear to control the location and trend of synrift structures.

Rift volcanism is concentrated along transfer zones, where volcanic vents appear to be, in part, controlled by a series of north-trending normal faults (Fig. 8.5; Taylor et al., 1990b). The Central Ridge transfer zone separates the Sumisu Rift into two depocenters, or sub-basins (Taylor et al., 1991; Klaus et al., 1992b). Smooth basin floors, from 2100 to 2275 m deep, shallow to 2000 m to the north and south, near the arc volcanoes that serve as sediment

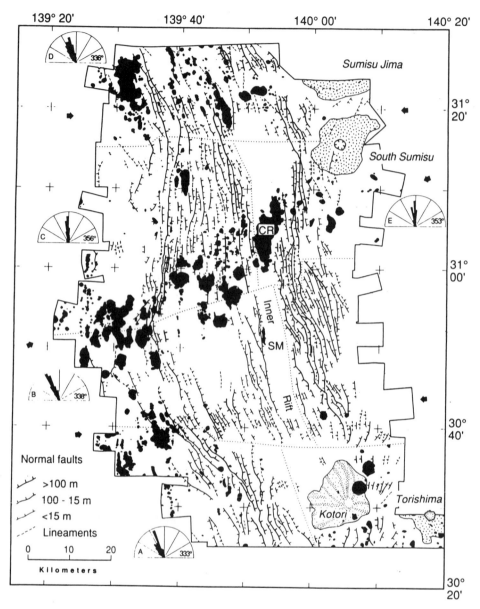

Fig. 8.5 Structural map of the Sumisu Rift area based on sidescan data. Normal faults indicated by ticked lines, volcanic vents and flows by dark shaded areas. Active arc calderas are dotted. Regional variations in fault strike and length are indicated by polar histograms (cumulative fault length vs. azimuth). The Central Ridge is indicated by CR and Shadow Mountain by SM. Trends of three transfer zones are indicated by arrows. From Klaus et al. (1992b).

sources (Taylor et al., 1991; Klaus et al., 1992b). Lavas of the Central Ridge (Fig. 8.5) are predominantly basaltic, with minor rhyolites and dacites (Fryer et al., 1990b; Hochstaedter et al., 1990a). Volcanic vents also occur along the flanks of arc volcanoes and on the proto-remnant margins (Fig. 8.5; Taylor et al., 1990b). Young, intra-rift, elongate dike-fed volcanic ridges may be spreading-center precursors (e.g., Shadow Mountain (SM) in Fig. 8.5; Taylor et al., 1990b). Multi-vent en-echelon ridges are better developed in Izu-Bonin backarc rifts with greater extension, and these may eventually coalesce into spreading systems

(Taylor, 1992). The lack of magnetic anomalies in the Sumisu Rift suggests the absence of sea-floor spreading and the presence of island-arc crust under the basin (Yamazaki et al., 1991). In the southern part of the Izu-Bonin arc system, the presence of linear magnetic anomalies suggests that the arc and parts of the backarc basin may be located on old oceanic crust (Yamazaki et al., 1991).

In addition to detailed studies in the Sumisu Rift, preliminary drilling results (Leg 135) are available from the Lau Basin (Fig. 8.1). These suggest that an early phase of crustal extension also occurred in this

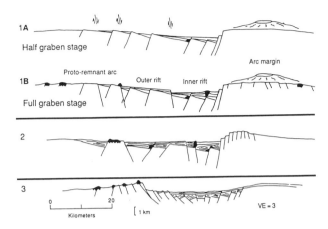

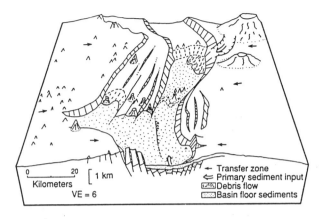

Fig. 8.6 A) Structural evolution of the Sumisu Rift from half-graben to full-graben stages (1A,B), and variations along strike (2 and 3).

B) Perspective view of the Sumisu Rift. From Taylor et al. (1991) and Klaus et al. (1992b).

basin prior to sea-floor spreading (Leg 135 Scientific Party, 1992). This early phase of crustal extension was associated with episodic volcanism and produced a series of graben-like sub-basins, which were rapidly filled with sediment (Leg 135 Scientific Party, 1992).

Seafloor-Spreading Phase of Intraoceanic Backarc Basins
Mature backarc basins range from rectangular to arcuate in plan view (Fig. 8.1). To some extent, basin width is a function of spreading rate and duration of spreading. The average lifespan of backarc-basin spreading systems is less than 25 my (Fig. 8.7; Tamaki and Honza, 1991).

Once seafloor-spreading has commenced, basin relief is more pronounced, and the basin is bounded by a remnant-arc ridge and a magmatic arc (Fig. 8.2). The remnant arc is a linear to arcuate fault-bounded aseismic ridge with flanking volcaniclastic aprons (Karig, 1972; White et al., 1980). The vol-caniclastic aprons on either side of the remnant arc ridge are of different origin, with younger syn- to postrift debris on the backarc-basin side, and older pre-rift arc-related debris on the opposite side of the ridge (Karig, 1972; White et al., 1980). The remnant arc may also be segmented into multiple en-echelon segments (Karig, 1972). Remnant-arc width and shape are functions of the locus of frontal-arc rifting (Karig, 1972); in extreme instances, where rifting initiates within the debris apron on the rear flank of the frontal arc, no remnant arc is produced (Karig, 1972).

The position of arc rifting, forearc versus backarc, determines the presence and nature of the remnant arc (Taylor and Karner, 1983). In the case of forearc rifting, the remnant arc consists of arc-axis volcanoes, which may be preserved as subcircular seamounts along the remnant arc (e.g., western Mariana Ridge; Karig, 1972). Forearc spreading is thought to have produced the Mariana and Lau Basins (Taylor and Karner, 1983; Hawkins et al., 1984; Parson et al.,

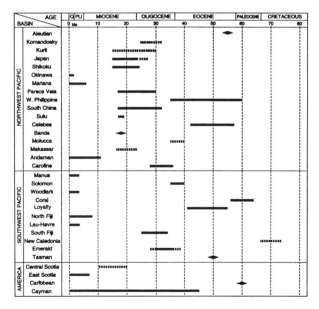

Fig. 8.7 Ages of backarc-basin formation in the Pacific region (Fig. 8.1) (from Tamaki and Honza, 1991). Estimated ranges are indicated by broken lines and time of entrapment is indicated by rhombic marks. See Tamaki and Honza (1991) for specific references.

1992; Taylor, 1992). In the case of backarc initiation of rifting, a remnant arc may not be produced (Karig, 1972; Taylor and Karner, 1983). Temporal and spatial variations in arc volcanism and spreading may result in overprinting of arc magmatism on backarc-basin crust (Taylor and Karner, 1983). Thus, the prerift arc axis may be incorporated into a remnant arc in the case of forearc rifting and spreading, or be over-printed by postrift arc volcanism in the case of backarc rifting and spreading.

Continental Backarc Basins

Except perhaps for the Andaman Sea, the Okinawa Trough (Figs. 8.1, 8.8) is the only modern example of an incipient extensional continental backarc basin (Letouzey and Kimura, 1986). The axis of opening within the Okinawa Trough is parallel to the Ryukyu arc axis (Letouzey and Kimura, 1985). It is a rela-tively shallow, southward deepening (<2,300 m deep) backarc depression flanked to the south and east by the partially emergent Ryukyu island arc, and to the north and west by the shallow East China Sea shelf, terminating to the south, against the northeast coast of Taiwan (Fig. 8.8; Lee et al., 1980). Geophysi-cal studies and dredged samples suggest that base-ment consists of slightly thinned continental crust, with oceanic crust limited to the southern axis of the trough (Lee et al., 1980; Furukawa et al., 1991; Hirata et al., 1991; Ishikawa et al., 1991; Kimura et al., 1991). This continental crust has been offset by a series of normal faults, forming well defined graben structures with tilted blocks draped by Pleistocene sedimentary sequences (Letouzey and Kimura, 1985). Geophysical data also indicate the presence of igneous intrusions within the basin (Letouzey and Kimura, 1985).

Early rift structure of continental backarc basins may be modified during sea-floor spreading. The Japan Sea is an example of a mature continental-margin backarc basin formed by multiaxial backarc spreading (Tamaki et al., 1990). The basin consists of deep basinal regions (2-3.7 km bsl) floored by oceanic crust and sill-sediment complexes, separated by nor-mal-fault-bounded ridges, thought to be foundered and rifted continental fragments (Ingle et al., 1990; Tamaki et al., 1990). Unlike the Okinawa Trough, the

axis of spreading in the northern Japan Sea (Japan Basin) was perpendicular to the modern arc axis, and this basin may have initiated as a pull-apart along a strike-slip margin (e.g., Tamaki et al., 1992).

Basin Subsidence

Rift Phase of Intraoceanic Backarc Basins

Structural provinces in the Sumisu Rift include the arc margin, the inner rift, the outer rift, and the proto-remnant arc (Figs. 8.5, 8.6; Taylor et al., 1991; Klaus et al., 1992). These structural provinces are, in turn, subdivided by oblique transfer zones, along which volcanism is concentrated; subprovinces show changes in fault trends and uplift/subsidence history (Taylor et al., 1991; Klaus et al., 1992).

The arc margin is characterized by fault-bounded flank uplifts, locally draped by volcaniclastic aprons adjacent to arc volcanoes. These uplifts are composed of volcaniclastic sequences probably deposited in intra-arc basins between volcanic centers. Rapid uplift concurrent with rift inception can be documented by seismic studies and the presence of unconformities and faunal changes (deep to shallow) within rift-flank sedimentary sequences (e.g., 0.6 to 1.6 km of uplift of strata deposited between 2.35 and 0.275 Ma at Site 788 on the Sumisu Rift flank; Fig. 8.9; Taylor et al., 1990a; Taylor, 1992). Taylor (1992) interpreted this as footwall uplift due to isostatic rebound during removal of material from the hanging wall of the main bounding faults during extension.

The inner rift is the major locus of sedimentation, showing maximum subsidence and sediment thick-ness near the arc margin (Fig. 8.6); the outer rift shows less subsidence and a thinner sedimentary cover (Taylor et al., 1991; Klaus et al., 1992). Variable sediment thickness and sedimentation rates, like those documented at Sites 790 and 791 within the inner rift, can be attributed to differential subsidence (Fig. 8.9), but benthic-foraminiferal evidence for deposition at relatively uniform water depths (Kaiho, 1992) suggests that sedimentation has approximately matched subsidence (Taylor et al., 1991; Klaus et al., 1992). In general, sediment accumulation rates have increased through time at Sites 790 and 791 in the

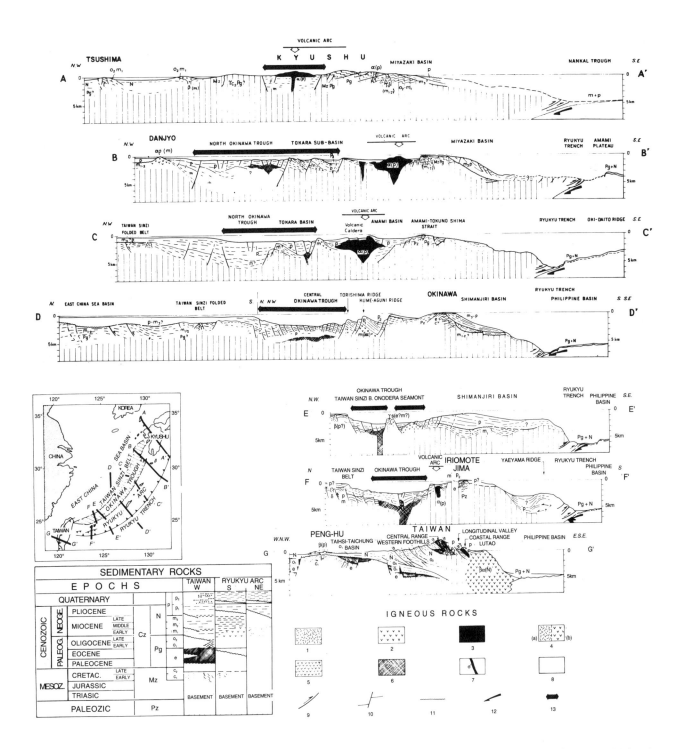

Fig. 8.8 Location map, representative stratigraphic sections and interpretive cross sections for the Okinawa Trough (from Letouzey and Kimura, 1985). Basement indicated by vertical lines. (1) Intrusive rocks; (2) extrusive rocks; (3) Pleistocene to Holocene igneous rocks of the Ryukyu volcanic arc; (4) Miocene igneous rocks; (5) Miocene to Pleistocene igneous rocks of the Taiwan-Luzon volcanic arc; (6) Inferred Pleistocene mafic rocks; (7) Lichi melange with ultramafic rocks in the Eastern Taiwan suture zone; (8) oceanic crust, ridge or plateau of the Philippine Sea plate; (9) thrust fault; (10) normal fault; (11) unconformity; (12) subduction; (13) Okinawa Trough.

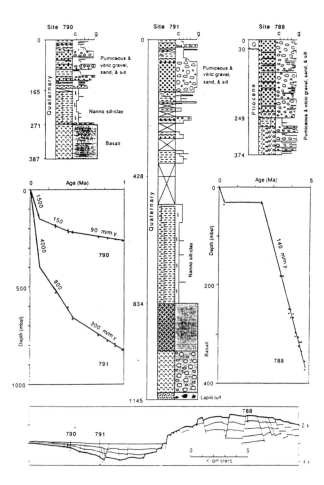

Fig. 8.9 Simplified cross section, stratigraphic columns and age-versus-depth plots for ODP Leg 126 Sites 788, 790, and 791 in the Sumisu Rift. Site 788 is located on the rift flank, and Sites 790 and 791 in the inner rift. Cross section based on interpretation of an east-west seismic profile at approximately latitude 31°N. Stratigraphic columns indicate lithology (dash, clay to silt; barbed dash, nannofossils; and black dots, sand/sandstone and gravel/conglomerate). Patterns at base of Sites 790 and 791 refer to basaltic breccias and flows. Xs in columns mark intervals with no recovery. Grain size ranges from clay(c) to gravel(g). Wavy vertical lines indicate bioturbation. Slope of age-versus-depth plots gives apparent sediment accumulation rates in meters/million years (m/my). Data points based on paleontologic data. Wavy line in Site 788 plot indicates unconformity. From Taylor et al. (1991), Klaus et al. (1992) and Taylor (1992).

Sumisu Rift, where recent accumulation rates are extremely high, up to 4000 m/my (Fig. 8.9; Taylor et al., 1991). Extrapolation of these results on seismic profiles extending into the basin suggests that progressive tilting of sedimentary sequences from nearly horizontal to 45° at 800 mbsf (meters below seafloor) at Site 791 is a result of syndepositional faulting and subsidence (Taylor et al., 1991; Klaus et al., 1992).

Sedimentation rates within the inner rift, to the east of Site 791, may reach 6000 m/my (Klaus et al., 1992). Some faults appear to be listric at depth, apparently soling on relatively shallow detachment surfaces (Klaus et al., 1992; Taylor, 1992).

The proto-remnant arc is structurally higher than the adjacent rift basin. This margin presumably consists of pre-rift, arc-related volcaniclastic sediments overlain by a thin cover of synrift sediments (Taylor et al., 1991; Klaus et al., 1992).

Seafloor-Spreading Phase of Intraoceanic Backarc Basins
The large-scale architecture of backarc basins is primarily controlled by subsidence rates. Relative subsidence rates can be determined by comparing heat flow versus sea-floor depth (Watanabe et al., 1977). As with oceanic lithosphere, mean heat-flow values and basement elevation decrease with age in backarc basins due to thermal subsidence (Watanabe et al., 1977). Kobayashi (1984) showed an age-depth correlation in backarc basins, and found that backarc-basin depth is generally 1000 m greater than expected for equivalent-aged crust in the major oceans. In a more detailed study of the Philippine Basin, Park et al. (1990) included corrections for sediment loading and found a similar correlation and depth disparity (800 m). They computed age-depth relationships for other Pacific backarc basins, compared them to curves determined for normal ocean and Philippine Sea crust, and found that older (>15 my old) basins have predictable subsidence (Fig. 8.10). Younger basins, in contrast, show highly variable average basement depths, with a positive correlation between basement depth and dip angle of subducting slab.

Mature backarc basins are generally asymmetrical, with greater depth adjacent to the remnant arc. This asymmetry may be a function of arc development on backarc-basin crust on the trenchward side of the basin, effects of the subducting slab, or differential subsidence of the remnant arc due to gravitational adjustment (Weissel, 1977; Taylor and Karner, 1983; Park et al., 1990).

After spreading has commenced, the isolated remnant arc, which may have been originally emergent, isostatically subsides, forming a submerged chain of seamounts or a ridge (Karig, 1972; Kobayashi, 1983). This subsidence (e.g., 325 m/my at Site 296 on the

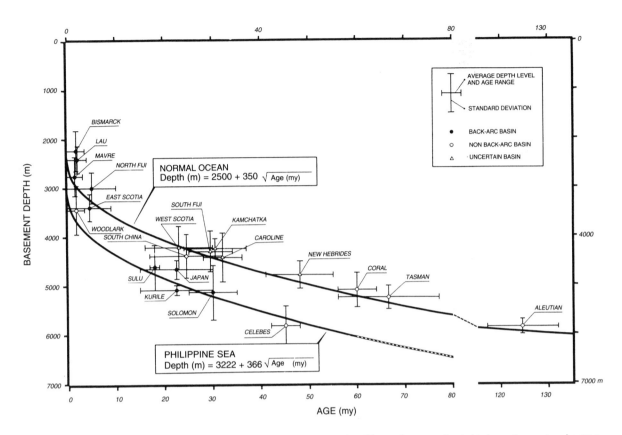

Fig. 8.10 Age-versus-depth plot, comparing data for selected marginal basins to curves for Philippine Sea backarc basins and normal ocean crust. "Non back-arc basins" are those floored by oceanic crust not produced by sea-floor spreading behind an active arc-trench system. From Park et al. (1990).

Palau-Kyushu Ridge) can be an order-of-magnitude greater than for normal oceanic crust (Kobayashi, 1983). The region of highest relief within nonactive backarc basins marks the final position of the spreading zone, where backarc crust is youngest, thinnest, hottest, and most buoyant (Karig, 1975).

Intraoceanic backarc rifts are characterized by periods of rapid subsidence, but sedimentation rates drop as the basin widens during sea-floor spreading (Karig, 1975). Park et al. (1990) found that in the Philippine Sea backarc basins, estimated sediment thickness is proportional to basement age, with thicker sediments occurring over older crust.

Continental Backarc Basins
The major structural elements of continental extensional backarc basins are the magmatic arc, an oceanic basin produced by sea-floor spreading, and a passive continental margin. Similar to intraoceanic backarc rifts, the rift phase of continental backarc-basin formation can result in extremely high subsidence rates (e.g., over 1 km/my during Miocene rifting in northern Japan; Yamaji, 1990; Urabe and Marumo, 1991).

What is known of the early subsidence history of continental-margin backarc basins is based on geophysical and sedimentological data from the Okinawa Trough, the Tyrrhenian Sea, and the Japan Sea region. Based on seismic stratigraphy across the Okinawa Trough, Letouzey and Kimura (1985) suggested that Late Miocene extension and subsidence along the arc axis was accompanied by rift flank uplift (Fig. 8.11). This rift stage was then followed by a Plio-Pleistocene drift or a passive-margin stage, where

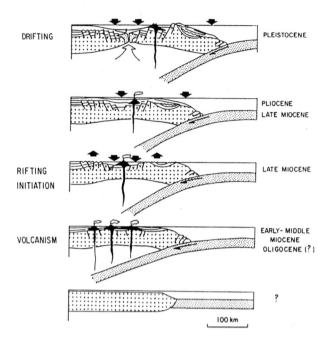

Fig. 8.11 Tectonic evolution of the Okinawa Trough. From Letouzey and Kimura (1985).

tilted fault blocks were draped by prograding sediment wedges off the continental shelf to the west and by frontal-arc volcanic sequences to the east (Letouzey and Kimura, 1985).

The subsidence history of the Sardinian passive margin adjacent to the Tyrrhenian Sea backarc basin was determined during Ocean Drilling Program Leg 107 (Trincardi and Zitellini, 1987; Kastens et al., 1988). Extension and subsidence of this passive margin are recorded in synrift transgressive sequences with abundant microfaults and slumps (Kastens et al., 1988).

In instances of incomplete or dispersed extension, a horst-and-graben structure may be produced, where basins floored by sediment-sill complexes alternate with basement blocks. For example, the Japan Sea backarc basin is subdivided into sub-basins by normal-fault-bounded continental blocks, isolated during rifting (Ingle et al., 1990; Tamaki et al., 1990). Sediment accumulation on these basement highs is minimal in comparison to the adjacent basins, where sediment up to 3000 m thick is present (Ingle et al., 1990; Tamaki et al., 1990). The subsidence history of the arc margin of the Japan Sea is preserved in the now-uplifted sequences in northeastern Honshu and within the

Yamato Basin. In northeastern Honshu, Sato and Amano (1991) documented the following stages of backarc-basin formation: 1) a rift stage associated with half-graben formation, intermediate to felsic volcanism, and fluvial to lacustrine sedimentation; 2) an opening stage associated with rapid subsidence due to crustal stretching, marine transgression to middle-bathyal depths, and voluminous bimodal volcanism; and 3) a thermal-subsidence stage characterized by hemipelagic sedimentation and decrease in backarc volcanism. Results from recent ocean drilling in the Yamato Basin suggest a similar history (Tamaki et al., 1990). Ingle (1992) analyzed basin subsidence throughout the Japan Sea region, incorporating data from ODP sites, offshore wells, and onshore sequences. In his model, the rift stage is characterized by slow thermal subsidence, followed by alternating periods of rapid mechanical and slow thermal subsidence, culminating in compressional uplift. This compression is related to incipient subduction of the Japan Sea along its eastern margin (Kikuchi et al., 1991).

Composition and Distribution of Volcanic Rocks

Volcanic components in backarc basins include lava flows, breccias, pyroclastic rocks, and reworked volcaniclastic materials. Possible sources of these materials in the rift stage include arc-axis volcanoes, intra-rift volcanism, and the proto-remnant arc. Volcanic sources for mature backarc basins are limited to arc-axis volcanoes and volcanoes associated with spreading.

Rift Phase of Intraoceanic Backarc Basins

The most complete picture of backarc-rift volcanism comes from studies of the Sumisu Rift. In this area, there are 200 or more volcanic centers that show an average relief of 100-200 m and an average diameter of 1-3 km (Fig. 8.5; Fryer et al., 1985b). In general, volcanism is bimodal, with felsic lava flows and calderas, and basaltic magmatic centers, lava flows and cones (Fryer et al., 1985b, 1990b; Hochstaedter et al., 1990a; Taylor et al., 1990b). The felsic centers are concentrated along the main arc axis, whereas the mafic centers are concentrated in transfer zones (Taylor et al., 1990b). Arc volcanoes lie at the northern (Sumisu and South Sumisu Calderas) and south-

ern (Torishima Caldera) limits of the rift (Fig. 8.5). Basaltic intrusions along fault zones disrupt synrift sediments and are associated with mound-like extrusive volcanic piles where they breach the surface (Taylor et al., 1991; Klaus et al., 1992). These basaltic volcanic centers are flanked by coarse talus slopes and debris aprons (Taylor et al., 1990b).

At about 2 Ma, a tensional regime developed within the Izu-Bonin arc, leading to the development of submarine calderas and the eruption of dacitic to rhyolitic magmas differentiated from subcrustal basalt (Gill et al., 1992). Backarc-basin basalt (BABB) and related rhyolite were present before rift formation, implying that the BABB source was either recently intruded and extension was the result of intrusion, or that the BABB source was always present and BABB extrusion was a passive result of extension (Hochstaedter et al., 1990b; Gill et al., 1992). The recycled slab component (Cs, U, Pb, Ba and ^{87}Sr) has increased in the arc volcanic rocks during rifting (Gill et al., 1992).

Basement rocks penetrated at Sites 790 and 791 within the Sumisu Rift formed during the first 0.5-1.0 my of rift development, and consist of subaqueous explosion breccia, apparently produced by deep-sea pyroclastic fountaining during a Strombolian eruption (Gill et al., 1990, 1992; Koyama et al., 1992). Their composition is similar to younger Sumisu Rift basalts recovered by dredging and submersible (Gill et al., 1992). Young basalts in the Sumisu Rift are nearly identical in composition to backarc-basin basalts from the Mariana Trough spreading axis (Taylor et al., 1990b), and similar, in terms of their major- and trace-element geochemistry, to backarc-basin basalts from mature backarc basins (Fryer et al., 1990b).

Seafloor-Spreading Phase of Intraoceanic Backarc Basins
Most igneous rocks recovered from within backarc basins are olivine-normative tholeiites with mineralogy and major-element chemistry similar to ocean-ridge tholeiites; they may be distinguished by their trace-element and isotopic signatures (Saunders and Tarney, 1984). With increasing age of the subduction zone and successive backarc spreading episodes, magmas erupted in intraoceanic backarc basins are enriched in large-ion lithophile elements; this enrichment is probably the dehydration product of sub-

ducted oceanic crust (Saunders and Tarney, 1984). Shikoku Basin basalts (DSDP Sites 442, 443 and 444) differ from MORB in that they may have high water contents, as determined by vesicularity and mineralogy, somewhat sodic plagioclase, and slight enrichment in alkalis (Dick, 1982).

As backarc basins evolve, so do basaltic magma compositions (Taylor and Karner, 1983). Tarney et al. (1981) and Saunders and Tarney (1984) suggested that backarc-basin basalt (BABB) is initially similar to island-arc basalt (IAB), but with time, becomes more like mid-ocean ridge basalt (MORB). Data from the Sumisu Rift and the Lau Basin imply that this change may be surprisingly abrupt: in each case, a MORB-type magma source with little slab-derived material (mostly water) was immediately tapped during rift inception (Gill et al., 1992). The Lau basin is more mature than the Sumisu Rift in that rifting began in the Lau forearc at approximately 5 Ma and progressed to spreading at about 3 Ma, now concentrated along an axial rift in the northern Lau Basin (Parson et al., 1990; Gill et al., 1992; Leg 135 Scientific Party, 1992).

Volcano spacing may also change with arc maturity and development, from many closely spaced volcanic centers to larger, widely spaced (50-70 km) volcanoes (Bloomer et al., 1989). Thus, volcano development may vary in situations where forearc rifting has occurred versus backarc rifting; in the former, widely spaced volcanic centers may survive rifting, whereas in the latter, the arc has to reestablish itself, resulting in many closely spaced volcanoes. Also, volcano development may change along basin strike as a function of basin age; larger, more widely spaced volcanoes occur adjacent to older parts of the backarc basin, from which rifting propagates (e.g., Bloomer et al., 1989). Finally, as in the Sumisu Rift, well defined structural controls exist (e.g., bounding faults and cross-arc fractures) in more mature backarc basins, such as the eastern Mariana Trough (Bloomer et al., 1989).

Continental Backarc Basins
Compositional data for continental backarc basins are sparse (Saunders and Tarney, 1984), but recent drilling in the Japan Sea during ODP Legs 127 and 128 has provided some new information. Analysis of

basaltic sills from the Yamato basin in the Japan Sea suggests a transition from arc-related calc-alkaline to spreading-related tholeiitic magmatism during basin evolution (Thy, 1992). The chemistry of igneous rocks recovered at Leg 127/128 sites in the Sea of Japan is transitional among arc basalts, intraplate basalts, and MORB (Allan and Gorton, 1992). According to Allan and Gorton (1992) and Cousens and Allan (1992), mantle heterogeneity beneath the Japan Sea is reflected in basalt compositional diversity within and among sites, with some subduction-related contamination, but there is little evidence for interaction with continental crust. Variations in basalt geochemistry can also be explained by a temporal change from enriched to depleted mantle sources (Nohda et al., 1992; Pouclet and Bellon, 1992).

Samples collected by submersible from the central graben of the Okinawa Trough give some insight into the effects of backarc rifting along a continental margin. Overall, Okinawa Trough basalts are similar in composition to Mariana Trough basalts, consistent with a water-rich MORB-like mantle source with a subducted-slab component (Honma et al., 1991). The $^{87}Sr/^{86}Sr$ and $^{143}Nd/^{144}Nd$ ratios for Okinawa Trough basalts are distinctly higher than those for Mariana Trough basalts, reflecting a continental component (Honma et al., 1991). This is perhaps one means of distinguishing intraoceanic and continental-margin backarc basins.

Sediments and Sedimentary Rocks

Facies
The major sediment sources for backarc basins are pelagic fallout, airborne ash (following hydraulic settling), and submarine gravity flows (some of which may initiate as pyroclastic flows and be reworked from shallow water surrounding the arc).

Characteristic lithofacies for backarc basins are outlined in the numerous reports of the Deep Sea Drilling Project and the Ocean Drilling Program. These lithofacies and controls on their distribution are summarized by Klein (1985a). Klein (1985a) found that backarc-basin sequences are highly variable. He documented nine major sediment types in DSDP cores from backarc sites in the western Pacific:

debris flows, submarine-fan depositional systems, silty basinal turbidites, hemipelagic clays, pelagic clays, biogenic pelagic silica sediments, biogenic pelagic carbonates, resedimented carbonates, and pyroclastites. The characteristics of each of these facies types are summarized in Table 8.1. Cores described on Leg 126 in the Sumisu Rift contain similar sediment types, but also include massive units of vitric silt and pumiceous gravel (Fig. 8.9). The latter may be characteristic of the earliest phases of backarc rifting, but were not cored on previous ocean drilling legs (Nishimura et al., 1991).

Nontectonic Controls on Sedimentation
According to Klein (1985a), the distribution of backarc-basin facies commonly is independent of time or tectonic processes. Important processes affecting the composition and distribution of sediments within backarc basins include: ocean-current circulation, latitudinal controls on biogenic (carbonate vs. silica) productivity and deposition, and sediment input from continental land masses (e.g., hemipelagic clay). For example, ocean circulation patterns may be responsible for periods of submarine erosion or nondeposition, especially on topographic highs. Widespread development of deep-marine hiatuses has been linked to periods of more intense circulation during the production of cold bottom water (e.g., Barron and Keller, 1982).

According to Klein (1985a), eustatic changes in sea level have little effect on clastic input into backarc basins, even where the arc is largely emergent. Recent work by Betzler et al. (1991) in the Celebes and Sulu backarc basins, however, suggests that in addition to a strong tectonic control on sedimentation, global sea-level changes may explain the synchronous breaks in turbidite sedimentation observed in both basins. In the case of continental-margin basins such as the Japan Sea, which are semi-isolated from the open ocean by shallow sills, eustatic fluctuations in sea level and tectonic uplift can result in periods of basin isolation and stagnation (Ingle et al., 1990). Pelagic sedimentation within this basin was affected by fluctuations in climate, sea level and tectonics, which altered water temperature and circulation patterns within the basin (e.g., Ingle et al., 1990; Tamaki et al., 1990; Alexandrovich, 1992; Koizumi,

Table 8.1   Sediment Facies in Backarc Basins after Klein (1985a)

1) Debris flows (1.2% of DSDP cores examined by Klein)
    A) Conglomerates
        1) Thick- to medium-bedded
        2) Massive except for local parallel lamination, reverse graded bedding, and clast imbrication
        3) Clasts primarily composed of basalt and andesitic rock fragments; range in size from granules to small cobbles, averaging pebble-size
        4) Matrix of volcanic sand and silt, glass shards
        5) Locally contain fragments of shallow-water fauna
    B) Sandy debris flows (or fluidized flows) with dish structures

2) Submarine-fan depositional systems (20.0% of DSDP cores examined by Klein)
    A) Inner and Midfan subsystem
        1) Interbedded thin- to medium-bedded sandstones and mudstones with local pebble horizons
        2) Coarse- to medium-grained sandstones, with common subrounded to angular volcanic rock fragments, dense minerals and plagioclase, and rarer shallow-water carbonate lithic clasts and bioclasts
        3) Sedimentary structures include flame structures, load casts, microfaults, slump folds, micro-cross-laminae, in-phase waves, dewatering pipes, multiple graded bedding, and gradational tops of sandstone beds; partial Bouma sequences and various trace fossils are common
        4) Massive to parallel-laminated mudstones with silty laminae
    B) Outer-Fan subsystem
        1) Interbedded thin-bedded sandstones, siltstones and mudstones
        2) Medium- to fine-grained sandstones, with common subrounded to well rounded volcanic rock fragments and plagioclase
        3) Sedimentary structures include parallel lamination, micro-cross-lamination, graded bedding, and sharp to gradational tops of sandstone beds; partial Bouma sequences and graded cycles of sand-silt-clay are present

3) Silty basinal turbidites (5.7% of DSDP cores examined by Klein)
    A) Thin-bedded silts and siltstones ranging to clays and claystones
    B) Sedimentary structures include graded bedding (silt to clay), parallel lamination, micro-cross-lamination and bioturbation

4) Hemipelagic clays (21.8% of DSDP cores examined by Klein)
    A) Interbedded and mixed silt (e.g., quartz, feldspar and zeolites) and clay (various types)
    B) Mostly structureless, but some sedimentary structures present, such as bioturbation, mottling, parallel lamination, and graded bedding

5) Pelagic clays (4.2% of DSDP cores examined by Klein)
    A) Primarily composed of red to dark reddish brown structureless clay with iron-manganese micronodules, and some silt/sand, volcanic ash and biogenic debris

6) Biogenic pelagic silica sediments (4.3% of DSDP cores examined by Klein)
    A) Fine-grained siliceous oozes or chert (layers or nodules)
    B) Main components are radiolarians, diatoms, and silicoflagellates

7) Biogenic pelagic carbonates (23.8% of DSDP cores examined by Klein)
    A) Fine-grained ooze, chalk, and limestone, principally composed of nannofossils and foraminifera, with lesser hemipelagic clay, siliceous microfauna, et al.
    B) Generally lack distinctive sedimentary structures, with some minor parallel lamination, and biogenic mottling

8) Resedimented carbonates (9.5% of DSDP cores examined by Klein)
    A) Composed of nannoplankton and foraminiferal ooze, chalk, or limestone
    B) Contains both reworked deep-water and shallow-water biogenic components, including foraminifera, pelecypod, coralline alga, coral, and bryozoan fragments
    C) Well developed sedimentary structures include graded bedding, parallel bedding, microfaults, distorted burrows, slump folds and blocks, load casts, and micro-cross laminae; partial Bouma sequences are also present

9) Pyroclastics (9.5% of DSDP cores examined by Klein)
    A) Volcanic ash and tuff composed of coarse silt to medium sand grains of vesicular glass, glass shards, quartz, feldspar, pyroxene, bioclasts, et al.
    B) Exhibit grading, parallel lamination, mottling and bioturbation
    C) Ash layers may be present in association with other lithofacies

1992; Muza, 1992). The effects of eustatic changes in sea level may be more apparent along the passive margins of continental backarc basins.

Tectonic Controls on Sedimentation

Plate tectonics dictates the overall basin architecture, and so, influences the distribution of sediments within backarc basins. For example, within incipient backarc basins such as the Sumisu Rift, sediment transport and deposition are affected by volcanic ridges and variable subsidence of rift blocks, and sediment depocenters are largely fault-controlled (Taylor et al., 1991; Klaus et al., 1992). Tectonics also determines water depth, which is especially important in the distribution of pelagic facies. Klein (1985a) observed that in many backarc-basin sequences, an upward (or younging) transition from biogenic pelagic carbonates to pelagic clays occurs, which he attributed to post-spreading, thermal subsidence of the basin floor below the CCD.

Plate reorganizations and changes in subduction mode along the arc (e.g., Chough and Barg, 1987) also influence backarc-basin sedimentation. Arc uplift may isolate continental-margin backarc basins, thus restricting circulation and affecting pelagic sedimentation. In situations such as the Sea of Japan, where the arc is emergent and drained by major rivers, submarine fans may form. Klein (1984, 1985a,b) found a correlation between tectonic uplift rate in intraoceanic arcs and the presence of submarine-fan systems and debris-flow deposits in adjacent backarc basins. According to Klein (1985a), the minimum uplift rate needed to support submarine-fan development is 400 m/my. Turbidite frequency may be controlled by the tectonic history of adjacent source terranes and/or relatively high stands in sea level (Klein, 1984; Betzler et al., 1991). Island-arc volcanism without associated uplift, exposure, and stream development will only produce volcano-proximal subaqueous aprons (or blanket-like deposits) of volcaniclastic sediment, rather than canyon-fed submarine-fan systems (Klein, 1985a).

Peripheral input of clastic sediment to backarc basins may reflect plate-boundary tectonics; for example, during the Pliocene, sediments derived from the southern Japan margin of the Shikoku basin, a strike-slip zone, were deposited in the adjacent Shikoku Basin (White et al., 1980; Marsaglia et al., 1992).

Volcanic Controls on Sedimentation

Sedimentation in backarc basins is complex because the production and composition of volcaniclastic products are highly variable (Fisher, 1984). Volcanic products from arc volcanoes and intrabasinal volcanic centers can range from felsic to mafic. In addition to lava flows, subaerial eruptions may introduce fallout ash and pyroclastic flows into water; submarine eruptions may produce hydroclastic and pyroclastic debris, including pumice (Cashman and Fiske, 1991); pyroclastic, epiclastic and hydroclastic materials may be reworked from land (Fisher, 1984). The types of submarine volcaniclastic deposits include olistostrome (slump and slide), mass-flow, turbidite and suspension-fallout (Fig. 8.12; Fisher, 1984), and these can be composed of pyroclastic and/or epiclastic debris. These materials, if deposited on steep slopes, may be remobilized and mixed by gravity-flow processes, and redeposited in backarc basins, especially in seismically active regions with high sedimentation rates (Fisher, 1984). Ash layers can be produced by submarine and subaerial eruptions, with distribution dependent on prevailing currents in the former and seasonal atmospheric patterns in the latter (Fisher, 1984; Klein, 1985a).

The arc is the major source of volcaniclastic debris carried into the backarc basin. The volume of volcaniclastic material supplied to the backarc basin is primarily a function of arc magmatism and eruption history. Relative intensity of arc volcanism, as measured by input of pyroclastic and epiclastic sediment into the basin, varies throughout backarc-basin history. A universal temporal relationship has yet to be demonstrated for rifting and arc volcanism (Taylor, 1992), in that backarc spreading has been linked to both volcanic minima and maxima (Karig, 1975, 1983b; Scott and Kroenke, 1980; Kroenke et al., 1981; Taylor, 1992). Gill et al. (1992) pointed out that apparent discrepancies in eruption frequency and intensity in ancient versus recent rift systems may reflect finer time resolution in younger rift sequences. Taylor (1992) proposed that many arc segments may exhibit

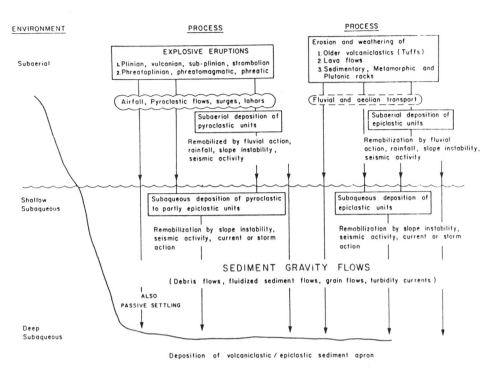

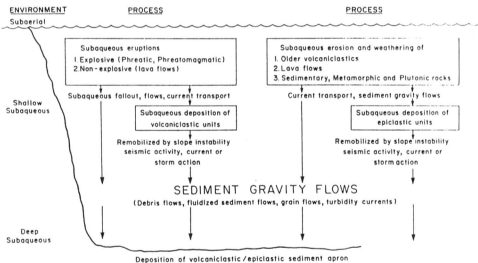

Fig. 8.12 Transport processes and sediment production associated with subaerial (Top) and subaqueous (Bottom) volcanic arcs. From Carey and Sigurdsson (1984).

volcanic cycles in which volcanism is common before and during rifting, reduced during latest rifting and early backarc spreading, and increased during middle and late-stage backarc spreading. Due to diachronous variations from rifting to spreading along an arc, however, volcanic minima can be masked by along-strike transport of volcanic material (Taylor, 1992). In addition, the volcanic history of a given arc segment may be variably preserved across the irregular submarine topography of a backarc basin due to prefer-

ential accumulation of volcaniclastic material in topographic lows (Karig, 1983b). Because of these and other factors, Karig (1983b) suggested that arc volcanic history would be best preserved in the midsection of the volcanic aprons flanking volcanic arcs.

Changes may occur in the composition of arc magmatism during backarc-basin formation, as reflected in backarc-basin stratigraphy. For example, immediately before and during backarc rifting in the Izu-Bonin system, arc volcanism was dominantly silicic

(subalkaline andesite-rhyolite), with minor backarc basaltic activity (Fujioka et al., 1992; Gill et al., 1992; Rodolfo et al., 1992; Taylor, 1992). In the Mariana arc/backarc system, which has evolved into the spreading phase, arc volcanism is dominated by basalt and basaltic andesite, with lesser andesite and minor dacite and rhyolite (Bloomer et al., 1989).

Volcanic influence on sedimentation can be great during both the early-rifting and sea-floor-spreading stages of backarc-basin formation. For example, the initiation of backarc rifting in the Izu-Bonin arc can be correlated with an increase in magma eruption rate in submarine calderas along the Izu-Bonin arc (Gill et al., 1992; Taylor, 1992). Arc calderas at the northern and southern ends of the Sumisu Rift provide the majority of sediment, which is distributed throughout the basin by volcanic eruption and posteruption mass-flow processes (Brown and Taylor, 1988; Nishimura and Murakami, 1988; Klaus et al., 1992b). The early fill of Sumisu Rift is nannofossil-rich clay, silty clay and clayey silt, with thin ash beds and scattered pumice and scoria clasts; this sediment is overlain by units of coarse pumiceous gravel and vitric silt episodically deposited within the basin by mass-flow and ashfall processes (Fig. 8.9; Nishimura et al., 1991, 1992). A dramatic increase in sediment accumulation rates was brought about by cyclic (every 30×10^3 y) rhyolitic eruption of the arc volcanoes (Nishimura et al., 1992). The pumice-producing eruptions were likely submarine, in that fine ash was incorporated into the pumiceous flows and not redistributed by wind (Nishimura et al., 1992). Quiescent periods were marked by fine-grained hemipelagic sedimentation (Nishimura et al., 1992). These sediments were, to a large extent, redistributed by submarine mass-flow processes.

Massive debris-flow units up to 100 m thick can be traced on seismic profiles from their source calderas down valleys into adjacent rift basins, over 20 km^2 across the basin floor (Klaus et al., 1992b). The lack of channelization and continuity of reflectors in seismic lines across the Sumisu Rift indicate basin-wide transport in unconfined mass flows (Klaus et al., 1992b).

The Grenada basin, a backarc basin within the Caribbean Sea behind the Lesser Antilles arc, is a well studied example of the volcanic influence on sedimentation in mature backarc basins. Sigurdsson et al. (1980) studied piston cores of Quaternary sediments in this basin, which has an average depth of 2800 m, is 150 km wide, and contains 4 to 7 km of sediment. Most of the volcanic sources are subaerial, silicic and explosive, producing Plinian pumice falls and pyroclastic flows (including ignimbrites and block-and-ash flows). Minor basaltic to basaltic-andesite volcanic centers produce some lava flows and air-fall deposits. Pyroclastic materials can enter the backarc basin by direct fallout, pyroclastic flow from land into the sea, or onshore erosion and offshore redistribution of loose, subaerial pyroclastic deposits (Sigurdsson et al., 1980). Individual subaqueous pyroclastic flows can be basinwide, forming the most voluminous deposits in the Granada basin (Sigurdsson et al., 1980). These deposits are composed of coarse ash, lapilli and pumice, with the only sedimentary structures being inverse and normal grading. Coarse-grained deposits are preferentially dispersed down the steeper backarc slope of the volcanic arc (Fig. 8.13); on the forearc slope, pyroclastic flows decelerate more quickly, and are reworked by westerly flowing ocean currents. Volcaniclastic turbidites and grain-flow deposits are also common but limited to localized fans proximal to arc sources (Sigurdsson et al., 1980). These coarse-grained facies are interbedded with gray, hemipelagic silty clay. Coring of proximal, nearshore deposits has not been successful because of the unconsolidated nature and coarse grain size; however, uplifted sequences suggest that they are composed of interbedded coarse volcanogenic sand and conglomerate. Submarine fans composed of these sediments extend seaward from the volcanic centers (Sigurdsson et al., 1980). Prevailing upper-level wind distributes air-borne ash to the east, and strong ocean currents transport volcaniclastic sand to the west. Thus, there is an asymmetric distribution of volcaniclastic material, with coarser-grained pyroclastic debris-flow deposits and ash turbidites preferentially accumulating in the backarc basin, and finer-grained ashfall layers preferentially deposited in the forearc basin (Fig. 8.13).

Controls of Rift Location on Sedimentation
Backarc-basin tectonic and depositional history appears to be especially sensitive to the locale of rift

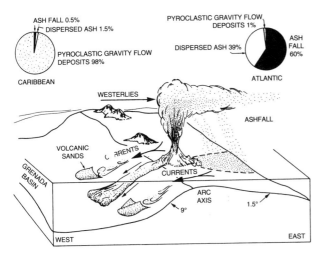

Fig. 8.13 Asymmetric distribution of volcaniclastic material across the Lesser Antilles arc, controlled by bathymetry and marine and atmospheric currents. Fine-grained ash is carried by the tropospheric westerlies to the east, whereas coarse-grained proximal sediments are preferentially distributed to the west. See text for discussion. From Sigurdsson et al. (1980).

inception, behind or in front of the arc axis. In the case of backarc rifting, arc volcanism may persist, continuously supplying volcaniclastic sediment to first the rift basin and then the backarc margin of the backarc basin; in contrast, forearc rifting may be characterized by a gap in volcanism, such as that proposed for the Miocene of the Philippine Sea. For example, at DSDP Site 451 on the West Mariana Ridge, basement consists of pillow-basalt rubble overlain by 930 m of sediment (Scott and Kroenke, 1980); sediment accumulation rates were very high from 11 to 9 Ma, and then decreased from 9 to 5 Ma. Scott and Kroenke (1980) associated the first pulse of volcaniclastic sedimentation with arc volcanism, which waned and then terminated at 5 Ma as spreading began. This sedimentation pattern may be a result of forearc rifting, whereas the sedimentary history recorded within the Sumisu Rift, namely basaltic rubble overlain by a rapidly accumulated pile of sediment, is typical of basins initiated by backarc rifting. The differences in sedimentation rate are apparently tied to the foundering of a volcanic arc during the transfer of active volcanism from the old, now remnant, arc to a "new" arc, assuming that intrabasinal rift volcanism does not supply a large quantity of sediment.

Depositional Models

Several models for backarc sedimentation have been proposed (e.g., Karig and Moore, 1975; Klein, 1975, 1985a; Carey and Sigurdsson, 1984; Busby-Spera, 1988a). Karig and Moore's (1975) model was an effort to help distinguish mid-oceanic from backarc-basin ophiolites. Karig and Moore (1975) reported that backarc basins young progressively away from continental margins in the western Pacific; basins adjacent to continents are strongly influenced by terrigenous sources. They focused on intraoceanic backarc basins isolated from terrigenous influx, with primarily volcanic sedimentary input from the volcanic arc and lesser montmorillonitic clay, biogenic debris, and continentally derived dust. Progressive extension within backarc basins produces an asymmetric distribution of sediment, with general thickening toward the arc and thinning over the spreading center (Fig. 8.14). In this model, sedimentation changes across the basin as it evolves. Initial sedimentary input is primarily volcanic; then as the basin widens, volcaniclastic input is limited to the frontal-arc side of the basin, and hemipelagic and pelagic sediment accumulates across the spreading center and adjacent to the remnant arc. Volcaniclastic input is a function of arc activity.

Klein (1975) also proposed a simple model for backarc-basin sedimentation that differed from Karig and Moore's model in several aspects. Most notably, it assumed basin initiation in oceanic crust and included sediment aprons on either side of the basin. Klein's (1975) basin configuration and sedimentation model are probably more appropriate for continental-margin backarc basins such as the Japan Sea. Continental backarc basins receive sediment from both the passive and active sides of the basins.

These early models were later refined and expanded by Carey and Sigurdsson (1984) and Klein (1985a), who incorporated more details on sediment distribution and facies, based on additional drilling. Carey and Sigurdsson (1984) based their model predominantly on the Karig and Moore (1975) model, but included some aspects of the Klein (1975) model. Carey and Sigurdsson (1984) added details on facies distribution and sedimentary architecture, emphasizing the volcaniclastic apron and subdividing basin

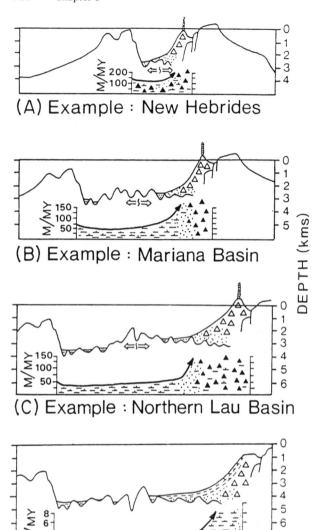

(A) Example : New Hebrides

(B) Example : Mariana Basin

(C) Example : Northern Lau Basin

(D) Example : Parece Vela Basin

Fig. 8.14 Model for backarc-basin evolution, after Karig and Moore (1975). From Carey and Sigurdsson (1984).

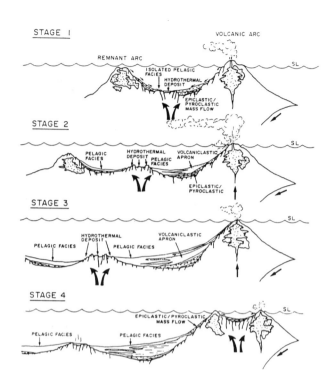

Fig. 8.15 Evolutionary model of backarc-basin sedimentation. During early-rifting stage (1) and backarc-spreading stage (2), there is a high influx of volcaniclastic material. The volcaniclastic apron produced during Stages 1 and 2 is draped by pelagic sediments as the basin matures and volcanism wanes in Stage 3. In Stage 4, the spreading axis becomes inactive and extension within the volcanic arc signals a second phase of backarc-basin formation. From Carey and Sigurdsson (1984).

evolution into four stages (Fig. 8.15). Stage 1 is characterized by a steep-sided rifted basin and the accumulation of gravity-flow volcaniclastic deposits on the basin floor. Stage 2 is characterized by basin widening, subsidence, smoother basin margins, and development of a volcaniclastic apron adjacent to the arc, which grades into pelagic facies deposited across the spreading ridge and the remnant-arc side of the basin. Stage 3 is characterized by cessation of backarc spreading and either continued volcanic input or the accumulation of brown clays. Stage 4 is characterized by a new cycle of rifting and backarc spreading along the arc axis, with pelagic sedimentation in the inactive backarc basin. Klein's (1985a) model (Fig. 8.16) focused on the definition, and spatial and temporal distribution of facies. Klein (1984, 1985a) linked tectonic uplift in source terranes to the distribution of submarine-fan systems and debris flows, and proposed that common upsection transitions from biogenic pelagic carbonates to pelagic clay record thermal subsidence of the basin floor below the CCD. He could not, however, relate the overall diversity of sediment types to the evolutionary history of a given basin because ocean circulation and regional wind patterns dominated. Together, Carey and Sigurdsson's (1984) and Klein's (1985a) models provide a clear picture of intraoceanic-arc sedimentation and evolution in space and time.

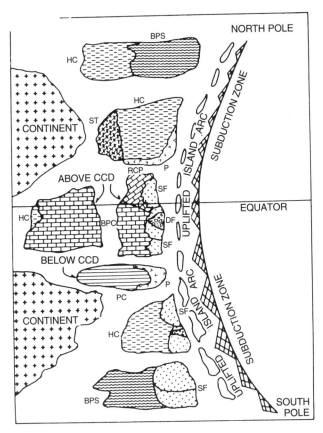

Fig. 8.16 Synthetic model of sediment distribution along the backarc region of a hypothetical subduction zone extending from pole to pole. DB is debris flow; SF is submarine fan; BPC is biogenic pelagic carbonate; RPC is resedimented pelagic carbonate; BPS is biogenic pelagic silica; HC is hemipelagic clay; PC is pelagic clay; ST is silty turbidites; and P is pyroclastic. See text for discussion. After Klein (1985a).

The above models focus on large-scale facies changes across backarc basins. Small-scale sedimentary facies architecture of backarc basins is not tightly constrained, in part because these depositional environments lack the two- to three-dimensional exposures upon which to build such models. Volcanic aprons are especially important because they may be the most diagnostic features of backarc-basin sedimentation (Carey and Sigurdsson, 1984). Arc-proximal facies of modern volcanic aprons have been studied least due to coring difficulties and poor recovery. Carey and Sigurdsson (1984) determined from piston-core studies in the Granada Basin that arc-apron facies are characterized by multiple volcanic sources, massive influxes of sediment, and few systematic facies transitions within gravity-flow deposits.

The most detailed picture of backarc-apron facies architecture is given by Busby-Spera (1987, 1988a), based on exposures of the Middle Jurassic Gran Cañon Formation on Cedros Island, Baja California (Fig. 8.17). The Gran Cañon Formation consists largely of a single, 500 to 1200 m-thick upward coarsening sequence of tuff, lapilli tuff, tuff breccia, and dacite pyroclastic flows (Fig. 8.17B). This is interpreted to record progradation of a deep-marine apron across rifted arc basement onto backarc ophiolitic basement, contemporaneously with the growth and gradual emergence of intraoceanic-arc volcanoes (Fig. 8.17A). Basalt lava flows were fed from fissures that extended down the backarc apron along the frontal-arc side of the backarc basin. Pyroclastic rocks of the backarc apron were blanketed with volcaniclastic sandstone and siltstone; they record abrupt cessation of volcanism and erosion of the arc within 10 my of ophiolite generation, reflecting the temporal episodicity of backarc basins. Busby-Spera's (1988a) model for deposition of the Jurassic Gran Cañon Formation is similar to that proposed by Bloomer et al. (1989, Fig. 8, p. 220) for development of some modern Mariana arc volcanoes along the contact between rifted frontal arc and backarc crust.

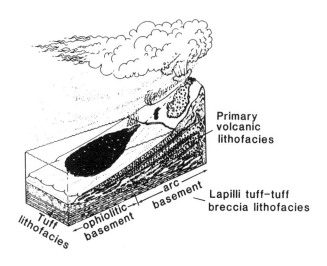

Fig. 8.17A Progradational-backarc-apron model for the Gran Cañon Formation, Cedros Island. From Busby-Spera (1988a).

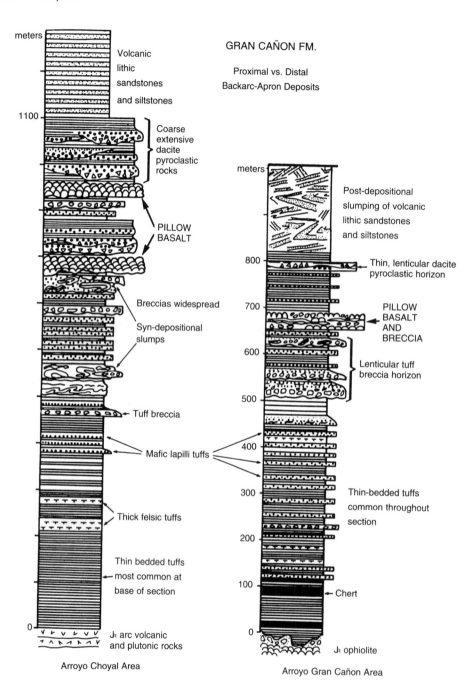

Fig. 8.17B Representative measured sections for the Gran Cañon Formation, Cedros Island. From Busby-Spera (1988a).

Sand Petrofacies and Diagenesis

Sand composition in backarc basins is variable, but dominantly volcaniclastic (see Gergen and Ingersoll, 1986; Packer and Ingersoll, 1986; Marsaglia and Ingersoll, 1992; Marsaglia et al., 1992). In the case of intraoceanic backarc basins, the volcanic arc supplies the majority of coarse-grained sediments throughout the history of the basin, but other sources, such as the proto-remnant arc and intrarift volcanic centers, may contribute during rifting (Marsaglia, 1992). In general, the volcanic component in intraoceanic backarc basins consists of mixtures of basaltic to felsic

volcanic clasts, with most beds showing mixed volcanic provenance (e.g., Marsaglia, 1992). In contrast, a rifted continental backarc basin may receive sediment from the frontal continental arc and the continental margin (Packer and Ingersoll, 1986; Marsaglia et al., 1992). Within continental backarc basins, such as the Sea of Japan, bimodal to polymodal provenance is possible, with volcaniclastic petrofacies characteristic of the arc margin of the basin and plutoniclastic petrofacies characteristic of the rifted continental-margin side of the basin (Boggs and Seyedolali, 1992a; Marsaglia et al., 1992). Similar trends are also evident in geochemical analyses of sand and mud (Maynard et al., 1982; Roser and Korsch, 1986).

Backarc basins formed by entrapment of old oceanic crust and basins produced along transitional plate boundaries such as the Gulf of California (see below) also exhibit polymodal provenance (e.g., Gergen and Ingersoll, 1986; Marsaglia, 1991; Marsaglia and Ingersoll, 1992). In general, the proportion of continental-derived material increases away from the arc across the basin towards the continental margin (Gergen and Ingersoll, 1986; Marsaglia, 1989; Marsaglia and Ingersoll, 1992).

Post-depositional alteration and dissolution of volcanic components is commonplace within backarc basins. Early diagenetic changes documented in the Plio-Pleistocene sediments at intra-arc and forearc sites suggest that dissolution of glassy fragments, ferromagnesian minerals, and calcareous microfossils precedes cementation (Marsaglia and Tazaki, 1992). Cementation, grain alteration, and grain replacement can be significant in Eocene to Miocene backarc-basin sequences, due perhaps, to high geothermal gradients (Lee and Klein, 1986; Marsaglia, 1989; R.B. Smith, 1991; Boggs and Seyedolali, 1992b). The major authigenic phases in these older sequences include zeolites and clay minerals, with lesser carbonate, and in rare instances, quartz.

Hydrothermal Processes

Actively spreading backarc basins are associated with volcanic intrusions, high heat flow and hydrothermal circulation. Mean heat-flow values gradually decrease with basin age (Watanabe et al., 1977).

Hydrothermal flow is probably the dominant mode of heat transfer in backarc basins. Water circulates through overlying sedimentary piles and through underlying, less permeable crust. Very low heat-flow values measured in active basins may delineate zones of recharge, and hot hydrothermal vents, zones of expulsion. Faults within the basin may serve as magma conduits, water recharge zones, and discharge zones for hydrothermal fluids. In the Sumisu Rift, hydrothermal vents are not consistently associated with faults, perhaps due to the permeable nature of the undercompacted, coarse volcaniclastic basin fill (Taylor et al., 1990b). Active hydrothermal circulation within the Sumisu Rift is indicated by high manganese and nickel concentrations within surficial sediments, water-column bacterial biomass anomalies, nonlinear thermal gradients, and high but variable (38 to 700 mW/m^2) heat-flow values (Mita et al., 1988; Nishimura and Murakami, 1988; Nishimura et al., 1988; Yamazaki, 1988; Nakao et al., 1990; Taylor et al., 1990b). Heat flow within the southern Okinawa Trough is also highly variable (30 to 180 mW/m^2), with extremely high heat flow (600 to 700 mW/m^2) along the axis of the trough (Kinoshita et al., 1991). The thermal history of the Japan Sea is partially reflected in the diagenetic state of siliceous sedimentary sequences within the basin; oxygen-isotopic data from pore waters and authigenic phases in Leg 127 cores indicate higher geothermal gradients immediately following rifting (Pisciotto et al., 1992).

Economic Deposits

Oil and Gas

Backarc basins contain some hydrocarbon potential because high heat flow and steep geothermal gradients during rifting and sea-floor spreading should enhance oil and gas generation (Schlanger and Combs, 1975). Other determinants of hydrocarbon potential include the presence of reservoir sand (turbidites) and the organic content of source rocks, which is dependent on the paleoproductivity of the marginal sea and the input of terrestrial plant matter from rivers draining surrounding land masses (Schlanger and Combs, 1975). Structural or stratigraphic trapping mechanisms are also needed.

The eastern side of the Japan Sea is one location where all these requirements are met. Economic accumulations of oil and gas have been found within Neogene onshore basins on Honshu within the "Green Tuff" region (Kikuchi et al., 1991). These are backarc basins formed during the opening of the Japan Sea, which now have been folded and faulted by compressional stresses related to incipient subduction along that margin (Kikuchi et al., 1991). The source rocks are likely middle Miocene shales, whereas the main reservoirs are Neogene sandstone turbidites, pyroclastic rocks and volcanic rocks (Kikuchi et al., 1991). The stratigraphic sequences in these basins are thicker than, but similar to, those cored at ODP sites within the Japan Sea (Ingle et al., 1990).

Minerals

Backarc basins are characterized by "polymetallic" or "kuroko"-type massive-base-metal (Pb-Zn-Cu: AgAu) deposits (Hutchison, 1980). These deposits are found in Lower Proterozoic and Phanerozoic successions, such as the Ordovician of the central Norwegian Caledonides (Roberts et al., 1984), and most notably, the Miocene Green Tuff region of Japan (Hutchison, 1980). The Kuroko massive-sulfide ore deposits in the "Green Tuff" region of eastern Japan have been well studied because they are young, relatively undeformed, and only weakly metamorphosed (Scott, 1980). The Hokuroku ore field, the most important Kuroko deposit in Japan, contains an estimated 140 million tons of ore, with an average grade of 1.6% copper, 3.0% zinc, and 0.8% lead (Tanimura et al., 1983). The Kuroko deposits formed within a backarc or intra-arc rift basin characterized first by intermediate and then by bimodal volcanism. The mineralizing hydrothermal fluids are thought to have been associated with aluminous felsic magma bodies (Urabe and Marumo, 1991). The Huroko region was a rapidly subsiding sedimentary basin or caldera filled by volcanic debris (Scott, 1980). Modern equivalents of these deposits have been described recently in the Okinawa Trough (Halbach et al., 1989; Sakai et al., 1990) and the Sumisu Rift region of the Izu-Bonin arc (Urabe and Kusakabe, 1990), where hydrothermal vents are found within these active rift systems.

Nonextensional and Hybrid Backarc Basins

Nonextensional intraoceanic backarc basins may form by entrapment of pre-existing ocean basins during plate reorganizations, which cause sudden shifts of subduction zones (Ben-Avraham and Uyeda, 1983; Tamaki and Honza, 1991). These shifts may occur when buoyant oceanic plateaus collide with subduction zones, or when subduction is initiated along a transform fault (Ben-Avraham and Uyeda, 1983). This group of basins includes the Aleutian Basin in the Bering Sea, part of the Caribbean Sea, the West Philippine Basin, and perhaps, part of the Okhotsk Sea north of the Kuril Basin (Fig. 8.1; Uyeda and Ben-Avraham, 1972; Scholl et al., 1975a; Ben-Avraham and Uyeda, 1983). In addition to old oceanic crust, these basins contain oceanic plateaus consisting of detached and submerged continental fragments or thick piles of oceanic basalt (Ben-Avraham and Uyeda, 1983). More recent studies indicate that the tectonic history of the Bering Sea is more complex than previously thought. In addition to entrapment of an oceanic plate, the evolution of this basin may have involved localized backarc spreading along the Vitus Arch and within the Bowers Basin (Cooper et al., 1992).

Backarc spreading may also be initialized by interaction with a spreading ridge along a continental margin. These "hybrid" backarc basins are produced along transitional plate boundaries. One example of this type of backarc basin is the Andaman Sea (Fig. 8.1), where oblique extensional rifting has resulted from ridge subduction (aseismic Ninetyeast Ridge) and oblique subduction associated with a nearby continental collision (Eguchi et al., 1979, 1980). Ridge subduction was also proposed for formation of the Japan Sea (Uyeda and Miyashiro, 1974) and the Gulf of California (Uyeda, 1977). The Japan Sea is considered by other workers to have formed as a pull-apart basin in response to plate reorganization and strike-slip movement (Lallemand and Jolivet, 1985; Tamaki, 1988; Jolivet and Tamaki, 1992; Tamaki et al., 1992), perhaps in response to microplate reorganization after India-Eurasia collisional events (Kimura and Tamaki, 1986).

Sedimentary sequences within these basins are complex; their sediment distribution and character are most similar to continental backarc basins. Clastic sediment sources include plateaus, plate margins, volcanic arc systems, and suture zones.

Ancient Examples

Ancient ophiolite-related sequences are often differentiated by means of their basalt geochemistry (e.g., Swinden et al., 1990), rather than their sedimentary facies. Many ophiolitic sequences, including the Bay of Islands complex in Newfoundland (Suen et al., 1979) and the Trinity Ophiolite in California (Brouxel and Lapierre, 1988), have been interpreted as obducted floors of backarc basins, rather than mid-oceanic crust (Wilson, 1989). Although both mid-ocean ridges and backarc basins exhibit variable basalt types, the latter tend to be less depleted in incompatible elements (Tarney et al., 1981). In addition, backarc basalts tend to be more vesicular, and basaltic glasses from backarc basins contain higher water contents and H_2O/CO_2 ratios than do MORB (Tarney et al., 1981). These compositional differences can be attributed to the addition of a subducted-slab component in backarc basins (Tarney et al., 1981).

Differentiating backarc basins and intra-arc basins is not straightforward (see Chapter 7). Cas and Wright (1987) suggested the following criteria for recognition of backarc basins: 1) ophiolitic basement; 2) volcaniclastic component in overlying pelagic sedimentary section; and 3) association with a contemporaneous arc and forearc succession. These criteria also characterize intra-arc basins. In addition, they proposed that ancient arc systems must be on a comparable scale with modern systems, possibly extending onland to equivalents floored by continental crust (e.g., the Havre intraoceanic backarc basin extends into the ensialic Taupo backarc basin; Cole, 1984; Lewis and Pantin, 1984) (see Figure 7.16 for alternative interpretation).

Backarc basins may be partially or wholly consumed by subduction after plate reorganizations; for example, an incipient subduction zone has formed in the eastern Japan Sea (Ingle et al., 1990; Tamaki et al., 1990, 1992), and the northern Philippine Sea basin is presently being subducted along the Nankai

Trough (Tamaki and Honza, 1991). Ancient examples of backarc basins are difficult to recognize in that they usually are preserved as tectonostratigraphic units (Leitch, 1984) in highly deformed suture belts (e.g., Zhijin, 1984; Floyd et al., 1992). In these tectonically telescoped suture zones, ophiolitic basement, volcaniclastic pelagic units, and arc and forearc successions (criteria outlined by Cas and Wright, 1987) may be juxtaposed or translated. Paleogeography can be palinspastically reconstructed by delineating arc terranes and adjacent backarc-basin facies (e.g., Guoqiang, 1984; Ingersoll and Schweickert, 1986).

The origin and mode of emplacement of a series of Jurassic ophiolites (Josephine, Smartville, and Coast Range) within the collage of terranes of southern Oregon and northern California have been tied to backarc-basin evolution (Schweickert and Cowan, 1975; Harper and Wright, 1984; Ingersoll and Schweickert, 1986). A backarc rather than an open-ocean setting for these rocks has been inferred from their close association with volcanic arc sequences (e.g., Schweickert and Cowan, 1975). Ingersoll and Schweickert (1986) combined and expanded previous models (i.e., Schweickert and Cowan, 1975; Harper and Wright, 1984) into a model for the Middle to Late Jurassic evolution of northern California (Fig. 8.18). This integrated model proposes that the Jurassic ophiolites of northern California formed during two phases of intra-arc to backarc spreading along an east-facing intraoceanic arc (Coast Range and Smartville ophiolites), and continental-margin backarc spreading (Josephine ophiolite) north of the collision zone between this intraoceanic arc and the west-facing continental-margin arc.

One of the best-documented and better-studied ancient examples of a continental-margin backarc basin occurs in southernmost Chile (Figs. 8.19 to 8.21). During the Late Jurassic to Early Cretaceous, a backarc basin formed behind the Andean arc (now represented by the Patagonian batholith), following a rifting phase associated with silicic volcanism. This basin then closed during the middle Cretaceous (Dalziel et al., 1974; Dalziel, 1981; De Wit and Stern, 1981; Tarney et al., 1981). The distribution and mode of occurrence of ophiolites throughout the region suggest that, from north to south, the basin was

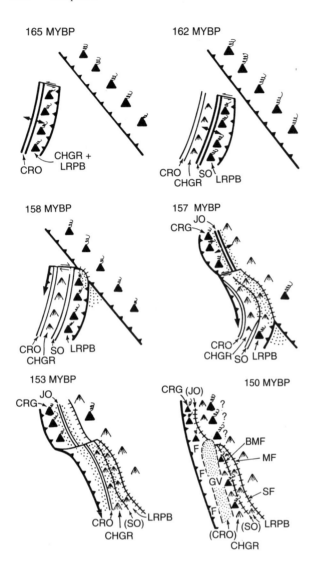

Fig. 8.18 Model of Middle to Late Jurassic tectonic evolution of northern California. Ophiolites generated in intraoceanic backarc basins (SO, Smartville ophiolite, and CRO, Coast Range ophiolite) are sequentially sutured to the North American continent during the Sierran phase of the Nevadan orogeny, and ophiolite generated in continental backarc basin (JO, Josephine ophiolite) later closed during the Klamath phase of the Nevadan orogeny. Ophiolite symbols in parentheses indicate partial preservation within fault zones. Symbols: volcanoes with smoke, active magmatic arcs; volcanoes without smoke, inactive magmatic arcs; barbed symbols, active subduction zones; suture pattern, suture zones; hachured line, rifted continental margin; double lines with arrows, active spreading centers; double lines without arrows, inactive spreading centers; thin arrows, transform faults; large arrow, southward propagating trench; stippled pattern, sites of deposition of Mariposa and Galice formations, and the Great Valley (GV) forearc basin. Other abbreviations: CRG, Chetco, Rogue, Galice arc complex; F, Franciscan Complex; LRPB, Logtown Ridge/Peñon Blanco arc complexes; CHGR, Cooper Hill/Gopher Ridge arc complex; BMF, Bear Mountain fault; MF, Melones fault; SF, Sonoran fault. From Ingersoll and Schweickert (1986).

The transitional continental/intraoceanic backarc basin represented by the modern Havre Trough may have an ancient counterpart in the Siluro-Devonian Hill End Trough of New South Wales, Australia (Cas and Jones, 1979). Cas and Jones (1979) proposed that this basin was an interarc basin, which merged southward into a continental block. They based their interpretation on the distribution and composition of volcanic sequences, basin physiography, the likely continental/ transitional basement of the basin, the presence of shallow-water carbonate and volcaniclastic detritus on flanking highs, the wedge-shaped

floored by transitional continental crust to oceanic crust assoicated with sea-floor-spreading (Fig. 8.20; De Wit and Stern, 1981). The Lower Cretaceous Yahgan Formation consists of a volcaniclastic, arc-derived turbidite sequence over 3000 m thick (Fig. 8.21; Katz and Watters, 1965; Winn, 1978; Winn and Dott, 1978; Andrews-Speed, 1980). Farther south, along the northern Antarctic Peninsula, the Cretaceous submarine-fan and slope-apron deposits of west James Ross Island are interpreted to record arc-proximal backarc deposition in a fault-bounded basin formed by oblique extension of a continental-margin arc (Farquharson et al., 1984; Ineson, 1989).

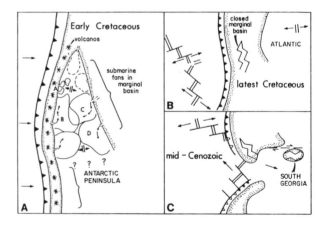

Fig. 8.19 (A) Schematic reconstruction of the Early Cretaceous backarc basin in southernmost South America, showing bilateral infilling by submarine fans. The basin was later closed (B) and partially dismembered (C). From Winn (1978).

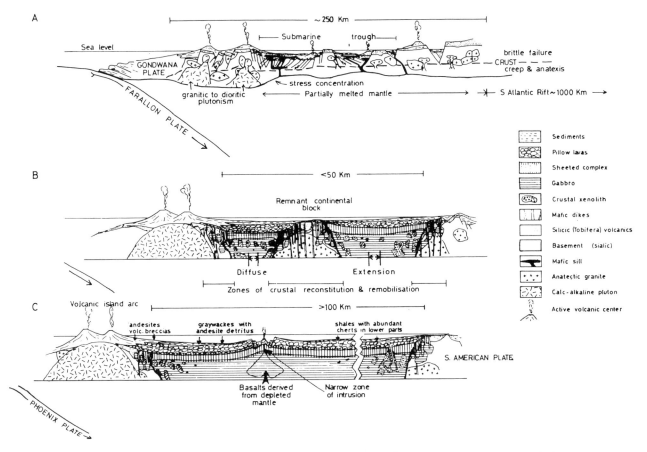

Fig. 8.20 North (A) to south (C) schematic cross sections through marginal basin pictured in Figure 8.19A. These sections represent the evolutionary stages of Mesozoic backarc-basin formation, showing the transition from continental (A) to oceanic (C) backarc crust. From De Wit and Stern (1981).

asymmetrical basin fill, metamorphic evidence for high heat flow along the basin axis, overall facies distribution, and the composition and dispersal pattern of sediments within the basin.

Interpretive Summary

Extensional backarc basins form as a result of crustal extension and rifting leading to sea-floor spreading; this process has been attributed to several mechanisms, including slab rollback, mantle convection and diapirism. These basins may develop on oceanic, continental, or transitional crust, commonly with initiation along or near active magmatic arcs. Using the Sumisu Rift as a model, early rifting may be characterized by normal faulting, high sedimentation rates, bimodal volcanism and hydrothermal circulation. In mature intraoceanic backarc basins, sea-floor spreading produces basins flanked by remnant arcs and magmatic arcs. During this stage, mean heat flow and basement elevation gradually decrease with time, and primary sources of sediment are magmatic arcs.

Continental-margin backarc basins also may form by crustal extension and rifting, followed by sea-floor spreading. The Okinawa Trough, an incipient continental-margin backarc basin, is characterized by normal faulting and thinned continental crust. It is bordered by a rifted continental margin and an active magmatic arc. The Japan Sea, a presently inactive, mature continental backarc basin, is floored by oceanic crust (formed by backarc spreading), sediment-sill complexes (incomplete/dispersed rifting), and continental blocks (isolated during rifting). Unlike intraoceanic backarc basins, continental-margin

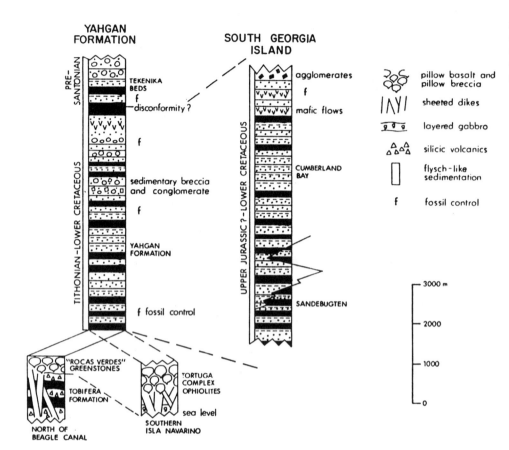

Fig. 8.21 Mesozoic stratigraphy of Tierra del Fuego and possible equivalent sections on South Georgia Island. From Winn (1978).

backarc basins may initially be subaerial and later marine, and receive sediments from both the passive (continental) and active (magmatic-arc) margins of the basin.

Depositional models for backarc basins need to address: 1) the tectonic controls on sedimentation, such as faulting and subsidence, which influence depocenters and water depth; 2) the volcanic controls on sedimentation, such as the distribution, composition, and eruption rates of arc-axis and intrabasinal volcanoes; and 3) other controls on sedimentation, such as ocean currents, productivity, and terrigenous input, as influenced by eustatic sea level. Models proposed by Klein (1985a) and Carey and Sigurdsson (1984) adequately cover these points, and are generally applicable to both extensional and trapped backarc basins.

Recent ODP drilling has provided a more detailed picture of early rifting of intraoceanic backarc basins (Leg 126—Sumisu Rift), and has underscored the importance of sill dynamics (i.e., barriers to circulation) in controlling sedimentation in continental-margin backarc basins (Legs 127 and 128 - Japan Sea). Future drilling efforts and more detailed studies of ancient equivalents will likely produce refinements of the backarc tectonic and sedimentological models presented here.

Acknowledgments

I thank Cathy Busby, Raymond Ingersoll and George Klein for their thorough reviews of the text.

Further Reading

Brooks DA, 6 coauthors, 1984, *Characteristics of backarc regions*: Tectonophysics, v. 102, p. 1-16.

Carey SN, Sigurdsson H, 1984, *A model of volcanogenic sedimentation in marginal basins*: Geological Society of London Special Publication 16, p. 37-58.

Karig DE, 1972, *Remnant arcs*: Geological Society of America Bulletin, v. 83, p. 1057-1068.

Klaus A, 6 coauthors, 1992b, *Structural and stratigraphic evolution of Sumisu Rift, Izu-Bonin Arc*: Proceedings of the Ocean Drilling Program, Scientific Results, v. 126, p. 555-574.

Klein GdeV, 1985a, *The control of depositional depth, tectonic uplift, and volcanism on sedimentation processes in the back-arc basins of the western Pacific Ocean*: Journal of Geology, v. 93, p. 1-25.

Taylor B (ed), in press, *Back-arc basins*: tectonics and magmatism: Plenum, New York.

Taylor B, Karner GD, 1983, *On the evolution of marginal basins*: Reviews of Geophysics and Space Physics, v. 21, p. 1727-1741.

Taylor B, Natland J (eds), in press, *Active margins and marginal basins: western Pacific drilling results*: American Geophysical Union Geophysical Monograph.

Retroarc Foreland and Related Basins **9**

Teresa E. Jordan

General Characteristics and Behavior of Foreland Basins

Retroarc foreland basins form along the continental-interior flanks of continental-margin orogenic belts. Commonly, subduction of an oceanic plate produces a volcanoplutonic arc, behind which may lie a "retroarc" foreland basin (Dickinson, 1974b) (even if the volcanoplutonic arc is inactive for millions of years, the foreland basin, if present, is referred to as "retroarc") (see Chapters 1 and 8 for discussion of backarcs). Collisional orogenic belts produce "peripheral" foreland basins, which have somewhat different facies and subsidence histories (see Chapter 11). Structural thickening of the upper crust drives tectonic subsidence in foreland basins. Retroarc foreland basins are large-scale, long-lived features (100s of kilometers wide, 1000s of kilometers long, with many kilometers of strata, and 10 to 100 million years in duration). The two best known examples of retroarc foreland-basin systems formed during mountain building of Mesozoic to early Cenozoic age in the North American Cordillera, and during Jurassic to Recent growth of the Andes Mountains in South America. Case studies used in this chapter are drawn from those areas. There are fewer examples of modern retroarc foreland thrust belts and basins than of peripheral ones, but nevertheless, they are large-scale, long-lived features.

The majority of foreland-basin development is related to thin-skinned thrust belts (i.e., where sedimentary cover rocks are shortened by folding and thrusting above undeformed basement). Less commonly, upper-crustal shortening that triggers subsidence occurs in zones of thick-skinned basement uplifts (i.e., crystalline basement rocks translate along reverse faults). The thrust belt typically is wedge-shaped in cross section, thickest adjacent to the core of the orogenic system, and tapering toward the continental interior (Fig. 9.1). The foreland basin is also wedge-shaped in cross section, being thickest on the margin that is most proximal to the thrust belt. Where basement uplifts drive subsidence, a suite of smaller, partially separated basins forms, referred to as "broken foreland basins".

The necessary conditions for thrust belts to form and widen are the existence of topographic highs and horizontal shortening (Davis et al., 1983; Molnar and Lyon-Caen, 1988). Horizontal compression may be generated either by coupling at the contact between subducting and overriding plates (e.g., Kanamori, 1986) or by traction at the base of the lithosphere due to circulation of mantle above the subducting plate (e.g., Wdowinski et al., 1989). However, those conditions are not sufficient to form a thrust belt: a thin-skinned thrust belt can only form where there is a pre-existing, thick pile of layered strata (e.g., Allmendinger et al., 1983a).

The spectacular examples of basement uplifts and broken foreland basins discussed in this chapter are known or inferred to have developed while they were underlain by subducted plates with very low angles of descent into the mantle ("flat-subduction") (Figs. 9.1, 9.2) (e.g., Jordan et al., 1983b; Cross, 1986). However, some regions with basement uplifts are not underlain by flat subduction (Allmendinger et al., 1983a). A cause-and-effect relationship between angle of subduction and style of foreland shortening (thin-skinned versus thick-skinned) is uncertain (e.g., Molnar and Lyon-Caen, 1988). What controls the subduction angle? The angle of descent of a subducted plate

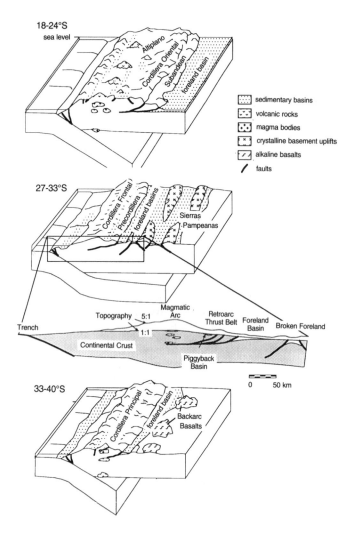

Fig. 9.1 Schematic block diagrams of lithosphere, and an accurately scaled cross section of crust, for a convergent continental margin, showing three different foreland configurations. These three cases coexist along the western (Andean) margin of South America (latitudes to which they correspond are indicated [see Fig. 9.14]). The block diagrams show that the subducting oceanic lithosphere descends beneath the continental lithosphere at variable angles. In South America, along-strike variability in the structure of the Andes generates along-strike variations in the geometry of foreland basins. In the northern area (18-24°S), a simple thin-skinned thrust belt borders an extensive, asymmetric, foreland basin. In the central area with flat subduction (27-33°S), foreland basins flank the thrust belt and occur among reverse-fault-bounded basement uplifts. In the southern area (33-40°S), thrust systems of the Andes are paired with backarc alkaline basalts on a slightly extensional or neutral plateau (Muñoz and Stern, 1989). For cross section of the crust in the 27-33°S segment, there is no vertical exaggeration, except that the upper topographic profile is portrayed both at 1:1 scale, and vertically exaggerated to 5:1 scale, which better portrays the character of the earth's surface. Block diagrams modified from Jordan et al. (1983a); cross section adapted from Allmendinger et al. (1990).

is influenced by three factors: regional mantle circulation, subducted plate buoyancy, and local shape of the margin of the overriding plate. The angle of descent is shallower for cases of high convergence rate and "absolute" trenchward overriding by the upper plate. A subducting plate may be relatively buoyant, and thus, subduct at a shallow trajectory, if it is either young or contains bathymetric features such as aseismic ridges or seamount chains (e.g., Pilger, 1981; Cross and Pilger, 1982; Cloos, 1993). The map-view shape of the margin of the overriding plate is important because it forces the descending oceanic lithosphere, which had the shape of a thin, spherical shell before entering the trench, to bend in three dimensions. Cahill and Isacks (1992) demonstrate that seaward convex parts of a continental margin can produce lessening in the angle of descent downdip of the trench, whereas seaward concave segments of a margin promote steepening. As a result, flat-subduction segments that alternate with steep-subduction segments are one of the stable configurations that accommodate the three dimensional shape of a lithospheric slab descending around the complex shape of a continental margin (Cahill and Isacks, 1992). The shape factor means that the pre-existing tectonic history of an orogenic belt and its plate margin play an active role in plate interactions. Furthermore, if in the course of an orogeny, the map-view shape of the continental margin changes, the dips of the subducted plate may change (Isacks, 1988).

Because the theory of basin formation is relatively well established, foreland basins are used to investigate the deformational history and rheological properties of the continental lithosphere. The fact that foreland basins typically contain strata deposited in nonmarine to shallow-water depositional environments provides excellent opportunities to combine geomorphological, sedimentological and mechanical studies of these basins.

Retroarc and peripheral foreland basins combined constitute the principal global occurrence of nonmarine strata, and contain large volumes of marine strata. Strata accumulate at rates ranging from 0.1 to 1 m/1000 years. Foreland basins contain large resources of oil, gas and coal.

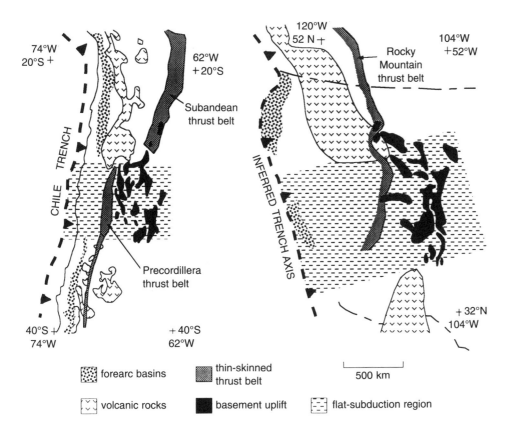

Fig. 9.2 Comparison of the late Cenozoic Andean orogen to earliest Eocene Cordillera of western North America, at the same scale. The region underlain by a plate with nearly horizontal subduction is indicated. North American example (after Dickinson, 1979b) includes palinspastic restoration of zones of Cenozoic extension and strike-slip displacement. Volcanic units shown for South America are largely younger than 10 Ma. The two systems are of quite similar scale and tectonic style, although the North American example is broader than is its South American counterpart, possibly reflecting insufficient palinspastic restoration (e.g., Wernicke et al., 1988). (Modified from Jordan et al., 1983b; Jordan and Allmendinger, 1986.)

:::: legend
- ⣿ forearc basins
- ⌄⌄ volcanic rocks
- ▓ thin-skinned thrust belt
- ■ basement uplift
- ⌶ flat-subduction region

500 km
::::

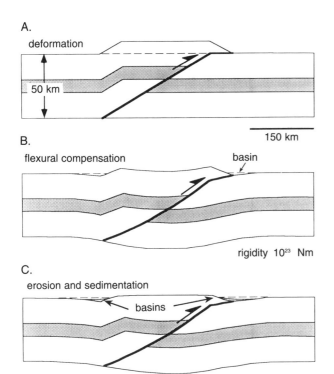

A.
deformation
50 km
150 km

B.
flexural compensation   basin
rigidity 10²³ Nm

C.
erosion and sedimentation
basins

Fig. 9.3 Three conceptual steps in formation of sedimentary basins flanking a thrust belt (vertical exaggeration 3×). A) Shortening causes thickening above a ramp, which acts as a load. B) Regional compensation of the load deflects lithosphere (note smooth arc at base of profile), producing basins at flanks of mountain range. C) Surface processes strip material from the mountain range and transfer it to the tectonic basins. Flexural compensation for erosion and deposition causes the basins in C to be wider than in B. Simplified model treats thickening in an orogenic belt as the result of shortening along one master fault that penetrates the crust; middle patterned layer is a marker horizon. Dashed line shows initial elevation of upper surface of crust (after Flemings and Jordan, 1990).

Physical Principles of Subsidence and Fill of Retroarc Foreland Basins

Foreland basins, whether retroarc or peripheral, largely result from deformation of a foreland fold-thrust belt. In thrust belts, horizontal shortening produces crustal thickening (Fig. 9.3A). This lateral transfer of upper-crustal mass triggers a series of responses. The first response is isostatic adjustment to the mass transfer, which produces high topography within the thickened region and a down-flexed "moat" adjacent to this region (Fig. 9.3B). This moat is the foreland basin. The second response is an additional transfer of mass, by simultaneous erosion of the topographic highs and sedimentary filling of the depression (Fig. 9.3C).

A genetic link between foreland thrust systems and basin subsidence was described first by J. Barrell, who realized that the thick nonmarine strata of the Gangetic plains accumulated in space made available by subsidence of the Indian crust beneath the mass of thrust plates of the Himalayan Range (Barrell, 1917, p. 787). Considerable time passed before Price (1973) stimulated renewed interest in the concept of thrust-load-driven subsidence and Beaumont (1981) and Jordan (1981) independently established a quantitative relationship between the mass of strata in foreland basins and the flexural compensation of thrust-plate mass. Below, we examine local isostatic compensation and flexure.

Local isostatic equilibrium governs vertical movements; the summed masses of any column of the earth above an equipotential surface in the asthenosphere must equal that of all other columns. The equipotential surface exists in the asthenosphere because the asthenosphere has low enough viscosity that it flows. Although differences may exist between two localities in the mass of a layer that overlies the asthenosphere, those mass differences must be balanced in another layer. In other words, if two locations are compared, the sum for all the layers of the individual mass differences must be zero (Suppe, 1985):

$$\sum_{0}^{i} \Delta (\rho h)_i = 0 \qquad (1)$$

$0 \rightarrow i$ are levels of differing densities
Δ indicates the difference between columns
ρ = density
h = thickness of each level

The direct application to stratigraphy of the principle of isostatic equilibrium is that differences in thicknesses of layers in a column produce differences in elevations. In a comparison between two columns,

$$\sum_{0}^{i} \Delta h_i = \Delta \text{ elevation.} \qquad (2)$$

In the same fashion, one can compare the masses and elevations of a single location at two times, such as before and after a tectonic event that affects the thickness or density of layers above the asthenosphere. If crustal mass moves from one column to another, as it does during thrusting, then there is a tendency for asthenospheric material to flow to compensate for the crustal changes. If the crust of density (ρ_c) thickens by an amount X, then its mass is compensated if a layer of high-density asthenosphere (ρ_m) that is of thickness $(\rho_c/\rho_m)(X)$ flows out from beneath the thrust belt. The thrusting and compensation generate a thicker column above the compensation level, and therefore, a topographic high.

This description of isostatic compensation is strictly true only for a weak lithosphere, in which each column behaves independently of its neighbors (i.e., *local* compensation). In fact, typical continental lithosphere has lateral strength, and individual columns do not bob up and down independently of one another. Rather, isostatic disequilibrium caused by shortening is compensated over a broad region (i.e., *regional* or *flexural* compensation), rather than just beneath the thrust belt. Integrated over a large region, the masses above the equipotential surface sum to the equilibrium quantity, but locally, some regions have greater mass and others have less mass. This response of the lithosphere approximates the bending of an elastic beam under a vertical load (Fig. 9.4A). Horizontal variations in the amount of tectonic subsidence within the basin can be visualized by considering an elastic beam which rests above a resisting fluid. The equation in two dimensions for the deflection, z, of an originally horizontal surface is:

$$T\rho_T g = D\frac{d^4z}{dx^4} + (\rho_A - \rho_i + \rho_T)gz, \text{ where} \qquad (3)$$

ρ_A and ρ_i = densities below and above plate (above the plate the downflexed area might be infilled with water or air or sediment; below is asthenosphere)
ρ_T = density of rock thickened by shortening
T = increased thickness of crust caused by shortening
X = horizontal position
g = gravity = 9.81 m/sec^2
D = flexural rigidity, where

$$D = \frac{Eh^3}{12(1-v^2)}$$

  h = effective elastic thickness (meters)
  E = Young's modulus ($\sim 10^{11}$ kg/m-s^2)
  v = Poison's ratio (0.25)

The solution to equation 3 describes a profile in which subsidence is greatest beneath the load (Fig. 9.4) and diminishes outward; there is an upward deflection at a distance from the load (called the peripheral bulge or forebulge). Beyond the peripheral bulge are subsiding and uplifting zones of outward-diminishing amplitude. The wavelength of the tectonic basin is a function of the strength of the lithosphere, which is expressed as the flexural rigidity. A beam of greater flexural rigidity (approximating a cool, thick lithosphere) distributes the isostatic compensation over a broader distance; hence, it causes less local subsidence but a broader basin (Fig. 9.4B). Because of this dependence on strength of the lithosphere, even if all thrust belts were identical, not all foreland basins would have equal profiles of tectonic subsidence. This is because the flexural rigidity of the lithosphere varies depending upon the geologic history of the continent, which determines the specific material properties of the continent and its thermal state (Karner and Watts, 1983). Furthermore, if the lithosphere is especially rigid, subsidence in the foreland basin may, in part, be driven by thickening in the interior part of the mountain belt (the hinterland).

The physics of isostatic equilibration and the modification by flexure apply equally to the sedimentary load that accumulates in the tectonically initiated basin. Erosion that transfers mass from the thrust belt to the basin causes uplift in the mountains and subsidence in the basin (Fig. 9.3C). This sediment-load subsidence amplifies and modifies the tectonically driven subsidence.

Elastic materials deform instantaneously in response to applied stress, and then maintain the

deformation until the stress is changed. However, lithospheric flexure is not strictly instantaneous because the asthenosphere is viscous and a finite amount of time is needed for isostatic compensation to be achieved by asthenospheric flow. Nevertheless, in only 10,000 years since melting of the large Pleistocene ice sheets, 90% of the isostatic readjustment is complete (Turcotte and Schubert, 1982). Therefore, if thrust thickening and mass transfer by erosion are incremental (i.e., much slower than deglaciation), then foreland subsidence keeps pace with shifting of masses without any perceptible lag time.

In detail, behavior of the lithosphere is not well understood. Consequently, some questions remain about subsidence histories of foreland basins. Beaumont and coworkers have investigated the

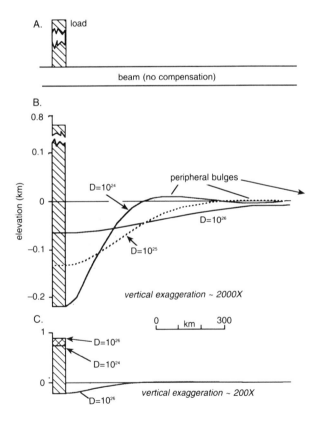

Fig. 9.4 Cross section of the deformation of an elastic beam (after Beaumont, 1981). A) Configuration of load and beam, in situation with no isostatic compensation. B) Regional compensation of a 50 km-wide by 1 km-thick load, showing that the deflected profile of the basin depends on the flexural rigidity, D, of the plate (units of D are Newton-meters). C) Same deformation as in B, but with less vertical exaggeration, showing that the elevation of the mountain formed from the load is much greater than the depth of subsidence of the adjacent basin.

consequences of progressive weakening of the lithosphere over long time periods, a phenomenon commonly referred to as viscoelasticity (e.g., Beaumont, 1981; Quinlan and Beaumont, 1984). Such weakening is postulated because, on a microscale, crystals at high temperatures are known to relieve stress by creep processes (e.g., dislocation creep, diffusion creep). The macroscopic consequence of these processes is that lithosphere exhibits a finite "effective viscosity"; in other words, bending stresses associated with flexure are relaxed as if by viscous flow of the lithosphere. As weakening occurs, a load that was originally supported by regional flexure increasingly is supported by local flexure (i.e., the flexural rigidity diminishes through time) (Fig. 9.5). Effective viscosity governs the time constant of relaxation. If the relaxation time is very long (100s of millions of years), then it may not be an important factor in basin evolution. However, if the relaxation time is in the range of 1 to 10s of millions of years, then it would have a distinctive impact on basin history. Quinlan and Beaumont (1984) suggested that, if weakening occurs on a time scale of a few million years, then a foreland basin will have a stratigraphic record of first broadening (due to elastic flexure) and then narrowing through time (due to viscous relaxation). This change in basin wavelength would be due to the properties of the lithosphere, rather than due to thrusting, erosion, or sediment loading. Despite the theoretical justification, it remains unclear whether changes in flexural response occur at a sufficiently large vertical scale at short enough time scales to be a factor in basin dynamics (e.g., Turcotte, 1979; Karner et al., 1983). Most stratigraphic tests designed thus far (e.g., Beaumont, 1981; Quinlan and Beaumont, 1984) considered flexural response only to tectonic loads, whereas erosion and sedimentation are dynamic processes that also govern loads and that operate on the time scale of viscoelastic behavior (Flemings and Jordan, 1989, 1990). Until more comprehensive tests are completed, it remains unclear whether or not viscoelasticity is a significant factor in producing foreland-basin stratigraphy.

The models of flexure are also unrealistic because they are designed for homogeneous materials, whereas

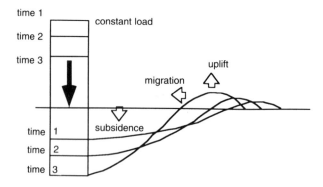

Fig. 9.5 Cross section of a viscoelastic beam, adjusting to a constant load over a long period of time. The initial response (time 1) is regional flexure, as in Fig. 9.4, but the basin narrows progressively through time (after Quinlan and Beaumont, 1984).

lithosphere and crust have laterally variable compositions and inherited zones of weakness. Therefore, true subsidence and uplift patterns in and adjacent to basins differ from the models. For example, Waschbusch and Royden (1992a) have shown, through mechanical analysis of plate flexure, that old weaknesses in the lithosphere may cause forebulges to remain fixed in location even while the loads shift in position, and to have larger amplitudes than illustrated in Fig. 9.4.

In view of the hypothesized mechanical relationships of thrust systems to foreland basins, the subsidence history of a foreland basin should reflect the history of thrust loading. Lateral variations in a thrust belt should produce lateral variations of subsidence of the basin, and episodicity or cyclicity in thrust movements should be expressed as variations in the tempo of subsidence. A major tendency of thrust systems is that the edge of the deformed belt advances toward the foreland through time (Armstrong and Oriel, 1965; Dahlstrom, 1970). In turn, early-deposited proximal strata are deformed in the leading edge of the thrust belt.

Despite their theoretical simplicity, it is not easy to quantify patterns of tectonic subsidence in a real foreland basin, and thus, interpret the history of thrusting or of weakening of the elastic plate. This is because geologists measure rock thickness, not tectonic subsidence. Under what conditions would trends in rock thickness be straightforward reflections of trends in tectonic subsidence? This would be true if every location in the basin accumulated

sediment without changing its surface elevation. In theory, one can use backstripping to calculate tectonic subsidence from rock thickness (see Chapter 4). Sclater and Christie (1980) demonstrated that the set of techniques referred to as "backstripping" can extract from measured sections the thicknesses that were controlled by the amplitude of eustatic sea-level change, elevation change, compaction, and the isostatic load of the sedimentary fill, leaving only tectonic subsidence. One can also improve backstripping techniques to treat regional rather than local isostatic compensation. However, the subsidence history at any location on an elastic plate is affected by sediment accumulation at all other sites along strike, and up and down dip. To successfully extract the tectonic component of accumulation, one must work with two- or three-dimensionsal backstripping of several sites or cross sections and incorporate progressive changes in topography of the basin.

Surface Modification and Sediment Fill

Long-term accumulation in a foreland basin requires tectonic subsidence. Nonetheless, surface processes impart such a strong signature on the stratigraphy that the pure tectonic signal is obscured. The general categories of controls that are independent of local tectonic activity are bedrock lithology, climate, eustatic sea level, and geologic age (age may influence erosion and deposition, which are affected by evolutionary changes in land plants). Sediment supply may also be independent of local tectonics, if a continental drainage pattern funnels sediment toward the basin. These controls operate through several interdependent processes. For instance, bedrock lithology, climate, and geologic age influence sediment supply to the basin, as does tectonic uplift. Bedrock lithology and climate influence grain size, which in combination with climate, geologic age, and eustatic sea level, influence depositional environments.

Underfilled and Overfilled Conditions

Sediment supply and depositional environments actively influence large-scale characteristics of fore-

land basins. A sedimentary basin is not necessarily a topographic basin. If sediment does not completely fill the tectonically subsided basin, then a topographic trough will be produced between the thrust belt and the forebulge. The trough may be subaerial or marine. If subaerial, then the valley may be internally drained, or at least may have drainage parallel to the thrust front. This trough is evidence of an "underfilled basin" (Fig. 9.6A) (e.g., Covey, 1986). An underfilled basin might form if there is little sediment supply from the mountain system, or if the transport processes are too inefficient to distribute the sediment across the tectonic basin, or if there is a baselevel outside of the basin that captures rivers in the basin. Alternatively, sediment supply may exceed subsidence and the basin may "overfill" (Fig. 9.6B), resulting in a sedimentary plain that extends across the forebulge. In this case, the sedimentary basin is a lowland region, but it is not a valley. Overfilling produces a foreland basin that is wider than the zone of tectonic subsidence. In two cases with equal tectonic subsidence and lithospheric strength, underfilling versus overfilling produces very different facies, basin architecture, and rates of accumulation.

In practice, position of a forebulge in an active foreland basin can only be confirmed by study of the gravity field. Position of the forebulge is indicated by high gravity values, reflecting upward deflection of the base of the crust (Karner and Watts, 1983). In modern retroarc foreland basins, forebulges rarely form topographic highs; therefore, they are not expressed in surface facies patterns. This implies that the forebulge is commonly capped by strata derived from the thrust belt, which is the case expected for overfilled basins. However, facies patterns in ancient foreland basins commonly parallel the forebulge, which is expected of underfilled basins. This apparent contradiction suggests that real foreland basins do not fit the overfilled and underfilled models in a simple manner. One modern example of this enigma is expressed by the Ganges basin (a peripheral foreland basin; see Chapter 11), in which the forebulge is overlapped by strata derived from the Himalayan thrust belt, and the Ganges River runs parallel to the mountain system on the south (distal) side of the gravity high that defines the forebulge (Karner and

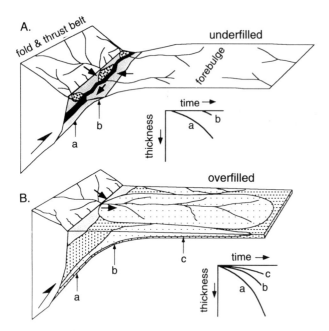

Fig. 9.6 Comparison of the geometry and facies of underfilled and overfilled foreland basins. Based on steady-state model of Flemings and Jordan (1989). A) The underfilled basin forms a valley, and receives sediment from both the thrust belt and peripheral bulge (forebulge). B) In the overfilled basin, the peripheral bulge receives sediment from the thrust belt and is expressed only by slower accumulation and thinner strata than in the proximal zone. Insets show the accumulation history at increasing distances from the thrust front, revealing that the accumulation rate is slower in distal sites compared to proximal sites. Although the tectonic-subsidence histories may be identical for these two basins, the accumulation histories of the underfilled and overfilled cases are quite different (see text).

Watts, 1983; Burbank, 1992). The enigma may reflect pre-existing topography and continental-scale drainage patterns. These two factors are largely independent of the orogenic belt and its tectonic subsidence, but they strongly control the orientation of drainage and positions of topographic lows in a foreland basin. Alternatively, the enigma may demonstrate that modern foreland basins, which have experienced accelerated sediment supply during deglaciation and have formed during low eustatic sea level of the late Cenozoic, are not ideal analogs for most ancient basins.

Flemings and Jordan (1989) evaluated the potential importance of those major stratigraphic controls that are intrinsic to the working of an orogenic system and tectonic subsidence for a nonmarine foreland basin. They used a forward model that incorporates steady deformation, flexural compensation, erosion, and deposition. (A forward model uses mathematical expressions describing the physics of geological processes to predict the stratigraphy of the basin.) The central question is whether surface processes (e.g., sediment supply and transport efficiency) are as important as tectonic controls (e.g., thrust history and lithospheric strength) on stratigraphy. The results indicate that, at a theoretical level, the answer is a firm "yes"; the values of sediment supply and transport efficiency within the sedimentary basin can control whether a basin is underfilled or overfilled. With the other three factors held constant, overfilling may result from any one of the following factors: 1) high flux of sediment to the basin, 2) efficient transport of sediment across the basin, 3) a weak lithosphere, or 4) a slow rate of advance of the thrust front. In reality, the efficiency of sediment transport depends on sedimentological factors (e.g., whether transport is by efficient rivers and turbidity currents or by inefficient shallow-marine processes) and discharge (e.g., humid climate causes more efficient transport of sediment than does arid climate). Weak lithosphere causes overfilling because it generates a narrow basin, which a given quantity of sediment fills more readily than a wide basin. Similarly, a slow rate of thrust advance implies a slow rate of addition of new tectonic subsidence, which a given quantity of sediment fills more readily than if subsidence were rapid.

The model results also predict that surface processes control facies characteristics, with a high rate of sediment supply to the basin or inefficient transport within the basin enhancing aggradation of basin-margin facies (e.g., facies deposited on steep topographic gradients) at the mountain/basin contact. If sediment flux is low or transport is especially efficient, then there will be a higher proportion of distal facies (e.g., facies deposited on gentle gradients).

These theoretical results suggest that the stratigraphic history of a foreland basin could include major changes in basin width, rate of accumulation, and facies successions that are due to climate change or drainage reorganization, rather than to tectonic activity or viscoelastic relaxation (Flemings and Jordan, 1989). However, the natural range of values for the key parameters (both surface processes and tec-

tonic controls) are not well known, and little work has been done to verify which factors were important to the evolution of specific basins. Furthermore, many complexities and interdependencies of uplift, climate, erosion, and drainages are not included in the forward model.

Depositional Environments and Facies Trends

Nonmarine and shallow-marine siliciclastic strata dominate retroarc foreland basins. Long-term progradation is due to progressive migration of the fold-and-thrust belt, and large regions of the basin eventually are incorporated into the thrust system. Progradation places shoreline deposits over marine strata, fluvial deposits over shoreline facies, and coarse-grained over fine-grained fluvial facies; long-term progradation is interrupted by short-term retrogradation.

Eustatic sea level is the prime control on whether a retroarc foreland basin is marine or nonmarine. This is because thrust-load-driven subsidence is not adequate to submerge normal-thickness continental lithosphere during low eustatic sea level. This contrasts with the case in peripheral foreland basins, for which subsidence commonly places the surface of attenuated continental crust beneath sea level. This illustrates that there exist differences in the causes of tectonic subsidence in retroarc versus collisional tectonic settings (see Chapter 11). At times of high eustatic sea level (e.g., the Cretaceous), retroarc foreland basins were marine, with maximum water depths on the order of 500 m (see below). At times of lowstand (e.g., today), comparable tectonic subsidence generates a nonmarine basin. Superposed upon these long-term trends, however, is the record of short-term sea-level fluctuation.

The character of prograding facies is governed by whether the basin is underfilled or overfilled, and marine or nonmarine. In an underfilled basin (Fig. 9.6A), initial distal facies are overlain progressively by basin-axis facies and then proximal facies. Provenance and paleocurrents reflect transition in sediment supply from the forebulge, to axial transport, to provenance in the adjoining thrust belt. Progradation implies that sections in the underfilled nonmarine

basin may fine upward from fluvial sandstones to lacustrine facies or coarsen upward to gravelly deposits of a high-discharge axial river, before changing upward to proximal alluvial-fan facies. In an overfilled, steadily prograding basin (Fig. 9.6b), both distal and proximal deposits show provenance and paleoflow from the thrust belt, and sections become coarser-grained upward. Furthermore, strata architecture may reflect the higher rate of preservation that occurs in proximal positions compared to distal sites (e.g., in proximal positions channel sandstones may be single-storied and isolated within overbank facies, whereas in distal positions, they may be interconnected in sheets and be multistoried) (Bridge and Leeder, 1979; Gardner et al., 1992).

Sediment Sources

Another key variable in foreland systems is the geometry of sediment supply to the basin. One aspect of the geometry is governed by the large-scale tectonic setting: sediment supply is strongly asymmetric. Much more sediment is supplied by the thrust belt than by the low-relief distal margin. In the case of an overfilled basin, there is no topographic expression of the forebulge, and the only detrital sediment supply is from the orogenic belt. For underfilled basins, this asymmetry insures that locations on the proximal side will be more completely filled than will locations on the distal side, and thus that stratal thickness systematically misrepresents tectonic subsidence patterns (compare accumulation curves a and b in Fig. 9.6A). An underfilled basin may receive additional sediment from along strike, through axial drainage.

Sediment may be supplied by a few point sources or the supply may approximate a line source, constituting a second geometrical variable (Fig. 9.7). This variable depends on the drainage pattern in the thrust belt. Commonly, river systems draining the interior of a mountain system pass through the marginal thrust belt along a few major antecedent valleys (Fig. 9.7b) (e.g., Eisbacher et al., 1974). In basins with major antecedent streams feeding the basin at point sources, the sediment supply varies greatly along strike because carrying capacity of major rivers differs greatly from the capacity of small rivers that

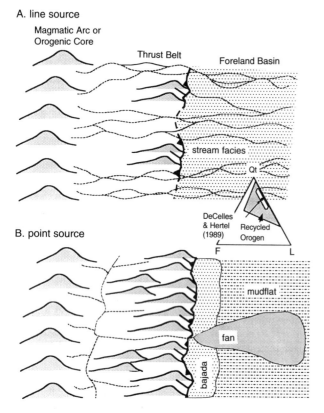

A. line source

Magmatic Arc or
Orogenic Core

Thrust Belt

Foreland Basin

stream facies

Qt

DeCelles
& Hertel
(1989)   Recycled
           Orogen

B. point source

F   L

mudflat

fan

bajada

Fig. 9.7 The drainage network in an orogenic belt is a primary control on facies in the foreland basin and on compositions of sandstones. A) Numerous minor rivers traverse the thrust belt, delivering a mixture of orogenic-core and thrust-belt sands to the foreland basin in a line-source configuration. B) A single major antecedent river cuts the thrust belt, concentrating an unusual volume of sediment in the foreland basin, where it empties, as a point source. Facies differ along strike in the basin, depending upon location relative to the mouth of the major river. Triangle (Qt = total quartz, F = feldspar, L = lithics, except polycrystalline quartz) shows that average sandstone composition lies in the recycled-orogen field of Dickinson and Suczek (1979). Holocene sands from rivers in the foreland basin of Peru and Bolivia are typical (DeCelles and Hertel, 1989). (See Fig. 9.14 for additional sand compositions.)

drain only the thrust belt. At a point source, a sedimentary lobe with facies corresponding to a delta or a large alluvial fan accumulates. Between point sources, linear coastlines or bajadas (coalesced small alluvial fans) may form. Structural geometry, thrust rate, bedrock lithology and climate probably control positions of major rivers and whether those positions remain fixed or vary through time. If the mountain drainage network reorganizes, then large shifts in points of sediment supply might generate locally complex stratigraphic patterns in the foreland basin

that could be confused with results of episodic tectonic activity or sea-level change (Fig. 9.7) (Damanti, 1989; Milana, 1991).

The nature of the drainage system also is a key factor in establishing whether a basin is overfilled or underfilled. Although tectonic subsidence usually largely results from activity in the thin-skinned thrust belt at the edge of an orogenic system, river systems commonly deliver sediment from the interior of the mountain system (Damanti, 1989; Schmitt and Steidtmann, 1990). Depending on the drainage system, a particular river entering the foreland basin may drain a very large region and transport sufficient sediment to overfill the basin, or it may drain only the thrust belt or a part of the thrust belt, and not carry enough sediment to adequately fill the basin. Along-strike transport of sediment within the foreland basin smooths out some volumetric irregularities; the irregularities are expressed in "depocenters" and in contrasting facies patterns along strike.

The nature of the drainage system is reflected in the compositions of sandstones in foreland basins (Fig. 9.7). The composition of the average foreland-basin sandstone lies in the recycled-orogen category of Dickinson and Suczek (1979); it is rich in quartz, has very little feldspar, and contains abundant lithic fragments (commonly sedimentary or metasedimentary, but locally volcanic), reflecting provenance in the thrust belt (e.g., Lawton, 1986a; Ingersoll et al., 1987; DeCelles and Hertel, 1989; Johnsson, 1990). Other workers (Mack and Jerzykiewicz, 1989; Schmitt and Steidtmann, 1990; Jordan et al., 1993) have found that the metamorphic and igneous interior of an orogenic belt may contribute large volumes of sediment where major river systems cut across the thrust belt. At other times and locations in these same basins, sandstone compositions reflect recycling of grains from older sedimentary rocks that typically comprise the thrust belt (Ingersoll et al., 1987; DeCelles and Hertel, 1989; Mack and Jerzykiewicz, 1989; Jordan et al., 1993).

Stratigraphic Completeness

Ample subsidence and sediment supply to proximal margins of foreland basins produce an exceptionally complete stratigraphic record (Fig. 9.8) (i.e., geologic

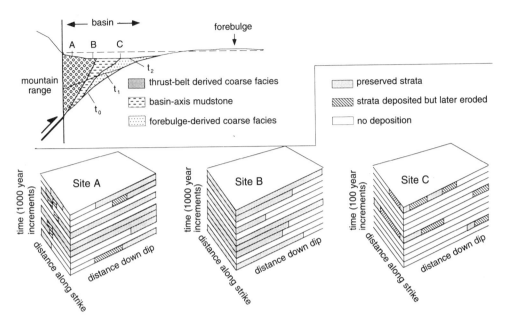

Fig. 9.8 Foreland basins are unusually complete registers of geologic time in nonmarine environments. Block diagrams compare the record of time (in 1000-year increments) at three positions in a cross section of a foreland basin. Site A is in a proximal zone, with ample subsidence and sediment supply to capture most depositional events, but where a channelized depositional environment tends to limit the lateral distribution of depositional events and to subject those strata to subsequent erosion. Site B is slightly farther down depositional slope. It contains the most complete temporal record: there are ample subsidence and sediment supply, coupled with unchannelized deposition (hence little erosional power) in distal alluvial fans. Site C is a distal site, with a very incomplete stratigraphic record. The primary factor is the slow subsidence, which reduces the chance of deposition and increases the time during which a stratum remains at the surface to be subject to erosion by subsequent events. Based on Barrell (1917), Wheeler (1958), and McRae (1990), with time increment consistent with work of Beer (1990). t_0, t_1, and t_2 in reference cross section are successive time lines.

time is represented primarily as accumulated strata, rather than unconformities). Even in the case of a nonmarine foreland basin, in which deposition during unusual floods alternates with decades or more of exposure, sediment accumulation tends to be rapid. N.M. Johnson et al. (1986) and Reynolds et al. (1990) document accumulation rates in proximal and medial positions of an Andean nonmarine foreland basin, averaged over intervals of several million years, up to about 1.0 mm/y and up to 1.4 mm/y over shorter time spans. Beer (1990) and McRae (1990) used magnetostratigraphy and physical stratigraphy to show that, for the foreland basins of the Andes and the Himalayas, strata representing as little time as 30,000 to 10,000 years, respectively, can be traced readily along strike for tens of kilometers. The degree of steadiness (i.e., constant in time) and lateral continuity (i.e., constant in space) of accumulated strata differs somewhat between the two study sites because the depositional environments differ. The less steady and less continuous case (Himalayas) is a region of channelized rivers, and the steadier and more continuous case (Andes) has sandflat and mudflat deposits. Steadiness and lateral continuity probably also differ down dip (Fig. 9.8); Beer's and McRae's studies were of proximal to medial positions in the respective basins, where a greater degree of stratigraphic completeness is expected than on the distal flank.

Stratigraphic Sequences

Natural basins are rarely steady over long time intervals. In real cases, a change in one of the controls causes the stratigraphy of the basin to deviate from the simple pattern described above. Short-term variability, expressed in reversing vertical sequences of facies or compositions, has been studied and interpreted by generations of geologists. A central goal of stratigraphic studies is to discern which of the possible controls (e.g., thrusting, climate, sea level) is responsible for vertical trends. In many cases, the primary

342   Chapter 9

interest has been to determine the time of deformation in the mountain belt based on strata in the foreland basin (Jordan et al., 1988). Traditionally, a phase of active tectonism was presumed to cause high relief, and thus an immediate flux of coarse sediment into the basin, which produced conglomerate (e.g., Longwell, 1937; Wiltschko and Dorr, 1983). Interbedded mudrocks were considered to indicate times of tectonic quiescence. Subsequently, reasoning based on the mechanics of subsidence led to the hypothesis that upward coarsening sections are evidence of pulses of active deformation *only* in proximal parts of the basin, whereas coarse-grained strata in distal positions indicate times of tectonic quiescence (Fig. 9.9) (Heller et al., 1988). Tectonic quiescence was also anticipated to produce unconformities within the basin, because the basin would rebound isostatically as erosion thinned the mountain belt (Beaumont, 1981; Heller et al. 1988).

It should be possible to extract from the sedimentary fill of a foreland basin the history of thrusting, relative sea level, climate, and drainage. In order to do this, it is necessary to determine thickness changes, patterns of stratal lap-out and truncation,

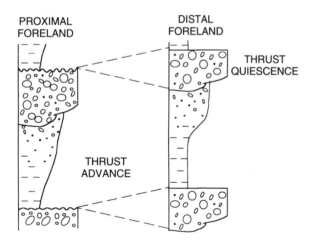

Fig. 9.9 Upward coarsening clastic sections in proximal and distal parts of foreland basin may be out-of-phase with one another (after Heller et al., 1988). Preserved strata in proximal part of basin correspond to time of thrust onset (fine-grained) through climax (coarse-grained), and unconformity represents time that thrust system is inactive. In distal part of basin, coarsest strata accumulate during time that thrusts are inactive. No specific correlations are implied between two columns during the interval of active thrusting.

and facies distributions within successive packages of rock that represent short increments of time. Appropriate packages of rock, "stratigraphic sequences,"[1] are defined such that their boundaries are mappable time surfaces. Subdivision into sequences facilitates basin analysis because it emphasizes time relationships among various sedimentary facies *within the same sequence*.

Although sequence stratigraphy is a powerful tool in the analysis of foreland-basin history, few detailed studies of the three-dimensional architecture of those sequences have been published. In contrast, stratigraphic sequences are well studied in passive-margin settings. In passive-margin basins, unconformities with considerable erosion within the nonmarine to shallow-marine zones are the hallmark of boundaries between depositional sequences (see Chapter 4). In passive-margin basins, sequence-bounding unconformities can form because subsidence rates are low on the landward margin of the basin. Conversely, the rate of subsidence in foreland basins is highest adjacent to the zone of principal sediment supply (the thrust belt). As a result of adequate subsidence and abundant sediment, unconformities do not appear to play a major role on the proximal foreland-basin margin (Swift et al., 1987; Jordan and Flemings, 1991). Instead, the dominant pattern on the proximal margin of the basin is that the sedimentary succession is punctuated by a series of marine flooding surfaces (i.e., laterally traceable, abrupt, vertical passages from nonmarine to marine strata) that separate upward shoaling progradational facies successions (e.g., Leckie, 1986; Swift et al., 1987; Plint, 1988; Van Wagoner et al., 1990; Gardner et al. 1992). As discussed above, the temporal record of the proximal margin of the foreland basin is relatively complete, with little time represented by the marine flooding surfaces.

More pronounced unconformities and erosion are seen at some distance from the thrust front. For instance, the Alberta foreland basin contains a series of lenticular conglomerates and sandstones, bounded by unconformities, and encased in marine mudstones, that lie at least 100 to 200 km from the contemporaneous thrust front. These coarse-grained units are interpreted by Plint (1988) as the erosional remnants of shoreface sandstones that originally pro-

graded from the thrust belt. In even more distal positions, shoreline facies that prograde *toward* the thrust belt and that are beveled by unconformities that cut deeper *away from* the thrust belt, may indicate uplift of the forebulge (Plint et al., 1993).

These observations reveal that several kinds of surfaces are common in foreland basins and that they express deviations from slow, steady progradation: marine flooding surfaces, unconformities in the middle of the basins, and distal unconformities. The stratigraphic sequences bounded by these surfaces span a broad range of time scales, from a few hundred-thousand years to several million years. At present, we do not have enough detailed documentation of these sequences to decide whether they reflect tectonic causes, eustatic causes, climate change, or combinations of causes.

Because the basic shape of a foreland basin and its asymmetric sediment supply can be readily portrayed by forward modeling, the models can be used to study stratigraphic sequences. As described below, the models show that stratigraphic sequences generated by episodic tectonism differ from those produced by eustatic sea-level change.

To model stratigraphic sequences in a retroarc foreland basin, a depositional model is incorporated with a basin-subsidence model. One approach is to simulate sedimentary processes in two dimensions by diffusion (Flemings and Jordan, 1990; Jordan and Flemings, 1991; Sinclair et al., 1991). Diffusion treats the rate of aggradation or erosion at each site as a linear function of the local curvature of topography; areas which are concave-up accumulate sediment and those that are convex-up erode. For cases of mixed marine and nonmarine deposition, nonmarine transport is treated as more efficient diffusion than is marine transport. Below, we compare the modelled stratigraphy for cases of variable thrusting and constant sea level (Figs. 9.10, 9.11a,b) to the stratigraphy of a case with variable sea level and constant thrusting (Fig. 9.11c,d).

In cases of constant sea level (or completely nonmarine basins) but variable thrusting, stratigraphic sequences are generated that are bounded by important unconformities in the distal part of the basin (Flemings and Jordan, 1990; Sinclair et al., 1991). The model shows that subaerial unconformities are

cut locally on the proximal and regionally on the distal margins of the basin (Figs. 9.10, 9.11a). Yet, ages of proximal-margin and distal-margin unconformities are out of phase with each other: erosion is most pronounced during tectonic quiescence on the proximal side and during thrusting on the distal side (Flemings and Jordan, 1990).

One might wonder why deposition occurs at all during tectonic quiescence, since thrust-derived tectonic loading and tectonic subsidence are absent. The answer is twofold. First, erosion of the highlands continues with or without additional uplift, and the eroded sediment has mass; wherever the sediment pauses, it causes flexural subsidence. Second, in a basin that was underfilled during thrust-load subsidence, the topographic trough or marine embayment is unused subsided space, which can be filled with sediment later. As a result of these two factors, subsidence occurs on a large scale during tectonic quiescence, but it is focused beneath the sediments (which are deposited in the previously formed trough) rather than beneath the integretated load of active thrusts and sediments.

In fact, in the thrust zone, the erosional removal of sediment reduces the crustal column and causes flexural uplift (Beaumont, 1981; Heller et al., 1988). Therefore, during thrust quiescence, the area immediately adjacent to the thrust belt, which was once the proximal zone of most rapid sedimentation, lies at a hinge between uplift of the thrust belt and subsidence of the region with greatest remnant space in the basin. This converts the proximal zone to a region of either minor uplift, bypass, or at most, minimal subsidence; accumulation rates in such a quiescent basin are greatest near the center and decrease in both proximal and distal directions (Fig. 9.10a,e). The shape of depositional packages in a quiescent-phase basin is lens-like (Fig. 9.10c), broader than it was during active thrusting. Whereas a forebulge still exists, it is a broader, less pronounced uplift, which is farther from the thrust belt. During filling of the tectonically quiescent basin, the depocenter shifts progressively to more distal positions as the asymmetric sediment supply continues to cause progradation.

Renewed thrusting reverts the nature of load and subsidence to their original form (Fig. 9.10c), causing

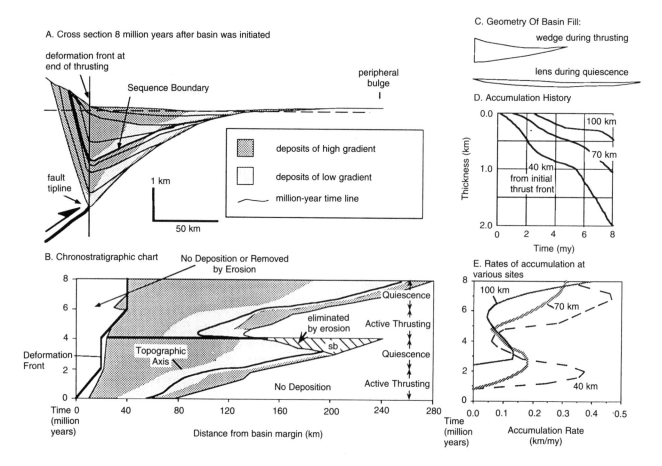

Fig. 9.10 Modeled stratigraphic sequences in a foreland basin are predicted to form as a result of episodic thrusting (here thrust periodicity is 4 million years). A) After two cycles of thrusting and quiescence. One-million-year time lines are shown, plus two facies categories. Sequence boundary caps the quiescent-phase strata. B) Chronostratigraphic chart of strata shown in A. Note that erosion during the initiation of thrusting removes quiescent-phase strata from the distal margin. Deposits of low gradient reach closest to thrust belt at onset of each phase of thrusting, whereas deposits of high gradients are most widespread in basin during quiescence. C) Cross sections for one-million-year intervals (from A) highlight the geometric contrast between wedges of sediment during intervals of thrusting and lenses during times of quiescence. D) History of accumulation of foreland-basin sections at increasing distances from the mountain front. E) Rates of accumulation (first derivative of curves in D). At the most proximal site (40 km), the rate is highest during times of thrusting, and declines slowly during times of quiescence. At the most distal site (100 km), the rate is highest during times of quiescence. Based on Flemings and Jordan (1990).

subsidence in the proximal region and drawing the forebulge toward the thrust belt. This mountainward shift of the forebulge erodes strata deposited in what was the distal quiescent basin (or at least slows accumulation), and produces a distal unconformity. Thus, tectonic variation is typified by a distal unconformity initiated by renewed thrusting, and locally, by a proximal unconformity that is initiated during quiescence (Fig. 9.10B, 9.11b).

In the absence of a proximal unconformity, a stratigraphic sequence would be defined by the distal unconformity and a correlative facies shift (upward fining) in the proximal region (Fig. 9.10A,B) (Flemings and Jordan, 1990). Where a proximal unconformity develops, architecture of strata (Fig. 9.11a) may suggest that stratigraphic sequences be defined, whose boundaries are the out-of-phase proximal and distal unconformities. Such a sequence boundary would be slightly diachronous down dip (Fig. 9.11b) (Jordan and Flemings, 1991).

For steady thrusting and variable sea level in a marine foreland basin, unconformities are more likely to form on the distal or forebulge side of the basin than on the proximal margin (Fig. 9.11c,d).

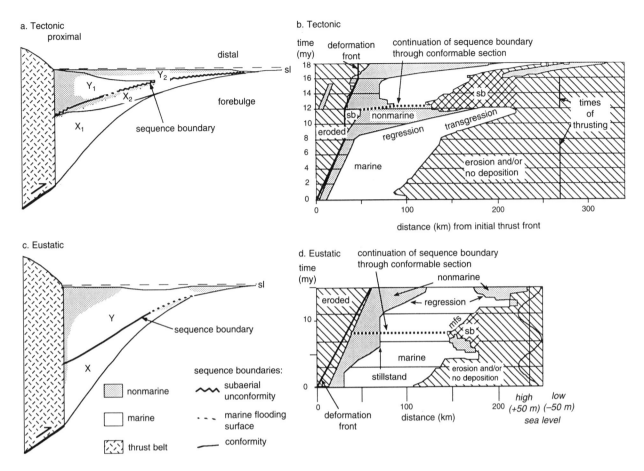

Fig. 9.11 Comparison of marine foreland-basin stratigraphic sequences predicted in response to (A, B) episodic thrusting but constant sea level, and (C, D) eustatic sea-level change but steady thrusting. Two models use identical parameters, except for thrust timing and eustatic variation. Eustatic variation generates an unconformable sequence boundary on the distal basin margin, which is traced into the proximal zone along conformable surfaces, and separates sequences X and Y. Eustatic variation causes stillstands and regressions (and with other values of the parameters, transgressions) but little erosion or progradation on the proximal margin. mfs = marine flooding surface; sb = sequence boundary (time and area that are incorporated in sequence-bounding unconformity shown by crossing-diagonal pattern). Episodic tectonism generates an unconformable sequence boundary across much of the basin (see text). X1 and Y1 are the syndeformational hemicycles, during whose deposition the basin is narrow and facies first retrograde and then stack vertically; X2 and Y2 are the quiescent hemicycles, during whose deposition considerable progradation occurs. A greater width is shown in B than in A because, although distal strata are widespread, they are too thin to resolve at scale of A. In cross sections, left margin of basin (located at position of deformation front at end of time interval) is shown as vertical line, to avoid the visual complications of deformed sequences. Two cycles are shown for each case, comprising 18 million years in A, B, and 16 million years in C, D. After Jordan and Flemings (1991).

Eustasy controls processes on the distal margin by alternately flooding and uncovering the forebulge region. Falling sea level creates an unconformity above a shoreline that progrades toward the thrust belt. In addition, in an entirely clastic system (which need not be true of many distal foreland basins), marine flooding adds to the distal unconformity by slowing erosion of the forebulge and thus cutting off sediment supply. Physically, much of the unconfor-mity is a surface of submarine erosion and nondeposition. The majority of the time recorded by the unconformity surface correlates to rising sea level (Fig. 9.11d). If the basin is sufficiently narrow, then the distal unconformity is capped by onlapping strata derived from the proximal side of the basin. The marine flooding surface above the distal-margin shorelines and its continuation through conformable strata toward the proximal margin define sequence

boundaries. The ages of the sequence boundaries correlate primarily to times of rising sea level (Fig. 9.11d) (Jordan and Flemings, 1991). If the amplitude of sea-level change is sufficiently large, then more prominent facies shifts will occur on the proximal basin margin than shown in Figure 9.11c, but the large degree of tectonic subsidence nevertheless impedes development of unconformities (T.E. Jordan and P.B. Flemings, unpublished model results).

In practice, we would like to differentiate between tectonic and eustatic stratigraphic sequences in marine clastic foreland basins (Fig. 9.11). Thickness patterns are useful to differentiate tectonic and eustatic cases. Tectonic cycles cause the pulses of major progradation of the depocenter during quiescence (subsequences X2 and Y2, Fig. 9.11a), whereas even large changes in sea level have comparatively little impact on the depocenter in the basin (Fig. 9.11c). Accurate chronostratigraphic correlation would also show that, for tectonically controlled stratigraphic sequences, transgressive-regressive cycles of the proximal margin are 180° out of phase with those of the distal margin.

Structural controls at local scale

The structural geometry and kinematic history of the thrust belt are important controls on the details of the stratigraphy of a foreland basin, especially in proximal regions. The characteristics of retroarc foreland basins described above are generalities that apply to large-scale trends in basins bounded by simple thrust belts. A "simple" thrust belt would be one whose structures produce a wedge-shaped zone of thickening (Davis et al., 1983), with emergent thrusts, and whose frontal thrust steps progressively toward the foreland basin (Armstrong and Oriel, 1965; Dahlstrom, 1970). In reality, because thrust zones commonly are not so simple, the subsidence, sediment supply, and therefore, proximal stratigraphy may deviate considerably from these generalities.

Temporally irregular thrusting is capable of generating regional unconformity-bounded stratigraphic sequences (see above). Episodic thrust history is suggested by some available data (e.g., Burbank et al., 1986; Jordan et al., 1988), but the information is

incomplete. Even if individual thrusts move episodically, the system of linked thrusts might operate continuously (Molnar and Lyon-Caen, 1988). If so, regional tectonic subsidence in the basin could be steady even while proximal zones or drainage patterns were affected by local episodicity.

Spatially discontinuous development of thrust belts may be equally important. At times, a new frontal thrust may form at a position kilometers to tens of kilometers farther into the basin than the former basin margin. As the new thrust advances, the proximal margin of the basin may be folded even while sediment continues to accumulate above the growing structures (Medwedeff, 1989; Suppe et al., 1992). The advance of a new fault into the basin also affects the regional stratigraphy, by moving canyon mouths rapidly into medial positions in the basin. If the thrust-front position changes little when a new phase of thrusting begins, then syndeformational basin-axis facies may overlie proximal facies of the quiescent phase (Fig. 9.10) (e.g., Heller et al., 1988; Flemings and Jordan, 1990). In contrast, if the thrust front moves significantly into the basin when a new thrusting phase begins, then proximal facies of the active phase may overlie proximal or distal facies of the quiescent phase (e.g., Burbank et al., 1988).

The pattern of erosional unconformities, conformities, and onlap is vastly more complicated on a local scale in the region where deformation spatially overlaps with deposition (Fig. 9.12) (Medwedeff, 1989). These conditions are common in the boundary between thrust belts and foreland basins, and in local synformal regions within thrust belts that are called "piggyback basins" (Ori and Friend, 1984; see Chapter 11). For example, Medwedeff (1989) showed that a simple history of shortening on a fault-bend fold (i.e., a fold that forms as strata move up over a ramp in the thrust plane) could generate syndeformational strata on the back limb that are folded conformably with pre-deformation strata, while syndeformational strata on the forelimb onlap the fold (Fig. 9.12). In cases for which sediment is supplied from a neighboring region (e.g., a principal drainage from the thrust belt) and can bury the growing local structure, the shape of syndeformational strata depends on the ratio

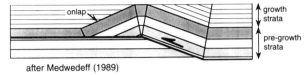

A. Rate of Accumulation = Rate of Uplift Above Ramp

onlap

growth strata

pre-growth strata

after Medwedeff (1989)

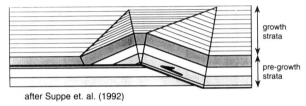

B. Rate of Accumulation > Rate of Uplift Above Ramp

growth strata

pre-growth strata

after Suppe et. al. (1992)

Fig. 9.12 Architecture of syndeformational strata deposited adjacent to and overlying a ramp in an active fault-bend fold. Two cases illustrate differing relative rates of uplift and aggradation; other cases (not shown) can involve erosion above fold. In case A, forelimb strata (left margin) appear to post-date the fold, whereas backlimb strata of identical age appear to predate the fold. In case B, equivalent-age strata accumulate more slowly over the crest of the anticline than on the flanks. The kink axes must converge upward to accommodate the changes in thickness.

of sediment accumulation rates to the rate of uplift of local structures (Fig. 9.12) (Suppe et al., 1992). The architecture of strata deposited over these growing structures varies considerably and depends on the detailed kinematics of the folds and on rates of accumulation. Although complicated, these stratal patterns offer outstanding opportunities to determine detailed motion histories of the folds and faults (DeCelles et al., 1991; Suppe et al., 1992).

Structural geometry and kinematic history also affect compositions of detrital sediments in foreland basins. On one hand, it is attractive to use the compositions of sands and gravels as clues to the times of motion of the thrust faults that most closely underlie their source units (e.g., Armstrong and Oriel, 1965; Wiltschko and Dorr, 1983). A comparison of predicted detrital compositions derived from a particular structure to the compositions of proximal strata can reveal details of structural motion (Graham et al., 1986; Ingersoll et al., 1987; DeCelles, 1988; DeCelles et al., 1991). However, exposure of a given source unit in the thrust belt is controlled by *all* of the thrust

sheets that underlie it, and a pulse of sediment derived from a particular source unit may reflect motion on any of the underlying thrusts (Fig. 9.13) (Steidtmann and Schmitt, 1988) or may represent drainage-network changes that are indirectly related to thrust motion (Damanti, 1989). Therefore, the three-dimensional structural geometry of the thrust belt and the drainage pattern must be determined before detrital compositions in most of the foreland basin can be interpreted in detail.

Nonmarine Examples: Andean Foreland Basins

The part of the Andes Mountains that lies south of the equator is a noncollisional mountain belt, consisting of a core zone of a late Cenozoic magmatic arc and high plateau, flanked on its eastern margin by Miocene to Recent retroarc thrust belts and foreland basins (Fig. 9.14) (Jordan et al., 1983a; Jordan and Gardeweg, 1989). The shape of the subducted plate varies along strike, with some segments of the plate descending at an angle of about 30° beneath the South American lithosphere, and other segments subhorizontal (Fig. 9.1) (Cahill and Isacks, 1992).

This foreland-basin system stretches across climates ranging from tropical to arid to glacial, creating wide variation in depositional and erosional conditions. In addition, the foreland compressional structures vary along strike, and consequently, the character of the foreland basins varies (Fig. 9.14). For

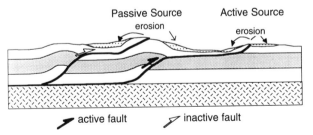

Passive Source   Active Source
erosion   erosion

active fault   inactive fault

Fig. 9.13 Source areas within the thrust belt may be uplifted by motion of the active fault immediately beneath them (active source) or may be rejuvenated by folding above a ramp in a deeper active fault (passive source). As shown, uplifted sources do not necessarily record history of motion of the nearest fault, which may be inactive. Sediments shed from the passive sources may be interpreted in terms of history of movement on a given fault only if the geometry of the set of faults is known. After Steidtmann and Schmitt (1988).

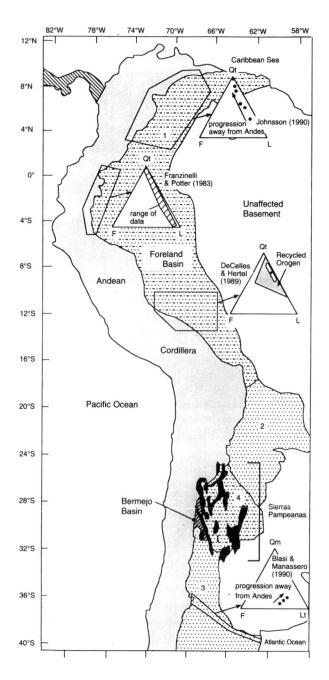

Fig. 9.14 Distribution of Quaternary foreland basins of the Andes Mountains and Holocene sand compositions (in triangles). Map distinguishes foreland basins with contrasting drainage styles. 1. Foreland basin in a tropical climate drains through Orinocco and Amazon rivers. 2. An overfilled foreland basin in a drier climate, in which nearly all of the sediment appears to be trapped in the basin, although some of water drains to Paranà River. 3. Drainage basin in dry climate, whose rivers traverse the short distance to the Atlantic margin. In northern part of 3, most of the sediment is retained in the foreland basin, but in the southern part of 3, there is little documentation of foreland basins along the edge of the Andes. 4. Internally drained basins, receiving sediment from the Sierras Pampeanas broken foreland (uplifts shown in black) and the Andes Mountains. The Bermejo basin is an internally drained basin, receiving majority of its sediment from Andes, but whose structure is affected by the Sierras Pampeanas (see Fig. 9.15). Map based on DeCelles and Hertel (1989) and J. Damanti (unpublished drainage map). Sand composition data derived from areas enclosed in polygons. Compositions of modern sands for northern localities fall in the recycled-orogen category of Dickinson and Suczek (1979) (plotted for reference on composition triangle near 8° S latitude). Sand compositions for 36° S region are strongly influenced by volcanic arc. Methods varied among the four datasets that are available. DeCelles and Hertel (1989), Johnsson (1990), and Franzinelli and Potter (1983) counted thin sections, whereas Blasi and Manassero (1990) counted grain mounts. DeCelles and Hertel (1989) and Johnsson (1990) stained for calcium and potassium feldspars; Franzinelli and Potter (1983) stained for only potassium feldspar. DeCelles and Hertel (1989) used the Gazzi-Dickinson point-count method (Ingersoll et al. 1984); Johnsson (1990) and Franzinelli and Potter (1983) did not.

there are basement blocks bounded by reverse faults (Sierras Pampeanas) east of a thin-skinned thrust belt (Precordillera). The Sierras Pampeanas produce a set of broken-foreland basins and the Precordillera produces a standard foreland (Bermejo) basin. From 33 to 40° S, shortening is confined to a narrow thrust belt on the eastern flank of the Andes, and there is a transition in the foreland from basins in the north to a plateau further south, with backarc alkaline basalts on the plateau, and little deformation or very minor normal faulting (Muñoz and Stern, 1989). All of these Central Andean basins are now nonmarine and were nonmarine during most of their Mio-Pliocene development (N.M. Johnson et al., 1986; Jordan and Alonso, 1987).

Compositions of sand and sandstone in the Andean foreland basins reflect a mixed provenance, including the Cenozoic volcanic chain of the Andes, Paleozoic and Mesozoic sedimentary, volcanic, and plutonic rocks of the Andes, and reworked older foreland-basin deposits. Most sands in modern river systems in the foreland basins of the northern half of the Andes fall in the recycled-orogen composition field of

example (Figs. 9.1, 9.14), from 18 to 24° S, the foreland basin is simple, flanking the Subandean thrust belt, with maximum subsidence along its western flank, and a source area to the west (Jordan and Alonso, 1987). Along much of its length, this basin is currently overfilled (Flemings and Jordan, 1989), with alluvial fans extending 600 km perpendicular to the mountain front (Iriondo, 1990). From 27 to 33° S,

Dickinson and Suczek (1979) (Fig. 9.14). Although weathering in a hot, humid climate tends to destroy unstable grains and alter the average composition, Johnsson (1990) showed that the primary compositional character is retained in sands of proximal parts of the foreland basin (Fig. 9.14). Only in the southern Andes are late Cenozoic volcanic arc rocks large components of the source area, producing sands in the Rio Colorado that plot in Dickinson and Suczek's magmatic-arc field (Fig. 9.14). Blasi and Manassero (1990) reason that Rio Colorado sands are not of recycled-orogen composition in part because of the high percentage of volcanic rocks in the drainage area within the Andes, plus the fact that volcaniclastic rocks are exposed close to the river in the foreland (Blasi and Manassero, 1990).

The Bermejo basin is a particularly well studied Miocene to Recent nonmarine foreland basin. It formed during the shortening of the Precordilllera thin-skinned thrust belt (Fig. 9.14). Up to 6000 m of strata accumulated in the principal depocenter between about 18 and 2.5 Ma; another 2000 to 3000 m of sediment have accumulated during the last 2.5 my, in a tectonic environment modified by the Sierras Pampeanas basement structures (see below). The Precordillera deformation front has migrated more than 100 km (Allmendinger et al., 1990) after thrusting began at approximately 20 Ma (Jordan et al., 1993), and consequently, the zone of maximum accumulation has migrated eastward through time (Fig. 9.15). The unusually great tectonic subsidence in the Bermejo basin may have resulted from greater than average tectonic thickening in the thrust belt (~10 km of thickening) (Allmendinger et al., 1990).

In western parts of the Bermejo basin, sediment which now occurs as isolated slivers in the thrust belt, coarsened upward to proximal gravels some 10 million years before similar lithologies were deposited in more distal parts of the basin (Fig. 9.16). Measured rates of accumulation increase abruptly upward, suddenly jumping a factor of 3 to 5 fold (Fig. 9.16, inset). Furthermore, this jump occurred later in the east than in the west, probably in conjunction with west-to-east migration of the deformation front. The rates, measured using magnetic-polarity stratigraphy, are low (~0.1–0.2 mm/y) for the basal part and high (>0.5 mm/y) in the main part of each section.

The major stages of shortening in the Precordillera thrust belt can be inferred from cross-cutting relations between folds and faults with strata within the thrust belt, and from foreland-basin strata (Jordan et al., 1993). The western thrusts first were active at about 20 Ma, and there may have been a stage of inactivity between about 19 and 15 Ma. Most thrust activity occurred in the interval from about 15 to 2 Ma. There is little evidence of strata deposited during the first phase of thrust activity (Jordan et al., 1993). Sediments that accumulated in the foreland basin during the time of likely tectonic quiescence (18–15 Ma) were dominantly eolian, grading eastward into playa deposits, and reach thicknesses of several-hundred meters (Fig. 9.16) (Milana, 1991; Jordan et al., 1993). The second phase of thrusting triggered more conventional nonmarine foreland-basin deposition, dominated by fluvial and alluvial units (Fig. 9.16). Sandstones older than about 16 Ma largely reflect provenance in the high Andes west of the thrust belt; the high Andes are composed of Paleozoic volcanic and plutonic rocks and produce volcaniclastic sandstones (Fig. 9.17a) (Jordan et al., 1993). Conglomerates immediately adjacent to the early western thrusts are composed of Precordilleran bedrock clasts. In most locations, sandstones younger than 16 Ma contain clasts eroded from the thrust belt (Reynolds et al., 1990; Jordan et al., 1993). By ~10 Ma, the topography of the western thrust belt was sufficiently high and continuous along strike that the drainage pattern changed (Fig. 9.17); minor rivers with sources in the Andean interior were integrated west of the Precordillera and funneled into one or two major antecedent rivers that cut the thrust belt (Damanti, 1989). Beginning at that time, these few rivers were the main sediment and water sources to the foreland basin. This may also have been the time that accumulation began in the Iglesia piggyback basin (Figs. 9.15, 9.17) (Beer et al., 1990). Shortening subsequently progressed eastward in the Precordillera, resulting in initiation of the easternmost thrust at about 5 Ma, which is, at least locally, active today (N.M. Johnson et al., 1986; Jordan et al., 1990).

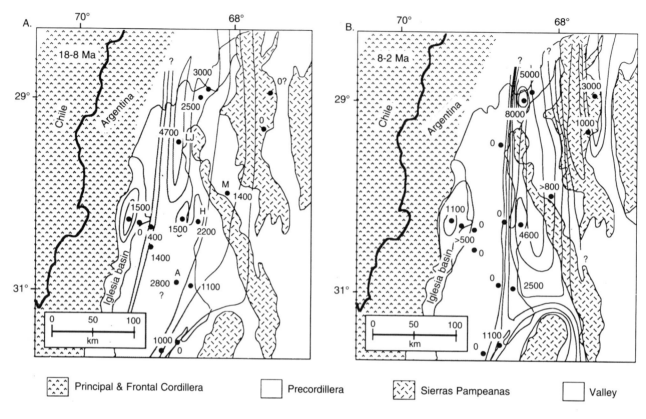

Fig. 9.15 Isopach maps of the Bermejo foreland basin (see location on Fig. 9.14) and Iglesia piggyback basin during two periods, showing progressive eastward migration of depocenter. Contour interval 1000 m. Modified from Jordan et al. (1989) based on new information in Beer et al. (1990), Milana (1991), and Jordan et al. (1993). Locations LJ (Las Juntas), M (Rio Mañero), A (Rio Azul), and H (Huaco) referred to in Fig. 9.16.

Basement shortening in the Sierras Pampeanas began ~ 8 Ma (Reynolds, 1987), which led eventually to disruption of what had been the distal part of the Bermejo basin (Fig. 9.15). Beginning at roughly 3 to 2 Ma, basement deformation just a few kilometers east of the easternmost thin-skinned thrust converted the main part of the Bermejo foreland basin to a broken-foreland basin (Fielding and Jordan, 1988; Milana and Jordan, 1989).

The Bermejo basin exceeded 220 km in width early in its history and narrowed through time (Fig. 9.15), although probably not steadily. Facies patterns suggest that it was underfilled through most or all of its history, with the topographic axis (a valley) about 100 to 160 km from the contemporaneous thrust front. Specifically, prior to 13 Ma, lacustrine facies in distal exposures (M, Fig. 9.15) show that the basin topographic axis was at Rio Mañero (Fig. 9.16). Subsequently, the topographic axis lay to the west of that

location (paleoflow depositing fluvial strata at Rio Mañero was to the west), which might indicate that the forebulge was elevated and shedding sediment (Malizzia, 1987). However, it is uncertain whether the topographic high on the distal (eastern) side of the valley was the peripheral bulge, as there are no data that directly define the history of positions of the forebulge.

Arid conditions dominated the Bermejo depositional regime throughout its history (N.M. Johnson et al., 1986; Malizzia, 1987; Beer and Jordan, 1989), although the climate in the headwater regions may have changed through time, causing variable discharge into and through the basin (Milana, 1990). Nearly all of the strata are products of ephemeral discharge. Proximal depositional environments vary along strike from narrow bajadas with flashy, short-lived discharge from thrust-belt drainages, to large alluvial fans fed by permanent antecedent rivers

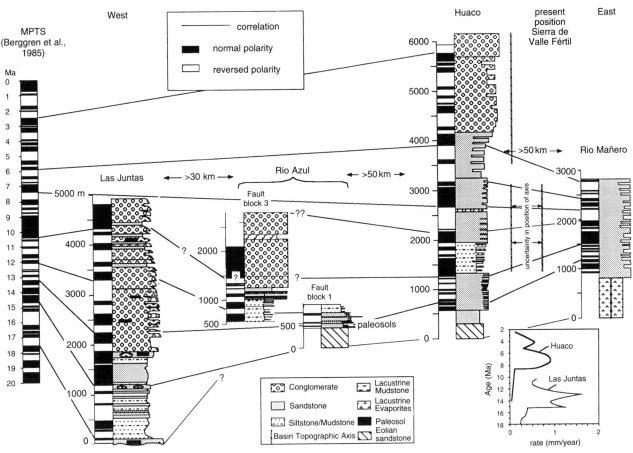

Fig. 9.16 Comparison of sections in Bermejo Basin, perpendicular to structural and depositional strike, for rocks spanning 18 to ~2 Ma. Correlations among sections, and of sections to global magnetic-polarity time scale (left column), are based on magnetic-polarity stratigraphy and radiometric dates of intercalated ashes (not shown). Locations shown in Fig. 9.15. Conservative estimates of pre-shortening distances perpendicular to strike between sections indicated. Units at base of Huaco section determined from borehole cuttings; thicknesses of those units are highly uncertain, due to structural complexity. Top of Río Mañero section apparently substantially truncated by faulting. Note that all of the sections began to accumulate sediment early in the basin history; western sections became involved in deformation at 9 or 8 Ma, while the eastern sections still accumulated strata; conglomerates in the western sections are nearly 10 my older than those in eastern sections. Inset compares rates of accumulation at Las Juntas, a proximal site, and Huaco, a more distal site, showing the diachroneity in their accumulation histories. Note that Las Juntas lies substantially north of the other sections, and may respond to a thrust history that differs somewhat from that of the southern sections. Based on Malizzia and Villanueva Garcia (1984), N.M. Johnson et al. (1986), Malizzia (1987), Reynolds (1987), Jordan et al. (1990), Reynolds et al. (1990), J.H. Reynolds (personal communication, 1991) and Jordan et al. (1993).

draining the interior parts of the Andes (Jordan et al., 1990, 1993). More distal environments typically were sandflats and mudflats of the major alluvial fans, with meter-thick sand bodies that can be traced many kilometers along strike (Beer and Jordan, 1989). The most distal facies exposed are thick clastic and evaporite lacustrine facies (reflecting times of closed drainage), intercalated with very fine-grained sheet-flood sandstones (Malizzia, 1987).

The Bermejo basin may contain stratigraphic sequences at several scales. This is likely for two reasons: 1) we know that marine foreland basins display stratigraphic sequences, and 2) tectonic activity, discharge, sediment supply, and drainage organization almost certainly varied through time in the Precordillera thrust belt-Bermejo basin system. However, good-quality seismic lines and surface exposures reveal only rare geometrically defined unconformities (primarily in growth structures related to the youngest basement structures that disrupted the basin). During the interval from ~14 to 2.5 Ma for the proximal margin of the basin, clear stratigraphic sequences are

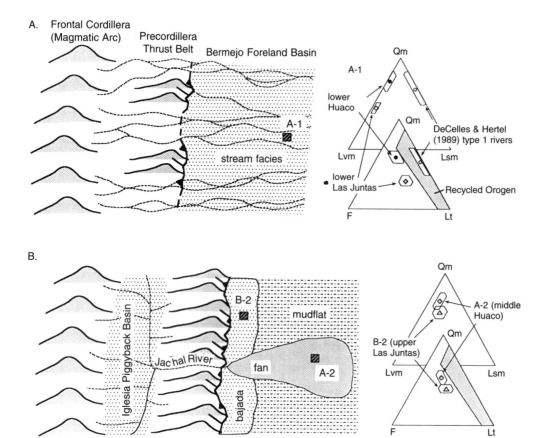

Fig. 9.17 The drainage network that fed the Bermejo basin evolved as deformation progressed in the Precordillera thrust belt. A) At early stages, in the deformation, minor rivers traversed the thrust belt and delivered a mixture of Frontal Cordillera and thrust-belt sands to site A-1 in the foreland basin. Because much of the bedrock of the Frontal Cordillera is Paleozoic and Mesozoic volcanic and plutonic rocks, these sandstones are rich in volcanic lithic grains. B) At later stages in deformation, a piggyback basin formed that was drained by a single major antecedent, Jàchal River. As a result, facies and sandstone compositions differ along strike in the basin. Site B-2 (bajada fringing the thrust belt) receives sediment from the Precordillera thrust belt only, which is composed of sedimentary and very low-grade metasedimentary rocks. However, erosion of deformed, slightly older foreland basin strata with the composition of A-1 recycles volcanic lithic grains. Sediment Site A-2 (alluvial fan fed by Jáchal River) is still dominated by the Frontal Cordillera provenance. Maps based on Jordan et al. (1990). Data for

Bermejo basin examples from Damanti (1989). Site A-1 "lower Huaco" example is the Jarillal Formation, and "lower Las Juntas" example is units older than 16.5 Ma from Las Juntas. Site A-2 ("middle Huaco") is based on the Huachipampa and Quebrada del Cura formations. Site B-2 ("upper Las Juntas") based on units younger than 16 Ma at Las Juntas. Because it proved difficult to differentiate devitrified rhyolites and chert in thin section with complete confidence (devitrified rhyolites are a major component of the source area and cherts minor), Qm is reported rather than Qp. For comparison, the upper pair of triangles also shows compositions of Holocene sands from rivers in Peru and Bolivia that drain the orogenic core zone, a situation similar to site A-1 ("type 1 rivers" of DeCelles and Hertel (1989)), and shows the recycled-orogen field of Dickinson and Suczek (1979). In Peru and Bolivia, the bedrock of the orogenic core and thrust belt both are largely sedimentary (and metasedimentary) rocks. DeCelles and Hertel's data and Dickinson and Suczek's field are recalculated for QmLvmLsm poles.

lacking in the extensive surface data. Their absence emphasizes the subtlety of depositional sequences in the proximal margins of foreland basins. Few surface data exist for the distal and forebulge regions of the basin, where unconformities might be best expressed.

In the absence of data for the distal part of the basin, and lacking seismic-stratigraphic analyses with which one might trace distal unconformities into conformable

proximal sections, what is the evidence for probable stratigraphic sequences in proximal and medial areas of the basin? The best evidence should be found in unconformities, facies changes, and accumulation rates.

At what scale might sequence-bounding unconformities occur in Bermejo basin sections? Beer (1990) showed, for strata that were several tens of kilometers from the contemporaneous thrust front (Fig. 9.8,

site B), that strata with rates of accumulation ranging from 0.24 to 0.88 mm/y are temporally complete at a resolution of less than 30,000 years (the best resolution of magnetic stratigraphy for late Miocene time [Berggren et al., 1985]). This rate implies that the section is complete (i.e., lacks unconformities) at a thickness resolution of less than 7.2 to 26.4 m. Thus, erosion represented by unconformities would remove less than this thickness of section, and would be resolvable only with high-resolution data. At what scale do unconformities occur in Bermejo basin alluvial and fluvial sections? Erosional relief between successive beds is extremely rare, and where present, is almost always less than 2 to 3 m. At this scale, it is difficult to differentiate sequence boundaries from boundaries between autocyclic repetitions.

Facies discontinuities occur at many levels in each stratigraphic section, but whether or not a given facies offset is related to a sequence boundary is difficult to determine. This uncertainty exists partly because down-dip continuity of exposure is usually much less than 50 m, and magnetostratigraphic correlations between sections separated by tens of kilometers are apparently too coarse for interpretations of subtle facies variations in terms of sequences.

Despite ambiguous evidence for unconformities or facies shifts, documented major changes in rates of accumulation through time (Fig. 9.16, inset) are almost certainly a signature of stratigraphic sequences; one of the controls on basin stratigraphy changed markedly. Rapid increases in accumulation rate upward in the section probably represent times of movement of thrust sheets located within the flexural wavelength of that site (N.M. Johnson et al., 1986; Reynolds et al., 1990; Milana, 1991; Jordan et al., 1993). If these pulses of high accumulation rate are responses to episodes of deformation, then the jump in the accumulation rate may reflect the conformable contact between an underlying stratigraphic sequence representing tectonic tranquility and an overlying sequence representing tectonic activity (if the sites are sufficiently close to the basin margin) (compare rates and sequence boundary in Fig. 9.10b,e to Fig. 9.16; Flemings and Jordan, 1990). However, the facies tend to be more proximal upward at the level where the rate increases (Beer and Jordan, 1989), in contrast to

models of tectonically controlled stratigraphic sequences, which predict that facies deposited on a low topographic gradient accompany increased tectonic loading (Figs. 9.9, 9.10) (Heller et al., 1988; Flemings and Jordan, 1990). Proximal shifts of facies may reflect an advance of the Precordillera thrust front many kilometers into the basin at the beginning of each deformation pulse (see above).

A noteworthy feature of these strata is the paucity of paleosols, which are indicators of hiatus (e.g., Kraus and Bown, 1988). Moderate pedogenesis is common in the basal parts of many sections, corresponding to times that magnetic stratigraphy reveals slow accumulation. Reddening and destruction of depositional laminae also exist in some proximal overbank facies, but soil profiles are poorly developed. Far more common is the preservation of fine-scale depositional laminae and lack of chemical alteration. Trace fossils in sheetflood sandstones are common, but they represent wet conditions immediately after flash-flood depositional events rather than long-term pedogenic conditions. Locally, roots extensively homogenized the depositional laminae, but without chemical alteration. Given the high rates of accumulation (which should be a common characteristic of nonmarine foreland basins), steadiness and lateral continuity of sedimentation (Beer, 1990), and arid climate (which need not be true of most foreland basins), the lack of soil development is expected.

In contrast to the foreland basin, the Iglesia piggyback basin on the western flank of the Precordillera has well defined stratigraphic sequences (Beer et al., 1990). The Iglesia basin is 70 km long by 35 km wide, lens-shaped in cross section, and has about 3500 m of Mio-Pliocene strata in the depocenter (Fig. 9.15). The structures that bound the basin at the surface are large folds rather than thrusts, and both margins were folded and uplifted simultaneously due to linking of subsurface structures (Allmendinger et al., 1990; Beer et al., 1990). As revealed on seismic data, the long-term trend was onlap of successively younger strata along the basin margins. This trend was interrupted by development of at least four sequence boundaries. At the time of formation of each sequence boundary, the region of accumulation contracted significantly toward the basin center,

forming unconformities along the basin margins that grade to apparent conformities at the basin center. Exposures near the basin margins reveal that the sequence boundaries range from surfaces across which provenance and drainages changed markedly, to gentle angular unconformities cut by channels, to smooth transitions with no local evidence of the unconformities (Beer et al., 1990). In contrast to Bermejo foreland strata, paleosols are common in the Iglesia basin. The Iglesia basin is a large-scale example of stratal accumulation on the flanks of a fold contemporaneous with the folding (e.g., Medwedeff, 1989). The intervals of accumulation represented by the stratigraphic sequences may reflect intervals of tectonic activity in the Precordillera, or they may have formed at times when no antecedent river drained through the Precordillera, or both (Beer et al., 1990).

Marine Example: Cretaceous Western Interior Seaway of North America

The Western Interior Seaway existed from Jurassic through early Tertiary time, spanning roughly 100 million years. During the Cretaceous, the Interior Seaway extended from Arctic Canada to the Gulf of Mexico, and was about 5000 km long by 1500 km wide during maximum flooding associated with Cretaceous eustatic highstand (Fig. 9.18) (Williams and Stelck, 1975; Kauffman, 1977, 1984). In the United States, the Sierra Nevada arc, Sevier thrust belt, and western part of the Western Interior Seaway constitute the magmatic arc, thrust belt, and retroarc foreland basin, respectively (Dickinson, 1976b). The Canadian part of the Cretaceous foreland basin may not be strictly a retroarc situation due to collisions of exotic terranes (Monger et al., 1982; Cant and Stockmal, 1989; see Chapter 11).

The width of the Western Interior Seaway greatly exceeds the widths of simple flexural moats related to tectonic subsidence driven by thrust belts (e.g., Fig. 9.4). The seaway is a composite of a western north-trending foreland basin with thick accumulations of subaerial and shallow-marine clastic strata (Figs. 9.19, 9.20), and a very broad eastern "axial basin" with thin accumulations, including extensive limestones, and greater water paleodepths (up to 500 m) (Fig. 9.20)

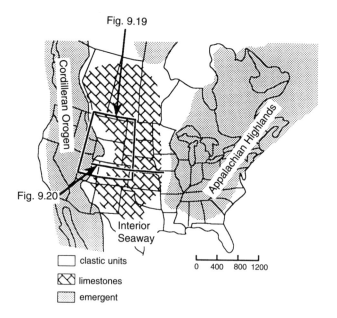

Fig. 9.18 Extent of the late early Turonian sea in the Western Interior Seaway near the time of maximum marine flooding (after Williams and Stelck, 1975), and distribution of clastic-rich and lime-rich facies (after Kauffman, 1984). The clastic units along the western margin of the basin constitute the principal foreland-basin fill; these grade eastward into thinner, deeper-water limestones.

(Kauffman, 1984). The mechanism responsible for subsidence in the "axial basin," at a great distance from the thrust belt, is controversial. The cause of the anomalously broad region of subsidence may be an underlying, shallowly dipping subducting plate: either loading due to the relatively high density of the cold subducting plate, or drag caused by flow in the asthenosphere induced by the temperature contrast between the subducting plate and the mantle (e.g., Cross and Pilger, 1978; Bird, 1984, 1988; Cross, 1986; Mitrovica et al., 1989).

The thrust belt that drove formation of the foreland basin of the Western Interior Seaway is typified by the Sevier thrust belt of Wyoming, Idaho, and Utah (Fig. 9.21c). The upper crust shortened by about 140 km (50%) and thickened by an average of 5 km (Royse et al., 1975; Jordan, 1981; Allmendinger, 1992). This thickening, if compensated by an elastic plate with a flexural rigidity of 10^{22} or 10^{23} Newton-meters, explains the general shape and volume of strata that accumulated in the western 600 km of the Western Interior Seaway (e.g., isopachs of Fig. 9.19b) (Jordan, 1981).

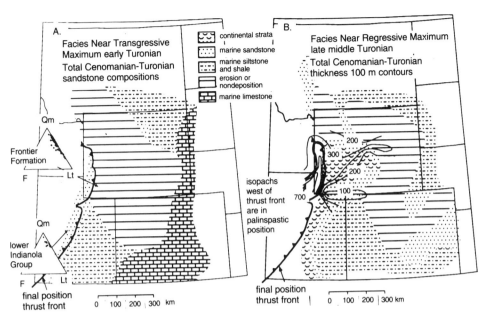

Fig. 9.19 Instantaneous distributions of facies and long-term nature of sandstone compositions and sediment accumulation, for ~7 my long Cenomanian and Turonian (location shown on Fig. 9.18). Distributions of dominant facies and regions of sediment bypass or erosion at times of A) high relative sea level and B) low relative sea level, mapped on a base of present positions (after Merewether and Cobban, 1986). In areas west of the easternmost thrust, facies patterns have been shortened by up to 150 km (e.g., Allmendinger, 1992). A is for the same time as shown in Fig. 9.18. Shown on A are data for compositions of nonmarine and marine sandstones, expressed as proportions of monocrystalline quartz (Qm), total feldspar (F), and total lithic grains (Lt) (Frontier data from Winn et al., 1984 [Cenomanian according to Winn et al., 1984, or Turonian according to Merewether and Cobban, 1986; Cenomanian through early Santonian Indianola data from Lawton, 1986a]). Superposed on B are isopachs of total Cenomanian and Turonian strata in basin adjacent to the Idaho-Wyoming salient of the thrust belt, mapped on a palinspastic base (after Jordan, 1981). The forebulge, defined by rapid thinning of isopachs to the east, is within the nonmarine zone at the time shown by the facies map, implying that the basin was overfilled, with a delta prograding across the forebulge.

The history of the forebulge is not well understood in the United States part of the Western Interior Seaway. Along much of the basin, the position of the forebulge (estimated only on the basis of thickness patterns during time intervals of several million years) appears to have been influenced by ancient zones of weakness in the crust (e.g., the Moxa and Sweetgrass arches); therefore, it may not have migrated in step with the shifting thrust and sediment loads (Jordan, 1981; Lorenz, 1982). Such long-term localization of a forebulge on a zone of weakness is consistent with the theoretical studies of Waschbusch and Royden (1992) (see above). According to Plint et al. (1993), a mobile forebulge is expressed in facies patterns and architecture of Cenomanian-Santonian marine units in the Canadian (Alberta) part of the foreland basin. There, stratal packages, whose average durations are 100,000 years, can be mapped, using detailed subsurface correlations; this permits recognition of short-term changes in bathymetry of the sea floor. The forebulge, located 200 to 500 km from the contemporaneous thrust front (Price, 1981), is expressed by four features (Plint et al., 1993): 1) Regional bevelling surfaces that are largely concordant in the west and erosionally truncate ~50 to 160 m of strata in the east. 2) East-facing notches in ravinement surfaces[2] that were eroded during east-to-west shoreline transgression, whose geometries indicate that the seafloor was varying through time in degree of eastward tilt. 3) Transgressive marine shales onlap from east to west above the beveling surfaces and ravinement surfaces described in (1) and (2). 4) Westward-prograding shoreline facies underlie the beveled surfaces of (1). Plint et al. (1993) suggested that the westward-prograding shorelines (4) formed on the western face of the forebulge, the eastward-tilting ravinement surfaces (2) and east-to-west onlap (3) formed on the

A.

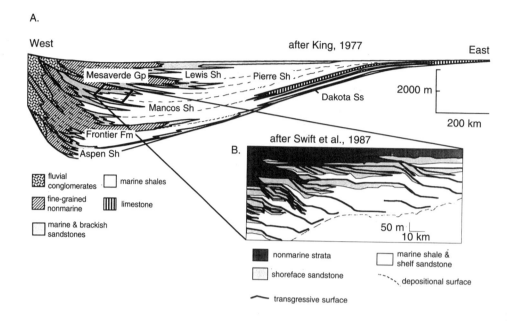

Fig. 9.20 A) Cross section of the Late Cretaceous foreland basin (location shown on Fig. 9.18). The Sevier thrust belt lay to the west, and strata thin at the western edge due to later erosion. The nonmarine clastic wedge prograded eastward into the seaway, generating deltaic deposits. At greater distances to the east, there was little clastic supply and limestones accumulated slowly. Note the complex pattern of transgression and regression of the western shoreline. B) Detail of the shoreline pattern in part of the Mesaverde Group, illustrating depositional topography of delta. Boundaries between depositional sequences are defined by slightly erosional transgressive unconformities. Modified from Jordan and Flemings (1991), based on King (1977) and Swift et al. (1987).

eastern face of the forebulge, and the beveling truncated the entire forebulge. At present, the historical positions of Plint et al.'s (1993) forebulge have not been reported.

Over long distances in the foreland basin, 3000 to 5000 m of strata accumulated in the western depocenter, and attempts have been made to determine the history of thrusting from the accumulation history (Fig. 9.21b) (e.g., Cross, 1986; Heller et al., 1986). Because of incomplete preservation, the long-term accumulation history can best be reconstructed in parts of the basin which were not originally the most proximal. Data from these locations show slow accumulation during the Late Jurassic, a decrease in rate of accumulation in the Early Cretaceous, followed by regional acceleration of accumulation near the beginning of the Late Cretaceous (Fig. 9.21b) (Cross, 1986; Heller et al., 1986). Slow accumulation during the Late Jurassic could be related to independently dated early deformation in the westernmost part of the thrust belt (Fig. 9.21b, "NW Utah - NE Nevada"). Rapid accumulation during the Late Cretaceous correlates to the time of motion of most of the eastern thrusts of the Sevier thrust belt (Fig. 9.21b,c, "Meade, Absaroka, Darby, Hogsback, Prospect").

Because the accumulation curves do not record the histories of proximal locations, they provide little or no information about the history of movement of the more western faults (Fig. 9.21, "Paris").

Foreland-basin facies and stratigraphic sequences in lower Upper Cretaceous (Cenomanian-Santonian) strata best illustrate response of the foreland basin to thrusting in the Sevier thrust belt (Figs. 9.18, 9.19) (see Fig. 9.22 for summary of chronology and stage names). Later in the basin history, beginning ~75 Ma, the basin ceased to be a simple retroarc foreland basin because of movement of a subhorizontal subducted plate beneath the region (Cross, 1986). After 75 Ma, superposed on thrust-belt-driven subsidence, there was an apparent long-wavelength subsidence that accommodated accumulation of 2 to 3 km of strata centered over 200 km east of the thrust belt (Cross and Pilger, 1978; Cross, 1986; Bird, 1984, 1988). Also, the region was broken by thick-skinned reverse faults and subdivided into local basins beginning at 75 Ma (see below).

The westernmost margin of the Interior Seaway consists of nonmarine clastic deposits, which grade eastward through large delta systems with sandy shoreline deposits, into muddy clastic marine shelf deposits (Fig. 9.20) (e.g., Weimer, 1970; McGookey,

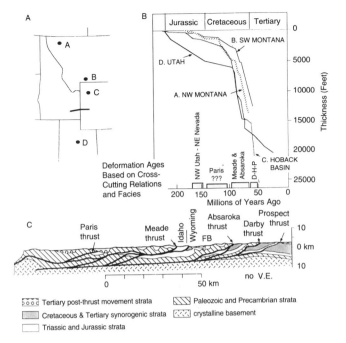

Fig. 9.21 Comparison of thrust history to accumulation history of sites in the foreland basin of the Western Interior Seaway. A) Location map: letters show sites used in B; line shows position of cross section illustrated in C. B) Accumulation histories in comparison to independently determined ages of deformation in the Sevier thrust belt. Accumulation histories (no backstripping corrections) after Cross (1986). Ages of deformation from Armstrong and Oriel (1965), Royse et al. (1975), Wiltschko and Dorr (1983), Allmendinger et al. (1984), and Heller et al. (1986). Age of Paris thrust shown as uncertain because of questions concerning paleontological samples (e.g., Heller et al., 1986). C) Structural geometry of Sevier thrust belt in southern Idaho and Wyoming, based on Royse et al. (1975), Bally (1984) and Allmendinger (1992). Heavy lines represent major thrust faults. Post-thrusting normal faults are not shown. FB indicates position of piggyback Fossil basin, projected along strike from the south.

1972; Ryer, 1977; Swift et al., 1987). Sandstones, conglomerates, and mudrocks of fluvial and fan-delta origin are the westernmost preserved facies (Lawton, 1986a; Pivnik, 1990). Deltas with little depositional relief (15 to 150 m) were common (Swift et al., 1987; Pivnik, 1990; Gardner et al., 1992). (In contrast, Campanian and Maastrichtian deltas had hundreds of meters of relief [e.g., Asquith, 1970; Perman, 1990]). Preservation of major accumulations of coal in deltas indicates that the stratigraphic record reflects long-term relative sea-level rise (Swift et al., 1987; Cross, 1988). These deltaic deposits consist of stacked series of offlapping progradational wedges; typically, each progradational unit is capped by a transgressive ravinement surface (Fig. 9.20, inset). Delta-front

deposits are dominantly shales and siltstones. Fine-grained sandstones in delta-front positions include turbidites (Pivnik, 1990; Van Wagoner et al., 1990) and hummocky cross-stratified units (Swift et al., 1987). The shales are commonly calcareous, suggesting slow accumulation.

Beyond the clastic prism, thin but regionally continuous limestones, interbedded with marine shales, are characteristic of the axial part of the Western Interior Seaway (Kauffman, 1977) (Figs. 9.18–9.20). Limestones were deposited primarily during times of high sea level, suggesting that transgression trapped clastic sediment close to source areas, causing sediment starvation across the rest of the Seaway. Bentonites are commonly concentrated in highstand strata of the axial basin (Kauffman, 1984), due to low rates of accumulation of other types of sediment. At times of lower sea level and less extreme sediment starvation, shales accumulated in the center of the seaway (Fig. 9.20).

Although it is clear that foreland-basin strata correlate in general to the time that the Sevier thrust belt was active, there is considerable uncertainty about how the characteristics of these strata reflect the thrust activity. This is largely because the ages of activity of the thrust faults are not known precisely. Methods of establishing times of motion on the major

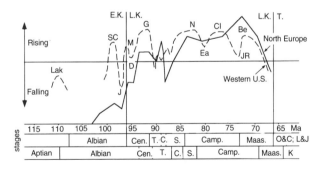

Fig. 9.22 Sea-level curves for the Western Interior Seaway and Europe. The similar sea-level history of the two locations leads to the interpretation that the large-scale transgressions and regressions of the marine foreland basin may be tied largely to eustatic variation. Abbreviations for stages: Cen. = Cenomanian; T. = Turonian; C. = Coniacian; S. = Santonian; Camp. = Campanian; Maas. = Maastrichtian; E.K. = Early Cretaceous; L.K. = Late Cretaceous; T. = Tertiary. Two commonly used time scales for the Western Interior are compared (O&C = Obradovich and Cobban [1975]; L&J = Lanphere and Jones [1978]; K = Kauffman [1977]). After Weimer (1986).

faults have included determining the ages of cross-cut stratigraphic units, determining the ages of appearance of diagnostic detrital populations in the basin, facies interpretations, and interpretations of accumulation rates (Fig. 9.21) (e.g., Armstrong and Oriel, 1965; Royse et al., 1975; Wiltschko and Dorr, 1983; Cross, 1986; Heller et al., 1986; Lawton, 1986a; Pivnik, 1990). The relationship between foreland-basin strata and thrust history is especially uncertain for strata deposited during the presumed early stages of deformation (e.g., Heller et al., 1986).

Links between thrust activity and depositional environments can be more readily demonstrated for extremely proximal parts of the basin, although proximal strata are rarely preserved (see above). In one case, a 500 m-thick transgressive unit, exposed within a few-hundred meters of the trace of the "old Absaroka" thrust fault, records active thrusting, cessation of thrusting, and subsequent quiescence of that thrust fault (Pivnik, 1990). At the scale of the entire Coniacian-Santonian Little Muddy Creek conglomerate, facies grade upward from subaerial fan-delta to shoreline, shelf and delta-slope deposits (although the shallow-marine facies define repeated progradational units at a smaller scale [5–25 m]). The age of "old Absaroka" thrust activity is bracketed, independent of the facies of the Little Muddy Creek conglomerate, to be Coniacian and/or Santonian: the conglomerate contains clasts derived from a Cenomanian-Turonian unit, and the conglomerate was cut by the "old Absaroka" thrust (Royse et al., 1975; Vietti, 1977; Jacobsen and Nichols, 1982; Merewether and Cobban, 1986; Pivnik, 1990). A comparison of source rocks uplifted above the thrust to the Little Muddy Creek clast compositions indicates that thrusting was active only during deposition of the lower part of the unit (Pivnik, 1990), which is the proximal progradational fan-delta deposit. Clast compositions also indicate that the overlying deposits accumulated during denudation of the quiescent highland, and thus that quiescence resulted in transgression immediately adjacent to the mountain front (Pivnik, 1990).

Several piggyback basins developed during progressive deformation of the Sevier thrust belt. The preserved piggyback basins are largely Cenozoic. The Fossil basin (Oriel and Tracey, 1970; Lamerson, 1982;

Steidtmann and Schmitt, 1988) and Axhandle basin of central Utah (Lawton and Trexler, 1991) are well preserved examples. The Fossil basin (Fig. 9.21C) accumulated 1700 m of fluvial and lacustrine strata during 15 million years, while being carried in the hanging wall of the Absaroka thrust. The Axhandle basin accumulated over 800 m of strata during ~25 million years in the Campanian to early Eocene, in the hanging wall of the Gunnison thrust. Repeated unconformities, which are best developed near the margins of the basin and pass into conformities toward the basin center, demonstrate that the underlying thrust experienced intermittent activity while the Axhandle basin filled (Lawton and Trexler, 1991).

Stratigraphic sequences of Cretaceous strata of the Western Interior Seaway are defined by unconformities and by vertical facies variations, and occur at various scales. Some of those sequences are probably due to eustatic sea-level fluctuation and some to tectonic episodicity; both interpretations have been made (e.g., Kauffman, 1984, 1985; Weimer, 1986; Swift et al., 1987). Other sequences may be due to variable sediment supply. Six large-scale "transgressive/regressive cycles," with periodicities averaging 6 million years, can be traced regionally (Kauffman, 1984). Small-scale progradational sequences, with durations as long as about 1 to 2 million years, are nested within those large cycles on the western margin of the Western Interior Seaway (e.g., Asquith, 1970; Gardner et al., 1992).

Commonly, the million-year-scale sequences (and similar shorter-scale sequences within them) in marine strata of the proximal margin are defined by progradation of clinoformal delta deposits, separated vertically by transgressive erosional surfaces (Fig. 9.20, inset) (e.g., Asquith, 1970; Swift et al., 1987; Gardner et al., 1992). Unconformities are poorly developed in the fluvial and deltaic strata of the proximal basin margin. Gardner et al. (1992) showed that the transgressive ravinement surfaces of the delta fronts merge landward with thick sections of delta-plain fluvial facies. Thus, fluvial sections accumulated sediment during times that transgression promoted erosion of the shorelines, whereas sediment bypassed the nonmarine realm (producing minor unconformities) at those times that accumulation in the delta

front and prodelta regions caused progradation. Swift et al. (1987) interpreted the transgressive parts of these sequences as evidence for pulses of accelerated relative sea-level rise, linked to pulses of accelerated thrusting; Van Wagoner et al. (1990) interpreted the same pulse-like pattern as evidence of eustatic sea-level change.

Plint (1991) described 100,000-year-duration sequences farther north, in the Alberta part of the foreland basin, that consist of progradational shelf successions, capped by erosion surfaces beneath shoreface sandstones, and terminated by marine flooding surfaces. In this case, the erosional surfaces were formed during falls of relative sea level. Plint (1991) reasoned that the relationships of these small-scale sequences to million-year-scale sequences, within which they are nested, is best explained if the short-term cycles reflect eustatic sea-level change.

Some of the most convincing clues to origin of the many sequences will come from high-resolution regional correlations which trace progradational units to the distal margin of the basin, and which define the architecture of those units (see above). Plint et al. (1993) provided examples, from Cenomanian-Santonian units in the Alberta basin, of lens-like units, with up to 150 m of regressive shoreline facies, alternating with westward-thickening wedges of coastal-plain strata. Those lenses and wedges represent much less than one million years. By comparison to the models (Fig. 9.11), Plint et al. (1993) attributed the interleaving to alternating times of increased (wedge-shaped accumulation) and decreased (lens-shaped accumulation) thrust loading. However, in the opinion of Plint et al. (1993), the vast majority of erosion surfaces are related to eustatic sea-level change, with a few exceptions that are due to tectonics.

Six major "transgressive/regressive cycles" of the Upper Cretaceous can be traced throughout the Western Interior Seaway (Weimer, 1960, 1986; Kauffman, 1984). Kauffman (1984, 1985) correlated these "cycles" to global eustatic cycles (Fig. 9.22), but also suggested that the transgressions correlate to times of possible thrust activity. Because the temporal resolution of thrust activity in the Sevier belt is much worse than the resolution of ages of the marine cycles, it is not yet possible to confidently differentiate between eustatic sea-level change and thrusting as causes for deposition of these "transgressive/regressive cycles."

Broken-Foreland Basins

Variants of typical foreland basins include "broken-foreland" basins (also referred to as "intraforeland", "Laramide", "tilted-block" or "intermontane foreland" basins (e.g., Chapin and Cather, 1981; Jordan and Allmendinger, 1986; Dickinson et al., 1988; Ingersoll, 1988b; Schwartz and DeCelles, 1988; McQueen and Beaumont, 1989). These basins form cratonward of thin-skinned thrusting, on both the footwalls and hanging walls of crustal-scale reverse faults that break the continental basement (Fig. 9.23). Basement uplifts and thin-skinned thrust belts commonly form contemporaneously (e.g., Fielding and Jordan, 1988). In retroarc forelands, this style of deformation is much less common than is thin-skinned thrusting (Rodgers, 1987). The typical result of deformation is a set of elongate, but variably oriented, nonmarine basins, many with closed drainages. The same areas used above as examples of simple foreland basins offer case studies of broken-foreland basins: the late Miocene to Recent Sierras Pampeanas (~10 my duration) of the Andean orogenic belt, and the latest Cretaceous-early Cenozoic Laramide province (~40 my duration) of the western United States (Fig. 9.2). In South America, the broken foreland has developed while the associated subducting plate has maintained an anomalously shallow angle beneath the continent (Jordan et al., 1983a; Jordan and Allmendinger, 1986). In North America, flat subduction is postulated to have coexisted with development of the broken foreland (Cross and Pilger, 1978; Dickinson and Snyder, 1978; Bird, 1984), although not all authors agree with this interpretation (Molnar and Lyon-Caen, 1988).

Because reverse faults of the broken foreland reach ductile levels deep in the crust or reach the mantle (e.g., Brewer et al., 1980), the crust is subdivided into short beams that can rotate about horizontal axes relative to one another when in-plane stress is applied (Fig. 9.24A) (McQueen and Beaumont, 1989). This rotation is a principal component of tectonic subsidence. In addition, as in the case of

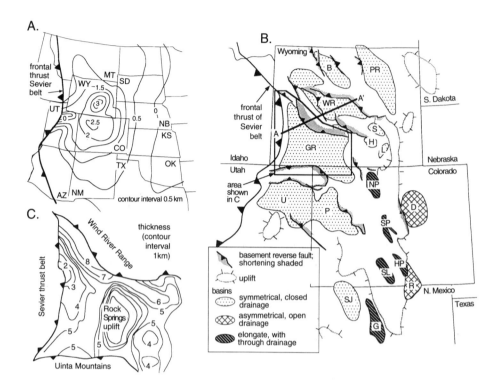

Fig. 9.23 Apparent subsidence patterns and sedimentary basins related to Rocky Mountain broken foreland. A) The thicknesses of uppermost Cretaceous (Campanian-Maastrichtian) units define a regional depocenter in eastern Wyoming and northern Colorado, too far from the thin-skinned thrust belt to reflect thrust-driven subsidence (after Cross, 1986). The thickness pattern implies that subsidence occurred on a scale exceeding individual fault-bounded structural blocks. B) Eocene basins and ranges, based on Chapin and Cather (1981). Amount of overhang on the basement reverse faults indicated by shading, from Gries (1983). "Uplifts," which formed contemporaneously with basin development, include monoclines and fault-bounded basement uplifts. (Basins are B=Big Horn, PR=Powder River, WR=Wind River, GR=Green River, H=Hanna, S=Shirley, L=Laramie, U=Uinta, P=Piceance, NP=North Park, SP=South Park & Echo Park, D=Denver, SL=San Luis, SJ=San Juan, HP=Huerfano Park, R=Raton, and G=Galisteo-El Rito.) Cross section A-A' shown in Fig. 9.24A. C) Thicknesses of units accumulated in the Green River basin during Maastrichtian, Paleocene and Eocene (from P.B. Flemings, 1985, unpublished compilation). Thickness increases toward the Wind River fault and the fault on north flank of the Uinta Mountains reflect tilting and bending of the crustal blocks.

simple foreland basins, the margin of each beam segment may also flex, in response to vertical loads (Hagen et al., 1985; McQueen and Beaumont, 1989). For plate-tectonic settings in which the subducted plate is nearly horizontal at a shallow depth beneath the continental lithosphere, there may be an additional regional-scale tectonic subsidence in broken-foreland provinces (Fig. 9.23A) due to coupling with the subducted lithosphere, flow of the lower crust (Cross and Pilger, 1978; Bird, 1984), or circulation of the asthenosphere above the subducted plate (Mitrovica et al., 1989). Typical dimensions of fault blocks and basins are 30 to 150 km wide by 100-300 km long (Fig. 9.23B), with structural relief across the

basin margins of as much as 12 km (Gries, 1983; Jordan and Allmendinger, 1986). Because the trend of major faults is quite variable and there is no systematic sense of vergence (Figs. 9.14, 9.23B) (Chapin and Cather, 1981; Jordan and Allmendinger, 1986), complex subsidence patterns emerge (Fig. 9.23C) that locally may involve strike-slip components (Chapin and Cather, 1981; McQueen and Beaumont, 1989).

Facies patterns are typified by rapid basinward changes from coarse basin-margin conglomerates to lacustrine facies in basin axes (Fig. 9.24C); whether the lacustrine units are evaporites, as in the Sierras Pampeanas (Malizzia, 1987) or deposits of permanent lakes, as in some Laramide basins (Surdam and Stan-

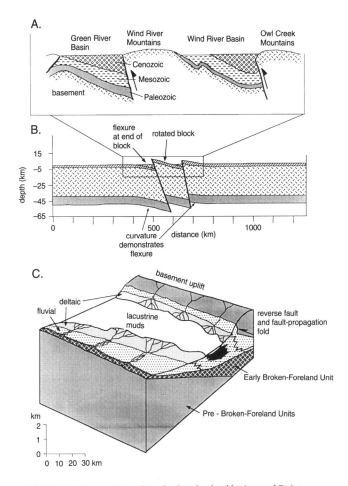

Fig. 9.24 A) Cross section of two broken foreland basins, and B) its interpretation in terms of a mechanical model. Modified from McQueen and Beaumont (1989). Approximate line of section shown in Fig. 9.23b. C) Facies mosaic in the Wind River basin during the Paleocene (after Flemings and Nelson, 1991).

ley, 1979; Dickinson et al., 1988) depends primarily on climate. Because the volume of potential sediment sources (uplifted basement blocks) is commonly small compared to areas of the basins, such basins are commonly underfilled (e.g., Crews and Angevine, 1988). Individual basins, however, may be linked to neighboring basins by river systems (Dickinson et al., 1988; Lillegraven and Ostresh, 1988), such that some underfilled basins are bypassed by sediment, while their downstream neighbors fill more completely. Similarly, rivers draining the main fold-thrust belt may fill some of the basins (Schmitt and Steidtmann, 1990). The erodability of the uplifts that flank a given basin control not only whether the basin fills, but

also the character of its facies. In the Laramide province, the ease of erosion of young mudstones during early stages of uplift of the basement blocks led to deposition of low-energy clastic facies reflecting little relief between ranges and basins, whereas subsequent exposure and erosion of more resistant crystalline basement rocks produced steep relief at the basin margins, which led to conglomeratic facies that are widespread within the basins (e.g., Flemings, 1987; DeCelles et al., 1991).

Due to limited sediment supply, broken-foreland basins may have more pronounced signatures of episodic tectonic activity than do simple foreland basins. Beck et al. (1988) reasoned that retrogradation of facies toward the faulted basin margin should accompany times of active deformation, with facies progradation during tectonic quiescence, essentially as illustrated in the theoretical models of simple foreland basins discussed above. In contrast, DeCelles et al. (1991) showed, for proximal regions in a Laramide basin, that local kinematics of faulting and folding overwhelm the regional trends of subsidence and uplift, producing proximal bypass of sediment during active deformation. In their example, sediment accumulates preferentially in distal parts of the basin during early episodes of shortening, and only accumulates in proximal areas after fault-propagation folds are broken by range-bounding reverse faults.

Challenges for Continued Research

The overall tectonic setting and mechanics of retroarc foreland basins are reasonably well understood. Continued studies of tectonics and mechanics are needed to refine rheological models of continental lithosphere and to measure more precisely the degree to which subsidence and uplift patterns are influenced by ductile flow of the lower crust or coupling to subducted lithosphere.

Given this basic understanding of the principal mechanical link between the thrust belt and basin, we are able to look critically at the details of the stratigraphic record in search of further information about deformation history and surface processes. Stratigraphic studies can make a major contribution to understanding of thrust belts, if the detailed move-

ment history of thrust faults can be deduced from foreland-basin strata. To measure subsidence based on the stratigraphic record requires regional three-dimensional description of thicknesses, ages, and facies. To fully understand the sedimentary basin requires that we partition the subsidence between thrust loads (true tectonic subsidence) and sedimentary loads that migrate in response to changing surface conditions. Whereas details of deformation at the frontal edge of the thrust belt can be resolved by study of detrital compositions and growth strata in proximal foreland-basin strata, these details may not reflect overall deformation in the thrust belt. Comparative studies of proximal trends and regional trends in foreland basins are required to resolve this problem.

Due to the natural combination of ample tectonic subsidence and a ready source of clastic sediment, proximal to intermediate positions in retroarc foreland basins have exceptionally complete stratigraphic records. Although intervals of nondeposition produce unconformities, the unconformities tend to be subtle. As a result, characteristics of stratigraphic sequences are markedly different from those of passive margins. A current challenge is to determine how episodic thrusting, eustatic sea-level variation, inplane-stress variations, climate change, or complex combinations of these phenomena affect the character of stratigraphic sequences .

This chapter paints an image of ideal behavior in retroarc foreland thrust belts and basins, and uses case histories largely to emphasize that real basins resemble the idealized models. However, the crust that predates deformation of any thrust belt and any foreland basin is non-ideal, with many inhomogeneities. These inhomogeneities play major roles in determining structural response and drainage patterns in the mountain belt, and subsidence patterns in the foreland basin. Any analysis of a foreland basin should be made with the likelihood of local or regional deviations from the idealized case firmly in mind.

Acknowledgments

I truly appreciate the guidance that Cathy Busby and Ray Ingersoll provided throughout the development and revision of this chapter. Critical reviews by Rick Allmendinger, Peter Flemings, Jim Beer, Dirk Vandervoort, Jeff Masek, Anibal Fernandez and Cathy Busby's class, coupled with discussion with Mike Gardner, greatly improved the clarity and content. I thank the National Science Foundation (EAR-8904537, EAR-9017168 and FAW EAR-9022811) for financial support during preparation of the manuscript.

Further Reading

Allen PA, Homewood P (eds), 1986, *Foreland basins*: International Association of Sedimentologists Special Publication 8, 453p.

Cross TA, 1986, *Tectonic controls of foreland basin subsidence and Laramide style deformation, western United States*: International Association of Sedimentologists Special Publication 8, p. 15-39.

Jordan TE, Allmendinger RW, 1986, *The Sierras Pampeanas of Argentina: a modern analogue of Rocky Mountain foreland deformation*: American Journal of Science, v. 286, p. 737-764.

Jordan TE, Flemings PB, Beer JA, 1988, *Dating of thrust-fault activity by use of foreland basin strata*, in Kleinspehn K, Paola C (eds), New perspectives in basin analysis: Springer-Verlag, New York, p. 307-330.

McGookey DP, 1972, *Cretaceous system*, in Geologic atlas of the Rocky Mountain region: Rocky Mountain Association of Geologists, Denver, p. 190-228.

Swift DJP, Hudelson PM, Brenner RL, Thompson P, 1987, *Shelf construction in a foreland basin: storm beds, shelf sandbodies, and shelf-slope depositional sequences in the Upper Cretaceous Mesaverde Group, Book Cliffs, Utah*: Sedimentology, v. 34, p. 423-457.

Endnotes

1. "Stratigraphic sequence" here refers to a packet of strata bounded above and below by a traceable, time-significant surface, across which the depositional system shifted laterally or changed. The term encompasses the concepts of "depositional sequences" (Mitchum et al., 1977) and "genetic sequences" (e.g., Busch and West, 1987; Galloway, 1989b). See also Walker (1990).

2. A ravinement surface is an erosional unconformity that is cut during transgression, by a shoreface high-energy wave regime that sweeps landward through time.

Remnant Ocean Basins

10

Raymond V. Ingersoll, Stephan A. Graham, and William R. Dickinson

Introduction

A *remnant ocean basin* is a shrinking ocean basin, which is flanked by at least one convergent margin and whose floor is typically covered by turbidites derived predominantly from associated suture zone(s). Graham et al. (1975) developed a general model for this class of basin; this model relates sediment dispersal to sequential suturing of orogenic belts (Fig. 10.1). Graham et al. (1975) used the Cenozoic development of the Himalayan-Bengal system as an analog for the late Paleozoic development of the Appalachian-Ouachita system (Fig. 10.2).

Rifted continental margins tend to be irregular in map view due to the alternation of coastal promontories and reentrants, the result of linked continental rifts, hot spots, transforms and failed rifts (Dewey and Burke, 1974; Şengör, 1976b). As a rifted continental margin approaches (generally obliquely) a subduction zone, coastal promontories collide first, resulting in diachronous orogenic uplift and erosion. Adjoining remnant ocean basins are the natural repositories for voluminous detritus eroded from the growing orogenic belts (Fig. 10.3). As sequential suturing progresses, the flux of sediment eroded from the growing accretionary orogen increases at the same time that the repository (remnant ocean basin) is shrinking as a result of continued plate convergence. The common result is extremely rapid sedimentation, followed closely by flexural loading to form peripheral foreland basins (see Chapter 11), terminal suturing, and ultimately, plate reorganization (e.g., Cloos, 1993). During the suturing of supercontinents (e.g., Laurasia and Gondwanaland), regionally diachronous and sequential closure of remnant ocean basins usually occurs (e.g., the Silurian through Permian Caledonide-Mauritanide-Appalachian-Ouachita Marathon

system) (e.g., Dewey and Kidd, 1974; Graham et al., 1975).

Details of collision processes and orogenesis are variable (Fig. 10.4). Major continent-continent collisions (e.g., Himalayan-Bengal and Appalachian-Ouachita systems) result in the largest orogenic and sediment-dispersal systems on Earth (e.g., Bouma et al., 1985; Dickinson, 1988). In contrast, collision of two intraoceanic arcs (e.g., Molucca Sea [Silver and Moore, 1978; Moore and Silver, 1983]) produces relatively small volumes of subaerially derived detritus. Nonetheless, significant accretionary wedges are formed and plate boundaries change as a result of any of these collisional processes. Between these end-member examples are several variants of collisional orogenic events involving rifted or transform margins and intraoceanic arcs (e.g., Taiwan [Teng, 1990] and Papua New Guinea [Crook, 1989]).

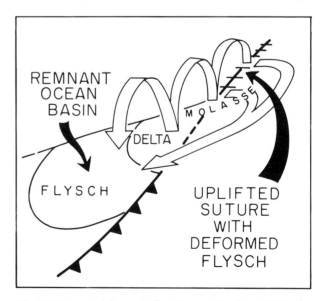

Fig. 10.1 Conceptual diagram to illustrate progressive incorporation of synorogenic flysch within an orogenic suture belt by sequential closure of remnant ocean basin. From Graham et al. (1975).

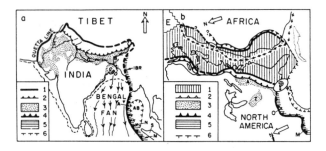

Fig. 10.2 Main tectonic elements of Himalayan-Bengal (A) and Appalachian-Ouachita (B) regions at same scale. Symbols: A. 1, Himalayan suture belt between India and Eurasia; 2, Main Boundary fault (thrust) of Himalayan foothills; 3. Quaternary alluvium of Indus (I), Ganges (G) and Brahmaputra (B) river systems; arrows denote channel trends on subsea Bengal Fan with head near Ganges-Brahmaputra Delta (GBD); 4, Indoburman-Sunda subduction zone, dashed where inactive; 5, melanges and deformed flyschoid rocks of Indoburman Ranges (IBR), Andaman (A) Nicobar (N) insular ridge, and Mentawai (M) Islands off Sumatra (S); 6, schematic margin of extensional Andaman basin (AB). Dashed line is 1,000 m isobath of continental slope off India. B. 1, Complex belt of multiple Paleozoic suturing between North America and West Africa cratons shown joined (E is Europe) prior to Mesozoic opening of the Atlantic Ocean approximately along the stitched line; 2, fronts of foreland fold-thrust belts of Appalachian Valley and Ridge province (American side) and Mauritanides (African side); 3, coarse Carboniferous clastic strata, terrestrial and littoral, of Appalachian (A) and Illinois (I) basins; 4, frontal thrusts and folds of Ouachita system; 5, outcrop and known subcrop of Ouachita system; 6, schematic extensional margin of Mesozoic Gulf of Mexico (present position of Cuba shown in dotted outline). Question mark shows location of Black Warrior basin at transition from Appalachian foreland to Ouachita remnant ocean basin. From Graham et al. (1975).

These examples have in common the rapid accumulation of turbidites (commonly called "flysch") immediately prior to accretion and suturing. Most such turbidites have been deposited in remnant ocean basins between colliding crustal components, and have been derived from orogenic belts along strike. All that usually remains of remnant ocean basins are the accretionary wedges formed by deformation of their sedimentary fill.

Sequential History

The early history of remnant ocean basins is as variable as the history of all oceanic crust. The age of oceanic lithosphere underlying remnant ocean basins is unrelated to the time of collision, so that very young or very old lithosphere may underlie the sedimentary cover. Rifted continental margins formed during the breakup of supercontinents (e.g., the eastern margin of India today and the southern margin [present orienta-

tion] of North America in the early Paleozoic) evolve from continental rifts through mature passive margins (see Chapters 1 through 4). Adjoining oceanic crust thermally subsides for approximately 80 m.y. to an average unsedimented depth of 5.5 km (Parsons and Sclater, 1977; Cloos, 1993). A complete thermal subsidence history characterizes mature passive margins drawn toward subduction zones (e.g., see discussion of Ouachita margin below).

As collision begins along one or more parts of the suture zone, clastic sediments eroded from uplifting areas flood into transversely adjoining peripheral foreland basins, transitional deltaic complexes and laterally adjoining remnant ocean basins (see Figs. 10.1, 10.5

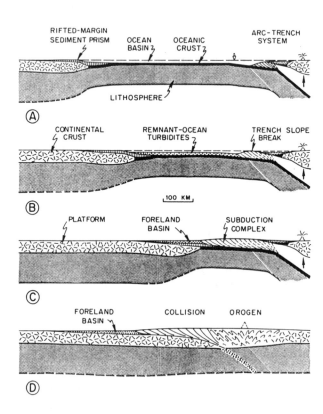

Fig. 10.3 Idealized true-scale diagrams showing inferred evolution (A to D) of sedimentary basins associated with crustal collision to form a cryptic intercontinental suture belt within a collisional orogen. Diagrams represent a sequence in time at one place along a developing collisional orogen, or coeval events at different places along a suture belt marked by diachronous closure. Hence, erosion in one segment (D), where suture has formed, provides sediment that is dispersed longitudinally through a peripheral foreland basin, past a migrating transition point (B to C), to feed subsea fans in a remnant ocean basin (B) along tectonic strike. (See Figs. 10.1 and 10.6.) From Dickinson (1976a).

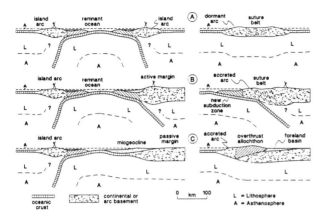

Fig. 10.4 Arc collision before (left) and after (right) suturing: A, intraoceanic arc-arc collision (new subduction zone may develop subsequent to collision on either flank of amalgamated arcs); B, intraoceanic-continental arc-arc collision, followed by initiation of subduction beneath flank of accreted arc (latter does not occur in all cases); C, intraoceanic-arc collision with passive continental margin (initiation of new subduction zone offshore, as in B, may reverse arc polarity). No vertical exaggeration; condition and thickness of lithosphere beneath active magmatic arcs uncertain (note queries). Remnant-ocean turbidites, limited in volume without influence of continent-continent collision (e.g., Fig. 10.3), not shown for reasons of scale.

and 10.6). Rates of subsidence and sedimentation increase rapidly as the tectonic and sedimentary loads of the growing accretionary wedge and sedimentary pile, respectively, affect the continental margin (Fig. 10.3). The transition from subduction of oceanic lithosphere (remnant ocean basin) to attempted subduction of continental crust (peripheral foreland basin) is complex in three dimensions. Transitional continental crust typically underlies large deltaic complexes formed in this setting (e.g., Ganges-Brahmaputra [Alam, 1989; Lindsay et al., 1991; Reimann, 1993]). Peripheral foreland basins receive voluminous detritus shed from the growing orogenic belt(s) (e.g., Tigris-Euphrates rivers entering the Persian Gulf); strike-slip zones may localize major drainages cutting through the orogenic belts. Big rivers (e.g., Ganges and Indus) drain along suture zones, across fold-thrust belts and longitudinally through foreland basins to build transitional deltas. The deltas, in turn, feed turbidite fans (e.g., Bengal and Indus fans) within remnant ocean basins. The four-dimensional nature of laterally suturing remnant ocean basins and peripheral foreland basins is illustrated by the fact that cross sections (e.g., Fig. 10.3) can be viewed as sequential

diagrams at any one location or as adjoining locations at one time; in either case, dominant sediment flux is perpendicular to the cross sections (e.g., Ganges-Brahmaputra-Bengal system).

Depositional Systems and Sediment Dispersal

Remnant ocean basins are dominated by turbidites and other submarine sediment gravitites. They clearly represent the largest sedimentary piles in the world today (e.g., Bengal and Indus fans); they are fed by the largest deltaic and fluvial systems, and derived from the greatest orogenic systems (e.g., Himalayas and Tibetan Plateau). Sediment in remnant ocean basins usually is destined to be incorporated into collision suture belts, where they comprise major segments of modern and ancient continental crust (e.g., Indoburman Range, Songpan-Ganzi terrane and Ouachita Mountains [Fig. 10.2 and below] and the Junggaro-Balkhash terrane of the Altaids [Şengör et al., 1993]). Most "flysch" deposits of orogenic belts were deposited as turbidites in remnant ocean basins (Graham et al., 1975). On the other hand, many other "flysch" deposits were deposited as turbidites during the early "starved" stages of peripheral foreland basins (e.g., Antler and Taconic flysch deposits) (see Chapter 11 for additional discussion). The diverse usage of the term "flysch" renders it almost useless in the absence of definition by the user; we recommend its use only as a local lithostratigraphic unit (e.g., specific Alpine Flysch units) or where its tectonic setting can be clearly specified (e.g., remnant ocean basin or initial stages of peripheral foreland basins) (also see Hsü, 1970; Miall, 1984). "Flysch" should not be used as a synonym for turbidites (e.g., forearc turbidites should not be called "flysch"). In the following discussion, we use "flysch" to refer to turbidites deposited in remnant ocean basins.

The dominant source areas for sediment fed into remnant ocean basins are rapidly uplifted and recycled sedimentary and metasedimentary strata of associated foreland foldthrust belts (Figs. 10.1 and 10.5) (Graham et al., 1976; Dickinson and Suczek, 1979; Ingersoll and Suczek, 1979). This recycled-orogenic provenance generates quartzolithic sand and sandstone, in strong

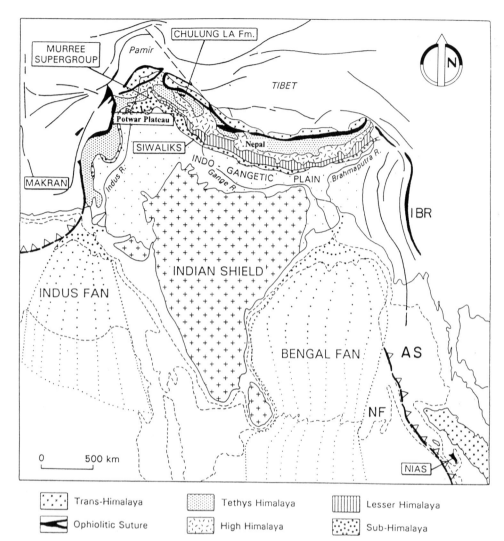

Fig. 10.5 Indian-Asian collision zone. Since initial collision during Eocene, enormous quantities of detritus have been shed from the Himalayas into ancient and modern peripheral foreland basins (Murees and Siwaliks, deformed and uplifted in Sub-Himalayan belt; modern Indo-Gangetic Plain) and remnant ocean basins (Indus and Bengal fans). Older parts of the Indus and Bengal fans have been (and are being) deformed at subduction zones, to form the Makran and Sunda accretionary complexes, respectively. Abbreviations: IBR, Indoburman Ranges; AS, Andaman Sea; NF, Nicobar Fan (subdivision of Bengal Fan). (Also see Figs. 10.8 and 11.21.) After Critelli and Garzanti (1994).

contrast with quartzofeldspathic compositions derived from basement uplifts and feldspatholithic compositions derived from magmatic arcs (Dickinson and Suczek, 1979; Dickinson, 1985). Basement uplifts and ancient magmatic arcs may locally contribute important components to remnant ocean basins, especially along long-lived and complex orogens. Nonetheless, on a regional scale, these sources are subordinate to sedimentary and metasedimentary sources.

Recycled-orogenic sand/sandstone composition varies due to the following factors: 1. larger cratons provide more quartzose compositions (Graham et al., 1976; Mack et al., 1983); 2. extreme uplift (as in the Himalayas) increases the proportion of feldspar, which is, nonetheless, still minor (Ingersoll and Suczek, 1979); 3. uplift of growing accretionary wedges results in unroofing, with increasing diagenetic/metamorphic grade through time (e.g., Taiwan; Dorsey, 1988); and 4. climatic effects can vary proportions of carbonate sediments and the overall compositional maturity (Suczek and Ingersoll, 1985). Some subduction complexes at trenches where remnant ocean basins are consumed also reflect this recycled-orogenic provenance, in spite of proximity of magmatic arcs (Moore, 1979; Critelli et al., 1990), because most magmatic-arc sand/sandstone is trapped within forearc basins (see Chapter 6).

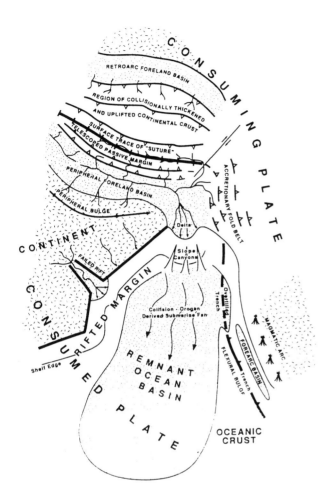

Fig. 10.6 Morphotectonic map showing most components of typical remnant ocean basins closing sequentially between colliding continents. Illustrated geography is analogous to the modern Bay of Bengal and surrounding areas (compare to Figs. 10.5 and 10.8), and to inferred Carboniferous of the Ouachita area (compare to Fig. 10.16). Details of components may be modified in many ways, as discussed in text (e.g., see Fig. 10.4).

Subsidence History and Basin Architecture

Remnant ocean basins generally begin their histories as deep oceanic basins, in contrast to thermally subsiding rifted margins and lithospherically loaded foreland basins. Initial depth of oceanic crust underlying remnant ocean basins is dependent on crustal age since formation at spreading ridges (e.g., Parsons and Sclater, 1977; Cloos, 1993) and this crust will still be thermally subsiding (very slowly if old) when it becomes a remnant ocean basin. Remnant ocean basins are preexisting depressions in Earth's surface

(relative to continental source areas); therefore, subsidence does not initiate due to tectonic processes (e.g., stretching or flexure of the lithosphere). Rather, subsidence is driven primarily by sedimentary loading as the oceanic lithosphere is confined between colliding buoyant crust and sediment flux to the basin increases. Determination of subsidence history is inherently difficult because of the likelihood that the sedimentary fill of remnant ocean basins will be deformed in accretionary wedges during suturing. Accurate chronostratigraphy is problematic in most deformed "flysch" sequences. Palinspastic reconstruction of remnant ocean basins is a daunting challenge!

Remnant ocean basins may be symmetric or asymmetric. Symmetric basins result where two subduction zones face each other (e.g., Molucca Sea) (Fig. 10.4). Asymmetric basins result where rifted or transform continental margins are drawn into subduction zones, either along continental margins (e.g., northeastern Bay of Bengal) or in front of intraoceanic arcs (e.g., Taiwan and northern Australia). In either case, longitudinal transport of sediment is the rule, with previously uplifted parts of the orogen providing the bulk of the detritus by way of rivers, deltas and submarine fans. Subduction zones seldom have surface expression as bathymetric trenches because sediment fills them faster than they subside. This is especially well illustrated by the Sunda arc. The Java Trench is deeper than 7000 m south of Java, where only oceanic sediments are being subducted/accreted; it shallows to 5000 m southwest of Sumatra, where the edge of the Bengal/Nicobar fan is being accreted (Hamilton, 1979) (Fig. 10.5). Northwest of Sumatra (west of the Andaman Islands), the bathymetric trench is replaced by a massive accretionary wedge, where the eastern edge of the Bengal fan is obliquely accreting. Major distributary channels of the Bengal/Nicobar fan system transport abundant sediment directly south, parallel to the Andaman Arc (Curray and Moore, 1971, 1974). A similar sedimentary symmetry combined with tectonic asymmetry characterizes the Indus fan, whose northern flank is being accreted along the Makran continental margin (Kolla and Coumes, 1987; Critelli et al., 1990) (Fig. 10.5).

Orogenic Development

Suture zones are inherently complex (Dewey, 1977; Dewey et al., 1989a; Şengör, 1990a, 1991d). Continental crust is compressed, thickened and shifted laterally as subduction of oceanic lithosphere forces nonsubductable blocks together (also see Cloos, 1993). Significant strike-slip motion is likely, within both continental and oceanic crust associated with suture zones. This is especially common where two continents collide (e.g., most of modern central to southern Asia). Where oblique oceanic convergence occurs (e.g., most of the Bengal Fan is converging obliquely with Southeast Asia) (Fig. 10.5), strike-slip motion is likely in intraarc and backarc segments of the overriding plate (e.g., Andaman Sea; Hamilton, 1979). During terminal suturing, deformation of remnant ocean turbidites may be extreme, to the point where detailed sedimentologic and paleoenvironmental studies are impossible.

During the evolution of complex suture zones, some tectonostratigraphic elements have higher preservation potential than others (Ingersoll, 1988b). Veizer and Jansen (1979, 1985) developed an empirical method of determining the "half life" of tectonostratigraphic elements; for "active-margin basins", the half life is only 30 my, but for "immature orogenic belts", it is 100 my. A major difficulty in applying the concept of "half life" to remnant ocean basins is that they evolve into "immature orogenic belts" during collision. A quantitative treatment of all types of sedimentary basins is needed before definitive statements can be made regarding their preservation potential. Nonetheless, our experience suggests that peripheral foreland basins are well represented in the ancient record (see Chapter 11), probably because they form late in the history of suturing and are found at the edges of highly uplifted orogenic belts. In contrast, remnant ocean basins are almost always obliterated during suturing, although the sedimentary fill commonly is preserved in the form of large accretionary wedges (e.g., Indoburman Range, Songpan-Ganzi and Junggaro-Balkhash terranes and Ouachita Mountains; see below, and Şengör and Okurogullari, 1991; Şengör et al., 1993). Transitional deltaic complexes have low preservation potential due to their locations at laterally migrating tectonic transition points (Fig. 10.1) immediately adjacent to evolving foreland foldthrust belts, within which sediments commonly are cannibalized. These deltaic complexes are most likely to be preserved at reentrants (commonly near syntaxes), where deltaic sedimentation may be concentrated for extended periods (e.g., Black Warrior basin; Fig. 10.2).

Variations

Remnant ocean basins develop between nonsubductable crustal blocks of any dimension (Fig. 10.7). At one extreme is the collision of major continents to form supercontinents (e.g., mid- to late Paleozoic suturing of Laurasia and Gondwanaland); major continents may also collide with subcontinents or intraoceanic arcs (e.g., modern Asia-India or Asia-Taiwan, respectively). At the other extreme are collisions involving two intraoceanic arcs (e.g., Molucca

MODERN	ANCIENT
Continent-Continent	
Bengal-Indus	Ouachita-Marathon (C - P)
	Southern Uplands (Scotland) (S - D)
Mediterranean Sea	Acadian Orogeny (S - D)
Gulf of Oman	Songpan-Ganzi (Ṱ)
Intraoceanic Arc-Continent	
	Lachlan Foldbelt (O - S)
	Alps-Carpathians (T)
NW Australia-Java	Apennines (T)
Sea of Japan	Persian Gulf (K - T)
Huon Gulf	Junggaro-Balkhash (D - P)
NE South China Sea	Taconic Orogeny (O)
	Antler Orogeny (D - M)
	NE Caribbean (T)
Intraoceanic-Continental Arc-Arc	
	Nevadan Orogeny (J)
Intraoceanic Arc-Arc	
Molucca Sea	

Fig. 10.7 Modern and ancient examples of continent-continent, intraoceanic arc-continent, intraoceanic-continental arc-arc, and intraoceanic arc-arc collisions. Examples are arranged approximately by size, with larger features at top and smaller features at bottom. See figure 10.4 and text for discussion.

Sea collision zone; Silver and Moore, 1978; Hamilton, 1979; Moore et al., 1981; Moore and Silver, 1983). The history of remnant ocean basins associated with continent-continent collisions may span many geological Periods (e.g., Silurian through Permian for Laurasian-Gondwanan system), although individual basins have shorter life spans. Most collisions involving intraoceanic arcs are shortlived (i.e., a few million years), so that associated remnant ocean basins are destroyed soon after their initiation. Sediment flux to remnant ocean basins is directly dependent on the rate and dimension of uplift in source areas in suture zones, so that overall sediment mass is huge for continent-continent collisions (e.g., Bengal and Indus fans) and relatively small for continent-arc collisions (e.g., Huon Gulf of the Solomon Sea; Crook, 1989; Silver et al., 1991). Accretionary wedges may grow to the point where their volume exceeds that of associated intraoceanic arcs (e.g., Taiwan orogen versus Luzon arc; Teng, 1990; Huang et al., 1992).

Collision of mid-sized continents (e.g., India) with large continents (e.g., Asia) is the ideal way to produce long-lived and voluminous sedimentary piles in remnant ocean basins fed by extremely indented and uplifted orogenic belts (e.g., Himalayas and Tibetan Plateau). The subcontinent of India is a relatively small part of the Indian oceanic plate, which is being pulled by subduction of dense lithosphere northward under Asia (e.g., Cloos, 1993). The small size of India means that its buoyancy cannot overcome the negative buoyancy of the Indian oceanic lithosphere, and it continues to be driven northward as an indentor (e.g., Tapponnier et al., 1986; Dewey et al., 1989a). In addition, the relatively modest size of India results in two large remnant ocean basins (Bay of Bengal and Arabian Sea) immediately adjacent to the rapidly growing orogen.

A further contributor to rapid erosion and production of voluminous detritus is the modifying effect of climate. There is growing evidence of accelerated uplift of the Himalayas and the Tibetan Plateau during the Miocene, which resulted in development of, or intensification of, the Asian monsoon system (e.g., Quade et al., 1989; Ruddiman and Kutzbach, 1989; Harrison et al., 1993). In effect, major collisional orogens create their own climates (as well as modifying

global climate); the greatest sediment accumulations probably form(ed) in tropical to subtropical settings.

The greatest accretionary bodies in the geologic record should have resulted from collision of a large continent with a moderate-sized continent, preferably in tropical to subtropical paleolatitudes. The Triassic Songpan-Ganzi terrane of central China is an excellent example (see Şengör and Okurogullari, 1991; Yin and Nie, 1993; and below). Collision of 2 moderate-sized continents also may create large accretionary bodies, especially when long-term complex plate interactions are involved (e.g., Altaids; Şengör et al., 1993). In contrast, collision of major continents (e.g., Appalachian-Ouachita system) results in terminal suturing, which eliminates most remnant ocean basins with less continental indentation and uplift, and shortens their life spans. Major remnant ocean basins, such as the Ouachita-Marathon example, are exceptions to this general rule. At the other extreme, collision of intraoceanic arcs and microcontinents results in rapid plate reorganization, without the long-term survival of remnant ocean basins.

Modern Examples

Bengal/Nicobar and Indus Submarine Fan Systems

Deposition of the immense Bengal/Nicobar and Indus submarine fans (Fig. 10.5) since the Eocene has been in direct response to the collision-induced uplift of the Himalayan suture belt (Curray and Moore, 1971, 1974). Recognition of this causal relationship between continental collision and synorogenic sedimentation in adjacent oceanic basins led to the first statement of the general model for remnant ocean basins (Graham et al., 1975). Remnant ocean basins both west and east of southern India are receiving voluminous sediment derived from the Himalayan suture belt, at the same time that their flanks are being accreted along subduction zones (Makran and Indoburman Ranges, respectively). Both systems are fed by immense fluvial and deltaic systems (Indus and Ganges-Brahmaputra, respectively). Some of the sediment derived from the Himalayas has accumulated in peripheral foreland basins of the Indo-Gangetic Plain

(e.g., Muree Supergroup and Siwalik Group), but most modern sediment bypasses these filled forelands and accumulates in the deltas and submarine fans along the continental margins. Much of the sediment has been recycled during various stages of erosion, transport and deposition.

The timing of continental collision between India and southern Asia is constrained by the following observations (Fig. 10.8): 1. The northward movement of India is tracked by magnetic anomalies of the Indian Ocean (e.g., McKenzie and Sclater, 1971; Johnson et al., 1976; Norton and Sclater, 1979; Patriat and Achache, 1984; Dewey et al., 1989a). 2. The passive margin of northern India experienced thermally influenced subsidence following rifting from Gondwanaland and prior to collision with Asia (Garzanti et al., 1987; Searle et al., 1987; Gaetani and Garzanti, 1991; Brookfield, 1993). 3. A peripheral foreland basin formed as the northern edge of India was pulled below the growing Himalayan orogen (Critelli and Garzanti, 1994). 4. The south-facing magmatic arc of southern Asia became inactive as the subduction zone was stifled by the attempted subduction of buoyant Indian continental crust (Graham et al., 1975). 5. Massive quantities of sediment were deposited along the flanks of India as the Himalayas were uplifted and the foreland basins filled (Graham et al., 1975; Curray, 1991).

The pre-Eocene history of southern Asia is complicated by the fact that several continental fragments collided to form the Cimmeride orogen as Paleotethys was closed during the Mesozoic (Şengör, 1984). From the Cretaceous to the Eocene, an arc-trench system characterized the southern Asian margin (Graham et al., 1975; Garzanti et al., 1987; Searle et al., 1987). The pre-Eocene history of northern India was dominated by passive-margin sedimentation during and following breakup of Gondwanaland during the late Mesozoic (Fig. 2 of Graham et al., 1975) (Garzanti et al., 1987; Gaetani and Garzanti, 1991; Brookfield, 1993). At the end of the Cretaceous, northern India was separated from southern Asia by the approximately 2,000km-wide Neotethys Ocean (Şengör, 1984; Garzanti et al., 1987).

By the early Eocene (approximately 55 Ma), the continental rise of northern India began to enter the Transhimalayan subduction zone, causing initial flexural bulging of the margin (Fig. 10.8) (Garzanti et al., 1987) (also see Chapter 11). The remnant ocean basin of Neotethys was bounded on the north by a trench, accretionary wedge and forearc basin (Garzanti et al., 1987), thus trapping most volcaniclastic detritus along the margin; to the south, Neotethys was bordered by the passive margin of northern India, which supplied limited quartzose detritus to the basin. As a result, no "flysch" analogous to the modern Bengal or Indus fans formed at this stage. Upon initial collision and uplift of the Himalayan margin in the Eocene, synorogenic sediment began to fill the evolving peripheral foreland basin and associated piggyback basins (Garzanti et al., 1987;

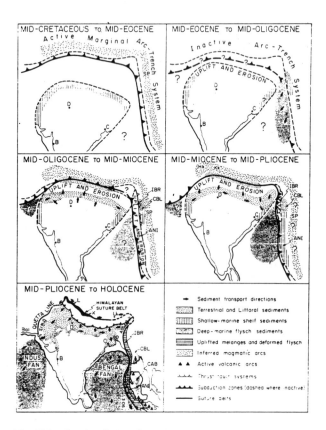

Fig. 10.8 Inferred evolution of Himalayan suture belt and surrounding areas. Abbreviations: ANI, Andaman-Nicobar islands; B, Bombay; Bal, Baluchistan; BD, Bangladesh; C, Calcutta; CAB, central Andaman basin; CBL, central Burmese lowland; D, Delhi; HK, Hindu Kush; IBR, Indo-burman Ranges; K, Kailas; Kk, Karakorum; L, Ladakh; Lh, Lhasa; 90E, Ninety-East Ridge; S, Sumatra; SP, Shan Plateau. Line off coast of India is present 1,000 m isobath. Central Andaman basin began to open in middle Miocene; early phases of opening not shown here. Modified from Graham et al. (1975).

Critelli and Garzanti, 1994); sediment was probably also transported east and west into remnant ocean basins away from the initial point of contact in the Ladakh area (e.g., Patriat and Achache, 1984). These early flysch deposits were quickly deformed and incorporated into the lengthening suture zone. Some of the older foreland deposits (fluvio-deltaic molasse) (e.g., Critelli and Garzanti, 1994), and (possibly) older flysch, are volcaniclastic, thus representing early recycling of forearc and subduction-complex detritus. Other than these oldest syn-collisional deposits, however, both molasse and flysch in the Himalayan system are overwhelmingly dominated by sedimentary and metasedimentary detritus, characteristic of recycled-orogenic provenance (Dickinson and Suczek, 1979; Ingersoll and Suczek, 1979; Suczek and Ingersoll, 1985; Garzanti et al., 1987; Critelli and Garzanti, 1994).

Once Neotethys was destroyed north of the Indian margin and peripheral foreland basins began to fill, the only remaining repositories for detritus eroded from the growing Himalayan suture were the remnant ocean basins west and east of India. Thus, the pre-Eocene continental-rise deposits of the rifted margins of west and east India are overlain by the Eocene to Holocene Indus and Bengal fans, respectively (and predecessors in Pakistan and the Indoburman Range) (Fig. 10.8) (Curray and Moore, 1971, 1974; Kolla and Coumes, 1987; Curray, 1991). Initial deposition was relatively slow, but by the Early Miocene, vigorous uplift of the Himalayas and Tibet resulted in rapid building of the submarine fans (Alam, 1989; Cochran, 1990; Copeland and Harrison, 1990; Klootwijk et al., 1992; Harrison et al., 1993). Since the Miocene, most sediment has been transported by the Indus and Ganges/Brahmaputra river systems to their respective deltas; much of the fluvial sediment bypasses the deltas and is transported, in places, over 3,000 km as turbidity currents to form the Indus and Bengal/Nicobar fans (Curray and Moore, 1971, 1974; Bowles et al., 1978; Kuehl et al., 1989; Cochran, 1990; Lindsay et al., 1991).

The northwestern edge of the older part of the Indus Fan is currently being deformed into the Makran accretionary wedge (Fig. 10.5) (Farhoudi and Karig, 1977; Critelli et al., 1990, and references

therein); almost symmetrically, the eastern edge of the Bengal/Nicobar Fan is being deformed into the Sunda accretionary wedge (Curray and Moore, 1971, 1974; Graham et al., 1975; Moore, 1979). The Makran is transitional westward into the continental-collision zone of the Zagros and the peripheral foreland basin of the Persian Gulf (Farhoudi and Karig, 1977), whereas the Sunda accretionary zone is transitional southeastward into the oceanic subduction zone of the Sunda arc (Hamilton, 1979). The Indoburman Ranges represent accreted proto-Bengal Fan (Graham et al., 1975). All evidence points to the ultimate derivation of most Indus and Bengal/Nicobar sediments from the Himalaya and bordering ranges (Ingersoll and Suczek, 1979; Suczek and Ingersoll, 1985). Most of the Makran accretionary wedge also has Himalayan sources, although younger shallower parts contain volcaniclastic detritus derived from the north (Critelli et al., 1990); Himalayan detritus constitutes most of the Neogene to Holocene Sunda accretionary wedge, although older parts of the accretionary wedge include considerable volcaniclastic detritus derived from the Sunda arc (Ingersoll and Suczek, 1979; Moore, 1979; Moore et al., 1982).

Both the Indus and Bengal/Nicobar fan systems are likely to continue receiving detritus from the Himalayas and Tibetan Plateau until their respective remnant ocean basins are closed by subduction. Final closure and destruction of these two remnant ocean basins is not likely to occur until additonal continent fragments or arcs move northward relative to India in the distant future. Alternatively, the Indian subcontinent could rotate clockwise to close the Arabian Sea, thus emplacing the Makran/Indus accretionary wedge over the west coast of India, or counterclockwise to close the Bay of Bengal, thus emplacing the Sunda/Bengal/Nicobar accretionary wedge over the east coast of India. The final result in any of these scenarios is the destruction of oceanic crust between colliding continents, and the creation of accretionary masses along suture zones (Graham et al., 1975; Şengör and Okurogullari, 1991). In rare cases, dormant ocean basins may persist, completely surrounded by collided continental blocks (e.g., North Caspian Depression) (see Chapter 1).

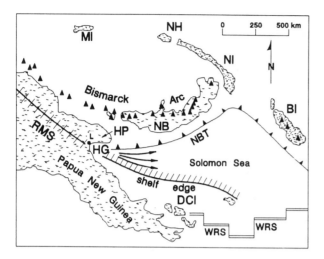

Fig. 10.9 Setting of Solomon Sea remnant ocean basin in Papua New Guinea. Barbed line is New Britain Trench (NBT) and crossed line is Ramu-Markham suture (RMS) formed by collision of Bismarck arc with continental Papua New Guinea; NBT-RMS transition point is near Lae (L) at head of Huon Gulf (HG) beside accreted Huon Peninsula (HP). Black triangles are arc volcanoes and arrows denote flow paths of turbidity currents transporting clastic detritus longitudinally to deep seafloor. Other symbols: BI, Bougainville I. (of Solomons chain); DCI, D'Entrecasteaux Is.; MI, Manus I.; NB, New Britain; NH, New Hanover; NI, New Ireland; WRS, Woodlark rift system. Note scale (distance from Lae to Bougainville is comparable to distance from London to Warsaw, or from San Francisco to Denver). (Also see Fig. 11.20.)

Huon Gulf

The Huon Gulf (HG) at the western extremity of the Solomon Sea (Papua New Guinea; Fig. 10.9) provides a modern example of clastic sedimentation in a remnant ocean basin being closed by the sequential collision of an intraoceanic island arc with a continental margin. Neogene collision of the western end of the Bismarck volcanic arc with the northern fringe of the Australian continental block has attached the Finisterre arc terrane of Huon Peninsula (HP) to mainland New Guinea (Jaques and Robinson, 1977; Cullen and Pigott, 1989). The suture zone lies along the longitudinal Ramu-Markham lowland that extends inland from the head of Huon Gulf (Crook, 1989; Silver et al., 1991). Active subduction continues beneath the New Britain segment of the Bismarck arc lying to the east of Huon Peninsula (Fig. 10.9).

The dispersal path for sediment derived from the erosion of rugged highlands flanking the Ramu-Markham suture follows the Markham River on land to the Markham submarine canyon, which leads down

Huon Gulf to the floor of the Solomon Sea. The onland Markham valley is flanked by conglomeratic alluvial fans and floored by the braided fluvial system of a trunk stream that passes into nonmarine and marine deltaic facies near Lae at the head of Huon Gulf. Syntectonic fluvio-deltaic deposition (of molasse) is coeval with turbidite deposition (of flysch) on the floor of the Solomon Sea and along the axis of the New Britain Trench (Crook, 1989). The Markham submarine channel system is incised partly into undeformed proximal turbidites and partly into deformed strata undergoing uplift near the longitudinal transition from molasse to flysch sedimentation (Silver et al., 1991). Undeformed turbidite sequences covering the deep seafloor near the western end of the Solomon Sea reach thicknesses in excess of 5000 m on seismic-reflection records.

As the collision suture propagates eastward, regions of flysch deposition are converted to molasse deposition as the turbidite fill of the Solomon Sea remnant ocean is deformed and succeeded by shallow-marine and nonmarine deposition of the Markham fluvio-deltaic system. Limited data suggest that the transition from flysch to molasse (i.e., remnant ocean basin to peripheral foreland basin) is migrating eastward as the suture evolves at a rate of 100-200 km/my (Crook, 1989; Silver et al., 1991). The facies change from flysch to molasse is thus time-transgressive in an orderly and monotonic fashion on a regional scale, but is complicated locally by multiple stratal discontinuities and interfingering lithosomes of disparate lithology.

Taiwan Arc-Continent Collision

Evolution of System

The late Cenozoic collision of Asia with the Luzon arc (Fig. 10.10) affords an opportunity to observe an arc-continent collision (Chai, 1972; Bowin et al., 1978) and to document the morphologic, structural and sedimentologic transitions along strike from pre-collisional settings to nascent remnant ocean basin to uplifted collisional orogen. Oceanic crust of the northeastern South China Sea was subducted obliquely eastward beneath the Luzon intraoceanic arc during the early Neogene until sometime in the late Miocene, when parts of the rifted/passive margin of Asia first encountered the Luzon convergent plate

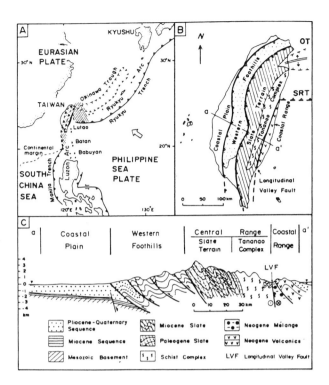

Fig. 10.10 Plate-tectonic setting (A) and geologic relations (B, C) of late Cenozoic Taiwan arc-continent collision. Northeasternmost part of South China Sea, shown in A, is a remnant ocean basin receiving detritus recycled from uplifted collision orogen on Taiwan, as well as from Asia. See text for discussion. Modified from Teng (1990).

boundary (Fig. 10.11; Teng, 1990). After about 5 Ma, a collisional orogenic belt developed in the area of initial contact at the northern tip of Taiwan, and because of the angle between the N-S Manila trench and the NE-SW Asian margin, propagated at least 400 km southward to its present ill-defined location near southern Taiwan (Teng, 1990). Plate convergence and consumption of South China Sea oceanic crust continue south of Taiwan, as demonstrated by sea-floor morphology, seismicity and volcanism of the Luzon-Manila arc-trench system (e.g., Suppe, 1984; Huang et al., 1992). To the north of Taiwan, the Okinawa trough has opened behind the Ryuku arc-trench system since 3 Ma, possibly as a consequence of the evolution of the collisional plate boundary to the south (Suppe, 1984).

The entire island of Taiwan may be envisioned as the collision zone (Fig. 10.12; Ernst et al., 1985). The rapid late Cenozoic uplift of Taiwan to present eleva-

tions as great as 4 km above sea level clearly is related to collisional orogenesis (Fig. 10.10). The geomorphology and basin development are direct reflections of the arc-continent collision and are reminiscent of other collision orogens. The fundamental petrotectonic boundary at the surface is the linear Longitudinal Valley that separates Neogene Luzon arc-affinity rocks of the Coastal Range of eastern Taiwan from rocks of Asian continental-margin affinity in the Central Range of Taiwan (Fig. 10.10; Page and Suppe, 1981). Synorogenic strata fill a peripheral foreland basin that is elongate parallel to the island beneath the Coastal Plain of western Taiwan and bounded on the eastern side by an active fold-thrust belt in the Western Foothills (Fig. 10.10; Covey, 1986). Synorogenic strata also were deposited in a "collisional basin" now deformed in the vicinity of the Longitudinal Valley (Figs. 10.10 and 10.11; Lundberg and Dorsey, 1988); this basin evolved as a successor to the northern continuation of the Luzon Trough forearc basin.

At the south end of the island, the Coastal Plain foreland basin, the Longitudinal Valley, the Coastal Range arc and other parallel elements of the orogenic system blend into corresponding, largely submarine morphotectonic features where collision has not yet occurred (Huang et al., 1992). The Coastal Range arc is contiguous with volcanically active islands of the northern Luzon arc and the Luzon forearc trough; the Hengchun ridge and Manila trench merge northward with the Longitudinal Valley-Coastal Range-Western Foothill collision complex (Fig. 10.13). To the southwest, South China Sea oceanic crust continues to subduct beneath the arc, flooring a remnant-ocean bight that narrows northward toward Taiwan between the continental rise of the Asian margin and the Manila Trench (Fig. 10.13).

The Taiwan collisional zone is an instructive example of a remnant ocean basin in terms of the limits it demonstrates: the geologically young collision has generated a remnant ocean basin, yet, the volume of synorogenic flysch either collapsed in the collision belt on Taiwan or in the modern ocean basin to the south is small (see below). Only recently has the system evolved enough to fully develop such facies. The critical issue is sediment supply. An aerially small, largely submergent arc system collided

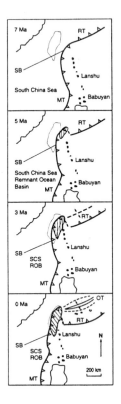

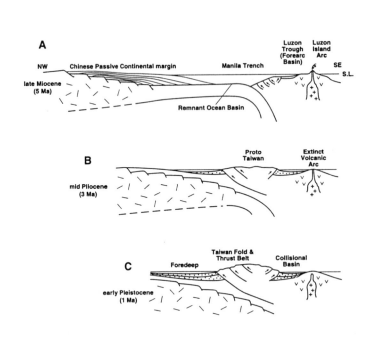

Fig. 10.11 Evolution of Taiwan arc-continent collision shown in plan view, at left, simplified from Teng (1990), and in cross section, at right, modified slightly from Lundberg and Dorsey (1988). MT: Manila Trench; OT: Okinawa Trough; RT: Ryukyu Trench; SB: shelf break; SCS ROB: South China Sea remnant ocean basin.

with a submerged rifted continental margin; only in the past 3 m.y. has the collision belt, the source for remnant-ocean-basin turbidites, emerged above sea level (Fig. 10.14; Teng, 1990). Uplift rates on Taiwan are high and have been increasing during the late Cenozoic (Teng, 1990), yet subaerial exposures are still only 36,000 km². Sediment is being dispersed generally southward parallel to tectonic grain into the northeastern South China Sea (Fig. 10.13), as demonstrated by provenance relations (Huang et al., 1992), but as yet, the supply has been insufficient to mask pre-collisional seafloor bathymetry.

The Taiwan example allows us to see sedimentary and structural relations at the onset of remnant-ocean-basin development, and suggests lower limits of sediment flux for remnant ocean basins. It suggests, for instance, that arc-continent collisions may not yield voluminous and petrologically distinct remnant-ocean-basin facies in ancient collision zones. This also presents a nomenclatural dilemma, discussed below, as to how best to name basin elements that are in transition, when no common label provides an entirely appropriate fit. Whether or not a

remnant-ocean-basin sequence develops in the next few million years as a voluminous turbidite apron blanketing the northeastern South China Sea and obscuring the Manila Trench will depend on continued collision, diachronous closure, uplift and erosion (enhanced by Taiwan's humid climate) as controlling factors on sediment supply.

Pre-Collisional Sequences on Taiwan

The arc-continent collision that has elevated Taiwan is sufficiently young and modest that it is still possible to discern the chief petrotectonic elements of the collisional orogen. The consumed margin of Asia, now represented in the orogen by the western three quarters of Taiwan, consists of a collapsed Paleogene-Miocene passive-margin sequence overlying Asian continental basement (Fig. 10.10; Ernst et al., 1985). The basement represents Mesozoic continental-margin arc batholiths and associated metamorphic rocks with metamorphic ages of 70 to 90 my (Ernst, 1983) that were rifted with the opening of the South China Sea early in the Paleogene (Taylor and Hayes, 1982).

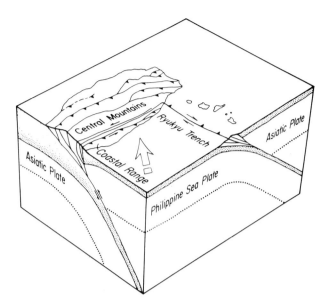

Fig. 10.12 Oblique block diagram showing Taiwan as an uplifted collisional orogen where Coastal Range arc of overriding Philippine Sea plate has encountered leading edge of Asian continental margin. Notice that details of structural style at boundary between Central Mountains and Coastal Range are complex (compare with Figs. 10.10 and 10.11). Components of both compression and strike slip are important in this zone. From Ernst et al. (1985).

This rifting event probably represented intra-arc and backarc spreading, in which case, the basement of western Taiwan is a continental remnant arc, which has subsided passively following rifting. A reversal of subduction polarity occurred in the late Paleogene, as shown in plate reconstructions (e.g., Karig, 1983a), in order to set up eastward subduction beneath the Luzon arc. In any event, the Paleogene-Miocene sedimentary cover of Asian basement exposed in western Taiwan is a siliciclastic, east-facing passive-margin sequence, including marginal-marine deposits in the Western Foothills to shelfal strata in the Central Range to slope strata up to 10 km thick in the east center of the island (Ernst et al., 1985). Detailed reconstruction of the passive margin is difficult because the entire sequence has been shortened by imbricate thrusting and is increasingly metamorphosed toward the east, where it is informally termed the "slate series" (Fig. 10.10; Ernst et al., 1985). The slate series is rootless in thrust sheets, which are para-autochthonous to the Asian margin based on provenance relations (Ernst et al., 1985).

A pre-collisional island-arc sequence, now represented by the eastern Coastal Range of Taiwan (Fig. 10.10), developed as the northernmost segment of the Luzon arc. Exposures of Lower Miocene, generally andesitic submarine volcanics of the Tuluanshan Formation comprise the arc basement (Fig. 10.14). This arc edifice is overlain locally by the thin reefal Kangkou Limestone and the upward deepening Takangkou Formation, which is characterized as "fly-

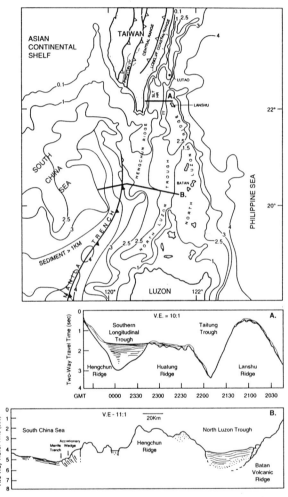

Fig. 10.13 Tectonic and morphologic relations between Taiwan and northeastern South China Sea remnant ocean basin. Note remnant-ocean-basin sediments ponded between Asian continental rise and Manila Trench. Older remnant-ocean-basin sediments and parts of Asian rise already have been offscraped and accreted to form uplifted Hengchun Ridge. HR: Huatung Ridge; SLT: southern longitudinal trough. Sources of map: bathymetry (km) and seismic line B simplified from Huang et al. (1992), thrust structures from Covey (1986), seismic line A from Chen et al. (1988), shaded area approximates region of sediment cover thicker than 1 km shown by Taylor and Hayes (1980).

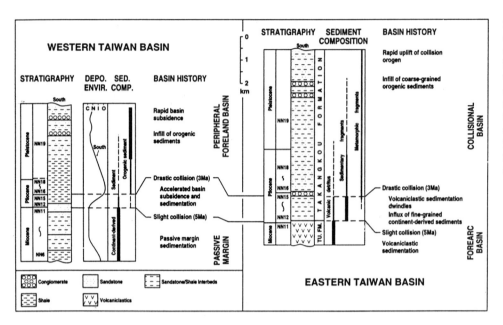

Fig. 10.14 Stratigraphic record of late Cenozoic Taiwan arc-continent collision, as reflected in peripheral foreland basin of western Taiwan, at left, and complex forearc-collisional basin of eastern Taiwan, at right. C: coastal; N: nearshore; I: inner offshore; O: outer offshore; TU.FM.: Tuluanshan Formation. Modified extensively from Teng (1990).

sch" by Ernst et al. (1985); the lower part of the Takangkou Formation contains magmatic-arc detritus (Fig. 10.14) and microfaunas indicative of water depths on the order of 1800 m (Page and Suppe, 1981). These biofacies, lithofacies and relationships to underlying arc rocks indicate that the lower part of the Takangkou Formation was deposited in a forearc basin along the west-facing northern Luzon arc (Fig. 10.11; Page and Suppe, 1981).

Remnants of oceanic facies that formerly lay between the Asian continental margin and the arc are limited to large blocks (up to 1 km² in outcrop) that were recycled into a syn-collisional olistostromal unit in the area of the Longitudinal Valley (Fig. 10.10). These blocks of ophiolite consist of upper parts of lower Miocene oceanic crust, including gabbro, peridotite, pillow basalt and red clay, which are of South China Sea affinity (Ernst et al., 1985), but which may be related to tranform faulting (Suppe et al., 1981), consistent with oblique subduction (e.g., Teng, 1990; Fig. 10.11).

Syn-Collisional Strata on Taiwan

The onset of collision occurred around 5 Ma (Teng, 1990); subsequently, oblique collision has sequentially closed the ocean from north to south by the length of the island, uplifted formerly submergent elements to 4 km above sea level, shortened the colli-

sional margin west of the Longitudinal Valley by at least 160 km (Fig. 10.11; Suppe, 1980a,b), induced at least 150 km of left slip along the Longitudinal Valley fault (Ernst et al., 1985), and caused clockwise rotation (Huang et al., 1992). Because uplift and exposure of the Asian margin did not occur prior to collision (Teng, 1990), the oldest distinctive Asian-derived recycled detritus is a significant indicator of collision (Fig. 10.14).

In the area of the Longitudinal Valley, where arc and Asian sequences are now juxtaposed, syn-collisional strata include upper elements of the Takangkou Formation turbidite sequence, which was deposited in an orogen-parallel trough-like basin (Chen and Wang, 1988). Volcaniclastic lower Takangkou sandstone is overlain by sedimentaclastic and metamorphiclastic detritus (Figs. 10.14 and 10.15). Teng et al. (1988) and Teng (1990) emphasized these differences in Takangkou provenance through alternative stratigraphic nomenclature that replaces "Takangkou" with "Fanshuliao Formation" for arc-derived strata and "Paliwan Formation" for strata derived from the Central Range (western source). Elevation and unroofing of the rising metamorphic core of the Taiwan orogen are recorded in sandstone compositions, whereby sedimentary lithic fragments are replaced upward by metasedimentary lithic fragments of increasing metamorphic grade (Fig. 10.15c; Dorsey, 1988) and detrital

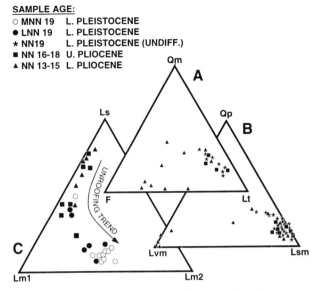

SAMPLE AGE:
○ MNN 19   L. PLEISTOCENE
● LNN 19   L. PLEISTOCENE
★ NN19   L. PLEISTOCENE (UNDIFF.)
■ NN 16-18 U. PLIOCENE
▲ NN 13-15 L. PLIOCENE

Fig. 10.15 Framework-grain composition of sandstone from late Cenozoic sedimentary basin of eastern Taiwan, revealing evolution of sand composition from volcaniclastic forearc basin to recycled-orogen provenance of younger, collisional phase (A and B). Detailed analysis of lithic fragments, shown in triangle C, reveals unroofing of metamorphic core of Taiwan collision orogen. Abbreviations: Qm=Monocrystalline quartz; F=total feldspar; Lt=total lithic fragments; Qp=polycrystalline quartz+chert; Lvm=volcanic+metavolcanic lithic fragments; Lsm= sedimentary+metasedimentary lithic fragments; Ls=sedimentary lithic fragments; Lm1=low-grade lithic fragments; Lm2=medium-grade lithic fragments. Redrawn from Dorsey (1988).

clay minerals show increasing levels of crystallinity (Buchovecky and Lundberg, 1988). Most spectacular is an olistostromal facies of the Takangkou Formation in the Pliocene section, the Lichi Melange (2.5 to 3.5 my old; Page and Suppe, 1981), which includes large blocks (in excess of 1 km) of ophiolite, sedimentary rocks of Asian affinity and Luzon arc volcanics (Page and Suppe, 1981; Barrier and Muller, 1984), reflecting the proximity of colliding elements in the final deep-marine phase of the collision. The sequence comprises principally turbidite deposits (Chen and Wang, 1988; Dorsey and Lundberg, 1988), but also includes shelf storm deposits (Dorsey and Lundberg, 1988) and is overlain unconformably by upper Pleistocene fluvial conglomerate of the Pinanshan Conglomerate (Page and Suppe, 1981). The Takangkou-Pinanshan basin subsequently was shortened by perhaps 70 km, and is being uplifted and recycled now along the flanks of the Longitudinal Valley (Dorsey and Lundberg, 1988; Lundberg and Dorsey, 1988).

These strata are transitional in tectonic setting, making their classification difficult. With forearc underpinnings (Fig. 10.11), Takangkou strata have variously been termed "forearc basin" (Teng, 1990), "remnant ocean basin" (Teng et al., 1988) and "defunct forearc basin" (Lundberg and Dorsey, 1988). Recognizing the collisional setting, Lundberg and Dorsey (1988) also applied "collisional basin" to the latest Neogene through Holocene phase of Takangkou-Pinanshan basin history. Simultaneous with uplift of the Asian passive margin during early collision was the development of a peripheral foreland basin in western Taiwan ("foredeep" of Fig. 10.11). In a manner parallel to the forearc/collisional basin of eastern Taiwan, the western Taiwan foreland basin includes a sedimentary sequence that reflects diachronous north-to-south basin development typified by shoaling from offshore marine to nonmarine environments, increasing rates of sediment accumulation and subsidence driven by flexural loading, and paleocurrent and provenance reversal from westward to eastward derived (Fig. 10.14) (Covey, 1986; Teng, 1990). These trends correspond to an increasing tempo of uplift of the Central Range inferred on the basis of mineral uplift ages (Teng, 1990).

South China Sea Remnant Ocean Basin
To the south of Taiwan, collision between Asia and the Luzon arc is just beginning, and the submarine morphotectonic elements can be traced southward into regions where South China Sea ocean crust continues to subduct beneath the Luzon arc via the Manila Trench (Fig. 10.13; Huang et al., 1992). Thus, the area south of Taiwan is a modern arc-continent remnant ocean basin; because of the modest surface area of the Taiwan orogen, it remains relatively free of thick turbidite sequences.

Bathymetry identifies morphotectonic elements of the system: the edge of the Asian continent is defined by the SE-facing shelf-slope-rise that trends NE to Taiwan (Suppe, 1984; Teng, 1990), with South China Sea crust occupying regions deeper than 3000 m between the Asian rise and the Manila Trench (Fig. 10.13). The trench and the parallel Luzon forearc basin and arc trend north to the orogen on Taiwan. Most authors show the Manila subduction front passing into the frontal thrusts of the western Taiwan foreland fold-

thrust belt (e.g., Page and Suppe, 1981), although some leave the transitional area blank, in recognition of the complexity of tectonic transition (e.g., Li, 1982). The northernmost active elements of the Luzon arc are the Batan-Babuyan Islands; Lutao Island and Lanshu Island expose arc volcanics as young as Pleistocene (Fig. 10.13; Huang et al., 1992).

Between 18 and 20 degrees N, a remnant ocean basin 100 to 200 km across and blanketed by more than 1 km of deep-marine strata exists (Taylor and Hayes, 1980; Huang et al., 1992). Sediment of recycled-orogen provenance is being supplied from the Central Range of Taiwan via the orogen-parallel Pingtun Valley (Huang et al., 1992) and from rivers that orthogonally drain westward across the Coastal Plain foreland basin to the Taiwan Straits. In addition, the coastal rivers of mainland China deliver sediment of Asian affinity to the shelf and rise, which comprises the western limit of the remnant ocean basin.

In detail, sea-floor morphology and underlying structure are complex in the remnant ocean basin for about 2 degrees of latitude south from Lutao Island to the pronounced kink in the Manila Trench at 20°N, where the first manifestations of collision probably occur (Figs. 10.10A and 10.13; Teng, 1990). The northernmost continuation of the northern Luzon trough and the southerly, offshore continuation of the Longitudinal Valley of Taiwan, termed the Southern Longitudinal Trough by Chen et al. (1988), are orogen-parallel troughs that pond turbidites between ridges of inferred accretionary complexes, the Hengchun and Huatung ridges (Huang et al., 1992). Sediment from Huatung Ridge has recycled-orogen provenance due to its derivation from Taiwan orogenic highlands (Huang et al., 1992). The ponded Southern Longitudinal Trough is filling with recycled-orogen detritus supplied by the longitudinally draining rivers of the Longitudinal Valley (Huang et al., 1992). Locally, the ponded turbidite fill has overtopped submarine ridges (Chen et al., 1988; Lundberg, 1988), smoothing seafloor topography (Fig. 10.13A) and apparently beginning to fill troughs of the Luzon forearc to the east and south (Lundberg, 1988), in a manner reminiscent of more voluminous remnant-ocean-basin systems. With continued collision, the ponded sequences likely will be deformed

along with intervening ridges (Lundberg, 1988).

The area immediately south of Taiwan is interesting because limited sediment supply and youthfulness of the orogen allow us to see the distribution of ponded remnant-ocean-basin turbidites and intervening uplifted structures of the most headward parts of the ocean basin. Not surprisingly, the structure of the remnant ocean basin in this focal region is complicated. If sediment supply is modest, as is the case in collision of intraoceanic arcs and rifted continental margins, then the distribution of remnant-ocean-basin facies is controlled by structure and resulting seafloor bathymetry. The result is orogen-parallel ponded areas, which with accretionary substrates in some areas, could be confused in the ancient record with trench-slope basins formed prior to any collision. Such details probably will not be easily decipherable in ancient systems (e.g., Chen and Wang [1988] inferred multiple late Cenozoic submarine-fan systems in the Longitudinal Valley region, but the evidence is equivocal).

Other Modern Examples

The only other examples of modern remnant ocean basins between major continents are the Mediterranean Sea between Europe and Africa, and the Gulf of Oman between the Arabian Peninsula and the Makran. The Mediterranean has an especially complicated history of collisions and microplate interactions between the major plates (Dewey et al., 1973, 1989b). All of the complexity of the Neogene to Holocene Mediterranean area (e.g., Boccaletti et al., 1990) is destined to be compressed into a suture zone when Africa terminally collides with Europe. Most of the eastern Mediterranean is a remnant ocean basin receiving sediment primarily from young suture zones (e.g., erosion of the Apennines provides sediment to the Ionian Sea; Critelli and Le Pera, 1994) and intraplate margins (e.g., north Africa, especially the Nile Delta; Bartolini et al., 1975). In contrast, the Tyrrhenian Sea is both a backarc/interarc basin relative to the Calabrian Arc and a remnant ocean basin relative to the European and African continents. Palinspastic reconstruction and paleotectonic interpretation of such an oceanic basin following terminal suturing would certainly be challenging!

Another example of possible ambiguity in interpreting backarc basins that become remnant ocean basins is the modern Sea of Japan. An incipient subduction zone is forming west of northern Japan (see Maruyama and Seno, 1986; Marsaglia et al., 1992; Hashimoto and Jackson, 1993); by this process, the backarc basin is being converted into a remnant ocean basin. Suturing of Japan to Asia would result in an orogenic belt intermediate in scale between the ongoing Himalayan and Taiwan collisions (see above), with a pre-collisional phase being dominated by backarc sediments being incorporated into growing accretionary wedges, and a syn-collisional phase being dominated by recycling of these uplifted wedges.

The extraordinarily complex area between southeast Asia and Australia includes several present or future remnant ocean basins (e.g., Hamilton, 1979). Subduction dominates the area, with many arc-related basins and accretionary wedges. Ultimately, however, as Australia continues to be pulled northward by subduction of the Indian plate, these basins, arcs and accretionary wedges are destined to be mashed between the continents in a complex suture zone. The accretion of Papua/New Guinea to northeastern Australia continues (see above) and the uplift of Timor as northwestern Australia is pulled into the Java Trench is just beginning (e.g., Veevers et al., 1978; Hamilton, 1979; Masson et al., 1991). Longitudinal transport of recycled orogenic detritus into remnant ocean basins is the inevitable result of these relatively modest collisions.

The only documented example of two intraoceanic arcs in the process of terminal suturing is the Talaud-Mindanao collision zone in the Molucca Sea, where the Sangihe and Halmahera arcs face each other and their accretionary wedges battle for supremacy (Silver and Moore, 1978; Moore et al., 1981; Moore and Silver, 1983; Silver et al., 1991). In this case, major accretionary wedges have formed despite the modest dimensions of associated magmatic arcs (also see Şengör and Okurogullari, 1991). The continued shrinking of this remnant ocean basin has resulted in ocean crust completely covered by accretionary wedges.

Ancient Examples

Ouachita Remnant Ocean

On the southern margin of the North American craton, the curvilinear Ouachita orogenic belt (Fig. 10.16) extends along tectonic trend for more than 2000 km, mostly in the subsurface, westward from the southern limit of the Appalachian orogen into northern Mexico (Arbenz, 1989; Viele, 1989). The chief lithic assemblage of the Ouachita system, as exposed in the Ouachita Mountains of Arkansas and Oklahoma, and in the Marathon region of west Texas, is a thick allochthonous succession of Paleozoic turbidites, chert, and mudrock that is thrust over coeval but contrasting platformal strata fringing the craton. The overthrust Ouachita-Marathon sedimentary assemblage is interpreted as the fill of a remnant ocean basin that was destroyed as the accreted assemblage was thrust cratonward over the continental margin during collisional orogenesis (Graham et al., 1975). The alternate hypotheses that the Ouachita succession was deposited in a backarc basin (Morris, 1974b) or within a failed rift trough (Lowe, 1985, 1989) do not account as well for overall geologic relationships (Viele and Thomas, 1989).

The southern or Ouachita margin of the North American craton was delineated by Cambrian rifting, related geometrically to the Precambrian rifting which established the Appalachian continental margin (Thomas, 1991). Several aulacogens (Fig. 10.16) extended into the adjacent craton as failed rifts oriented at high angles to the Ouachita continental margin. Sedimentation that initiated along the nascent continental margin, within the failed rifts, and on the floor of the adjacent ocean basin by latest Cambrian time continued without significant tectonic interruption through much of Paleozoic time. In the interval from mid-Pennsylvanian to earliest Permian time, however, diachronous thrusting, beginning first in the east and lasting longest in the west, carried the Ouachita-Marathon oceanic assemblage perhaps 75 km over shelf strata deposited along the continental margin. This terminal orogenesis reflected the attempted subduction of the continental margin beneath the flank of an arc-trench system that

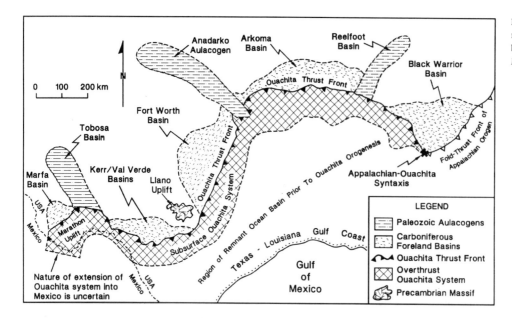

Fig. 10.16 Regional tectonic relations of Ouachita orogenic belt. See text for discussion. Modified after Thomas (1989b).

arrived from the south and faced the continental block (Graham et al., 1975; Wickham et al., 1976). The landmass upon which the arc was built is often termed Llanoria, and may be represented, in part, by the basement of Yucatan, rifted away from Texas in Mesozoic time to form the Gulf of Mexico.

Voluminous Carboniferous turbidites were transported as synorogenic flysch down the axial trough of the Ouachita remnant ocean basin from sources lying dominantly to the east near the syntaxis between the Appalachian and Ouachita orogenic belts. As thrusting proceeded, synorogenic basins that developed in front of advancing allochthonous masses evolved from a remnant ocean basin to a string of discrete foreland basins floored by continental crust (Fig. 10.16). As basin architecture changed, turbidite (flysch) sedimentation in the remnant ocean basin was succeeded by deltaic and nonmarine (molasse) deposition in the foreland basins. Intermediate phases of basin evolution and the transition from flysch to molasse deposition involved transient hybrid basins floored by partly oceanic and partly continental substratum (Houseknecht, 1986). The nature of basement beneath sediment deposited during the transitional stages of tectonic evolution is uncertain because synorogenic sedimentary sequences were detached from underlying basement as the Ouachita allochthon overrode the continental margin.

Summary accounts (McBride, 1970, 1989; Morris, 1974b, 1989; Ethington et al., 1989; Lowe, 1989) of the overthrust Ouachita-Marathon sedimentary assemblage permit the recognition of successive stratigraphic intervals that represent sequential phases of deposition during basin evolution (Fig. 10.17). Sub-Carboniferous strata are preorogenic deposits of an open ocean basin lying south of a passive Ouachita continental margin, whereas Carboniferous strata are synorogenic deposits of the remnant ocean basin, and related depocenters, that developed just before and during arc-continent collision in the Ouachita region. Interpretations of nearly all facies relationships are partly speculative because sedimentary packages were telescoped and shuffled as multiple thrust sheets during Ouachita orogenesis. The schematic geohistory diagram of Figure 10.18 depicts inferred patterns of tectonic and isostatic subsidence during evolution of the Ouachita system as exposed now in the Ouachita Mountains.

Preorogenic Sedimentary Succession
A lower Paleozoic (uppermost Cambrian to mid-Upper Ordovician) sequence of shaly hemipelagites and associated sandy turbidites was derived from the adjacent craton and deposited along or near the base of a newly formed continental slope and rise. Although less than 1000 m thick in the Marathon

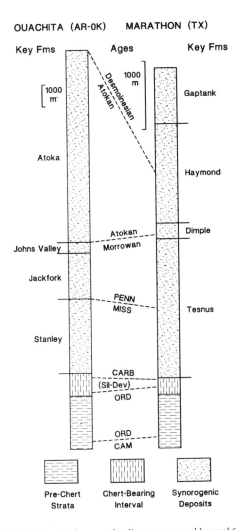

OUACHITA (AR-OK)   MARATHON (TX)

Key Fms   Ages   Key Fms

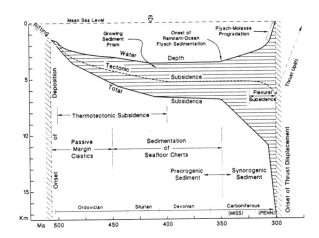

Fig. 10.17 Relative thicknesses of sedimentary assemblages of Ouachita system (see Fig. 10.16) in Ouachita Mountains (left) and Marathon region (right); note difference in scales of schematic stratigraphic columns. Pre-chert strata form post-rift succession deposited prior to accumulation of seafloor cherts. Synorogenic deposits filled remnant ocean basin during sequential crustal collision (transition from turbidite flysch to fluviodeltaic and shelfal molasse occurs in upper horizons of Atoka and Haymond formations; all synorogenic units are siliciclastic except Dimple Limestone of carbonate turbidites in Marathon region).

region (McBride, 1989), and no more than 2500 m thick in the Ouachita Mountains (Lowe, 1989), this basal interval includes multiple formational units of heterogeneous lithology. Sediment accumulation rates were 25 to 50 m/my. Sparse paleocurrent indicators, lenticular quartzose to quartzofeldspathic and locally conglomeratic sandstone units, calcarenitic turbidites derived from platform carbonate sources, and lensoid olistostromal beds all reflect transport of

debris downslope into deep-marine environments from the nearby passive continental margin (Lowe, 1989; McBride, 1989). The presence of coarse detritus within the dominantly shaly strata implies deposition within a proximal part of the nascent ocean basin that developed south of the rifted Ouachita continental margin.

A middle Paleozoic (uppermost Ordovician to lowest Mississippian) sequence, approximately 300 m thick in the Marathon region and 600 to 900 m thick in the Ouachita Mountains, includes prominent Upper Ordovician and Devonian units composed dominantly of bedded spicular and radiolarian chert with interstratified argillite (Lowe, 1989; McBride, 1989). Overall sediment accumulation rates were 5 to 10 m/my, but were less than 5 m/my for some of the chert intervals, which are regarded here as biogenic pelagites deposited on a deep seafloor formed by thermotectonic subsidence of oceanic or transitional

Fig. 10.18 Hypothetical geohistory diagram for Ouachita succession of Ouachita Mountains (see Fig. 10.16). Thicknesses from Lowe (1989) and Morris (1989), and ages from Ethington et al. (1989) using DNAG time scale (Palmer, 1983). Thermotectonic subsidence constrained to rates established for cooling of oceanic lithosphere (350 m times square root of elapsed time in million years after rifting) for a period of 100 my after rifting, with no tectonic subsidence thereafter until onset of flexural subsidence under influence of structural loading by thrust sheets of Ouachita allochthon. Flexural subsidence constrained with flexural geometry inferred by Goebel (1991), assuming allochthon movement of 10 km/my (faster rate of 100 km/my would confine flexural subsidence to last 2.5 my of depositional history, and would smooth elbow of water-depth curve at transition from aggradational to progradational phases of flysch sedimentation, but would also sharpen corresponding elbow in curve for total subsidence of substratum). Backstripping constrained by net sediment densities inferred from equations for depth-porosity relations given by Dickinson et al. (1987).

crust lying south of the shelf break along the passive Ouachita continental margin. In the Ouachita Mountains, 300 to 400 m of uppermost Ordovician and Silurian turbidites and associated hemipelagites occur between Upper Ordovician and Devonian chert formations. Paleocurrent indicators of westerly paleoflow and quartzolithic sandstone of orogenic derivation in the Lower Silurian Blaylock Sandstone of the Ouachita Mountains are suggestive of distal derivation from the Taconic orogen of the Appalachian region (Viele and Thomas, 1989).

Synorogenic Sedimentary Succession
Synorogenic Carboniferous strata of the Ouachita remnant ocean basin and the successor troughs that developed during the orogenic movements which closed it reach an aggregate thickness of 12,000 to 15,000 m in the Ouachita Mountains; analogous strata in the Marathon region are less than 5000 m thick (McBride, 1989; Morris, 1989). Facies relations in the two areas are similar, but different in detail. In general, the transition from flysch to molasse deposition occurred during mid-Pennsylvanian time. Deformation was complete before the end of Pennsylvanian time in the Ouachita Mountains (Viele and Thomas, 1989), and by the end of Wolfcampian (earliest Permian) time in the Marathon region (Ross, 1986). Sediment accumulation rates were perhaps 125 m/my during the initial phases of turbidite sedimentation, but increased to almost 1000 m/my before deposition was terminated by thrust-dominated compressional deformation. Accommodation space presumably was enhanced by the flexural effects of thrust loading as the accretionary Ouachita allochthon advanced toward the craton of the Ouachita foreland.

Mississippian depositional systems record the earliest delivery of sediment derived from Llanoria. Widespread turbidite paleocurrent indicators of northwesterly paleoflow are dominant in the Stanley Group of the Ouachita Mountains and in the Tesnus Formation of the Marathon region. Lateral facies patterns in both units are more proximal to the southeast and more distal to the northwest (Niem, 1976; Flores, 1977). Both units also contain ashfall and subaqueous ashflow tuffs, apparently derived from arc eruptions farther south (Niem, 1977; Imoto and McBride, 1990).

Paleocurrent indicators in Pennsylvanian flysch (Jackfork and Atoka formations) of the Ouachita Mountains record dominantly longitudinal paleoflow westerly along the axis of the trough that formed as the remnant ocean basin narrowed. Regional tectonic relationships (Graham et al., 1975) and the distribution of turbidite facies (Moiola and Shanmugam, 1984; Link and Roberts, 1986) imply that the main depositional system was an elongate submarine fan built westward from an initial apex near the southern end of the Appalachian orogenic belt. The Ouachita flysch fan was analogous to the modern Bengal fan built into the Bay of Bengal from an apex near the eastern end of the Himalayan orogen, and was of generally comparable size (Graham et al., 1975).

Facies relationships and paleocurrent trends are more complex, however, in flysch units deposited as the Ouachita allochthon began to override the continental margin. Thrust loading of the foreland induced normal faulting in the basement (Houseknecht, 1986), and locally caused segmentation of the residual Ouachita trough into parallel elongate sub-basins with turbidite paleocurrent indicators of local paleoflow to the east as well as the west (Ferguson and Suneson, 1988). Moreover, in the Marathon region, carbonate turbidites of the Dimple Limestone were transported toward the arriving allochthon from sources along a forebulge shelf, and local fan deltas containing coarse clasts derived from the advancing allochthon prograded over foredelta turbidites of the mid-Pennsylvanian Haymond Formation (Flores, 1978).

Ouachita Flysch Sources
The immense volume of the Ouachita flysch and molasse requires a provenance able to supply sediment rapidly to the remnant ocean basin and its derivative troughs. The quartzose to quartzolithic character of the bulk of the Carboniferous sandstones (Fig. 10.19) implies recycling of sedimentary and metasedimentary detritus without unroofing of deep crustal basement, although significant feldspathic components in some units reflect additional contributions from igneous sources. Given the regional tectonic setting of the Ouachita system, we inferred (Graham et al., 1976) that the source of the voluminous detritus was the vigorously uplifted interior of older increments of the

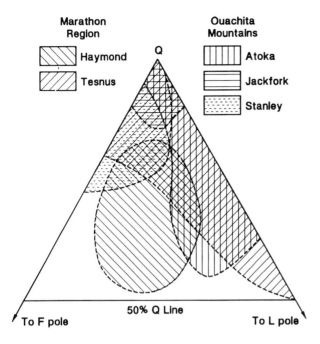

Fig. 10.19 QFL diagrams for compositions of sandstones in Carboniferous flysch and molasse of Ouachita sedimentary assemblages (data from McBride, 1966, 1970, 1989; Flores, 1974; Morris, 1974b, 1989; Graham et al., 1976; Morris et al., 1979; Lowe, 1985, 1989; Houseknecht, 1986; Wuellner et al., 1986). See Figure 10.17 for relative stratigraphic positions of five formational units for which compositional fields are plotted. Note that baseline of diagram denotes 50% monocrystalline quartz grains.

Appalachian-Ouachita orogen, which developed diachronously by sequential closure of the oceanic region that lay between Laurentian and Gondwanan segments of Pangea. In this view, the longitudinal dispersal of turbidites down the axis of a closing remnant ocean basin systematically preceded orogenic deformation along each increment of the evolving orogenic belt. On a regional scale, sequential initiation of coarse-clastic sedimentation within foreland depocenters along the cratonal flank of the Appalachian-Ouachita belt indicates diachronous suturing, from northeast to southwest, of Laurentian and Gondwanan continental blocks (Fig. 10.20).

East of the Ouachita Mountains of Arkansas and Oklahoma, subsurface trends of the Ouachita orogen meet the southern projection of the Appalachian orogen in the area south of the Black Warrior basin (Thomas, 1984, 1989a; Hale-Erlich and Coleman, 1993), a foreland basin nestled within the tectonic syntaxis where the two orogens join (Figs. 10.2 and

10.16). The Black Warrior basin is, thus, situated where the apex of the turbidite fan of the Ouachita remnant ocean basin would initially have been located as longitudinal progradation began. Its position in the Carboniferous was analogous to the present position of the Ganges-Brahmaputra fluviodeltaic plain of Bangladesh within the syntaxis where the Himalayan and Indoburman orogens join. In accord with expectation, the types of lithic fragments and the proportions of different kinds in Pennsylvanian flysch sandstones from the Ouachita Mountains and coeval molasse sandstones from the Black Warrior basin are indistinguishable (Graham et al., 1976). Sources in both the Appalachian and Ouachita orogens apparently fed detritus into the Black Warrior basin (Mack et al., 1981, 1983; Liu and Gastaldo, 1992), and down the Ouachita flysch trough to the west. In the Marathon region of west Texas, the combination of both longitudinal and transverse paleocurrent indicators in Carboniferous turbidites (McBride, 1970, 1989) suggests that detritus was also fed into the flysch trough from the evolving Ouachita orogen as continental collision proceeded.

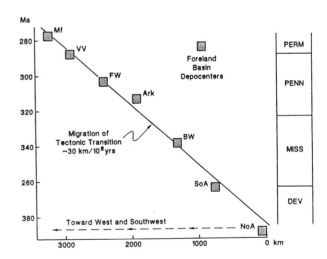

Fig. 10.20 Approximate ages (plotted boxes) of oldest post-suture clastic strata in foreland basins along southeastern flank of North American craton (adjacent to Appalachian-Ouachita orogen): Mf, Marfa (Luff and Pearson, 1988); VV, Val Verde (Wuellner et al., 1986); FW, Fort Worth (T.J. Bornhorst, 1977, personal communication); Ark, Arkoma (Houseknecht, 1986); BW, Black Warrior (Thomas, 1988); SoA and NoA, southern and northern Appalachian (Milici and deWitt, 1988). Indicated mean rate of migration of "flysch-molasse" transition is approximately 30 km/my as collisional Appalachian-Ouachita belt evolved. Modified after unpublished diagram by T.J. Bornhorst.

Songpan-Ganzi Complex

Evolution of System

Prominent on any geologic or tectonic map of Asia is a vast triangular region in central China underlain by deformed Triassic deep-marine strata (Zhang et al., 1984; Zhu, 1989). These rocks comprise the Songpan-Ganzi Complex, long recognized as an accretionary complex of deep-marine "flysch," trapped between cratonal blocks in the tectonic collage of Asia (Fig. 10.21) (Şengör, 1981, 1984; Klimetz, 1983; Şengör and Hsü, 1984; Ji and Coney, 1985; Watson et al., 1987). We believe, following Yin and Nie (1993), Nie et al. (1993) and Zhou and Graham (1993), that, with present surface area of 220,000 km² (Huang and Chen, 1987), the Songpan-Ganzi Complex may be the largest, best preserved record of sedimentation within, and tectonic accretion of the fill of, an ancient remnant ocean basin. If so, it may hold important answers about the evolution of remnant ocean basins in general, as well as the evolution of the Asian tectonic mosaic. Accordingly, we summarize what is known and reasonably inferred about the Songpan-Ganzi Complex.

The Songpan-Ganzi Complex is nestled among continental nuclei within the pan-Eurasian Cimmeride orogenic system (Şengör, 1981, 1984, 1987e; Klimetz, 1983). To the north of the Songpan-Ganzi Complex is an amalgam of cratonal elements, including Tarim, Qaidam and Sino-Korea, or following Yin and Nie (1993), the "North China Block" (NCB) (Fig. 10.21). The southern margin of the NCB has a complicated and incompletely understood Paleozoic history as a convergent margin (Ji and Coney, 1985; Mattauer et al., 1985; Şengör and Okurogullari, 1991; Eide, 1993; Yin and Nie, 1993). The contact between the Songpan-Ganzi Complex and the NCB is primarily tectonic, and where relatively well described, as in the Qinling Mountains, involves nappes of Triassic flysch in thrust contact with NCB basement and platform cover rocks (Hsü et al., 1987). To the southeast lies the South China Block (SCB), cored by the Yangtze craton. The SCB and its Paleozoic - lower Triassic carbonate platform cover (Yin and Nie, 1993) are in tectonic contact with the Songpan-Ganzi Complex across the Longmen Shan

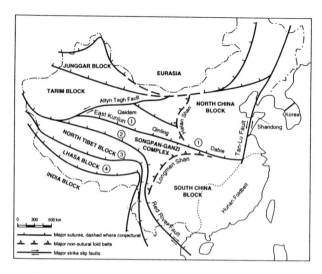

Fig. 10.21 Selected tectonic features of China, emphasizing major accreted blocks, sutures between those blocks, and fault and fold systems discussed in text. Note triangular Songpan-Ganzi Complex, interpreted here as the accreted fill of a Triassic remnant ocean basin, surrounded by North China, South China and North Tibet blocks. Sutures: 1 = Kunlun-Qinling-Dabie, 2 = Jinsha River, 3 = Bangong Lake and 4 = Yarlung River. Map modified and simplified from Watson et al. (1987).

fold-thrust belt (Fig. 10.21) (Şengör, 1984; Zhang et al., 1984; Klimetz, 1985), which evolved in late Triassic time as a peripheral foreland system during closure of the Songpan-Ganzi ocean basin (Huang and Chen, 1987; Lu et al., 1990). The North Tibet (Qiangtang) block and one to three accreted volcanic arcs and associated ophiolite belts of late Paleozoic to early Mesozoic age tectonically bound the Songpan-Ganzi Complex to the west and southwest (Şengör, 1984; Şengör and Hsü, 1984).

The timing and geometry of assembly of these continental blocks remain controversial. As the Songpan-Ganzi Complex is bounded on two sides by the NCB and SCB, and wedges out eastward between them (Fig. 10.21), it is the history of the collision of the NCB and SCB that is most significant. Conclusions regarding timing of this collision have revolved largely around the age and distribution of ophiolitic, fold-thrust and magmatic belts of inferred subduction origin; the timing, nature and distribution of metamorphic rocks; and paleomagnetic data. Disparate views have arisen about the closing of the ocean between the NCB and the SCB: mid-Paleozoic (Mattauer et al., 1985) and late Paleozoic suturing (Zhang et al., 1984) have been advocated largely on

geologic/geochronologic grounds, whereas Triassi closing has been inferred on the strength of paleo-magnetic and geologic arguments (e.g., McElhinney et al., 1981; Klimetz, 1983; Şengör, 1984).

These opposing perspectives are reconciled, to some extent, in recent studies (Watson et al., 1987; Eide, 1993; Yin and Nie, 1993) that argue for diachro-nous ocean closure between NCB and SCB from latest Paleozoic in easternmost China to Triassic in the west in the Qinling and Kunlun Mountains (Fig. 10.22). Diachronous closing of the NCB-SCB remnant ocean basin in late Paleozoic through early Mesozoic time explains the following: (1) paleomagnetic data place NCB and SCB apart in Permian time and together by Middle Jurassic time (McElhinny et al., 1981; Klimetz, 1983; Nie, 1991; Enkin et al., 1992), (2) coesite-and-diamond-bearing ultrahigh-pressure (UHP) metamor-phic terranes within the nominal NCB-SCB suture zone in the Dabie Mountains and offset of these ter-ranes along the Tan-Lu strike-slip fault to the Shan-dong Peninsula (Wang and Liou, 1987, 1991; Yin and Nie, 1993), (3) the curious distribution of strike-slip offset along the Tan-Lu fault, which is as great as 540 km, but dies to nothing within the SCB just south of the Dabie Mountains (Okay and Şengör, 1992; Yin and Nie, 1993), (4) a pronounced decrease in meta-morphic grade and probably age of metamorphism from the Dabie Mountains westward toward the Qin-ling Mountains (Wang and Liou, 1991; Eide, 1993;

Yin and Nie, 1993), (5) the limited time span (5 to 35 my) between peak UHP metamorphism and uplift cooling ages (Eide et al., 1992; Eide, 1993), (6) the absence of an early Mesozoic magmatic arc along the Dabie segment of the NCB (e.g., Zhang et al., 1985), (7) the apparent absence of Songpan-Ganzi flysch in the Dabie suture region (Zhang et al., 1984, 1985), although Hsü et al. (1987) noted that presumed lower Paleozoic metaflysch within the Qinling segment may be much younger, and (8) the westward younging of initiation of rapid clastic sedimentation in the SCB peripheral foreland basins (Yin and Nie, 1993).

Yin and Nie's (1993) hypothesis of diachronous ocean closure accounts for all of these phenomena and suggests that diachroneity of closure is related to irregularity of the northern passive margin of SCB as it collided with the smoothly arcuate convergent margin of NCB (Fig. 10.22), as commonly is the case (Dewey and Burke, 1974; Graham et al., 1975; Dewey, 1977). In this scenario, a northeastern penin-sula of SCB first impacted the NCB convergent mar-gin in the east during the Permian in Shandong and the Early Triassic in the Dabie region farther west, driving a crustal wedge that produced UHP metamor-phism, collisional strike-slip faulting and thrusting analogous to the effects within Asia of the late Ceno-zoic collision of India and Asia (e.g., Molnar and Tap-ponier, 1975, 1977). The NCB-SCB collision elevated and erosionally exhumed high-grade metamorphic

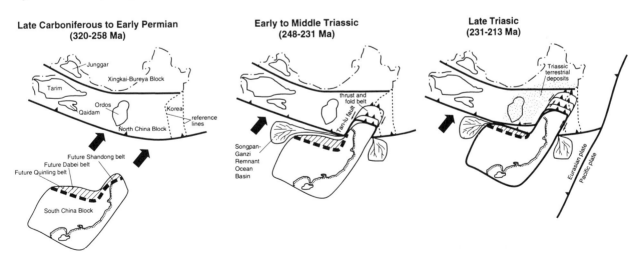

Fig. 10.22 Development of Songpan-Ganzi remnant ocean basin and its flysch fill as a consequence of diachronous collision of North China and

South China blocks, as envisioned in tectonic model of Yin and Nie (1993). Simplified and modified slightly from Yin and Nie (1993).

terranes; rapid erosion was enhanced by the tropical humid climate that prevailed by Late Triassic time (Wang, 1985; Huang and Chen, 1987). The diachronous collision also established the Songpan-Ganzi remnant ocean basin to the west as a receptacle for detritus eroded from the collision orogen (Fig. 10.22), which itself was deformed by continued ocean closure in the latest Triassic. A variant on this theme (Eide, 1993) emphasizes the 60 to 90-degree paleomagnetic azimuthal discrepancy between the NCB and SCB in the Late Permian (e.g., Lin et al., 1985; Zhao and Coe, 1987) and suggests that diachroneity was related to clockwise rotation of SCB relative to NCB after initial collision in the east (Ji and Coney, 1985), where the pivotal area became the focus of maximum crustal telescoping, metamorphism and strike-slip faulting (Eide et al., 1992; Eide, 1993).

Syncollisional Strata

The character, age and origins of the rocks of the Songpan-Ganzi Complex are poorly known (Huang and Chen, 1987), apparently due to geographic isolation, complex structure and monotonous bedding style, but the term "flysch" has been widely applied to sedimentary elements of the complex (e.g., Wang, 1985). Structural slivers of ultramafic rocks are rare (Zhang et al., 1989), and depending upon definitions of terrane boundaries, the arc assemblages of the southwest Songpan-Ganzi area have sometimes been included within the complex. Some authors (Zhang et al., 1984; Wang, 1985; Ren et al., 1987) have alluded to a microcontinent concealed beneath the flysch, but evidence is scant (e.g., Huang and Chen, 1987; Şengör and Okurogullari, 1991). The Songpan-Ganzi flysch is extensively folded, faulted and metamorphosed to varying degrees (Yang et al., 1986; Zhang et al., 1989). Reliable estimates of shortening across the complex are not available.

The flysch overlies Carboniferous pelagic limestone, which in turn, overlies oceanic basement within the Songpan-Ganzi Complex (Zhou and Graham, 1993). The stratigraphic column in the western Qinling Mountains generally consists of Lower Triassic pelagic carbonates and carbonate turbidites overlain by Middle Triassic siliciclastic turbidites (Zhou, 1987). The most striking deposits in the Middle Triassic section are olis-

tostromes containing large blocks of carbonate-platform facies, presumably derived from carbonate-dominated margins (Zhang et al., 1984; Zhou, 1987). Many blocks contain Devonian-Permian faunas that clearly predate the Triassic sediment-gravity flows in which they are entombed (Zhang et al, 1984; Zhou, 1987). Elsewhere in the Songpan-Ganzi Complex, the flysch sequence ranges into the Upper Triassic (Huang and Chen, 1987; Ren et al., 1987). The flysch sequence is overlain with angular unconformity by less deformed, nonmarine Jurassic strata (Huang and Chen, 1987). The total thickness of the section is poorly known, with intact sections greater than 5 km thick in the Qinling Mountains (Zhou, 1987), and total thickness variably estimated from 7 km (Yang et al., 1986) to 20 km (Huang and Chen, 1987)!

The source(s) of the Songpan-Ganzi flysch remain(s) poorly documented. Huang and Chen (1987, p. 32), for instance, suggested Qinling, Kunlun and Qilian sources, but stated that, in view of the massive volume of the flysch, the source "remains an enigma". Şengör (1984; Şengör and Okurogullari, 1991) suggested that the Songpan-Ganzi Complex is an accretionary wedge that accumulated in front of the eastern Kunlun arc segment of the NCB by off-scraping during head-on subduction, and Şengör (1984; his Fig. 16) depicted the Songpan-Ganzi Complex as bathymetrically isolated from sediment sources in the region of the NCB-SCB collisional orogen.

In contrast, reconstructions invoking east-to-west diachronous collision of the NCB and SCB predict that the voluminous flysch of the Songpan-Ganzi was the fill of a remnant ocean basin lying along tectonic strike to the west of the region of early collision (Fig. 10.22) (Watson et al., 1987; Yin and Nie, 1993; Nie et al., 1993; Zhou and Graham, 1993). Several lines of evidence favor, or are consistent with, this interpretation. First, the Triassic age of the flysch matches the timing of HP and UHP metamorphism (244–201 Ma; Eide, 1993) and uplift cooling (230–195 Ma; Eide et al., 1992) of the Dabie suture zone. Second, the current $2.2 \times 10^5 km^2$ outcrop area of the Songpan-Ganzi complex (Huang and Chen, 1987), even without palinspastic reconstruction, compares favorably with the area of the largest modern remnant-ocean-basin submarine-fan system, the Bengal fan, at 2.8-$3.0 \times 10^6 km^2$ (Emmel

and Curray, 1985). Furthermore, the $2.2 \times 10^6 km^3$ volume of the Songpan-Ganzi flysch (estimated by Huang and Chen, 1987, assuming 10 km average thickness) accommodates the $1.6 \times 10^6 km^3$ of material estimated by Okay and Şengör (1992) to have been eroded from the Dabie UHP terrane during the Triassic-Jurassic. Some of the sediment eroded from the collision orogen may have been transported to the Pacific basin (Okay and Şengör, 1992; Yin and Nie, 1993) and some is sequestered in Triassic-Jurassic nonmarine and shallow-marine siliciclastic sediments that supplanted carbonate deposition on the NCB-SCB platforms diachronously from east to west (Yin and Nie, 1993), but much detritus presumably passed through these foreland regions to the Songpan-Ganzi remnant ocean basin to the west (Fig. 10.22).

Detrital composition provides permissive support for a source in the collision belt to the east for at least some of the Songpan-Ganzi turbidites. In general, Lower Triassic pelagic carbonate strata are overlain by carbonate turbidites, consistent with proximity to low-latitude carbonate-dominated continental margins. Olistostromes containing blocks of Carboniferous platform facies are interlayered with, and overlain by, siliciclastic turbidites. Compositional data are limited, but we have petrographically confirmed the arkosic composition of turbidite sandstones reported by Zhou (1987) from the Qinling area. Thus, the upward variation in sandstone compositions from carbonatoclastic to plutoniclastic may reflect unroofing of the Dabie suture region. Arkosic compositions are not typical of remnant-ocean-basin sandstones (Graham et al., 1976; Dickinson and Suczek, 1979), but in this case, are consistent with the exposure of mid-crustal rocks during the Triassic in the focal region of collisional uplift (Fig. 10.22) (also see Ingersoll and Suczek, 1979; Suczek and Ingersoll, 1985). It should also be noted that volcanic sandstones have been described, but not documented, from the western Songpan-Ganzi triangle, that may have Kunlun arc sources (Zou et al., 1984).

Finally, although the eastward tectonic wedge-out of Triassic flysch of the Songpan-Ganzi Complex between the NCB and SCB strongly suggests sediment derivation from the suture (Figs. 10.21 and 10.22), paleocurrent and facies data are needed to document the westerly dispersal pathway that we infer from the Dabie suture to the Songpan-Ganzi remnant ocean basin. At present, paleocurrent data are available only from the western Qinling Mountains (Zhou, 1987); these are directed toward the west and southwest, consistent with derivation of detritus from the NCB and the suture zone. A good modern analog is the Indus Fan and the Makran (e.g., Critelli et al., 1990).

Songpan-Ganzi Basin Closure

The timing of closure and mode of deformation of the Songpan-Ganzi remnant ocean basin are known generally, but not well documented in detail. Some deformation of the Songpan-Ganzi flysch was related to ongoing accretion at the southern margin of Eurasia (e.g., Şengör and Okurogullari, 1991). However, principal deformation of the flysch must have been associated with terminal suturing of the Songpan-Ganzi ocean, which is constrained to have occurred between deposition of the Upper Triassic deformed flysch and the Jurassic nonmarine coal-bearing molasse deposits that overlie the flysch with angular unconformity in many locations (Şengör, 1981; Klimetz, 1983; Şengör and Hsü, 1984; Wang, 1985; Zhou and Graham, 1993). This deformation reflects not only the east-to-west diachronous closing of the remnant ocean basin, but later in the Triassic, the approach and collision of island arcs and the North Tibet terrane from the southwest (Fig. 10.21). Thus incorporated by middle Mesozoic into the growing collage of terranes along the southern margin of Eurasia, the collapsed Songpan-Ganzi basin was redeformed and structurally reorganized by folding, thrusting and strike-slip faulting during subsequent collisional events in the late Mesozoic, culminating in the collision of the Indian subcontinent and the Himalayan orogeny (e.g., Watson et al., 1987).

In addition, numerous granitic plutons intruded and locally metamorphosed the Songpan-Ganzi flysch during the middle and late Mesozoic (Zhang et al., 1984); the origin of these plutons is not fully understood, but they might be related to either crustal thickening or the last phases of subduction beneath the Songpan-Ganzi flysch (e.g., Şengör, 1984; Hsü et al., 1987; Şengör and Okurogullari, 1991). Thus, the Songpan-Ganzi remnant ocean

basin likely represents the complicated case of multiple colliding terranes and nonterminal suturing, followed by repeated tectonic overprinting.

Other Ancient Examples

Other workers have used relationships among the Himalayan suture, the Bengal Fan and the Sunda arc-trench system as modern analogs for Paleozoic systems. The Southern Uplands of Scotland have been interpreted as an accretionary mass resulting from northward subduction of Iapetus and culminating in suturing in the Devonian (Mitchell and McKerrow, 1975; Mitchell, 1978; Leggett et al., 1982, 1983). Earliest Paleozoic history is obscure, but Late Ordovician through Middle Devonian history of the northern margin of Iapetus seems to have consisted of:
1. accretion of primarily arc-derived detritus (Williams et al., 1992) following initiation of subduction;
2. increasing flux of nonvolcaniclastic detritus, probably derived from along strike where suturing had already occurred (e.g., Norway Caledonides); and 3. uplift of the accretionary wedge as the English continental margin was drawn into the subduction zone. Fergusson and Coney (1992) have proposed a similar analogy between the Bengal Fan and the Paleozoic Lachlan fold belt of southeastern Australia.

A likely example of flysch deposited in a remnant ocean basin between colliding arcs is the Mariposa Formation, which was deposited just before and during the Jurassic Nevadan orogeny in California (Schweickert and Cowan, 1975; Ingersoll and Schweickert, 1986) (see Fig. 8.18). Bradley (1983) proposed a similar model for the Siluro-Devonian Acadian orogeny in northern New England.

Cherven (1986) applied our model (Graham et al., 1975) to the Cretaceous-Paleogene remnant ocean basin of western Iran. Suturing proceeded from southeast to northwest, with recycling of detritus from sutured areas into the remnant ocean basin. Closing of this remnant ocean basin preceded terminal suturing of Arabia to Iran, which began in the Miocene (Şengör and Kidd, 1979; Şengör, 1990b).

Many of the flysch units of the Alps and Carpathians were probably deposited in remnant ocean basins (e.g., Schwab, 1981; Homewood and Caron, 1982;

Homewood, 1983; Caron et al., 1989). Paleogeographic and paleotectonic complexity is the rule in collisional orogens, especially those involving several arcs and microcontinents. The modern Mediterranean and Indonesian areas provide many modern examples of small ocean basins, continental fragments, arcs and accretionary wedges destined to be incorporated into orogenic belts similar to the Alpine system.

Prior to formation of the Apennine foreland basins (Miocene to Holocene; Ricci-Lucchi, 1986), a remnant ocean basin existed between the European margin (represented by Corsica, Sardinia and Calabria) and the Apulia margin (Adriatic foreland) (Boccaletti et al., 1990a; Critelli, 1993). Closing of this remnant ocean basin (Ligurian Ocean) occurred concurrently with opening of the Balearic and Tyrrhenian ocean basins, driven by slab pull (Malinverno and Ryan, 1986; Dewey et al., 1989b; Patacca et al., 1990). The Paleogene Liguride Complex of the southern Apennines is the accretionary wedge resulting from destruction of this remnant ocean basin (Critelli, 1993).

Diachronous uplift and recycling of sediment have characterized the northeastern Caribbean area from the latest Cretaceous to the present (Heubeck et al., 1991). A small remnant ocean basin (which originated as a backarc basin) has been closed by transpression that has progressed from northwest to southeast. During each stage of closing, uplifted older sediments have been the dominant sources for turbidites deposited in the remnant ocean basin (Heubeck et al., 1991).

Implications for Paleotectonic Reconstructions

Our actualistic model for the evolution of remnant ocean basins (Fig. 10.1) has significant implications for paleotectonic reconstruction and geodynamic modeling of orogenic sedimentary basins. Most importantly, remnant ocean basins are inevitably destroyed during suturing and their former presence must be inferred. Structurally complex accretionary wedges are all that remain following suturing; these accretionary wedges commonly contain deep-water turbidites (flysch) that have been thrust over rifted continental margins or other accretionary wedges. The emplacement of these accretionary wedges usually results in tectonic flexure

of adjoining continents, and the formation of peripheral foreland basins. The allochthonous nature of
highly deformed flysch renders detailed palinspastic
reconstruction difficult. Intricate structural relations
must be determined in conjunction with sedimentologic and paleoecologic studies; after approximately
200 years of study, the Alpine system is finally revealing much of its history (e.g., Homewood and Caron,
1982; Homewood, 1983; Caron et al., 1989).

Geosynclinal models for flysch and molasse (e.g.,
Aubouin, 1965) were essentially one-dimensional.
The ophiolitic suite was overlain by the flysch facies,
which was overlain by the molasse facies (Fig. 10.23).
We now interpret the same vertical sequence (rarely
completely preserved) as the result of seafloor spreading to form oceanic crust, followed by oceanic sedimentation, flysch sedimentation in a remnant ocean
basin, and collision to form a peripheral foreland
basin with molasse sedimentation. This one-dimensional view of sedimentation related to sutures is useful as a first approximation, but it fails to account for
the four-dimensional nature of these complex events.
Especially significant are the time-transgressive
nature of suturing (Fig. 10.20) and the lateral derivation of sediment derived from previously uplifted
suture zones (Fig. 10.1). Two-dimensional models for
foreland loading (Stockmal et al., 1986) provide
important insights regarding flexure of continental
margins as suturing occurs, but they fail to account
for the dominant longitudinal flux of sediment. Useful mechanical models for the complex interactions of
suturing, erosion, sedimentation and accretion will
require far more sophisticated treatment.

The most ambiguous aspect of our model (Fig.
10.1) is the nature of the deltaic transition. Modern
deltaic systems, such as the Ganges-Brahmaputra,
Indus and Markham deltas, are relatively clear examples of tectonic and environmental transitions. However, ancient examples of transitional deltas constitute volumetrically minor proportions of orogenic
sequences (Fig. 10.23). Sutures that form in sediment-starved settings (intraoceanic settings, in general, and possibly continental settings during sea-level
highstands) tend to have minor deltaic deposits. In
fact, the scarcity of sediment in such systems may
delay the sedimentologic transition from flysch to

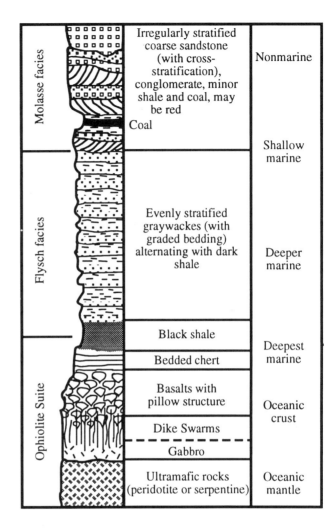

Fig. 10.23 Idealized vertical sequence from oceanic igneous rocks and
chert (ophiolite sequence) upward through graptolitic-graywacke turbidite (flysch) strata to red clastic (molasse) deposits. Such successions are commonly seen in orogenic belts and represent a progression from deep-
marine to nonmarine conditions. We now interpret this succession as
typically resulting from closing of remnant ocean basins, accretion of
turbidites into suture zones, and creation of foreland basins adjacent to the
orogens. See text for discussion. Modified from Dott and Batten (1988).

molasse beyond the tectonic transition from remnant
ocean basin to peripheral foreland basin. In this case,
early deposits within the peripheral foreland will be
turbidites (Fig. 10.23; also see Chapter 11); ambiguity
as to whether the terms "flysch" and "molasse" refer
to sedimentologic or tectonic features results.

An additional ambiguity results from the transition along tectonic strike of unfilled oceanic trenches
to submarine fans that cover trenches (e.g., Java-

Sumatra-Bay of Bengal transition, see above) (also see Chapter 5). Subduction complexes of arc-trench systems are transitional to accretionary wedges of suture zones. A usable distinction in modern examples is that subduction complexes form by accretion of distantly derived oceanic and turbidite sediments, and local arc-derived sediment, whereas accretionary wedges of suture zones form by accretion of longitudinally derived recycled-orogenic sediments. These distinctions are more difficult to make in highly deformed ancient examples, but provenance studies are the most promising method of making the distinction (e.g., Graham et al., 1976; Dickinson and Suczek, 1979; Ingersoll and Suczek, 1979; Dickinson, 1985; Critelli et al., 1990; Critelli, 1993).

Conclusions

Remnant ocean basins form during collisions between nonsubductable continental crustal elements of variable proportions. These elements include major and minor continents, small continental blocks and intraoceanic magmatic arcs. At least one of the elements must include an arc-trench system, along which oceanic crust is subducted. Collisions usually involve irregular continental margins and oblique convergence, so that events are diachronous and complex; the result is uplift of orogens immediately adjacent to remnant ocean basins, which receive the bulk of their detritus.

Sediment deposited in remnant ocean basins represents the greatest accumulations on Earth, both modern and ancient. Collisions involving one major continent and one moderate-sized continent (e.g., Asia and India) seem to produce the greatest volume of sediment, as a result of protracted collision and persistence of adjoining remnant ocean(s). Major sediment accumulations form rapidly at various stages during terminal suturing of supercontinents, but the remnant ocean basins are usually destroyed rapidly as suturing progresses. Collisions involving intraoceanic arcs produce less detritus.

Regardless of the dimensions of collision, most of the detritus eroded from collisional orogens accumulates in remnant ocean basins and is usually deformed in accretionary wedges during both oceanic subduction and terminal suturing. As a result, the sedimentary basins (floored by oceanic crust) are seldom preserved intact. Paleogeographic and paleotectonic reconstruction of ancient remnant ocean basins is difficult, but it must be attempted because they represent major ancient basins. Geophysical models for the evolution of remnant ocean basins and their accretionary remains are nonexistent. Detailed structural, stratigraphic, sedimentologic and petrologic studies are the primary methods by which remnant ocean basins must be reconstructed.

Much of the continental crust has been created through the processes of uplift, erosion, deposition, accretion and metamorphism during the evolution of collision orogens and remnant ocean basins. If we are to understand the development of sedimentary basins and the evolution and growth of continental crust, in general, we must understand remnant ocean basins and the accretion of their sedimentary fill during crustal collision.

Acknowledgments

We appreciate guidance to the geology of the Ouachita Mountains in Arkansas and Oklahoma from B.R. Haley, C.G. Stone, and G.W. Viele, and of the Marathon Basin in west Texas from E.F. McBride and W.R. Muehlberger; discussions of Ouachita-Marathon geology with D.W. Houseknecht and W.A. Thomas were also helpful.

We also thank S. Critelli, E. Garzanti, A. Yin and D. Zhou for illuminating discussions and correspondence about Eurasian remnant oceans. T.J. Bornhorst provided the original version of Figure 10.20, and we thank him for permission to modify and update it. Work on the Ouachitas was supported by NSF Grant EAR-9105479 to Dickinson.

We thank C. Busby, S. Critelli, S. Nie, and A.M.C. Şengör for reviewing the manuscript.

Further Reading

Crook KAW, 1989, *Suturing history of an allochthonous terrane at a modern plate boundary traced by flysch-to-molasse facies transitions*: Sedimentary Geology, v. 61, p. 49–79.

Dickinson WR, 1988, *Provenance and sediment dispersal in relation to paleotectonics and paleogeography of sedimentary basins, in Kleinspehn KL, Paola C (eds)*, New perspectives in basin analysis: Springer-Verlag, New York, p. 3–25.

Dorsey RJ, 1988, *Provenance evolution and unroofing history of a modern arc-continent collision: evidence from petrography of Plio-Pleistocene sandstones, eastern Taiwan*: Journal of Sedimentary Petrology, v. 58, p. 208–218.

Graham SA, Dickinson WR, Ingersoll RV, 1975, *Himalayan-Bengal model for flysch dispersal in the Appalachian-Ouachita system*: Geological Society of America Bulletin, v. 86, p. 273–286.

Şengör AMC, Okurogullari AH, 1991, *The role of accretionary wedges in the growth of continents: Asiatic examples from Argand to plate tectonics:* Eclogae Geologicae Helvetiae, v. 84, p. 535–597.

Teng LS, 1990, *Geotectonic evolution of late Cenozoic arc-continent collision in Taiwan:* Tectonophysics, v. 183, p. 57–76.

Yin A, Nie S, 1993, *An indentation model for the North and South China collision and the development of the Tan-Lu and Honam fault systems, eastern Asia:* Tectonics, v. 12, p. 801–813.

Collision-Related Foreland Basins

11

Andrew D. Miall

Introduction

The foreland or foredeep is defined as "a stable area marginal to an orogenic belt, toward which the rocks of the belt were thrust or overfolded. Generally the foreland is a continental part of the crust, and is the edge of the craton or platform area" (Bates and Jackson, 1987, p. 254). Forelands typically are depressed ensialic "moats" adjacent to the orogen (Figs. 11.1, 11.2), and may be filled with hundreds of meters of sediment (e.g., Fig. 11.3). Schematic models by Dewey and Bird (1970) and Dickinson (1974b, 1976a) illustrate the gross tectonic framework of forelands in collision zones as they were understood two decades ago (Fig. 11.4); these models still form the basis for our understanding of foreland basins. Early plate-tectonic interpretations of forelands at suture zones suggested that they form by flexural bending of continental crust adjacent to subduction zones (Dickinson, 1974b). However, Price (1973) pointed out that it is the regional isostatic subsidence beneath the supracrustal load of the fold-thrust belt marginal to the orogen that generates the moat in which sediments collect. Quantitative modeling of flexural loading was done by Beaumont (1981), Jordan (1981), and Quinlan and Beaumont (1984); this work provides the geodynamic framework for our current interpretations of this class of sedimentary basin.

Arc-related foreland basins are divided into two major classes. Retroarc foreland basins (Dickinson, 1974b) occur behind (on the continental side of) compressional arcs formed during subduction of oceanic plates. This type of basin is described by Jordan (see Chapter 9), where the development of quantitative models for foreland basins is discussed in some detail (e.g., Cordilleran foreland of North

America; Fig. 11.3). Collisional foreland basins result from arc-arc, arc-continent and continent-continent collision (see Chapter 10). These basins are described in the present chapter.

Collisional foreland basins are subdivided into two types. Collisional retroarc basins lie on the overriding plate, behind the arc; they are similar to those described in Chapter 9 (i.e., RAB in Fig. 11.4B). At certain stages of its development, the Alberta basin may be a good example, and some of the interior basins of western China, located behind (north of) the collision mountains of Tibet, are also partly of this type. Peripheral foreland basins (Dickinson, 1974b) are initiated by attempted subduction and

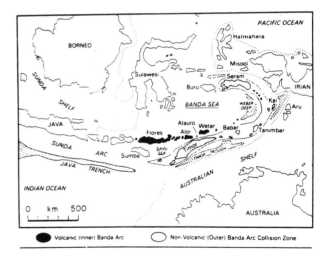

Fig. 11.1 Tectonic setting of Timor Trough, a peripheral foreland basin that has been actively forming since collision of Australian continental margin with Banda arc along a north-dipping subduction zone, beginning in Pliocene. Northern edge of Australian continental crust is shown by dotted line, which underlies Australian shelf and includes continental islands of Irian and Halmahera (Reproduced with permission from Audley-Charles, 1986).

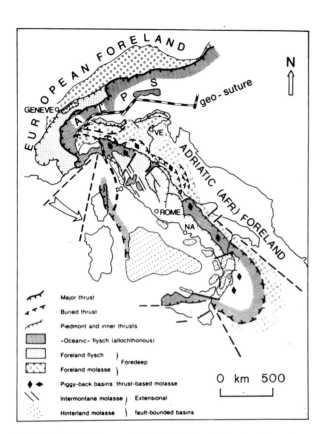

Fig. 11.2 Tectonic framework of Switzerland and Italy, showing setting of two major Alpine and Apennine collision-related foreland basins. Adriatic foreland is a peripheral basin formed by collision of Italy with Adriatic-Yugoslavian continent along a southwest-dipping subduction zone during Oligocene-Recent. European foreland, corresponding to Swiss molasse basin, is a peripheral basin formed where Italy overrode European margin along a south-dipping subduction zone during Eocene-Miocene. Backthrusting on south side of Alps (south of "geo-suture" and north of VE [Venice]), in part, preceded accretion of Apennines. Therefore, Po Valley (north and west of VE) began as a collisional retroarc foreland (relative to Alps), but is now influenced primarily by Apennine compression (peripheral foreland). Therefore, it is a hybrid collisional foreland (both retroarc and peripheral) (R.V. Ingersoll, personal communication, 1994). Terminology of legend can be translated into terminology of this book as follows: "Oceanic" flysch (allochthonous): remnant-ocean turbidites; Foreland flysch: peripheral foreland turbidites; Foreland molasse: peripheral foreland molasse; Piggyback basins: thrust-based molasse: piggy-back molasse; Intermontane molasse: successor molasse; Hinterland molasse: backarc turbidites. (Reproduced with permission from Ricci-Lucchi, 1986.)

loading of rifted continental margins in forearc settings (PFB in Fig. 11.4B). They overlie, adjoin, and may be overriden by subduction complexes and overlap marginal sediment wedges of subducting plates.

The western Pacific offers several examples of peripheral basins formed by arc-arc collision. For example, a series of small peripheral basins developed in central Honshu, Japan, as a result of the collision of the Izu-Bonin arc with Honshu during the late Cenozoic (Ito and Masuda, 1986).

Arc-continent collisions may lead to the development of significant peripheral basins. For example, during the late Cenozoic, the Luzon arc has collided with the Chinese mainland, creating the island of Taiwan, the western part of which constitutes a fold-thrust belt and associated peripheral basin (Covey, 1986; Teng, 1990; also see Chapter 10). Similarly, the Banda arc is colliding with the northern Australian continental margin, with the formation and uplift of the island of Timor. The Timor Trough, to the south, is a developing peripheral basin (Audley-Charles, 1986; Figs. 11.1, 11.3). Well known ancient examples are the Taconic foreland of the northern Appalachians (St. Julien and Hubert, 1975; Hiscott et al., 1986; Bradley, 1989; Bradley and Kidd, 1991), and the Antler foreland of Nevada (Speed and Sleep, 1982).

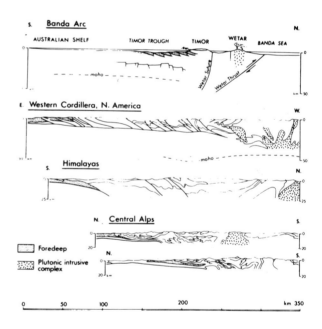

Fig. 11.3 Comparative cross sections across four collisional orogens and their flanking foreland basins. Banda Arc section shows arc at Wetar overriding Australian continental margin, which is subducting northward as part of Indian plate (structural interpretation is controversial [see text], see Fig. 11.1). Western Cordilleran foredeep of North America is an example of a primarily retroarc foreland basin. Himalayan and Alpine basins are peripheral foreland basins developed on partially subducted continental crust (see Fig. 11.2). (Reproduced with permission from Audley-Charles, 1986.)

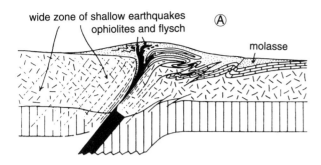

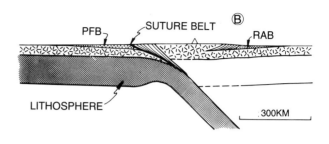

Fig. 11.4 Schematic diagrams illustrating early plate-tectonic interpretations of foreland basins. (A) Development of orogenic welt and molasse wedge in adjacent foreland basin (reproduced with permission from Dewey and Bird, 1970).

(B) Relationship of peripheral foreland basin (PFB) and retroarc collisional foreland basin (RAB) to suture (reproduced with permission from Dickinson, 1976a). These two types of foreland basin are the subject of this chapter.

The world's largest and best-known peripheral foreland basins (Himalayan foredeep, Appalachian basin, Swiss Flysch-Molasse basin, Adriatic foreland, and Persian Gulf; Figs. 11.2, 11.3) have resulted from continent-continent collision. Typically, these basins form above, and rest on, the extensional continental margins of downgoing plates that developed during preceding cycles of ocean opening and closing. For example, the Himalayan foredeep basin is formed above the Paleozoic through early Cenozoic Tethyan continental margin of the Indian subcontinent (Gansser, 1964). Development of the peripheral basin may involve several distinct but diachronous phases, as first the arc on the overriding plate undergoes compression, and later, the main mass of continental crust reaches the subduction zone.

Distinguishing among different types of foreland basins in the ancient record may be difficult because most orogens undergo several phases of accretion, changes in subduction polarity and changes in convergence angle, leading to such complications as strike-slip displacement of a basin and its source area(s), and superposition of basins controlled by different plate-tectonic mechanisms. Analysis of the Cordilleran and Appalachian foreland basins of North America is beset by such difficulties, as noted below.

Analysis of the tectonic and stratigraphic evolution of foreland basins can be structured into two or three hierarchical levels (Jordan et al., 1988). At the first level, there are regional or continental-scale plate-tectonic controls, which influence global sea level, the trajectories of plate convergence and their rates of movement. At the second level are variations in plate-margin geometry and age that govern details of diachronous tectonism, and variations in loading, subsidence history and large-scale basin architecture along strike within the foreland. At the third level is the response of the basin to individual loading events, such as movement on a specific thrust plate. The structural evolution of a basin, its stratigraphy, its facies, its petrographic characteristics, and the nature of the stratigraphic sequences within it are products of these three types of controls overlain and integrated within the final basin fill.

Geodynamics

Modeling studies have shown that when subjected to supracrustal loads, the lithosphere behaves approximately as a uniform elastic plate (Beaumont, 1981; Jordan, 1981). The configuration of the resulting moat, the composition of the basin fill, and the subsidence history of the basin depend on the mechanical properties of the loaded crust and the rate of emplacement of the load. The thickness and age of crust underlying the basin are important variables, determining its thermal properties, its buoyancy and flexural strength. Variations in these properties lead to wide variations in the paleogeographic evolution of the basin (Beaumont et al., 1982a; Stockmal et al., 1986; Stockmal and Beaumont, 1987). Figure 11.5 shows deformation of an elastic beam (beneath the foreland basin) with three conditions of flexural rigidity. For example, where collision closes a young

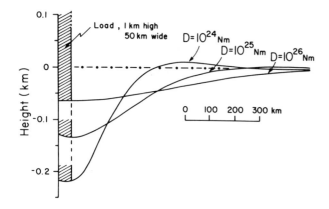

Fig. 11.5 Cross sections of a deformed elastic beam in response to a load 1 km high and 50 km wide, with a density of 2400 kg/m³. Sediments of same density fill each basin. Three cross sections are shown, corresponding to three values of flexural rigidity. Each basin has approximately the same cross-sectional area but differs in degree to which load is spread over greater surface areas of crust. (Reproduced with permission from Beaumont, 1981.)

ocean, the rifted margin that forms the substrate to the developing foreland basin is floored by relatively warm, buoyant and weak lithosphere with relatively low flexural rigidity. Where the lithosphere is weak, the basin tends to be narrow. At first, it will be relatively shallow because the underlying rifted margin will not have subsided to its maximum depth (see Chapter 4 for a discussion of the evolution of rifted continental margins). Given a low flexural strength of the basement (typically $D = <10^{24}$ Nm), the foreland basin increases in depth relatively rapidly, over periods of millions to tens of millions of years, and assumes a configuration comparable to the narrowest shown in Figure 11.5. Where a foreland basin develops over an old continental margin, it initially is relatively deep, but tends to deepen more slowly under the imposition of the initial overthrust load; the basin tends to be relatively wide because of the greater rigidity of the old basement ($D = 10^{26}$ curve in Fig. 11.5). Some of these effects were modelled by Stockmal et al. (1986) and Stockmal and Beaumont (1987) using the Alberta and Swiss basins as typical examples of basins formed over "old" and "young" continental lithosphere, respectively.

Flexural bending of the crust produces uplift on the distal margin of the basin (Fig. 11.5). This is termed the *outer peripheral high, peripheral bulge,* or

forebulge. The position and elevation of this uplift are linked mechanically to the supracrustal load; for example, the uplift migrates cratonward as the colliding terrane is thrust over the former rifted margin. Waschbusch and Royden (1992a,b) and Watts (1992) showed that variations in flexural strength of forelands can lead to variations in shapes of foreland basins as supracrustal loads migrate. They also discussed how the forebulge may remain fixed for millions of years if part of the migrating load is accommodated by additional subsidence over weaknesses in the underlying foreland. The forebulge uplift typically has an amplitude of up to a few tens of meters, and is located several-hundred kilometers from the thrust front. As discussed below, the stratigraphic record of the forebulge may yield a considerable amount of information about the tectonic history of the basin. The actual topographic expression of the uplift depends on the local rates of erosion and sedimentation. For example, in basins with a high sediment supply, the uplift may be buried by sediment and have no topographic expression (overfilled basin; see Chapter 9).

As summarized by Allen et al. (1986), there has been some debate regarding the rheology of the lithosphere beneath foreland basins. Jordan (1981) was able to explain the development of the foreland basin of the Rocky Mountains, Wyoming, using an elastic model, whereas Beaumont (1981) found that a viscoelastic model better explained the details of the Alberta and Appalachian basins. The main difference between the two interpretations is that a viscoelastic crust accommodates some of the loading stress by gradual permanent changes in the lithosphere, resulting in progressive weakening. This weakening leads to the formation of a basin that is deeper and narrower relative to one formed on elastic crust. The shape of the flexural wave thus changes through time, one of the more important implications of which is that the forebulge is uplifted and migrates basinward. As shown by Tankard (1986), careful stratigraphic studies of the margins of the basin, and its associated forebulge, may be used to investigate this process (see below). The reader is referred to Chapter 9 for additional discussion of crustal rheology.

The major evolutionary theme in the development of a peripheral foreland basin is its gradual shallowing, following initial subsidence. The oldest sediments are typically deep-water; commonly, the stratigraphic section is condensed, indicating sediment starvation. Most basins are then filled successively by deep-water clastics, such as turbidites, and then by shallow-water and nonmarine sediments. This type of stratigraphic succession was once termed the "geosynclinal cycle" (e.g., Clark and Stearn, 1960) (see Fig. 10.23). Plate-tectonic models for foreland basins provide a geodynamic framework for the cycle, as shown in Figure 11.6 for the case of a peripheral foreland basin. The initial loading by the overriding margin (typically the subduction complex of an arc, or an accreted terrane) onto the continental slope of a rifted continental margin (Fig. 11.6a) causes flexural depression of the foreland and uplift of a low-relief peripheral bulge, as recorded by the development of an unconformity. This collision generally takes place below sea level, so little sediment is available to be shed into the basin. Seaward of the peripheral bulge, the continental slope is steepened by flexural loading and may become unstable, with extensional faulting and the formation of slump and slide masses (e.g., olistostromes). As the overthrust mass builds in relief with continued shortening and begins to rise above sea level, increasing volumes of sediment are shed into the basin, forming a deep-marine clastic wedge (Fig. 11.6b). This is the so-called flysch phase of basin development, as discussed below. Continued shortening causes the accreted arc or continent to climb beyond the thinned edge of the rifted margin, resulting in greater topographic uplift and increased sediment supply (Fig. 11.6c). The basin fills with sediment and conditions change to shallow-marine and nonmarine (the molasse phase). With cessation of thrusting, erosion disperses the lithospheric load and the basin undergoes isostatic uplift and erosion (Fig. 11.6d).

The Coastal Plain of western Taiwan is a modern peripheral basin that has formed during the collision of the Luzon arc with China (see Chapter 10). Its sedimentary evolution provides a good illustration of the geodynamic cycle of Fig. 11.6 (Covey, 1986; Teng, 1990; Fig. 10.10). The North Alpine foreland basin of Switzerland is a well studied ancient example of a peripheral basin (Homewood et al., 1986; Allen et al.,

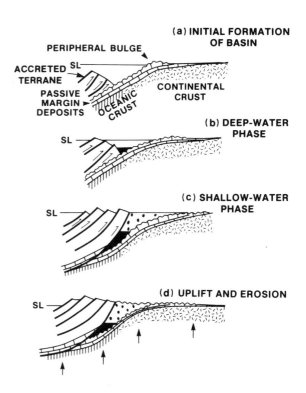

Fig. 11.6 Idealized model for terrane accretion, and development of sedimentary succession in a peripheral foreland basin. See text for discussion. (Reproduced with permission from Cant and Stockmal, 1989.)

1991; Sinclair and Allen, 1992; Fig. 11.7). Both examples exhibit the classic asymmetric basin architecture, flanked and overlain on one side by the fold-thrust belt and shallowing gently onto the continental interior on the other side. Figure 11.8 illustrates some of the details of the structural geology of this type of basin in an example from northern Italy.

Sinclair and Allen (1992) attempted to quantify the relationships between thrust advance rates, flexural loading, exhumation of the source area and sediment supply. In the North Alpine foreland basin of Switzerland, they recognized an accretionary-wedge phase, characterized by rapid advance of the accretionary prism and rapid basin subsidence, slow exhumation and sedimentation rates, and consequent deep-water and shelfal sedimentation. This was followed by a continental-wedge phase, during which backthrusting led to higher exhumation rates and higher sedimentation rates, so that the basin was filled to sea level, accumulating shallow-marine and nonmarine sediments.

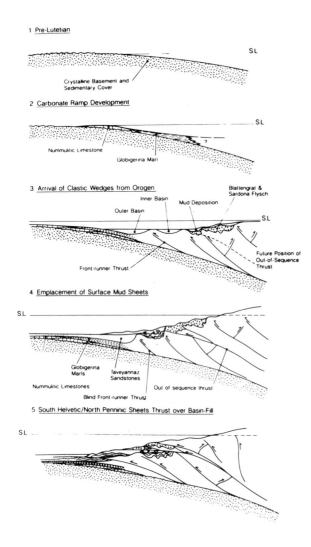

Fig. 11.7 Schematic evolution of North Alpine peripheral foreland basin, Switzerland, from initial uplift and erosion of forebulge on downgoing (European) plate in mid-Eocene (panel 1) to overthrusting from south in Oligocene (panel 5). (Reproduced with permission from Allen et al., 1991.)

As discussed below, the stratigraphic record of foreland basins may contain several clastic pulses. Each may constitute the sedimentary response to flexural loading following repeated thrusting or terrane-accretion episodes, as overriding crustal masses are driven onto the foreland. Similar cycles of loading and sedimentary response may develop in retroarc collisional basins, where terranes are emplaced over the foreland by delamination and obduction (e.g., Cant and Stockmal, 1989).

Variations in basin filling reflect variations in flexural response to loading, caused by differences in the type of crust, the type of foreland basin, or the age of the rifted margin underlying the foreland basin. For example, in the Alberta retroarc basin, the deep-water clastic phase (Fig. 11.6b) is represented by only a few hundred meters of beds (Fernie Formation), whereas in the Alpine peripheral foreland basin of Switzerland, several kilometers of deep-water clastics were deposited (the classic flysch). The Alberta basin developed over thick Precambrian crust and a mature Proterozoic-Paleozoic rifted-continental-margin wedge, characterized by high flexural rigidity, whereas the Swiss basin formed at the margin of a young (Mesozoic) continental plate margin, which underwent more rapid subsidence (Stockmal and Beaumont, 1987). Some, if not most, of the Swiss flysch was deposited on the oceanic crust of one or more remnant ocean basins preceding continental collision (see Chapter 10). Variations in the configuration and stratigraphic fill of the Appalachian basins, which reflect some of these tectonic variables, are described briefly below.

Transport of sediment within the basin can also alter the timing and relative importance of these phases of sedimentation, relative to the stages of collision at any given cross section of the basin. Thus, a large volume of turbidites may be transported away from the initial point of collision into remnant ocean basins along strike, where they are deposited on oceanic crust (Graham et al., 1975; see Chapter 10). If continued plate convergence consumes the remnant ocean, this flysch will then be incorporated into the subduction complex that develops in the early

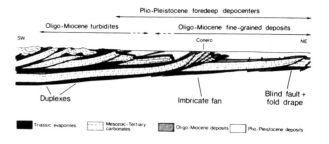

Fig. 11.8 Typical fold-thrust belt and associated foreland basin: Apenninic-Adriatic foredeep, Italy. Foreland basin is of Plio-Pleistocene age, and reaches a maximum depth of 1200 m. Width of section is approximately 100 km. (Reproduced with permission from Ori et al., 1986.)

stages of peripheral-basin development. Much of the Cenozoic flysch of the Swiss basin and the upper Paleozoic flysch of the Ouachita basin is thought to have formed in this way, in remnant ocean basins, rather than as early stages in the filling of the succeeding foreland basins (Graham et al., 1975; Chapter 10). Progressive closure of a remnant ocean basin will lead to diachronous sedimentation and tectonism, and overlap in ages of various phases along the final suture.

Many orogens develop by the amalgamation of terranes following the subduction of intervening oceans. The kinematic history of rifting, subduction, collision, rotation, and strike-slip displacement can be extremely complex (Dewey et al., 1991), and may involve multiple phases of basin formation. Specific basins may undergo phases of platform sedimentation, extensional faulting and subsidence, deep- to shallow-water sedimentation, deformation, and then renewed sedimentation related to plate-margin processes along a successor suture zone as another terane arrives. Much of the east Asian continental margin evolved in this way, with several separate phases of foreland-basin formation and deformation (e.g., south China: Hsü et al., 1990; Korea: Cluzel et al., 1990). The North American Cordillera underwent a similar evolution (e.g., Gabrielse and Yorath, 1992), with many of the tectonic events recorded by the sedimentary evolution of the major retroarc foreland basin occupying the Western Interior. Terrane amalgamation events within accretionary orogens may be dated by reference to overlap of sediments deposited on or close to the suture. For example, Bowser basin in British Columbia is a peripheral basin formed where oceanic rocks of the Cache Creek Terrane were thrust over Stickinia, a continental block, beginning in the Early Jurassic. The oldest sediments in the basin are shales of late Early Jurassic age, indicating the first flexural loading of Stickinia. These are overlain gradationally by Bajocian conglomerates containing radiolarian cherts derived from the Cache Creek terrane, and indicating the first emergence of the overriding fold-thrust belt as a significant subaerial sediment source (Ricketts et al., 1992).

It is now thought that plate-margin stresses may be transmitted for long distance across plate interiors, activating folding, faulting, basin formation and deformation thousands of kilometers from the plate margin (Cloetingh, 1988). Thus, the Cenozoic geology of east Asia is dominated by the effects of the India-Asia collision. Tapponnier et al. (1986) described the continental-scale accommodation to this collision by the movement of major blocks along strike-slip faults. In many areas, fold-thrust belts caused crustal loading and the development of retroarc foreland basins (Chen and Dickinson, 1986; Liu, 1986). Amongst these, in China, are the Qaidam, Junggar, Juiquan, and Tarim basins (Figs. 11.9, 11.10).

Cant and Stockmal (1989) and Stockmal et al. (1992) applied this reasoning to the development of the retroarc Alberta foreland basin, arguing that terrane-collision events hundreds of kilometers from the foreland basin generated pulses of shortening, uplift and progradation of molasse into the basin. However, loading models require that the supracrustal load be emplaced above a continuous plate. Where the continental margin is broken by steep faults that penetrate the lithosphere, the effects of the load may not be transmitted fully across the fault, with some of the displacement being accommodated by downfaulting rather than bending of the foreland. Terranes may be stripped from their lithospheric basement by the accretion process (lithospheric delamination of Price, 1986), and emplaced onto the continental margin above a major décollement surface, below which the lithosphere is flexurally continuous with the foreland basin (Stockmal and Beaumont, 1987; Cant and Stockmal, 1989; Stockmal et al., 1992). Multiple terrane collision events may then cause multiple pulses of thrusting and foreland-basin subsidence. The stratigraphy of the basin is not affected directly by a load positioned more than one flexural half-wavelength away from the basin (at least several-hundred kilometers), but terrane accretion may result in compression and renewed shortening of older, inboard terranes.

The pattern of subsidence at any single point in a foreland basin is distinctive (Burbank et al., 1986; Cross, 1986; Homewood et al., 1986; Johnson et al., 1986). The initial subsidence rate for a peripheral foreland is slow, as the supracrustal load is emplaced on the distal edge of the continental margin. It accelerates as the accreted wedge shortens, thickens and

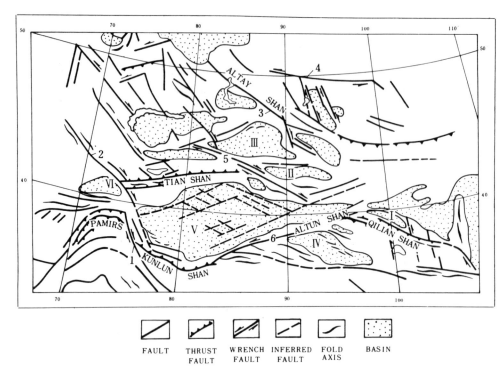

Fig. 11.9 Basins and major structures in west China and adjacent areas of Russia, Afghanistan and Indian subcontinent. Many basins are bounded by fold-thrust belts, and have character of foreland basins. Basins: I=Jiuquan, II=Turpan, III=Junggar, IV=Qaidam, V=Tarim, VI=Fergana. Faults: 1=Karakoram, 2=Talaso-Fergana, 3=Ertix, 4=Changajn, 5=Borohoro, 6=Altun Shan (see Fig. 11.10). (Reproduced with permission from Liu, 1986.)

climbs the underlying continental margin, with a consequently high rate of flexural bending, and reaches a local maximum when the basin is overridden by the emplaced load. The subsidence rate then decreases as the basin itself is uplifted by thrust faulting or post-tectonic erosional rebound. Among the most distinctive features of foreland-basin subsidence curves are sharp inflection points, indicating sudden increases in subsidence rate. They may indicate flexural response to specific thrust-loading events and commencement of uplift, and can therefore, be used in dating fold-thrust-belt tectonism (Burbank et al., 1986; Cross, 1986; Johnson et al., 1986; Jordan et al., 1988; Chapter 9).

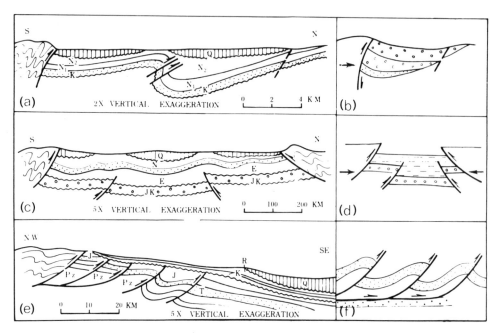

Fig. 11.10 Cross sections through some foreland basins in west China. (a,b) Jiuquan Basin, (c,d) Qaidam Basin and (e,f) Junggar Basin. Locations of basins are given in Fig. 11.9. Pz=Paleozoic, T=Triassic, J=Jurassic, JK=Jurassic-Cretaceous, K=Cretaceous, E=Paleogene, N=Neogene, Q=Quaternary. (Reproduced with permission from Liu, 1986.)

Some retroarc foreland basins are wider and deeper than can be accounted for by the flexural-loading model. Supracrustal loading typically can account for a basin up to about 400 km wide, whereas some basins are double this width. The Alberta basin, for example, is more than 1000 km wide in its central part (from the fold-thrust belt to the edge of the Canadian Shield), occupying most of Alberta, plus the southern parts of the adjacent provinces of Saskatchewan and Manitoba. Beaumont (1982) suggested that a regional basinward tilting of the crust occurred at the time of the flexural loading, and proposed that the tilting was a response of the lithosphere to convective flow coupled to a subduction zone. A pattern of secondary asthenospheric flow in the overriding lithospheric plate was suggested by Toksöz and Bird (1977). Flow takes place toward the subduction zone and is drawn down parallel to the cold, descending oceanic plate. The crust is tilted toward the subduction zone by the mechanical drag effects of the downgoing current. When subduction ceases, buoyancy forces, coupled with erosional unroofing of the accreted terrane, combine to reverse the tilting process, leading to uplift of the basin. Mitrovica and Jarvis (1985) and Mitrovica et al. (1989) modeled this process, and showed that the width of the crust affected by the tilting increases as the subduction angle decreases. At subduction angles of 20°, the tilt effect extends for more than 1500 km from the thrust front. The degree of tilting is also affected by the flexural rigidity of the overriding lithosphere. The cessation of crustal shortening accompanies the termination of subduction and its associated mantle convection currents, so that the mechanical down-drag effect ceases, and the basin then tends to rebound (Mitrovica et al., 1989).

In some modeling studies of foreland-basin dynamics (e.g., Karner and Watts, 1983; Royden and Karner, 1984), simulation of the flexural history has suggested that the calculated magnitude of the supracrustal load is inadequate to explain the total subsidence. In the case of peripheral basins, the negative buoyancy of the cold subducting lithosphere may provide extra "load" early during collision history, but this mechanism does not account for apparent discrepancies during intermediate to late stages of collision. These results of modeling have influenced the analysis of orogens, as means have been sought to recreate large loads by appropriate structural interpretations. For example, Pfiffner (1986) suggested that steep dips on some thrust faults at depth beneath the central Swiss Alps thickened the thrust loads. However, Stockmal et al. (1986) pointed out that such interpretations may be unnecessary if it is remembered that the supracrustal load is emplaced into a pre-existing "hole," which is the bathymetric expression of the pre-existing thinned continental margin. The apparent need for extra load disappears if the discrepancy is due simply to the inappropriate simplicity of the mechanical model.

Uplift and erosion of the fold-thrust belt begin during crustal shortening, and continue following the cessation of subduction, because of the effects of isostatic rebound resulting from the erosion (Beaumont, 1981). Both subsidence and uplift may be very rapid. Fortuin and de Smet (1991), based on geohistory analysis of stratigraphic sections, determined subsidence rates of up to 0.2 mm/y and uplift rates of up to 1 mm/y in the Timor Trough. Cerveny et al. (1988) determined present denudation rates of 5 mm/y for the Himalayas. Studies of thermal maturation of foreland-basin sediments can be used to reconstruct burial and unroofing histories, and they commonly reveal considerable vertical movements. At least 2 km of strata have been removed from proximal parts of the Alberta basin (Hacquebard, 1977; Beaumont, 1981), and at least 6.5 km of Carboniferous strata have been stripped from the Appalachian basin of New York (Friedman, 1987). Structural studies by Pfiffner (1986) indicate former burial of the present proximal edge of the Swiss Molasse basin to depths of at least 10 to 12 km. However, it is possible that not all of the uplift took place during isostatic rebound. Some may relate to later tectonic episodes.

Structural Evolution

The development of a peripheral foreland basin is preceded by the consumption of oceanic crust at a Benioff Zone (B-subduction of Bally, 1975). As continental crust enters the subduction zone, the accretionary wedge above the subduction zone begins to

incorporate rocks of the downgoing continental plate. This zone evolves into the distinctive fold-thrust belt that borders and overlies the basin (Fig. 11.4b). Similar fold-thrust belts occur in retroarc settings, where collisional orogens are thrust over interior parts of continents (e.g., this process affected some of the retroarc or hinterland basins of west China during the India-Asia collision: Figs. 11.9, 11.10). Bally (1975) referred to this retroarc shortening process as A-subduction, naming it (in Bally and Snelson, 1980) after the Swiss geologist Ampferer. Bally and Snelson (1980, p. 20) suggested that "some amounts of sialic crust may be disposed at intermediate depths below the megasuture." However, the fold-thrust process leads only to crustal thickening; sialic crust is not consumed by the mantle. To apply the term "A-subduction" to the shortening of the continental crust, therefore, seems misleading to this writer.

The leading edge of the downgoing continental crust at a collision zone is typically a rifted margin, one of the principal structural characteristics of which is the presence of numerous extensional faults (Figs. 11.11, 11.12, 11.13; see Chapters 3 and 4). Most of these are oceanward dipping listric faults. They include shallow growth faults in areas of rapid sedimentation, and deeper faults, some bottoming out at mid-crustal levels, and others possibly penetrating to the Moho, that developed during initial continental stretching and rifting. One effect of flexural bending of the downgoing plate by the emplacement of a supracrustal load may be reactivation of these faults. Later, collisional shortening reverses the displacement, and they may serve as the nuclei for thrusts (Jackson, 1980; Dewey et al., 1986; Price, 1986).

Typically, major shortening of thrust plates and nappes takes place above décollement surfaces, which are flat-lying thrusts, from which imbricate thrust stacks grow (Figs. 11.3, 11.7, 11.8, 11.10, 11.11, 11.12, 11.13). Initially, separation from basement may take place along reactivated extensional faults (e.g., those undergoing extension in Early Jurassic in Fig. 11.13; see also dotted lines showing future faults in Fig. 11.12), but, décollements may also occur at the basement-cover contact (Fig. 11.12), at the contact between upper and lower crust, at the Moho, or at anisotropies within the lithosphere (Mattauer, 1986).

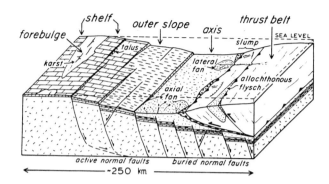

Fig. 11.11 Schematic block diagram of Taconic foredeep, New England, showing structural regime and facies belts shortly before plate convergence ended. (Reproduced with permission from Bradley and Kidd, 1991.)

As shortening proceeds, the level of detachment may shift, commonly biting deeper and farther into the foreland of the downgoing plate (Fig. 11.12), and displacing proximal parts of the foreland basin (Figs. 11.7, 11.8). Stratigraphic units of low strength on this plate may be exploited as decollements, along which several-hundred kilometers of shortening may occur. For example, in the Alps, décollements occur within Mesozoic carbonates and evaporites (Homewood et al., 1986); in the Apennine foreland basin, the décollement is along Triassic evaporites (Ricci Lucchi, 1986); and in Pakistan, below the Himalayan peripheral basin, the detachment is along Cambrian salt beds (G.D. Johnson et al., 1986).

Descriptions of the structural geology, and igneous and metamorphic history of fold-thrust belts and collisional core zones are beyond the scope of this book. These topics are, therefore, not discussed here, except where relevant to the depositional history of a foreland basin. For example, the sequence of faulting affects the growth of subbasins within forelands; the stacking of thrust and nappe slices, and their subsequent erosional unroofing control the nature of detritus shed into basins.

In general, deformation steps progressively into the foreland during collisional shortening (in-sequence thrusting; Dahlstrom, 1970; Fig. 11.8). (Also see the front-runner thrust illustrated in Fig. 11.7). Early basin-fill strata are, thus, involved in the deformation, and are cannibalized by uplift, erosion, and subsequent recycling into the basin (the migrating foredeep of Bally et al., 1966). The basinward progression of deformation

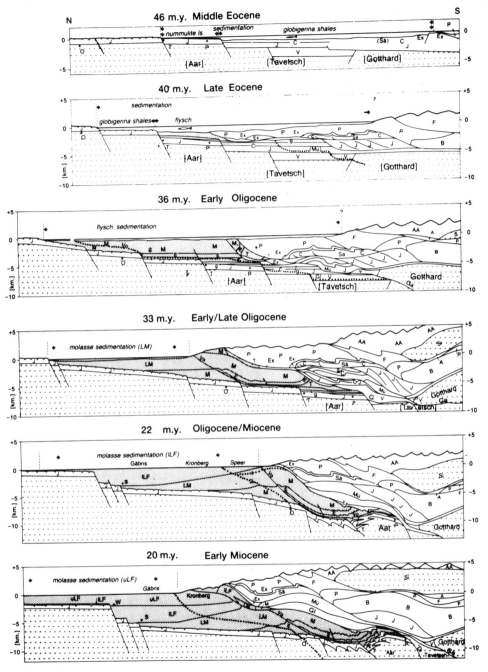

Fig. 11.12 Evolution of North Alpine fold belt and peripheral foreland basin. Dotted lines indicate traces of future faults. Crosses=crystalline basement, shaded=foreland sediments. Nappes, formations and thrusts: A=Arosa ophiolitic melange, AA=Austroalpine cover sediments, B=Bündnerschiefer, C=Cretaceous, Di=Disentis thrust, E=Engi Formation, Ex=exotic strip sheets, F=Falknis nappe, G=Globigerina shales, Ga=Garvera thrust, Gl=Glarus thrust, J=Mesozoic (mainly Jurassic), LF and LM=Lower Freshwater and Marine Molasse (u,l=upper, lower), M=Matt Formation, Mü=Mürtschen thrust, P=Penninic flysch, Sä=Säntis thrust, Si=Silvretta basement, T=Taveyannaz Formation, UF and UM=Upper Freshwater and Marine Molasse, V=Verrucano, Vo=Vorstegstock thrust (Pfiffner, 1986).

takes place, in part, because of the actual movement of displaced masses into the basin, and in part, because of the gradual expulsion of pore fluids and consequent increase in internal friction within inner parts of the fold-thrust belt. Some faults, particularly those formed farthest into the basin, may not reach the surface. They may or may not be detectable at the surface as anti-clines. Such structures are termed "blind thrusts"; examples are illustrated in Figs. 11.7 and 11.8. Their development is accompanied by a distinctive type of syntectonic sedimentation, as described below. Other thrusts cease displacement before sedimentation terminates, in which case, they may be blanketed by flat-lying strata. Such faults also leave no surface expression.

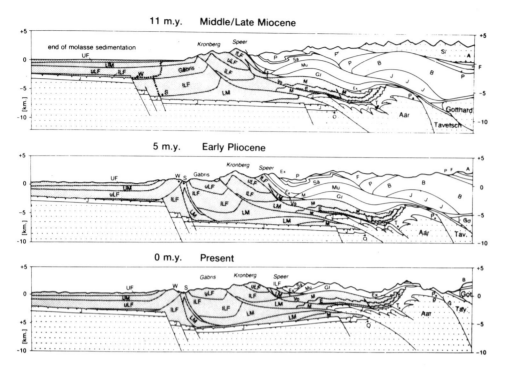

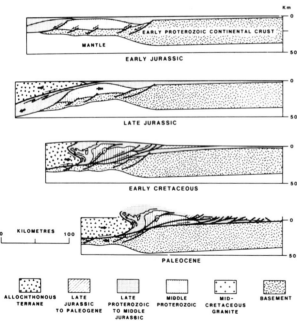

Fig. 11.13 Evolution of southeastern Canadian Cordillera, based on structural reconstructions by Price (1986). Foreland basin is Late Jurassic to mid-Cenozoic in age. Major décollement flooring the allochthonous block was initiated as an extensional detachment on pre-Jurassic rifted continental margin, and became a surface along which an accreted terrane was delaminated during Mesozoic. (Reproduced with permission from Price, 1986.)

Another important process is that of tectonic wedging and delamination, described by Price (1986). This process leads to the development of backthrusts and triangle zones (segments of the fold-thrust belt that are bounded in cross section by a triangular arrangement of faults: décollement, thrust and backthrust; Figs. 11.12, 11.13), and may also lead to out-of-sequence thrusts (e.g., Fig. 11.7), with consequences for the evolution of the foreland basin where these faults approach or actually cut the basin floor. Out-of-sequence faulting may also occur where preexisting structural weaknesses are exploited during loading and shortening. Examples of these syndepositional processes are also discussed below.

In the simplest case, a foreland basin is an asymmetric moat, deepest adjacent to the fold-thrust belt and shallowing toward the continental interior (Fig. 11.14A). However, as a result of the complexities in thrust kinematics noted above, the proximal edge of the basin may evolve into several separate sub-basins, isolated or semi-isolated during sedimentation by faults growing within the basin (Fig. 11.14B-E). Ori and Friend (1984) introduced the term "piggyback basin" for basins carried on the backs of

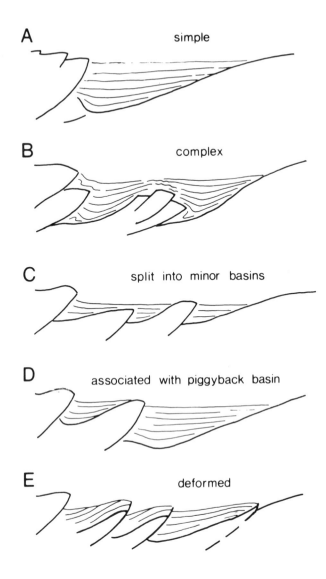

A simple

B complex

C split into minor basins

D associated with piggyback basin

E deformed

Fig. 11.14 Variations in relationship between fold-thrust belt and fore-land basin, based on seismic profiles. Minor basins and piggyback basins are varieties of satellite basins. (Reproduced with permission from Ricci Lucchi, 1986.)

thrust plates (Fig. 11.14D), but Ricci Lucchi (1986) suggested that the specific condition described by Ori and Friend is only one of several possibilities, which he preferred to include in a wider category of minor or satellite basins (Fig. 11.14C, D). The depositional style of these minor basins, and the intimate relationship between sedimentation and tectonics are discussed below.

Sedimentary Facies and Depositional Systems

Initial deposits within a peripheral foreland basin generally are deep-water, fine-grained clastic and associated facies. Hemipelagic shales predominate, and may be associated with various condensed facies, including carbonaceous shales, sapropels, silica-rich sediments, glauconite, and volcanogenic sediments, including tephras and bentonites. Ricci Lucchi (1986) briefly described these facies in the Apennine peripheral foreland basin.

Sand-dominated turbidites and associated facies typically denote the development of a significant subaerial sediment source (the uplifted fold-thrust belt of Fig. 11.6b) while the basin is still under predominantly deep water. Good examples are described from the Alps (the original flysch basins) by Homewood and Caron (1982), the Carpathians (Hesse, 1982), the Apennines (Ricci Lucchi, 1986), the Antler foreland basin of Nevada (Poole, 1974), and the Taconic basin of eastern North America (Hiscott et al., 1986; Fig. 11.11). In fact, flysch sedimentation commonly precedes the development of the foreland basin, as deposition occurs on oceanic crust within a remnant ocean basin (Graham et al., 1975; see Chapter 10).

There is not the space to dwell here on the details of submarine depositional systems. A review of deep-water clastic sedimentation in peripheral foreland basins and remnant ocean basins is very much the history of the evolution of our ideas regarding submarine-fan depositional models, and of the controversies regarding tectonic, eustatic and autogenic mechanisms for the development of fan channels, lobes, cycles and other depositional elements. Many of our ideas on these matters originated with the Italian school, led by Emiliano Mutti and Franco Ricci Lucchi, who collected most of their data in the Cenozoic deep-water deposits of the Apennine foreland basins (e.g., Mutti and Ricci Lucchi, 1972; Mutti, 1985; Mutti and Normark, 1987). As our understanding of allogenic depositional controls improves, there is increasing recognition of the inadequacy of existing fan models and a tendency even to avoid the use of the term "fan" because of wide variations in size, shape, physiography and composition of turbidite

depositional systems (Hiscott et al., 1986; Ricci Luc-chi, 1986; Reading, 1991). Use of the term is, of course, entirely appropriate for a depositional system consisting of feeder channels and outer-fan lobes, especially if it can be demonstrated to fan out from a point source (see discussion of fan models in Picker-ing et al., 1989).

At the proximal margins of nonmarine (molasse) foreland basins, there is commonly a zone of coarse alluvial-fan conglomerate (e.g., McLean, 1977; Puigde-fabregas et al., 1986). As discussed below, dating of these conglomerates may provide important informa-tion regarding timing of tectonism. Fan deposits pass downstream into sand-dominated fluvial assemblages (e.g., Hirst and Nichols, 1986; Homewood et al., 1986; Puigdefabregas et al., 1986), and these, in turn, may pass distally into shallow-marine successions (e.g., Homewood and Allen, 1981).

Important early work on the sedimentology of shallow-marine and nonmarine sediments was carried out in foreland basins. Bersier (1948) discussed possi-ble tectonic mechanisms for the origin of upward fin-ing cycles in the Alpine molasse, and Allen (1964, 1970) based his fluvial model for upward fining cycles largely on field studies of the Old Red Sandstone, the Devonian molasse-type fill of the Appalachian and Anglo-Welsh foreland basins. King (1959) coined the term "clastic wedge" for the sheets of detritus flanking mountain ranges, and gave as his initial example the largely nonmarine Catskill and associated deposits of Devonian age in the Appalachian foreland basin of New York and Pennsylvania (Fig. 11.15). He illus-trated this and other clastic wedges of various ages in an accompanying map (Fig. 11.16). More recently, the spectacular exposures of fluvial facies in the Ebro basin and related foreland basins in Spain have been the basis for useful advances in our understanding of fluvial-channel processes and architecture (e.g., Friend, 1983).

Sedimentation and Tectonics

Regional Tectonic-Sedimentary Relationships

The first evidence of an arc-arc or arc-continent colli-sion in the stratigraphic record may be the transport of sediments into a remnant ocean basin from a point of collision along strike. The first evidence that the continental margin is being affected by collision may be extensional faulting on the downgoing rifted con-tinental margin. These faults may be growth faults, and they may be associated with local wedges of coarse scarp breccias derived from precollision rocks (e.g., European margin of Swiss Molasse basin: Homewood et al., 1986; Pfiffner, 1986; Fig. 11.12; Taconic orogen of New England: Bradley and Kidd, 1991; Fig. 11.11). Active faulting may step inboard along the continental margin as the downgoing plate moves into the zone of flexural loading. This can be documented by tracking the age of scarp breccias and

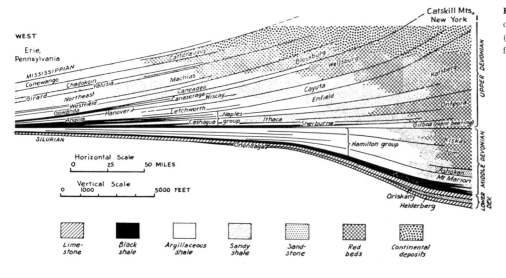

Fig. 11.15 Catskill clastic wedge of New York and Pennsylvania. (Reproduced with permission from King, 1959.)

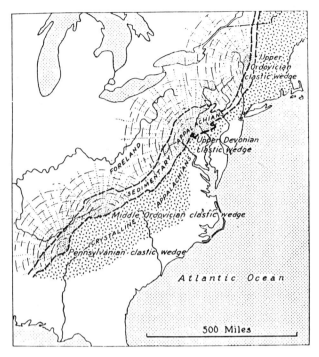

Fig. 11.16 Clastic wedges flanking Appalachian orogen in eastern North America, including two Taconic wedges (Middle and Upper Ordovician), Acadian Catskill wedge (Upper Devonian), and Alleghenian wedge (Pennsylvanian). Fan-shaped dispersal patterns are diagrammatic representations of several dispersal systems radiating from limited areas of active uplift. (Reproduced with permission from King, 1959.)

units that thicken across growth faults, and by onlap relationships (e.g., Taconic basin of New York: St. Julien and Hubert, 1975; Bradley and Kusky, 1986; Arkoma basin: Houseknecht, 1986).

As a peripheral basin develops and fills with sediment, the main theme is the progressive shallowing of the basin and coarsening of the sediment. These are the basic characteristics that suggested the old geosynclinal shale-flysch-molasse cycle (Fig. 11.6). However, in detail, the history of basin filling may be far more complex, as the basin evolves in response to local tectonics. Sediment transported longitudinally from older collisional mountains along strike may predominate in the basin fill (Figs. 11.17, 11.18). The dominant local sediment source in a mature basin is typically the adjacent uplifted, thrust-faulted terrane of the overriding plate, but significant volumes of sediment may also be derived from erosion of the downgoing plate. Uplift and migration of the peripheral bulge yield some sediment, but the cratonic interior, beyond this uplift, may also become an impor-

tant source, particularly if it is cut by basement structures that are reactivated. In a zone of multiple collisions, as in many major orogenic belts, the foreland of one developing suture may be the hinterland of another, so that there may be significant uplifted terranes on both sides of the peripheral basin, and along strike. For example, the Apennine peripheral foreland overlies the flank of the slightly older retroarc foreland of the Alpine belt to the north (Fig. 11.2). The latter was uplifted during Apennine-basin sedimentation, and acted as a significant sediment source (Fig. 11.17).

As noted under the heading Geodynamics, variations in basin geometry and composition of stratigraphic fill may be interpreted in geodynamic terms. Figure 11.19 illustrates three of the Appalachian clastic wedges. The Taconic wedge of Tennessee records the earliest approach of accretionary terranes, which loaded thin, weak lithosphere. The resulting peripheral foreland basin was deep and narrow, and was filled mainly with fine-grained, deep-water clastics (Sevier and Tellico units). It is not clear how the subsequent Appalachian foreland basins should be classified according to the three types described at the beginning of this chapter. During the Acadian orogeny, accretionary terranes amalgamated with North America, and were thickened by thrusting; the lithosphere beneath the foreland underwent broad flexure (Rast, 1989). The classic Catskill facies was

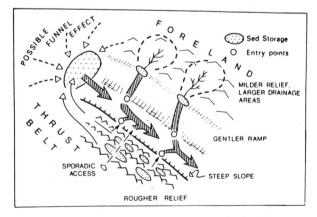

Fig. 11.17 Drainage pattern in an idealized Apennine foreland basin. Thrust belt to southwest verges northeastward and represents crustal shortening associated with northeastward overriding of Adriatic foreland, a spur of African plate. To northeast, this foreland is underlain by slightly older Alpine chain (see Fig. 11.2). (Reproduced with permission from Ricci Lucchi, 1986.)

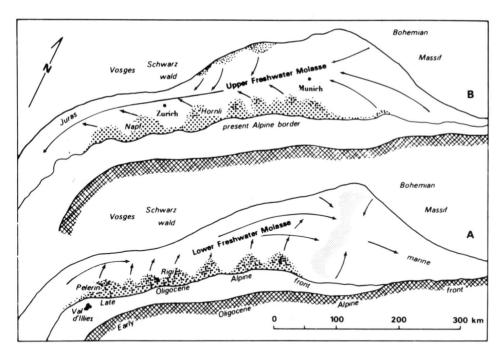

Fig. 11.18 Major alluvial fans and axial fluvial dispersal of detritus in north Alpine peripheral basin. A. Chattian-Aquitanian Lower Freshwater Molasse, B. Tortonian-Pontian Upper Freshwater Molasse. Note reversal in direction of axial transport. (Reproduced with permission from Van Houten, 1981.)

deposited at this time. Note that the basal shale facies and the nonmarine redbed facies are well represented, but the deep-water-clastics facies is absent (Fig. 11.15). Finally, during the Alleghenian orogeny, thin thrust sheets were emplaced over thick, strong lithosphere, with resulting widespread deposition of relatively coarse clastics in shallow-water to nonmarine environments (Tankard, 1986). Plate movements during the Alleghenian orogeny included a significant transpressional component (Rast, 1989).

Drainage patterns within foreland basins have been discussed by several authors (e.g., Eisbacher et al., 1974; Graham et al., 1975; Şengör, 1976b; Dewey, 1977; Miall, 1978, 1981; Van Houten, 1981). Typically, there are two basic patterns, both in subaqueous and subaerial settings. First, there is transverse flow, in which the primary paleoslope within the basin is oriented perpendicular to the fold-thrust belt, leading to radial sediment dispersal directly across the basin; and second, longitudinal or axial flow, in which drainage is funneled along a main channel parallel to the fold-thrust belt (Figs. 11.17, 11.20). In basins receiving a large sediment supply from the fold-thrust belt, transverse drainage may extend across the entire basin. Such basins are said to be overfilled (Sinclair and Allen, 1992; see Chapter

9). In contrast, axial drainage may take place along the deepest part of the basin. It collects and funnels sediment entering the basin along transverse drainage systems, and therefore, tends to be characterized by larger channels. Such basins are commonly underfilled, in the terminology of Jordan (see Chapter 9). Eisbacher et al. (1974) demonstrated that transverse drainage derives mainly from structural salients along fold-thrust belts, such as the central, most uplifted parts of thrust plates, where the belt tends to have a convex curvature facing the basin, whereas many major longitudinal streams enter the basin at reentrants, such as at the termination of thrust plates or at terrane sutures. They cited the modern Himalayan foredeep as an example. Major alluvial fans, such as that of the Kosi River, represent transverse drainage from structural salients, whereas the Brahmaputra River, one of the major axial drainage channels in the basin, enters at the major syntaxis that marks the end of the Himalayan ranges in Assam.

Hoffman and Grotzinger (1993) speculated that the climatic belt in which a rising orogen develops also influences the tectonic style of the orogen and the architecture of the adjacent foreland basin. Areas of high precipitation, as in monsoonal belts, are char-

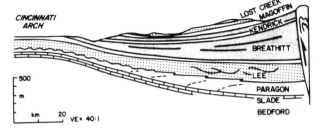

3. ACADIAN – ALLEGHENIAN TRANSITION
Late Mississippian – Early Pennsylvanian

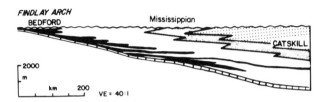

2. ACADIAN
Devonian

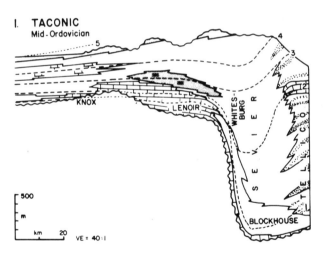

1. TACONIC
Mid-Ordovician

Fig. 11.19 Cross sections through three clastic wedges of Appalachian basin, illustrating varying stratigraphic response to dynamics of foreland-basin construction. Note various scales of these diagrams. Each wedge may reflect important differences in plate-tectonic evolution. Thus, Taconic wedge represents a peripheral foreland basin, Acadian wedge is a retroarc basin, and Alleghenian wedge may have been deposited under the influence of transpressional tectonism. (R.V. Ingersoll, personal communication, 1992; reproduced with permission from Tankard, 1986.)

acterized by rapid erosional unroofing, leading to rapid uplift, deep erosion, and the development of a foreland basin overfilled with nonmarine sediments (but of low preservation potential). The Grenville orogen (1.1 Ga) of eastern North America may be cited as a possible example. This contrasts with orogens such as the Alleghenian, which may have developed in more arid climates. Erosional unroofing would not compensate uplift, leading to preservation of the fold-thrust belt and an underfilled foreland basin. Sinclair and Allen (1992) pointed out the feedback effects of elevating a fold-thrust belt, resulting in greater orographic rainfall, increased rates of erosional unroofing, and consequently enhanced uplift rates.

Baltzer and Purser (1990) provided a good description of the relationship between drainage and tectonics in the Persian Gulf peripheral basin. The Tigris and Euphrates rivers are good examples of axial drainage, and the fans of the Karun and Fuka rivers in Iran represent the same type of transverse drainage as the Kosi of India. The Mehran fan delta is another depositional system that drains transversely into the Persian Gulf basin, although it is fed by a longitudinal river draining the Zagros Mountains of Iran (Baltzer and Purser, 1990). Many small transverse fans bordering the north side of the Persian Gulf (the marine part of the foreland basin) are deposited by consequent streams draining the flanks of anticlinal ridges (Baltzer and Purser, 1990). The Markham River in Papua New Guinea is another example of a major trunk drainage flowing along the axis of a peripheral foreland basin (Fig. 11.20). It is fed by transverse rivers and fans entering the basin from the fold-thrust belt to the north (also see Fig. 10.9).

In an ancient basin, detailed paleocurrent work may aid in tectonic analysis, such as a reconstruction of the subtle structural reentrants which occur where thrust plates terminated or overlapped. For example, Hirst and Nichols (1986) reconstructed radial dispersal patterns of two large fluvial-fan systems, and showed how fluvial entry points could be determined by projecting measured flow directions back to fan apexes.

The direction of axial drainage is away from the locus of maximum crustal shortening (the point of initial collision) and toward any remnant ocean (or lacustrine) basins located along strike, where the bulk of deep-water sediments are likely to be deposited (e.g., modern Bengal and Indus cones; Graham et al., 1975; Şengör, 1976b; see Chapter 10). Most suture zones form by the consumption of an ocean between

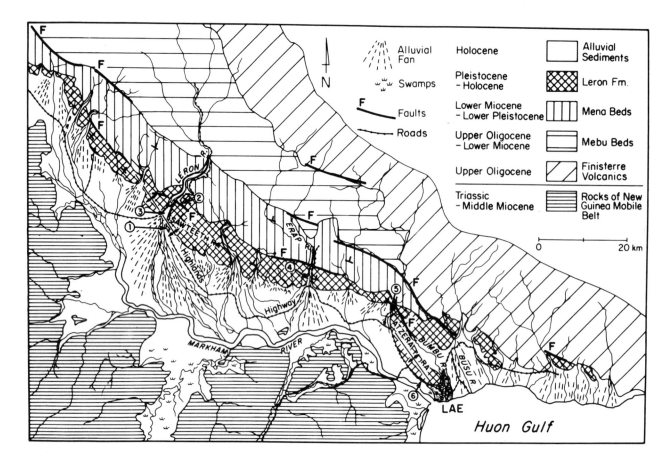

Fig. 11.20 Geological and drainage map of Huon Gulf area of Papua New Guinea. Markham River flows along axis of a peripheral basin developing between New Guinea mobile belt to south and Finisterre terrane to north. It is fed by minor rivers and their associated fans draining transversely from mountains to north. Faults oriented NW-SE along north side of Markham River represent south-verging fold-thrust belt bordering this basin. Nonmarine deposits are filling Markham Valley and prograding into Huon Gulf, where deep-water sediments are currently accumulating. Earlier stages in diachronous (west to east) suturing of these two plates are represented by uplifted Oligocene to Holocene beds exposed in fold-thrust belt. Leron Formation consists of deep-water clastics similar to those presently forming in Huon Gulf (also see Fig. 10.9). (Reproduced with permission from Crook, 1989.)

irregular continental margins that do not match in shape when they collide. The suturing process is, therefore, diachronous, involving progressive collision, uplift, and closure of remnant basins as the suture zips up along strike (see Chapter 10). Filling and uplift of the basin, therefore, also are diachronous, with the deep- and shallow-water depositional systems migrating along strike toward remnant ocean basins. Axial progradation into the basin emphasizes and amplifies this lateral shifting of depositional systems. As noted by numerous authors (see in particular Graham et al., 1975; Miall, 1984), an important implication of this is that the deep-water facies (flysch) may form contemporaneously with along-strike shallow-water to nonmarine facies (molasse), as the basin fills along strike, contrary to the simple time sequence implicit in the geosynclinal cycle. Again, the Himalayan foredeep and adjacent remnant ocean basins provide an excellent modern example. The nonmarine, fluvial deposits presently being deposited by the Ganga, Indus and other major rivers of the north-India plains are feeding major submarine depositional systems in adjacent remnant ocean basins of the Arabian Sea (Indus Cone) and Bay of Bengal (Bengal Fan) (Figs. 10.5 and 11.21). Another excellent example of diachronous, along-strike basin evolution was described by Crook (1989). Huon Gulf and Markham Valley are the deep-water and nonmarine

components, respectively, of a peripheral basin currently developing by diachronous closure of a small ocean basin adjacent to Papua New Guinea (Figs. 10.9 and 11.20). Similarly, the Tigris-Euphrates system is progressively filling the Persian Gulf with fluvial-deltaic sediments, while marine sediments are accumulating in the gulf itself (Baltzer and Purser, 1990). Uplift of Taiwan is shedding sediment from the fold thrust belt across the coastal-plain peripheral basin and into the remnant ocean basin to the south (Teng, 1990; see Chapter 10).

As noted above, the forebulge at the edge of a foreland basin is sensitive to the flexural loading, relaxation and unloading history of the basin (Figs.

11.11, 11.22). Detailed studies of the stratigraphic onlap/offlap, wedging and pinchout relationships within and adjacent to the forebulge can, therefore, provide a great deal of information with regard to basin history (Figs. 11.22, 11.23). The principles of forebulge evolution were elucidated by Quinlan and Beaumont (1984), with particular reference to the Appalachian foreland basin. Tankard (1986) provided

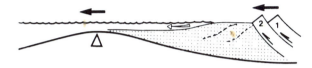

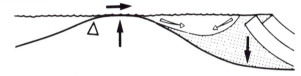

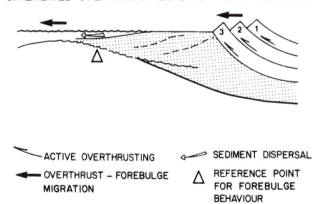

Fig. 11.22 Development and migration of a forebulge in response to cycles of deformation and tectonic quiescence. Forebulge develops as part of the flexural wave generated in the crust by loading (see Fig. 11.5). Following cessation of thrusting, there may be a viscoelastic relaxational response to loading, which leads to deepening and narrowing of the basin. In the case of an elastic crust this phase would not occur. The forebulge may be exposed to erosion, or become a shoal area within an otherwise deeper-marine basin. With renewed crustal shortening, the forebulge migrates cratonward. Subsidence and uplift of the forebulge produce onlap and wedge-out unconformities, respectively, on the flanks of the uplift. (Reproduced with permission from Tankard, 1986.)

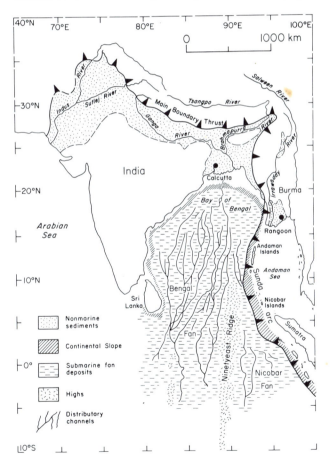

Fig. 11.21 Axial flow of Indus, Ganga, and Brahmaputra rivers is overfilling the peripheral foreland basin of the northern Indian subcontinent. Much of this detritus reaches the Bay of Bengal (a remnant ocean basin), where it contributes to the growth of the giant Bengal fan (the world's largest depositional system; also see Fig. 10.5). (Reproduced with permission from Miall, 1990.)

EARLY ALLEGHENIAN — THRUST FLEXURAL PHASE

OVERFILLED BASIN

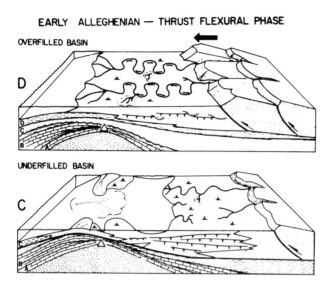

UNDERFILLED BASIN

LATE ACADIAN — RELAXATION PHASE

ARCH UPLIFT — ONLAP — EROSION

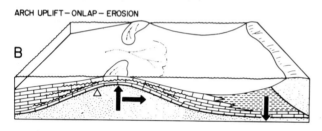

BLANKET SEDIMENTATION

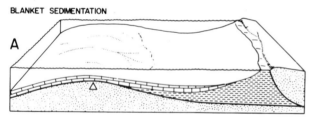

Fig. 11.23 Evolution of Appalachian landscapes during the Acadian (Middle Devonian to Early Mississippian) and Alleghenian (Early Pennsylvanian to Permian) orogenic episodes. (Reproduced with permission from Tankard, 1986.)

a stratigraphic-sedimentologic study of parts of this basin, which explored these ideas in some detail. Another useful study is that of Lash (1987b). These workers focused on a basin where the crustal response to loading was apparently viscoelastic. Flemings and Jordan (1990) discussed a foreland-basin model developed under conditions of an elastic crust.

If the basin is underfilled, then the forebulge develops as a low-relief topographic barrier. The Taconic foreland of New England was an area of platform-carbonate sedimentation, which underwent uplift and karstification (Fig. 11.11). In the case of a subaerial, fluvial basin, the forebulge may divert flow, as may have occurred during the Early Cretaceous in the Alberta basin. There, a major conglomerate unit, the Cadomin Formation, formed by transverse drainage of the fold-thrust belt, which fed a longitudinal river system flowing northwest, parallel to the mountain front. McLean (1977) termed this the Spirit River Channel. In the case of a subaqueous basin, the forebulge may serve as a shoal area, with deposition of offshore barrier-type islands, or as a region of enhanced wave and tide reworking, or as a carbonate bank. The Waverley Arch in Kentucky was interpreted by Tankard (1986) as a forebulge that was an active shoal area during the Carboniferous (Fig. 11.23). At times, it served as the foundation for a major suite of offshore sand bars and "barrier" complexes.

Foreland basins that develop within complex orogens, such as those that undergo changes in convergence vectors or multiple terrane-accretion events, may be characterized by extremely complex histories. Tectonic styles may change during basin development, and the result may be a hybrid basin that is difficult to classify by plate-tectonic origin. The progressive (diachronous) collision of irregular margins and the reactivation of inherited basement structures may lead to repeated pulses of loading and subsidence. The simple rheological responses predicted by the geophysical models are then overprinted by the local responses to inherited structural anisotropies. Relative sea level and sediment source relief are affected by resulting tectonism on both local and regional scales, with the possibility for developing very complex patterns of intersecting clastic wedges (e.g., Ricci Lucchi, 1986; Cluzel et al., 1990; Hsü et al., 1990) and tectonic-eustatic stratigraphic sequences (e.g., Mutti, 1985). Within large orogens, such as the Mediterranean-Alpine region and the Himalayan orogen of west China, basins may become isolated from the world ocean system within surrounding mountain ranges as a result of regional collisional events (e.g., Tarim, Jiuquan, Qaidam basins: Fig. 11.9), leading to centripetal drainage, changes from marine to fresh water, lowering of sea level, anoxic events, or basin evaporation causing salinity crises (Ricci Lucchi, 1986). Paleoflow patterns may undergo major changes, even

reversals, as a response to basin tilting, diachronous closure or the influence of reactivated basement structures. For example, in the Swiss Molasse basin, longitudinal drainage patterns reversed from northeast to southwest during the Miocene (Fuchtbauer, 1967; Van Houten, 1981; Fig. 11.18); a similar axial reversal (from northward to southward longitudinal flow) occurred in the Alberta basin between Early and Late Cretaceous time (compare McLean, 1977, with Rahmani and Lerbekmo, 1975). In detail, the Taconic clastic wedge of Quebec and adjacent areas also shows local axial paleoflow reversals, because of changes in position along strike of major sediment entry points (Hiscott et al., 1986). The effects of reactivation of transverse basement structures on basin tectonics and sedimentary patterns have been described from the Marathon basin (Wuellner at al., 1986), the Apennines (Ricci Lucchi, 1986) and the Himalayan foredeep of Pakistan (G.D. Johnson et al., 1986).

Oblique collision may lead to the development of peripheral basins dominated by strike-slip faulting. For example, in the western Swiss Alps, anomalous thickening in the Molasse has been attributed to the presence of strike-slip faults in the Jura, the massif which underlies the foreland basin on its northwest flank (Homewood et al., 1986). Collision of small terranes with larger continental masses may lead to strike-slip faulting adjacent to the collision, where a region undergoing crustal shortening passes along strike into a region not affected by shortening. For example, collision of the Izu-Bonin arc with the main Japanese island of Honshu in the late Cenozoic led to the development of a series of small peripheral basins (Ito and Masuda, 1986), several of which, adjacent to the point of collision, have all the characteristics of basins on transform plate margins (Ito, 1989). In larger orogens, crustal shortening may be accommodated by sideways squeezing of major crustal blocks, with the development of strike-slip faults in the foreland, the hinterland or both (Burke and Şengör, 1986; Tapponnier et al., 1986).

Local Tectonic-Sedimentary Relationships

Minor satellite or piggyback basins develop when fault ramps that cut the foreland basin are uplifted,

isolating part of the foreland and acting as sediment barriers (Fig. 11.14). Ponding of sediment takes place within the minor basins, with diversion of flow along the basin axes toward reentrants in the main thrust front. Such reentrants commonly form at terminations or overlaps of thrust plates. Careful stratigraphic work on the fill of these satellite basins may assist in unravelling the thrust kinematics of the fold-thrust belt (e.g., Ori and Friend, 1984; Johnson et al., 1986; Ricci Lucchi, 1986; Bentham et al., 1992). Continued crustal shortening may culminate in the uplift of these minor basins, so that only remnants may be preserved within the eroded foldbelt, as in the Swiss Alps (Homewood et al., 1986; Pfiffner, 1986) and the Appalachian orogen (Lash, 1990).

The intimate association of active tectonism with contemporaneous sedimentation leads to the development in many foreland basins of a distinctive class of structures termed *syndepositional* or *intraformational unconformities*. These were first described in detail using examples from the Ebro basin, Spain (Fig. 11.24; Riba, 1976; Anadon et al., 1986; Puigdefabregas et al., 1986), but have also been observed in many other foreland basins (e.g., Miall, 1978; Ori et al., 1986). For example, satellite basins may fill by onlap of the rising upper plates of the thrust ramps that underlie them (Figs. 11.14C, D). Continued deformation of these ramps may produce folded onlap unconformities (Fig. 11.24II, III). Blind thrusts may be overlain by anticlines, the rise of which may also generate unconformities (Fig. 11.24IV). Fanglomerate deposited against a rising (growing) fold or fault will be uplifted and deformed by continued movement of the structure. In many young foreland basins, neotectonic observations document the development of such structures. For example, modern alluvium can be observed dipping at high angles close to basin-margin faults, and incised by channel downcutting and lateral migration (Miall, 1978, and personal observations in west China; Trifonov, 1978). In Iraq, ancient drainage canals have been deformed by the growth of anticlines beneath them (Lees, 1955). In central Asia, growing anticlines are bringing about diversions in drainage patterns (Trifonov, 1978). Syndepositional structures tend to be laterally restricted. The unconformities described by Anadon et al.

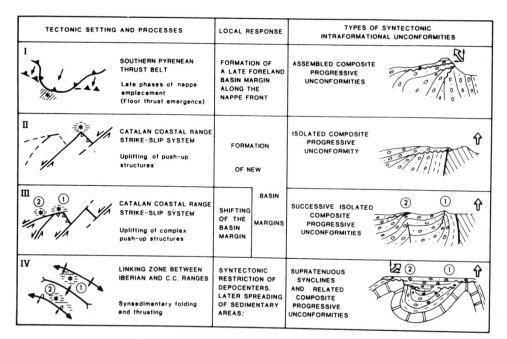

(1986) typically die into the basin within 500 m and extend along strike for as little as 5 to 10 km.

Analogous structures within foreland basins are *denudation complexes*. These are stratigraphically complex units that develop by erosion and resedimentation of rocks at the crests of actively growing blind structures on the basin floor. Blind structures generate their own topography that influences local depositional and erosional patterns. Denudation complexes have been particularly well described by Ori et al. (1986), based on superb seismic-reflection data obtained from the Apennine foredeep beneath the Adriatic Sea (Fig. 11.25). Chaotic slump deposits and progradational clinoform units are characteristic features of the seismic facies in this basin.

Megabeds are another type of syntectonic deposit in many foreland basins. These are unusually thick and areally extensive units, typically recording catastrophic sediment-gravity flows, probably triggered by basin-margin seismicity. They include basinwide turbidite units several meters thick (e.g., Taconic basin: Hiscott et al., 1986; Apennine basin: Ricci Lucchi, 1986), and subaerial debris-flow deposits extending for tens of kilometers across alluvial-fan surfaces (e.g., Swiss Molasse basin: Burgisser, 1984). Hiscott and Pickering (1984) showed that some turbidite megabeds contain evidence of repeated reversals of flow, and suggested

that within ponded basins, major turbidity currents may be reflected from the basin margins several times before their momentum is depleted.

Rates of Regional Tectonism

Structural and stratigraphic data can be used to reconstruct rates and styles of plate convergence and crustal shortening. Veevers et al. (1978) and Fortuin and de Smet (1991) noted an interval of subsidence, followed by uplift, and migration of facies belts in the Timor Trough following initiation of collision of the Banda arc with the Australian continental margin (Fig. 11.1). The average rate of facies migration in this trough is 2.6 cm/y. Sea-floor spreading evidence indicates that, in the vicinity of Taiwan, plate convergence has a rate of 7 cm/y, with a minimum of 160 km of shortening having occurred during the last 4 my (4 cm/y; Covey, 1986). Jordan et al. (1988) compiled structural evidence on rates of shortening in several foreland fold-thrust belts in various plate-tectonic settings. They range from 0.2 to 3 cm/y. The minimum value seems low compared to many plate convergence rates, but shortening in the fold-thrust belt may, of course, be only part of the total shortening that takes place across a plate margin during foreland-basin development.

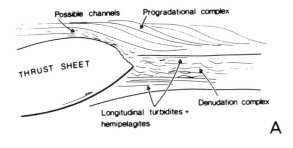

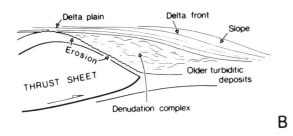

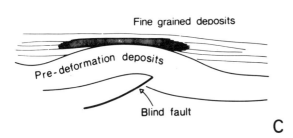

Fig. 11.25 Three examples of denudation complexes developed by erosion and resedimentation over blind structures in a foredeep, Apennine basin. The diagrams are simplified and idealized from seismic-reflection sections. The deposits formed in a deep-marine environment, and include turbidite channels and progradational depositional systems. Short, irregular reflections are coarse-grained, chaotic, slump deposits. (Reproduced with permission from Ori et al., 1986.)

Crook (1989) addressed a different problem, that of the rate at which suturing progresses along strike as two plates converge along oblique trajectories. He used the lateral shift in facies from deep-marine to shallow-marine to nonmarine as indicators of convergence. The data from his project area, the Huon Gulf of Papua New Guinea, are incomplete, but they permit an estimate of between 7 and 18 cm/y for the rate of lateral "zipping-up" of a suture (see Chapter 10).

Reconstruction of rates of convergence, shortening and crustal thickening in ancient orogens is complicated by the destruction of evidence that is an inte-gral part of the orogenic process. Foreland basins have been described as "migrating foredeeps" (Bally et al., 1966). The rate of migration of depocenters should provide an estimate of shortening and assist in estimating rates of plate convergence. Ricci Lucchi (1986) showed that, in the Apennine foreland basin, this rate varied between 0.3 and 0.8 cm/y. In the western part of the Swiss Molasse basin, the migration rate was 0.6 cm/y (Homewood et al., 1986), whereas structural reconstructions by Pfiffner (1986) indicate a range of 0.2 to 0.3 cm/y for the central part of this basin.

It has been demonstrated that arc-continent collision produces a succession of datable structural and stratigraphic events that can be used to reconstruct the rate of plate convergence (Bradley and Kusky, 1986; Bradley, 1989; Bradley and Kidd, 1991). The first indication of an approaching arc-continent collision is the initiation of a broad uplift, which records the passage of the forebulge. This is marked by low-angle subregional unconformities in the stratigraphic record. Following this uplift, activation of extensional faults occurs on the rifted margin of the downgoing plate, as a result of flexural extension (Fig. 11.11). This event can be dated by the age of fault breccias and sedimentary thickening. The margin then subsides under the supracrustal load, and its gradual subsidence may be measurable using biostratigraphic and biofacies data relating to onlap and deepening. The sediment transport directions then undergo reversal as the colliding terrane or arc becomes the principal sediment source, and the sediments change facies from deep-water mud to sand-dominated turbidites and related facies. Figure 11.26 is a time-distance plot for some of these events as tracked by Bradley and Kusky (1986) in the Taconic foreland of New York (also see Fig. 10.20). Figure 11.27 illustrates some of the variations in the collision patterns between irregular continental margins that can be reconstructed, complete with plate convergence trajectories and rates, using this method. Bradley and Kusky (1986) were able to obtain reasonably consistent data demonstrating a convergence rate of 2 to 3 cm/y for the Taconic foreland of New York, and Bradley (1989) applied the same methods to the Taconic rocks over the entire length of the Appalachian orogen. He

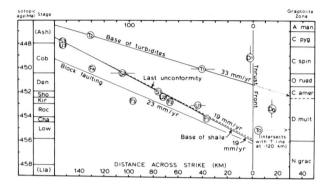

Fig. 11.26 Time-distance plots for events in the Taconic peripheral basin of New York. Circled numbers and letters are data points. Block faulting is the normal faulting of the continental margin as it approaches subduction zone. The last unconformity is that associated with the final passage of migrating forebulge. Turbidites are those associated with the development of a deep-water basin and sediment derivation from the allochthon. (Reproduced with permission from Bradley and Kusky, 1986.)

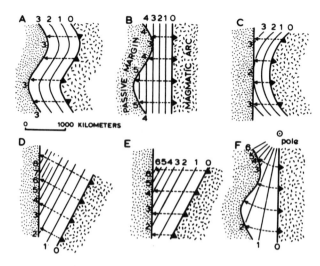

Fig. 11.27 Variations in the shapes and trajectories of approaching continents that can lead to diachronous collision. Numbered lines refer to datable structural-stratigraphic events that can be documented in order to reconstruct collision; dashed lines are regional transects through the orogen (e.g., Fig. 11.26). (Reproduced with permission from Bradley, 1989.)

showed how the same methods could be used to reconstruct the amount of shortening of major thrust allochthons. Once a regional convergence rate is established, it can be used in the development of palinspastic reconstructions of fold-thrust belts. Unfolding of folds and restoration of faulted units to pre-faulting positions can be carried out using the restored positions to calculate rates of shortening for comparison with regional rates. This serves as a check on the validity of the palinspastic reconstruction, and may assist in determining the total amount of local shortening during orogeny.

Analysis of rates of movement of individual thrust faults and analysis of specific tectonic and depositional episodes are more difficult problems (Burbank and Raynolds, 1988; Jordan et al., 1988). Structural and stratigraphic cross-cutting data have been used to estimate rates of thrust-tip migration, but these yield average rates, such as the rate of 0.7 cm/y calculated for the western Swiss Molasse basin (Homewood et al., 1986). Burbank and Raynolds (1984, 1988) used magnetostratigraphic techniques to date fluvial deposits of the Himalayan foredeep of Pakistan. The technique permits a precision of dating to within the nearest 10^5 y (Jordan [Chapter 9] discusses examples of studies that obtained a precision of 30×10^3 y). These results show that, in the Himalayan foredeep, thrusting and uplift episodes were very rapid and

spasmodic, and that they did not occur in sequence into the basin, as the classic model (e.g., Dahlstrom, 1970) would predict (Fig. 11.28). In one case, Burbank and Raynolds (1988) were able to demonstrate the uplift and removal of 3 km of sediment over the crest of an anticline within the basin over a period of 200,000 y, an average uplift and erosion rate of 1.5 cm/y. In most basins, this precision of dating cannot be attempted, and this, therefore, suggests that caution should be used in assessing published reconstructions of rates of convergence and crustal shortening, except where these are offered as long-term (>1 my) averages.

Petrology, and Its Relationship to Local Tectonic Evolution

The detrital petrology of sandstone and conglomerate within foreland basins has long been used as an aid to tectonic interpretations. At the regional to continental scale, it has been shown that modal (average) values of sandstone composition can be used to discriminate among broad classes of source-terrane type, with the sediments of foreland basins falling chiefly into a class of "recycled-orogen" petrofacies (Dickinson and Suczek, 1979). Ingersoll (1990) pointed out that at

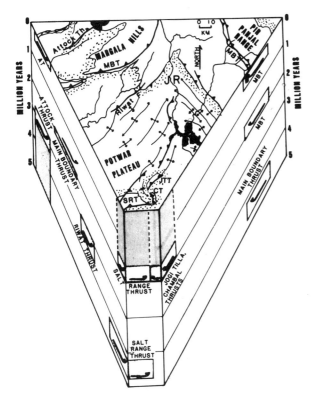

Fig. 11.28 Chronology of fault motions within part of the Himalayan foredeep of Pakistan. The shaded boxes indicate the time and space domain over which each thrust is interpreted to have been active. (Reproduced with permission from Burbank and Raynolds, 1988.)

the local level (his first- and second-order sampling scales), this type of analysis may break down because of small-scale variations in source-area geology. On the other hand, foreland basins have uniform recycled-orogen compositions at all sampling scales as a result of their uniform sedimentary and metasedimentary source rocks (Ingersoll et al., 1993). Local studies can be extremely useful in analyzing the tectonic history of a specific basin and its source terranes. Basin-margin conglomerates have long been interpreted as synorogenic deposits, indicating the timing of thrust-faulting and uplift events; the sequence of detrital species preserved within the basin-fill succession records the unroofing history of the source terrane (e.g., Spieker, 1946; Armstrong and Oriel, 1965; Price and Mountjoy, 1970; Rahmani and Lerbekmo, 1975; Graham et al., 1986; Lawton, 1986a; Ingersoll et al., 1987; Burbank and Raynolds, 1988; Jordan et al., 1988).

In the simplest case, unroofing of a flat-lying sequence of rocks and transport of the detritus into an adjacent basin produces an *inverted stratigraphy* of detrital particles (Fig. 11.29A). Numerous studies of detrital petrology, particularly of syntectonic conglomerates, have been based on this premise (e.g., Graham et al., 1986; Lawton, 1986a; Ingersoll et al., 1987). However, in many cases, structural development of the fold-thrust belt does not permit such a simple analysis (Steidtmann and Schmitt, 1988). In the first instance, strata in the source terrane may be steeply dipping, or uplifted along steep faults, so that detritus is blended in the basin fill (Fig. 11.29B). Secondly, the moving thrust plate may carry with it inactive, older thrust plates that had been uplifted and eroded at earlier stages of basin evolution (terrane B in Fig. 11.30). Detritus from these older sources will then be reintroduced into the basin. Schmitt and Steidtmann (1990) proposed the term *interior ramp-supported uplift* for such structures. A third complication is that, where there are interior uplifts at the margins of the basin (satellite or piggyback-basin configurations: Fig. 11.14), sediment may be shed toward, and trapped within, the core of the fold-thrust belt, rather than toward the main foreland basin, as is usually assumed (antithetic dispersal, D in Fig. 11.30).

Basin-margin conglomerates are traditionally regarded as synorogenic (e.g., Graham et al., 1986; Burbank and Raynolds, 1988; Jordan et al., 1988), and there is ample evidence to support this interpretation where proximal facies belts are preserved, as in the deposits containing syndepositional unconformities

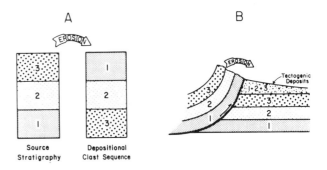

Fig. 11.29 Composition of basin fill in terms of detrital source petrography: A. classic inverted stratigraphy, B. blended compositions. (Reproduced with permission from Steidtmann and Schmitt, 1988.)

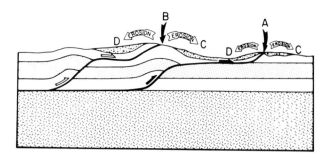

Fig. 11.30 Structural setting and drainage patterns at a basin margin. A and B are active source terrains. Fault plate A carries older plate (B) as an uplifted ramp. Drainage may be toward the basin (C), the "normal" or "expected" case, or it may be away from uplifts toward the interior of the fold belt (antithetic drainage, D). (Reproduced with permission from Steidtmann and Schmitt, 1988).

described by Riba (1976) and Anadon et al. (1986) (see above). However, recent field studies in parts of the Rocky Mountains by various workers (e.g., Beck et al., 1988; Heller et al., 1988; Heller and Paola, 1989), a review of "tectonic cyclothems" by Blair and Bilodeau (1988), and quantitative modeling of foreland-basin subsidence and stratigraphy by Heller et al. (1988) have suggested alternative detailed models for gravel transport and its relationship to thrust-faulting activity. The data of Beck et al. (1988) and Heller et al. (1988) suggest that the elastic reponse to loading of a thrust sheet results in proximal subsidence that outpaces the rate of gravel supply. Coarse-grained detritus is thus confined to a narrow belt immediately adjacent to the thrust front (possible modern analogs are the small alluvial fans flanking the Zagros Mountains at the edge of the modern Persian Gulf; Baltzer and Purser, 1990), and the adjacent moat may be filled with marine, lacustrine or fluvial sediments, in which paleoflow patterns may be toward or parallel to the thrust rather than away from it. The basin-margin deposits contemporaneous with the thrusting may be fine-grained (an extremely narrow fanglomerate belt could be developed immediately adjacent to the fault), with upward coarsening to gravel following the cessation of fault-induced loading, when erosion and isostatic rebound occurs. Heller and Paola (1989) showed that Lower Cretaceous gravels flanking the Sevier orogen of the United States Western Interior are too areally extensive to be explained on the basis of the uplift and paleoslope

associated with a flexural thrust-loading model. They suggested that widespread uplift caused by regional thermal doming associated with arc magmatism is the most likely cause for dispersal of these gravels. Gravels derived from the fold-thrust belt may be distinguished from those derived from regional uplifts by their isopach patterns (Yingling and Heller, 1992). Many additional case studies are needed to determine the local validity of these alternative models. For example, Burbank et al. (1988) studied the Plio-Pleistocene gravels of the Himalayan foredeep of Pakistan. Their estimates of the age and progradation rates of the gravel, based on magnetostratigraphic data, support the original syntectonic-conglomerate model. The Himalayan case may not be typical of all foreland basins because of the presence of a particularly rigid underlying crust, which affects the flexural behavior of the basin, and because of the presence of very large, vigorous antecedent rivers capable of carrying huge volumes of detritus into the basin. Heller et al. (1989) and Burbank et al. (1988, 1989) debated these and other points in a discussion (going beyond the scope of this book) that reveals the complexity of the problem of modeling basin formation, subsidence, and filling mechanisms.

Sequence Concepts

It has long been known that foreland-basin strata are characterized by the intertonguing of marine and nonmarine strata, indicating episodic regression and transgression. For example, Weimer (1960) recognized four regional regressive-transgressive cycles in his classic work on the Upper Cretaceous of the Western Interior of the United States, plus many minor events that he did not, at that time, attempt to correlate. Many nonmarine (molasse) successions consist of several separate wedges of coarse detritus that formed by episodic progradation into the basin (e.g., Miall, 1978; Van Houten, 1981; Blair and Bilodeau, 1988). In recent years, a revival of interest in regional and intercontinental stratigraphic sequences (e.g., Vail et al., 1977) has led to a renewed search for sequence-forming mechanisms, and to a vigorous debate regarding the relative importance of regional and intercontinental tectonism versus eustatic sea-level

change as mechanisms for sedimentary episodicity. This debate is of particular importance, and particularly difficult to resolve, in foreland basins, which, by their very nature, owe their origins to regional tectonic activity. It has long been accepted that episodic cycles of uplift, related to crustal shortening, are responsible for large-scale pulses of molasse deposition (Miall, 1978), although as noted above, the relationship between thrust loading and gravel progradation may be more complex than was once thought. It is becoming possible to correlate individual pulses with specific tectonic events, such as times of terrane accretion (Cant and Stockmal, 1989; Stockmal et al., 1992), and thrust faulting (Burbank and Raynolds, 1988). More difficult to resolve is the question of causality for intertonguing fine-grained marine and nonmarine clastic successions far from sediment sources, where the influence of basin-margin tectonism is not obvious. Alternative models of sequence development, relating transgression and regression primarily to eustatic changes in sea-level, have received considerable attention in recent years (e.g., Vail et al., 1977; Wilgus et al., 1988). However, these models were developed primarily for rifted continental margins (Posamentier et al., 1988) and their applicability to foreland basins remains unproven. As pointed out by Swift et al. (1987), patterns of subsidence in foreland basins are quite different from those of rifted continental margins (also see Chapter 9). In both cases, the basins undergo differential subsidence around a hinge line. Along rifted continental margins, the hinge is located on the edge of the basin, with maximum subsidence in areas distant from the sediment source. In foreland basins, the hinge is also located on the edge of the basin, but with maximum subsidence adjacent to the source area (Fig. 11.31).

It is now known that regional allogenic sequences fall into several classes, differentiated in the first instance by their time span (Vail et al., 1977; Miall, 1990). Of concern here are the so-called third-, fourth- and fifth-order sequences, which have indicated durations of 1–10, 0.2–0.5, and 0.01–0.2 million years, respectively. Longer-term sequences (>10 my) are related to global plate-tectonic controls and are beyond the concern of this chapter (see review in Miall, 1990). The relative importance of tectonism and eustasy in

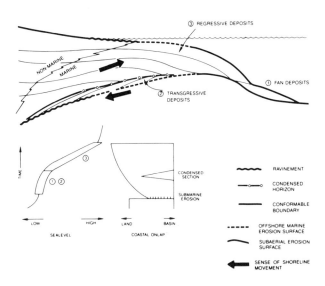

Fig. 11.31 Sequence architecture within a foreland basin, showing the succession formed during an episode of rising and falling sea level. (Reproduced with permission from Swift et al., 1987.)

generating these sequences is by no means clear, either in general global terms or in the case of specific sequences in foreland basins. These problems can be exemplified by brief reviews of Cenozoic sequences of two European peripheral basins, and Cretaceous cyclicity of the North American Western Interior Seaway. The latter classifies as a retroarc foreland basin, but the processes of flexural loading and tectonic control of sequence development are considered to be similar for all classes of foreland basin (see Chapter 9).

Sequence concepts have been employed in the analysis of several foreland-basin stratigraphic successions in the Alpine belt of Europe. For example Crumeyrolle et al. (1991) discussed a sequence analysis of the Alpine molasse succession in the Digne basin, France. This basin constitutes one of the sub-basins within the Cenozoic Alpine foreland basin of France, Switzerland and Austria. The Oligocene-Miocene succession is more than 2 km thick and has been subdivided lithostratigraphically, in ascending order (Fig. 11.32) into:

1. Red Molasse (Oligocene): fluvial channel sandstones and channelized fanglomerates.
2. Intermediate Molasse (Aquitanian): thin fluvial and deltaic sandstones and marls.
3. Lower Marine Molasse (Burdigalian): transgressive, wave-dominated deltaic clastics.

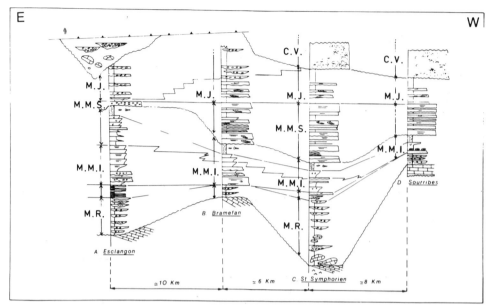

Fig. 11.32 Stratigraphic cross-section through the Digne molasse basin. M.R.=Red Oligocene Molasse, M.M.I.= Lower Marine Molasse, M.M.S.= Upper Marine Molasse, M.J.= Yellow Molasse, C.V.=Valensole Conglomerate. Unconformities constitute the main sequence boundaries. (Reproduced with permission from Crumeyrolle et al., 1991.)

4. Upper Marine Molasse (Langhian): wave- and tide-dominated sandstones and marls.
5. Yellow Molasse (Serravallian): prograding succession of channelized conglomerates grading distally into alluvial-plain and shallow-marine deposits.

The Lower Marine Molasse constitutes two third-order stratigraphic sequences. The total section is more than 800 m thick. The lowermost sequence boundary is a ravinement surface. It is overlain by a succession of landward stepping, wave-dominated, deltaic cycles constituting a "transgressive systems tract." A condensed interval consists of bored and glauconitized beds, and is overlain by progradational deltaic cycles corresponding to the "highstand systems tract." The base of the second sequence contains a bioclastic tidal-bar deposit resting on an erosion surface that corresponds to the sequence boundary. The top of the tidal-bar deposit is marked by a "maximum-flooding surface," and is overlain by shelf mudstones. These, in turn are overlain by prograding highstand deltaic deposits. The uppermost of these consists of brackish lagoonal marls containing oysters and pelecypods.

Crumeyrolle et al. (1991) suggested that the overall molasse succession is a long-term (second-order?) trangressive-regressive cycle induced by foreland-basin tectonism. The succession is punctuated by unconformities and is subdivisible into third-order sequences, which they suggested may be related to eustatic events and can be correlated with the Haq et al. (1987, 1988) global cycle chart. Recent debate (Miall, 1992, 1993; Dickinson, 1993; Snyder and Spinosa, 1993) indicates that this type of correlation is speculative and should be viewed with caution.

The South Pyrenean foreland basin of northeast Spain contains a mixed carbonate-siliciclastic Paleocene-Eocene succession 3 km thick that has been subdivided into nearly twenty third-order sequences. Contrasting analyses of different, but overlapping parts of this succession were provided by Puigdefabregas et al. (1986) and Luterbacher et al. (1991). In the first of these papers, the authors attributed development of the sequences to tectonism, the gradual southward overthrusting of the fold-thrust belt leading to a migration and offlapping of depocenters (Fig. 11.33). However, in the second paper, it is claimed that correlation of the sequences with the Haq et al. (1987, 1988) chart can be carried out, and the authors interpreted eustasy as the main driving mechanism.

An idealized cross section of the Eocene-Oligocene section is shown in Figure 11.33. The overall structure of the sequences is that of a coarse, nonmarine clastic wedge prograding southward from the thrust front into basin-center marine marls. Shallow-water carbonate deposits are formed on the distal ramp and

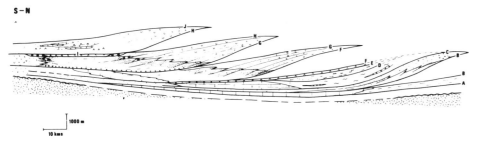

Fig. 11.33 Idealized cross section of the Eocene-Lower Oligocene sequence stratigraphy of the South Pyrenean basin. (Reproduced with permission from Puigdefabregas et al., 1986.)

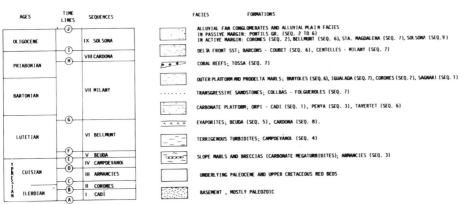

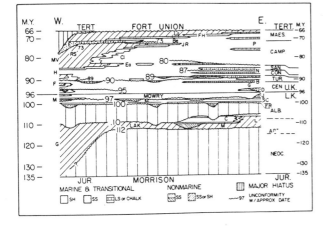

Fig. 11.34 Diagrammatic cross section through the Western Interior Seaway of the Rocky Mountains, showing stratigraphic positions and approximate dates of major transgressive units and interregional unconformities. Formations or groups to the west are: G = Gannett; SC = Skull Creek; M = Mowry; F = Frontier; H = Hilliard; MV = Mesaverde; RS = Rock Springs; E = Ericson; Ea = Eagle; Cl = Claggett; JR = Judith River; Be = Bearpaw; FH = Fox Hills; La = Lance. To the east, formations are L = Lytle; LAK = Lakota; FR = Fall River; SC = Skull Creek; J and D sands of Denver basin; G = Greenhorn; B = Benton; N = Niobrara; P = Pierre; M and C = McMurray and Clearwater of Canada. (Reproduced with permission from Weimer, 1986.)

forebulge of the foreland basin, away from the influence of the fold-thrust-belt clastic source. Carbonates are found in the transgressive stage of each sequence. They may be associated with evaporites formed during the subsequent regressive stage. Careful dating of these sequences by Burbank et al. (1992, p. 1118) using magnetostratigraphy indicated "that progradation was controlled primarily by subsidence, rather than sediment supply or base-level changes."

Ten major third-order transgressive-regressive cycles have been recognized in the Western Interior basin of the United States (Weimer, 1960, 1986; Kauffman, 1984). Most of these are shown in Figs. 11.34 and 11.35. High sea level was characterized by the development of thick and areally extensive mudstone units (e.g., Mancos Shale), and by fine-grained limestone (including chalk) in areas distant from sediment sources (e.g., Niobrara Formation). Lower sea level gave rise to extensive clastic wedges, in which nonmarine sandstone and conglomerate passed basinward into shoreline and shelf sandbodies. The Indianola and Mesaverde groups of Utah and Colorado are good examples of major regressive stratigraphic packages, which contain numerous minor transgressive-regressive cycles (Weimer, 1960; Fouch et al., 1983;

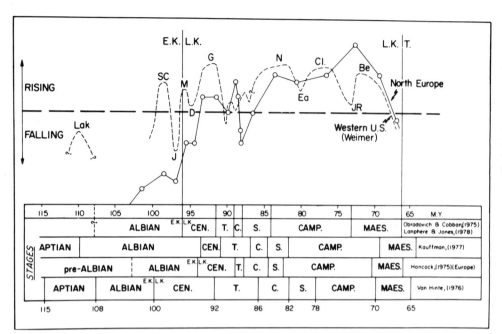

Fig. 11.35 Sea-level curves for United States and Europe. Letters designate same formations/groups as in Fig.11.34. Abbreviations for stages: Cen = Cenomanian; T = Turonian; C = Coniacian; S = Santonian; Camp = Campanian; Maes = Maastrichtian; EK = Early Cretaceous; LK = Late Cretaceous; T = Tertiary. (Reproduced with permission from Weimer, 1986.)

Lawton, 1986a,b; Swift et al., 1987; Van Wagoner et al., 1990; Fig. 11.36).

Some of the third-order cycles may be global in origin, and related to eustatic changes in sea level. As shown in Fig. 11.35, there is a fair degree of correlation between the sea-level curve for the western United States and that for northern Europe, although correlations are bedeviled by difficulties in reconciling various chronostratigraphic time scales. Arguments regarding the validity of the type of correlation shown in Fig. 11.35, and the precision of the dating on which such correlation depends, are beyond the scope of this chapter (see Miall, 1991, 1992, 1993; Dickinson, 1993; Snyder and Spinosa, 1993). Kauffman (1984) claimed that there is a consistent correlation among transgression, thrusting (in Wyoming and Utah), and volcanism in the Western Interior, but his data do not bear this out. However, there is no doubt that some of the transgressions were contemporaneous with thrust-faulting episodes within the Sevier orogen of Utah, possibly including episodes in the Albian, Santonian and Maastrichtian (compare Kauffman, 1984; Lawton, 1986a,b; Weimer, 1986). These transgressions were regional in scope, and cannot, therefore, be attributed to flexural loading by individual thrust plates. However, where thrusting is part of regional compression, possibly related to terrane

accretion or variations in subduction parameters, the mechanical-drag effect described by Mitrovica et al. (1989; see above) may be invoked as a cause of regional basin subsidence and transgression (although it is not yet clear whether the rate of change induced by this process is that of a third-order stratigraphic cycle). This does not explain how such transgressions might be correlated with eustatic change (such as the events suggested by the correlation of the two curves in Fig. 11.35), unless changes in subduction rates were part of an oceanwide or global reordering of plate-rotation vectors large enough to impact worldwide sea level. Cloetingh (1988) demonstrated that plate-margin stresses can be transmitted for thousands of kilometers across plate interiors, and this might explain the relationship between continent-wide third-order changes in sea level and regional tectonic events (some possible examples of such a relationship are summarized by Miall, 1991).

Other thrusting events within the Sevier orogen are not clearly correlated with regional changes in sea level in the basin, but are correlated with the development of major clastic wedges that prograded across the basin margins. Lawton (1986a,b) indicated a link between thrusting and clastic-wedge formation in Utah between the mid-Albian and the late Campan-

ian, although it is not possible to provide the tight correlation between individual tectonic and stratigraphic events that is now available for parts of the Himalayan foreland.

Are individual clastic tongues within major wedges tectonic or eustatic in origin? Figure 11.36 illustrates the broad stratigraphic architecture of part of the Upper Cretaceous clastic wedge of Utah, and Fig. 11.37 is a sequence model for a fourth-order cycle of a type that is common in parts of Alberta and British Columbia. There is no doubting an overriding tectonic control of sedimentation for many of these successions. Paleocurrent and petrologic evidence indicates shifting sediment sources and changes in regional paleoslope during deposition of the Mesaverde Group and associated units in Utah. Lawton (1986a,b) documented unroofing of intrabasin uplifts, from which early basin-fill sediments were cannibalized, and changes in dispersal patterns related to basin tilting. The volume of sediment within the wedge is too great to have been controlled by passive changes in sea level (also see Galloway, 1989b). But the question remains whether tectonically induced sediment input was modulated by sea-level control, leading to fourth- and higher-order cyclicity along the fringes of the clastic wedge (such

as in the Book Cliffs of Utah: see Swift et al., 1987; Van Wagoner et al., 1990). Posamentier et al. (1988) argued that fluvial coastal-plain progradation is switched on during initial sea-level fall after a time of highstand, as a result of lateral shift in stream profiles and creation of sedimentary accommodation space. However, Miall (1991) argued that this concept is faulty on several grounds (e.g., base-level fall normally leads to incision), and maintained that variations in sediment supply, driven by tectonism, is the major cause of coastal-plain progradation. The sequence models applied by Van Wagoner et al. (1990) to parts of the Mesaverde Group of Utah are those developed for passive margins; these models, as noted above, and in Chapter 9, may be unsuitable for foreland basins. Their successful comparison of a model to a specific field case does not prove that the preferences of the authors with regard to mechanisms are correct. There is room for considerable additional field work, including detailed mapping of the sequences in order to permit a reconstruction of their architectural evolution, and detailed dating, to permit a test of the correlation with regional or global (eustatic) cycles of sea-level change. As noted above, problems of time resolution make such correlations questionable at present.

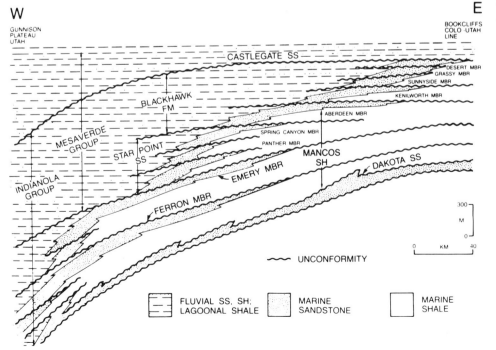

Fig. 11.36 Regional stratigraphic framework of Upper Cretaceous strata of east-central Utah. (Reproduced with permission from Swift et al., 1987.)

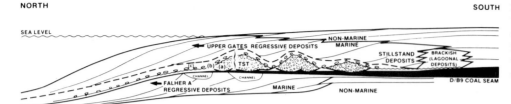

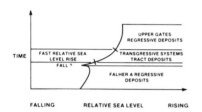

Fig. 11.37 Sequence model for Upper Gates Formation (Lower Cretaceous), northeastern British Columbia. (Reproduced with permission from Carmichael, 1988.)

Other examples of fourth-order cycles dominated by shelf and shoreline depositional systems are abundant farther north within the Western Interior Seaway, particularly in Alberta and British Columbia (Leckie, 1986; Plint et al., 1986; Leggitt et al., 1990; Fig. 11.37). In all these cases, appeal was made to tectonic mechanisms to explain short-term changes in relative sea level. Leggitt et al. (1990) described beveling of an erosion surface, the geometry and orientation of which appears to indicate intrabasin tilting between intervals of sedimentation.

Conclusions

Increasing refinement in the study of the structural geology, stratigraphy and sedimentology of foreland basins in recent years has been coupled with an increasing sophistication in interpretation, as computer-aided models of geodynamics, sediment dispersal and sequence development permit the testing of ever more complex ideas regarding crustal evolution. These trends in research have, as such trends commonly do, raised as many questions as they have answered. In the case of foreland basins, refinements in litho- and allostratigraphic correlation (e.g., Cardium studies of Alberta: Plint et al., 1986; Walker and Eyles, 1991) and in chronostratigraphic correlation (e.g., magnetostratigraphic studies of the Himalayan foredeep: Burbank et al., 1986) have revealed a hitherto unsuspected abundance of short-term sedimentary and tectonic events. This is bring-

ing about a reexamination of existing geodynamic models to focus on short-term crustal processes.

Particularly complex questions relate to the mechanisms controlling third-, fourth- and fifth-order stratigraphic sequences. As discussed in this chapter, there is some evidence that suggests a global correlation of some of these events, and at the same time, there may be a correlation with localized events, such as thrust-movement pulses. The complete chain of cause and effect, from global plate movements on the large scale, to intraplate structural readjustments at the other end of the scale (with corresponding stratigraphic consequences), has yet to be modeled successfully, but promises to be one of the most fruitful and exciting lines of research to be carried out in foreland basins in the next few years.

Further Reading

Allen PA, Homewood P (eds), 1986, *Foreland basins:* International Association of Sedimentologists Special Publication 8, 453 p.

Beaumont C, 1981, Foreland basins: *Geophysical Journal of the Royal Astronomical Society*, v. 65, p. 291–329.

Jordan TE, Flemings PB, Beer JA, 1988, Dating thrust-fault activity by use of foreland-basin strata, in Kleinspehn KL, Paola C (eds), *New perspectives in basin analysis:* Springer Verlag, New York, p. 307–330.

Quinlan GM, Beaumont C, 1984, Appalachian thrusting, lithospheric flexure, and the Paleozoic stratigraphy of the eastern interior of North America: *Canadian Journal of Earth Sciences*, v. 21, p. 973–996.

Stockmal GS, Beaumont C, Boutilier R, 1986, Geodynamic models of convergent margin tectonics: transition from rifted margin to overthrust belt and consequences for foreland-basin development: *American Association of Petroleum Geologists Bulletin*, v. 70, p. 181–190.

Strike-Slip Basins

12

Tor H. Nilsen and Arthur G. Sylvester

Introduction

Various types of basins, previously grouped together by many workers as "pull-apart basins" (Burchfiel and Stewart, 1966), commonly form along major strike-slip faults. We refer to these basins herein as *strike-slip basins,* following Mann et al. (1983), because "pull-apart basins" are only one of a range of basin types that may develop as a result of strike-slip faulting.

Strike-slip basins range in size from small sag ponds to rhombochasms as wide as 50 km (Fig. 12.1). Basin length-to-width ratios are typically 4:1, with a range from 1 to 10:1 (Aydin and Nur, 1982). The surface dimensions of a strike-slip basin may be measured either structurally by bounding faults or flexures, or physiographically by the extent of the area of subsidence. The basin margins, however, are commonly deformed by younger folds and faults, or have been displaced from the basin depocenter by continued strike slip. Bounding faults may dip in varying directions and be of various types. As a result, definition of basin size and shape may be a difficult task because bounding faults may dip either more steeply or more gently with depth, and may merge at depth into a single master fault; some basin margins may sag rather than be faulted. Some strike-slip basins

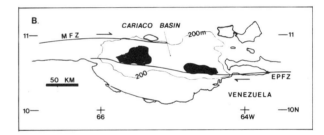

Fig. 12.1 Examples of range of scales of stepover basins. A) Photograph of sag pond on central San Andreas fault, California. B) Fault and bathymetric map of Cariaco basin at right step between dextral Moron (MFZ) and El Pilar (EPFZ) fault zones, Venezuelan borderland (Reproduced with permission from Mann, et al., 1983).

may also be bounded at their bases by sub-horizontal faults.

Strike-slip basins are common and have been best studied along continental transform boundaries, such as the San Andreas fault of California, where extensive hydrocarbon exploration has yielded abundant three-dimensional control. The processes leading to crustal extension and subsidence in strike-slip settings are generally not as well understood as they are in other tectonic settings, however, and they may be extremely complex at different scales. Basins that develop along and adjacent to strike-slip faults are generally small and short-lived compared to those that develop in other plate-tectonic settings. Thermo-mechanical models for their formation, as well as structural and stratigraphic models of their evolution, are generally poorly developed. Each basin must be examined in its temporal, spatial, and tectonic setting within a very broad framework of paradigms that have been collected from strike-slip basins around the world. New concepts are required to model their behavior and resolve their sedimentary and tectonic histories.

Classification of Faults and Basins

The plate-tectonic settings of strike-slip faults and strike-slip basins, as developed by Woodcock (1986) and Reading (1980), respectively, reflect the complex settings and histories of horizontal slip along plate margins and provide a framework for fault classification (Fig. 12.2). Strike-slip faults may be divided into transform and transcurrent faults (Table 12.1; Sylvester, 1988).

The term "transform fault" is used for strike-slip faults that define plate boundaries, penetrate the crust, and include: (1) ridge transforms that displace spreading ridges; (2) boundary transforms that accommodate horizontal displacement between plates; and (3) trench-linked transforms sub-parallel to trenches in convergent settings.

The term "transcurrent fault" is restricted to intraplate faults that penetrate only the upper crust and include: (1) indent-linked strike-slip faults in zones of convergence and tectonic escape; (2) intra-continental strike-slip faults that are unrelated to

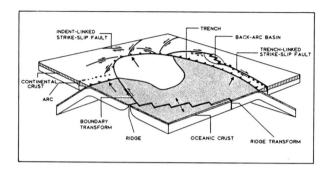

Fig. 12.2 Plate-tectonic settings of strike-slip faults (Reproduced with permission from Woodcock, 1986). Diagram is a schematic representation of India-Asia collision, looking northeast from southern Indian Ocean (compare with Fig. 10.5).

indentation and typically separate distinct regional tectonic domains; (3) tear faults that accommodate the differential displacement within or at the margin of an allochthon; and (4) transfer faults connecting overstepping segments of parallel or en-echelon strike-slip faults (Allen and Allen, 1990).

We adopt a six-fold classification of strike-slip basins based on the geometry of bounding faults and the kinematic setting of the basin (Fig. 12.3):
1. fault-bend basins
2. stepover basins
3. transrotational basins
4. transpressional basins
5. polygenetic basins
6. polyhistory basins

This classification permits easy recognition of basins in their initial stages of formation. Either part or all of a strike-slip basin may undergo stages of growth and development, however, that could fit into any of these basin types. As strike-slip basins evolve through time and space, they may undergo significant transla-tion along principal strike-slip zones. Changes in their plate-tectonic framework, even including minor changes in direction and rate of plate motion, may radically affect basin structure and history. The basins are commonly subjected to multiple cycles of subsi-dence and uplift while being translated laterally (the "porpoising" effect), and in general, are partly to wholly uplifted.

Fault-bend basins generally develop at releasing bends along strike-slip faults (Figs. 12.3, 12.4). Exten-sion results as one fault block slides past and away

Table 12.1 Classification of strike-slip faults by Sylvester (1988), based partly on Woodcock's (1986) relation of strike-slip faults to plate-tectonic setting (see Fig. 12.2)

INTERPLATE (deep-seated)	INTRAPLATE (thin-skinned)
TRANSFORM faults (delimit plates, cut lithosphere, fully accommodate motion between plates)	TRANSCURRENT faults (confined to the crust)
Ridge Transform faults* • Displace segments of oceanic crust having similar spreading vectors • Present examples: Owen, Romanche, and Charlie Gibbs fracture zones	Indent-linked strike-slip faults* • Separate continent-continent blocks which move with respect to one another because of plate convergence • Present examples: North Anatolian fault (Turkey); Karakorum, Altyn Tagh, and Kunlun fault (Tibet)
Boundary transform faults* • Join unlike plates which move parallel to the boundary between the plates • Present examples: San Andreas fault (California), Chaman fault (Pakistan), and Alpine fault (New Zealand)	Tear faults • Accommodate differential displacement within a given allochthon, or between the allochthon and adjacent structural units (Biddle and Christie-Blick, 1985) • Present examples: northwest- and northeast-striking faults in Asiak fold-thrust belt (Canada)
Trench-linked strike-slip faults* • Accommodate horizontal component of oblique subduction; cut and may localize arc intrusions and volcanic rocks; located about 100 km inboard of trench • Present examples: Semanko fault (Burma), Atacama fault (Chile), and Median Tectonic Line (Japan)	Transfer faults • Transfer horizontal slip from one segment of a major strike-slip fault to its overstepping or en-echelon neighbor • Present examples: Lower Hope Valley and Upper Hurunui Valley faults between the Hope and Kakapo faults (New Zealand), and Southern and Northern Diagonal faults (eastern Sinai)
	Intracontinental transform faults • Separate allochthons of different tectonic styles • Present example: Garlock fault (California)

*See Woodcock (1986, p. 20) for additional examples, both ancient and modern, and for their geometric and kinematic characteristics.

from the other. Sharply defined clefts may form at the mesoscopic scale along releasing fault bends. At larger and longer scales, the sliding block may sag into the extended zone. More commonly, parts of both blocks sag toward the fault to form an elongate zone of subsidence along the fault bend, referred to as a lens-shaped basin (Crowell, 1974b) or a "lazy pull apart" (Mann et al., 1983). How the walls of a fault-bend basin converge at depth and merge with the master fault is largely unknown. Fault-bend basins may also form near restraining bends, where one of the fault blocks extends differentially as it slides around the bend, out of the zone of restraint.

Fault-bend basins are strongly asymmetric, have prominent coarse-grained aprons along their principal displacement zones, are commonly lens-shaped in map view, and generally develop in transtensional set-

tings, subsequently undergoing inversion in transpressional settings. These basins may resemble other rift basins (see Chapter 3), especially in two-dimensional seismic-reflection profiles.

Stepover basins (Fig. 12.3B) form between the ends of two parallel to sub-parallel strike-slip faults that are not connected (Aydin and Nur, 1985; Schubert, 1986; Sarewitz and Lewis, 1991). Because the zone between two fault segments in en-echelon arrangement is termed a "stepover," basins that form as a result of extension between the two faults are referred to as "stepover basins." The bounding faults may merge at depth into a single master fault. At basement level, the extended domain between the two en-echelon faults generally forms a mesh-like arrangement of normal and strike-slip faults that are steep at depth. Stepover basins typically

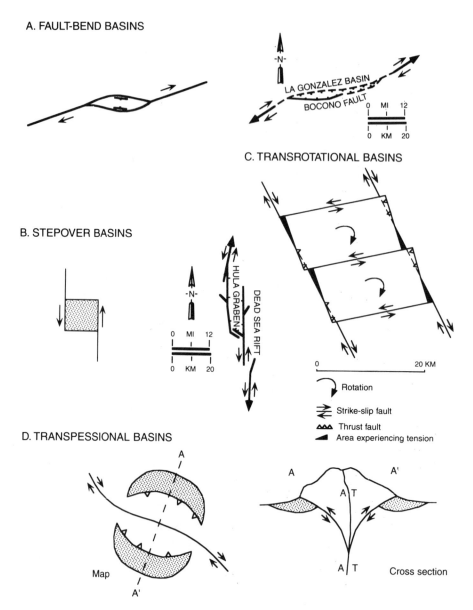

Fig. 12.3 Diagrammatic maps of six strike-slip basin types: A) fault-bend basin (left) with map of La Gonzalez basin, Venezuela (right); B) stepover basin (left) with map of part of Dead Sea rift (right); C) transrotational basins (black areas); D) transpressional basins (dot pattern) in map view (left) and cross section (right); E) polygenetic basins (dot pattern) in regional extension (left) and in regional shortening (right); and F) polyhistory basins initiated as rift basin.

form between left-stepping faults in left slip and between right-stepping faults in right slip (Fig. 12.3B).

Stepover basins may be more symmetric than fault-bend basins, with coarse-grained aprons shed basinward from all faulted margins. Depocenters may not lie preferentially adjacent to one of the marginal faults. Transverse structures may be common, seg-

menting the basin into separate subbasins. Continued transtension may extend and rupture the crust, producing magmatic activity, high heat flow, and in extreme cases, generation of new crust that may be younger than the overlying sedimentary succession. The floors of other stepover basins may consist of gently dipping faults at the basement-cover interface or deeper in the basement.

E. POLYGENETIC BASINS

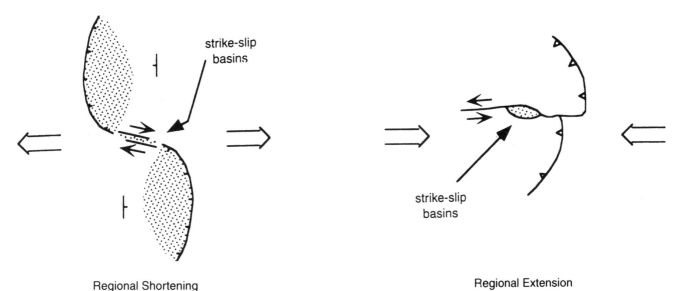

Regional Shortening

Regional Extension

F. POLYHISTORY BASINS

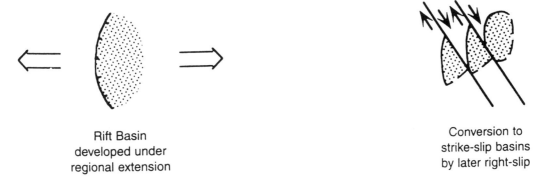

Rift Basin
developed under
regional extension

Conversion to
strike-slip basins
by later right-slip

Fig. 12.3 Continued.

Transrotational basins (Ingersoll, 1988b) develop as a result of continued shear strain that causes the extension fractures and the blocks between them to rotate about a subvertical axis in the same direction as the direction of principal shear strain, clockwise in right simple shear and counterclockwise in left simple shear (Fig. 12.3C). The rate and magnitude of rotation depend on the rate of shear strain. Triangular gaps or transrotational basins form among irregularly shaped, rotated blocks (Fig. 12.5). Detachment faults within the crust may floor the basins and separate upper rotated blocks from underlying unrotated blocks. The upper block may undergo rotation and strike-slip deformation during and following deposition. Major amounts of slip along basin-bounding faults are not necessary.

Fig. 12.4 Small fault-bend basin from Superstition Hills earthquake zone. Fault strikes away from viewer, from bottom of photograph to tower on horizon. Axis of basin is about 25° to fault strike. Net right-lateral strike-slip at this site was 1.5 m (photograph by A. G. Sylvester).

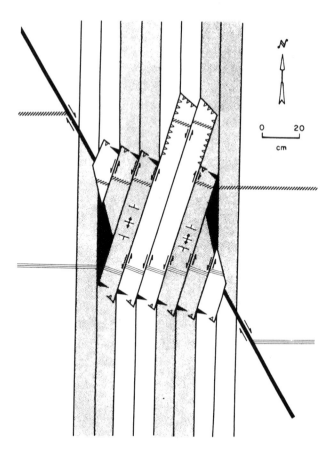

Fig. 12.5 Idealized diagram of gaps or basins (black areas) among rotated blocks in a right-simple-shear couple (Reproduced with permission from Terres and Sylvester, 1981). Hachures are on overthrust parts of blocks.

Transpressional basins (Fig. 12.3D) are generally long and narrow structural depressions, parallel to regional faults and folds, and are commonly bounded by underlying thrust faults and flanking strike-slip, or reverse faults. These basins may be dominated by axial transport of sediments parallel to structural depressions that develop in several ways, generally in response to flexural subsidence. They form next to uplifted and overthrust fault blocks that dip back into the principal displacement zone or other strike-slip faults (Fig. 12.6), typically adjacent to positive "flower" or "palm tree" structures in zones of trans-pression along strike-slip faults. Subsidence results from flexural loading of the marginal crust, forming mini-foreland basins adjacent to the uplifted blocks.

Piggyback basins, forebulges, and related structures may develop in the transpressive regions. Paired transpressional basins may form on opposite sides of transpressive uplifts. With continued strike-slip along the principal displacement zone, patterns of sedimentation and subsidence can be very complex. Other transpressive basins may develop in response to synclinal downfolding or paired reverse faulting within transpressive regions.

Polygenetic basins (Fig. 12.3E) as defined herein develop in local regimes of strike slip in generally convergent or divergent tectonic settings. In rift settings, transfer faults or accommodation zones may link border faults of normal displacement and bound small strike-slip basins. In convergent settings, strike-slip faults and strike-slip basins may be confined to

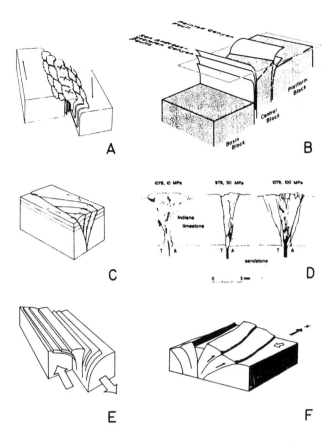

A

B

C

D

E

F

Fig. 12.6 Conceptual diagrams of flower or palm-tree structures in right simple shear (from Sylvester, 1988): A) from Lowell (1972, p. 3099); B) from Sylvester and Smith (1976); C) from Woodcock and Fisher (1986); D) from Bartlett et al. (1981); E) adapted with modifications from Ramsay and Huber (1987, p. 529); and F) with axial graben from Steel et al. (1985).

the upper plate of allochthonous thrust sheets, yielding a type of piggyback basin. Polygenetic strike-slip basins are also common in accommodation zones, where they may be confined to upper structural plates or hanging walls.

Polyhistory basins (Fig. 12.3F) are those in which episodes of pure extensional rifting or pure compressional thrusting alternate with episodes of strike slip, generating complex and commonly multicyclic basins. They can be of almost any size and shape, and they generally record pulses of subsidence caused by varying mechanisms. Many "successor basins" in complex orogenic belts are of this type; they may be long-lasting, with multiple histories of uplift and subsidence related to shifting tectonic settings. Polyhistory strike-slip basins may also have the characteristics of many of the other types of strike-slip basins,

but they tend to be even more complicated because of their varying tectonic framework.

Structural Framework

Fault Pattern

The structural patterns that develop along strike-slip fault systems depend on four principal factors (Christie-Blick and Biddle, 1985): 1) the kinematics (convergent, divergent, or parallel) of the fault system; 2) the magnitude of the displacement; 3) the material properties of the rocks and sedimentary infills in the deforming zone; and 4) the configuration of pre-existing structures. These factors may yield a curviplanar principal displacement zone with subsidiary faults having widely divergent and changing strikes. Extension occurs at releasing bends (Fig. 12.3B), and shortening takes place at restraining bends (Fig. 12.6); thus, basins and uplifts may be produced by curviplanar faults as well as overstepping faults.

Because basins along and adjacent to strike-slip fault systems may change both gradually and abruptly through time and space from divergence (transtension) to convergence (transpression), their tectonic and sedimentary histories may be extraordinarily complex. In simple strike-slip fault systems, the strike of the fault relative to the block or plate-motion vectors determines whether displacement along the fault has a component of transpression or transtension.

The upward branching "flower structures," characteristic in cross sections of strike-slip faults, range from simple, single-strand features with uniform dips toward the principal displacement zone to complex, upward convex, multi-strand faults that dip in variable directions (Harding et al., 1983; Harding, 1985; Fig. 12.6). In transpressional zones, most faults that make up "positive flower structures" have the cross-sectional appearance of reverse faults; in transtensional zones, most faults that make up "negative flower structures" appear to be normal faults (Fig. 12.7). Positive flower structures have an overall antiformal character with abundant folds as a result of net shortening, whereas negative flower structures have an overall synformal character with folds that result from net extension.

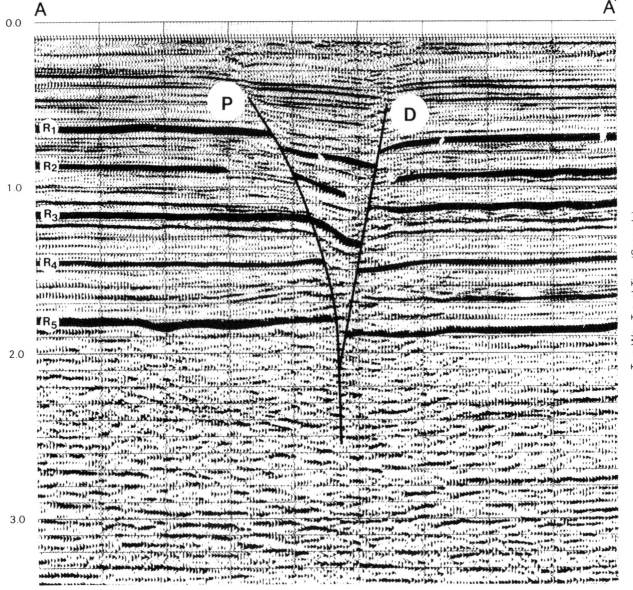

Fig. 12.7 Seismic-reflection image of negative flower structure (Reproduced with permission from D'Onfro and Glagola, 1983). P and D indicate fault names.

Where subparallel to the principal displacement zone, individual faults may have various dip directions, dip amounts, senses of displacement, and separation of rocks of different ages. Lateral changes from transpressional to transtensional settings cause gradual to abrupt changes in senses of displacement of either parts of or all of the flower structure. Because the overall strike-slip displacements include regional to local oblique as well as dip-slip displacements, individual faults must be studied in four dimensions.

Useful criteria for the correct interpretation of positive and negative flower structures were described and discussed by Harding (1985, 1990).

Rotation of blocks about vertical to subvertical axes within strike-slip fault zones may also produce areas of shortening and extension (e.g., Luyendyk and Hornafius, 1987). Rotation of large blocks may result in formation of basins of regional extent. Some strike-slip basins have subhorizontal detachment surfaces that separate structurally higher, rotated crustal

fragments characterized by brittle deformation from structurally lower crustal units characterized by more ductile deformation.

Basin Geometry

Strike-slip basins develop by various processes related to local transtensional and transpressive regimes. Oversteps and bent fault segments (releasing bends) commonly yield strike-slip basins that become more elongate parallel to the principal displacement zone through time. Some of the parameters that exert control on the length, width and depth of strike-slip basins are:

1. the degree of lateral and vertical curvature of the principal displacement zone and other fault surfaces
2. the depth to the brittle-ductile transition in the crust
3. the amount of displacement along the principal displacement zone and other faults
4. the age of the basin relative to the age of the principal displacement zone
5. the spacing between the overstepping faults
6. the length that the overstepping faults extend beyond one another

None of these parameters has been sufficiently quantified, unfortunately, to yield a predictive model that fits all strike-slip basins.

The type of and depth to basement rocks exert strong controlling influences on strike-slip-basin geometry. Where basement rocks are shallow, consist of massive and rigid plutono-metamorphic rocks, and are overlain by a thin cover of sedimentary strata, strike-slip basins are generally sharply defined and bounded by steeply dipping faults (Ben Avraham, 1985). Where basement consists of thick, easily deformed sedimentary or layered metamorphic rocks, overlain by a thick cover of sedimentary strata, strike-slip basins are generally poorly defined and bounded by low-angle listric transpressional or transtensional faults. As an example, the sedimentary cover on the Bering Shelf is more than 5,000 m thick and the basement rocks consist of a succession of sedimentary and layered metamorphic rocks (Worrall, 1991). Because the sedimentary cover is so thick, the strike-slip basins are less well-defined at the surface than

they are at depth. Faults do not propagate from the basement all the way to the surface, and folds die out upward in the stratigraphic succession. Thus, recognition of the basin form and genesis requires three-dimensional definition and analyses at several different structural levels.

Most basins in their embryonic stages appear to be related to either releasing-bend geometries or non-parallel master faults. The development of basins along releasing bends of strike-slip faults yields s-shaped or z-shaped basins that, with continued displacement, form rhomb-shaped basins, locally referred to as "rhombochasms" (e.g., Mann et al., 1983). With continued elongation, continental crust may thin and extend sufficiently to rupture, yielding mantle uplift and generation of an oceanic spreading ridge, such as the Cayman trough (Perfit and Heezen, 1978; Rosencrantz et al., 1988). In these types of basins, length is commonly more than three times greater than width.

At the continental scale, strike-slip basins may develop between tectonic plates as they diverge in transtension or converge in transpression. Individual basins may develop by one or several mechanisms; because of their scale, the basins may be quite complex and involve diverse mixtures of normal, lateral and reverse faults. In continental interiors, equally complex basin types and fault patterns are possible (e.g., Zolnai, 1991). Basins along intraoceanic transform faults appear to be complex structurally, but simpler depositionally, as a result of lower sedimentation rates and less diverse source areas (e.g., Barany and Karson, 1989).

Extension, Subsidence and Thermal History

The formation of strike-slip basins depends largely on the orientation of the principal direction of extension relative to the direction of bulk shear strain, on the overstepping arrangement of discontinuous and discrete fault segments, and on the bending geometry of the fault. In instantaneous, homogeneous, bulk simple-shear (Fig. 12.8), the principal direction of extension is 45° to the mean shear direction. The brittle manifestations of bulk simple shear in rocks are extension fractures (designated T), shear fractures (R, R' and P), and faults that are parallel to the principal displacement zone. The T and R fractures

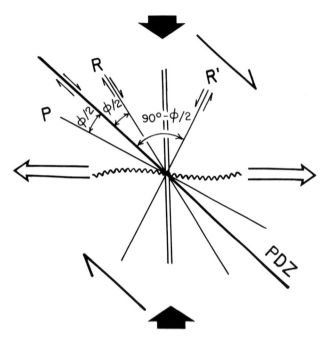

Fig. 12.8 Geometric relations among associated structures formed in right simple shear (Reproduced with permission from Sylvester, 1988). Plan view of Riedel model with orientations of structures formed in incipient right simple shear along a southeast-striking vertical fault. Double parallel line represents orientation of extension (T) fractures; wavy line represents orientation of fold axes. Abbreviations: P, P fractures; R and R', synthetic and antithetic shears, respectively; PDZ, principal displacement zone; ø, angle of internal friction. Stubby black arrows = axis of principal shortening; open arrows = principal axis of lengthening.

develop commonly in en-echelon arrays with a self similarity from the scale of micro- and mesoscopic laboratory experiments, to surficial fault ruptures with dimensions measurable in meters, to fault zones from 1 to 2 km wide (Tchalenko, 1970). The development of R and T fractures at larger scales is problematic and depends on the width of the zone of strike-slip shearing.

Extension may form obliquely across T or R fractures, parallel to the entire shear zone, yielding a sag within the zone, or between pairs or sets of overstepping R fractures, or as an array of en-echelon, external faults (Fig. 12.8). In the latter instance, the sense of overstepping is opposite to the arrangement of stepover basins.

Two major strike-slip basin types may be distinguished on the basis of mantle involvement (Allen and Allen, 1990), although basin histories may be difficult to summarize in simple terms or to model effec-

tively because source areas and depocenters migrate laterally through time: 1) Those in which the mantle has been involved, that is, "hot basins;" and 2) those in which the mantle has not been involved, and are thus, generally thin-skinned, that is, "cold basins."

Because strike-slip basins evolve so rapidly and complexly, most writers have had limited success in modeling their thermal evolution and subsidence/uplift histories. In "hot" basins, uniform-extension models with modifications for lateral loss of heat have been applied with some success. In "cold" basins, post-extensional thermal subsidence is insignificant. Transtension during formation of some strike-slip basins may yield extraordinarily high rates of subsidence and sedimentation; equally high rates of uplift and erosion may occur during transpression. Many strike-slip basins pass through multiple phases of subsidence and uplift during the complex evolution of the basin, and different parts of the same basin may undergo rapid uplift and rapid subsidence at the same time.

Continental-margin and intracontinental strike-slip basins may receive huge volumes of sediment, whereas intraoceanic basins may be starved of sediment. Because narrow basins cool rapidly by lateral heat conduction, they appear to undergo rapid early subsidence to form deep basins with associated sediment starvation during early stages of development (Allen and Allen, 1990). As a result of greater lateral heat loss, however, narrower basins subside at faster rates and to greater depths than do wider basins; this relation has profound implications for hydrocarbon generation and trapping in strike-slip basins. As the initial subsidence rate decreases and pathways are established for sediment to enter the basin, sedimentation rates will ultimately exceed subsidence rates, leading to basin filling.

Depositional Framework

The sedimentary fill of strike-slip basins may be extraordinarily variable and complex, depending on whether the basins are submarine, sublacustrine, subaerial, combinations of these, or variable through time and space. Nilsen and McLaughlin (1985) summarized the stratigraphic and sedimentological aspects of three well exposed strike-slip basins, the

Little Sulphur Creek (McLaughlin and Nilsen, 1982) and Ridge (Link, 1982) basins of California, and the Hornelen basin (Steel and Gloppen, 1980) of western Norway, which are principally of nonmarine origin and of varying size and character, as follows (Fig. 12.9):

1. The basins are asymmetric, with their structurally deepest parts close to and subparallel to the syndepositionally most active strike-slip margins.

2. The basins are characterized by diverse depositional facies, including talus, landslide, alluvial-fan, braided- and meandering-fluvial, deltaic, fan-delta, shoreline, shallow- and deep-lacustrine, turbidite, chemical-precipitate and algal-limestone deposits.

3. The basin fill is characterized dominantly by axial infilling, subparallel to the principal displacement zones.

4. The basin-margin deposits are distinctive. Small debris-flow-dominated alluvial fans containing coarse sedimentary breccia and conglomerate formed along the syndepositionally active principal displacement zones. Along the inactive or less active margins are larger streamflow-dominated alluvial fans and fluvial deposits that contain finer-grained conglomerate and little breccia (compare with rifts; see Chapter 3).

5. The basin fill is characterized by abrupt facies changes.

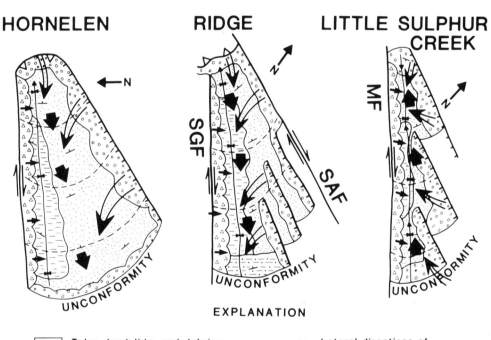

Fig. 12.9 Tectonic and sedimentary comparison of Hornelen basin, Norway, Ridge basin, southern California, and Little Sulphur Creek basin, northern California (Reproduced with permission from Nilsen and McLaughlin, 1985). Orientations and scales of basins vary considerably but are shown here at approximately the same orientation and size for comparison. Length of each basin is approximately: Hornelen, 50 km; Ridge, 25 km; Sulphur, 10 km. Abbreviations: MF, Maacama fault; SAF, San Andreas fault; SGF, San Gabriel fault.

EXPLANATION

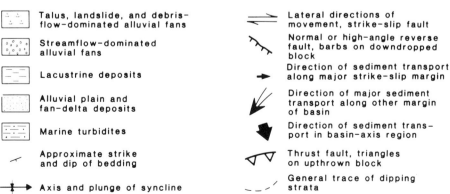

- Talus, landslide, and debris-flow-dominated alluvial fans
- Streamflow-dominated alluvial fans
- Lacustrine deposits
- Alluvial plain and fan-delta deposits
- Marine turbidites
- Approximate strike and dip of bedding
- Axis and plunge of syncline
- Lateral directions of movement, strike-slip fault
- Normal or high-angle reverse fault, barbs on downdropped block
- Direction of sediment transport along major strike-slip margin
- Direction of major sediment transport along other margin of basin
- Direction of sediment transport in basin-axis region
- Thrust fault, triangles on upthrown block
- General trace of dipping strata

6. The basin fill was derived from multiple basin-margin sources that changed through time as a result of continued lateral movement along basin-margin faults.

7. The basin fill may be petrographically diverse and complex as a result of the multiple sources that changed through time and space.

8. The basins contain very thick sedimentary sections compared to their areas.

9. The basin fill is characterized by high rates of sedimentation, roughly 2.5–3.0 mm/y.

10. The basins are characterized by depocenters that migrated in the same directions as source terranes along the principal displacement zones; these migration directions are generally opposite to the directions of axial sediment transport.

11. The basin fill is characterized by abundant synsedimentary slumping and deformation, possibly in response to basinwide shaking from earthquakes along basin-margin faults.

Many of these attributes are also characteristic of other generally elongate and restricted types of basins, especially rift basins, foreland basins and trench-slope basins. Strike-slip faults may form significant components, in fact, of these three basin types, as well as of many others. Recognition of strike-slip basins on the basis of structural criteria is probably a more straightforward process than is their recognition on the basis of stratigraphic and sedimentologic criteria. The most useful criterion may simply be the recognition of source areas that have been laterally displaced from sedimentary successions that contain unique detritus, especially if coarse-grained; however, even this criterion may be difficult to apply in areas of excellent exposure because: 1) the uniqueness of the source area must be proven convincingly; 2) the possible removal of similar rocks from areas that have undergone subsequent uplift and erosion must be convincingly disproved; 3) the transport paths of the unique detritus into the basin must be demonstrable; and 4) fault displacement contemporaneous with basin filling, rather than postdepositional displacement, must be demonstrable. These relations have been difficult to prove in most ancient basins, especially where strike-slip faults are no longer active or where erosion has stripped away most of the source area(s) or basin(s).

The sedimentary fill of many strike-slip basins is, in general, dominated by repetitive, basinwide, upward coarsening sequences that cut across all facies boundaries. These sequences seem to be tectonically controlled by basin-wide changes in base level that induce progradation of marginal and axial coarse-grained facies over axial finer-grained marine, lacustrine and related facies. Upward coarsening sequences are especially well developed in the Little Sulphur Creek basins, Hornelen basin and Ridge basin (Steel and others, 1977; Steel and Gloppen, 1980; Nilsen and McLaughlin, 1985). The sequences may result from periodic major tectonic subsidence of the basin, uplift of the basin margins, lowering of sea level and (or) lake level, climatic changes, or combinations of all these factors.

Recognition of Ancient Strike-Slip Basins

Several factors make recognition of ancient strike-slip basins difficult:

1. Their depositional and structural histories are complex.

2. Lateral movements along principal displacement zones and other faults commonly detach, rotate and translate basins or parts of basins from their places of origin, making paleogeographic reconstructions tenuous.

3. Polycyclic episodes of subsidence and uplift commonly remove major parts of the stratigraphic and structural record.

4. Most easily studied late Cenozoic strike-slip basins are nonmarine, resulting in poor intrabasinal and interbasinal age control and stratigraphic correlations.

5. Many cross-sectional and plan-view features of strike-slip basins resemble those of rift basins and forelands, making their recognition problematic; detailed three- and four-dimensional studies are required, and reliance upon two-dimensional seismic-reflection lines is commonly misleading.

6. Strike-slip faults are commonly reactivated as normal, reverse, oblique, thrust and more complex zones of faulting; changes in tectonic framework may yield preservation of only parts of the original strike-slip basins.

7. Few ancient deep-sea strike-slip basins survive subsequent episodes of subduction intact enough to provide suitable models for their structural and depositional evolution.

Reliable criteria for recognition of strike-slip basins need to be developed from the sedimentary pattern, stratal succession and filling mechanisms of the basins. Of the depositional criteria outlined by Nilsen and McLaughlin (1985), lateral migration of depocenters parallel to principal displacement zones appears to us to be the most useful. Depocenters in most other basins tend to migrate either transversely away from or toward the principal bounding faults, or to remain in more or less the same location through time. In strike-slip basins, however, depocenters migrate laterally, parallel to the principal bounding faults. By this process, basins lengthen through time without excessive widening, accumulate extraordinary thicknesses of fill, and contain abrupt facies changes and petrographic variations throughout their history. Polycyclic histories, changing structural settings, and repeated episodes of rapid subsidence and uplift characterize long-lived strike-slip zones, such as many of those along the San Andreas fault (Crowell, 1974a,b), even though the basins themselves are commonly short-lived.

Other useful structural and stratigraphic criteria for recognizing strike-slip basins include:

1. lateral displacement of related depocenters
2. lateral offset of matched source areas and deposits
3. presence of coarse-grained alluvial-fan, fan-delta, submarine/and sublacustrine-apron, and related deposits along the flanks of basins that contain principal displacement zones (these depocenters typically alternate along strike in rift basins from one flank to the other, reflecting activity of border faults [see Chapter 3])
4. presence of thick, but laterally restricted sedimentary sequences characterized by high sedimentation rates
5. localized uplift and erosion, resulting in unconformities of the same age as thick nearby sedimentary fill
6. abrupt lateral facies variations

7. presence of strike-slip faults on one or more sides of the basin
8. development of basinwide upward coarsening sequences that develop in response to tectonically induced basin deepening

Examples of Strike-Slip Basins

Fault-Bend Basins

Most modern and ancient strike-slip basins are probably fault-bend basins (Fig. 12.3A) because bends in strike-slip faults are common. Although modern Death Valley in California (Burchfiel and Stewart, 1966) is generally regarded as a classic example of a "pull-apart basin," we consider it and several adjacent basins as polygenetic basins (Fig. 12.3E), because they have formed within a large region dominated principally by extension.

Ridge Basin, Southern California

Ridge basin is one of the best studied fault-bend basins in the world (Fig. 12.10). It developed during Miocene and Pliocene time adjacent to the right-lateral San Gabriel fault (Crowell and Link, 1982). The basin is 30 to 40 km in length, 6 to 15 km wide, and about 400 km² in areal extent; its cumulative fill of 7,000 to 13,000 m was deposited at a rate of about 3 mm/y (Nilsen and McLaughlin, 1985).

Ridge basin developed as a stretched and sagged crustal wedge northeast of the San Gabriel fault in the area where the fault had a curvilinear trace. Crowell (1982a) inferred formation of the basin along a restraining bend, whereas May et al. (1993) inferred formation of the basin along a releasing bend of the San Gabriel fault (Fig 12.11A). The San Gabriel fault was the principal zone of right slip during most of the history of the basin, and its curved trace may have been the reason for uplift of a source area along the southwestern margin of the basin that migrated northwestward relative to stable North America through time, shedding detritus that formed the Violin Breccia (Crowell, 1974a,b, 1982b). The Clearwater and Liebre fault zones became active strike-slip faults as the basin widened and lengthened.

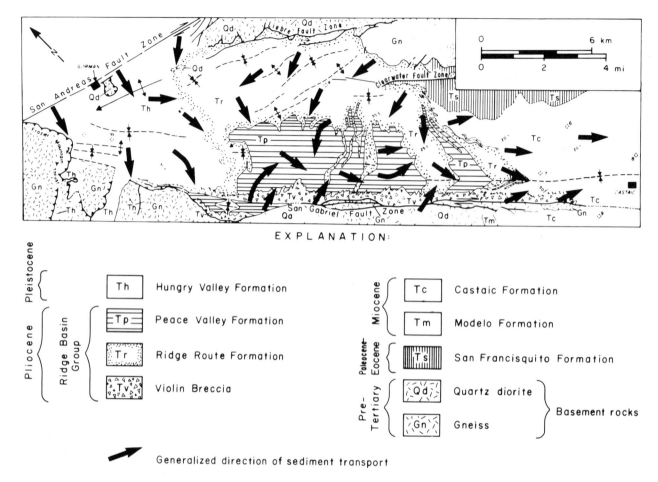

EXPLANATION:

Th	Hungry Valley Formation
Tp	Peace Valley Formation
Tr	Ridge Route Formation
Tv	Violin Breccia

Pleistocene / Pliocene — Ridge Basin Group

Tc	Castaic Formation
Tm	Modelo Formation
Ts	San Francisquito Formation
Qd	Quartz diorite
Gn	Gneiss

Miocene — Pre-Tertiary / Paleocene-Eocene — Basement rocks

Generalized direction of sediment transport

Fig. 12.10 Generalized map (left) and schematic composite stratigraphy (right) of Ridge basin (Reproduced with permission from Crowell and Link, 1982). Note that strata are shingled, such that no one vertical section contains all units (see Fig. 12.11C).

Most of the sediment filling the basin was carried by rivers draining source areas located to the northeast (Fig. 12.10). The basin depocenter migrated northwestward as the source of the Violin Breccia moved northwestward along the southwest side of the San Gabriel fault (Fig. 12.11B); as a result, the Violin Breccia is progressively younger northwestward, forming shingled sediment lobes that record northwestward migration of the depocenter. The Violin Breccia formed as a coarse-grained apron of sediment deposited transverse to the basin axis. It consists of fan-shaped bodies that were moderately deep marine during the early history of the basin, and were fan deltas and alluvial fans during the later history of the basin. These fan-delta and alluvial-fan deposits interfinger basinward with fluvial and lacustrine strata that record longitudinal basin filling,

southward and southeastward down the axis of the basin concurrently with northwestward migration of the depocenter and deposition of the Violin Breccia (Fig. 12.10). Deep- to shallow-lacustrine facies were deposited in the depocenter area adjacent to the San Gabriel fault (Fig. 12.10) (Link and Osborne, 1978). The total stratigraphic thickness is much greater than the accumulated thickness at any particular locality in the basin (Fig. 12.11C). Abundant synsedimentary deformational structures in the basin fill may record penecontemporaneous seismicity. When the principal locus of right slip transferred to the San Andreas fault about 5 Ma, the basin-margin faults became inactive, and the basin was folded, uplifted, and transported northwestward about 220 to 240 km relative to stable North America.

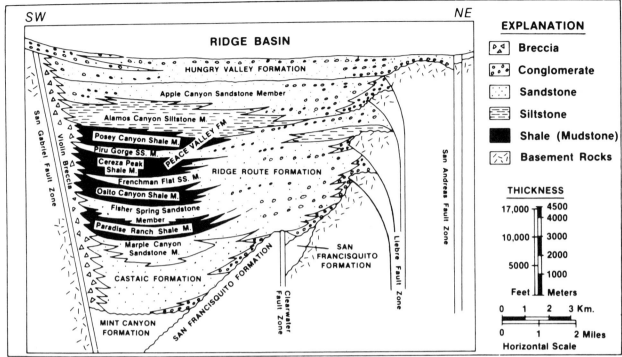

SW · NE

RIDGE BASIN

HUNGRY VALLEY FORMATION

Apple Canyon Sandstone Member

Alamos Canyon Siltstone M.

Posey Canyon Shale M.

Piru Gorge SS. M.

Cereza Peak Shale M.

Frenchman Flat SS. M.

Osito Canyon Shale M.

Fisher Spring Sandstone Member

Paradise Ranch Shale M.

Marple Canyon Sandstone M.

PEACE VALLEY FM.

RIDGE ROUTE FORMATION

CASTAIC FORMATION

San Gabriel Fault Zone

Violin Breccia

MINT CANYON FORMATION

SAN FRANCISQUITO FORMATION

SAN FRANCISQUITO FORMATION

Clearwater Fault Zone

Liebre Fault Zone

San Andreas Fault Zone

EXPLANATION

Breccia

Conglomerate

Sandstone

Siltstone

Shale (Mudstone)

Basement Rocks

THICKNESS

17,000	4500
	4000
10,000	3000
	2000
5000	1000
Feet	Meters

0 1 2 3 Km.

0 1 2 Miles

Horizontal Scale

Fig. 12.10 Continued.

Fig. 12.11 Diagrammatic sketches of Ridge basin showing: A) its tectonic origin (after May et al., 1993); B) sedimentation model that gives rise to shingled arrangement of basin-margin strata (Violin Breccia)(Reproduced with permission from Crowell, 1982b); and C) arrangement of basin axial strata, parallel to depositional trough. Depocenter remained fixed relative to source area(s) across from San Gabriel fault as previously deposited strata were rotated and carried relatively south. Upper and center cross sections depict scheme at early and late stages, whereas lower cross section is diagrammatic depiction of present pattern of exposures after uplift and erosion. Total stratigraphic thickness (A+B+C . . .) is much greater than maximum thickness in any single well (from Crowell, 1982a).

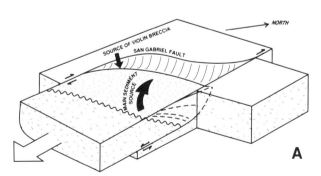

A

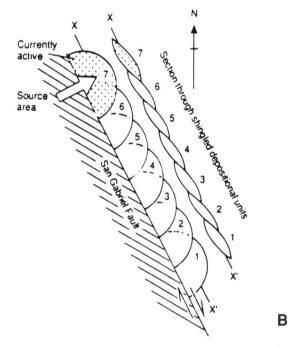

B

La González Basin, Venezuela

La González basin is located along the Boconó fault, a late Tertiary (probably Pliocene) to recently active right-slip fault that is 500 km long and has a displacement of less than 100 km (Schubert, 1980,

1984, 1988). The fault-bend basin is about 40 km long and 5 km wide; 7 to 9 km of right-lateral slip and at least 1 km of vertical separation in middle to late Pleistocene time caused the basin to form (Fig. 12.12). The upper Pliocene to upper Pleistocene

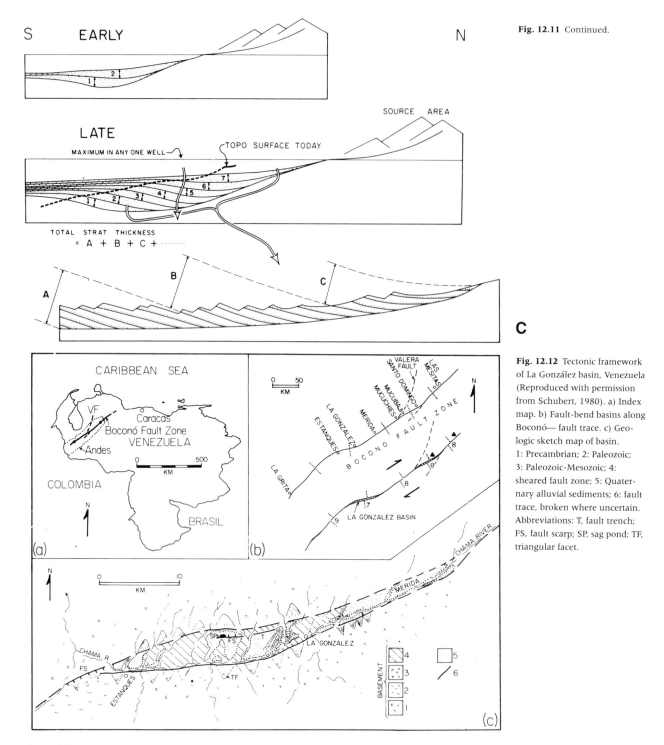

Fig. 12.11 Continued.

Fig. 12.12 Tectonic framework of La González basin, Venezuela (Reproduced with permission from Schubert, 1980). a) Index map. b) Fault-bend basins along Boconó— fault trace. c) Geologic sketch map of basin. 1: Precambrian; 2: Paleozoic; 3: Paleozoic-Mesozoic; 4: sheared fault zone; 5: Quaternary alluvial sediments; 6: fault trace, broken where uncertain. Abbreviations: T, fault trench; FS, fault scarp; SP, sag pond; TF, triangular facet.

fluvial basin fill overlies pre-Tertiary rocks, and the modern basin floor lies more than 2 km below the surrounding mountain ranges (Schubert, 1984). The fluvial basin fill can be divided into four climatic-tectonic sequences and is at least several hundred meters thick (Schubert, 1992); it has been transported into the basin from both flanks. The modern drainage is primarily axial to the west-southwest.

Stepover Basins

The geometric patterns, structural axes, and depocenters of stepover basins depend on the spacing between overstepping or en-echelon strike-slip faults, the amount of overlap between the faults, and the depth to basement (Rodgers, 1980). Where basement is shallow, small subbasins develop. Where basement is deep, a single, central basin may form where the overlap is slight; with increasing overlap, two or more subbasins may form. Several individual basins between overstepping strike-slip faults may coalesce into a single, larger basin, such as Binchuan basin in Yunnan province of southern China (Deng et al., 1986).

North China Basin

The North China basin is a large ($\sim 2 \times 10^5$ km^2) strike-slip basin that formed transtensionally in an intracratonic setting at an overstep in a large-scale

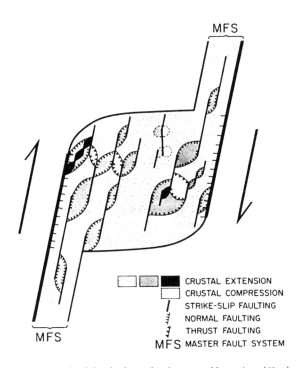

Fig. 12.13 Idealized sketch of postulated process of formation of North China basin (Reproduced with permission from Nábelek et al., 1987). Basic geometry of basin is repeated on several scales within basin. Regions of uplift are present locally, either as a result of small bends along strike-slip faults or due to interaction between neighboring strike-slip faults. Complex block rotations may have also occurred. East-west width of North China basin is approximately 600 km.

right-slip system of nonparallel, right-stepping, northeast-striking faults (Nábelek et al., 1987; Chen and Nábelek, 1988). Localized and rapid post-mid-Pliocene subsidence resulted from right slip on the master bounding faults, as well as interplay of numerous other types of faults of different scales in the interior of the basin (Fig. 12.13).

The basin began to form in early to middle Eocene time. Several fault sets divide the extensively fractured underlying basement rocks into rhombic and triangular blocks; these pre-existing faults strongly influenced the locus and sequence of numerous large earthquakes along the north margin of the basin in 1978 (Nábelek et al., 1987).

Three major earthquakes occurred in the 1978 Tangshan sequence. The epicenter of the main earthquake was at the junction of two strike-slip faults, a northnortheast-striking right-slip fault and an east-northeast-striking left-slip fault having a significant component of thrusting. One of the two strongest aftershocks occurred at the northeast end of the main fault in a right stepover in the northnortheast-striking fault system; its nodal planes strike east-west, its focal mechanism was almost purely normal, and almost 1 m of subsidence occurred here during the earthquake. The earthquake sequence demonstrated that pull-apart subsidence caused by large faults can overprint displacements on smaller ones, producing subsidence on a regional scale.

Dead Sea Basin

The Dead Sea basin is a large (132 x 16 km) and deep (8–10 km) symmetric graben (Fig. 2.14A) formed by divergent left slip along an active fault system that extends from the Gulf of Aqaba to southeastern Turkey (tenBrink and Ben-Avraham, 1989). The basin is located along the transform plate boundary that separates Arabia from Africa, and it opened in response to Miocene differential extension between the Gulf of Suez and the northern Red Sea. The total of about 105 km of left slip along the Dead Sea transform probably started in middle Miocene time (15.5–11.5 Ma); slip of only 30 km has occurred in the last 5 my, implying a low relative plate velocity (6 mm/y) between Arabia and Africa.

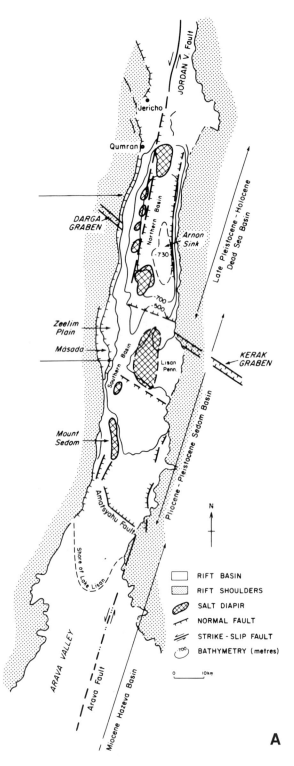

Fig. 12.14 Geologic setting of Dead Sea basin. A) Geologic map of Dead Sea-Arava depression, illustrating tectonic elements, physiographic features, northward migration of depocenter, and basin asymmetry expressed by bathymetry (Reproduced with permission from Manspeizer, 1985). Early Miocene (25–14 Ma) strike slip of about 60–65 km opened Arava basin that filled with about 2 km of red beds during a pause in strike-slip displacement. Later displacement in last 4.5 my allowed deposition of more than 4 km of marine to lacustrine rock salt of the Sedom Formation, overlain by 3.5 km of lacustrine evaporitic carbonate and clastic sedimentary strata. B) Schematic block diagram of Dead Sea region, viewed to north, with generalized rift physiography, tectonic framework and depositional domains (Reproduced with permission from Manspeizer, 1985). Bedrock of different ages on rift shoulders, basin bathymetry and eastward thickening of strata illustrate tectonic asymmetry.

(Fig. 12.14B). The basin is bounded longitudinally by an overstepping pair of vertical left-slip faults, and transversely by several sharply defined faults that structurally separate the basin into several 20 to 30 km segments (Fig. 12.14A). Deformation occurs mostly along the strike-slip faults, whose surface traces extend discontinuously along the length of the basin, and to a lesser degree on transverse faults. The intervening sedimentary strata are relatively undeformed due partly to the presence of a salt layer beneath much of the basin that decouples the sedimentary fill from the basement (tenBrink and Ben-Avraham, 1989).

Sediment accumulation and subsidence rates of the Dead Sea basin are high, with more than 7 km of upper Cenozoic marine, lacustrine, evaporite and fluvial-deltaic deposits (Manspeizer, 1985; tenBrink and Ben-Avraham, 1989). Subsidence of the Dead Sea basin probably began in Pliocene time, when much of the evaporitic succession was deposited, together with local marine strata. Subsidence accelerated in Pleistocene time when several kilometers of lacustrine, fluvial and other terrestrial sediments accumulated. The depocenter has migrated northward, accompanied by diapiric intrusion of evaporites, generating an asymmetric basin with extraordinarily abrupt facies changes.

The basin is narrowest and shallowest toward its northern and southern ends, and widest and deepest in the central part. Gravity data suggest that the basin sags toward its center and lacks northern and southern boundary faults (tenBrink et al., in press). The basin formed and continues to form by crustal extension along its long axis (Fig. 12.14B); maximum extension is concentrated in the central part, where it widens by passive collapse of its flanks between the two overstepping strike-slip faults that bound the

The shallow basement rocks consist of massive and rigid crystalline plutonic and metamorphic rocks that are overlain by a thin cover of sedimentary strata

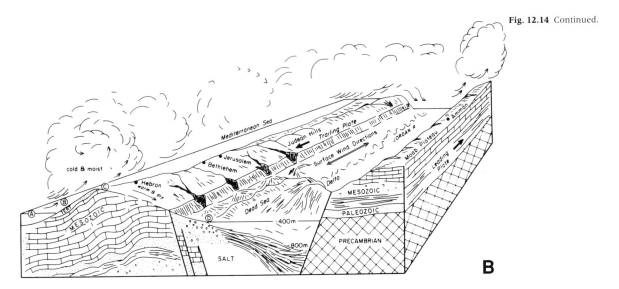

Fig. 12.14 Continued.

basin. The basin lengthened northward by simultaneous northward propagation of the southern strand of the Dead Sea fault and a retardation of the northern strand (tenBrink and Ben-Avraham, 1989; tenBrink et al., in press).

Transrotational Basins

Fault-bounded transrotational basins develop between blocks that rotate differentially in a shear zone rather than along a single fault strand (Fig. 12.3C). Rotation of blocks about vertical or subvertical axes yields prominent triangular or rhomb-shaped basins that may be quite large in very wide zones of distributed

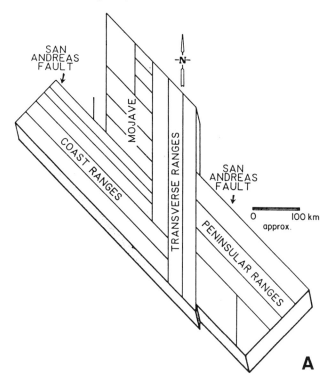

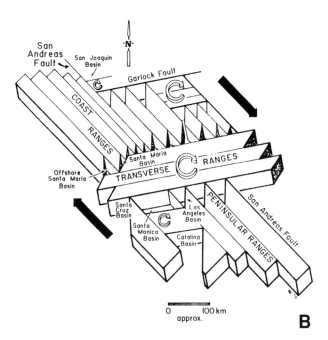

Fig. 12.15 Rotation model for southern California. See Luyendyk and Hornafius (1987) for discussion and details of faults used in model. A) Inferred initial fracture pattern and prerotation geometry during Oligocene. B) Late Miocene geometry after clockwise rotation, which occurred mainly during Middle Miocene. Sinistral and dextral slip has occurred, and deltoid basins have opened at joins of rotated and unrotated blocks. (Reproduced with permission from Luyendyk and Hornafius, 1987.)

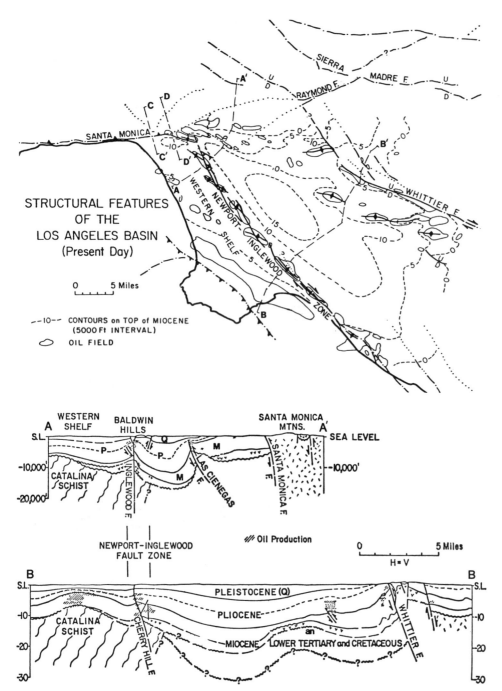

Fig. 12.16 Generalized structural map (A) and regional cross sections (B) of Los Angeles basin. Abbreviations: M, upper and upper-middle Miocene; P, Pliocene; Q, Quaternary; an, Anaheim nose. Vertical scale of (B) is in kilofeet. (Reproduced with permission from Wright, 1987.)

shear, such as the rhomb-shaped Bir Zreir graben in eastern Sinai, formed by rotation and consequent subsidence between two bent strike-slip faults of the Dead Sea Rift zone (Eyal and Eyal, 1986). Several southern California Neogene basins appear to have formed during clockwise rotation of as much as 90° during Neogene time near the margins of the east-trending Transverse Ranges and in the Mojave Desert (Fig. 12.15) (Luyendyk and Hornafius, 1987; Luyendyk, 1989).

Los Angeles Basin, Southern California

The Los Angeles basin (Fig. 12.16) is regarded by some writers as a transrotational basin, formed by differential clockwise rotation of blocks in a wide zone of right simple shear (Luyendyk et al., 1980, 1985; McKenzie and Jackson, 1986; Luyendyk and Hornafius, 1987; Luyendyk, 1991). Other writers (Crouch and Suppe, 1993) have proposed that it formed by large-magnitude crustal extension in early Miocene time (25–22 Ma), and after its Miocene and Pliocene sedimentary succession accumulated during rapid rifting from 16 to 5 Ma, it was deformed by late Pliocene and younger right-slip faults.

The roughly triangular basin is approximately 40 km wide and 60 km long (Fig. 12.16A). It consists of several triangular sub-blocks and basins bounded by fault zones that also control the main hydrocarbon-bearing structural traps in the basin. Many of the faults are active, judging from historic earthquakes, as well as from abundant geologic evidence of recent deformation. Along its north flank, a series of north-dipping Quaternary thrust faults carries crystalline basement of the Transverse Ranges southward over the basin (Dibblee, 1982; U.S. Geological Survey, 1987). The east and southeast flanks of the basin consist of a faulted shelf of deformed Upper Cretaceous and lower Tertiary strata that rest on Mesozoic sialic basement overlain by Oligocene, Miocene and Pliocene sedimentary rocks, some of which are faulted and intruded by Miocene igneous rocks. The Elsinore-Whittier right-slip fault system extends northwestward from the Peninsular Ranges into the eastern part of the Los Angeles basin, where it must intersect somehow in the subsurface with the northern basin-margin thrust faults. The southwest side of the basin is bounded by the northwest-striking Newport-Inglewood fault zone. Crouch and Suppe (1993) suggested that the Newport-Inglewood fault zone was the locus of the main extensional breakaway for the Early Miocene rifting that formed the basin. Interpretations of geologic, seismic and seismic-reflection data indicate that the entire basin is underlain by an active, blind, south-dipping regional décollement, which with its imbricate thrusts, ramps and folds, has accommodated 15 to 30 km of north-south shortening in the last 2 to 4 my (Davis and Namson, 1989). Many modern topographic highs and lows in the basin correspond to structural highs and lows, and, together with deep thrust earthquakes, attest to continuing deformation.

The Tertiary sedimentary succession in the Los Angeles basin is dominated by Neogene deep-marine turbidites and hemipelagic units that contain the greatest concentration of petroleum per unit area of any basin in the world (Biddle, 1991). Miocene strata include organic-rich, siliceous shale of the Monterey Formation and related units, which are the primary source rocks. Miocene volcanic rocks are present at depth nearly everywhere in the central part of the Los Angeles basin, but crop out only locally along its edges. This implies that magma intruded in response to cracking of the crust and opening of the basin, and it rose nearly hydrostatically to a compensating level within the sedimentary fill (Wright, 1991). Some of the volcanic rock units are also rotated clockwise, based on interpretations of paleomagnetic vectors (Hornafius et al., 1986), supporting the conclusion that transrotational deformation, in part, followed Early Miocene formation of the basin by rifting and detachment faulting (Crouch and Suppe, 1993).

Thick strata consisting of thin-bedded Pliocene marine turbidites, derived largely from sources north and east of the basin, are the principal reservoir rocks for petroleum. The Pliocene and Quaternary clastic marine and nonmarine strata in the central part of the basin are nearly 4 km thick, based on drilling and seismic-reflection studies (Blake, 1991; Mayer, 1991; Wright, 1991) (Fig. 12.16B).

Mid-Pliocene (younger than 5 Ma) transpression may have caused the Newport-Inglewood fault zone to be reactivated as a right-slip fault (Crouch and Suppe, 1993). On the sea floor above the fault, right-stepping, en-echelon, anticlinal hills formed (Yeats, 1973) that guided turbidity currents down the axes of synclinal troughs. Continued folding of the anticlines tilted the edges of the sand-filled synclines and formed some oil traps (Harding, 1973). Extensive folding and faulting also generated abundant structural traps in and along the margins of the basin (Wright, 1991). Now the basin is filled so that alluvial fans stretch from the northern bounding mountains across the basin almost to the sea, and surface streams carry sedimentary debris to offshore basins.

Many modern topographic highs and lows correspond to structural highs and lows, attesting to the recency of deformation.

Mojave Desert Basins, Southern California

The Mojave region experienced a major episode of uplift, extensional rifting, and detachment faulting during Early Miocene time, followed by regional, northwest-southeast simple shear that caused irregularly shaped blocks to rotate differentially about subvertical axes (Dokka and Travis, 1990). Several basins formed between misfit areas among the faulted pieces, edges and corners of the rotated blocks. More than 1000 m of Miocene and Pliocene alluvial, fluvial, lacustrine, volcaniclastic and evaporite deposits accumulated in the largest of the basins, the Barstow basin, which is 100 km long and 20 km wide (Link, 1980; Woodburne et al., 1990). These deposits are moderately folded and faulted, but little rotated (MacFadden et al., 1990). Early Middle Miocene evaporite deposits in Kramer basin (Dibblee, 1980) contain 100 m of sodium borate (borax), one of the largest deposits in the world. It was deposited mostly from hot springs into an ephemeral playa lake (Siefke, 1980) that formed along a minor right-slip fault (Dibblee, 1967).

Transpressional Basins

Narrow basins that are marginal to transpressional principal displacement zones or positive flower structures may result from flexural subsidence induced by loading from branching reverse faults. This flexurally induced subsidence is similar to the mechanisms of foreland-basin development, and these marginal strike-slip basins are, in fact, miniforeland basins that may be overidden by advancing allochthons. Small piggy-back basins are also common in these settings (e.g., Namson and Davis, 1988). Rapid sedimentation in transpressional basins induces additional subsidence by sediment loading.

Sediments deposited in transpressional basins in deep-marine settings along transform faults may consist mostly of pelagic or hemipelagic mudstone with some proximal landslide, debris-flow, volcaniclastic or talus deposits. Where deep-marine transpressional basins are situated close to larger landmasses, thick turbidite successions may be deposited. In nonmarine settings, alluvial-fan, lacustrine, fan-delta, delta, fluvio-deltaic and turbidite facies are most common. Most sediment is derived from the adjacent transpressional uplift blocks, but sediment may also be derived from more distal sources.

In more complex transpressional regimes, marked by multiple faults, fault strands and interfault basins, transpressional basins may originate from synclinal downfolding or fault-dominated subsidence marginal to reverse faults. These types of transpressional basins have more complex patterns of sedimentation and structural evolution. Individual transpressional basins may have multiple sources of sediment; longitudinal, rather than transverse sediment transport directions, may predominate. We discuss two California basins as examples of transpressional basins, the Ventura and San Joaquin basins. It is clear, however, that both basins have had long and complex histories and should be grouped with polyhistory basins. Here, we focus only on their last stages of evolution, which have been transpressional.

Ventura Basin, Southern California

The Ventura basin of southern California is located in the southwestern part of the Transverse Ranges province (Fig. 12.17) (Crowell, 1976). The west-trending and west-plunging narrow basin developed in Early Miocene time (about 22 Ma) in response to 90° clockwise rotation of the Transverse Ranges province (Luyendyk et al., 1980, 1985; Luyendyk, 1989). Although early evolution of the basin was transtensional or transrotational, its later history has been dominated by transpression (Crowell, 1976). As the province underwent rotation, extensive left-lateral faults sliced the block into a series of subblocks, some of which underwent uplift, and others of which underwent subsidence in response to rotation-induced shear.

Like the nearby Los Angeles basin, deposition commenced in Miocene time (14–6 Ma) with diatomaceous and phosphatic mud (now the Monterey Formation) in protected and moderately deep water. These deposits lapped across a subsea shoulder or ridge with seaknolls on it that was uplifted into an

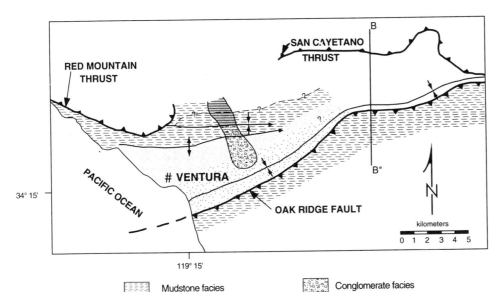

Fig. 12.17 Generalized map, cross section and depositional model of Ventura basin, California. A) Facies distribution of Pliocene and Pleistocene deep-sea sedimentary rocks (modified from Hsü et al., 1980); B) Structure section B-B' (from Nagle and Parker, 1971); C) Process model of turbidity-current deposition in deep-sea basin of Ventura type (modified from Hsü et al., 1980).

Mudstone facies	Conglomerate facies
Thin-bedded sand facies	Eroded slope
Graded sand facies	

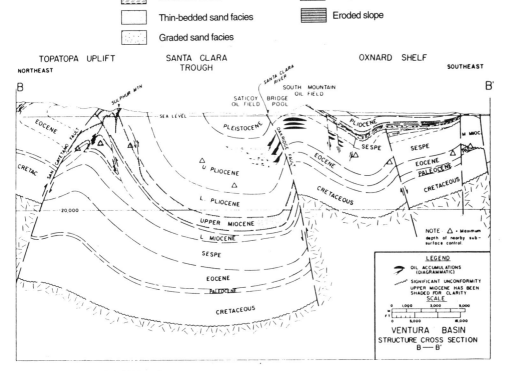

VENTURA BASIN MODEL

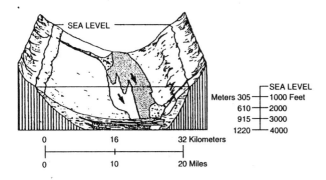

antiformal arch along the length of the basin. Major subsidence occurred in Early Pliocene time, and one of the thickest Pliocene bathyal marine sequences in the world was deposited, with sedimentation rates exceeding 2 mm/y. Locally, basin inversion has occurred during the past one-half million years.

The thick Pliocene deposits consist of sandy siliciclastic turbidites transported axially down the basin from basement source areas more than 50 km to the east (Hsü, 1977; Hsü et al., 1980). These deposits form spectacular petroleum reservoirs in prolific oil fields along the axes of inversion anticlines. Coarser-grained submarine debris flows were shed into the basin from the subparallel north and south flanks of the basin. Basin inversion also caused uplift of older rocks along basinward directed reverse faults along the north and south flanks of the basin (Fig. 12.17B).

The entire western Transverse Ranges province has been interpreted as a fold-and-thrust belt by some workers (Namson and Davis, 1988). The abundant left-lateral faults, rotation history, and spatial association of the basin with the San Andreas fault suggest to us that it forms a type of fold-and-thrust belt within a wide zone of transpressional deformation. The Pliocene to Holocene Ventura basin formed within this zone as a complex transpressional basin characterized by rapid subsidence and high rates of sedimentation typical of strike-slip basins.

San Joaquin Basin, Central California
The late Cenozoic San Joaquin basin in central California (Bartow, 1991) developed in the southern part of the Great Valley, a forearc basin that originated in Mesozoic time as a result of lithospheric plate convergence (see Chapter 6). The San Joaquin basin is a deep (12 km), irregular-shaped basin that began to subside rapidly in Late Oligocene and Early Miocene time in response to transpression extension (Goodman and Malin, 1992). Subsequent middle Miocene to Holocene subsidence and basin-margin uplift have been partly in response to associated with the San Andreas fault (Wilcox et al., 1973). Subsidence increased at about 16 Ma, with deposition of phosphatic shale and diatomaceous strata, the principal source rocks for petroleum.

The southern and southwestern flanks of the San Joaquin basin are bounded by transpressional regions of the San Andreas fault (Fig. 12.18; Wilcox et al., 1973; Harding, 1976). The San Emigdio Range and Tehachapi Mountains along the south flank of the basin are being thrust northward over the basin along the Pleito thrust and along several buried thrusts. The Temblor Range along the southwestern flank is being thrust northeastward over the basin along mostly buried thrust faults (Webb, 1981; Namson and Davis, 1988; Medwedeff, 1989; Ryder and Thomson, 1989). Marginal to the thrusts is a series of recent lacustrine basins, including, from southeast to northwest, Kern, Buena Vista, and Tulare lakes.

Structural mapping of basinwide lacustrine and volcanic ash units within the Tulare Formation, a widespread and thick Pliocene to Holocene siliciclastic unit in the San Joaquin Valley and surrounding San Emigdio and Temblor Ranges, reveals that these units have been structurally and topographically depressed adjacent to the allochthonous masses. The lacustrine transpressional basins form mini-foreland basins depressed by structural loading and resultant flexural bending. Some workers attribute the convergent foldthrust belt entirely to thrust tectonics (Namson and Davis, 1988; Medwedeff, 1990). Together with Harding (1973), however, we maintain that the transpressional setting is a consequence of partitioned convergent strike slip.

The San Joaquin basin originally opened westward toward the Pacific Ocean in early Miocene time (20 to 16 Ma); it was subsequently truncated, probably rotated and offset about 300 km by the San Andreas fault. The offset western part of the Early Miocene basin lies in the Santa Cruz Mountains (Bartow, 1991). During Miocene time (12–7 Ma), the block southwest of the San Andreas fault was uplifted and provided detritus to fan deltas and turbidite systems into the deep basin (Ryder and Thomson, 1989; Fig. 12.18). Prior to that, most of the sediment came into the basin through submarine canyons from the Sierra Nevada region to the east and flowed northward down the axis of this mid-Miocene trough. At the same time, however, en-echelon folds were growing and forming submarine hills along the west edge of the basin, partly as a result of shear along the San Andreas fault. The growing folds focused turbidity currents along synclinal valleys and through saddles to basin plains. The

basin became isolated from the Pacific at the beginning of the Pliocene and has filled with nonmarine deposits of Pliocene and Quaternary age.

About 4 Ma, motion along the San Andreas transform changed from predominately transtensional to transpressional. The consequent shortening caused blind thrusts to develop along the west edge of the basin, further deforming the early en-echelon anticlines. The thrust faults and anticlines are active today.

Polygenetic Basins

In divergent settings, strike-slip basins may develop along transfer or accommodation zones that link major loci of normal faults. They may also form along the flanks of individual detached crustal segments, especially in horseshoe faults, where low-angle strike-slip faults mark the margins of detached segments (Fig. 12.3E). In convergent settings, strike-slip basins develop along transfer zones that link intact individual segments of larger thrust or underthrust crustal segments (Fig. 12.3E).

Death Valley Basin, Eastern California
Death Valley is a north-trending, asymmetric half graben that has formed as a result of regional extension (Fig. 12.19). Although extension commenced about 15 Ma, the present physiography of the valley has developed in the last 4 my. The east side of the half graben is bounded by a major range-front fault zone with oblique normal slip, the Black Mountains fault zone. Structural relief exceeds 6,000 m across the Black Mountains fault zone, as measured from the base of the sedimentary section in the deepest part of the half graben to the top of basement rocks in the adjacent bounding mountain range to the east.

Alluvial, fluvial, and lacustrine sediments exceed 3000 m in thickness on the east side of the basin next to the Black Mountains fault, and thin westward to less than a few-hundred meters. Some sediment is derived from the bounding mountains and transported into the half graben as debris-flow-dominated alluvial fans that are small and steep on the east side, and by larger, gentler-sloping and regionally extensive stream-flow-dominated fans on the west side (see also Fig. 3.32). The axial and ephemeral Amargosa River transports northward-fining sediment into

the valley from a larger area surrounding Death Valley. Evaporitic playa deposits and wind-blown dune fields are also present in the axial parts of the valley (close to the east side).

Northwest-striking right-slip faults at the north and south ends of central Death Valley are tear faults or transfer faults that separate the half graben from differentially extending tracts to the north and south (Burchfiel and Stewart, 1966). At the north end, the Furnace Creek fault extends southeastward across the north end of the valley to the Black Mountains fault zone. At the south end, the Death Valley fault zone extends northwest, cutting the southern end of the Black Mountains fault zone and the southern end of the graben, there bending northward into minor normal faults along the west edge of the valley. The amount of displacement is about 25 km along both of these fault zones.

In addition to being a pull-apart between two fault systems, as proposed originally by Burchfiel and Stewart (1966), Death Valley has been interpreted as one of several half grabens that developed locally in a regionally extended crustal slab (Wernicke, 1985; Wernicke et al., 1988; Hodges et al., 1989). The graben formed as the hanging wall extended northwestward, parallel to the strike-slip transfer faults that accommodate or transfer the extension from one half graben to another (Burchfiel et al., 1987).

Vienna Basin, Austria
The Vienna basin, one of several Miocene basins in the Carpathian Mountains of Austria and Czechoslovakia, is a rhombohedral basin along an active left-stepping, left-lateral tear fault between two thrust blocks of the Carpathian thrust belt (Fig. 12.20).

The basin is about 200 km long and 60 km wide. It is filled with Miocene shallow-marine to locally brackish clastic strata (as thick as 6 km) that were deposited during two episodes of basin subsidence; Pliocene and Quaternary fluvial deposits, as thick as 200 m, cap the Miocene sequence (Royden, 1985).

Three overstepping and northeast-striking left-slip faults controlled the early and late episodes of development of the two rhombohedral subbasins that make up the Vienna basin. The northernmost fault has accumulated about 80 km of Cenozoic left slip.

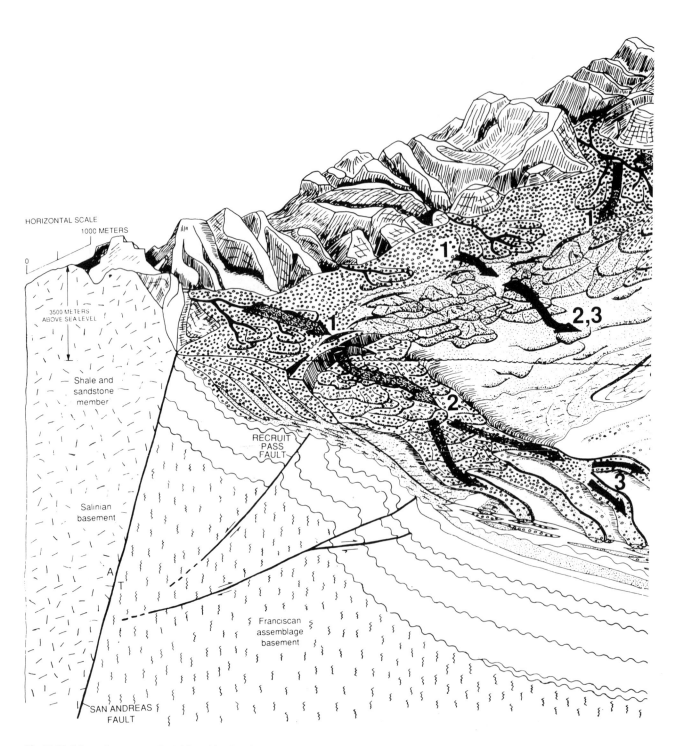

Fig. 12.18 Schematic reconstruction of depositional environments, depositional processes and paleotectonic setting of Santa Margarita Formation, southern San Joaquin basin, California. Water depth is greatly exaggerated to allow depiction of fan deltas and submarine fans.

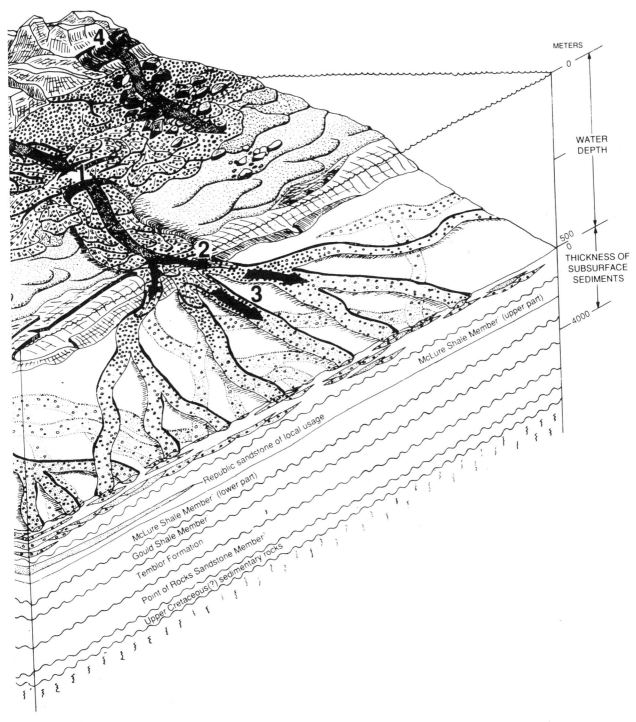

METERS

0

WATER
DEPTH

500
0

THICKNESS OF
SUBSURFACE
SEDIMENTS

4000

McLure Shale Member (upper part)

Republic sandstone of local usage

McLure Shale Member (lower part)

Gould Shale Member

Temblor Formation

Point of Rocks Sandstone Member

Upper Cretaceous(?) sedimentary rocks

Numerals: 1) subaerial debris-flow deposit; 2) subaqueous debris-flow deposit; 3) high-concentration turbidity-current deposit; and 4) debris-avalanche deposit. (Reproduced with permission from Ryder and Thomson, 1989.)

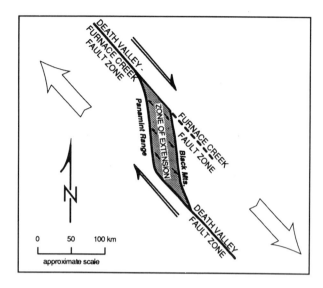

Fig. 12-19 Diagrammatic sketch map of Death Valley "pull-apart." (Modified from Burchfiel and Stewart, 1966.)

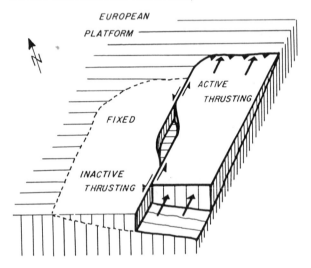

Fig. 12.20 Schematic diagram of opening of Vienna basin as a fault-bend basin at a left step in a left-slip tear fault within an allochthon thrust northeastward onto the European platform (Reproduced with permission from Royden, 1985).

The faults are roughly parallel to structural trends in the underlying nappes.

The depocenters of the two episodes of subsidence do not coincide. Subsidiary synsedimentary faults along the northwest margin of the basin dip 40° to 50° basinward and have several kilometers of normal separation at the surface. The basin fill thickens northwestward across the basin, and the older strata dip more steeply than the younger.

Left-slip faults developed contemporaneously with thin-skinned thrusts (Royden, 1985). The Vienna basin is confined entirely to the allochthon and formed along a releasing bend; therefore, in this type of polygenetic basin, mantle was not involved and a thermal anomaly is not present beneath the basin.

Polyhistory Basins

Polyhistory basins have complex structural patterns that result from alternating episodes of strike-slip, compressional, and extensional faulting. These basins commonly form in areas of changing plate-tectonic framework. They may develop as polygenetic basins along thrust-sheet boundaries in collisional orogens that are later modified by post-collisional extensional faults; in this type of setting, strike-slip basins along tear faults confined to the upper plate of thrust sheets may undergo reversal of movement along tear faults during subsequent extensional faulting. Some collapse basins, such as the strike-slip-bounded, Devonian Hornelen basin of western Norway (Steel and Gloppen, 1980), may have resulted from this process.

Bowser Basin, Western Canada
The northwestern part of the North American Cordillera contains Jurassic, Cretaceous and Cenozoic "successor" basins that record complex tectonic and depositional adjustments to the accretion and juxtaposition of tectonostratigraphic terranes. The Tyaughton-Methow, Bowser and Gravina-Nutzotin basins are three such complex successor basins that developed along the continental margin in response to accretionary events (Fig. 12.21). Oblique convergence and transform motion of oceanic and arc-type lithospheric terranes has generated a wide region of generally right-lateral displacement (Eisbacher, 1981, 1985).

The Bowser basin originally developed in Late Jurassic time above and marginal to the belt of faults along which the Stikine terrane accreted to North America. The Stikine terrane, primarily made up of an early Mesozoic calc-alkaline magmatic-arc complex, subsided by Late Jurassic time to form the basement of the Bowser basin. Initially, Bowser basin sediments were transported westward into the basin from uplifted, previously accreted terranes to the

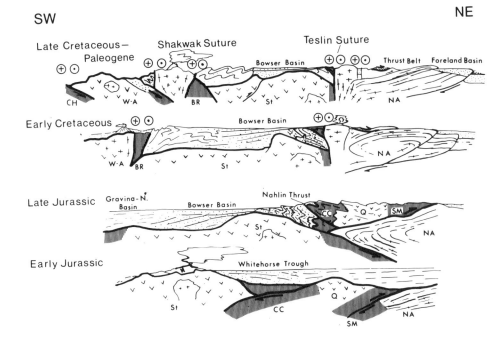

SW

NE

Late Cretaceous–
Paleogene

Shakwak Suture

Teslin Suture

Bowser Basin

Thrust Belt   Foreland Basin

CH   W-A   BR   St   NA

Early Cretaceous

Bowser Basin

W-A   BR   St   NA

Late Jurassic

Gravina-N.
Basin

Bowser Basin

Nahlin Thrust

St   CC   Q   SM

NA

Early Jurassic

Whitehorse Trough

St   CC   Q   SM   NA

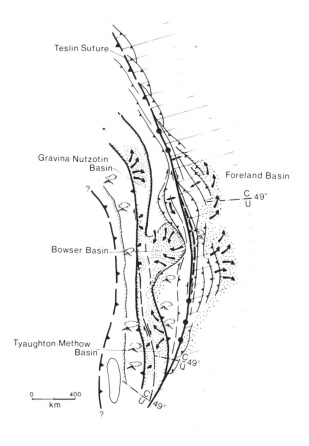

Teslin Suture

Gravina-Nutzotin
Basin

Foreland Basin

C/U 49°

Bowser Basin

Tyaughton-Methow
Basin

C/U 49°

0   400
km

C/U 49°

Fig. 12.21 Cross sections and map illustrating tectonic evolution of Bowser basin. A) Schematic evolution of peri-collisional clastic basins and strike-slip faults in a southwest-northeast transect through north-central Canadian Cordillera. Vertical ruling indicates oceanic-type terranes and ophiolitic mélanges (SM, Slide Mountain terrane; CC, Cache Creek terrane; BR, Bridge River terrane; CH, Chugach terrane). V-pattern indicates island arcs and oceanic plateaus (Q, Quesnel terrane; St, Stikine terrane; W-A, Wrangell-Alexander terrane). NA, North American craton. Cretaceous-Paleogene right slip developed along zones of complex crustal convergence characterized by multiple changes in structural vergence and subsequent pluton activity. B) Paleogeographic model of Middle Jurassic to Early Cretaceous setting of clastic basins near Teslin suture. Diagonal ruling indicates extent of North American continental crust locally overlapped in west by ophiolitic thrust masses of closing Teslin suture, and characterized by uplift of metamorphic core complexes. Accreted terranes west of thrust belt are shown approximately 1500 to 2000 km south of their present positions with respect to North America (49th parallel is used as displaced marker line). Strike-slip faults and arc polarity west of marginal Gravina-Nutzotin-Bowser-Tyaughton-Methow troughs are speculative. Areas of volcanic activity and generalized sediment dispersal depicted by cones and arrows, respectively. (Reproduced with permission from Eisbacher, 1985.)

east; more than 3,000 m of mid-Jurassic to Upper Cretaceous submarine-fan, slope and fluvial-dominated deltaic deposits accumulated in the basin. There is little direct evidence during this early history of the basin for major strike-slip involvement; basin subsidence and basin-margin uplift appear to have resulted primarily from convergence.

Later in the Early Cretaceous, however, right-slip faulting began to dominate. The eastern flank of the basin underwent uplift and transpression along the Teslin suture, an active zone of right slip that widened eastward through time. The eastern Bowser basin was tectonically reactivated, and 500 to 1000 m of generally coarse-grained fluvial and lacustrine strata were deposited unconformably on the older

marine fill. Right slip and uplift along the Shakwak suture on the western flank of the basin resulted in eastward transport of nonmarine clastic detritus into the basin. As deformation continued during the Tertiary, the basin narrowed to a trough that filled with 300 to 800 m of uppermost Cretaceous to Eocene alluvial-fan and fluvial units deposited chiefly by east-flowing systems.

Although much detailed sedimentologic and stratigraphic work remains to be done in the basin, the framework established by Eisbacher (1981, 1985), as well as by workers in related basins, including the outboard Queen Charlotte basin, reflects the shifting influences of convergent and strike-slip tectonics on depositional systems and basin morphology.

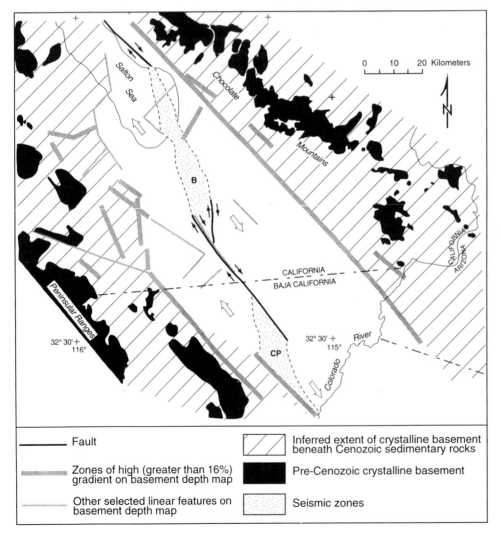

Fig. 12.22 Structural and tectonic map of Salton Trough. Seismic zones (stippled) are spreading centers (B, Brawley; CP, Cerro Prieto) that connect right-stepping right-slip faults and are probable locations of incipient stepover basins in San Andreas strike-slip system that trends obliquely up axis of Salton Trough. Direction of spreading in each stepover is indicated by open arrows. Sediment is carried into basin by Colorado River and by ephemeral drainages from surrounding mountains. (Modified from Fuis and Kohler, 1984.)

Salton Trough, Southern California

The Salton trough is an active strike-slip basin with an active oceanic spreading ridge constituting the southern part of its floor (Fig. 12.22). The trough is an elongate, northnorthwest-trending sedimentary basin that has formed adjacent to the southern terminus of the intracontinental San Andreas fault; the plate boundary passes southeastward across a series of short spreading ridges and long oceanic transform faults to the Rivera triple junction and the East Pacific Rise (see Chapter 2). Many aspects of the basin are well understood as a result of extensive subsurface geothermal exploration, frequent and large earthquakes, excellent exposures of rocks along both basin margins, and numerous studies of modern depositional systems.

Although initial formation of the basin during Middle Miocene time appears to have been the result of extensional tectonics related to formation of the Gulf of California and the Basin and Range (Crowell, 1987), the later and modern history of the basin has reflected right-slip along the San Andreas fault and related transform faults to the south. The initial record of extensional opening is well preserved in outcrops along the western margin of the trough, where pre-rift, braided-stream deposits are overlain in angular unconformity by synrift alluvial-fan, landslide-breccia, lacustrine, fluviodeltaic and evaporitic deposits. The basin in this area appears to have been a half graben, with its principal north-striking border fault to the west. Subsequent increased subsidence allowed the terrestrial succession to be overlain by widespread shallow- to deep-marine, late synrift or postrift strata of the Imperial Formation.

During deposition of the Imperial Formation, the transform boundary jumped eastward into the rift valley, thus initiating activity along the modern San Andreas fault. The later history of the basin has been dominated by strike-slip tectonism associated with several right-slip faults of the San Andreas system, including the San Andreas, Imperial, San Jacinto and Elsinore faults. Coarse-grained and very thick alluvial-fan deposits grade axially into lacustrine mud, similar to the modern setting surrounding the Salton Sea. The later history is dominated by deposi-

tion of huge volumes of sediment by the Colorado River, whose delta separates the south end of the Salton trough from the north end of the Gulf of California (Winker and Kidwell, 1986).

The Salton Trough has reached a stage in its tectonic history that involves uplift and arching above an en-echelon chain of hot mantle domes, judging from Quaternary volcanism within the trough, high heat flow, high Bouguer gravity values, and greenschist-facies metamorphism of Pliocene and Pleistocene sediments at depths reached by geothermal drill holes (4000–6000 m). These phenomena are associated with oblique sea-floor spreading and opening of the Gulf of California. Uplift of the mountainous margins may be attributed to thermal expansion of the lithosphere beneath a region wider than the Salton trough itself. Oblique stretching of the region follows the strike of transform faults and agrees with the plate motion between the Pacific and North American lithospheric plates. Faults and deformation patterns outline uplifted blocks and subbasins nested within the Salton Trough. The subbasins are similar to those strike-slip basins in the Gulf of California that have not filled with sediments; orientation of the small basins relative to the strikes of the major transform faults and the plate boundary suggests that their direction of stretching corresponds to the direction of extension in right simple shear (Crowell, 1985).

The Salton Trough reflects the complex evolution of a polyhistory basin with an initial rift origin, an irregularly stretched floor, strike slip, development of en-echelon subbasins, and rerouting of a major fluvial-dominated delta into the widening trough. Tectonic mobility of the Salton trough began about 12 Ma and has intensified during the last 5 my. Modern strike-slip faulting is superposed on Miocene detachment faults which were superposed on Mesozoic and early Tertiary thrusts.

Conclusions

We have described six major types of strike-slip basins based on their fault patterns and mechanisms of formation; however, strike-slip basins constitute the most complicated of basin types. They form in diverse tectonic settings, and are deformed and

reformed as fault blocks rise, fall, converge, and diverge in space and time. Their form, structure, depositional setting, and history depend upon the complex interplay of structural, depositional, climatic, and paleogeographic factors.

Most long-lived strike-slip basins are polycyclic and undergo repeated episodes of generally transtensive and transpressive subsidence and uplift within continuing strike-slip settings. Repeated subsidence and uplift commonly yield complex basins that may be difficult to reconstruct, however, and each basin must be evaluated separately.

Laterally displaced source areas, facies, and depocenters provide general criteria for the recognition of areas of strike-slip faulting. Complex and abrupt facies changes within narrow, generally elongate basins characterized by high sedimentation rates, very thick stratigraphic successions, and abundant evidence of paleoseismic activity are typical of the fill of strike-slip basins. Half-graben patterns in transverse cross sections are common. Coarse-grained debris-flow-dominated aprons characterize active strike-slip basin margins, whereas fluvio-deltaic facies characterize the opposite margins, and submarine or sublacustrine turbidite, fluviodeltaic, eolian and playa-evaporite facies typify axial deposystems. Depocenters migrate parallel to the most active strike-slip margin, resulting in accumulation of extraordinarily thick sedimentary sections in small basins. Basinwide upward coarsening sequences record progradational sedimentary systems that result from forced regressions caused by periodic lowering of base level induced by tectonism, sea/lake-level changes and climatic changes.

From our perspective, an understanding of strike-slip basins is best enhanced by a thorough understanding of how other types of basins develop. Strike slip is a significant component of almost all plate-bounding and intraplate settings; thus, to a greater or lesser degree, their understanding helps the understanding of all other tectonic settings. Continued fault displacement and polycyclic episodes of sedimentation and uplift commonly place most long-lived strike-slip basins in the general category of "complex basins." Their original paleotectonic framework may be difficult to recognize and restore.

John C. Crowell (1987), in characterizing Tertiary basins in southern California, wrote: "complexity is the hallmark of their tectonic history," and by inference, their sedimentation history as well. He wrote (p. 240):

> Crustal mobility and sedimentation have gone on together across the whole region, but with belts and areas where, for a time, either deformation or sedimentation has dominated. Cross sections are bound to get more complicated at depth, inasmuch as older beds have been subject to more deformation than younger. Geologists must abandon simplicity as a main guiding principle, and expect complexity instead. Each new bit of information, such as that obtained from a new drill hole or a new seismic line, will help in elucidating the history, but it will also add complexity to the cross section already drawn. No longer does the Principle of Simplicity, so aptly exploited in physics, for example, apply unqualified to our task. We must substitute the Principle of Complexity: in historical or configurational science, new details add complications.

Acknowledgments

The authors thank the editors for helpful reviews of the manuscript and for their patience and understanding. THN thanks Greg Zolnai, consulting geologist from Pau, France, for helpful discussions. THN also thanks Elf Aquitaine Production for abruptly terminating his one-year contract, providing him with the time to complete his part of the manuscript.

Further Reading

Ballance PF, Reading HG (eds), 1980, *Sedimentation in oblique-slip mobile zones*: International Association of Sedimentologists Special Publication 4, 265p.

Biddle KT, Christie-Blick N (eds), 1985, *Strike-slip deformation, basin formation, and sedimentation*: Society of Economic Paleontologists and Mineralogists Special Publication 37, 386p.

Christie-Blick N, Biddle KT, 1985, *Deformation and basin formation along strike-slip faults*: Society of Economic Paleontologists and Mineralogists Special Publication 37, p.1–34.

Reading HG, 1980, *Characteristics and recognition of strike-slip fault systems*: International Association of Sedimentologists Special Publication 4, p.7–26.

Sylvester AG (compiler), 1984, *Wrench fault tectonics*: American Association of Petroleum Geologists Reprint Series 28, 374p.

Sylvester AG, 1988, Strike-slip faults: *Geological Society of America Bulletin*, v.100, p.1666–1703.

Wallace RE (ed), 1990, *The San Andreas fault system*: United States Geological Survey Professional Paper 1515, 283p.

Woodcock NH, 1986, The role of strike-slip fault systems at plate boundaries: *Philosophical Transactions of the Royal Society of London*, v.A317, p.13–29.

Zolnai G, 1991, *Continental wrench-tectonics and hydrocarbon habitat* [revised]: American Association of Petroleum Geologists Continuing Education Course Note Series 30, variably paginated.

Intracratonic Basins **13**

George D. Klein

Introduction

Intracratonic basins occur within continental interiors away from plate margins. They are oval in plan and saucer-shaped in cross section. Intracratonic basins are floored with continental crust, and in most instances, are also underlain by failed or fossil rifts (see Chapters 2 and 3). Their evolution involves a combination and succession of basin-forming processes, which include continental extension, thermal subsidence over a wide area, and later isostatic readjustments.

Controversy has surrounded understanding of the origin and evolution of intracratonic basins. Eleven hypotheses (Klein, 1991) for the formation of intracratonic basins have been proposed and include: (1) increase in density of the crust by an eclogite phase transformation (Michigan basin, Haxby et al., 1976; Williston basin, Fowler and Nisbet, 1985), (2) rifting associated with impingement of a thermal plume at the base of the lithosphere (Illinois basin, Burke and Dewey, 1973; McGinnis et al., 1976), (3) thermal metamorphism of the lower crust to boundary conditions of the greenschist and amphibolite facies (Australian intracratonic basins, Middleton, 1980), (4) mechanical subsidence caused by isostatically uncompensated excess mass of igneous intrusions (Williston basin, DeRito et al., 1983), (5) tectonic reactivation along older structures (Michigan basin, Howell, 1989; Parana, Maranhao, and Amazonas basins, Brazil, DeBrito Neves et al., 1984), (6) thermal subsidence (Illinois basin, Sleep, 1976; Sleep and Snell, 1976; Sleep et al., 1980; Heidlauf et al., 1986; Williston basin, Ahern and Mrkvicka, 1984; Ahern and Ditmars, 1985; Michigan basin, Sleep et al., 1980; Nunn et al., 1984; Nunn, 1986; Hudson Bay basin, Quinlan, 1987; Amadeus basin, Australia,

Lindsay and Korsch, 1989), (7) partial melting in lower crust and drainage of resulting igneous melt to mid-ocean ridges by volcanism, resulting in basin subsidence above the zone from which igneous melt was removed (Sloss and Speed, 1974), (8) changes in intraplate stress, assuming a visco-elastic plate (Australian intracratonic basins, Lambeck, 1983a,b; Shaw et al., 1991) or an elastic plate (all intracratonic basins, Karner, 1986), (9) thermal subsidence following intrusion of anorogenic granites in response to changing heat flow during supercontinent breakup (all intracratonic basins, Klein and Hsui, 1987), (10) thermal subsidence followed by subsidence caused by an isostatically uncompensated excess mass of cooling igneous intrusions (Illinois basin, Treworgy et al., 1989; Kolata and Nelson, 1990a,b; Hamdani et al., 1991), and (11) subsidence caused by tectonic events at adjacent plate margins (Howell and van der Pluijm, 1990; Leighton and Kolata, 1990).

In North America (Fig. 13.1), subsidence analysis has shown that initiation of subsidence of the Illinois, Michigan, and Williston basins, and initiation of subsidence of latest Precambrian and earliest Paleozoic passive margins were coeval with the age of breakup of a late Precambrian supercontinent (Bond and Kominz, 1984, 1991; Bond et al., 1984; Klein and Hsui, 1987). In Australia, initial formation of late Proterozoic cratonic basins was coeval with the breakup of a supercontinent (Lindsay et al., 1987; Korsch and Lindsay, 1989; Lindsay and Korsch, 1989). Similar timing of subsidence followed breakup of Pangea in the Paris basin (Brunet and LePichon, 1982; Perrodon and Zabek, 1990), the North Sea (Ziegler, 1977, 1982, 1987, 1988), and Gondwana basins of India (Kailasham, 1976). Sloss (1972, 1979) and Sloss and Scherer (1975) observed that systematic synchronous

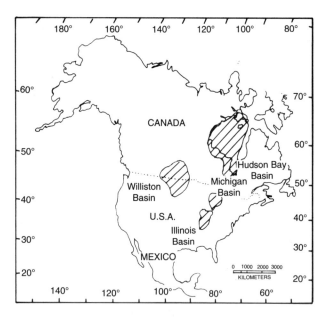

KLEIN - Intracratonic Figure 13.1

Fig. 13.1 Map of North America, showing location of major intracratonic basins.

changes in sediment volume were characteristic of intracratonic basins in North America, Brazil (Soares et al., 1978; DeBrito Neves et al., 1984), the Russian platform, and Africa (Peters, 1979). Moreover, continentwide interregional unconformities characteristic of Sloss's (1963, 1988a) cratonic sequence boundaries established in North America are found in other intracratonic settings, including eastern Europe and the former Soviet Union (Sloss, 1972, 1976), Brazil (Soares et al., 1978; DeBrito Neves et al., 1984; Zalan et al., 1990), and Africa (Peters, 1979). These stratigraphic observations indicate a commonality in time and space of formation of intracratonic basins in response to a global process, which Sloss (1988a,b) emphasized is tectonic.

This chapter focuses on some of these problems and suggests an hypothesis not only for the origin of intracratonic basins, but also for their global commonality. This chapter summarizes recent observations that clarify some of the earlier explanations for mechanisms of formation of intracratonic basins listed above, as well as new data that permit resolution of much of the controversy regarding the formation of intracratonic basins. This chapter also summarizes what is known about North American Paleozoic

intracratonic basins, including the Illinois basin (Sleep, 1971, 1976; Sleep et al., 1980; Heidlauf et al., 1986; Treworgy et al., 1989; Kolata and Nelson, 1990), the Michigan basin (Sleep and Sloss, 1978; Nunn et al., 1984; Nunn, 1986; Ahern and Dikeou, 1989; Howell, 1989), and the Williston basin (Gerhard et al., 1982, 1990; DeRito et al., 1983; Ahern and Mrkvicka, 1984; Ahern and Ditmars, 1985; Kent, 1987; Gerhard and Anderson, 1988), as well as selected intracratonic basins, such as the Amadeus basin of Australia, and the Parana basin of Brazil.

Intracratonic Stratigraphic Sequences

Within intracratonic basins and adjacent platforms, stratigraphic subdivision differs from international stratigraphic subdivisions. Intracratonic stratigraphic subdivision follows the now well established concept of "Cratonic Sequences" of Sloss (1963, 1988a; also see Leighton and Kolata, 1990). Sloss (1963) recognized that the stratigraphy of the North American craton is subdivided into six cratonic sequences, which are bounded by major interregional (continent-wide) unconformities. These cratonic sequences are (from oldest to youngest): Sauk (590 to 488 Ma), Tippecanoe (488 to 401 Ma), Kaskaskia (401 to 330 Ma), Absaroka (330 to 186 Ma), Zuni (186 to 60 Ma), and Tejas (60 Ma to present). Each cratonic sequence records a nearly complete trangressive-regressive sedimentary cycle. Internally, these cratonic sequences are subdivided into subsequences separated by lesser regional unconformities; these sequences include basal onlap and are capped by minor offlapping sequences below the unconformities. It is the writer's view that Sloss's North American cratonic sequences are comparable to the classic European geological systems and are unique to intracratonic settings in other regions of the world, including the Russian platform (Sloss, 1972, 1976, 1979), Brazil (Soares et al., 1978; DeBrito Neves et al., 1984), and Africa (Peters, 1979). Intracratonic areas outside North America show coeval, continentwide interregional unconformities, and parallel temporal trends to areal distribution and sediment volume (Sloss, 1972, 1976, 1979). Although the cratonic sequences were defined from a single tectonic setting (intracratonic) in North America, they

are observed globally (Soares et al., 1978; Peters, 1979; Sloss, 1979). By contrast, the classic European type sections were recognized and established in differing tectonic settings (e.g., active margins, rift basins, platforms, intracratonic basins, and foreland basins), a fact which has contributed to never-ending arguments over global correlation of system boundaries in different tectonic domains.

Summary of North American Intracratonic Basins

Four major intracratonic basins are known from North America (Fig. 13.1). Of these, the Illinois, Michigan, and Williston basins are summarized below.

Illinois Basin

The Illinois basin is oval in plan (Figs. 13.1, 13.2). Six-thousand meters of sedimentary strata accumulated within this basin during Paleozoic time (Figs. 13.3, 13.4; Kolata and Nelson, 1990a,b; Nelson, 1990a,b). Initially, sediments accumulated on a south-facing embayed ramp (Reelfoot aulacogen; McGinnis, 1970; McGinnis et al., 1976; Kolata and Nelson, 1990a,b; Nelson, 1990a,b). Uplift of the Pascola arch on its south side closed this embayment between Pennsylvanian and Cretaceous time (Sterns, 1957; Marcher, 1961; Bethke, 1985).

Controversy exists concerning basement structure. Magnetic and gravity anomalies and the location of Holocene earthquake epicenters (e.g., McGinnis, 1970; McGinnis et al., 1976; Braille et al., 1982, 1986; Hildenbrand, 1985) suggest that the Reelfoot aulacogen (Fig. 13.2) was part of a three-arm rift system. Moreover, seismic data suggest that the crust is thin below the depocenter and the Reelfoot aulacogen (Braille et al., 1986). More recent seismic surveys (e.g., Pratt et al., 1989; Bretagne and Leising, 1990; Heigold, 1990; Heigold and Oltz, 1990) and structural mapping (e.g., Nelson, 1990a,b) show that the basement consists of bedded volcanics; only one of the proposed three arms, the Rough Creek graben, could be identified as extending east of the depocenter from the Reelfoot aulacogen (Fig. 13.2). To what extent

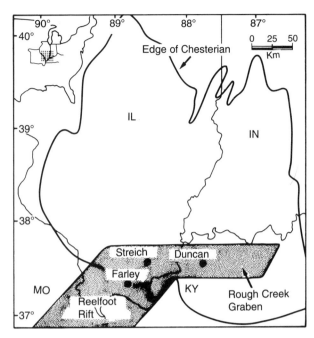

KLEIN - Intracratonic - Figure 13.2

Fig. 13.2 Map showing location of three wells used for tectonic-subsidence analysis (Fig. 13.5) in Illinois basin (basin boundary marked by edge of Chesterian Series). Location of Reelfoot rift and Rough Creek graben, both stippled, after Nelson (1990b). IL = Illinois; IN = Indiana; KY = Kentucky; MO = Missouri.

the boundaries of the other two arms (mapped by earlier geophysical studies) represent localized faults remains problematic. Even the origin of the Rough Creek "graben" now appears to be in dispute because recent work by Goetz et al. (1992) shows it to have been formed in a strike-slip zone involving transpression throughout its history.

I recomputed the tectonic-subsidence history of the Illinois basin using stratigraphic data from Heidlauf et al. (1986; their tables 1–3) and the Harland et al. (1990) time scale. Methods used for this computation were those summarized in Sclater and Christie (1980) and Bond and Kominz (1984). Tectonic-subsidence analysis of three deep wells in the basin depocenter (Fig. 13.5) shows that the basin formed by initial, rapid fault-controlled subsidence associated with rifting and crustal thinning, followed by a period of rapid thermal subsidence that lasted about 60 my (shown by an upward concave profile beyond the initial steep

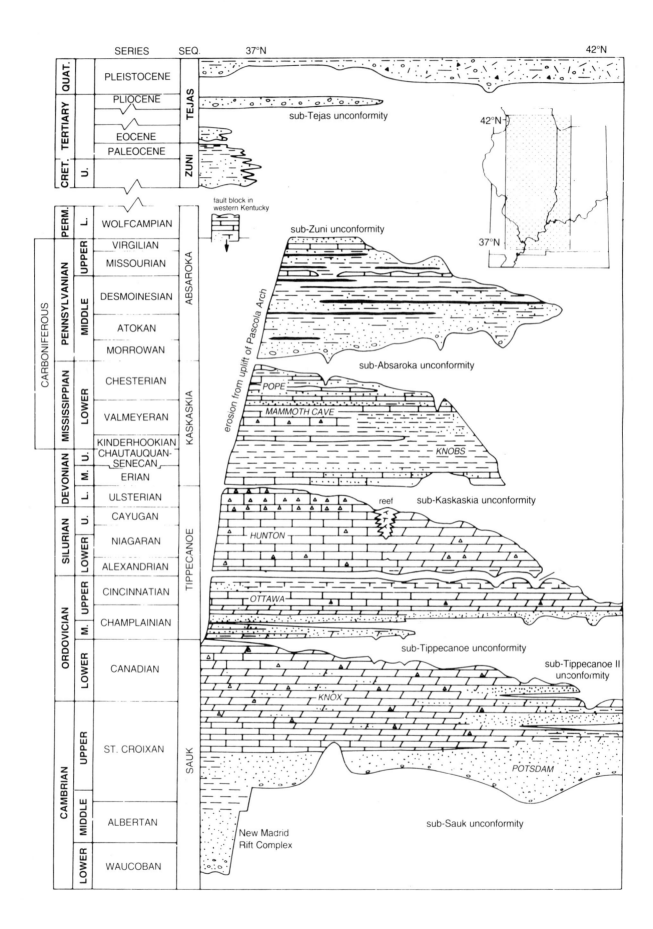

Fig. 13.3 Schematic stratigraphic section showing relationship of stratigraphic megagroups, lithofacies, series and cratonic sequences of Sloss (1963) in the Illinois basin. (Redrawn after Buschbach and Kolata, 1990.)

line). Thermal subsidence was verified by comparing the elevation of basement depth with t^{-2} (Fig. 13.5); the linear trend that was established showed strong positive correlation coefficients ranging from 0.95 to 0.98 (Heidlauf et al., 1986), as outlined by McKenzie (1978a). This phase of thermal subsidence began around 510 Ma and continued until 450 Ma.

Heidlauf et al. (1986) observed that, during this 60 my thermal-subsidence phase, subsidence rates suddenly increased at various times. These changes in subsidence rates were attributed originally to changes in sea level (e.g., Sleep and Snell, 1976; Sleep et al., 1980; Heidlauf et al., 1986) or inaccuracies in earlier time scales (e.g., Heidlauf et al., 1986). However, sea level was rising continuously during this time interval (Watso and Klein, 1989), so sea-level changes appear not to have caused these departures from subsidence rate. Davis (1987, 1990), Clendenin (1989), and Watso and Klein (1989) suggested, instead, that during the initial phase of rapid thermal subsidence, regional transfer faulting (Clendenin, 1989) and associated normal faulting contributed to additional subsidence. During the earliest phase of thermal subsidence, an interval of overlapping, but diminished, mechanical subsidence occurred; this faulting may be attributed to progressively increasing lithospheric rigidity with cooling (Karner et al., 1983). Quinlan (1987) reported a similar combination of mechanical and thermal subsidence.

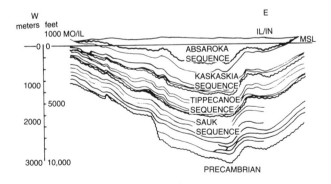

Fig. 13.4 West-east stratigraphic cross section of Illinois basin, approximately 100 km north of depocenter. (Redrawn after Collinson et al., 1988.)

Rapid thermal subsidence was followed by slower subsidence, which Heidlauf et al. (1986) also interpreted as thermal subsidence (cf., McKenzie, 1978a; Korsch and Lindsay, 1989). More recently, Treworgy et al. (1989) and Kolata and Nelson (1990a,b) suggested that this slower phase of subsidence, which lasted from approximately 450 to 360 Ma, was caused by an isostatically uncompensated excess mass (cf., DeRito et al., 1983). This model of DeRito et al. (1983) requires concurrent large-scale tectonic stresses. Tectonic events known in the Illinois basin started during latest Mississippian and Pennsylvanian time (Nelson and Krause, 1981; Nelson and Lumm, 1984; Kolata and Nelson, 1990a,b). Alternatively, in my view, after the relatively rapid phase of thermal subsidence lasting 60 my, progressive cooling of crustal magma would cause a phase change below the basin and subsidence would occur in response to a different style of isostatic compensation (cf., Hamdani et al., 1991), particularly as lithospheric rigidity increased (Karner et al., 1983); it would be accompanied also by a concurrent slower rate of thermal subsidence and decay (cf., McKenzie, 1978a). Thus, I conclude that this period of subsidence (450–360 Ma) was caused by a combination of late stages of decaying thermal subsidence and an isostatically uncompensated excess mass caused by a phase change associated with rift-related magmatic cooling (e.g., Hamdani et al., 1991), independent of large-scale compression.

During Carboniferous time, the rate and magnitude of basin subsidence increased a second time (Heidlauf et al., 1986). Collision tectonics of the Alleghenian-Hercynian-Ouachita orogeny occurred along the eastern and southern margins of North America and caused widespread regional compressive stresses (Craddock and van der Pluijm, 1989), which yielded a consistent regional fracture trend, associated strike-slip faulting in the Illinois basin (Nelson and Krause, 1981; Nelson and Lumm, 1984; Nelson, 1990a,b), and altered sedimentation patterns (Greb, 1989). Subsidence was reactivated by a far-field tectonic effect involving tectonic forces at the margin, which influenced a widespread area in the interior of the continent (e.g., Kluth and Coney, 1981; Kluth, 1986; Ziegler, 1987, 1988; Leighton and Kolata, 1990); in this case, thrust loading on eastern and

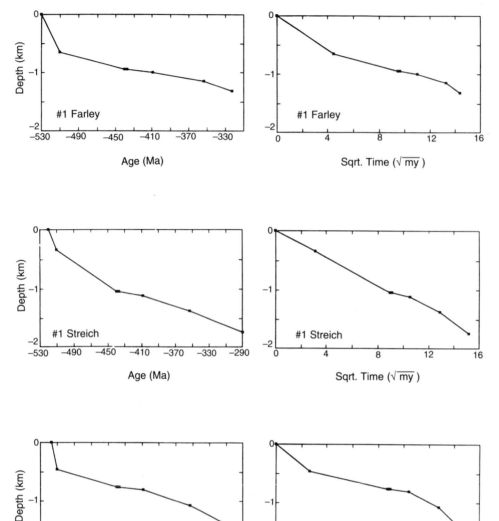

Fig. 13.5 Left: decompacted and backstripped tectonic subsidence for Texas Pacific Co. No. #1 Farley Well (upper), Texas Pacific Co. No. #1 Streich Well (middle), and Exxon Corp. No. #1 Choice Duncan Well (lower), Illinois basin. Data from Heidlauf et al. (1986), recomputed according to time scale of Harland et al. (1990). Right: Comparison of tectonic-subsidence with respect to t^2 for Farley (upper), Streich (middle), and Duncan (lower) wells, Illinois basin. (See Fig. 14.2 for locations.)

southern Laurussia caused major tectonic deformation and subsidence in the continental interior of North America (Quinlan and Beaumont, 1984; Beaumont et al., 1987). Thrust loading caused widespread flexural stress (cf., Craddock and van der Pluijm, 1989), which was transmitted through the lithosphere; the associated flexural wave reactivated and focused basin subsidence along older trends in the Illinois and Michigan basins, as well as in the central Appalachian foreland basin (cf., also Karner et al., 1983; Lambeck, 1983a,b; Quinlan and Beaumont,

1984; Karner, 1986). This thrust loading may have caused uplift and inversion of a Precambrian rift basin on the Cincinnati arch (Shrake et al., 1991). Widespread intracratonic basin subsidence in North America occurred during three periods of subsidence associated with Late Ordovician, Late Silurian-Early Devonian, and Carboniferous orogenic events (Willard and Klein, 1990). The thermal age of the crust below the Appalachian foeland basin is approximately 600 Ma (Bond et al., 1984; Willard and Klein, 1990); lithospheric rigidity increased gradually with

time and during each successive orogenic event, as confirmed by decreasing rates of subsidence throughout Paleozoic time (Willard and Klein, 1990). Consequently, foreland-basin subsidence was shallower and broader during the Alleghenian-Hercynian orogeny (see Chapters 9 and 11). Distribution of sedimentary facies in the central Appalachian basin fits such a geodynamic model, inasmuch as the basin was filled with submarine turbidites during the Late Ordovician, turbidites overlain by deltas during the Late Devonian, and fluvio-deltaic sediments during the Pennsylvanian (Tankard, 1986; Willard and Klein, 1990).

Such shallower and broader subsidence favored lateral transmission of compressive stresses, which reactivated subsidence in the Illinois basin during Middle and Late Carboniferous time, and reactivated Precambrian faults, causing uplift along the eastern boundary of the basin, the Cincinnati arch (Shrake et al., 1991). This subsidence was associated also with deposition of Pennsylvanian cyclothems (Tankard, 1986), which were influenced by concurrent tectonic and climatically induced cyclic processess (Klein and Willard, 1989; Klein, 1992a; Klein and Kupperman, 1992). This second phase of subsidence in the Illinois basin was associated with intrusion of Late Pennsylvanian and Permian alnoites, derived by partial melting of the lower crust (Zartman et al., 1967). Their presence suggests renewed extension in the Reelfoot aulacogen (Klein and Hsui, 1987).

Sedimentary systems in the Illinois basin range from fluvial to coastal to shelf, with local turbidite units below deltaic strata (see Heidlauf et al., 1986, their Fig. 6; Kolata and Nelson, 1990a, their Fig. 18-1; Leighton et al., 1990, p. 75–164, for recent summaries and syntheses). Cambrian fluvial and coastal deposits occur in the Reelfoot aulacogen, and onlapping strata occur immediately above each of the cratonic sequences reported by Sloss (1963, 1988), Collinson et al. (1988), and Leighton et al. (1990, p.75–164). Cambrian fluvial-coastal deposition was associated with rifting and thermal subsidence of the Illinois basin (Heidlauf et al., 1986); as thermal subsidence diminished, deposition of carbonate platform facies was dominant. The source of lower and middle Paleozoic clastic deposits appears to have been the Canadian shield (Potter and Pryor, 1960).

During Silurian time, extensive pinnacle reefs developed; these are major reservoirs for petroleum and form a linear trend along the basin axis (Whitaker, 1988). Carbonate-platform deposition persisted until latest Mississippian time. During Alleghenian-Hercynian orogenic movements in the Appalachian and Ouachita regions, the Illinois basin was linked with the continental margin by flexural tectonics (Quinlan and Beaumont, 1984; Beaumont et al., 1987) and the depositional motif was dominated by coal-bearing cyclothems, particularly during Pennsylvanian time (Klein and Willard, 1989). Deposition of these cyclothems was influenced concurrently by regional flexural tectonics, glacial eustasy, and long-term climate change (Klein and Willard, 1989; Cecil, 1990; Klein, 1992a,b; Klein and Kupperman, 1992). Fluvial systems controlled by concurrent faulting (e.g., Greb, 1989) and extensive deltas characterized deposition of these cyclothems. The principal source of clastic components of these cyclothems appears to have been the Appalachians (Potter and Pryor, 1960; Greb, 1989).

The Illinois basin has been a major oil-producing province since oil was discovered in 1886 (Leighton et al., 1990). Principal source beds are in the uppermost Devonian to lowest Mississippian New Albany Shale (Cluff and Byrnes, 1990) and the Ordovician Maquoketa Shale, both of anoxic relatively deepwater origin. Other sources include Pennsylvanian coals, and the Cambrian intertidal Eau Claire Shale.

Reservoirs occur primarily in the Silurian (primarily reefs), Devonian (shelf carbonates), Mississippian and Pennsylvanian systems (Howard, 1990; Seyler and Cluff, 1990). The Mississippian reservoirs are diverse, ranging from oolitic shoals in the Salem Limestone, to tidal sand bodies of the Aux Vases Sandstone. Pennsylvanian channel and delta-distributary deposits comprise remaining reservoir types.

Timing and patterns of oil migration have been modelled relatively recently (Bethke, 1985; Bethke and Marchak, 1990; Bethke et al., 1991). Because the Illinois basin was a cratonic embayment through most of Paleozoic time (Kolata and Nelson, 1990a,b), it did not assume its present shape until the Pascola arch on its southern boundary was uplifted as early as Permian time (Bethke, 1985; Bethke et al., 1991).

With this uplift, topographically driven gravity flow of basin fluids was initiated to the north. These fluids were heated during migration into the basin depocenter, where temperatures were higher, and this thermal effect caused maturation of oil from the principal source beds. Lateral migration into reservoirs occurred over a wider region (Fig. 13.6).

Trapping mechanisms show considerable variability (Seyler and Cluff, 1990), with structural traps (i.e., anticlines and faults) being dominant; facies changes and lateral pinchouts provide important stratigraphic traps, particularly in the Pennsylvanian and the Mississippian, and also in Silurian pinnacle reefs. Seals consist mostly of overlying shale, or shale bounding one of the walls of a fault-bounded trap.

Williston Basin

The Williston basin is also saucer-shaped, with an oval plan view (Fig. 13.1). It is filled with 3700 m of sedimentary strata, which are also subdivided according to Sloss's (1963, 1988a) cratonic sequences (Fig. 13.7). Strata consist mostly of carbonate rocks and evaporites, and smaller volumes of sandstone and

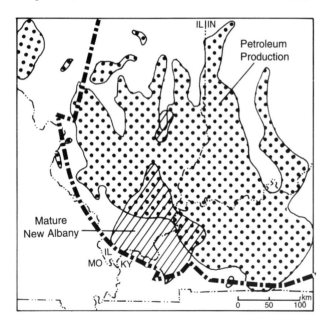

Fig. 13.6 Areal distribution of petroleum production and thermally mature, oil-prone New Albany Shale (Upper Devonian - Lower Mississippian), Illinois basin. Edge of Illinois basin represented by heavy black line with white markings. (Redrawn from Bethke and Marshak, 1990).

shale, representing all periods of the Phanerozoic (e.g., Gerhard et al., 1982, 1990; Peterson and Mac-Cary, 1987; Gerhard and Anderson, 1988). These strata include coastal to shallow-shelf deposits, and Devonian pinnacle reefs. Drilling to basement indicates that the basin is underlain by Archean basement (see Hoffman, 1988), Proterozoic suture zones, a major transform fault, and a province of anorogenic granite (Kent, 1987; Gerhard et al., 1990).

The Williston basin underwent thermal subsidence (Fig. 13.8) starting from about 520 to 500 Ma (Gerhard et al., 1982, 1990; Ahern and Mrkvicka, 1984; Ahern and Ditmars, 1985; Fowler and Nisbet, 1985; Gerhard and Anderson, 1988). This rapid phase of thermal subsidence was followed by both a lower rate of thermal subsidence coupled with subsidence due to an isostatically uncompensated excess mass during middle Paleozoic time (e.g., DeRito et al., 1983). A Proterozoic failed rift underneath the Williston basin has been inferred both by Stewart (1972) and Kent (1987); this rift was reactivated (DeRito et al., 1983; Quinlan, 1987). A transtensional basin precursor may also have been associated with reactivation of Precambrian orogenic zones (Gerhard et al., 1990). The reactivation that caused the thermal event at about 500 Ma appears to have been associated either with a phase change at the base of the crust (Fowler and Nisbet, 1985) or with a mantle intrusion into the crust (Ahern and Ditmars, 1985). Both these interpretations are based on analysis of tectonic subsidence. Foreland flexural subsidence, (e.g., see Chapters 9 and 11) similar to Carboniferous subsidence in the Illinois and Michigan basins, is inferred to have developed from post-Jurassic time through the Late Eocene, presumably in response to convergent tectonics in the North American Cordillera.

The Williston basin is one of North America's major petroleum provinces (Gerhard et al., 1990). The principal source bed is the Upper Devonian to Lower Mississippian Bakken Formation, which formed during a transgressive anoxic episode (Gerhard et al., 1990). Subsidiary source beds include the Ordovician Winnipeg Group (also of anoxic, relatively deep-water conditions) and parts of the Pennsylvanian Tyler

Fig. 13.7 Generalized stratigraphy, Williston basin. (Redrawn after Gerhard et al., 1990).

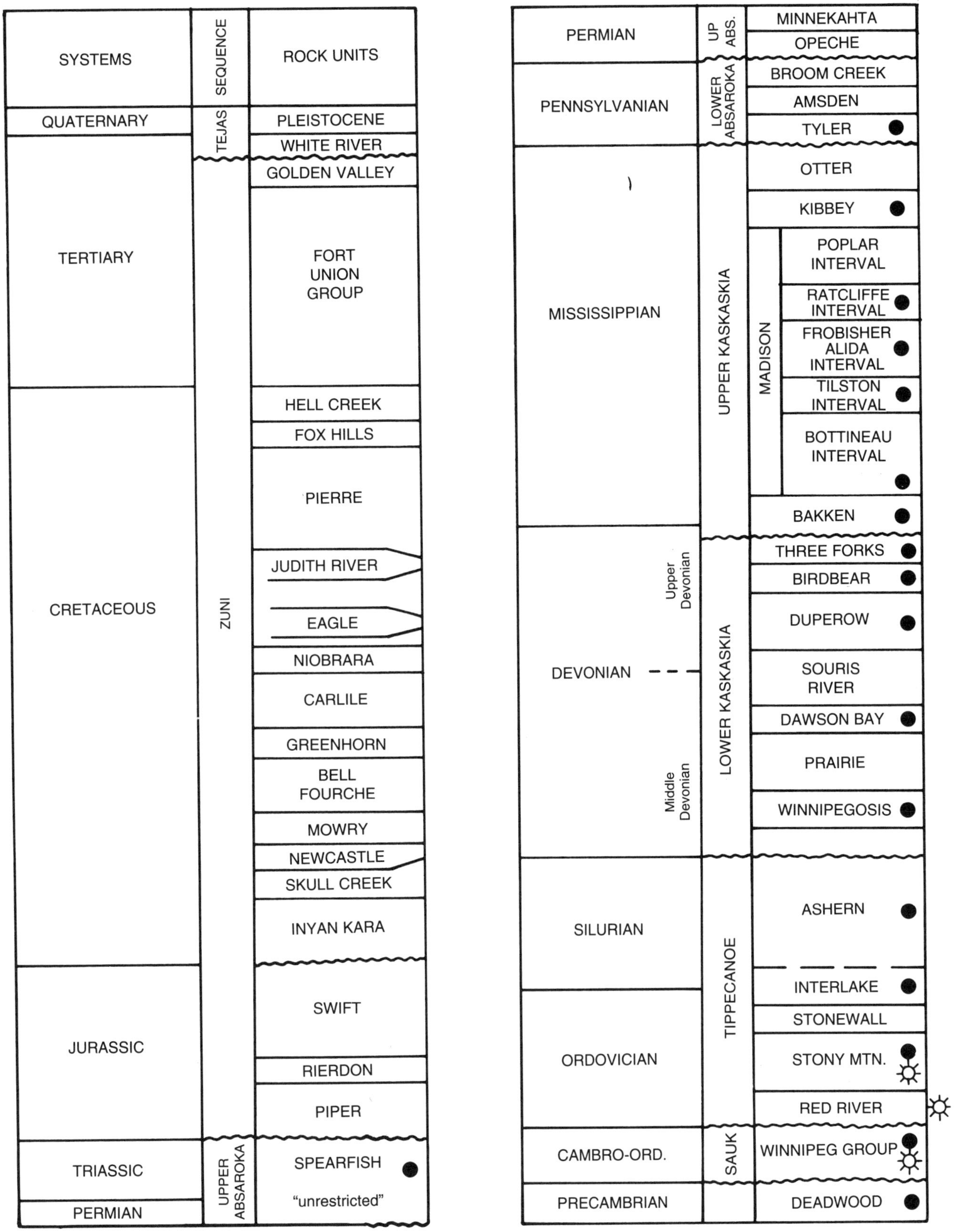

Left Column

SYSTEMS	SEQUENCE	ROCK UNITS
QUATERNARY	TEJAS	PLEISTOCENE
		WHITE RIVER
		GOLDEN VALLEY
TERTIARY		FORT UNION GROUP
		HELL CREEK
		FOX HILLS
		PIERRE
	ZUNI	JUDITH RIVER
CRETACEOUS		EAGLE
		NIOBRARA
		CARLILE
		GREENHORN
		BELL FOURCHE
		MOWRY
		NEWCASTLE
		SKULL CREEK
		INYAN KARA
		SWIFT
JURASSIC		RIERDON
		PIPER
TRIASSIC	UPPER ABSAROKA	SPEARFISH ● "unrestricted"
PERMIAN		

Right Column

SYSTEMS	SEQUENCE		ROCK UNITS
PERMIAN	UP ABS.		MINNEKAHTA
			OPECHE
PENNSYLVANIAN	LOWER ABSAROKA		BROOM CREEK
			AMSDEN
			TYLER ●
			OTTER
			KIBBEY ●
MISSISSIPPIAN	UPPER KASKASKIA	MADISON	POPLAR INTERVAL
			RATCLIFFE INTERVAL ●
			FROBISHER ALIDA INTERVAL ●
			TILSTON INTERVAL ●
			BOTTINEAU INTERVAL ●
			BAKKEN ●
		Upper Devonian	THREE FORKS ●
			BIRDBEAR ●
DEVONIAN	LOWER KASKASKIA		DUPEROW ●
			SOURIS RIVER
			DAWSON BAY ●
		Middle Devonian	PRAIRIE
			WINNIPEGOSIS ●
SILURIAN	TIPPECANOE		ASHERN ●
			INTERLAKE ●
			STONEWALL
ORDOVICIAN			STONY MTN. ● ☼
			RED RIVER ☼
CAMBRO-ORD.	SAUK		WINNIPEG GROUP ● ☼
PRECAMBRIAN			DEADWOOD ●

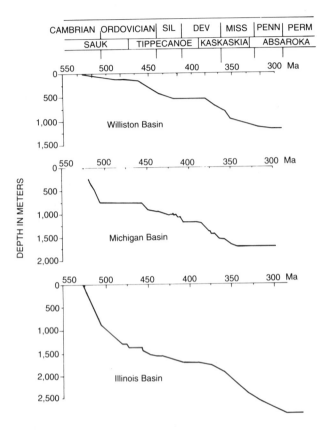

Fig. 13.8 Backstripped and decompacted tectonic subsidence for the Williston, Michigan, and Illinois basins, showing similar time of basin formation around 550 to 530 Ma, and initiation of thermal subsidence between 530 and 500 Ma. Stratigraphic data for these curves came from McClure-Sparks Well in Michigan basin (Hinze et al., 1978), and regional stratigraphic summaries for other two basins (see Bond and Kominz, 1991, for details). (Redrawn from Bond and Kominz, 1991, Fig. 2E.)

Formation (mostly coastal deposits). Reservoirs are primarily Devonian and include the Winnipegosis Formation (major reef trend), and the Duperow and Dawson Bay formations (both stromatoporoid banks). The Pennsylvanian Tyler Formation contains productive reservoirs, which are mostly barrier-island and deltaic-distributary reservoirs. A recent play has focused on the Sherwood Formation (Mississippian), where oil has accumulated in stratigraphic traps at carbonate facies pinchouts into mudstones and evaporites (Shirley, 1991).

Petroleum migration was mostly vertical, along regional lineaments and vertical faults (Gerhard et al., 1990). Most of the traps are structural, and even the reefs and stromatoporoid reservoirs are fault-bounded.

Michigan Basin

The Michigan basin (Fig. 13.1) also is saucer-shaped in cross section and oval in plan view. It is filled with 4500 m of sedimentary strata, which consist dominantly of carbonate rocks and evaporites, with subordinate shale and sandstone (Fisher et al., 1988; Catacosinos et al., 1990; Howell and van der Pluijm, 1990). These strata represent coastal to shallow-shelf depositional settings. Silurian evaporites and fringing pinnacle reefs represent one of the most unique set of facies in this basin and serve as major economic resources (see Fisher et al., 1988 and Catacosinos et al., 1900 for detailed reviews). Sedimentary strata renge in age from Cambrian through Carboniferous, and are capped with Jurassic red shale (Fig. 13.9). All six cratonic sequences of Sloss (1963, 1988a) occur in this basin.

The Michigan basin is underlain by an arm of the Midcontinent rift system, whose age is 1.1 Ga (Van Schmus and Hinze, 1985). The relationship between mechanical subsidence associated with this rifting and the onset of thermal subsidence from around 520 to 500 Ma (e.g., Sleep and Sloss, 1978; Nunn et al., 1984; Nunn, 1986; Ahern and Dikeou, 1989) remains a problem because of the 600 my interval between the two tectonic events. This problem was resolved recently by Howell (1989) and Howell and van der Pluijm (1990), who proposed that intracratonic-basin formation and subsidence, especially in the Michigan basin, were associated with reactivation of earlier rift systems caused by decoupling of the upper crust and upper mantle because of weakening of the lower crust. Howell's (1989) and Howell and van der Pluijm's (1990) model is supported by discovery of approximately 670 Ma thermal metamorphism (McCallister et al., 1978) of Keweenawan (1.1 Ga) basalts recovered by drilling from the Michigan basin depocenter, and by secondary magnetization of this basement following deposition of overlying sediments, which retained their primary magnetism (Van der Voo and Watts, 1978). It is suggested, therefore, that earliest Paleozoic tectonic reactivation of subsidence in the Michigan basin was controlled in part by late Proterozoic structural trends of the Midcontinent rift system (cf., DeBrito Neves et al., 1984; Leighton and Kolata, 1990). Thermal subsidence followed this

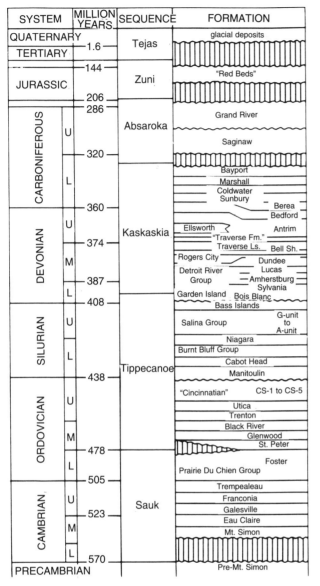

SYSTEM	MILLION YEARS	SEQUENCE	FORMATION
QUATERNARY		Tejas	glacial deposits
TERTIARY	1.6		
JURASSIC	144	Zuni	"Red Beds"
	206		
CARBONIFEROUS U	286	Absaroka	Grand River
			Saginaw
CARBONIFEROUS L	320		Bayport
		Kaskaskia	Marshall
	360		Coldwater
			Sunbury
DEVONIAN U			Berea
			Bedford
	374		Ellsworth Antrim
			"Traverse Fm."
DEVONIAN M			Traverse Ls. Bell Sh.
	387		Rogers City Dundee
			Detroit River Lucas
			Group Amherstburg
DEVONIAN L	408		Sylvania
			Garden Island Bois Blanc
			Bass Islands
SILURIAN U		Tippecanoe	Salina Group G-unit to A-unit
			Niagara
SILURIAN L			Burnt Bluff Group
			Cabot Head
	438		Manitoulin
ORDOVICIAN U			"Cincinnatian" CS-1 to CS-5
			Utica
			Trenton
ORDOVICIAN M			Black River
			Glenwood
	478		St. Peter
ORDOVICIAN L			Foster
	505		Prairie Du Chien Group
			Trempealeau
CAMBRIAN U		Sauk	Franconia
	523		Galesville
CAMBRIAN M			Eau Claire
			Mt. Simon
CAMBRIAN L	570		Pre-Mt. Simon
PRECAMBRIAN			

Fig. 13.9 Stratigraphic subdivision of Michigan basin. (Redrawn from Catacosinos et al., 1990.)

reactivation rifting; it began coevally with thermal subsidence in the Illinois and Williston basins (Fig. 13.8). This observation is consistent with the suggestion of Sloss (1988a) and Howell and van der Pluijm (1990) that around 500 Ma, the Michigan basin was a northern extension of the Reelfoot aulacogen underlying the Illinois basin; thus, thermal subsidence of the Michigan basin was linked to thermal subsidence in the Illinois basin.

The Michigan basin was characterized by renewed subsidence during the middle Devonian (Fig. 13.8),

and also, during the late Mississippian and Pennsylvanian (Middle and Late Carboniferous) due to far-field flexural subsidence in response to the Alleghenian-Hercynian orogeny (Quinlan and Beaumont, 1984; Beaumont et al., 1987; Leighton and Kolata, 1990). The flexural stress associated with that orogeny caused not only shallow and broad subsidence of the Appalachian basin, but also transmission of these stresses into the intracratonic interior, reactivating subsidence of intracratonic basins (cf., Ziegler, 1987). Howell and van der Pluijm (1990) also provided evidence suggesting that subsidence rates increased during Late Ordovician and Late Silurian to Early Devonian time because basin subsidence was renewed by reactivation along inherited crustal trends due to similar far-field influences during early and middle Paleozoic orogenic movements in the Appalachians (Taconic and Acadian orogenies, respectively).

Although petroleum is produced from the Michigan basin, source and reservoir units appear to be less favorable than in either the Illinois or the Williston basins. Principal source beds are the Precambrian lacustrine Nonesuch Shale, several Ordovician through Mississippian shales of anoxic origin (Nunn et al., 1984; Catacosinos et al., 1990), and possibly Silurian horizons. Reservoirs are confined to marine carbonate-platform beds of the Traverse Formation (Devonian) and Niagaran (Silurian) pinnacle reefs. Migration of petroleum appears to have been lateral (Catacosinos et al., 1990), although modeling studies by Nunn et al. (1984) suggest a vertical component of oil migration. Principal traps and seals appear to be carbonate mudstone enclosing pinnacle reefs, and unconformities and fault-bounded features associated with the Traverse Formation.

Other Intracratonic Basins

Amadeus Basin

The Amadeus basin of Australia is a late Proterozoic to early Paleozoic intracratonic basin (Fig. 13.10) that shows an evolutionary history comparable to other intracratonic basins. The basin is an elongate ellipse in plan, with dimensions of 800 x 300 km; strata

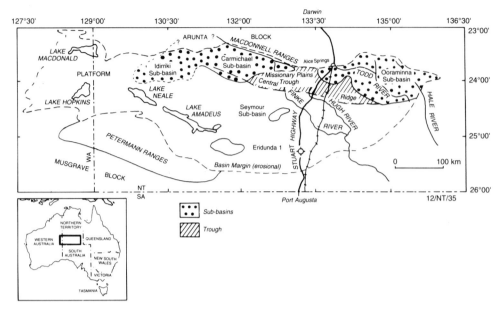

Fig. 13.10 Location map, Amadeus basin, central Australia. Peterman Ranges on southwest side of basin were principal source terrane for clastic sediments during basin filling. "Trough" refers to basin depocenter. (Redrawn after Lindsay and Korsch, 1989.)

reach a maximum thickness of 14 km. The discussion that follows is based on recent work by Lindsay (1987), Lindsay et al. (1987), Korsch and Lindsay (1989), and Shaw et al. (1991).

The stratigraphy of the basin (Fig. 13.11) is dominated by clastic rocks, mostly of coastal-plain, deltaic and continental-shelf facies, and some of tidal origin. Locally, salt deposits are known. Basal Upper Proterozoic units are mostly sandstone with interbedded basalt; they appear to represent a rift sequence. Overlying widespread strata contain quartz arenite, carbonate rocks, and evaporites. This sequence is continuous in the basin depocenter and overlaps basement; the boundary represents an equivalent to a breakup unconformity of a passive margin (see Chapter 4). This sequence is overlain unconformably by fluvial, glacial, and shelf facies. The basal marine onlapping beds of this third unit are arkosic and phosphatic; locally, a deep-water facies occurs at its base.

Latest Proterozoic and Cambrian sedimentation (Stage 2 in Fig. 13.11) occurred in major subbasins of the Amadeus basin. Sedimentation began with deposition of deltaic, coastal-plain and shelfal sandstone, followed by evaporites. The entire succession is capped by a widespread Ordovician to Silurian unconformity. Sequence 2 in the Amadeus basin is coeval with the Sauk Sequence (cf., Sloss, 1963, 1988a). It is overlain by marine Devonian strata throughout the basin; these are capped by a fluvial conglomerate.

Onlap curves were derived from field stratigraphic relations, and a relative sea-level curve was constructed (Fig. 13.11). Several regular cycles are evident. Drops in relative sea level were observed and found to correlate to subsidence (Fig. 13.12) and tectonic evolution of the basin. Lindsay and Korsch (1989) discriminated between larger-order cycles controlled by regional tectonics, and smaller-order cycles controlled by both climate and local tectonics. The climatic-eustatic sequences may have been influenced by global tectonic patterns (Lindsay and Korsch, 1989) during evolution of the Amadeus basin.

Subsidence analysis was completed in three locations, one from a deep well in the platform zone adjoining the basin, and the other two from within the basin (Fig. 13.12). The subsidence curves show an early history of crustal extension and thermal subsidence (Stages 1 and 2; Fig. 13.11) followed by regional thrusting (Stage 3). The subsidence history is similiar to that reported from the Illinois basin (Heidlauf et al., 1986; Treworgy et al., 1989; Kolata and Nelson, 1990b), the Michigan basin and the Williston basin (Howell and van der Pluijm, 1990; Bond and Kominz, 1991). Lambeck (1983a,b) and Shaw et al. (1991) argued, however, from model calculations and unconformity mapping, that the principal driving mechanisms for formation of the Amadeus basin

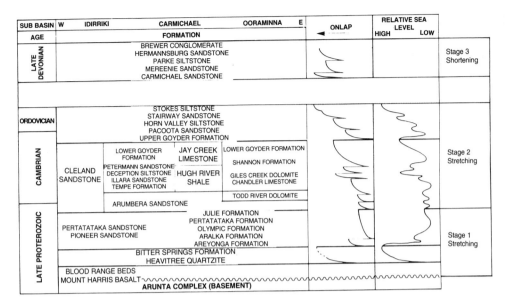

Fig. 13.11 Stratigraphic synthesis, onlap curves, interpreted relative sea level and crustal evolution, Amadeus basin, Australia. (Redrawn after Lindsay and Korsch, 1989.)

were regional intraplate stresses caused by compressional tectonic activity at continental margins (cf., also Leighton and Kolata, 1990), and not thermal subsidence. The presence of early extensional faults (Korsch and Lindsay, 1989; Lindsay and Korsch, 1989) argues against this interpretation. Devonian subsidence, associated with regional thrust faulting, may have been more widespread (also see, Quinlan and Beaumont, 1984; Shaw et al., 1991).

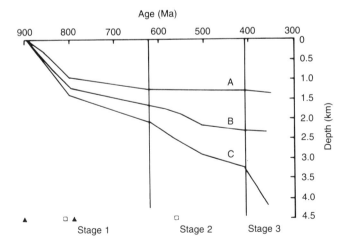

Fig. 13.12 Tectonic subsidence from (A) platform setting, (B) Missionary plains trough, and (C) Carmichael subbasin, Amadeus basin, Australia (see Fig. 13.10). (Redrawn after Lindsay and Korsch, 1989.)

Parana Basin

The intracratonic Parana basin incorporates an area of 1,400,000 sq. km in southern Brazil, Paraguay, northeast Argentina, and Uruguay (Figs. 13.13, 13.14). The discussion that follows is derived from DeBrito Neves et al. (1984), Stanley et al. (1985), and Zalan et al. (1985, 1990).

The basin is filled with sedimentary and volcanic strata nearly 6000 m thick; 1700 m of the basin section consist of Jurassic and Cretaceous volcanic rocks, which cover two-thirds of the basin's outcrop area.

Most of the basin sedimentary rocks are siliciclastic, and are arranged into five stratigraphic sequences bounded by unconformities (Fig. 13.15). The lower two sequences are equivalent to Sloss's (1963, 1988a) Tippecanoe sequence, whereas the remainder are equivalent to the Absaroka and Zuni sequences (also see, Soares et al., 1978). This dominance of clastics is attributed to the basin's paleoposition in polar latitudes (Zalan et al., 1990). The Paleozoic sequences represent transgressive-regressive cycles, whereas the Mesozoic sequences are all nonmarine. The Mesozoic also contains a thick (1700 m) volcanic succession, which appears to be the thickest volcanic sequence reported from an intracratonic basin (Zalan et al., 1990). These lavas range in age from 147 to 119 Ma (Zalan et al., 1990). Each of the sequences can be correlated to major regional tectonic events

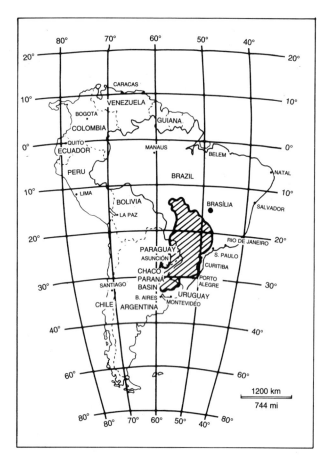

Fig. 13.13 Index map of Brazil, showing location of Parana basin. (Redrawn after Zalan et al., 1990.)

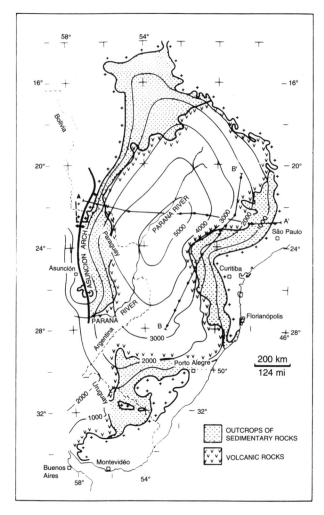

Fig. 13.14 Isopach map, incorporating both sedimentary and volcanic basin fill, Parana basin, Brazil. (Redrawn after Zalan et al., 1990.)

(Fig. 13.15). Eruption of the lavas was associated with rifting of South America from Africa.

The basin is criss-crossed by regional normal faults and lineaments, some of which are known to extend into basement (DeBrito Neves et al., 1984; Zalan et al., 1985, 1990). These are oriented primarily northeast-southwest, and northwest-southeast. All tend to follow older (Proterozoic) tectonic trends within the basin and probably were formed by propagation from and reactivation of basement faults. Reactivation of these basement-controlled features may have controlled formation and evolution of the Parma basin (deBrito Neves et al., 1984). Reactivation by extension facilitated recurring basin subsidence (Fig. 13.16). These subsidence events coincided with accumulation of sedimentary sequences during Silurian-Devonian, Permo-Carboniferous, and Late Jurassic-Early Cretaceous times.

Silurian-Devonian subsidence was thermal and was associated with a westward facing passive margin (Zalan et al., 1990). Permo-Carboniferous reactivation of subsidence was associated with rifting (Zalan et al., 1990). Jurassic-Cretaceous subsidence appears to have been caused entirely by volcanic loading. This subsidence accounts for the low slope at this time in the tectonic-subsidence curve (Fig. 13.16) and was critical for hydrocarbon formation, inasmuch as Permian source beds underwent thermal maturation during this volcanic loading event (Zalan et al., 1990). The presence of these thick volcanics contrasts with other intracratonic basins.

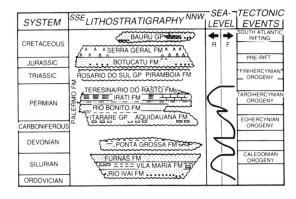

Fig 13.15 Chronostratigraphic diagram of Parana basin compared to regional tectonic events and to interpreted changes in sea level. (Redrawn after Zalan et al., 1990.)

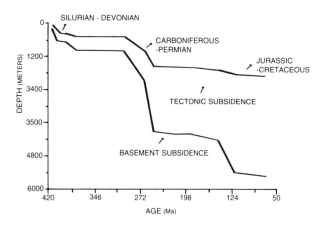

Fig 13.16 Tectonic and basement subsidence, north-central Parana basin, Brazil. (Redrawn after Zalan et al., 1990.)

No commercial petroleum exists in this basin, although drilling has found gas in the basin center and oil in the southern part of the basin. Tar sands and seeps are known in the eastern part of the basin. Two shale beds, the Permian Irati Formation and the Devonian Ponta Grossa Formation constitute the best source beds. Reservoirs are known from multiple pay zones, including Permian fluvial sand, and Triassic fluvial and eolian sands. A Silurian prospect exists in marine sandstones (Zalan et al., 1990). Vertical migration (both up and down) of hydrocarbons has occurred along faults and lineaments. Trapping geometries are poorly known, but include fault boundaries, and basaltic sills and dikes as boundaries (Zalan et al., 1990). The role of late Mesozoic igneous activity on limiting the petroleum potential of this basin remains unknown.

Mechanisms of Formation of Intracratonic Basins

The above summary of the geology and geodynamics of intracratonic basins demonstrates that the Illinois, Williston, Michigan, Amadeus, and Parana basins formed by a similar sequence of processes. These processes, in order of occurrence were: 1) lithospheric stretching, 2) mechanical, fault-controlled subsidence, 3) thermal subsidence and contraction, and 4) merging of slower thermal subsidence with reactivated subsi-

dence due to isostatically uncompensated excess mass (e.g., Sleep, 1976; Sleep and Snell, 1976; Sleep and Sloss, 1978; Sleep et al., 1980; DeRito et al., 1983; Ahern and Mrkvicka, 1984; Nunn et al., 1984; Ahern and Ditmars, 1985; Heidlauf et al., 1986; Klein and Hsui, 1987; Ahern and Dikeou, 1989). These events were followed by flexural foreland subsidence in some of the basins (e.g., Quinlan and Beaumont, 1984; Beaumont et al., 1987; Quinlan, 1987; Howell and van der Pluijm, 1990; Leighton and Kolata, 1990). Moreover, timing of initiation of thermal subsidence in the Illinois, Michigan, and Williston basins is remarkably close (Fig. 13.8; 530 to 500 Ma). Mesozoic and Cenozoic intracratonic basins of western Europe and India also share a common timing of initiation of thermal subsidence (see summary in Klein and Hsui, 1987).

To explain these basin-forming processes, concurrent initiation of thermal subsidence of North American Paleozoic intracratonic basins, and concurrent initiation of thermal subsidence of Mesozoic/Cenozoic intracratonic basins of Eurasia, several common characteristics of the basins must be considered. First and foremost are the cause and localization of rifting and reactivation of the basins (Ahern and Dikeou, 1989; Howell, 1989). Secondly, major Paleozoic intracratonic basins of North America, Europe, Africa, and South America appear to have been initiated during breakup of major supercontinents, either during the Latest Proterozoic/Early Cambrian (Bond et al., 1984;

Klein and Hsui, 1987) or during the Late Triassic/Early Jurassic (Ziegler, 1977, 1978, 1982, 1987, 1988). Thirdly, sedimentary sequences of intracratonic basins in North America, Africa, Asia, Europe, and South America show similar ages for interregional unconformities that separate intracratonic sedimentary sequences, and similar trends in thickness and volume (Sloss, 1963, 1972, 1976, 1979; Sloss and Scherer, 1975; Soares et al., 1978; Peters, 1979; DeBrito Neves et al., 1984; Zalan et al., 1990). These observations suggest a common global explanation, and the near synchroneity of the ages of formation of both North American (earliest Paleozoic) and Eurasian (early Mesozoic) intracratonic basins implies a large-scale process that caused both rifting and thermal subsidence.

Since 2 Ga, four major episodes of supercontinent accretion have been followed by supercontinent breakup (Condie, 1982b; Bond et al., 1984; Worsley et al., 1984, 1986). During periods when continents were dispersed, heat loss from the earth's interior was dissipated through mid-ocean ridges (Fischer, 1984; Worsley et al., 1984), whereas during times when continents were merged into a supercontinent, ridge volcanism and associated heat loss diminished considerably, partially beacuse spreading rates may have been reduced as well (Fischer, 1984; Worsley et al., 1984). Consequently, supercontinents may have acted as heat lenses (Anderson, 1982; Worsley et al., 1984, 1986). As a consequence of developing supercontinents, the rate of partial melting of continental lithosphere would have increased in response to focused heat flow, particularly in the lower crust and upper mantle. Breakup of a supercontinent into smaller plates would have changed the heat-flow regime to one more like the present. This cyclic repetition would also have been influenced by the changing convective history of the mantle, as plates shifted from separate geoid highs to merge over geoid lows (Gurnis, 1988, 1990a,b).

Research concerning the origin and nature of Proterozoic granite of the North American craton has documented two contrasting modes of granitic crustal formation (e.g., Anderson and Cullers, 1978; Emslie, 1978; Anderson, 1980, 1983; Anderson et al., 1980). During plate convergence and associated mountain building, crustal formation included emplacement of orogenic granite. Intrusion of anorogenic granite postdated these orogenic intrusive events by up to several hundred-million years (Anderson and Cullers, 1978; Emslie, 1978; Anderson 1980, 1983; Anderson et al., 1980). Extensive chemical petrologic analysis of anorogenic granite from several continents demonstrates that they formed by partial melting of older orogenic granitic crust (Anderson and Cullers, 1978; Emslie, 1978; Anderson et al., 1980; Nurmi and Haapala, 1986). Continent-wide intrusion of anorogenic granite (e.g., 1.5 to 1.4 Ga) may have been associated with supercontinent development and regional extension (Silver et al., 1977; Anderson and Cullers, 1978; Emslie, 1978; Anderson, 1983; Bickford et al., 1986; Nurmi and Haapala, 1986; Haapala and Ramo, 1987). The simultaneous partial melting of such scattered coeval anorogenic granite requires a common mechanism (Fig. 13.17A), postulated to be the buildup of mantle-derived heat below a supercontinent (Anderson, 1982; Worsley et al., 1984; Anderson, 1987). Alternatively, coeval development of a mafic igneous body in the upper mantle could also produce partial melting in the lower crust (Emslie, 1978; Anderson, 1983; Nurmi and Haapala, 1986).

One consequence of partial melting is that the remainder of the crust must have been thinned (cf., Howell, 1989; Howell and van der Pluijm, 1990). Intrusion of anorogenic granite on a supercontinent scale would have created major lateral physical discontinuities in properties of the continental crust. When stress fields changed from compressional to extensional, or vice versa, or in direction and magnitude, major ruptures and associated surficial deformation would be expected on supercontinents, especially at crustal inhomogeneities. Such changes in stress also would have favored the reactivation of basin subsidence (DeRito et al., 1983; Lambeck, 1983a,b; DeBrito Neves et al., 1984; Quinlan, 1987; Howell, 1989; Howell and van der Pluijm, 1990; Leighton and Kolata, 1990; Zalan et al., 1990; Shaw et al., 1991). Extension may have focused on zones within the lithosphere where anorogenic granite intruded (Anderson and Cullers, 1978; Emslie, 1978; Nurmi and Haapala, 1986; Haapala and Ramo, 1987), thus causing both thermal doming and weakening of

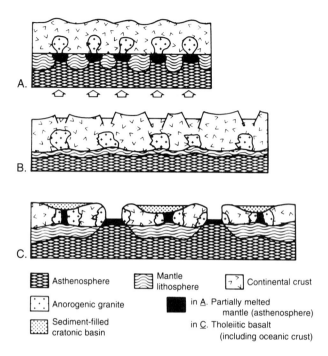

Fig 13.17 Evolution and simultaneous formation of intracratonic basins as crust of supercontinent is partially melted and anorogenic granite intrudes crust during stage when supercontinent acted as heat lens (A), followed by rifting and associated fault-controlled, mechanical subsidence (B). These rift zones underwent thermal subsidence, forming saucer-shaped intracratonic basins as anorogenic bodies cooled (C). Consequently, intracratonic basins occur above anorogenic-granite bodies. This process accounts for nearly simultaneous initiation of intracratonic basins and their global regional unconformities, parallel tectonic-subsidence histories, and parallel changing sediment volumes. Schematic representation superposed, in part, on continental-heat-lens and breakup model of Worsley et al. (1984, Fig. 8).

crust (Fig. 13.17B). Crust immediately above partially melted zones would have been thinned, thus favoring localized and regional stretching. This configuration would have accounted for stretching within the crust above areas of anorogenic granite intrusion and partial melting during supercontinent breakup. Because of increased viscosity of these granite magmas, the partially melted bodies should have split, separated, and accelerated uplift on the flanks of the rifted basin (Neugebauer and Reuther, 1987) during continued extension.

With continued stretching, the magma bodies themselves may have fractured; such fractures would have become conduits for subsequent intrusion of upwelled igneous materials derived from the upper mantle (Fig. 13.17C). It would be expected that asso-

ciated rift valleys that formed in response to stretching associated with supercontinent breakup would be floored by anorogenic granite intruded by tholeiitic basalt derived from the upper mantle. Such an association of anorogenic granite and tholeiitic basalt of approximately coeval age (577 to 500 Ma) has been documented in the southern Oklahoma aulacogen (e.g., Gilbert, 1983; Lanbert et al., 1988). Several younger Phanerozoic examples of rift-related, paired, anorogenic granite and tholeiitic basalt are known, including the Jurassic-Cretaceous White Mountain magma series of New Hampshire, Jurassic granite of Nigeria, and Miocene-Pliocene volcanics of the Aden Sea region, the latter being involved with relatively recent breakup and rifting of Africa and the Arabian Peninsula (Anderson and Cullers, 1978, p. 311; Emslie, 1978). In the Precambrian of south Africa, Cooper (1990) documented several cyclic recurrences of paired anorogenic granite and tholeiitic basalt associated with earliest stages of Precambrian intracratonic basin evolution. A similar succession of intracratonic basin development after intrusion of anorogenic granite has been reported from the Precambrian of Finland by Gaal and Gorbatschev (1987).

Three North American intracratonic basins, the Illinois, Williston, and Michigan basins, formed following rifting (Howell, 1989; Gerhard et al., 1990) during a period of Late Precambrian breakup of a supercontinent, which was followed by thermal basin subsidence (Sleep and Sloss, 1978; Braille et al., 1982; Keller et al., 1983; Ahern and Mrkvicka, 1984; Fowler and Nisbet, 1984; Nunn et al., 1984; Ahern and Ditmars, 1985; Heidlauf et al., 1986; Ahern and Dikeou, 1989). Thus, following Klein and Hsui (1987), the location of intracratonic basins was controlled by rifting above partial melting of lower crust and intrusion of anorogenic granite during Late Precambrian breakup of a supercontinent (Fig. 13.17C; cf., also Cooper, 1990). Intracratonic basins display both a common time of formation, and synchroneity of interregional unconformities and sediment volume changes. Crustal zones which were domed upward above anorogenic granite are susceptible to rupture by rifting, which is why intracratonic basins occur in these zones. Cooling of anorogenic granite, basalt, and asthenosphere to form lithosphere caused subsequent thermal subsidence.

476   Chapter 13

Bally (1989) and Sloss (1988b) objected to this model on several grounds. Bally (1989) rejected the concept of supercontinent accretion and breakup cycles, and questioned the existence of anorogenic granite bodies below intracratonic basins (also see Sloss, 1988b). Moreover, Bally (1989) argued that subsidence histories of intracratonic basins are independent, and cited Quinlan's (1987) discussion of subsidence history of the Hudson Bay basin, which clearly subsided thermally at least 100 my after the other three North American basins. Both Sloss (1988b) and Bally (1989) also called attention to a 600my delay between Keweenawan rifting below the Michigan basin and early Paleozoic thermal subsidence. They failed to consider arguments concerning reactivation of such rift systems (Quinlan, 1987), thermal metamorphism of the Keweenawan basalts in the depocenter (McCallister et al., 1978), and secondary magnetization of the basalts in the depocenter (Van der Voo and Watts, 1978), and were unaware of subsequent work establishing reactivation in that basin by Howell (1989). Moreover, Bond and Kominz (1991) demonstrated that subsidence histories of the Michigan, Williston, and Illinois basins (Fig. 13.12) are nearly synchronous, with minor deviations due to differential subsidence rates in each basin (also see Sloss, 1988a).

It should be remembered that supercontinent breakup is not an instantaneous process. The ages of rifting associated with late Proterozoic and early Paleozoic aulacogens and interior rifts differ as one moves away from continental margins where initial breakup occurred. Klein and Willard (1989, Fig. 2) compared subsidence histories of the Illinois and central Appalachian basins, and showed, as did Bond et al., (1984), that the late Proterozoic-early Paleozoic passive margin of eastern North America rifted around 620 Ma (Willard and Klein, 1990), whereas rifting below the Illinois basin started around 530 Ma. It is my interpretation that as a supercontinent breaks up, the areas of greatest heat flow migrate away from supercontinents both to mid-ocean ridges (also see Sloss and Speed, 1974) and toward interiors of remaining continents. It is, for instance, well known that the breakup of Pangea occurred in successive stages over a period in excess of 120 my (e.g., Watts

et al., 1982; Ziegler, 1982, 1988, among many others), and some of these areas of later breakup contained anorogenic granite and tholeiitic basalt (Anderson and Cullers, 1978). Therefore, later thermal subsidence of the most interior intracratonic basins would be expected. This change in timing may account also for evidence of localized differential subsidence reported by Sloss (1988a).

The proposed basin-forming mechanism for the origin of intracratonic basins poses two consequences. First, below older Precambrian intracratonic basins, where the root zones are exposed, anorogenic granite should occur. Gaal and Gorbatschev (1987) demonstrated that in the Jotnian intracratonic basin of southwestern Finland, the basal Jotnian Sandstone overlies anorogenic granite (Laitakari, 1983; Nurmi and Haapala, 1986). In an unnamed intracratonic basin in northeastern central Finland along the Gulf of Bothnia, the basal arkosic Muhos Formation overlies anorogenic granite (Veltheim, 1969; Paakola, 1983). Recent seismic surveys have demonstrated that the crust is thin below the anorogenic granite (Gaal and Tuomi, in press). The basement-sediment contact in the Williston basin (Kent, 1987) shows anorogenic granite below part of the basal Phanerozoic succession. In southern Africa, Cooper (1990) reported paired anorogenic granite and tholeiitic basalt below successive Precambrian intracratonic basins.

A second consequence of this hypothesis is that formation of intracratonic basins follows major collisional orogenic events. Both examples from Finland are postorogenic basins (Nurmi and Haapala, 1986). Intracratonic basins were observed to have developed in India starting during latest Permian time following the breakup of Gondwana (Kailasham, 1976). The Paris basin and the Viking graben of the North Sea show subsidence patterns that are similar to those of North American Paleozoic intracratonic basins; thermal subsidence of the former started during latest Permian and Triassic time (Sclater and Christie, 1980; Brunet and LePichon, 1982; Ziegler, 1982, 1988; Perrodon and Zabek, 1990). These findings support the interpretation advanced in this chapter.

Although the stages of subsidence of intracratonic basins are sequential, overlapping of subsidence mechanisms has occurred (Fig. 13.18). In the Illinois

basin, Davis (1987), Clendenin (1989), and Watso and Klein (1989) recognized that Cambrian and Ordovician normal faulting occurred when thermal subsidence was dominant. Following a 60my period of rapid thermal subsidence, a period of slower thermal subsidence overlapped a time of increasing subsidence due to isostatically uncompensated excess mass (cf., DeRito, et al., 1983; Treworgy et al., 1989; Kolata and Nelson, 1990). This slow thermal subsidence was associated with Devonian faulting (Davis, 1987). Late phases of subsidence are documented in the three US Paleozoic intracratonic basins (Illinois, Williston, and Michigan) in response to foreland tectonics and lateral transmission of intraplate stress during times of orogenesis along continental margins of North America (Quinlan and Beaumont, 1984; Beaumont et al., 1987; also see Craddock and van der Pluijm, 1989; Klein and Willard, 1989; Willard and Klein, 1990). During times of orogenesis, partial melting may occur locally, giving rise to igneous intrusions, such as the late Paleozoic alnoites reported from the Illinois basin (Zartman et al., 1967). Such localized partial melting may have caused shifting of depocenters, as reported from the Illinois, Williston, and Michigan basins by Heidlauf et al. (1986) and Sloss (1987). Partial melting is likely to be associated with reactivation of older rifts

and tectonic belts beneath intracratonic basins, such as those reported by DeBrito Neves et al. (1984), Quinlan (1987), Howell (1989), Howell and van der Pluijm (1990), and Leighton and Kolata (1990).

Finally, it must be emphasized that this model of basin formation of intracratonic basins requires further testing. Since the original proposal of the model (Klein and Hsui, 1987), additional data add plausability; recurring cycles of intrusions of anorogenic granite preceding intracratonic-basin evolution in southern Africa (Cooper, 1990), and the importance and verification of reactivation along older tectonic trends have been demonstrated (DeBrito Neves et al., 1984; Quinlan, 1987; Howell, 1989; Howell and van der Pluijm, 1990; Leighton and Kolata, 1990). The major difficulty with the hypothesis is that Phanerozoic intracratonic basins have not been drilled to basement and inadequate sampling leaves the problem unresolved. I predict that if ultradeep scientific drilling is undertaken in the Illinois basin, for instance, then anorogenic granite ranging in age from 600 to 510 Ma should be recovered, below the layered crustal structure reported by Pratt et al. (1989). The proposed model for intracratonic basins accounts for all the common properties and features of intracratonic basins.

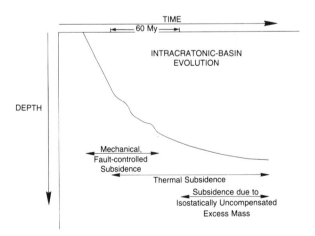

Fig. 13.18 Generalized backstripped and decompacted tectonic subsidence, showing sequential stages of subsidence processes for intracratonic basins.

Acknowledgments

My research in cratonic basins has been supported by various sources, including the National Science Foundation (EAR-90-01448), the University of Illinois Research Board, and the Department of Geology at the University of Illinois. Discussions with J.J. Eidel, L.C. Gerhard, P.D. Howell, A.T. Hsui, D.R. Kolata, M.W. Leighton, W.J. Nelson and B.J. van der Pluijm have proved helpful. In particular, I thank M.W. Leighton for making arrangements for me to obtain preprints of chapters in AAPG Memoir 51. Michael R. Leeder, Jeffrey A. Nunn, and the editors, Cathy Busby and Raymond V. Ingersoll, are thanked for their constructive comments on earlier manuscript versions of this chapter.

Further Reading

Bond GC, Kominz MA, 1991, Disentangling middle Paleozoic sea level and tectonic events in cratonic margins and cratonic basins of North America: *Journal of Geophysical Research*, v. 94, p. 6619–6639.

Hamdani Y, Mareschal JC, Arkani-Hamed J, 1991, Phase changes and thermal subsidence in intracontinental sedimentary basins: *Geophysical Journal International*, v. 106, p. 657–665.

Heidlauf DT, Hsui AT, Klein GdeV, 1986, Tectonic subsidence analysis of the Illinois basin: *Journal of Geology*, v. 94, p. 779–794.

Klein, GdeV, 1991, Origin and evolution of North American cratonic basins: *South African Journal of Geology*, v.94, p. 3–18.

Leighton MW, Kolata DR, Oltz DF, Eidel JJ (eds), 1990, *Interior cratonic basins*: American Association of Petroleum Geologists Memoir 51, 819p.

Sleep NH, Nunn JA, Chou L, 1980, Platform basins: *Annual Review of Earth and Planetary Sciences*, v. 8. p. 17–34.

Sloss LL, 1988a, Tectonic evolution of the craton in Phanerozoic time, in Sloss LL (ed), *Sedimentary cover - North American craton*: U.S. [The Geology of North America, v. D-2]: Geological Society of America, Boulder, p. 25–51.

References

Aalto KR, 1976, Sedimentology of a melange: Franciscan of Trinidad, California: Journal of Sedimentary Petrology, v. 46, p. 913–929.

Aalto KR, 1982, The Franciscan Complex of northernmost California: sedimentation and tectonics: Geological Society of London Special Publication 10, p. 419–432.

Aalto KR, 1989, Franciscan Complex olistostrome at Crescent City, northern California: Sedimentology, v. 36, p. 471–495.

Abbott DH, Menke W, Hobart MA, Anderson RN, 1981, Evidence for excess pore pressures in southwestern Indian Ocean sediments: Journal of Geophysical Research, v. 86, p. 1813–1827.

Abbott DH, Menke W, Morin R, 1983, Constraints upon water advection in sediments of the Mariana Trough: Journal of Geophysical Research, v. 88, p. 1075–1093.

Adams BF, Busby CJ, 1994, Stratigraphic evolution of a Cretaceous oceanic arc terrane, Baja California, Mexico: Abstracts of the 14th International Association of Sedimentologists Congress, Recife, p. F-1.

Ahern JL, Dikeou PJ, 1989, Evolution of the lithosphere beneath the Michigan basin: Earth and Planetary Science Letters, v. 95, p. 73–84.

Ahern JL, Ditmars RC, 1985, Rejuvenation of continental lithosphere beneath an intracratonic basin: Tectonophysics, v. 120, p. 21–35.

Ahern JL, Mrkvicka SR, 1984, A mechanical and thermal model for the evolution of the Williston basin: Tectonics, v. 3, p. 79–102.

Ahorner L, 1970, Seismo-tectonic relations between the graben zones of the Upper and Lower Rhine Valley, in Illies JM, Mueller S (eds.), Graben problems: E. Schweizerbart'sche Verlagsbuchandlung, Suttgart, p. 155–166.

Ahorner L, Schneider G, 1974, Herdmechanismen von Erdbeben im Oberrhein Graben und in seinen Randgebirgen, in Illies JM, Fuchs K (eds.), Approaches to taphrogenesis: E. Schweizerbart'sche Verlagsbuchandlung, Suttgart, p. 104–117.

Ahr WM, 1973, The carbonate ramp: an alternative to the shelf model: Transactions of the Gulf Coast Association of Geological Societies, v. 23, p. 221–225.

Alam M, 1989, Geology and depositional history of Cenozoic sediments of the Bengal basin of Bangladesh: Palaeogeography, Palaeoclimatology, Palaeoecology, v. 69, p. 125–139.

Albers JP, Bain JHC, 1985, Regional setting and new information on some critical geologic features of the west Shasta district, California: Economic Geology, v. 80, p. 2072–2091.

Aldiss DT, Ghazali SA, 1984, The regional geology and evolu-

tion of the Toba volcano-tectonic depression, Indonesia: Journal of the Geological Society of London, v. 141, p. 487–500.

Aleinikov AL, 8 coauthors, 1980, Dynamics of the Russian and west Siberian platforms: American Geophysical Union and Geological Society of America Geodynamics Series 1, p. 53–71.

Alexander JA, Leeder MR, 1987, Active tectonic control of alluvial architecture: Society of Economic Paleontologists and Mineralogists Special Publication 39, p. 243–252.

Alexander JA, Bridge JS, Leeder MR, Collier REL, Gawthorpe RL, 1994, Holocene meander-belt evolution in an active extensional basin, Southwestern Montana: Journal of Sedimentary Research, v. B64, p. 542–559

Alexandrovich JM, 1992, Radiolarians from Sites 794, 795, 796, and 797 (Japan Sea): Proceedings of the Ocean Drilling Program, Scientific Results, v. 127/128, Pt. 1, p. 291–307.

Allan JF, Gorton MP, 1992, Geochemistry of igneous rocks from Legs 127 and 128, Sea of Japan: Proceedings of the Ocean Drilling Program, Scientific Results, v. 127/128, Pt. 2, p. 905–929.

Allen JRL, 1964, Studies in fluviatile sedimentation: six cyclothems from the Lower Old Red Sandstone, Anglo-Welsh basin: Sedimentology, v. 3, p. 163–198.

Allen JRL, 1970, Studies in fluviatile sedimentation: a comparison of fining-upward cyclothems, with special reference to coarse-member composition and interpretation: Journal of Sedimentary Petrology, v. 40, p. 298–323.

Allen JRL, 1985, Principles of physical sedimentology [Chapter 12]: George Allen and Unwin, London, p. 223–242.

Allen PA, Allen JR, 1990, Basin analysis principles and applications: Blackwell Scientific, London, 451 p.

Allen PA, Homewood P (eds.), 1986, Foreland basins: International Association of Sedimentologists Special Publication 8, 453 p.

Allen PA, Homewood P, Williams GD, 1986, Foreland basins: an introduction: International Association of Sedimentologists Special Publication 8, p. 3–12.

Allen PA, Crampton SL, Sinclair HD, 1991, The inception and early evolution of the North Alpine foreland basin, Switzerland: Basin Research, v. 3, p. 143–163.

Allmendinger RW, 1992, Fold and thrust tectonics of the western United States exclusive of the accreted terrains, in Burchfiel BC, Lipman PW, Zoback ML (eds.), The Cordilleran orogen: conterminous U.S. [The geology of North America, v. G-3]: Geological Society of America, Boulder, p. 583–607.

Allmendinger RW, Ramos VA, Jordan TE, Palma M, Isacks BL, 1983a, Paleogeography and Andean structural geometry, north-

west Argentina: Tectonics, v. 2, p. 1–16.

Allmendinger RW, 7 coauthors, 1983b, Cenozoic and Mesozoic structure of the eastern Basin and Range province, Utah, from COCORP seismic-reflection data: Geology, v. 11, p. 532–536.

Allmendinger RW, Miller DM, Jordan TE, 1984, Known and inferred Mesozoic deformation in the hinterland of the Sevier belt, northwest Utah, in Kerns GJ, Kerns RL (eds.), Geology of northwest Utah, southern Idaho and northeast Nevada: Utah Geological Association Publication 13, p. 21–34.

Allmendinger RW, 6 coauthors, 1986, Phanerozoic tectonics of the Basin and Range—Colorado Plateau transition from COCORP data and geologic data: a review: American Geophysical Union and Geological Society of America Geodynamics Series 14, p. 257–267.

Allmendinger RW, 5 coauthors, 1987, Deep seismic reflection characteristics of the continental crust: Geology, v. 15, p. 304–310.

Allmendinger RW, 5 coauthors, 1990, Foreland shortening and crustal balancing in the Andes at 30°S Latitude: Tectonics, v. 9, p. 789–809.

Almgren AA, 1978, Timing of Tertiary submarine canyons and marine cycles of deposition in the southern Sacramento Valley, California, in Stanley DJ, Kelling G (eds.), Sedimentation in submarine fans, canyons, and trenches: Dowden, Hutchinson and Ross, Stroudsburg, p. 276–291.

Almogi-Labin A, Hemleben C, Meischner D, Erlenkeuser H, 1991, Palaeoenvironmental events during the last 13000 years in the Central Red Sea as recorded by pteropoda: Paleooceanography, v. 6, p. 83–98.

Alonso RN, Jordan TE, Tabbutt KT, Vandervoon D, 1991, Giant evaporite belts of the Neogene central Andes: Geology, v. 19, p. 401–404.

Ammon WL, 1981, Geology and plate tectonic history of the Marfa basin, Presidio County, Texas, in Pearson BT (ed.), Marathon-Marfa region of west Texas [Symposium and Guidebook]: Permian Basin Section, Society of Economic Paleontologists and Mineralogists, Midland, p. 75–101.

Anadon P, Cabrera L, Colombo F, Marzo M, Riba O, 1986, Syntectonic intraformational unconformities in alluvial fan deposits, eastern Ebro basin margins (NE Spain): International Association of Sedimentologists Special Publication 8, p. 259–271.

Andersen DW, Rymer MJ (eds.), 1983, Tectonics and sedimentation along faults of the San Andreas System: Pacific Section, Society of Economic Paleontologists and Mineralogists, Los Angeles, 110 p.

Anderson DL, 1982, Hotspots, polar wander, Mesozoic convection and the geoid: Nature, v. 297, p. 391–393.

Anderson DL, 1994, Superplumes or supercontinents?: Geology, v. 22, p. 39–42.

Anderson EM, 1951, The dynamics of faulting (2nd edition): Oliver and Boyd, Edinburgh, 206 p.

Anderson JL, 1980, Mineral equilibria and crystallization conditions in the late Precambrian Wolf River rapakivi massif, Wisconsin: American Journal of Science, v. 280, p. 289–332.

Anderson JL, 1983, Proterozoic anorogenic granite plutonism of North America: Geological Society of America Memoir 161, p. 134–154.

Anderson JL, 1989, Proterozoic anorogenic granites of the southwestern United States: Arizona Geological Society Digest, v. 17, p. 211–238.

Anderson JL, Cullers RL, 1978, Geochemistry and evolution of the Wolf River batholith, a late Precambrian rapakivi massif in north Wisconsin: Precambrian Research, v. 7, p. 287–324.

Anderson JL, Cullers RL, VanSchmus WR, 1980, Anorogenic metaluminous and peraluminous granite plutonism in the Mid-Proterozoic of Wisconsin, US: Contributions to Mineralogy and Petrology, v. 74, p. 311–328.

Anderson RE, 1989, Tectonic evolution of the intermontane system, Basin and Range, Colorado Plateau, and High Lava Plains: Geological Society of America Memoir 172, p. 163–177.

Andrews-Speed CP, 1980, The geology of central Isla Hoste, southern Chile: sedimentation, magmatism and tectonics in part of a Mesozoic back-arc basin: Geological Magazine, v. 117, p. 339–349.

Angelier J, 1985, Extension and rifting: the Zeit region, Gulf of Suez: Journal of Structural Geology, v. 7, p. 605–612.

Apperson KD, 1991, Stress fields of the overriding plate at convergent margins and beneath active volcanic arcs: Science, v. 254, p. 670–678.

Aramaki S, 1984, Formation of the Aira caldera, southern Kyushu, ~22,000 years ago: Journal of Geophysical Research, v. 89, p. 8485–8501.

Arbenz JK, 1989, The Ouachita system, in Bally AW, Palmer AR (eds.), The geology of North America: an overview [The Geology of North America, v. A]: Geological Society of America, Boulder, p. 371–396.

Armijo R, Tapponnier P, Mercier JL, Man TL, 1986, Quaternary extension in southern Tibet: field observations and tectonic implications: Journal of Geophysical Research, v. 91, p. 13,803–13,872.

Armstrong FC, Oriel SS, 1965, Tectonic development of Idaho-Wyoming thrust belt: American Association of Petroleum Geologists Bulletin, v. 49, p. 1847–1866.

Armstrong RL, 1974, Magmatism, orogenic timing and orogenic diachronism in the Cordillera from Mexico to Canada: Nature, v. 247, p. 348–351.

Armstrong RL, 1982, Cordilleran metamorphic core complexes: from Arizona to southern Canada: Annual Review of Earth and Planetary Sciences, v. 10, p. 129–154.

Arthaud F, Matte P, 1977, Late Paleozoic strike-slip faulting in southern Europe and northern Africa: result of a right-lateral shear zone between the Appalachians and the Urals: Geological Society of America Bulletin, v. 88, p. 1305–1320.

Arthur MA, von Huene R, Adelseck CG, 1980, Sedimentary evolution of the Japan fore-arc region off northern Honshu, Legs 56 and 57, Deep Sea Drilling Project: Initial Reports of the Deep Sea Drilling Project, v. 56/57, Part 1, p. 521–568.

Artyushkov EV, 1973, Stresses in the lithosphere caused by crustal thickness inhomogeneities: Journal of Geophysical Research, v. 78, p. 7675–7708.

Ashley GM, Wellner RW, Esker D, Sheridan RE, 1991, Clastic sequences developed during late Quaternary glacio-eustatic sea-level fluctuations on a passive margin: example from the inner continental shelf near Barnegat Inlet, New Jersey: Geological Society of America Bulletin, v. 103, p. 1607–1621.

Asmus HE, Ponte FC, 1973, The Brazilian marginal basins, in Nairn AEM, Stehli FG (eds.), The ocean basins and margins: Plenum Press, New York, p. 87–133.

Asquith GO, 1970, Depositional topography and major marine environments, Late Cretaceous, Wyoming: American Association of Petroleum Geologists Bulletin, v. 54, p. 1184–1224.

Athy LF, 1930, Density, porosity, and compaction of sedimentary rocks: American Association of Petroleum Geologists Bulletin, v. 14, p. 1–35.

Aubouin J, 1965, Geosynclines [Developments in Geotectonics 1]: Elsevier, New York, 335 p.

Aubouin J, 1989, Some aspects of the tectonics of subduction zones: Tectonophysics, v. 160, p. 1–21.

Aubouin J, 1990, The west Pacific geodynamic model: Tectonophysics, v. 183, p. 1–7.

Aubouin J, Stephan J F, Roump J, Renard V, 1982a, The Middle America Trench as an example of a subduction zone: Tectonophysics, v. 86, p. 113–132.

Aubouin J, 14 coauthors, 1982b, Initial reports of the Deep Sea Drilling Project [v. 67]: United States Government Printing Office, Washington, 799 p.

Aubouin J, Bourgois J, Azema J, 1985a, Two types of active margins: convergent-compressional margins and convergent-extensional margins, in Nasu N, Uyeda S, Kushiro, I, Kobayashi K, Kagami H (eds.), Formation of active ocean margins: Terra Scientific Publishing, Tokyo, p. 109–129.

Aubouin J, Bourgois J, Azema J, von Huene R, 1985b, Guatemala margin: model of convergent extensional margin: Initial Reports of the Deep Sea Drilling Project, v. 84, p. 911–917.

Audley-Charles MG, 1986, Timor-Tanimbar Trough: the foreland basin of the evolving Banda orogen: International Association of Sedimentologists Special Publication 8, p. 91–102.

Audley-Charles MG, Curray JR, Evans G, 1977, Location of major deltas: Geology, v. 5, p. 341–344.

Avila-Salinas W, 1992, Petrologic and tectonic evolution of the Cenozoic volcanism in the Bolivian western Andes: Geological Society of America Special Paper 265, p. 245–257.

Axen GJ, Taylor WJ, Bartley JM, 1993, Space-time patterns and tectonic controls of Tertiary extension and magmatism in the Great Basin of the western United States: Geological Society of America Bulletin, v. 105, p. 56–76.

Aydin A, Nur A, 1982, Evolution of pull-apart basins and their scale independence: Tectonics, v. 1, p. 91–105.

Aydin A, Nur A, 1985, The types and roles of stepovers in strike-slip tectonics: Society of Economic Paleontologists and Mineralogists Special Publication 37, p. 35–44.

Bachman SB, 1982, The Coastal belt of the Franciscan: youngest phase of northern California subduction: Geological Society of London Special Publication 10, p. 401–418.

Bachman SB, Lewis SD, Schweller WJ, 1983, Evolution of a forearc basin, Luzon Central Valley, Philippines: American Association of Petroleum Geologists Bulletin, v. 67, p. 1143–1162.

Badley ME, Price JD, Rambach Dahl D, Agdestein T, 1988, The structural evolution of the northern Viking graben and its bearing upon extensional models of basin formation: Journal of the Geological Society of London, v. 145, p. 455–472.

Bahlburg H, Brietkreuz C, 1991, Paleozoic evolution of active margin basins in the southern central Andes (northwestern Argentina and northern Chile): Journal of South American Earth Science, v. 4, p. 171–188.

Bailey EH, Everhart DL, 1964, Geology and quicksilver deposits of the New Almaden district, Santa Clara County, California: United States Geological Survey Professional Paper 360, p. 1–206.

Bailey EH, Blake MC, Jr, Jones DL, 1970, On-land Mesozoic oceanic crust in California Coast Ranges: United States Geological Survey Professional Paper 700-C, p. C70–C81.

Baldridge WS, Olsen KH, Callender JF, 1984, Rio Grande rift: problems and perspectives: New Mexico Geological Society Guidebook 35, p.1–12.

Baldridge WS, Eyal Y, Bartov Y, Steinitz G, Eyal M, 1991, Miocene magmatism of Sinai related to the opening of the Red Sea: Tectonophysics, v. 197, p. 181–202.

Baldwin SL, Harrison TM, Burke K, 1986, Fission track evidence for the source of accreted sandstones, Barbados: Tectonics, v. 5, p. 457–468.

Ball MM, 8 coauthors, 1985, Seismic structure and stratigraphy of northern edge of Bahaman-Cuban collision zone: American Association of Petroleum Geologists Bulletin, v. 69, p. 1275–1294.

Ballance PF, 1976, Evolution of the upper Cenozoic magmatic arc and plate boundary in northern New Zealand: Earth and Planetary Science Letters, v. 28, p. 356–370.

Ballance PF, 1991, Gravity flows and rock recycling on the Tonga landward trench slope: relation to trench–slope tectonic processes: Journal of Geology, v. 99, p. 817–828.

Ballance PF, Gregory MR, 1991, Parnell Grits—large subaqueous volcaniclastic gravity flows with multiple particle-support mechanisms: Society for Sedimentary Geology Special Publication 45, p. 189–200.

Ballance PF, Reading HG (eds.), 1980, Sedimentation in oblique–slip mobile zones: International Association of Sedimentologists Special Publication 4, 265 p.

Ballance PF, 5 coauthors, 1989, Subduction of a Late Cretaceous seamount of the Louisville Ridge at the Tonga Trench: a model of normal and accelerated tectonic erosion: Tectonics, v. 8, p. 953–962.

Ballesteros MW, Moore GF, Taylor B, Ruppert S, 1988, Seismic stratigraphic framework of the Lima and Yaquina forearc basins, Peru: Proceedings of the Ocean Drilling Program, v. 112, p. 77–90.

Bally AW, 1975, A geodynamic scenario for hydrocarbon occurrences: Proceedings of the 9th World Petroleum Congress, Tokyo [v. 2 (Geology)]: Applied Science Publishers, Essex, p. 33–44.

Bally AW (chair), 1979, Continental margins, geological and geophysical research needs and problems [National Research Council Report]: National Academy of Sciences, Washington, 302 p.

Bally AW, 1984, Tectogenese et sismique reflexion: Bulletin Societe Geologique France, v. 26, p. 279–285.

Bally AW, 1989, Phanerozoic basins of North America, in Bally AW, Palmer AR (eds.), The Geology of North America: an overview [The Geology of North America, v. A]: Geological Society of America, Boulder, p. 397–447.

Bally AW-Snelson S, 1980, Realms of subsidence: Canadian Society of Petroleum Geologists Memoir 6, p.9–94.

Bally AW, Gordy PL, Stewart GA, 1966, Structure, seismic data and orogenic evolution of southern Canadian Rocky Mountains: Bulletin of Canadian Petroleum Geology, v. 14, p. 337–381.

Baltuck M, vonHuene R, Arnott RJ, 1985, Sedimentology of the western continental slope of Central America: Initial Reports of the Deep Sea Drilling Project, v. 84, p. 921–937.

Baltz EM, 1978, Résumé of Rio Grande depression in north-central New Mexico: New Mexico Bureau of Mines and Mineral Resources Circular 163, p. 210–228.

Baltzer F, Purser BH, 1990, Modern alluvial fan and deltaic sedimentation in a foreland tectonic setting: the lower Mesopotamian Plain and the Arabian Gulf: Sedimentary Geology, v. 67, p. 175–197.

Barany I, Karson JA, 1989, Basaltic breccias of the Clipperton fracture zone (east Pacific): sedimentation and tectonics in a fast-slipping oceanic transform: Geological Society of America Bulletin, v. 101, p. 204–220.

Baranyi I, 1974, The subvolcanic breccias of the Kaiserstuhl volcano, in Illies JH, Fuchs K (eds.), Approaches to taphrogenesis: E. Schweizerbart'sche Verlagsbuchhandlung, Stuttgart, p. 226–230.

Barka AA, Kadinsky-Cade K, 1988, Strike-slip fault geometry in Turkey and its influence on earthquake activity: Tectonics, v. 8, p. 663–684.

Barker F, Farmer GL, Ayuso RA, Plafker G, Lull JS, 1992, The 50 Ma granodiorite of the eastern Gulf of Alaska: melting in an accretionary prism in the forearc: Journal of Geophysical Research, v. 97, p. 6757–6778.

Barker PF, 1972, A spreading centre in the east Scotia Sea: Earth and Planetary Science Letters, v. 15, p. 123–132.

Barker PF, Hill I, 1981, Back-arc extension in the Scotia Sea: Philosophical Transactions of the Royal Society of London, v. A300, p. 249–262.

Barnard WD, 1978, The Washington continental slope: Quaternary tectonics and sedimentation: Marine Geology, v. 27, p. 79–114.

Barnes DA, 1984, Volcanic arc derived, Mesozoic sedimentary rocks, Vizcaino Peninsula, Baja California Sur, Mexico, in Frizzell VA, Jr (ed.), Geology of the Baja California Peninsula: Pacific Section, Society of Economic Paleontologists and Mineralogists, Los Angeles, p. 119–130.

Barnes NE, Normark WR, 1985, Diagnostic parameters for comparing modern submarine fans and ancient turbidite systems, in Bouma AH, Normark WR, Barnes NE (eds.), Submarine fans and related turbidite systems: Springer-Verlag, New York, p. 13–14.

Barr D, 1987a, Structural/stratigraphic models for extensional basins of halfgraben type: Journal of Structural Geology, v. 9, p. 491–500.

Barr D, 1987b, Lithospheric stretching, detached normal faulting and footwall uplift: Geological Society of London Special Publication 28, p. 75–94.

Barrell, J., 1917, Rhythms and the measurements of geologic time: Geological Society of America Bulletin, v. 28, p. 745–904.

Barrier E, Muller C, 1984, New observations and discussion on the origin and age of the Lichi Melange: Geological Society of China Memoir 6, p. 303–326.

Barrier E, Huchon P, Aurelio M, 1991, Philippine fault: a key for Philippine kinematics: Geology, v. 19, p. 32–35.

Barron JA, Keller G, 1982, Widespread Miocene deep-sea hiatuses: coincidence with periods of global cooling: Geology, v. 10, p. 577–581.

Bartlett WL, Friedman M, Logan JM, 1981, Experimental folding and faulting of rocks under confining pressure: part IX. wrench faults in limestone layers: Tectonophysics, v. 79, p. 255–277.

Bartolini C, Malesani PG, Manetti P, Wezel FC, 1975, Sedimentology and petrology of Quaternary sediments from the Hellenic Trench, Mediterranean Ridge and the Nile Cone from DSDP, Leg 13, cores: Sedimentology, v. 22, p.205–236.

Barton MD, Hanson RB, 1989, Magmatism and the development of low-pressure metamorphic belts: implications from the western United States and thermal modeling: Geological Society of America Bulletin, v. 101, p. 1051–1065.

Barton P, Wood R, 1983, Tectonic evolution of the North Sea basin: crustal stretching and subsidence: Geophysical Journal of the Royal Astronomical Society, v. 79, p. 987–1022.

Bartow JA, 1991, The Cenozoic evolution of the San Joaquin Valley, California: United States Geological Survey Professional Paper 1501, 40 p.

Bartow JA, Nilsen TH, 1990, Review of the Great Valley sequence, eastern Diablo Range and northern San Joaquin Valley, central California, in Kuespert JG, Reid SA (eds.), Structure, stratigraphy, and hydrocarbon occurrences of the San Joaquin basin, California [Book 64]: Pacific Section, Society of Economic Paleontologists and Mineralogists, Los Angeles, p. 253–265.

Baş H, 1979, Petrologische und Geochemische Untersuchungen an Subrezenten Vulkaniten der nordanatolischen Storungszone (Abschnitt: Erzincan-Niksar), Turkei [PhD thesis]: University of Hamburg, Hamburg, 177p.

Bates RL, Jackson JA, 1987, Glossary of geology [third edition]: American Geological Institute, Alexandria, 788 p.

Beach A, Bird T, Gibbs A, 1987, Extensional tectonics and crustal structure: deep seismic reflection data from the northern North Sea Viking graben: Geological Society of London Special Publication 28, p. 467–476.

Beauchamp J, 1988, Triassic sedimentation and rifting in the High Atlas (Morocco), in Manspeizer W (ed.), Triassic-Jurassic rifting. Part A: Elsevier, Amsterdam, p. 477–497.

Beauchamp J, Petit JP, 1983, Sédimentation et taphrogénèse triasique au Maroc: l'exemple du Haut Atlas de Marrakech: Bulletin de Centres de Recherche, Exploration et Production d'Elf-Aquitaine, v. 7, p. 389–397.

Beaudry D, Moore GF, 1981, Seismic-stratigraphic framework of the forearc basin off central Sumatra, Sunda arc: Earth and Planetary Science Letters, v. 54, p. 17–28.

Beaudry D, Moore GF, 1985, Seismic stratigraphy and Cenozoic evolution of west Sumatra forearc basin: American Association of Petroleum Geologists Bulletin, v. 69, p. 742–759.

Beaumont C, 1981, Foreland basins: Geophysical Journal of the Royal Astronomical Society, v. 65, p. 291–329.

Beaumont C, 1982, Platform sedimentation, in 11th International Sedimentology Congress, Abstracts: International Association of Sedimentologists, Hamilton, p. 132.

Beaumont C, Keen CE, Boutilier R, 1982a, A comparison of foreland and rifted margin sedimentary basins: Philosophical Transactions of the Royal Society of London, v. A305, p. 295–317.

Beaumont C, Keen CE, Boutilier R, 1982b, On the evolution of rifted continental margins: comparison of models and observations for the Nova Scotian margin: Geophysical Journal of the Royal Astronomical Society, v. 70, p. 667–715.

Beaumont C, Quinlan GM, Hamilton J, 1987, The Alleghenian orogeny and its relationship to the evolution of the eastern interior, North America: Canadian Society of Petroleum Geologists Memoir 12, p.425–446.

Beck ME, Jr, 1988, Analysis of Late Jurassic-Recent paleomagnetic data from active plate margins of South America: Journal of South American Earth Sciences, v. 1, p. 39–52.

Beck RA, Vondra CF, Filkins JE, Olander JD, 1988, Syntectonic sedimentation and Laramide basement thrusting, Cordilleran foreland; timing of deformation: Geological Society of America Memoir 171, p. 465–487.

Becker DG, Cloos M, 1985, Melange diapirs into the Cambria slab: a Franciscan trench slope basin near Cambria, California: Journal of Geology, v. 93, p. 101–110.

Becker H, 1934, Die Beziehungen zwischen Felsengebirge und Großen Becken im westlichen Nordamerika: Zeitschrift der Deutschen Geologischen Gesellschaft, v. 86, p. 115–120.

Beer JA, 1990, Steady sedimentation and lithologic completeness, Bermejo basin, Argentina: Journal of Geology, v. 98, p. 501–517.

Beer JA, Jordan TE, 1989, The effects of Neogene thrusting on deposition in the Bermejo basin, Argentina: Journal of Sedimentary Petrology, v. 59, p. 330–345.

Beer JA, Allmendinger RW, Figueroa DE, Jordan TE, 1990, Seismic stratigraphy of a Neogene piggyback basin, Argentina: American Association of Petroleum Geologists Bulletin, v. 74, p. 1183–1202.

Beets DJ, 6 coauthors, 1984, Magmatic rock series and high-pressure metamorphism as constraints on the tectonic history of the southern Caribbean: Geological Society of America Memoir 162, p. 95–130.

Beggs JM, 1984, Volcaniclastic rocks of the Alisitos Group, Baja California, Mexico, in Frizzell VA, Jr (ed.), Geology of the Baja California Peninsula [Book 39]: Pacific Section, Society of Economic Paleontologists and Mineralogists, Los Angeles, p. 43–52.

Bell JW, 1984, Quaternary fault map of Nevada [Reno Sheet]: Nevada Bureau of Mines and Geology Map 79, 1:250,000.

Bell JW, Katzer T, 1990, Timing of late Quaternary faulting in the 1954 Dixie Valley earthquake area, central Nevada: Geology, v. 18, p. 622–625.

Bell RE, Karner GD, Steckler MS, 1988, Early Mesozoic rift basins of eastern North America and their gravity anomalies: the role of detachments during extension: Tectonics, v. 7, p. 447–462.

Beloussov VV, 1962, Basic problems in geotectonics: McGraw-Hill, New York, 809 p.

Beloussov VV, 1980, Geotectonics: Mir Publishers, Moscow, 330 p.

Ben-Avraham Z, 1985, Structural framework of the Gulf of Elat (Aqaba), northern Red Sea: Journal of Geophysical Research, v. 90, p. 703–726.

Ben-Avraham Z, 1994, Development of asymmetric basins along continental transform faults: Tectonophysics, in press.

Ben-Avraham Z, Lyakhovsky V, 1992, Faulting processes along the northern Dead Sea transform and the Levant margin: Geology, v. 20, p. 1139–1142.

Ben-Avraham Z, Uyeda S, 1983, Entrapment origin of marginal seas: American Geophysical Union Geodynamics Series, v. 11, p. 91–104.

Ben-Avraham Z, Zoback MD, 1992, Transform-normal extension and asymmetric basins: an alternative to pull-apart models: Geology, v. 20, p. 423–426.

Benkhelil J, Dainelli P, Ponsard JF, Popoff M, Sangy L, 1988, The Benue trough—wrench-fault related basin on the border of the equatorial Atlantic, in Manspeizer W (ed.), Triassic-Jurassic rifting. Part B: Elsevier, Amsterdam, p. 787–819.

Benson RN, Doyle RG, 1984, Inner margin of Baltimore Canyon trough: future exploration play: American Association of Petroleum Geologists Bulletin, v. 68, p. 454.

Benson RN, Doyle RG, 1988, Early Mesozoic rift basins and the development of the United States middle Atlantic continental margin, in Manspeizer W (ed.), Triassic-Jurassic rifting. Part B: Elsevier, Amsterdam, p. 99–127.

Bentham P, 5 coauthors, 1991, Tectono-sedimentary development of an extensional basin: the Neogene Megara basin, Greece: Journal of the Geological Society of London, v. 148, p. 923–934.

Bentham PA, Burbank DW, Puigdefabregas C, 1992, Temporal and spatial controls on the alluvial architecture of an axial drainage system: late Eocene Escanilla Formation, southern Pyrenean foreland basin, Spain: Basin Research, v. 4, p. 335–352.

Berger WH, 1973, Cenozoic sedimentation in the eastern tropical Pacific: Geological Society of America Bulletin, v. 84, p.1941–1954.

Berger WH, Winterer EL, 1974, Plate stratigraphy and the fluctuating carbonate line: International Association of Sedimentologists Special Publication 1, p. 11–48.

Bergerat F, 1983, Paléocontraintes et évolution tectonique paléogène du Fossé Rhénan: Comptes Rendus hébdomadaires de l'Academie des Sciences de Paris, v. 297, p. 77–80.

Bergerat F, 1987, Stress fields in the European platform at the time of Africa-Eurasia collision: Tectonics, v. 6, p. 99–132.

Berggren WA, VanCouvering JA (eds.), 1984, Catastrophes and Earth history: the new uniformitarianism: Princeton University Press, Princeton, 464 p.

Berggren WA, Kent DV, Flynn JJ, VanCouvering JA, 1985, Cenozoic geochronology: Geological Society of America Bulletin, v. 96, p. 1407–1418.

Beridze MA, 1984, Geosynklinalny vulkanogenno osadochni litogenez: Akademii Nauk, Gruzinsky SSR, Tbilisi, 182 p.

Bernstein-Taylor BL, Kirchoff-Stein KS, Silver EA, Reed DL, 1992a, Large-scale duplexes within the New Britain accretionary wedge: a possible example of accreted ophiolitic slivers: Tectonics, v. 11, p. 732–752.

Bernstein-Taylor BL, Brown KM, Silver EA, Kirchoff-Stein KS, 1992b, Basement slivers within the New Britain accretionary wedge: implications for emplacement of some ophiolitic slivers: Tectonics, v. 11, p. 753–765.

Berry JP, Wilkinson BH, 1994, Paleoclimatic and tectonic control on the accumulation of North American cratonic sediment: Geological Society of America Bulletin, v. 106, p. 855–865.

Bersier A, 1948, Les sedimentations rhythmiques synorogeniques dans l'avant–fosse molassique alpine: 18th International Geological Congress, Part IV, p. 83–93.

Bertucci PF, 1983, Petrology and provenance of the Stony Creek Formation, northwestern Sacramento Valley, California, in Bertucci PF, Ingersoll RV (eds.), Guidebook to the Stony Creek Formation, Great Valley Group, Sacramento Valley, Cali-

fornia: Pacific Section, Society of Economic Paleontologists and Mineralogists, Los Angeles, p. 1–16.

Bessis F, 1986, Some remarks on the study of subsidence of sedimentary basins. Application to the Gulf of Lions margin (western Mediterranean): Marine and Petroleum Geology, v. 3, p. 37–63.

Bethelsen A, Şengör AMC, 1990, How was Europe put together, in Calder N (ed.), Scientific Europe: research and technology in 20 countries: Nature and Technology Scientific Publishers Ltd., Maastricht, p. 108–121.

Bethke CM, 1985, A numerical model of compaction driven groundwater flow and heat transfer and its application to the paleohydrology of intracratonic sedimentary basins: Journal of Geophysical Research, v. 90, p. 6817–6828.

Bethke CM, Marshak S, 1990, Brine migrations across North America—the plate tectonics of groundwater: Annual Review of Earth and Planetary Sciences, v. 18, p. 287–315.

Bethke CM, Reed JD, Oltz DF, 1991, Long-range petroleum migration in the Illinois basin: American Association of Petroleum Geologists Bulletin, v. 75, p. 925–945.

Betzler C, Nederbragt AJ, Nichols GJ, 1991, Significance of turbidites at Site 767 (Celebes Sea) and Site 768 (Sulu Sea): Proceedings of the Ocean Drilling Program, Scientific Results, v. 124, p. 431–446.

Bhatia MR, 1985, Composition and classification of Paleozoic flysch mudrocks of eastern Australia: implications in provenance and tectonic setting interpretation: Sedimentary Geology, v. 41, p. 249–268.

Bickford ME, VanSchmus WR, Zietz I, 1986, Proterozoic history of the midcontinent region of North America: Geology, v. 14, p. 492–496.

Biddle KT, 1991, The Los Angeles basin—an overview: American Association of Petroleum Geologists Memoir 52, p. 5–24.

Biddle KT, Christie-Blick N (eds.), 1985, Strike-slip deformation, basin formation, and sedimentation: Society of Economic Paleontologists and Mineralogists Special Publication 37, 386 p.

Biju-Duval B, 14 coauthors, 1984, Initial reports of the Deep Sea Drilling Project [v. 78A]: United States Government Printing Office, Washington, 621 p.

Bird JM, Dewey JF, 1970, Lithosphere plate—continental margin tectonics and the evolution of the Appalachian orogen: Geological Society of America Bulletin, v. 81, p.1031–1060.

Bird P, 1984, Laramide crustal thickening event in the Rocky Mountain foreland and Great Plains: Tectonics, v. 3, p. 741–758.

Bird P, 1988, Formation of the Rocky Mountains, western United States: a continuum computer model: Science, v. 239, p. 1501–1507.

Bird P, 1991, Lateral extrusion of lower crust from under high topography, in the isostatic limit: Journal of Geophysical Research, v. 96, p. 10275–10286.

Blair TC, Bilodeau WL, 1988, Development of tectonic cyclothems in rift, pull-apart, and foreland basins: sedimentary response to episodic tectonism: Geology, v. 16, p. 517–520.

Blake GH, 1991, Review of the Neogene biostratigraphy and stratigraphy of the Los Angeles basin and implications for basin evolution: American Association of Petroleum Geologists Memoir 52, p. 135–184.

Blake MC, Jr, Jones DL, Landis CA, 1974, Active continental margins: contrasts between California and New Zealand, in Burk CA, Drake CL (eds.), The geology of continental margins:

Springer-Verlag, New York, p. 853–872.

Blarez E, Mascle J, 1988, Shallow structures and evolution of the Ivory Coast and Ghana transform margin: Marine and Petroleum Geology, v. 5, p. 54–64.

Blasi A, Manassero MJ, 1990, The Colorado River of Argentina: source, climate, and transport as controlling factors on sand composition: Journal of South American Earth Sciences, v. 3, p. 65–70.

Block L, Royden LH, 1990, Core complex geometries and regional scale flow in the lower crust: Tectonics, v. 9, p. 557–567.

Bloomer SH, Stern RJ, Smoot NC, 1989, Physical volcanology of the submarine Mariana and Volcano arcs: Bulletin of Volcanology, v. 51, p. 210–224.

Blundell DJ, Gibbs AD, 1990, Tectonic evolution of the North Sea rifts: Clarendon, Oxford, 272 p.

Blundell D, Meissner R, Scott-Robinson R, Thomas S, the BABEL Working Group, 1991, Deep structure of the Tornquist zone: a Caledonian plate boundary revealed?: XXth General Assembly, IUGG, Vienna, IASPEI Program and Abstracts, p. 58.

Boccaletti M, 10 coauthors, 1990a, Palinspastic restoration and paleogeographic reconstruction of the peri-Tyrrhenian area during the Neogene: Palaeogeography, Palaeoclimatology, Palaeoecology, v. 77, p. 41–50.

Boccaletti M, Nicolich R, Tortorici L, 1990b, New data and hypothesis on the development of the Tyrrhenian basin: Palaeogeography Palaeoclimatology Palaeoecology, v. 77, p. 15–40.

Bodine JH, Watts AB, 1979, On lithospheric flexure seaward of the Bonin and Mariana trenches: Earth and Planetary Science Letters, v. 43, p. 132–148.

Boehlke JE, Abbott PL, 1986, Punta Baja Formation, a Campanian submarine canyon fill, Baja California, Mexico, in Abbott PL (ed.), Cretaceous stratigraphy western North America [Book 46]: Pacific Section, Society of Economic Paleontologists and Mineralogists, Los Angeles, p. 91–102.

Bogdanov AA, 1962, O nekotorich problemach tektoniki Europi: Vestnik Moskovskogo Universiteta, Series IV (Geologii), v. 5, p. 46–66.

Bogdanov AA, Muratov MV, Khain VE, 1963, Osnovnie strukturnie elementi zemnoi kori: Bjulleten Moskovskogo Obshchestva Ispytatelej Prirody, Otdel Geologichesky, v. 38, p. 3–32.

Boggs S, Jr, 1984, Quaternary sedimentation in the Japan arc-trench system: Geological Society of America Bulletin, v. 95, p. 669–685.

Boggs S, Jr, Seyedolali A, 1992a, Provenance of Miocene sandstones from Sites 796, 797, and 799, Japan Sea: Proceedings of the Ocean Drilling Program, Scientific Results, v. 127/128, Pt. 1, p. 99–113.

Boggs S, Jr, Seyedolali A, 1992b, Diagenetic albitization, zeolitization, and replacement in Miocene sandstones, Sites 796, 797, and 799, Japan Sea: Proceedings of the Ocean Drilling Program, Scientific Results, v. 127/128, Pt. 1, p. 131–151.

Bohannon RG, 1986a, How much divergence has occurred between Africa and Arabia as a result of the opening of the Red Sea?: Geology, v. 14, p. 510–513.

Bohannon RG, 1986b, Tectonic configuration of the western Arabian continental margin, southern Red Sea: Tectonics, v. 5, p. 477–499.

Bohannon RG, Eittreim SL, 1991, Tectonic development of passive continental margins of the southern and central Red

Sea with a comparison to Wilkes Land, Antarctica: Tectonophysics, v. 198, p. 129–154.

Bohannon RG, Naeser CW, Schmidt DL, Zimmermann RA, 1989, The timing of uplift, volcanism, and rifting peripheral to the Red Sea: a case for passive rifting?: Journal of Geophysical Research, v. 94, p. 1683–1701.

Boles JR, 1986, Mesozoic sedimentary rocks in the Vizcaino Peninsula-Isla de Cedros area, Baja California, Mexico, in Abbott, P.L. (ed.), Cretaceous stratigraphy, western North America [Book 46]: Pacific Section, Society of Economic Paleontologists and Mineralogists, Los Angeles, p. 63–77.

Boles JR, Coombs DS, 1977, Zeolite facies alteration of sandstones in the Southland syncline, New Zealand: American Journal of Science, v. 277, p. 982–1012.

Boles JR, Landis CA, 1984, Jurassic sedimentary melange and associated facies, Baja California, Mexico: Geological Society of America Bulletin, v. 95, p. 513–521.

Bonatti E, 1985, Punctiform initiation of seafloor spreading in the Red Sea during transition from a continental to an oceanic rift: Nature, v. 316, p. 33–37.

Bonatti E, Seyler M, 1987, Crustal underplating and evolution in the Red Sea rift: uplifted gabbro/gneiss crustal complexes on Zabargad and Brothers islands: Journal of Geophysical Research, v. 92, p. 12803–12821.

Bond GC, Kominz MA, 1984, Construction of tectonic subsidence curves for the early Paleozoic miogeocline, southern Canadian Rocky Mountains: implications for subsidence mechanisms, age of breakup, and crustal thinning: Geological Society of America Bulletin, v. 95, p. 155–173.

Bond GC, Kominz MA, 1991, Disentangling middle Paleozoic sea level and tectonic events in cratonic margins and cratonic basins of North America: Journal of Geophysical Research, v. 94, p. 6619–6639.

Bond GC, Nickeson PA, Kominz MA, 1984, Breakup of a supercontinent between 625 Ma and 555 Ma: new evidence and implications for continental histories: Earth and Planetary Science Letters, v. 70, p. 325–345.

Borissjak A, 1903, Ueber die Tektonik des Donez-Höhenzuges in seinen nordwestlichen Ausläufern: Centralblatt für Mineralogie, Geologie und Paläontologie 1, p. 644–649.

Bosworth W, 1985, Geometry of propagating continental rifts: Nature, v. 316, p. 625–627.

Bott MHP, 1981, Crustal doming and the mechanism of continental rifting: Tectonophysics, v. 73, p. 1–9.

Bott MHP, Dean DS, 1972, Stress systems at young continental margins: Nature, v. 235, p. 23–25.

Bottjer DJ, Link MH, 1984, A synthesis of Late Cretaceous southern California and northern Baja California paleogeography, in Crouch JK, Bachman SB (eds.), Tectonics and sedimentation along the California margin [Book 38]: Pacific Section, Society of Economic Paleontologists and Mineralogists, Los Angeles, p. 79–90.

Bouma AH, Smith LB, Sidner BR, McKee TR, 1978, Intraslope basin in northwest Gulf of Mexico: American Association of Petroleum Geologists Studies in Geology 7, p. 289–302.

Bouma AH, Normark WR, Barnes NE (eds.), 1985, Submarine fans and related turbidite systems: Springer-Verlag, New York, 351 p.

Bouysse P, 1984, The Lesser Antilles island arc: structure and geodynamic evolution: Initial Reports of the Deep Sea Drilling Project, v. 78, p. 83–103.

Bowin C, Lu RS, Lee C-S, Schouten H, 1978, Plate convergence and accretion in Taiwan-Luzon region: American Association of Petroleum Geologists Bulletin, v. 62, p. 1645–1672.

Bowin C, 6 coauthors, 1980, Arc-continent collision in Banda Sea region: American Association of Petroleum Geologists Bulletin, v. 64, p. 868–915.

Bowles FA, Ruddiman WF, Jahn WH, 1978, Acoustic stratigraphy, structure, and depositional history of the Nicobar Fan, eastern Indian Ocean: Marine Geology, v. 26, p. 269–288.

Boyer SE, Elliot D, 1982, Thrust systems: American Association of Petroleum Geologists Bulletin, v. 66, p. 1196–1230.

Boynton CH, Westbrook GK, Bott MHP, Long RE, 1979, A seismic refraction investigation of crustal structure beneath the Lesser Antilles island arc: Geophysical Journal of the Royal Astronomical Society, v. 58, p. 371–393.

Bradley DC, 1983, Tectonics of the Acadian orogeny in New England and adjacent Canada: Journal of Geology, v. 91, p. 381–400.

Bradley DC, 1989, Taconic plate kinematics as revealed by foredeep stratigraphy, Appalachian orogen: Tectonics, v. 8, p. 1037–1049.

Bradley DC, Kidd WSF, 1991, Flexural extension of the upper continental crust in collisional foredeeps: Geological Society of America Bulletin, v. 103, p. 1416–1438.

Bradley DC, Kusky TM, 1986, Geologic evidence for rate of plate convergence during the Taconic arc-continent collision: Journal of Geology, v. 94, p. 667–681.

Braille LW, Keller GR, Hinze WJ, Lidiak EG, 1982, An ancient rift complex and its relation to contemporary seismicity in the New Madrid seismic zone: Tectonics, v. 1, p. 225–237.

Braille LW, Hinze WJ, Keller GR, Lidiak EG, Sexton JL, 1986, Tectonic development of the New Madridr rift complex, Mississippi embayment, North America: Tectonophysics, v. 131, p. 1–21.

Brandon MT, 1989, Deformational styles in a sequence of olistostromal melanges, Pacific Rim Complex, western Vancouver Island, Canada: Geological Society of America Bulletin, v. 101, p. 1520–1542.

Bray CJ, Karig DE, 1985, Porosity of sediments in accretionary prisms and some implications for dewatering processes: Journal of Geophysical Research, v. 90, p. 768–778.

Bray CJ, Karig DE, 1988, Dewatering and extensional deformation of the Shikoku basin hemipelagic sediments in the Nankai Trough: Pure and Applied Geophysics, v. 128, p. 725–747.

Breen N, 1989, Structural effect of Magdalena fan deposition on the northern Colombia convergent margin: Geology, v. 17, p. 34–37.

Breitkreuz C, 1991, Fluvio-lacustrine sedimentation and volcanism in a Late Carboniferous tensional intra-arc basin, northern Chile: Sedimentary Geology, v. 74, p. 173–187.

Bretagne AJ, Leising TC, 1990, Interpretation of seismic data from the Rough Creek graben of western Kentucky and southern Illinois: American Asociation of Petroleum Geologists Memoir 51, p. 199–208

Brewer JA, Smithson SB, Oliver JE, Kaufman S, Brown LD, 1980, The Laramide orogeny: evidence from COCORP deep crustal seismic profiles in the Wind River Mountains, Wyoming: Tectonophysics, v. 62, p. 165–189.

Bridge JS, Leeder MR, 1979, A simulation model of alluvial stratigraphy: Sedimentology, v. 26, p. 617–644.

Brookfield ME, 1993, The Himalayan passive margin from Precambrian to Cretaceous times: Sedimentary Geology, v. 84, p. 1–35.

Brookfield ME, Andrews-Speed CP, 1984, Sedimentology, petrography and tectonic significance of the shelf, flysch and molasse clastic deposits across the Indus suture zone, Ladakh, NW India: Sedimentary Geology, v. 40, p. 249–286.

Brooks DA, 6 coauthors, 1984, Characteristics of backarc regions: Tectonophysics, v. 102, p. 1–16.

Brooks M, Ferentinos G, 1984, Tectonics and sedimentation in the Gulf of Corinth and the Zakynthos and Kefallinia channels, western Greece: Tectonophysics, v. 101, p. 25–54.

Brouxel M, Lapierre H, 1988, Geochemical study of an early Paleozoic island-arc–back-arc basin system. Part 1: the Trinity Ophiolite (northern California): Geological Society of America Bulletin, v. 100, p. 1111–1119.

Brown C, Girdler RW, 1980, Structure of the Red Sea at 20N from gravity data and its implications for continental margins: Nature, v. 298, p. 51–53.

Brown G, Taylor B, 1988, Sea-floor mapping of the Sumisu Rift, Izu-Ogasawara (Bonin) island arc: Bulletin of the Geological Survey of Japan, v. 39, p. 23–38.

Brown KM, 1990, The nature and hydrologic significance of mud diapirs and diatremes for accretionary systems: Journal of Geophysical Research, v. 95, p. 8969–8982.

Brown LF, Jr, Fisher WL, 1977, Seismic-stratigraphic interpretation of depositional systems: example from Brazilian rift and pull–apart basins: American Association of Petroleum Geologists Memoir 26, p. 213–248.

Browne PRL, Graham IJ, Parker RJ, Wood CP, 1992, Subsurface andesite lavas and plutonic rocks in the Rotokawa and Ngatamariki geothermal systems, Taupo volcanic zone, New Zealand: Journal of Volcanology and Geothermal Research, v. 51, p. 199–215.

Bruce RM, Nelson EP, Weaver SG, Lux DR, 1991, Temporal and spatial variations in the southern Patagonian batholith: constraints on magmatic arc development: Geological Society of America Special Paper 265, p. 1–12.

Bruhn RL, Yusas MR, Huertos F, 1982, Mechanics of low-angle normal faulting: an example from Roosevelt Hot Springs geothermal area, Utah: Tectonophysics, v. 86, p. 243–361.

Brunet MF, LePichon X, 1982, Subsidence of the Paris basin: Journal of Geophysical Research, v. 87, p. 8547–8560.

Bruns TR, Cooper AK, Mann DM, Vedder JG, 1986, Seismic stratigraphy and structure of sedimentary basins in the Solomon Islands region: Circum-Pacific Council for Energy and Mineral Resources Earth Science Series, v. 4, p. 177–214.

Bruns TR, von Huene, R, Culotta RC, Lewis SD, Ladd JW, 1987, Geology and petroleum potential of the Shumagin margin, Alaska: Circum-Pacific Council for Energy and Mineral Resources Earth Science Series, v. 6, p. 157–189.

Buchovecky EJ, Lundberg N, 1988, Clay mineralogy of mudstones from the southern Coastal Range, eastern Taiwan: unroofing of the orogen versus in-situ diagenesis: Acta Geologica Taiwanica, n. 26, p. 247–262.

Buck WR, 1986, Small-scale convection induced by passive rifting: the cause for uplift of rift shoulders: Earth and Planetary Science Letters, v. 77, p. 362–372.

Buck WR, 1988, Flexural rotation of normal faults: Tectonics, v. 7, p. 959–973.

Buck WR, 1991, Modes of continental lithospheric extension: Journal of Geophysical Research, v. 96, p. 20161–20178.

Buck WR, Martinez F, Steckler MS, Cochran JR, 1988, Thermal consequences of lithospheric extension: pure and simple: Tectonics, v. 7, p. 213–233.

Burbank DW, 1992, Causes of recent Himalayan uplift deduced from deposited patterns in the Ganges basin: Nature, v. 357, p. 680–683.

Burbank DW, Raynolds RGH, 1984, Sequential late Cenozoic structural disruption of the northern Himalayan foredeep: Nature, v. 311, p. 114–118.

Burbank DW, Raynolds RGH, 1988, Stratigraphic keys to the timing of thrusting in terrestrial foreland basins: aplications to the northwestern Himalaya, in Kleinspehn KL, Paola C (eds.), New perspectives in basin analysis: Springer-Verlag, New York, p. 331–351.

Burbank DW, Tahirkheli RAK, 1985, The magnetostratigraphy, fission-track dating, and stratigraphic evolution of the Peshawar intermontane basin, northern Pakistan: Geological Society of America Bulletin, v. 96, p. 539–552.

Burbank DW, Raynolds RGH, Johnson GD, 1986, Late Cenozoic tectonics and sedimentation in the northwestern Himalayan foredeep: II. Eastern limb of the northwest syntaxis and regional synthesis: International Association of Sedimentologists Special Publication 8, p. 293–306.

Burbank DW, Beck RA, Raynolds RGH, Hobbs R, Tahirkheli RAK, 1988, Thrusting and gravel progradation in foreland basins: a test of post-thrusting gravel dispersal: Geology, v. 16, p. 1143–1146.

Burbank DW, Beck RA, Raynolds RGH, 1989, Reply to comment by Heller et al. on "Thrusting and gravel progradation in foreland basins: a test of post-thrusting gravel dispersal": Geology, v. 17, p. 960–961.

Burbank DW, Puigdefabregas C, Munoz JA, 1992, The chronology of the Eocene tectonic and stratigraphic development of the eastern Pyrenean foreland basin, northeast Spain: Geological Society of America Bulletin, v. 104, p. 1101–1120.

Burchfiel BC, 1966, Tin Mountain landslide, southeastern California, and the origin of megabreccia: Geological Society of America Bulletin, v. 77, p. 95–100.

Burchfiel BC, Davis GA, 1972, Structural framework and evolution of the southern part of the Cordilleran orogen, western United States: American Journal of Science, v. 272, p. 97–118.

Burchfiel BC, Davis GA, 1975, Nature and controls of Cordilleran orogenesis, western United States: extension of an earlier synthesis: American Journal of Science, v. 275-A, p. 363–396.

Burchfiel BC, Royden LH, 1991, Tectonics of Asia 50 years after the death of Emile Argand: Eclogae Geologicae Helvetiae, v. 84, p. 599–629.

Burchfiel BC, Stewart JH, 1966, "Pull-apart" origin of the central segment of Death Valley, California: Geological Society of America Bulletin, v. 77, p. 439–441.

Burchfiel BC, Hodges KV, Royden LH, 1987, Geology of Panamint Valley–Saline Valley pull–apart system, California: palinspastic evidence for low-angle geometry of a Neogene range–bounding fault: Journal of Geophysical Research, v. 92, p. 10422–10426.

Burg JP, Teyssier C, Lespinasse M, Etchecopar A, 1982, Direction de contraintes et dynamique du bassin de Saint-Flour-Saint-Alban (Massif Central français) à l'Oligocéne: Comptes Rendus hébdomadaires de l'Academie des Sciences de Paris, v. 294, p. 1021–1024.

Burgess CF, 6 coauthors, 1988, The structural and stratigraphic evolution of Lake Tanganyika: a case study of continental rifting, in Manspeizer W (ed.), Triassic-Jurassic rifting. Part B: Elsevier, Amsterdam, p. 859–881.

Burggraf DR, Vondra CF, 1982, Rift valley facies and paleoenvironments: an example from the East African rift system of Kenya and southern Ethiopia: Zeitschrift für Geomorphologie Neue Folge, Supplement, v. 42, p. 43–73.

Burgisser HM, 1984, A unique mass flow marker bed in a Miocene streamflow molasse sequence, Switzerland: Canadian Society of Petroleum Geologists Memoir 10, p. 147–163.

Burk CA, Drake CL (eds), 1974, The geology of continental margins: Springer-Verlag, New York, 1009 p.

Burkart B, Self S, 1985, Extension and rotation of crustal blocks in northern Central America and effect on the volcanic arc: Geology, v. 13, p. 22–26.

Burke K, 1972, Longshore drift, submarine canyons, and submarine fans in development of Niger delta: American Association of—Petroleum Geologists Bulletin, v. 56, p.1975–1983.

Burke K, 1975, Atlantic evaporites formed by evaporation of water spilled from Pacific, Tethyan and southern oceans: Geology, v. 3, p. 613–616.

Burke K, 1976, Development of graben associated with the initial ruptures of the Atlantic Ocean: Tectonophysics, v. 36, p. 93–112.

Burke K, 1977a, Aulacogens and continental breakup: Annual Review of Earth and Planetary Sciences, v. 5, p. 371–396.

Burke K, 1977b, Are lakes George and Champlain in Neogene graben reactivating early Paleozoic rifts?: Geological Society of America Abstracts with Programs, v. 9, p. 247–248.

Burke K, 1978, Evolution of continental rift systems in the light of plate tectonics, in Ramberg IB, Neumann ER (eds.), Tectonics and geophysics of continental rifts: D. Reidel Publishing Company, Dordrecht, p. 1–9.

Burke K, 1980, Intracontinental rifts and aulacogens [Continental Tectonics]: National Academy of Sciences, Washington, p. 42–49.

Burke K, Dewey JF, 1973, Plume-generated triple junctions: key indicators in applying plate tectonics to old rocks: Journal of Geology, v. 81, p. 406–433.

Burke K, Dewey JF, 1974, Two plates in Africa during the Cretaceous?: Nature, v. 249, p. 313–316.

Burke K, Şengör AMC, 1986, Tectonic escape in the evolution of the continental crust: American Geophysical Union Geodynamics Series 14, p. 41–53.

Burke K, Whiteman AJ, 1973, Uplift, rifting and the break-up of Africa, in Tarling DH, Runcorn SK (eds.), Implications of continental drift to the earth sciences: Academic Press, London, p. 735–755.

Burke K, Wilson JT, 1972, Is the African plate stationary?: Nature, v. 239, p. 387–390.

Burke K, Wilson JT, 1976, Hot spots on the earth's surface: Scientific American, v. 235, p. 46–57.

Burke KC, Dessauvagie TFJ, Whiteman AJ, 1972, Geological history of the Benue Valley and adjacent areas, in Dessauvagie TFJ, Whiteman AJ (eds.), African geology: Department of Geology, University of Ibadan, Ibadan, p. 187–205.

Burke K, Kidd WSF, Wilson JT, 1973, Relative and latitudinal motion of Atlantic hot spots: Nature, v. 245, p. 133–137.

Burke K, Dewey JF, Kidd WSF, 1977, World distribution of sutures—the sites of former oceans: Tectonophysics, v. 40, p. 69–99.

Burke K, 7 coauthors, 1978, Rifts and sutures of the world [Report to Geophysics Branch, Earth Survey Applications Division]: NASA Goddard Space Flight Center, Greenbelt.

Burke K, Cooper C, Dewey JF, Mann P, Pindell JL, 1984, Caribbean tectonics and relative plate motions: Geological Society of America Memoir 162, p. 31–63.

Burke K, Kidd WSF, Kusky TM, 1985, The Pongola structure of southeastern Africa: the world's oldest preserved rift?: Journal of Geodynamics, v. 2, p. 35–49.

Busby-Spera CJ, 1984a, The lower Mesozoic continental margin and marine intra-arc sedimentation at Mineral King, California, in Crouch JK, Bachman SB (eds.), Tectonics and sedimentation along the California margin: Pacific Section, Society of Economic Paleontologists and Mineralogists Publication 38, p.135–155.

Busby-Spera CJ, 1984b, Large-volume rhyolite ash flow eruptions and submarine caldera collapse in the lower Mesozoic Sierra Nevada, California: Journal of Geophysical Research, v. 89, p. 8417–8427.

Busby-Spera CJ, 1985, A sand-rich submarine fan in the lower Mesozoic Mineral King caldera complex, Sierra Nevada, California: Journal of Sedimentary Petrology, v. 55, p. 376–391.

Busby-Spera CJ, 1986, Depositional features of rhyolitic and andesitic volcaniclastic rocks of the Mineral King submarine caldera complex, Sierra Nevada, California: Journal of Volcanology and Geothermal Research, v. 27, p. 43–76.

Busby-Spera CJ, 1987, Lithofacies of deep marine basalts emplaced on a Jurassic backarc apron, Baja California (Mexico): Journal of Geology, v. 95, p. 671–686.

Busby-Spera CJ, 1988a, Evolution of a Middle Jurassic back-arc basin, Cedros Island, Baja California: evidence from a marine volcaniclastic apron: Geological Society of America Bulletin, v. 100, p. 218–233.

Busby-Spera CJ, 1988b, Speculative tectonic model for the early Mesozoic arc of the southwest Cordilleran United States: Geology, v. 16, p. 1121–1125.

Busby-Spera CJ, Boles JR, 1986, Sedimentation and subsidence styles in a Cretaceous forearc basin, southern Vizcaino Peninsula, Baja California (Mexico), in Abbott PL (ed.), Cretaceous stratigraphy, western North America [Book 46]: Pacific Section, Society of Economic Paleontologists and Mineralogists, Los Angeles, p. 79–90.

Busby-Spera CJ, Saleeby JB, 1990, Intra-arc strike-slip fault exposed at batholithic levels in the southern Sierra Nevada, California: Geology, v. 18, p. 225–259.

Busby-Spera CJ, White JDL, 1987, Variation in peperite textures associated with differing host–sediment properties: Bulletin of Volcanology, v. 49, p. 765–775.

Busby-Spera CJ, Schermer ER, Riggs NR, Mattinson JM, 1990a, Controls on basin subsidence within continental magmatic arcs: International Association of Sedimentologists 13th Congress Abstracts, p. 70–71.

Busby-Spera CJ, Mattinson JM, Riggs NR, Schermer ER, 1990b, The Triassic-Jurassic magmatic arc in the Mojave-Sonoran Deserts and the Sierran-Klamath region: similarities and differences in paleogeographic evolution: Geological Society of America Special Paper 255, p. 325–338.

Busch RM, West RR, 1987, Hierarchical genetic stratigraphy: a framework of paleoceanography: Paleoceanography, v. 2, p. 141–164.

Buschbach TC, Kolata DR, 1990, Regional setting of Illinois basin: American Asociation of Petroleum Geologists Memoir 51, p. 29–55.

Byrne DE, Wang W-H, Davis DM, 1993, Mechanical role of backstops in the growth of forearcs: Tectonics, v. 12, p. 123–144.

Byrne T, 1982, Structural evolution of coherent terranes in the Ghost Rocks Formation, Kodiak Island, Alaska: Geological Society of London Special Publication 10, p. 229–242.

Byrne T, 1986, Eocene underplating along the Kodiak shelf, Alaska: implications and regional correlations: Tectonics, v. 5, p. 403–421.

Cabrera J, Sebrier M, Mercier JL, 1987, Active normal faulting in high plateaus of central Andes: the Cuzco region (Peru): Annales Tectonicae, v. 1, p. 116–138.

Cadet JP, 17 coauthors, 1987, The Japan Trench and its juncture with the Kuril Trench: cruise results of the Kaiko project, Leg 3: Earth and Planetary Science Letters, v. 83, p. 267–284.

Cahill T, Isacks BL, 1992, Seismicity and shape of the subducted Nazca plate: Journal of Geophysical Research, v. 97, p. 17503–17529.

Cande SC, Leslie RB, 1986, Late Cenozoic tectonics of the southern Chile trench: Journal of Geophysical Research, v. 91, p. 471–496.

Cande SC, Leslie RB, Parra JC, Hobart M, 1987, Interaction between the Chile Ridge and Chile Trench: geophysical and geothermal evidence: Journal of Geophysical Research, v. 92, p. 495–520.

Cant DJ, Stockmal GS, 1989, The Alberta foreland basin: relationship between stratigraphy and Cordilleran terrane-accretion events: Canadian Journal of Earth Sciences, v. 26, p. 1964–1975.

Carey SN, Sigurdsson H, 1984, A model of volcanogenic sedimentation in marginal basins: Geological Society of London Special Publication 16, p. 37–58.

Carey SW, 1958, The tectonic approach to continental drift, in Carey SW (ed.), Continental drift, a symposium: Geology Department, University of Tasmania, Hobart, p. 177–356.

Carlé W, 1950, Erläuterungen zur Geotektonischen Übersichtskarte der südwestdeutschen Großscholle: Geologische Abteilung des Württembergischen Statistischen Landesamts, Stuttgart, 31p.

Carlisle D, 1963, Pillow breccias and their aquagene tuffs, Quadra Island, British Columbia: Journal of Geology, v. 71, p. 48–71.

Carlson C, 1984, Stratigraphic and structural significance of foliate serpentinite breccias, Wilbur Springs: Society of Economic Paleontologists and Mineralogists Field Trip Guidebook 3, 1984 Midyear Meeting, p. 108–112.

Carlson PR, Nelson CH, 1969, Sediments and sedimentary structures of the Astoria submarine canyon-fan system, northeast Pacific: Journal of Sedimentary Petrology, v. 39, p. 1269–1282.

Carlson PR, Nelson CH, 1987, Marine geology and resource potential of Cascadia Basin: Circum-Pacific Council for Energy and Mineral Resources Earth Science Series, v. 6, p. 523–535.

Carmalt SW, St. John B, 1986, Giant oil and gas fields: American Association of Petroleum Geologists Memoir 40, p. 11–53.

Carmichael SMM, 1988, Linear estuarine conglomerate bodies formed during a mid-Albian marine transgression: "Upper Gates" Formation, Rocky Mountain Foothills of northeastern British Columbia: Canadian Society of Petroleum Geologists Memoir 15, p. 49–62.

Caron C, Homewood P, Wildi W, 1989, The original Swiss flysch: a reappraisal of the type deposits in the Swiss Prealps: Earth-Science Reviews, v. 26, p. 1–45.

Carr MJ, 1976, Underthrusting and Quaternary faulting in northern Central America: Geological Society of America Bulletin, v. 87, p. 825–829.

Carson B, 1977, Tectonically induced deformation of deep-sea sediments off Washington and northern Oregon: mechanical consolidation: Marine Geology, v. 24, p. 289–308.

Carson B, 6 coauthors, 1986, Modern sediment dispersal and accumulation in Quinault submarine canyon—a summary: Marine Geology, v. 71, p. 1–13.

Carter RM, Hicks MD, Morris RJ, Turnbull IM, 1978, Sedimentation patterns in an ancient arc-trench-ocean basin complex: Carboniferous to Jurassic Rangitata orogen, New Zealand, in Stanley DJ, Kelling G (eds.), Sedimentation in submarine canyons, fans, and trenches: Dowden, Hutchinson and Ross, Stroudsburg, p. 340–361.

Cas R, Busby-Spera C (eds.), 1991, Volcaniclastic sedimentation: Sedimentary Geology, v. 74, 362 p.

Cas RAF, Jones JG, 1979, Paleozoic interarc basin in eastern Australia and a modern New Zealand analogue: New Zealand Journal of Geology and Geophysics, v. 22, p. 71–85.

Cas RAF, Wright JV, 1987, Volcanic successions modern and ancient: Allen and Unwin, Boston, 528 p.

Cas RAF, Wright JV, 1991, Subaqueous pyroclastic flows and ignimbrites: an assessment: Bulletin of Volcanology, v. 53, p. 357–380.

Cas RAF, Landis CA, Fordyce RE, 1989, A monogenetic, Surtla-type, surtseyan volcano from the Eocene-Oligocene Waiareka-Deborah volcanics, Otago, New Zealand: a model: Bulletin of Volcanology, v. 51, p. 281–298.

Cashman KV, Fiske RS, 1991, Fallout of pyroclastic debris from submarine volcanic eruptions: Science, v. 253, p. 270–280.

Catacosinos PA, Daniels PA, Jr, Harrison WB, III, 1990, Structure, stratigraphy, and petroleum geology of the Michigan basin: American Association of Petroleum Geologists Memoir 51, p. 561–601.

Cather SM, Chapin CE, 1990, Paleogeographic and paleotectonic setting of Laramide sedimentary basins in the central Rocky Mountain region: alternative interpretation: Geological Society of America Bulletin, v. 102, p. 256–258.

Cawood PA, 1983, Modal composition and detrital clinopyroxene geochemistry of lithic sandstones from the New England fold belt (east Australia): a Paleozoic forearc terrane: Geological Society of America Bulletin, v. 94, p. 1199–1214.

Cecil CB, 1990, Paleoclimate contols on stratigraphic repetiton of chemical and siliciclastic rocks: Geology, v. 18, p. 533–536.

Cerveny PF, Naeser ND, Zeitler PK, Naeser CW, Johnson NM, 1988, History of uplift and relief of the Himalaya during the

past 18 million years: evidence from fission-track ages of detrital zircons from sandstones of the Siwalik Group, in Kleinspehn KL, Paola C (eds.), New perspectives in basin analysis: Springer-Verlag, New York, p. 43–61.

Chai BHT, 1972, Structure and evolution of Taiwan: American Journal of Science, v. 272, p. 389–442.

Chamot-Rooke N, Renard V, LePichon X, 1987, Magnetic anomalies in the Shikoku Basin: a new interpretation: Earth and Planetary Science Letters, v. 83, p. 214–228.

Channell JET, D'Argenio B, Horvath F, 1979, Adria, the African promontory, in Mesozoic Mediterranean palaeogeography: Earth Science Reviews, v. 15, p. 213–292.

Channell JET, Winterer EL, Jansa LF (eds.), 1991, Palaeogeography and paleoceanography of Tethys: Palaeogeography, Palaeoclimatology, Palaeoecology, v. 87, p. 1–493.

Chapin CE, 1979, Evolution of the Rio Grande rift: a summary, in Riecker RE (ed.), Rio Grande rift: tectonics and magmatism: American Geophysical Union, Washington, p.1–6.

Chapin CE,-Cather SM, 1981, Eocene tectonics and sedimentation in the Colorado Plateau—Rocky Mountain area: Arizona Geological Society Digest, v. 14, p.173–198.

Charlton TR, Barber AJ, Barkham ST, 1991, The structural evolution of the Timor collision complex, eastern Indonesia: Journal of Structural Geology, v. 13, p. 489–500.

Chase CG, 1978, Extension behind island arcs and motion relative to hotspots: Journal of Geophysical Research, v. 83, p. 5385–5387.

Chatalov GA, 1991, Triassic in Bulgaria: a review: Bulletin of the Technical University of Istanbul, v. 44, p. 103–135.

Cheadle MJ, 6 coauthors, 1986, Geometries of deep crustal faults: evidence from the COCORP survey: American Geophysical Union and Geological Society of America Geodynamics Series 14, p. 305–312.

Cheadle MJ, McGeary S, Warner MR, Matthews DH, 1987, Extensional structures on the western UK continental shelf: a review of evidence from deep seismic profiling: Geological Society of London Special Publication 28, p. 445–465.

Chen MP, Shieh YT, Shyr CS, 1988, Seafloor physiography and surface sediments off southeastern Taiwan: Acta Geologica Taiwanica n. 26, p. 333–353.

Chen Q, Dickinson WR, 1986, Contrasting nature of petroliferous Meszoic-Cenozoic bains in eastern and western China: American Association of Petroleum Geologists Bulletin, v. 70, p. 263–275.

Chen W-P, Molnar P, 1983, Focal depths of intracontinental and intraplate earthquakes and their implications for the thermal and mechanical properties of the lithosphere: Journal of Geophysical Research, v. 88, p. 4183–4214.

Chen W-P, Nábelek J, 1988, Seismogenic strike-slip faulting and the development of the North China basin: Tectonics, v. 7, p. 975–989.

Chen WS, Wang Y, 1988, Development of deep-sea fan systems in Coastal Range basin, eastern Taiwan: Acta Geologica Taiwanica, n. 26, p. 37–56.

Cherven VB, 1983, A delta-slope-submarine fan model for Maestrichtian part of Great Valley sequence, Sacramento and San Joaquin basins, California: American Association of Petroleum Geologists Bulletin, v. 67, p. 772–816.

Cherven VB, 1986, Tethys-marginal sedimentary basins in western Iran: Geological Society of America Bulletin, v. 97, p. 516–522.

Chijiwa K, 1988, Post-Shimanto sedimentation and organic metamorphism: an example of the Miocene Kumano Group, Kii Peninsula: Modern Geology, v. 12, p. 363–387.

Chorowicz J, 1989, Transfer and transform fault zones in continental rifts: examples in the Afro-Arabian rift system. Implications of crust breaking: Journal of African Earth Sciences, v. 8, p. 203–214.

Chorowicz J, Sorlien C, 1992, Oblique extensional tectonics in the Malawi rift, Africa: Geological Society of America Bulletin, v. 104, p. 1015–1023.

Chorowicz J, Deffontaines B, Villemin T, 1989, Interprétation des structures transverses NE-SW du fossé rhénan en termes de failles de transfert. Apport de données multisources: Comptes Rendus hébdomadaires de l'Academie des Sciences de Paris, v. 309, p. 1067–1073.

Chough SK, Barg E, 1987, Tectonic history of Ulleung basin margin, East Sea (Sea of Japan): Geology, v. 15, p. 45–48.

Christensen MN, 1963, Structure of metamorphic rocks at Mineral King, California: University of California Publications in Geological Sciences, v. 42, p. 159–198.

Christie-Blick N, Biddle KT, 1985, Deformation and basin formation along strike-slip faults: Society of Economic Paleontologists and Mineralogists Special Publication 37, p. 1–34.

Christie-Blick N, Mountain GS, Miller KG, 1990, Seismic stratigraphic record of sea-level change in Sea-level change [Studies in Geophysics]: National Research Council, Washington, p. 116–140.

Clague DA, 1981, Linear island and seamount chains, aseismic ridges and intraplate volcanism: results from DSDP: Society of Economic Paleontologists and Mineralogists Special Publication 32, p. 7–22.

Clark TH, Stearn CW, 1960, The geological evolution of North America: Ronald Press, New York, 434 p.

Clarke SH, Jr, 1987, Geology of the California continental margin north of Cape Mendocino: Circum-Pacific Council for Energy and Mineral Resources Earth Science Series, v. 6, p. 337–351.

Clarke SH, Jr, 1992, Geology of the Eel River basin and adjacent region: implications for late Cenozoic tectonics of the southern Cascadia subduction zone and Mendocino triple junction: American Association of Petroleum Geologists Bulletin, v. 76, p. 199–224.

Cleary WJ, Curran HA, Thayer PA, 1984, Barbados Ridge: inner trench slope sedimentation: Journal of Sedimentary Petrology, v. 54, p. 527–540.

Clendenin WC, 1989, Influence of a rigid block on rift margin evolution: Geology, v. 17, p. 412–415.

Cloetingh S, 1988, Intraplate stress: a new element in basin analysis, in Kleinspehn KL, Paola C (eds.), New perspectives in basin analysis: Springer-Verlag, New York, p. 205–230.

Cloetingh S, McQueen H, Lambeck K, 1985, On a tectonic mechanism for regional sealevel variations: Earth and Planetary Science Letters, v. 75, p. 157–166.

Cloos H, 1936, Einführung in die Geologie: Gebrüder Borntraeger, Berlin, 503 p.

Cloos H, 1939, Hebung-Spaltung-Vulkanismus: Geologische Rundschau, v. 30, p. 405–525.

Cloos M, 1982, Flow melanges: numerical modeling and geologic constraints on their origin in the Franciscan subduction

complex, California: Geological Society of America Bulletin, v. 93, p. 330–345.

Cloos M, 1984, Flow melanges and the structural evolution of accretionary wedges: Geological Society of America Special Paper 198, p. 71–79.

Cloos M, 1993, Lithospheric buoyancy and collisional orogenesis: subduction of oceanic plateaus, continental margins, island arcs, spreading ridges, and seamounts: Geological Society of America Bulletin, v. 105, p. 715–737.

Cloos M, Shreve RL, 1988a, Subduction-channel model of prism accretion, mélange formation, sediment subduction, and subduction erosion at convergent plate margins: 1. background and description: Pure and Applied Geophysics, v. 128, p. 455–500.

Cloos M, Shreve RL, 1988b, Subduction-channel model of prism accretion, mélange formation, sediment subduction, and subduction erosion at convergent plate margins: 2. implications and discussion: Pure and Applied Geophysics, v. 128, p. 501–545.

Cluff RM, Byrnes, AP, 1990, Lopatin analysis of maturation and petroleum generation in the Illinois basin: American Association of Petroleum Geologists Memoir 51, p. 425–454.

Cluzel D, Cadet J-P, Lapierre H, 1990, Geodynamics of the Ogcheon belt (South Korea): Tectonophysics, v. 183, p. 41–56.

Cochran JR, 1983a, A model for development of Red Sea: American Association of Petroleum Geologists Bulletin, v. 67, p.41–69.

Cochran JR, 1983b, Effects of finite rifting times on the development of sedimentary basins: Earth and Planetary Science Letters, v. 66, p.289–302.

Cochran JR, 1990, Himalayan uplift, sea level, and the record of Bengal Fan sedimentation at the ODP Leg 116 Sites: Proceedings of the Ocean Drilling Program, Scientific Results, v. 116, p. 397–414.

Cohen CR, 1980, Plate tectonic model for the Oligo-Miocene evolution of the western Mediterranean: Tectonophysics, v. 68, p. 283–311.

Colbert EH, 1968, Men and dinosaurs: the search in field and laboratory: E.P. Dutton and Company, New York, 283 p.

Colchen M, Fort M, Freytet P, 1980, Évolution paléogéographique et structurale du fossé de la Thakkhola-Mustang (Himalaya du Népal), implications sur l'histoire récente de la chaine himalayenne: Comptes Rendus hébdomadaires de l'Academie des Sciences de Paris, v. 290, p. 311–314.

Cole JW, 1979, Structure, petrology, and genesis of Cenozoic volcanism, Taupo volcanic zone, New Zealand—a review: New Zealand Journal of Geology and Geophysics, v. 22, p. 631–657.

Cole JW, 1984, Taupo-Rotorua depression: an ensialic marginal basin of the North Island, New Zealand: Geological Society of London Special Publication 16, p. 109–120.

Cole JW, 1986, Distribution and tectonic setting of Late Cenozoic volcanism in New Zealand: Royal Society of New Zealand Bulletin 23, p. 7–20.

Cole JW, Lewis KB, 1981, Evolution of the Taupo-Hikurangi subduction system: Tectonophysics, v. 72, p. 1–21.

Colella A, D'Alessandro A, De Rosa R, 1992, Deep-water trace fossils and their environmental significance in forearc and backarc Cenozoic successions around Izu-Bonin arc, Leg 126 ODP: Proceedings of the Ocean Drilling Program, Scientific Results, v. 126, p. 209–229.

Coleman JM, Prior DB, Lindsay JR, 1983, Deltaic influences on shelf edge instability processes: Society of Economic Paleontologists and Mineralogists Special Publication 33, p. 121–137.

Coleman RG, 1984, The Red Sea: a small ocean basin formed by continental extension and sea floor spreading: Proceedings of the 27th International Geological Congress, Moscow, v. 23, p. 93–121.

Coleman RG, 1993, Geologic evolution of the Red Sea: Oxford University Press, New York, 186 p.

Colletta B, Angelier J, 1982, Sur les systèmes de blocs faillés basculés associés aux fortes extensions: étude préliminaire d'exemples ovest-américains (Nevada, U.S.A. et Basse Californie, Mexique): Comptes Rendus hébdomadaires de l'Academie des Sciences de Paris, v. 294, p. 467–469.

Collier REL, 1990, Eustatic and tectonic controls upon Quaternary coastal sedimentation in the Corinth basin, Greece: Journal of the Geological Society of London, v. 147, p. 301–314.

Collier REL, Thompson J, 1991, Transverse and linear dunes in an Upper Pleistocene marine sequence, Corinth basin, Greece: Sedimentology, v. 38, p. 1021–1040.

Collinson C, Sargeant ML, Jennings JR, 1988, Illinois basin region, in Sloss, L.L. (ed.), Sedimentary cover—North American craton: US [The Geology of North America, v. D-2]: Geological Society of America, Boulder, p. 383–426.

Collot J-Y, 27 coauthors, 1992, Proceedings of the Ocean Drilling Program, Initial Reports [v. 134]: Ocean Drilling Program, College Station, 1136 p.

Condie KC, 1982a, Plate-tectonics model for Proterozoic continental accretion in the southwestern United States: Geology, v. 10, p. 37–42.

Condie KC, 1982b, Plate tectonics and crustal evolution: Pergamon Press, New York, 310 p.

Coney PJ, 1979, Tertiary evolution of Cordilleran metamorphic core complexes, in Armentrout JM, Cole MR, TerBest H, Jr (eds.), Cenozoic paleogeography of the western United States [Pacific Coast Paleogeography Symposium 3]: Pacific Section, Society of Economic Paleontologists and Mineralogists, Los Angeles, p. 15–28.

Coney PJ, 1980a, Cordilleran metamorphic core complexes: an overview: Geological Society of America Memoir 153, p. 7–31.

Coney PJ, 1980b, Introduction: Geological Society of America Memoir 153, p. 3–6.

Coney PJ, 1987, The regional tectonic setting and possible causes of Cenozoic extension in the North American Cordillera: Geological Society of London Special Publication 28, p. 177–186.

Coney PJ, Harms TA, 1984, Cordilleran metamorphic core complexes: Cenozoic extensional relics of Mesozoic compression: Geology, v. 12, p. 550–554.

Coney PJ, Reynolds SJ, 1977, Cordilleran Benioff zones: Nature, v. 270, p.403–406.

Conrey RM, 1990, Olivine analcite in the Cascade Range of Oregon: Journal of Geophysical Research, v. 95, p. 19639–19650.

Cook HE, 1975, North American stratigraphic principles as applied to deep-sea sediments: American Association of Petroleum Geologists Bulletin, v.59, p.817–837.

Cook HE, Mullins HT, 1983, Basin margin environment:

American Association of Petroleum Geologists Memoir 33, p. 539–617.

Coombs DS, 1954, The nature and alteration of some Triassic sediments from Southland, New Zealand: Royal Society of New Zealand Transactions, v. 82, p. 65–109.

Coombs DS, 1961, Some recent work on the lower grades of metamorphism: Australia Journal of Science, v. 24, p. 203–215.

Coombs DS, 5 coauthors, 1976, The Dun Mountain ophiolite belt, New Zealand, its tectonic setting, constitution, and origin, with special reference to the southern portion: American Journal of Science, v. 276, p. 561–603.

Cooper AK, Bruns TR, Wood RA, 1986, Shallow crustal structure of the Solomon Islands intra-arc basins from sonobuoy seismic studies: Circum Pacific Council for Energy and Mineral Resources Earth Science Series, v. 4, p. 135–156.

Cooper AK, Marlow MS, Scholl DW, 1987, Geologic framework of the Bering Sea crust: Circum-Pacific Council for Energy and Mineral Resources Earth Science Series, v. 6, p. 73–122.

Cooper AK, Marlow MS, Scholl DW, Stevenson AJ, 1992, Evidence for Cenozoic crustal extension in the Bering Sea region: Tectonics, v. 11, p. 719–731.

Cooper MR, 1990, Tectonic cycles in southern Africa: Earth Science Reviews, v. 28, p. 321–364.

Copeland P, Harrison TM, 1990, Episodic rapid uplift in the Himalaya revealed by 40Ar/39Ar analysis of detrital K-feldspar and muscovite, Bengal fan: Geology, v. 18, p. 354–357.

Coppens Y, 1991, The origin and evolution of man: Diogenes 155, p. 111–134.

Costain JK, 6 coauthors, 1989, Geophysical characteristics of the Appalachian crust, in Hatcher RD Jr, Thomas WA, Viele GW (eds.), The Appalachian-Ouachita orogen in the United States [The Geology of North America, v. F-2]: Geological Society of America, Boulder, p. 385–416.

Couch RW, Pitts GS, Gemperle M, Braman DE, Veen CA, 1982, Gravity anomalies in the Cascade Range in Oregon: structural and thermal implications: Oregon Department of Geology and Mineral Industries Open File Report O-82-9, 66 p.

Coulbourn WT, 1981, Tectonics of the Nazca plate and the continental margin of western South America: Geological Society of America Memoir 154, p. 587–618.

Coulbourn WT, 1986, Sedimentologic summary, Nankai Trough Sites 582 and 583, and Japan Trench Site 584: Initial Reports of the Deep Sea Drilling Project, v. 87, p. 909–926.

Coulbourn WT, Moberly R, 1977, Structural evidence of the evolution of fore-arc basins off South America: Canadian Journal of Earth Sciences, v. 14, p. 102–116.

Courtillot V, 1982, Propagating rifts and continental breakup: Tectonics, v. 1, p. 239–250.

Courtillot V, Vink GE, 1983, How continents break-up: Scientific American, v. 249, p. 43–49.

Cousens BL, Allan JF, 1992, A Pb, Sr, and Nd isotopic study of basaltic rocks from the sea of Japan, Legs 127/128: Proceedings of the Ocean Drilling Program, Scientific Results, v. 127/128, Pt. 2, p. 805–817.

Covey M, 1986, The evolution of foreland basins to steady state: evidence from the western Taiwan foreland basin: International Association of Sedimentologists Special Publication 8, p. 77–90.

Cowan DS, 1974, Deformation and metamorphism of the Franciscan subduction zone complex, Pacheco Pass, California:

Geological Society of America Bulletin, v. 85, p. 1623–1634.

Cowan DS, 1985, Structural styles in Mesozoic and Cenozoic melanges in the western Cordillera of North America: Geological Society of America Bulletin, v. 96, p. 451–462.

Cowie PA, Scholz CH, 1992, Growth of faults by accumulation of seismic slip: Journal of Geophysical Research, v. 97, p. 11085–11095.

Cox A, Hart RB, 1986, Plate tectonics: how it works: Blackwell Scientific Publications, Palo Alto, 392 p.

Craddock JP, vanderPluijm BA, 1989, Late Paleozoic deformation of the cratonic carbonate cover of eastern North America: Geology, v. 17, p. 416–419.

Crawford AJ, Greene HG, Exon NF, 1988, Geology, petrology and geochemistry of submarine volcanoes around Epi Island, New Hebrides Island arc: Circum Pacific Council for Energy and Mineral Resources Earth Science Series, v. 8, p. 301–328.

Crews SG, Angevine CL, 1988, Modelling syntectonic sediment supply and subsidence adjacent to active thrust belts: Geological Society of America Abstracts with Programs, v. 20, p. A179.

Critelli S, 1993, Sandstone detrital modes in the Paleogene Liguride Complex, accretionary wedge of the southern Apennines (Italy): Journal of Sedimentary Petrology, v. 63, p. 464–476.

Critelli S, Garzanti E, 1994, Provenance of the lower Tertiary Murree redbeds (Hazara-Kashmir syntaxis, Pakistan) and initial rising of the Himalayas: Sedimentary Geology, v. 89, p. 265–284.

Critelli S, Le Pera E, 1994, Detrital modes and provenance of Miocene sandstones and modern sands of the southern Apennines thrust-top basins (Italy): Journal of Sedimentary Research, v. A64, p. 824–835.

Critelli S, DeRosa R, Platt JP, 1990, Sandstone detrital modes in the Makran accretionary wedge, southwest Pakistan: implications for tectonic setting and long-distance turbidite transportation: Sedimentary Geology, v. 68, p. 241–260.

Crone AJ, Haller KM, 1989, Segmentation of basin-and-range faults: examples from east-central Idaho and southwestern Montana: United States Geological Survey Open-File Report 89-315, p. 110–130.

Cronin DS, 5 coauthors, 1984, Hydrothermal and volcaniclastic sedimentation on the Tonga-Kermadec Ridge and in its adjacent basins: Geological Society of London Special Publication 16, p. 137–149.

Crook KAW, 1989, Suturing history of an allochthonous terrane at a modern plate boundary traced by flysch-to-molasse facies transitions: Sedimentary Geology, v. 61, p. 49–79.

Cross TA, 1986, Tectonic controls of foreland basin subsidence and Laramide style deformation, western United States: International Association of Sedimentologists Special Publication 8, p. 15–39.

Cross TA, 1988, Controls on coal distribution in transgressive-regressive cycles, Upper Cretaceous, western interior, U.S.A.: Society of Economic Paleontologists and Mineralogists Special Publication 42, p. 371–380.

Cross TA, Pilger RH, Jr, 1978, Tectonic controls of Late Cretaceous sedimentation, western interior, USA: Nature, v. 274, p. 653–657.

Cross TA, Pilger RH, 1982, Controls of subduction geometry, location of magmatic arcs, and tectonics of arc and back-arc

regions: Geological Society of America Bulletin, v. 93, p. 545–562.

Crossley DR, 1984, Controls of sedimentation in the Malawi rift valley, central Africa: Sedimentary Geology, v. 40, p. 33–50.

Crouch JK, Suppe J, 1993, Late Cenozoic tectonic evolution of the Los Angeles basin and inner California borderland: a model for core complex-like crustal extension: Geological Society of America Bulletin, v. 105, p. 1415–1434.

Crough ST, 1983, Rifts and swells: geophysical constraints on causality: Tectonophysics, v. 94, p. 23–37.

Crough ST, Jurdy DM, 1980, Subducted lithosphere, hotspots, and the geoid: Earth and Planetary Science Letters, v. 48, p. 15–22.

Crowell JC, 1974a, Sedimentation along the San Andreas fault, California: Society of Economic-Paleontologists and Mineralogists Special Publication 19, p.292–303.

Crowell JC, 1974b, Origin of late Cenozoic basins in southern California: Society of Economic Paleontologists and Mineralogists Special Publication 22, p.190–204.

Crowell JC, 1976, Implications of crustal stretching and shortening of coastal Ventura basin, California, in Howell DG (ed.), Aspects of the geologic history of the California borderland: Pacific Section, American Association of Petroleum Geologists Miscellaneous Publication 24, p. 365–382.

Crowell JC, 1982a, The tectonics of Ridge basin, southern California, in Crowell JC, Link MH (eds.), Geologic history of Ridge basin southern California [Book 22]: Pacific Section, Society of Economic Paleontologists and Mineralogists, Los Angeles, p. 25–41.

Crowell JC, 1982b, The Violin Breccia, Ridge basin, southern California, in Crowell JC, Link MH (eds.), Geologic history of Ridge basin southern California [Book 22]: Pacific Section, Society of Economic Paleontologists and Mineralogists, Los Angeles, p. 89–97.

Crowell JC, 1985, The recognition of transform terrane dispersion within mobile belts: Circum-Pacific Council for Energy and Mineral Resources Earth Science Series, v. 1, p. 51–62.

Crowell JC, 1987, Late Cenozoic basins of onshore southern California: complexity is the hallmark of their tectonic history, in Ingersoll RV, Ernst WG (eds.), Cenozoic basin development of coastal California [Rubey Volume VI]: Prentice-Hall, Englewood Cliffs, p. 207–241.

Crowell JC, Link MH (eds.), 1982, Geologic history of Ridge basin southern California [Book 22]: Pacific Section, Society of Economic Paleontologists and Mineralogists, Los Angeles, 304 p.

Crumeyrolle P, Rubino J-L, Clauzon G, 1991, Miocene depositional sequences within a tectonically controlled transgressive-regressive cycle: International Association of Sedimentologists Special Publication 12, p. 373–390.

Cullen AB, Pigott JD, 1989, Post-Jurassic tectonic evolution of Papua New Guinea: Tectonophysics, v. 162, p. 291–302.

Cunningham AB, Abbott PL 1986, Sedimentology and provenance of the Upper Cretaceous Rosario Formation south of Ensenada, Baja California, Mexico, in Abbott PL (ed.), Cretaceous stratigraphy western North America [Book 46]: Pacific Section, Society of Economic Paleontologists and Mineralogists, Los Angeles, p. 103–118.

Curray JR. 1991, Possible greenschist metamorphism at the base of a 22-km sedimentary section, Bay of Bengal: Geology, v. 19, p. 1097–1100.

Curray JR, Moore DG, 1971, Growth of the Bengal deep-sea fan and denudation in the Himalayas: Geological Society of America Bulletin, v. 82, p.563–572.

Curray JR, Moore DG, 1974, Sedimentary and tectonic processes in the Bengal deep-sea fan and geosyncline in Burk CA, Drake CL (eds.), The geology of continental margins: Springer-Verlag, New York, p.617–627.

Curray JR, Shor GG, Jr, Raitt RW, Henry M, 1977, Seismic refraction and reflection studies of the eastern Sunda and western Banda arcs: Journal of Geophysical Research, v. 82, p. 2479–2489.

Curray JR, Moore DG, Kelts K, Einsele G, 1982, Tectonics and geological history of the passive continental margin at the tip of Baja California: Initial Reports of the Deep-Sea Drilling Project, v. 64, Part 2, p. 1089–1116.

Dahlen FA, Suppe J, Davis D, 1984, Mechanics of fold-and-thrust belts and accretionary wedges: cohesive Coulomb theory: Journal of Geophysical Research, v. 89, p. 10087–10101.

Dahlstrom CDA, 1970, Structural geology in the eastern margin of the Canadian Rocky Mountains: Bulletin of Canadian Petroleum Geology, v. 18, p. 332–406.

Dalmayrac B, Molnar P, 1981, Parallel thrust and normal faulting in Peru and constraints on the state of stress: Earth and Planetary Science Letters, v. 55, p. 473–481.

Daly MC, Chorowicz J, Fairhead JD, 1989, Rift basin evolution in Africa: the influence of reactivated steep basement shear zones: Geological Society of London Special Publication 44, p. 309–334.

Dalziel IW, 1981, Back-arc extension in the southern Andes: a review and critical reappraisal: Philosophical Transactions of the Royal Society of London, v. A300, p. 319–335.

Dalziel IWD, 1991, Pacific margins of Laurentia and East Antarctica-Australia as a conjugate rift pair: evidence and implications for an Eocambrian supercontinent: Geology, v. 19, p. 598–601. (Also, see Comments and Replies, v. 20, p. 87–88 and 190–191.)

Dalziel IW, DeWit MJ, Palmer KF, 1974, Fossil marginal basin in the southern Andes: Nature, v. 250, p. 291–294.

Damanti JF, 1989, Evolution of the Bermejo foreland basin: provenance, drainage development, and diagenesis [PhD thesis]: Cornell University, Ithaca, 176 p.

Damuth JE, 1977, Late Quaternary sedimentation in the western equatorial Atlantic: Geological Society of America Bulletin, v. 88, p. 695–710.

Damuth JE, 1979, Migrating sediment waves created by turbidity currents in the northern South China Basin: Geology, v. 7, p. 520–523.

Damuth JE, Embley RW, 1979, Upslope flow of turbidity currents on the northwest flank of the Ceara Rise: western equatorial Atlantic: Sedimentology, v. 26, p. 825–834.

Damuth JE, 7 coauthors, 1983, Distributary channel meandering and bifurcation patterns on the Amazon deep-sea fan as revealed by long-range side scan sonar (GLORIA): Geology, v. 11, p. 94–98.

Davey FJ, 5 coauthors, 1986, Structure of a growing accretionary prism, Hikurangi margin, New Zealand: Geology, v. 14, p. 663–666.

Davies GF, 1981, Regional compensation of subducted lithos-

phere: effects on geoid, gravity, and topography from a preliminary model: Earth and Planetary Science Letters, v. 54, p. 431–441.

Davies GF, Richards MA, 1992, Mantle convection: Journal of Geology, v. 100, p. 151–206.

Davis D, Suppe J, Dahlen FA, 1983, Mechanics of fold-and-thrust belts and accretionary wedges: Journal of Geophysical Research, v. 88, p. 1153–1172.

Davis EE, Hyndman RD, 1989, Accretion and recent deformation of sediments along the northern Cascadia subduction zone: Geological Society of America Bulletin, v. 101, p. 1465–1480.

Davis EE, 6 coauthors, 1987, Massive sulfide in a sedimented rift valley, northern Juan de Fuca Ridge: Earth and Planetary Science Letters, v. 86, p. 49–61.

Davis EE, Hyndman RD, Villinger H, 1990, Rates of fluid expulsion across the northern Cascadia accretionary prism: constraints from new heat flow and multichannel seismic reflection data: Journal of Geophysical Research, v. 95, p. 8869–8889.

Davis GA, 1977, Tectonic evolution of the Pacific Northwest: Precambrian to present [Subappendix 2R C, Preliminary Safety Analysis Report, Amendment 23, Nuclear Project No. 1]: Washington Public Power Supply System, p. i 2Rc–46.

Davis GA, Monger JWH, Burchfiel BC, 1978, Mesozoic construction of the Cordilleran "collage", central British Columbia to central California, in Howell DG, McDougall KA (eds.), Mesozoic paleogeography of the western United States [Pacific Coast Paleogeography Symposium 2]: Pacific Section, Society of Economic Paleontologists and Mineralogists, Los Angeles, p. 1–32.

Davis HG, 1987, Pre-Mississippian hydrocarbon potential of Illinois basin: American Association of Petroleum Geologists Bulletin, v. 71, p. 546–547.

Davis HG, 1990, Pre-Mississippian hydrocarbon potential of the Illinois basin: American Association of Petroleum Geologists Memoir 51, p. 473–489.

Davis PM, 1991, Continental rift structures and dynamics with reference to teleseismic studies of the Rio Grande and East African rifts: Tectonophysics, v. 197, p. 309–325.

Davis RA, Jr, 1983, Depositional systems: a genetic approach to sedimentary geology: Prentice-Hall, Englewood Cliffs, 669 p.

Davis TL, Namson J, 1989, A cross section of the Los Angeles area: seismically active fold and thrust belt, the 1987 Whittier Narrows earthquake, and earthquake hazard: Journal of Geophysical Research, v. 94, p. 9644–9664.

DeBoer JZ, Clifford AE, 1988, Mesozoic tectonogenesis: development and deformation of "Newark" rift zones in the Appalachians (with special emphasis on the Hartford Basin, Connecticut), in Manspeizer W (ed.), Triassic-Jurassic rifting. Part A: Elsevier, Amsterdam, p. 275–306.

DeBoer J, Odom LA, Ragland PC, Snider FS, Tilford NR, 1980, The Bataan orogene: eastward subduction, tectonic rotations, and volcanism in the western Pacific (Philippines): Tectonophysics, v. 67, p. 251–282.

DeBrito Neves BB, Fuck RA, Cordani UG, Thomaz A, 1984, Influence of basement structures on the evolution of the major sedimentary basins of Brazil: a case of tectonic heritage: Journal of Geodynamics, v. 1, p. 495–510.

DeCelles PG, 1988, Lithologic provenance modeling applied to the Late Cretaceous synorogenic Echo Canyon Conglomerate, Utah: a case of multiple source areas: Geology, v. 16, p. 1039–1043.

DeCelles PG, Hertel F, 1989, Petrology of fluvial sands from the Amazonian foreland basin, Peru and Bolivia: Geological Society of America Bulletin, v. 101, p. 1552–1562.

DeCelles PG, 6 coauthors, 1991, Kinematic history of a foreland uplift from Paleocene synorogenic conglomerate, Beartooth Range, Wyoming and Montana: Geological Society of America, v. 103, p. 1458–1475.

DeCharpal O, Guennoc P, Montadert L, Roberts DG, 1978, Rifting, crustal attenuation and subsidence in the Bay of Biscay: Nature, v. 275, p. 706–711.

Defant MJ, DeBoer JZ, Oles D, 1988, The western central Luzon volcanic arc, the Philippines: two arcs divided by rifting?: Tectonophysics, v. 145, p. 305–317.

DeGraciansky PC, 14 coauthors, 1985, The Goban Spur transect: geologic evolution of a sediment-starved passive continental margin: Geological Society of America Bulletin, v. 96, p. 58–76.

DeHeinzelin J, 1982, The lower Omo and Turkana basin in Plio- lower Pleistocene times: Zeitschrift fur Geomorphologie, Supplementband 42 (Graben, Geology and Geomorphogenesis), p. 35–41.

DeKock GS, 1992, Forearc basin evolution in the Pan-African Damara belt, central Namibia: the Hureb Formation of the Khomas Zone: Precambrian Research, v. 57, p. 169–194.

DeLong SE, Dewey JF, Fox PJ, 1979, Topographic and geologic evolution of fracture zones: Journal of the Geological Society of London, v. 136, p. 303–310.

DeMets C, 1992, Oblique convergence and deformation along the Kuril and Japan trenches: Journal of Geophysical Research, v. 97, p. 17615–17625.

Deng Q, Wu D, Zhang P, Chen S, 1986, Structure and deformation character of strike-slip fault zones: Pure and Applied Geophysics, v. 124, p. 204–223.

Dengo G, Bohnenberger O, Bonis S, 1970, Tectonics and volcanism along the Pacific marginal zone of Central America: Geologische Rundschau, v. 59, p. 1215–1232.

Dennis JG, 1967, International tectonic dictionary: American Association of Petroleum Geologists Memoir 7, 196 p.

DePolo CM, Clark DG, Slemmons DB, Aymard WH, 1989, Historical Basin and Range province surface faulting and fault segmentation: United States Geological Survey Open-File Report 89-315, p. 131–162.

Dercourt J, 17 coauthors, 1986, Geologic evolution of the Tethys belt from the Atlantic to the Pamirs since the Lias: Tectonophysics, v. 123, p. 241–315.

DeRito RF, Cozzarelli FA, Hodge DS, 1983, Mechanism of subsidence of ancient cratonic rift basins: Tectonophysics, v. 94, p.141–168.

Dewey JF, 1971, A model for the lower Palaeozoic evolution of the southern margin of the early Caledonides of Scotland and Ireland: Scottish Journal of Geology, v. 7, p. 219–240.

Dewey JF, 1977, Suture zone complexities: a review: Tectonophysics, v. 40, p.53–67.

Dewey JF, 1980, Episodicity, sequence and style at convergent plate boundaries: Geological Association of Canada Special Paper 20, p. 553–573.

Dewey JF, 1988, Extensional collapse of orogens: Tectonics, v. 7, p. 1123–1139.

Dewey JF, Bird JM, 1970, Mountain belts and the new global

tectonics: Journal of Geophysical Research, v. 75, p. 2625–2647.

Dewey JF, Burke K, 1974, Hot spots and continental break-up: implications for collisional orogeny: Geology, v.2, p.57–60.

Dewey JG, Kidd WSF, 1974, Continental collisions in the Appalachian-Caledonian orogenic belt: variations related to complete and incomplete suturing: Geology, v. 2, p. 543–546.

Dewey, J.F., and Lamb, S.H., 1991, Active tectonics of the Andes: Bulletin of the Technical University of Istanbul, v. 44, p. 497–522.

Dewey JF, Şengör AMC, 1979, Aegean and surrounding regions: complex multiplate and continuum tectonics in a convergent zone: Geological Society of America Bulletin, v. 90, Part 1, p. 84–92.

Dewey JF, Windley BF, 1988, Palaeocene-Oligocene tectonics of NW Europe: Geological Society of London Special Publication 39, p. 25–31.

Dewey JF, Pitman WC, III, Ryan WBF, Bonnin J, 1973, Plate tectonics and the evolution of the Alpine system: Geological Society of America Bulletin, v. 84, p. 3137–3180.

Dewey JF, Hempton MR, Kidd WSF, Saroğlu F, Şengör AMC, 1986, Shortening of continental lithosphere: the neotectonics of Eastern Anatolia–a young collision zone: Geological Society of London Special Publication 19, p. 3–36.

Dewey JF, Shackleton RM, Chang C, Sun Y, 1988, The tectonic evolution of the Tibetian Plateau: Philosophical Transactions of the Royal Society of London, v. 327A, p. 379–413.

Dewey JF, Cande S, Pitman WC, III, 1989a, Tectonic evolution of the India/Eurasia collision zone: Eclogae Geologicae Helvetiae, v. 82, p. 717–734.

Dewey JF, Helman ML, Turco E, Hutton DHW, Knott SD, 1989b, Kinematics of the western Mediterranean: Geological Society of London Special Publication 45, p. 265–283.

Dewey JF, Gass IG, Curry GB, Harris NBW, Şengör AMC (eds.), 1991, Allochthonous terranes: Cambridge University Press, Cambridge, 199 p.

DeWit MJ, Stern CR, 1981, Variations in the degree of crustal extension during formation of a back-arc basin: Tectonophysics, v. 72, p. 229–260.

Diament M, 9 coauthors, 1992, Mentawai fault zone off Sumatra: a new key to the geodynamics of western Indonesia: Geology, v. 20, p. 259–262.

Dibblee TW, Jr, 1967, Areal geology of the western Mojave Desert, California: United States Geological Survey Professional Paper 522, 152 p.

Dibblee TW, Jr, 1977, Strike-slip tectonics of the San Andreas fault and its role in Cenozoic basin development, in Nilsen TH (ed.), Late Mesozoic and Cenozoic sedimentation and tectonics in California [Short Course Notes]: San Joaquin Geological Society, Bakersfield, p. 26–38.

Dibblee TW, Jr, 1980, Cenozoic rocks of the Mojave Desert, in Fife DL, Brown AR, (eds.), Geology and mineral wealth of the California Desert: South Coast Geological Society, Santa Ana, p. 41–68.

Dibblee TW, Jr, 1982, Regional geology of the Transverse Ranges province of southern California, in Fife DC, Minch JA (eds.), Geology and mineral wealth of the California Transverse Ranges [Mason Hill Volume]: South Coast Geological Society, Santa Ana, p. 7–26.

Dick HJ, 1982, The petrology of two back-arc basins of the northern Philippine Sea: American Journal of Science, v. 282,

p. 644–700.

Dickinson WR, 1968, Sedimentation of volcaniclastic strata of the Pliocene Koroimavua Group in northwest Viti Levu, Fiji: American Journal of Science, v. 266, p. 440–453.

Dickinson WR, 1970a, Relations of granites, andesites, and derivative sandstones to arc-trench tectonics: Reviews of Geophysics and Space Physics, v. 8, p. 813–862.

Dickinson WR, 1970b, Interpreting detrital modes of graywacke and arkose: Journal of Sedimentary Petrology, v. 40, p. 695–707.

Dickinson WR, 1971a, Clastic sedimentary sequences deposited in shelf, slope, and trough settings between magmatic arcs and associated trenches: Pacific Geology, v. 3, p. 15–30.

Dickinson WR, 1971b, Detrital modes of New Zealand graywackes: Sedimentary Geology, v. 5, p. 37–56.

Dickinson WR, 1973, Widths of modern arc-trench gaps proportional to past duration of igneous activity in associated magmatic arcs: Journal of Geophysical Research, v. 78, p. 3376–3389.

Dickinson WR, 1974a, Sedimentation within and beside ancient and modern magmatic arcs: Society of Economic Paleontologists and Mineralogists Special Publication 19, p. 230–239.

Dickinson WR, 1974b, Plate tectonics and sedimentation: Society of Economic Paleontologists and Mineralogists Special Publication 22, p.1–27

Dickinson WR, 1975, Potash-depth (K-h) relations in continental margin and intra-oceanic magmatic arcs: Geology, v. 3, p. 53–56.

Dickinson WR, 1976a, Plate tectonic evolution of sedimentary basins: American Association of Petroleum Geologists Continuing Education Course Notes Series 1, 62 p.

Dickinson WR, 1976b, Sedimentary basins developed during evolution of Mesozoic-Cenozoic arc-trench system in western North America: Canadian Journal of Earth Sciences, v. 13, p. 1268–1287.

Dickinson WR, 1978, Plate tectonic evolution of north Pacific rim: Journal of the Physics of the Earth, v. 26, Supplement, p. S1–S19.

Dickinson WR, 1979a, Mesozoic forearc basin in central Oregon: Geology, v. 7, p. 166–170.

Dickinson WR, 1979b, Cenozoic plate tectonic setting of the Cordilleran region in the United States, in Armentrout JM, Cole MR, TerBest H, Jr (eds.), Cenozoic paleogeography of the western United States [Pacific Coast Palegeography Symposium 3]: Pacific Section, Society of Economic Paleontologists and Mineralogists, Los Angeles, p. 1–13.

Dickinson WR, 1980, Plate tectonics and key petrologic associations: Geological Association of Canada Special Paper 20, p.341–360.

Dickinson WR, 1981, Plate tectonics and the continental margin of California, in Ernst WG (ed.), The geotectonic development of California [Rubey Volume I]: Prentice-Hall, Englewood Cliffs, p. 1–28.

Dickinson WR, 1982, Compositions of sandstones in circum-Pacific subduction complexes and fore-arc basins: American Association of Petroleum Geologists Bulletin, v. 66, p. 121–137.

Dickinson WR, 1985, Interpreting provenance relations from detrital modes of sandstones, in Zuffa GG (ed.), Provenance of arenites: D. Reidel, Dordrecht, p. 333–361.

Dickinson WR, 1988, Provenance and sediment dispersal in relation to paleotectonics and paleogeography of sedimentary basins, in Kleinspehn KL, Paola C (eds.), New perspectives in basin analysis: Springer-Verlag, New York, p. 3–25.

Dickinson WR, 1990, Paleogeographic and paleotectonic setting of Laramide sedimentary basins in the central Rocky Mountain region: reply: Geological Society of America Bulletin, v. 102, p. 281–282.

Dickinson WR, 1991, Tectonic setting of faulted Tertiary strata associated with the Catalina core complex in southern Arizona: Geological Society of America Special Paper 264, 106 p.

Dickinson WR, 1993a, Exxon global cycle chart: an event for every occasion: Comment: Geology, v. 21, p. 282–283.

Dickinson WR, 1993b, Basin geodynamics: Basin Research, v. 5, p. 195–196.

Dickinson WR, Coney PJ, 1980, Plate tectonic constraints on the origin of the Gulf of Mexico, in Pilger, RH, Jr (ed.), The origin of the Gulf of Mexico and the early opening of the central North Atlantic Ocean: Louisiana State University, Baton Rouge, p. 27–36.

Dickinson WR, Rich EI, 1972, Petrologic intervals and petrofacies in the Great Valley sequence, Sacramento Valley, California: Geological Society of America Bulletin, v. 83, p. 3007–3024.

Dickinson WR, Seely DR, 1979, Structure and stratigraphy of forearc regions: American Association of Petroleum Geologists Bulletin, v. 63, p. 2–31.

Dickinson WR, Snyder WS, 1978, Plate tectonics of the Laramide orogeny: Geological Society of America Memoir 151, p. 355–366.

Dickinson WR, Suczek CA, 1979, Plate tectonics and sandstone compositions: American Association of Petroleum Geologists Bulletin, v.63, p.2164–2182.

Dickinson WR, Ingersoll RV, Graham SA, 1979, Paleogene sediment dispersal and paleotectonics in northern California: Geological Society of America Bulletin, v. 90, Part I, p. 897–898, Part II, p. 1458–1528.

Dickinson WR, Ingersoll RV, Cowan DS, Helmold KP, Suczek CA, 1982, Provenance of Franciscan graywackes in coastal California: Geological Society of America Bulletin, v. 93, p. 95–107.

Dickinson WR, Harbaugh DW, Saller AH, Heller PL, Snyder WS, 1983, Detrital modes of upper Paleozoic sandstones derived from Antler orogen in Nevada: implications for nature of Antler orogeny: American Journal of Science, v. 283, p. 481–509.

Dickinson WR, 8 coauthors, 1987, Geohistory analysis of rates of sediment accumulation and subsidence for selected California basins, in Ingersoll RV, Ernst WG (eds.), Cenozoic basin development of coastal California [Rubey Volume VII]: Prentice-Hall, Englewood Cliffs, p. 2–23.

Dickinson WR, 6 coauthors, 1988, Paleogeographic and paleotectonic setting of Laramide sedimentary basins in the central Rocky Mountain region: Geological Society of America Bulletin, v. 100, p. 1023–1039.

Dickinson WR, 6 coauthors, 1990, Paleogeographic and paleotectonic setting of Laramide sedimentary basins in the central Rocky Mountain region: reply: Geological Society of America Bulletin, v. 102, p. 259–260.

Dietz RS, Holden JC, 1966, Miogeoclines (miogeosynclines) in space and time: Journal of Geology, v.74, p.566–583. (Also, see two discussions and replies, v.76, p.111–121.)

Dietz RS, Holden JC, 1970, Reconstruction of Pangaea: breakup and dispersion of continents, Permian to present: Journal of Geophysical Research, v. 75, p. 4939–4956.

Dingle RV, 1976, A review of the sedimentary history of some post-Permian continental margins of Atlantic-type: Anais da Academia Brasileira de Ciências, v. 48 (Suplemento), p. 67–80.

Dixey F, 1956, The East African rift system: Colonial Geology and Mineral Resources Bulletin, Supplement 1, p. 1–7.

Dixon JE, Fitton JG, Frost RTC, 1981, The tectonic significance of post-Carboniferous igneous activity in the North Sea basin, in Illing LV, Hobson GD (eds.), Petroleum geology of the continental shelf of north west Europe: Heydon and Sons, London, p. 121–137.

Dixon TH, Ivins ER, Franklin BJ, 1989, Topographic and volcanic asymmetry around the Red Sea: constraints on rift models: Tectonics, v. 8, p. 1193–1216.

Dobson MR, Scholl DW, Stevenson AJ, 1991, Interplay between tectonics and sea-level changes as revealed by sedimentation patterns in the Aleutians: International Association of Sedimentologists Special Publication 12, p. 151–163.

Doebl F, 1970, Die Tertiären und Onartären sedimente des südlichen Rheingrabens, in Illies JM, Mueller S (eds.), Graben problems: E.Schweizerbart'sche Verlagsbuchhandlung, Stuttgart, p. 56–66.

Dokka RK, Travis CJ, 1990: Late Cenozoic strike-slip faulting in the Mojave Desert, California: Tectonics, v. 9, p. 311–340.

Dolan JF, Beck C, Ogawa Y, 1989, Upslope deposition of extremely distal turbidites: an example from the Tiburon Rise, west-central Atlantic: Geology, v. 17, p. 990–994.

Dolan JF, Beck C, Ogawa Y, Klaus A, 1990, Eocene-Oligocene sedimentation in the Tiburon Rise/ODP Leg 110 area: an example of significant upslope flow of distal turbidity currents: Proceedings of the Ocean Drilling Program, Scientific Results, v. 110, p. 47–84.

D'Onfro P, Glagola P, 1983, Wrench fault, southeast Asia: American Association of Petroleum Geologists Studies in Geology Series 15, v. 3, p. 4.2-9–4.2-12.

Dorsey RJ, 1988, Provenance evolution and unroofing history of a modern arc-continent collision: evidence from petrography of Plio-Pleistocene sandstones, eastern Taiwan: Journal of Sedimentary Petrology, v. 58, p. 208–218.

Dorsey RJ, 1992, Collapse of the Luzon volcanic arc during onset of arc-continent collision: evidence from a Miocene-Pliocene unconformity, eastern Taiwan: Tectonics, v. 11, p. 177–191.

Dorsey RJ, Lundberg N, 1988, Lithofacies and basin reconstruction of the Plio-Pleistocene collisional basin, Coastal Range of eastern Taiwan: Acta Geologica Taiwanica, n. 26, p. 57–132.

Dott RH, Jr, 1978, Tectonics and sedimentation a century later: Earth-Science Reviews, v. 14, p. 1–34.

Dott RH, Jr, Batten RL, 1988, Evolution of the Earth [Fourth Edition]: McGraw-Hill, New York, 643 p.

Dott RH, Jr, Shaver RH (eds.), 1974, Modern and ancient geosynclinal sedimentation: Society of Economic Paleontologists and Mineralogists Special Publication 19, 380 p.

Dow WG, 1979, Petroleum source beds on continental slopes and rises: American Association of Petroleum Geologists Memoir 29, p. 423–442.

Duane DB, Stubblefield WL, 1988, Sand and gravel resources: U.S. Atlantic continental shelf, in Sheridan RG, Grow SA (eds.),

The Atlantic continental margin: U.S. [The Geology of North America, v. I–2]: Geological Society of America, Boulder, p. 481–500.

Dubois J, Launay J, Recey J, 1975, Some new evidence on lithospheric bulges close to island arcs: Tectonophysics, v. 26, p. 189–196.

DuDresnay R, 1979, Sédiments jurassiques du domaine des chaîne atlasiques du Maroc: Publication spéciale de l'Association Sedimentologique de France 1, p. 345–365.

Duffield WA, Bacon CR, Delaney PT, 1986, Deformation of poorly consolidated sediment during shallow emplacement of a basalt sill, Coso Range, California: Bulletin of Volcanology, v. 48, p. 97–107.

Dumitru TA, 1988, Subnormal geothermal gradients in the Great Valley forearc basin, California, during Franciscan subduction: a fission-track study: Tectonics, v. 7, p. 1201–1221.

Dunbar JA, Sawyer DS, 1988, Continental rifting at pre-existing lithospheric weaknesses: Nature, v. 333, p. 450–452.

Dunbar JA, Sawyer DS, 1989, How pre-existing weaknesses control the style of continental breakup: Journal of Geophysical Research, v. 94, p. 7278–7292.

Dunkelman TJ, Karson JA, Rosendahl BR, 1988, Structural style of the Turkana rift, Kenya: Geology, v. 16, p. 258–261.

Durkee EF, Pederson SL, 1961, Geology of northern Luzon, Philippines: American Association of Petroleum Geologists Bulletin, v. 45, p. 137–168.

Eardley AJ, 1962, Structural geology of North America (2nd edition): Harper and Row, New York, 743 p.

Eaton GP, 1979, A plate-tectonic model for late cenozoic crustal spreading in the western United States, in Riecker RE (ed.), Rio Grande rift: tectonics and magmatism: American Geophysical Union, Washington, p. 6–32.

Ebinger CJ, 1989a, Tectonic devolopment of the western branch of the East African rift system: Geological Society of America Bulletin, v. 101, p. 885–03.

Ebinger CJ, 1989b, Geometric and kinematic development of border faults and accommodation zones, Kivu-Rusizi rift, Africa: Tectonics, v. 8, p. 117–33.

Ebinger CJ, Bechtel TD, Forsyth DW, Bowin CO, 1989a, Effective elastic plate thickness beneath the East African and Afar plateaus and dynamic compensation of the uplifts: Journal of Geophysical Research, v. 94, p. 2883–901.

Ebinger CJ, Deino AL, Drake RE, Tesha AL, 1989b, Chronology of volcanism and rift basin propagation: Rungwe volcanic province, East Africa: Journal of Geophysical Research, v. 94, p. 15785–5803.

Ebinger CJ, Karner GD, Weissel JK, 1991, Mechanical strength of extended continental lithosphere: constraints from the western rift system, East Africa: Tectonics, v. 10, p. 1239–256.

Eckel EB, Meyers WB, 1946, Quicksilver deposits of the New Idria district, San Benito and Fresno counties, California: California Journal of Mines and Geology, v. 42, p. 81–24.

EEZ SCAN 84 Scientific Staff, 1986, Atlas of the Exclusive Economic Zone, western conterminous United States: United States Geological Survey Miscellaneous Investigations Series I 1792, 1:500,000.

Egloff F, 8 coauthors, 1991, Contrasting structural styles of the eastern and western margins of the southern Red Sea: the 1988 SONNE experiment: Tectonophysics, v. 198, p. 329–354.

Eguchi T, Uyeda S, Maki T, 1979, Seismotectonics and tectonic history of the Andaman Sea: Tectonophysics, v. 57, p. 35–51.

Eguchi T, Uyeda S, Maki T, 1980, Seismotectonics and tectonic history of the Andaman Sea, in Toksöz MN, Uyeda S, Francheteau J (eds.), Oceanic ridges and arcs: Elsevier, New York, p. 425–441.

Eide EA, 1993, Petrology, geochemistry, and structure of high-pressure metamorphic rocks in Hubei Province, east central China, and their relationship to continental collision [PhD thesis]: Stanford University, Stanford, 235p.

Eide EA, Liou JG, McWilliams MD, 1992, Continent-continent collision and the tectonic history of the eastern Qinling-Dabie orogenic belt, east-central China: EOS, v. 73, p. 653.

Einsele G, 1982, Mechanism of sill intrusion into soft sediment and expulsion of pore water: Initial reports of the Deep-Sea Drilling Project, v. 64, Pt. 2, p. 1169–1176.

Einsele G, 1986, Interaction between sediments and basalt injections in young Gulf of California type spreading centers: Geologische Rundschau, v. 75, p. 197–208.

Einsele G, 1992, Sedimentary basins: evolution, facies, and sediment budget: Springer-Verlag, Berlin, 628p.

Eisbacher GH, 1974, Evolution of successor basins in the Canadian Cordillera: Society of Economic Paleontologists and Mineralogists Special Publication 19, p. 274–291.

Eisbacher GH, 1981, Late Mesozoic-Paleogene Bowser basin molasse and Cordilleran tectonics, western Canada: Geological Association of Canada Special Paper 23, p. 125–151.

Eisbacher GH, 1985, Pericollisional strike-slip faults and synorogenic basins, Canadian Cordillera: Society of Economic Paleontologists and Mineralogists Special Publication 37, p. 265–282.

Eisbacher GH, Carrigy MA, Campbell RB, 1974, Paleodrainage pattern and late-orogenic basins of the Canadian Cordillera: Society of Economic Paleontologists and Mineralogists Special Publication 22, p. 143–166.

Elder WP, Gustason ER, Sageman BB, 1994, Correlation of basinal carbonate cycles to nearshore parasequences in the Late Cretaceous Greenhorn seaway, Western Interior, U.S.A.: Geological Society of America Bulletin, v. 106, p. 892–902.

Ellis PG, McClay KR, 1988, Listric extensional fault systems - results of analogue model experiments: Basin Research, v. 1, p. 55–70.

Embley RW, Langseth MG, 1977, Sedimentation processes on the continental rise of northeastern South America: Marine Geology, v. 25, p. 279–297.

Emery KO, 1977, Structure and stratigraphy of divergent continental margins: American Association of Petroleum Geologists Education Course Note Series 5, p. B1–B20.

Emmel FJ, Curray JR, 1985, Bengal fan, Indian Ocean, in Bouma AH, Normark WR, Barnes NE (eds.), Submarine fans and related turbidite systems: Springer-Verlag, New York, p. 107–112.

Emslie RF, 1978, Anorthosite massifs, rapakivi granites, and Late Proterozoic rifting of North America: Precambrian Research, v. 7, p. 61–98.

Enfield M, Coward MP, 1987, The structures of the West Orkney basin, northern Scotland: Journal of the Geological Society of London, v. 144, p. 871–884.

Engebretson DA, Cox A, Gordon RG, 1985, Relative motions between oceanic and continental plates in the Pacific Basin:

Geological Society of America Special Paper 206, 59p.

England PC, Houseman GA, 1988, The mechanics of the Tibetan Plateau: Philosophical Transactions of the Royal Society of London, v. 326A, p. 301–320.

England PC, Houseman GA, 1989, Extension during continental convergence, with application to the Tibetan Plateau: Journal of Geophysical Research, v. 94, p. 17561–17579.

Enkin RJ, Yang Z, Chen Y, Courtillot V, 1992, Paleomagnetic constraints on the geodynamic history of the major blocks of China from the Permian to the present: Journal of Geophysical Research, v. 97, p. 13953–13989.

Erlich EN, 1968, Recent movements and Quaternary volcanic activity within the Kamchatka territory: Pacific Geology, v. 1, p. 23–39.

Erlich EN, 1979, Recent structure of Kamchatka and position of Quaternary volcanoes: Bulletin Volcanique, v. 42, p. 13–42.

Ernst WG, 1970, Tectonic contact between the Franciscan melange and the Great Valley sequence—crustal expression of a late Mesozoic Benioff zone: Journal of Geophysical Research, v. 75, p. 886–901.

Ernst WG, 1971, Metamorphic zonations on presumably subducted lithospheric plates from Japan, California, and the Alps: Contributions to Mineralogy and Petrology, v. 34, p. 43–59.

Ernst WG, 1975, Systematics of large-scale tectonics and age progressions in Alpine and circum-Pacific blueschist belts: Tectonophysics, v. 26, p. 229–246.

Ernst WG, 1983, Mountain building and metamorphism: a case history from Taiwan, in Hsu KJ (ed.), Mountain building: Academic Press, London, p. 247–256.

Ernst WG, 1984, Californian blueschists, subduction, and the significance of tectonostratigraphic terranes: Geology, v. 12, p. 436–440.

Ernst WG, Ho CS, Liou JG, 1985, Rifting, drifting, and crustal accretion in the Taiwan sector of the Asiatic continental margin: Circum-Pacific Council for Energy and Mineral Resources Earth Science Series, v. 1, p. 375–390.

Etheridge MA, 1986, On the reactivation of extensional fault systems: Philosophical Transactions of the Royal Society of London, v. 317, p. 179–194.

Etheridge MA, Symonds PA, Lister GS, 1989, Application of the detachment model to reconstruction of conjugate passive margins: American Association of Petroleum Geologists Memoir 46, p. 21–40.

Ethington RL, Finney SC, Repetski JE, 1989, Biostratigraphy of the Paleozoic rocks of the Ouachita orogen, Arkansas, Oklahoma, west Texas, in Hatcher RD, Jr, Thomas WA, Viele GW (eds.), The Appalachian-Ouachita orogen in the United States [The Geology of North America, v. F-2]: Geological Society of America, Boulder, p. 563–574.

Evans JW, 1925, Regions of tension: Proceedings of the Geological Society of London, v. lxxxi, p. lxxix–cxxii.

Evarts RC, 1977, The geology and petrology of the Del Puerto ophiolite, Diablo Range, central California Coast Ranges: Oregon Department of Geology and Mineral Industries Bulletin 95, p. 121–139.

Everhart DL, 1950, Skaggs Springs quicksilver mine, Sonoma County, California: California Journal of Mines and Geology, v. 46, p. 385–394.

Ewart A, Brothers RN, Mateen A, 1977, An outline of the geology and geochemistry and the possible petrogenetic evolution of the volcanic rocks of the Tonga-Kermadec-New Zealand island arc: Journal of Volcanology and Geothermal Research, v. 2, p. 205–250.

Exon NF, Marlow MS, 1988, Geology and offshore resource potential of the New Ireland-Manus region—a synthesis: Circum-Pacific Council for Energy and Mineral Resources Earth Science Series, v. 9, p. 241–262.

Eyal Y, Eyal M, 1986, The origin of the Bir Zreir rhomb-shaped graben, eastern Sinai: Tectonics, v. 5, p. 267–277.

Eyidogan H, Jackson JA, 1985, A seismological study of normal faulting in the Demirci, Alasehir and Gediz earthquakes of 1969-1970 in western Turkey: implications for the nature and geometry of deformation in the continental crust: Geophysical Journal of the Royal Astronomical Society, v. 81, p. 569–607.

Fairhead D, 1978, A gravity link between the domally uplifted Cainozoic volcanic centres of North Africa and its similarity to the East African rift system anomaly: Earth and Planetary Science Letters, v. 42, p. 109–113.

Fairhead JD, 1988, Late Mesozoic rifting in Africa, in Manspeizer, W. (ed.), Triassic-Jurassic rifting. Part B: Elsevier, Amsterdam, p. 821–831.

Fairhead JD, Stuart GW, 1982, The seismicity of the East African rift system and comparison with other continental rifts: American Geophysical Union and Geological Society of America Geodynamics Series 8, p. 41–61.

Falvey DA, 1974, The development of continental margins in plate tectonic theory: Australian Petroleum Exploration Association Journal, v. 14, p. 95–106.

Farhoudi G, Karig DE, 1977, Makran of Iran and Pakistan as an active arc system: Geology, v. 5, p. 664–668.

Farquharson GW, Hamer RD, Ineson JR, 1984, Proximal volcaniclastic sedimentation in a Cretaceous back-arc basin, northern Antarctic Peninsula: Geological Society of London Special Publication 16, p. 219–229.

Farre JA, McGregor BA, Ryan WBF, Robb JM, 1983, Breaching the shelfbreak: passage from youthful to mature phase in submarine canyon evolution: Society of Economic Paleontologists and Mineralogists Special Publication 33, p. 25–39.

Ferentinos G, Papatheodorou G, Collins MB, 1988, Sediment transport processes on an active submarine fault escarpment: Gulf of Corinth, Greece: Marine Geology, v. 83, p. 43–61.

Ferguson CA, Suneson NH, 1988, Tectonic implications of early Pennsylvanian paleocurrents from flysch in the Ouachita Mountains frontal belt, southeast Oklahoma, in Johnson KS (ed.), Shelf-to-basin geology and resources of Pennsylvanian strata in the Arkoma basin and frontal Ouachita Mountains of Oklahoma: Oklahoma Geological Survey Guidebook 25, p. 49–61.

Fergusson CL, 1985, Trench-floor sedimentary sequences in a Paleozoic subduction complex, eastern Australia: Sedimentary Geology, v. 42, p. 181–200.

Fergusson CL, Coney PJ, 1992, Implications of a Bengal Fan-type deposit in the Paleozoic Lachlan fold belt of southeastern Australia: Geology, v. 20, p. 1047–1049.

Fielding EJ, Jordan TE, 1988, Active deformation at the boundary between the Precordillera and Sierras Pampeanas, Argentina, and comparison with ancient Rocky Mountain deformation: Geological Society of America Memoir 171, p. 143–163.

Fischer AG, 1984, The two Phanerozoic supercycles: in

Berggren WA, VanCouvering JA (eds.), Catastrophes in Earth history: the new uniformitarianism: Princeton University Press, Princeton, p. 129–150.

Fisher AT, Hounslow MW, 1990, Heat flow through the toe of the Barbados accretionary complex: Proceedings of the Ocean Drilling Program, Scientific Results, v. 110, p. 345–364.

Fisher D, Byrne T, 1987, Structural evolution of under-thrusted sediments, Kodiak Islands, Alaska: Tectonics, v. 6, p. 775–793.

Fisher JH, Barratt MW, Droste JB, Shaver RH, 1988, Michigan basin in Sloss LL (ed.), Sedimentary cover—North American craton: U.S. [The Geology of North America, v. D-2]: Geological Society of America, Boulder, p. 361–382.

Fisher MA, 1986, Tectonic processes at the collision of the D'Entrecasteaux zone and the New Hebrides island arc: Journal of Geophysical Research, v. 91, p. 10470–10486.

Fisher MA, 1988, Petroleum geology of the central New Hebrides arc: Circum-Pacific Council for Energy and Mineral Resources Earth Science Series, v. 8, p. 279–283.

Fisher MA, Magoon LB, 1978, Geologic framework of lower Cook Inlet, Alaska: American Association of Petroleum Geologists Bulletin, v. 62, p. 373–402.

Fisher MA, Detterman RL, Magoon LB, 1987, Tectonics and petroleum geology of the Cook-Shelikof basin, southern Alaska: Circum-Pacific Council for Energy and Mineral Resources Earth Science Series, v. 6, p. 213–228.

Fisher MA, Falvey DA, Smith GL, 1988, Seismic stratigraphy of the summit basins of the New Hebrides island arc: Circum Pacific Council for Energy and Mineral Resources Earth Science Series, v. 8, p. 201–223.

Fisher MA, Collot J-Y, Geist E, 1991, Structure of the collision zone between Bougainville guyot and the accretionary wedge of the New Hebrides island arc, southwest Pacific: Tectonics, v. 10, p. 887–903.

Fisher RV, 1961, A proposed classification of volcaniclastic sediments and rocks: Geological Society of America Bulletin, v. 72, p. 1409–1414.

Fisher RV, 1984, Submarine volcaniclastic rocks: Geological Society of London Special Publication 16, p. 5–27.

Fisher RV, Schmincke H-U, 1984, Pyroclastic rocks: Springer-Verlag, New York, 472 p.

Fisher RV, Smith GA (eds.), 1991, Sedimentation in volcanic settings: SEPM (Society for Sedimentary Geology) Special Publication 45, 257 p.

Fiske RS, 1963, Subaqueous pyroclastic flows in the Ohanapecosh Formation, Washington: Geological Society of America Bulletin, v. 74, p. 391–406.

Fiske RS, Tobisch OT, 1978, Paleogeographic significance of volcanic rocks of the Ritter Range pendant, central Sierra Nevada, California, in Howell DG, McDougall KA (eds.), Mesozoic paleogeography of the western United States [Pacific Coast Paleogeography Symposium 2]: Pacific Section, Society of Economic Paleontologists and Mineralogists, Los Angeles, p. 209–221.

Fitch TJ, 1972, Plate convergence, transcurrent faults, and internal deformation adjacent to southeast Asia and the western Pacific: Journal of Geophysical Research, v. 77, p. 4432–4460.

Fitton JG, 1983, Active versus passive continental rifting: evidence from the West African rift system: Tectonophysics, v. 94,

p. 473–481.

Flemings PB, 1987, Paleogeography of the Maastrichtian and Paleocene Wind River basin: a record of Laramide deformation [MS thesis]: Cornell University, Ithaca, 175 p.

Flemings PB, Jordan TE, 1989, A synthetic stratigraphic model of foreland basin development: Journal of Geophysical Research, v. 94, p. 3851–3866.

Flemings PB, Jordan TE, 1990, Stratigraphic modelling of foreland basins: interpreting thrust deformation and lithospheric rheology: Geology, v. 18, p. 430–435.

Flemings PB, Nelson SN, 1991, Paleogeographic evolution of the latest Cretaceous and Paleocene Wind River basin: The Mountain Geologist, v. 28, p. 37–52.

Flint S, Turner P, Jolley EJ, Hartley AJ, 1993, Extensional tectonics in convergent margin basins: an example from the Salar de Atacama, Chilean Andes: Geological Society of America Bulletin, v. 105, p. 603–617.

Florensov NA, 1965, The Baikal rift zone: Geological Survey of Canada Paper 66-14, p. 172–180.

Flores RM, 1974, Characteristics of the Pennsylvanian lower-middle Haymond delta-front sandstones, Marathon Basin, west Texas: Geological Society of America Bulletin, v. 85, p. 709–716.

Flores RM, 1977, Marginal marine deposits in the upper Tesnus Formation (Carboniferous), Marathon Basin, Texas: Journal of Sedimentary Petrology, v. 47, p. 582–592.

Flores RM, 1978, Braided fan delta deposits of the Pennsylvanian upper Haymond Formation in the northeastern Marathon basin, Texas, in Mazzullo SJ (ed.), Tectonics and Paleozoic facies of the Marathon geosyncline, west Texas [Publication 78-17]: Permian Basin Section, Society of Economic Paleontologists and Mineralogists, Midland, p. 149–159.

Floyd PA, Kelling G, Gökçen SL, Gökçen N, 1992, Arc-related origin of volcaniclastic sequences in the Misis Complex, southern Turkey: Journal of Geology, v. 100, p. 221–230.

Forsyth DW, 1980, Comparison of mechanical models of the oceanic lithosphere: Journal of Geophysical Research, v. 85, p. 6364–6368.

Forsyth DW, 1992, Finite extension and low angle normal faulting: Geology, v. 20, p. 27–30.

Forsyth DW, Uyeda S, 1975, On the relative importance of the driving forces of plate motion: Geophysical Journal of the Royal Astronomical Society, v. 43, p. 163–200.

Forsythe RD, Nelson EP, 1985, Geological manifestations of ridge collision: evidence from the Golfo de Penas-Taitao basin, southern Chile: Tectonics, v. 4, p. 477–495.

Forsythe RD, 7 coauthors, 1986, Pliocene near-trench magmatism in southern Chile: a possible manifestation of ridge collision: Geology, v. 14, p. 23–27.

Fort M, 1980, Les formations quaternaires lacustres de la Basse Thakkhola (Himalaya du Népal): intérêt paléogéographique, néotectonique et chronologique: Comptes Rendus hébdomadaires de l'Academie des Sciences de Paris, v. 290, p. 171–174.

Fort M, 1987, Sporadic morphogenesis in a continental subduction setting: an example from the Annapurna Range, Nepal Himalaya: Zeitschrift für Geomorphologie Neue Folge, Supplement, v. 63, p. 9–36.

Fortuin AR, deSmet MEM, 1991, Rates and magnitudes of late Cenozoic vertical movements in the Indonesian Banda Arc

and the distinction of eustatic effects: International Association of Sedimentologists Special Publication 12, p. 79–89.

Fouch TD, Lawton TF, Nichols DJ, Cashion WB, Cobban WA, 1983, Patterns and timing of synorogenic sedimentation in Upper Cretaceous rocks of central and northeast Utah, in Reynolds MW, Dolly ED (eds.), Mesozoic paleogeography of the west-central United States [Rocky Mountain Paleogeography Symposium 2]: Rocky Mountain Section, Society of Economic Paleontologists and Mineralogists, Denver, p. 305–336.

Foucher J-P, Le Pichon X, Sibuet JC, 1981, The ocean-continent transition in the uniform lithosphere stretching model: role of partial melting in the mantle: Philosophical Transactions of the Royal Society of London, v. A305, p. 27–43.

Foucher JP, 7 coauthors, 1990, Heat flow, tectonics, and fluid circulation at the toe of the Barbados Ridge accretionary prism: Journal of Geophysical Research, v. 95, p. 8859–8867.

Fowler CMR, Nisbet EG, 1985, The subsidence of the Williston basin: Canadian Journal of Earth Sciences, v. 22, p. 48–415.

Fox KF, Jr, Fleck RJ, Curtis GH, Meyer CE, 1985, Implications of the northwestwardly younger age of the volcanic rocks of west-central California: Geological Society of America Bulletin, v. 96, p. 647–654.

Franzinelli E, Potter PE, 1983, Petrology, chemistry, and texture of modern river sands, Amazon River system: Journal of Geology, v. 91, p. 23–39.

Friedman GM, 1987, Deep-burial diagenesis: its implications for vertical movements of the crust, uplift of the lithosphere and isostatic unroofing—a review: Sedimentary Geology, v. 50, p. 67–94.

Friend PF, 1983, Towards the field classification of alluvial architecture or sequence: International Association of Sedimentologists Special Publication 6, p. 345–354.

Fritz WJ, Howells MF, 1991, A shallow marine volcaniclastic facies model: an example from sedimentary rocks bounding the subaqueously welded Ordovician Garth Tuff, North Wales, UK: Sedimentary Geology, v. 74, p. 217–240.

Fritz WJ, Howells MF, Reedman AJ, Campbell SDG, 1990, Volcaniclastic sedimentation in and around an Ordovician subaqueous caldera, Lower Rhyolitic Tuff Formation, north Wales: Geological Society of America Bulletin, v. 102, p. 1246–1256.

Frost CD, Coombs DS, 1989, Nd isotopic character of New Zealand sediments: implications for terrane concepts and crustal evolution: American Journal of Science, v. 289, p. 744–770.

Frostick LE, Reid I, 1987, Tectonic control of desert sediments in rift basins ancient and modern: Geological Society of London Special Publication 35, p. 53–68.

Frostick LE, Reid I, 1990, Structural control of sedimentation patterns and implications for the economic potential of the East African Rift basins: Journal of African Earth Sciences, v. 10, p. 307–318.

Frostick LE, Steel RJ, 1993, Sedimentation in divergent plate-margin basins: International Association of Sedimentologists Special Publication 20, p. 111–128.

Fryer P, 1990, Recent marine geological research in the Mariana and Izu-Bonin island arcs: Pacific Science, v. 44, p. 95–114.

Fryer P, Ambos EL, Hussong DM, 1985a, Origin and emplacement of Mariana forearc seamounts: Geology, v. 13, p. 774–777.

Fryer P, Langmuir C, Taylor B, Zhang Y, Hussong D, 1985b, Rifting of the Izu Arc, III: relationship of chemistry to tectonics:

EOS, v. 66, p. 421.

Fryer P, 26 coauthors, 1990a, Proceedings of the Ocean Drilling Program, Initial Reports [v. 125]: Ocean Drilling Program, College Station, 1092 p.

Fryer P, Taylor B, Langmuir CH, Hochstaedter AG, 1990b, Petrology and geochemistry of lavas from the Sumisu and Torishima backarc rifts: Earth and Planetary Science Letters, v. 100, p. 161–178.

Fuchtbauer H, 1967, Die sandsteine in der Molasse nordlich der Alpen: Geologische Rundschau, v. 56, p. 226–300.

Fuis GS, Kohler WM, 1984, Crustal structure and tectonics of the Imperial Valley region, California, in Rigsby CA (ed.), The Imperial basin—tectonics, sedimentation and thermal aspects [Book 40]: Pacific Section, Society of Economic Paleontologists and Mineralogists, Los Angeles, p. 1–13.

Fujioka K, Nishimura A, Matsuo Y, Rodolfo KS, 1992, Correlation of Quaternary tephras throughout the Izu-Bonin areas: Proceedings of the Ocean Drilling Program, Scientific Results, v. 126, p. 23–46.

Fujita K, Cook DB, 1990, The Arctic continental margin of eastern Siberia, in Grantz A, Johnson L, Sweeney JF (eds.), The Arctic Ocean region [The Geology of North America, v. L]: Geological Society of America, Boulder, p. 289–304.

Fukushima Y, Parker G, Pankin HM, 1985, Prediction of ignitive trbidity currents in Scripps submarine canyon: Marine Geology, v. 67, p. 55–81.

Fulford MM, Busby CJ, 1993, Tectonic controls on nonmarine sedimentation in a Cretaceous fore-arc basin, Baja California, Mexico: International Association of Sedimentologists Special Publication 20, p. 301–333.

Furlong KP, Chapman DS, Alfeld PW, 1982, Thermal modeling of the geometry of subduction with implications for the tectonics of the overriding plate: Journal Geophysical Research, v. 87, p. 1786–1802.

Furukawa M, Tokuyama H, Abe S, Nishizawa A, Kinoshita H, 1991, Report on DELP 1988 cruises in the Okinawa Trough - Part 2: seismic reflection studies in the southwestern part of the Okinawa Trough: Bulletin of the Earthquake Research Institute, University of Tokyo, v. 66, p. 17–36.

Gaal G, Gorbatschev R, 1987, An outline of the Precambrian evolution of the Baltic shield: Precambrian Research, v. 35, p. 15–52.

Gaal G, Tuomi A, 1994, Geology of the Baltic DDS profile, Finland, in Green A (ed.), XV General Assembly, European Geophysical Society, Proceedings, in press.

Gabrielse H, Yorath CJ (eds.), 1992, Geology of the Cordilleran orogen in Canada [Geology of Canada, v. 4]: Geological Survey of Canada, Ottawa, 844 p.

Gaetani M, Garzanti E, 1991, Multicyclic history of the northern India continental margin (northwestern Himalaya): American Association of Petroleum Geologists Bulletin, v. 75, p. 1427–1446.

Gallagher K, 1989, An examination of some uncertainties associated with estimates of sedimentation rates and tectonic subsidence: Basin Research, v. 2, p. 97–114.

Galloway WE, 1974, Deposition and diagenetic alteration of sandstone in northeast Pacific arc-related basins: implications for graywacke genesis: Geological Society of America, v. 85, p. 379–390.

Galloway WE, 1979, Diagenetic control of reservoir quality in

arc-derived sandstones: implications for petroleum exploration: Society of Economic Paleontologists and Mineralogists Special Publication 26, p. 251–262.

Galloway WE, 1989a, Genetic stratigraphic sequences in basin analysis I: architecture and genesis of flooding-surface bounded depositional units: American Association of Petroleum Geologists Bulletin, v. 73, p. 125–142.

Galloway WE, 1989b, Genetic stratigraphic sequences in basin analysis II: application to northwest Gulf of Mexico Cenozoic basin: American Association of Petroleum Geologists Bulletin, v. 73, p. 143–154.

Gans PB, 1987, An open-system, two-layer crustal stretching model for the eastern Great Basin: Tectonics, v. 6, p. 1–12.

Gans PB, Miller EL, McCarty J, Ouldcott ML, 1985, Tertiary extensional faulting and evolving ductile-brittle transition zones in the northern Snake Range and vicinity: new insights from seismic data: Geology, v. 13, p. 189–193.

Gans PB, Mahood GA, Schermer E, 1989, Synextensional magmatism in the Basin and Range Province: a case study from the eastern Great Basin: Geological Society of America Special Paper 233, p. 1–53.

Gansser A, 1964, Geology of the Himalayas: Interscience, London, 289 p.

Garcia MO, 1992, Volcanic sands from ODP Site 842, 300 km west of Hawaii: turbidites from giant debris flows: EOS, v. 73, p. 513–514.

Garcia MO, Hull DM, 1994, Turbidites from giant Hawaiian landslides: results from Ocean Drilling Program Site 842: Geology, v. 22, p. 159–162.

Gardner MH, Barton MD, Tyler N, Fisher RS, 1992, Architecture and permeability structure of fluvial-deltaic sandstones, Ferron Sandstone, east central Utah, in Flores RM (ed.), Mesozoic of the western interior [Field Guidebook]: Society of Economic Paleontologists and Mineralogists, Tulsa, p. 5–21.

Garfunkel Z, 1988, Relation between continental rifting and uplifting: evidence from the Suez rift and northern Red Sea: Tectonophysics, v.150, p. 33–49.

Garzanti E, van Haver T, 1988, The Indus clastics: forearc basin sedimentation in the Ladakh Himalaya (India): Tectonophysics, v. 59, p. 237–249.

Garzanti E, Baud A, Mascle G, 1987, Sedimentary record of the northward flight of India and its collision with Eurasia (Ladakh Himalaya, India): Geodinamica Acta (Paris), v. 1, p. 297–312.

Gass IG, 1970, Tectonic and magmatic evolution of the Afro-Arabian dome, in Clifford TN, Gass IG (eds.), African magmatism and tectonics: Oliver and Boyd, Edinburgh, p. 285–300.

Gass IG, 1980, Red Sea case study: The Open University Coursebook S336 RS, 142 p.

Gastil RG, 1985, Terranes of Peninsular California and adjacent Sonora: Circum-Pacific Council for Energy and Mineral Resources Earth Sciences Series, v. 1, p. 273–283.

Gastil RG, Allison EC, 1966, An Upper Cretaceous fault-line coast: American Association of Petroleum Geologists Bulletin, v. 50, p. 647–648.

Gastil RG, Phillips RC, Allison EC, 1975, Reconnaissance geology of the state of Baja California: Geological Society of America Memoir 140, 170 p.

Gastil RG, Morgan GJ, Krummenacher D, 1978, Mesozoic history of Peninsular California and related areas east of the Gulf of California, in Howell DG, McDougall KA (eds.), Mesozoic paleogeography of the western United States [Pacific Coast Paleogeography Symposium 2]: Pacific Section, Society of Economic Paleontologists and Mineralogists, Los Angeles, p. 107–115.

Gawthorpe RL, 1986, Sedimentation during carbonate ramp-to-slope evolution of the Bowland basin (Dinantian), northern England: Sedimentology, v. 33, p. 185–206.

Gawthorpe RL, Colella A, 1990, Tectonic controls on coarse-grained delta depositional systems in rift basins: International Association of Sedimentologists Special Publication 10, p. 113–128.

Gawthorpe RL, Hurst JM, Sladen CP, 1990, Evolution of Miocene footwall-derived coarse-grained deltas, Gulf of Suez, Egypt: implications for exploration: American Association of Petroleum Geologists Bulletin, v. 74, p. 1077–1086.

Gealey WK, 1980, Ophiolite obduction mechanism, in Panayiotou A (ed.), Ophiolites: Cyprus Geological Survey Department, Nicosia, p. 228–243.

Geist EL, Scholl DW, 1992, Application of continuum models to deformation of the Aleutian Island arc: Journal of Geophysical Research, v. 97, p. 4953–4967.

Geist EL, Childs JR, Scholl DW, 1988, The origin of summit basins of the Aleutian Ridge: implications for block rotation of an arc massif: Tectonics, v. 7, p. 327–341.

George AD, 1992, Deposition and deformation of an Early Cretaceous trench-slope basin deposit, Torlesse terrane, New Zealand: Geological Society of America Bulletin, v. 104, p. 570–580.

George PG, Dokka RK, 1994, Major Late Cretaceous cooling events in the eastern Peninsular Ranges, California, and their implications for Cordilleran tectonics: Geological Society of America Bulletin, v. 106, p. 903–914.

Gergen LD, Ingersoll RV, 1986, Petrology and provenance of Deep Sea Drilling Project sand and sandstone from the North Pacific Ocean and the Bering Sea: Sedimentary Geology, v. 51, p. 29–56.

Gerhard LC, Anderson SB, 1988, Geology of the Williston basin (United States portion), in Sloss LL (ed.), Sedimentary cover –North American craton: US [The Geology of North America, v. D-2]: Geological Society of America, Boulder, p. 221–241.

Gerhard LC, Anderson SB, Lefever JA, Carlson CG, 1982, Geological development, origin, and energy mineral resources of Williston basin, North Dakota: American Association of Petroleum Geologists Bulletin, v. 68, p. 989–1010.

Gerhard LC, Anderson SB, Fischer DW, 1990, Petroleum geology of the Williston basin: American Association of Petroleum Geologists Memoir 51, p. 507–559.

Gibbs A, 1983, Balanced cross-section construction from seismic sections in areas of extensional tectonics: Journal of Structural Geology, v. 5, p. 153–160.

Gibbs A, 1984, Structural evolution of extensional basin margins: Journal of the Geological Society of London, v. 141, p. 609–620.

Gibbs A, 1987, Development of extension and mixed-mode sedimentary basins: Geological Society of London Special Publication 28, p. 19–33.

Gibson JR, Walsh JJ, Watterson J, 1989, Modelling of bed contours and cross-sections adjacent to planar normal faults:

Journal of Structural Geology, v. 11, p. 317–328.

Gieskes JM, Vrolijk P, Blanc G, 1990, Hydrogeochemistry of the northern Barbados accretionary complex transect: Ocean Drilling Project Leg 110: Journal of Geophysical Research, v. 95, p. 8809–8818.

Gilbert D, Cunningham R, 1985, Organic facies variations in sediments from Leg 84, off Guatemala: Initial Reports of the Deep Sea Drilling Project, v. 84, p. 739–742.

Gilbert GK, 1928, Studies of Basin-Range structure: United States Geological Survey Professional Paper 153, 80 p.

Gilbert MC, 1983, Timing and chemistry of igneous events associated with the southern Oklahoma aulacogen: Tectono-physics, v. 94, p. 439–455.

Gill J, 1981, Orogenic andesites and plate tectonics: Springer-Verlag, Berlin, 390 p.

Gill J, 16 coauthors, 1990, Explosive deep water basalt in the Sumisu backarc rift: Science, v. 248, p. 1214–1217.

Gill JB, Seales C, Thompson P, Hochstaedter AG, Dunlap C, 1992, Petrology and geochemistry of Pliocene-Pleistocene vol-canic rocks from the Izu Arc, Leg 126: Proceedings of the Ocean Drilling Program, Scientific Results, v. 126, p. 383–404.

Girdler RW, 1958, The relationship of the Red Sea to the East African rift system: Journal of the Geological Society of London, v. 114, p. 79–105.

Girdler RW, 1962, Initiation of continental drift: Nature, v. 194, p. 521–524.

Girdler RW, 1991, The Afro-Arabian rift system—an overview: Tectonophysics, v. 197, p. 139–153.

Glatzmaier GA, Schubert G, Bercovici D, 1990, Chaotic, sub-duction-like downflows in a spherical model of convection in the Earth's mantle: Nature, v. 347, p. 274–277.

Glazner AF, 1991, Plutonism, oblique subduction and conti-nental growth: an example from the Mesozoic of California: Geology, v. 19, p. 784–786.

Glazner AF, Bartley JM, 1984, Timing and tectonic setting of Tertiary low-angle normal faulting and associated magmatism in the southwestern United States: Tectonics, v. 3, p. 385–396.

Goebel KA, 1991, Paleogeographic setting of Late Devonian to Early Mississippian transition from passive to collisional margin, Antler foreland, eastern Nevada and western Utah, in Cooper JD, Stevens CH (eds.), Paleozoic paleogeography of the western United States—II, [Book 67]: Pacific Section, Society of Eco-nomic Paleontologists and Mineralogists, Los Angeles, p. 401–418.

Goetz CW, 1989, Geology of the Rancho El Rosarito area: evi-dence for latest Albian, east over west, ductile thrusting in the Peninsular Ranges [MS thesis]: San Diego State University, San Diego, 134 p.

Goetz CW, Girty GH, Gastil RG, 1988, East over west ductile thrusting along a terrane boundary in the Peninsular Ranges: Rancho El Rosarito, Baja California, Mexico: Geological Society of America Abstracts with Programs, v. 20, n. 3, p. 164.

Goetz LK, Tyler JG, Macarevich RL, Brewster DL, Sonnad JR, 1992, Tectonic setting and hydrocarbon potential of the Pro-terozoic to early Paleozoic section in the Rough Creek graben, Illinois and Kentucky: American Association of Petroleum Geoloists Bulletin, v. 76, p. 1275.

Gold DP, 1980, Structural geology, in Siegal BS, Gillespie AR (eds.), Remote sensing in geology: John Wiley, New York, p. 419–484.

Goldfarb RJ, Leach DL, Miller ML, Pickthorn WJ, 1986, Geol-ogy, metamorphic setting, and genetic constraints of epigenetic lode-gold mineralization within the Cretaceous Valdez Group, south-central Alaska: Geological Association of Canada Special Paper 32, p. 87–105.

Goldfarb RJ, Leach DL, Pickthorn WJ, Paterson CJ, 1988, Ori-gin of lode-gold deposits of the Juneau gold belt, southeastern Alaska: Geology, v. 16, p. 440–443.

Goldfarb RJ, Snee LW, Miller LD, Newberry RJ, 1991, Rapid dewatering of the crust deduced from ages of mesothermal gold deposits: Nature, v. 354, p. 296–298.

Goldfinger C, 5 coauthors, 1992, Transverse structural trends along the Oregon convergent margin: implications for Cascadia earthquake potential and crustal rotations: Geology, v. 20, p. 141–144.

Goodman ED, Malin PE, 1992, Evolution of the southern San Joaquin basin and mid-tertiary `transtensional' tectonics, cen-tral California: Tectonics, v. 11, p. 478–498.

Gordon MB, Hempton MR, 1986, Collision-induced rifting: the Grenville orogeny and the Keweenawan rift of North America: Tectonophysics, v. 127, p. 1–25.

Gorsline DS, 1984, A review of fine-grained sediment origins, characteristics, transport and deposition, in Stow DAV, Piper DJW (eds.), Fine-grained sediments: deep water processes and facies: Blackwell Scientific Publications, Oxford, p. 17–34.

Gould SJ, 1989, Wonderful life. The Burgess Shale and the nature of history: W. W. Norton and Co., New York, 347 p.

Graham SA, 1987, Tectonic controls on petroleum occurrence in central California, in Ingersoll RV, Ernst WG (eds.), Cenozoic basin development of coastal California [Rubey Volume VI]: Prentice-Hall, Englewood Cliffs, p. 47–63.

Graham SA, Dickinson WR, Ingersoll RV, 1975, Himalayan-Bengal model for flysch dispersal in Appalachian-Ouachita sys-tem: Geological Society of America Bulletin, v.86, p.273–286.

Graham SA, Ingersoll RV, Dickinson WR, 1976, Common provenance for lithic grains in Carboniferous sandstones from Ouachita Mountains and Black Warrior basin: Journal of Sedi-mentary Petrology, v.46, p.620–632.

Graham SA, McCloy C, Hitzman M, Ward R, Turner R, 1984, Basin evolution during change from convergent to transform margin in central California: American Association of Petro-leum Geologists Bulletin, v. 68, p. 233–249.

Graham SA, 14 coauthors, 1986, Provenance modelling as a technique for analysing source terrane evolution and controls on foreland sedimentation: International Association of Sedi-mentologists Special Publication 8, p. 425–436.

Grant NK, 1971, South Atlantic, Benue trough, and Gulf of Guinea Cretaceous triple junction: Geological Society of Amer-ica Bulletin, v. 82, p. 2295–2298.

Greb SF, 1989, Structural controls on the formation of the sub-Absaroka unconformity in the US eastern interior basin: Geology, v. 17, p. 889–892.

Greene G, 27 coauthors, 1991, Documenting twin-ridge colli-sion and arc deformation: Geotimes, v. 36, n. 5, p. 23–25.

Greene HG, Johnson DP, 1988, Geology of the central basin region of the New Hebrides arc inferred from single-channel seismic-reflection data: Circum-Pacific Council for Energy and Mineral Resources Earth Science Series, v. 8, p. 177–199.

Greene HG, Wong FL (eds.), 1988, Geology and offshore resources of Pacific island arcs-Vanuatu region: Circum-Pacific

Council for Energy and Mineral Resources Earth Science Series, v. 8, 442 p.

Greene HG, Wong FL, 1989, Ridge collisions along the plate margins of South America compared with those in the southwest Pacific: Circum-Pacific Council for Energy and Mineral Resources Earth Science Series, v. 11, p. 39–57.

Greene HG, Macfarlane A, Johnson DP, Crawford AJ, 1988a, Structure and tectonics of the central New Hebrides arc: Circum-Pacific Council for Energy and Mineral Resources Earth Science Series, v. 8, p. 377–411.

Greene HG, Macfarlane A, Wong FL, 1988b, Geology and offshore resources of Vanuatu—introduction and summary: Circum-Pacific Council for Energy and Mineral Resources Earth Science Series, v. 8, p. 1–25.

Gregory JW, 1894, Contributions to the physical geography of British East Africa: Geographical Journal, v. 4, p. 293–297.

Gries R, 1983, North-south compression of Rocky Mountain foreland structures, in Lowell JD (ed.), Rocky Mountain foreland basins and uplifts: Rocky Mountain Association of Geologists, Denver, p. 9–32.

Griffith RC, 1987, Geology of the southern Sierra Calamuje area: structural and stratigraphic evidence for latest Albian compression along a terrane boundary, Baja California, Mexico [MS thesis]: San Diego State University, San Diego, 119 p.

Griffith RC, 5 coauthors, 1986, Stratigraphic and structural implications of metasedimentary and metavolcanic rocks of Arroyo Calamuje, Baja California, Mexico: Geological Society of America Abstracts with Programs, v.18, p.111.

Griffiths RW, Campbell IH, 1991, Interaction of mantle plume heads with the earth's surface and onset of small-scale convection: Journal of Geophysical Research, v. 96, p. 18295–18310.

Griggs GB, Kulm LD, 1970, Sedimentation in Cascadia deep-sea channel: Geological Society of America Bulletin, v. 81, p. 1361–1384.

Grindley GW, 1960, Geological map of New Zealand, sheet 8—Taupo: New Zealand Geological Survey, 1:250,000.

Grindley GW, 1965, The geology, structure, and exploitation of the Wairakei geothermal field, Taupo, New Zealand: New Zealand Geological Survey Bulletin 75, 131 p.

Grocott J, Brown M, Dallmeyer RD, Taylor GK, Trelor PJ, 1994, Mechanisms of continental growth in extensional arcs: an example from the Andean plate-boundary zone: Geology, v. 22, p. 391–394.

Groshong RH, Jr, 1989, Half-graben structures: balanced models of extensional fault-bend folds: Geological Society of America Bulletin, v. 101, p. 96–105.

Grossheim VA, Khain VE (eds.), 1967, Atlas of the lithological-paleogeographical maps of the USSR: Ministry of Geology of the USSR, 55 sheets.

Grow JA, 1973, Crustal and upper mantle structure of the central Aleutian arc: Geological Society of America Bulletin, v. 84, p. 2169–2192.

Grow JA, Sheridan RE, 1981, Deep structure and evolution of the continental margin off the eastern United States [Colloque C3: Geologie des marges continentales]: Oceanologica Acta, v. 4, Supplement, p. 11–19.

Guffanti M, Clynne MA, Smith JG, Muffler LJP, Bullen TD, 1990, Late Cenozoic volcanism, subduction, and extension in the Lassen region of California, southern Cascade Range: Journal of Geophysical Research, v. 95, p. 19453–19464.

Guilbert JM, Park CF, Jr, 1986, The geology of ore deposits: W.H. Freeman and Company, New York, 985 p.

Guiraud M, Seguret M, 1985, A releasing solitary overstep model for the Late Jurassic–Early Cretaceous (Wealdian) Soria strike-slip basin (northern spain): Society of Economic Paleontologists and Mineralogists Special Publication 37, p. 159–175.

Guoqiang P, 1984, The Late Precambrian and early Paleozoic marginal basin of south China: Geological Society of London Special Publication 16, p. 279–284.

Gurnis M, 1988, Large-scale mantle convection and the aggragation and dispersal of supercontinents: Nature, v. 332, p. 695–699.

Gurnis M, 1990a, Bounds on global dynamic topography from Phanerozoic flooding of continental platforms: Nature, v. 344, p. 754–756.

Gurnis M, 1990b, Plate-mantle coupling and continental flooding: Geophysical Research Letters, v. 17, p. 623–626.

Gussev GS, Koval'skii VV, Parfenov LM, Petrov AF, Fradkin GS, 1985, Evolution of the crust of northeastern Siberia: Geologiiya'i Geofizika (Soviet Geology and Geophysics), v. 26, n. 9, p. 3–11.

Gust DA, Biddle KT, Phelps DW, Uliana MA, 1985, Associated middle to late Jurassic volcanism and extension in southern South America: Tectonophysics, v. 116, p. 223–253.

Gwinn VE, 1964, Thin-skinned tectonics in the plateau and northwestern Valley and Ridge province of the central Appalachians: Geological Society of America Bulletin, v. 75, p. 863–900.

Haapala IJ, Ramo OT, 1987, Petrogenesis of the Rapakivi granites of Finland: Geological Society of America Abstracts with Programs, v. 19, p. 689.

Hack JT, 1957, Studies of longitudinal stream profiles in Virginia and Maryland: United States Geological Survey Professional Paper 294B, 97 p.

Hackett WR, Houghton BF, 1989, A facies model for a Quaternary andesitic composite volcano: Ruapehu, New Zealand: Bulletin of Volcanology, v. 51, p. 51–68.

Hacquebard PA, 1977, Rank of coal as an index of organic metamorphism for oil and gas in Alberta: Geological Survey of Canada Bulletin 262, p. 11–22.

Hagen ES, Schuster MW, Furlong KP, 1985, Tectonic loading and subsidence of intermontane basins, Wyoming foreland province: Geology, v. 13, p. 529–532.

Hagstrum JT, McWilliams M, Howell DG, Gromme S, 1985, Mesozoic paleomagnetism and northward translation of the Baja California Peninsula: Geological Society of America Bulletin, v. 96, p. 1077–1090.

Halbach P, 17 coauthors, 1989, Probable modern analogue of Kuroko-type massive sulphide deposits in the Okinawa Trough backarc basin: Nature, v. 338, p. 496–499.

Hale-Erlich WS, Coleman JL, Jr, 1993, Ouachita-Appalachian juncture: a Paleozoic transpressional zone in the southeastern U.S.A.: American Association of Petroleum Geologists Bulletin, v. 77, p. 552–568.

Hallam A, 1971, Mesozoic geology and the opening of the North Atlantic: Journal of Geology, v. 79, p. 129–157.

Halle TG, 1911, On the geological structure and history of the Falkland Islands: Bulletin of the Geological Institute of the University of Uppsala, v. XI, p. 117.

Ham WE, Wilson JL, 1967, Paleozoic epeirogeny and orogeny

in the central United States: American Journal of Science, v. 265, p. 332–407.

Hamdani Y, Mareschal JC, Arkani-Hamed J, 1991, Phase changes and thermal subsidence in intracontinental sedimentary basins: Geophysical Journal International, v. 106, p. 657–665.

Hamilton W, 1969, Mesozoic California and the underflow of Pacific mantle: Geological Society of America Bulletin, v. 80, p. 2409–2430.

Hamilton W, 1979, Tectonics of the Indonesian region: United States Geological Survey Professional Paper 1078, 345p.

Hamilton W, 1981, Crustal evolution by arc magmatism: Philosophical Transactions of the Royal Society of London, v. A301, p. 279–291.

Hamilton W, 1985a, Review of "Processes of continental rifting": Journal of Volcanology and Geothermal Research, v. 24, p. 362–364.

Hamilton W, 1985b, Subduction, magmatic arcs, and foreland deformation: Circum-Pacific Council for Energy and Mineral Resources Earth Science Series, v. 1, p. 259–262.

Hamilton W, 1987, Crustal extension in the Basin and Range Province, southwestern United States: Geological Society of London Special Publication 28, p. 155–176.

Hamilton WB, 1988a, Plate tectonics and island arcs: Geological Society of America Bulletin, v. 100, p. 1503–1527.

Hamilton WB, 1988b, Detachment faulting in the Death Valley region, California and Nevada: United States Geological Survey Bulletin 1790, p. 51–85.

Hamilton W, 1989, Convergent-plate tectonics viewed from the Indonesian region, in Sengor AMC (ed.), Tectonic evolution of the Tethyan region [NATO ASI Series]: Kluwer Academic Publishers, Dordrecht, p. 655–698.

Hammond PE, 1979, A tectonic model for evolution of the Cascade Range, in Armentrout JM, Cole MR, Terbest H, Jr, (eds.), Cenozoic paleogeography of the western United States [Pacific Coast Paleogeography Symposium 3]: Pacific Section, Society of Economic Paleontologists and Mineralogists, p. 219–237.

Han TL, coauthors, 1984, Le système tectonique actif du Tibet méridional, in Mercier JL, Li GC (eds.), Mission Franco-Chinoise au Tibet: Centre National de Recherche Scientifique, Paris, p. 393–412.

Hancock PL, Bevan TG, 1987, Brittle modes of foreland extension: Geological Society of London Special Publication 28, p. 127–137.

Hansen VL, Goodge JW, Keep M, Oliver DH, 1993, Asymmetric rift interpretation of the North American margin: Geology, v. 21, p. 1067–1070.

Hansen WR, 1990, Paleogeographic and paleotectonic setting of Laramide sedimentary basins in the central Rocky Mountain region: alternative interpretation: Geological Society of America Bulletin, v. 102, p. 280–282.

Hanson RE, 1991, Quenching and hydroclastic disruption of andesitic to rhyolitic intrusions in a submarine island-arc sequence, northern Sierra Nevada, California: Geological Society of America Bulletin, v. 103, p. 804–816.

Haq BU, Hardenbol J, Vail PR, 1987, The chronology of fluctuating sea levels since the Triassic (250 million years ago to present): Science, v. 235, p. 1156–1167.

Haq BU, Hardenbol J, Vail PR, 1988, Mesozoic and Cenozoic chronostratigraphy and cycles of sea-level change: Society of Economic Paleontologists and Mineralogists Special Publication 42, p. 71–108.

Harding TP, 1973, Newport-Inglewood trend, California—an example of wrenching style of deformation: American Association of Petroleum Geologists Bulletin, v. 57, p. 97–116.

Harding TP, 1976, Tectonic significance and hydrocarbon trapping consequences of sequential folding synchronous with San Andreas faulting, San Joaquin Valley, California: American Association of Petroleum Geologists Bulletin, v. 60, p. 356–378.

Harding TP, 1983a, Graben hydrocarbon plays and structural styles: Geologie en Mijnbouw, v. 62, p. 3–23.

Harding TP, 1983b, Divergent wrench fault and negative flower structure, Andaman Sea: American Association of Petroleum Geologists Studies in Geology Series 15, v. 3, p. 4.2-1–4.2-8.

Harding TP, 1985, Seismic characteristics and identification of negative flower structures, positive flower structures, and positive structural inversion: American Association of Petroleum Geologists Bulletin, v. 69, p. 582–600.

Harding TP, 1990, Identification of wrench faults using subsurface structural data: criteria and pitfalls: American Association of Petroleum Geologists Bulletin, v. 74, p. 1590–1609.

Harding TP, Gregory RF, Stephens LH, 1983, Convergent wrench fault and positive flower structure, Ardmore basin, Oklahoma: American Association of Petroleum Geologists Studies in Geology Series 15, v. 3., p. 4.2-13–4.2-17.

Harding TP, Vierbuchen RC, Christie-Blick N, 1985, Structural styles, plate-tectonic settings, and hydrocarbon traps of divergent (transtensional) wrench faults: Society of Economic Paleontologists and Mineralogists Special Publication 37, p. 51–77.

Harland WB, 1971, Tectonic transpression in Caledonian Spitsbergen: Geological Magazine, v. 108, p. 27–41.

Harland WB, 5 coauthors, 1990, A geological time scale, 1990: Cambridge University Press, Cambridge, 263 p.

Harmand C, Laville E, 1983, Magmatisme alcalin mésozoique et phénoménes thermiques associés dans le Haut Atlas central, Maroc: Bulletin de Centres de Recherche, Exploration et Production d'Elf-Aquitaine, v. 7, p. 967–976.

Harms TA, Jayko AS, Blake MC, Jr, 1992, Kinematic evidence for extensional unroofing of the Franciscan Complex along the Coast Range fault, northern Diablo Range, California: Tectonics, v. 11, p. 228–241.

Harper GD, 1989, Field guide to the Josephine ophiolite and coeval island arc complex, Oregon-California, in Aalto K, Harper GD (eds.), Geologic evolution of the northernmost Coast Ranges and western Klamath Mountains, California: 28th International Geological Congress Fieldtrip Guidebook T308, p. 2–20.

Harper GD, Wright JE 1984, Middle to Late Jurassic tectonic evolution of the Klamath Mountains, California-Oregon: Tectonics, v. 3, p. 759–772.

Harrison TM, Copeland P, Kidd WSF, Yin A, 1992, Raising Tibet: Science, v. 255, p. 1663–1670.

Harrison TM, 6 coauthors, 1993, Isotopic preservation of Himalayan/Tibetan uplift, denudation, and climatic histories of two molasse deposits: Journal of Geology, v. 101, p. 157–175.

Hart EW, 1966, Mines and mineral resources of Monterey County, California: California Division of Mines and Geology County Report 5, 142 p.

Harvey AM, 1989, The occurrence and role of arid zone alluvial fans, in Thomas DG (ed.), Arid zone geomorphology: Halstead Press, Belhaven, p. 136–158.

Hasebe K, Fujii N, Uyeda S, 1970, Thermal process under island arcs: Tectonophysics, v. 10, p. 335–355.

Hashimoto M, Jackson DD, 1993, Plate tectonics and crustal deformation around the Japanese Islands: Journal of Geophysical Research, v. 98, p. 16149–16166.

Hathon EG, Underwood MB, 1991, Clay mineralogy and chemistry as indicators of hemipelagic sediment dispersal south of the Aleutian Arc: Marine Geology, v. 97, p. 145–166.

Hawkins JW, Bloomer SH, Evans CA, Melchior JT, 1984, Evolution of intra-oceanic arc-trench systems: Tectonophysics, v. 102, p. 175–205.

Haxby WF, Turcotte DL, Bird JM, 1976, Thermal and mechanical evolution of the Michigan basin: Tectonophysics, v. 36, p. 57–75.

Hayes DE, Lewis SD, 1984, A geophysical study of the Manila Trench, Luzon, Philippines 1. Crustal structure, gravity, and regional tectonic evolution: Journal of Geophysical Research, v. 89, p. 9171–9195.

Hays JD, Pitman WC, III, 1973, Lithospheric plate motion, sea-level changes, and climatic and ecologic consequences: Nature, v. 246, p. 18–22.

Healy J, Schofield JC, Thompson BN, 1974, Geological map of New Zealand, sheet 5–Rotorua: New Zealand Geological Survey, 1:250,000.

Heckel PH, 1984, Changing concepts of Midcontinent Pennsylvanian cyclothems, Nroth America, in Congres International de Stratigraphie et de Geologie du Carbonifere, 9th [Compte Rendu, v. 3]: Southern Illinois University Press, Carbondale, p. 535–538.

Heezen BC, 11 coauthors, 1973, Diachronous deposits, a kinematic interpretation of the post-Jurassic sedimentary sequence on the Pacific Plate: Nature, v. 241, p. 25–32.

Hefferan KP, Karson JA, Saquaque A, 1992, Proterozoic collisional basins in a Pan-African suture zone, Anti-Atlas Mountains, Morocco: Precambrian Research, v. 54, p. 295–319.

Heidlauf DT, Hsui AT, Klein GD, 1986, Tectonic subsidence analysis of the Illinois basin: Journal of Geology, v. 94, p. 779–794.

Heigold PC, 1990, Crustal character of the Illinois basin: American Association of Petroleum Geologists Memoir 51, p. 247–261.

Heigold PC, Oltz DF, 1990, Seismic expression of the stratigraphic succession: American Association of Petroleum Geologists Memoir 51, p. 169–178.

Heller PL, 1983, Sedimentary response to Eocene tectonic rotation in western Oregon [PhD thesis]: University of Arizona, Tucson, 321 p.

Heller PL, Angevine CL, 1985, Sea-level cycles during the growth of Atlantic-type oceans: Earth and Planetary Science Letters, v. 75, p. 417–426.

Heller PL, Dickinson WR, 1985, Submarine ramp facies model for delta-fed, sand-rich turbidite systems: American Association of Petroleum Geologists Bulletin, v. 69, p. 960–976.

Heller PL, Paola C, 1989, The paradox of Lower Cretaceous gravels and the initiation of thrusting in the Sevier orogenic belt, United States western interior: Geological Society of America Bulletin, v. 101, p. 864–875.

Heller PL, Paola C, 1992, The large-scale dynamics of grain-size variation in alluvial basins, 2: application to syntectonic conglomerate: Basin Research, v. 4, p. 91–102.

Heller PL, Ryberg PT, 1983, Sedimentary record of subduction to forearc transition in the rotated Eocene basin of western Oregon: Geology, v. 11, p. 380–383.

Heller PL, 7 coauthors, 1986, Time of initial thrusting in the Sevier orogenic belt, Idaho-Wyoming and Utah: Geology, v. 14, p. 388–391.

Heller PL, Tabor RW, Suczek CA, 1987, Paleogeographic evolution of the United States Pacific Northwest during Paleogene time: Canadian Journal of Earth Sciences, v. 24, p. 1652–1667.

Heller PL, Angevine CL, Winslow NS, Paola C, 1988, Two-phase stratigraphic model of foreland-basin sequences: Geology, v. 16, p. 501–504.

Heller PL, Angevine CL, Paola C, 1989, Comment on "Thrusting and gravel progradation in foreland basins: a test of post-thrusting gravel dispersal": Geology, v. 17, p. 959–960.

Hellinger SJ, Sclater JG, 1983, Some comments on two-layer extensional models for the evolution of sedimentary basins: Journal of Geophysical Research, v. 88, p. 8251–8269.

Hempton MR, 1983, The evolution of thought concerning sedimentation in pull-apart basins, in Boardman SJ (ed.), Revolution in the earth sciences: Kendall/Hunt, Dubuque, p. 167–180.

Herzer RH, Exon NF, 1985, Structure and basin analysis of the southern Tonga forearc: Circum-Pacific Council for Energy and Mineral Resources Earth Science Series, v. 2, p. 55–73.

Hesse R, 1982, Cretaceous-Paleogene flysch zone of the eastern Alps and Carpathians: identification and plate-tectonic significance of 'dormant' and 'active' deep-sea trenches in the Alpine-Carpathian Arc: Geological Society of London Special Publication 10, p. 471–494.

Heubeck C, Mann P, Dolan J, Monechi S, 1991, Diachronous uplift and recycling of sedimentary basins during Cenozoic tectonic transpression, northeastern Caribbean plate margin: Sedimentary Geology, v. 70, p. 1–32.

Hibbard JP, Karig DE, 1990, Structural and magmatic responses to spreading ridge subduction: an example from southwest Japan: Tectonics, v. 9, p. 207–230.

Hibbard JP, Karig DE, Taira A, 1992, Anomalous structural evolution of the Shimanto accretionary prism at Murotomisaki, Shikoku Island, Japan: The Island Arc, v. 1, p. 133–147.

Hickey JJ, 1984, Stratigraphy and composition of a Jura-Cretaceous volcanic arc apron: Punta Eugenia, Baja California Sur, Mexico, in Frizzell VA (ed.), Geology of the Baja California Peninsula: Pacific Section, Society of Economic Paleontologists and Mineralogists, Los Angeles, p. 149–160.

Hilde TWC, 1983, Sediment subduction versus accretion around the Pacific: Tectonophysics, v. 99, p. 381–397.

Hilde TWC, Uyeda S, 1983, Trench depth: variation and significance: American Geophysical Union Geodynamics Series, v. 11, p. 75–89.

Hildebrand RS, Bowring SA, 1984, Continental intra-arc depressions: a nonextensional model for their origin, with a Proterozoic example from Wopmay orogen: Geology, v. 12, p. 73–77.

Hildenbrand TG, 1985, Rift structure of the northern Mississippi embayment from the analysis of gravity and magnetic data: Journal of Geophysical Research, v. 90, p. 12607–12611.

Hill RI, 1991, Starting plumes and continental breakup: Earth and Planetary Science Letters, v. 104, p. 398–416.

Hill RI, Campbell IH, Davies GF, Griffiths RW, 1992, Mantle plumes and continental tectonics: Science, v. 256, p. 186–193.

Hinz K, 1981, A hypothesis on terrestrial catastrophes: wedges of very thick oceanward dipping layers beneath passive continental margins: Geologische Jahrbuch, Series E, v. 22, p. 3–28.

Hinze WJ, Bradley JW, Brown AR, 1978, Gravimeter survey in the Michigan basin deep borehole: Journal of Geophysical Research, v. 83, p. 5864–5868.

Hirata N, 7 coauthors, 1991, Report on DELP 1988 cruises in the Okinawa Trough - Part 3: crustal structure of the southern Okinawa Trough: Bulletin of the Earthquake Research Institute, University of Tokyo, v. 66, p. 37–70.

Hirst JPP, Nichols GJ, 1986, Thrust tectonic controls on Miocene alluvial distribution patterns, southern Pyrenees: International Association of Sedimentologists Special Publication 8, p. 247–258.

Hisatomi K, 1988, The Miocene forearc basin of southwest Japan and the Kumano Group of the Kii Peninsula: Modern Geology, v. 12, p. 389–408.

Hiscott RN, Pickering KT, 1984, Reflected turbidity currents on an Ordovician basin floor, Canadian Appalachians: Nature, v. 311, p. 143–145.

Hiscott RN, Pickering KT, Beeden DR, 1986, Progressive filling of a confined Middle Ordovician foreland basin associated with the Taconic orogeny, Quebec, Canada: International Association of Sedimentologists Special Publication 8, p. 309–325.

Hochstaedter AF, 6 coauthors, 1990a, Volcanism in the Sumisu Rift, I. Major element, volatiles and stable isotope geochemistry: Earth and Planetary Science Letters, v. 100, p. 179–194.

Hochstaedter AG, Gill JB, Morris JD, 1990b, Volcanism in the Sumisu Rift II: subduction and non-subduction related components: Earth and Planetary Science Letters, v. 100, p. 195–209.

Hodges KV, Wernicke BP, Walker JD, 1989, Middle Miocene (?) through Quaternary extension, northern Panamint Mountains area, California, in Wernicke BP, 6 coeditors (eds.), Extensional tectonics in the Basin and Range province between the southern Sierra Nevada and the Colorado Plateau [28th International Geological Congress Guidebook T138]: American Geophysical Union, Washington, p. 45–55.

Hoffman PF, 1973, Evolution of an early Proterozoic continental margin: the Coronation geosyncline and associated aulacogens of the NW Canadian shield: Philosophical Transactions of the Royal Society of London, v. 273A, p. 547–581.

Hoffman PF, 1987, Continental transform tectonics: Great Slave Lake shear zone (ca. 1.9 Ga), northwest Canada: Geology, v. 15, p. 785–788.

Hoffman PF, 1988a, United plates of America: the birth of a craton: early Proterozoic assembly and growth of Laurentia: Annual Review of Earth and Planetary Sciences, v. 16, p. 543–603.

Hoffman PF, 1988b, Belt basin: a landlocked remnant oceanic basin? (analogous to the south Caspian and Black seas): Geological Society of America Abstracts with Programs, v. 20, n. 7, p. A50.

Hoffman PF, 1991, Did the breakout of Laurentia turn Gondwanaland inside-out?: Science, v. 252, p. 1409–1412.

Hoffman PF, Grotzinger JP, 1993, Orographic precipitation, erosional unloading, and tectonic style: Geology, v. 21, p. 195–198.

Hoffman PF, Dewey JF, Burke K, 1974, Aulacogens and their genetic relation to geosynclines with a Proterozoic example from Great Slave Lake, Canada: Society of Economic Paleontologists and Mineralogists Special Publication 19, p. 38–55.

Holbrook WS, Mooney WD, 1987, The crustal structure of the axis of the Great Valley, California, from seismic refraction measurements: Tectonophysics, v. 140, p. 49–63.

Holmquist PJ, 1932, Über sog. Fiederspalten: Geologiska Föreningens i Stockholm Förhandlingar, v. 54, p. 99–118.

Homewood PW, 1977, Ultrahelvetic and north-Penninic flysch of the Prealps: Eclogae Geologicae-Helvetiae, v.70, p.627–641.

Homewood P, 1983, Palaeogeography of Alpine flysch: Palaeogeography, Palaeoclimatology, Palaeoecology, v.44, p.169–184.

Homewood P, Allen PA, 1981, Wave-, tide- and current-controlled sandbodies of Miocene Molasse, western Switzerland: American Association of Petroleum Geologists Bulletin, v. 65, p. 2534–2545.

Homewood P, Caron C, 1982, Flysch of the western Alps, in Hsu KJ (ed.), Mountain building processes: Academic Press, New York, p. 157–168.

Homewood P, Allen PA, Williams GD, 1986, Dynamics of the Molasse basin of western Switzerland: International Association of Sedimentologists Special Publication 8, p. 199–217.

Honma H, 6 coauthors, 1991, Major and trace element chemistry and D/H, $^{18}O/^{16}O$, $^{87}Sr/^{86}Sr$ and $^{143}Nd/^{144}Nd$ ratios of rocks from the spreading center of the Okinawa Trough, a marginal back-arc basin: Geochemical Journal, v. 25, p. 121–136.

Honza E, Tamaki K, 1985, The Bonin Arc, in Nairn AEM, Stehli FG, Uyeda S (eds.), The ocean basins and margins [v. 7A, The Pacific Ocean]: Plenum Press, New York, p. 459–502.

Hooke RL, 1972, Geomorphic evidence for late Wisconsinan and Holocene tectonic deformation, Death Valley, California: Geological Society of America Bulletin, v. 83, p. 2073–2098.

Hopson CA, Mattinson JM, Pessagno EA, Jr, 1981, Coast Range ophiolite, western California, in Ernst WG (ed.), The geotectonic development of California [Rubey Volume I]: Prentice-Hall, Englewood Cliffs, p. 418–510.

Horine RL, Moore GF, Taylor B, 1990, Structure of the outer Izu-Bonin forearc from seismic reflection profiling and gravity modeling: Proceedings of the Ocean Drilling Program, Initial Results, v. 125, p. 81–94.

Hornafius JS, Luyendyk BP, Terres RR, Kamerling MJ, 1986, Timing and extent of Neogene tectonic rotation in the western Transverse Ranges, California: Geological Society of America Bulletin, v. 97, p. 1476–1487.

Horváth F, Berckhemer H, Stegena L, 1981, Models of Mediterranean back-arc basin formation: Philosophical Transactions of the Royal Society of London, v. A300, p. 383–402.

Hou L, Luo D, Fu D, Hu S, Li K, 1991, Triassic sedimentary-tectonic evolution in western Sichuan and eastern Xizang region: People's Republic of China Ministry of Geology and Mineral Resources Geological Memoirs, Series 3, n. 13, 220p.

Houghton BF, Landis CA, 1989, Sedimentation and volcanism in a Permian arc-related basin, southern New Zealand: Bulletin of Volcanology, v. 51, p. 433–450.

Houseknecht DW, 1986, Evolution from passive margin to foreland basin: the Atoka Formation of the Arkoma basin,

south-central U.S.A.: International Association of Sedimentologists Special Publication 8, p. 327–345.

Howard RH, 1990, Hydrocarbon reservoir distribution in the Illinois basin: American Association of Petroleum Geologists Memoir 51, p. 299–327.

Howell DG, Jones DL, Schermer ER, 1985, Tectonostratigraphic terranes of the circum-Pacific region: Circum-Pacific Council for Energy and Mineral Resources Earth Sciences Series, v. 1, p. 3–30.

Howell PD, 1989, Epeirogeny, intracratonic basin reactivation and stress-induced weakening of the lower crust: Geological Society of America Abstracts with Programs, v. 21, n. 6, p. A81.

Howell PD, vanderPluijm BA, 1990, Early history of the Michigan basin: subsidence and Appalachian tectonics: Geology, v. 18, p. 1195–1998.

Hsü KJ, 1970, The meaning of the word flysch - a short historical search: Geological Association of Canada Special Paper 7, p. 1–11.

Hsü KJ, 1975, Catastrophic debris streams (Sturzstroms) generated by rock falls: Geological Society of America Bulletin, v. 86, p. 129–140.

Hsü KJ, 1977, Studies of Ventura field, California, I: facies geometry and genesis of lower Pliocene turbidites: American Association of Petroleum Geologists Bulletin, v. 61, p. 137–168.

Hsü KJ, Briegel U, 1991, Geologie der Schweiz: Birkhäuser, Basel, 219 p.

Hsü KJ, Kelts K, Valentine JW, 1980, Resedimented facies in Ventura basin, California, and model of longitudinal transport of turbidity currents: American Association of Petroleum Geologists Bulletin, v. 64, p. 1034–1051.

Hsü KJ, Wang Q, Li J, Zhou D, Sun S, 1987, Tectonic evolution of Qinling Mountains, China: Eclogae Geologicae Helvetiae, v. 80, p. 735–752.

Hsü KJ, 5 coauthors, 1990, Tectonics of South China: key to understanding west Pacific geology: Tectonophysics, v. 183, p. 9–39.

Hsui AT, Toksöz MN, 1981, Back-arc spreading: trench migration, continental pull, or induced convection?: Tectonophysics, v. 74, p. 89–98.

Huang C-Y, Shyu C-T, Lin SB, Lee T-Q, Sheu DD, 1992, Marine geology in the arc-continent collision zone off southeastern Taiwan: implications for late Neogene evolution of the Coastal Range: Marine Geology, v. 107, p. 183–212.

Huang J, Chen B, 1987, The evolution of the Tethys in China and adjacent regions: Geologic Publishing House, Beijing, 109 p.

Hubbard SS, Çoruh C, Costain JK, 1991, Paleozoic and Grenvillian structures in the southern Appalachians: extended interpretation of seismic reflection data: Tectonics, v. 10, p. 141–170.

Hubert JF, 1967, Sedimentology of Prealpine flysch sequences, Switzerland: Journal of Sedimentary Petrology, v. 37, p. 885–907.

Huchon P, LePichon X, 1984, Sunda Strait and the central Sumatra fault: Geology, v. 12, p. 668–672.

Hughes SS, 1990, Mafic magmatism and associated tectonism of the central High Cascade Range, Oregon: Journal of Geophysical Research, v. 95, p. 19623–19638.

Humphries CC, Jr, 1978, Salt movement on continental slope, northern Gulf of Mexico: American Association of Petroleum

Geologists Studies in Geology 7, p. 69–85.

Hussong DM, Fryer P, 1982, Structure and tectonics of the Mariana arc and fore-arc: drill site selection surveys: Initial Reports of the Deep Sea Drilling Project, v. 60, p. 33–44.

Hussong DM, Uyeda S, 1981, Tectonic processes and the history of the Mariana arc: a synthesis of the results of Deep Sea Drilling Project Leg 60: Initial Reports of the Deep Sea Drilling Project, v. 60, p. 909–929.

Hussong DM, Wipperman LK, 1981, Vertical movement and tectonic erosion of the continental wall of the Peru-Chile Trench near 11°30′S latitude: Geological Society of America Memoir 154, p. 509–524.

Hussong DM, 13 coauthors, 1981, Initial Reports of the Deep Sea Drilling Project [v. 60]: United States Government Printing Office, Washington, 929 p.

Hutchinson DR, Grow JA, 1985, New York Bight fault: Geological Society of America Bulletin, v. 96, p. 975–989.

Hutchinson DR, Klitgord KD, 1988, Evolution of rift basins on the continental margin off southern New England, in Manspeizer W (ed.), Triassic-Jurassic rifting. Part A: Elsevier, Amsterdam, p. 81–98.

Hutchinson DR, Grow JA, Klitgord KD, Swift BA, 1982, Deep structure and evolution of the Carolina trough: American Association of Petroleum Geologists Memoir 34, p. 129–152.

Hutchinson RW, 1980, Massive base metal sulphide deposits as guides to tectonic evolution: Geological Association of Canada Special Paper 20, p. 659–684.

Hutchinson RW, Engels GG, 1972, Tectonic evolution in the southern Red Sea and its possible significance to older rifted continental margins: Geological Society of America Bulletin, v. 83, p. 2989–3002.

Huyghe P, Mugnier J-L, 1992, The influence of depth on reactivation in normal faulting: Journal of Structural Geology, v. 14, p. 991–998.

Hyndman RD, Yorath CJ, Clowes RM, Davis EE, 1990, The northern Cascadia subduction zone at Vancouver Island: seismic structure and tectonic history: Canadian Journal of Earth Sciences, v. 27, p. 313–329.

Hynes A, Mott J, 1985, On the causes of back-arc spreading: Geology, v. 13, p. 387–389.

Ikeda Y, Yuasa M, 1989, Volcanism in nascent back-arc basins behind the Schichito Ridge and adjacent areas in the Izu-Ogasawara arc, northwest Pacific: evidence for mixing between E-type MORB and island arc magmas at the initiation of backarc rifting: Contributions to Mineralogy and Petrology, v. 101, p. 377–393.

Illies JH, 1962, Oberrheinisches Grundgebirge und Rheingraben: Geologische Rundschau, v. 52, p. 317–332.

Illies JH, 1967, Development and tectonic pattern of the Rhinegraben: Abhandlungen der geologischen Landesamt Baden Württemberg, v. 6, p. 7–12.

Illies JH, 1970, Graben tectonics as related to crust-mantle interaction, in Illies JH, Mueller S (eds.), Graben problems: E. Schweizerbart'sche Verlagsbuchhandlung, Stuttgart, p. 4–27.

Illies JH, 1972, The Rhine graben rift system-plate tectonics and transform faulting: Geophysical Surveys, v. 1, p. 27–60.

Illies JH, 1974a, Intra-Plattentektonik in Mitteleuropa und der Rheingraben: Oberrheinische Geologische Abhandlungen, v. 23, p. 1–24.

Illies JH, 1974b, Taphrogenesis and plate tectonics, in Illies JH,

Fuchs K (eds.), Approaches to taphrogenesis: E. Schweizer-bart'sche Verlagsbuchandlung, Stuttgart, p. 433–460.

Illies JH, 1975a, Rheingraben und Alpen: Wechselbeziehungen zwischen Taphrogenese und Orogenese: Mitteilungen aus dem Geologisch-Paläontologischen Institut der Universität Hamburg, v. 44, p. 403–409.

Illies JH, 1975b, Intraplate tectonics in stable Europe as related to plate tectonics in the Alpine system: Geologische Rundschau, v. 64, p. 677–699.

Illies JH, 1977, Ancient and recent rifting in the Rhinegraben: Geologie en Mijnbouw, v. 56, p.329–350.

Illies JH, Greiner G, 1976, Regionales stress-Feld und Neotektonik in Mitteleuropa: Oberrheinische Geologische Abhandlungen, v. 25, p. 1–40.

Illies JH, Greiner G, 1978, Rhinegraben and the Alpine system: Geological Society of America Bulletin, v. 89, p. 770–782.

Imoto N, McBride EF, 1990, Volcanism recorded in the Tesnus Formation, Marathon uplift, Texas, in LaRoche TM, Higgins L (eds.), Marathon thrust belt: structure, stratigraphy, and hydrocarbon potential [Field Seminar]: West Texas Geological Society, Midland, p. 93–98.

Ineson JR, 1989, Coarse-grained submarine fan and slope apron deposits in a Cretaceous back-arc basin, Antarctica: Sedimentology, v. 36, p. 793–819.

Ingersoll RV, 1978a, Paleogeography and paleotectonics of the late Mesozoic forearc basin of northern and central California, in Howell DG, McDougall KA (eds.), Mesozoic paleogeography of the western United States [Pacific Coast Paleogeography Symposium 2]: Pacific Section, Society of Economic Paleontologists and Mineralogists, Los Angeles, p. 471–482.

Ingersoll RV, 1978b, Submarine fan facies of the Upper Cretaceous Great Valley sequence, northern and central California: Sedimentary Geology, v. 21, p. 205–230.

Ingersoll RV, 1978c, Petrofacies and petrologic evolution of the Late Cretaceous fore-arc basin, northern and central California: Journal of Geology, v. 86, p. 335–352.

Ingersoll RV, 1979, Evolution of the Late Cretaceous forearc basin, northern and central California: Geological Society of America Bulletin, v. 90, part I, p.813–826.

Ingersoll RV, 1982a, Triple-junction instability as cause for late Cenozoic extension and fragmentation of the western United States: Geology, v. 10, p. 621–624.

Ingersoll RV, 1982b, Initiation and evolution of the Great Valley forearc basin of northern and central California, U.S.A.: Geological Society of London Special Publication 10, p.459–467.

Ingersoll RV, 1983, Petrofacies and provenance of late Mesozoic forearc basin, northern and central California: American Association of Petroleum Geologists Bulletin, v. 67, p. 1125–1142.

Ingersoll RV, 1988a, Development of the Cretaceous forearc basin of central California, in Graham SA (ed.), Studies of the geology of the San Joaquin basin [Book 60]: Pacific Section, Society of Economic Paleontologists and Mineralogists, Los Angeles, p. 141–155.

Ingersoll RV, 1988b, Tectonics of sedimentary basins: Geological Society of America Bulletin, v. 100, p. 1704–1719.

Ingersoll RV, 1990, Actualistic sandstone petrofacies: discriminating modern and ancient source rocks: Geology, v. 18, p. 733–736.

Ingersoll RV, 1992, Comment on "Plutonism, oblique subduction, and continental growth: an example from the Mesozoic of California": Geology, v. 20, p. 280–281.

Ingersoll RV, Dickinson WR, 1981, Great Valley Group (sequence), Sacramento Valley, California, in Frizzell V (ed.), Upper Mesozoic Franciscan rocks and Great Valley sequence, central Coast Ranges, California: Pacific Section, Society of Economic Paleontologists and Mineralogists, Los Angeles, p. 1–33.

Ingersoll RV, Nilsen TH (eds.), 1990, Sacramento Valley symposium and guidebook: Pacific Section, Society of Economic Paleontologists and Mineralogists, v. 65, 215 p.

Ingersoll RV, Schweickert RA, 1986, A plate-tectonic model for Late Jurassic ophiolite genesis, Nevadan orogeny and fore-arc initiation, northern California: Tectonics, v. 5, p. 901–912.

Ingersoll RV, Suczek CA, 1979, Petrology and provenance of Neogene sand from Nicobar and Bengal fans, DSDP sites 2ll and 218: Journal of Sedimentary Petrology, v.49, p.1217–1228.

Ingersoll RV, Bullard TF, Ford RL, Grimm JP, Pickle JD, Sares SW, 1984, The effect of grain size on detrital modes: a test of the Gazzi-Dickinson point-counting method: Journal of Sedimentary Petrology, v. 54, p. 103–116.

Ingersoll RV, Cavazza W, Graham SA, Indiana University Graduate Field Seminar Participants, 1987, Provenance of impure calclithites in the Laramide foreland of southwestern Montana: Journal of Sedimentary Petrology, v. 57, p. 995–1003.

Ingersoll RV, Cavazza W, Baldridge WS, Shafiqullah M, 1990, Cenozoic sedimentation and paleotectonics of north-central New Mexico: implications for initiation and evolution of the Rio Grande rift: Geological Society of America Bulletin, v. 102, p. 1280–1296.

Ingersoll RV, Kretchmer AG, Valles PK, 1993, The effect of sampling scale on actualistic sandstone petrofacies: Sedimentology, v. 40, p. 937–953.

Ingle JC, Jr, 1992, Subsidence of the Japan Sea: stratigraphic evidence from ODP sites and onshore sections: Proceedings of the Ocean Drilling Program, Scientific Results, v. 127/128, Pt. 2, p. 1197–1218.

Ingle JC, Jr, 27 coauthors, 1990, Introduction, background, and principal results of Leg 128 of the Ocean Drilling Program, Japan Sea: Proceedings of the Ocean Drilling Program, Initial Reports, v. 128, p. 5–38.

Inman DL, Nordstrom CE, Flick RE, 1976, Currents in submarine canyons: an air-sea-land interaction: Annual Review of Fluid Mechanics, v. 8, p. 275–310.

Iriondo MH, 1990, Map of the South American plains—its present state, in Rabassa J (ed.), Quaternary of South America and Antarctic Peninsula: A.A. Balkema, Rotterdam, p. 297–308.

Isacks BJ, 1988, Uplift of the central Andean plateau and bending of the Bolivian orocline: Journal of Geophysical Research, v. 93, p. 3211–3231.

Ishikawa M, 6 coauthors, 1991, Report on DELP 1988 cruises in the Okinawa Trough—Part 6: petrology of volcanic rocks: Bulletin of the Earthquake Research Institute, University of Tokyo, v. 66, p. 151–177.

Isler A, Pantiç N, 1980, Schistes-Iustrés-Ablagerungen der Tethys: Eclogae Geologicae Helvetiae, v. 73, p. 799–822.

Issler DH, McQueen H, Beaumont C, 1989, Thermal and isostatic consequences of simple shear extension of the continental lithosphere: Earth and Planetary Science Letters, v. 91, p. 341–358.

Ito M, 1989, The Itsukaichimachi Group: a Middle Miocene strike-slip basin-ill in the southeastern margin of the Kanto Mountains, central Honshu, Japan, in Taira A, Masuda F (eds.), Sedimentary facies in the active plate margin: Terra Scientific Publishing, Tokyo, p. 659–673.

Ito M, Masuda F, 1986, Evolution of clastic piles in an arc–arc collision zone: late Cenozoic depositional history around the Tanzawa Mountains, central Honshu, Japan: Sedimentary Geology, v. 49, p. 223–259.

Jackson JA, 1980, Reactivation of basement faults and crustal shortening in orogenic belts: Nature, v. 283, p. 343–346.

Jackson JA, 1987, Active normal faulting and crustal extension: Geological Society of London Special Publication 28, p. 3–17.

Jackson JA, Leeder MR, 1994, Drainage systems and the development of normal faults: an example from Pleasant Valley, Nevada: Journal of Structural Geology, v. 16, p. 1041–1059.

Jackson JA, McKenzie D, 1983, The geometrical evolution of normal fault systems: Journal of Structural Geology, v. 5, p. 471–482.

Jackson JA, White NJ, 1989, Normal faulting in the upper continental crust: observations from regions of active extension: Journal of Structural Geology, v. 11, p. 15–36.

Jackson JA, King G, Vita-Finzi C, 1982, The neotectonics of the Aegean: an alternative view: Earth and Planetary Science Letters, v. 61, p. 303–318.

Jackson JA, White NJ, Garfunkel Z, Anderson H, 1988, Relations between normal-fault geometry, tilting and vertical motions in extensional terrains: an example from the southern Gulf of Suez: Journal of Structural Geology, v. 10, p. 155–170.

Jacobi RD, 1981, Peripheral bulge—a causal mechanism for the Lower/Middle Ordovician unconformity along the western margin of the northern Appalachians: Earth and Planetary Science Letters, v. 56, p. 245–251.

Jacobi RD, 1984, Modern submarine sediment slides and their implications for mélange and the Dunnage Formation in north-central Newfoundland: Geological Society of America Special Paper 198, p. 81–102.

Jacobsen SR, Nichols DJ, 1982, Palynological dating of syntectonic units in the Utah-Wyoming thrust belt: the Evanston Formation, Echo Canyon Conglomerate and the Little Muddy Creek Conglomerate, in Powers RB (ed.), Geologic studies of the Cordilleran thrust belt: Rocky Mountain Association of Geologists, Denver, p. 735–750.

Jacobson RRE, MacLeod WN, Black N, 1958, Ring-complexes in the younger granite province of northern Nigeria: Geological Society of London Memoir 1, 72 p.

Jacobson RS, Shor GG, Jr, Kieckhefer RM, Purdy GM, 1979, Seismic refraction and reflection studies in the Timor-Aru trough system and Australian continental margin: American Association of Petroleum Geologists Memoir 29, p. 209–222.

Jaeger JC, Cook NGW, 1971, Fundamentals of rock mechanics: Chapman and Hall, London, 515 p.

James DE, 1971, Plate tectonic model for the evolution of the central Andes: Geological Society of America Bulletin, v. 82, p. 3325–2246.

Jansa LF, Wiedmann J, 1982, Mesozoic-Cenozoic development of the eastern North American and northwest African continental margins: a comparison, in VonRad U, Hinz K, Sarnthein M, Seibold E (eds.), Geology of the northwest African continental margin: Springer-Verlag, Berlin, p. 215–267.

Jaques AL, Robinson GP, 1977, The continent/island-arc collision in northern Papua New Guinea: BMR Journal of Australian Geology and Geophysics, v. 2, p. 289–303.

Jaritz W, 1987, The origin and development of salt structures in northwest Germany, in Lerche I, O'Brien JJ (eds.), Dynamical geology of salt and related structures: Academic Press, New York, p. 479–493.

Jarrard RD, 1986a, Relations among subduction parameters: Reviews of Geophysics, v. 24, p. 217–284.

Jarrard RD, 1986b, Terrane motion by strike-slip faulting of forearc slivers: Geology, v. 14, p. 780–783.

Jarvis GT, McKenzie DP, 1980, Sedimentary basin formation with finite extension rates: Earth and Planetary Science Letters, v. 48, p. 42–52.

Jayko AS, Blake MC, Jr, Harms T, 1987, Attenuation of the Coast Range ophiolite by extensional faulting, and nature of the Coast Range "thrust," California: Tectonics, v. 6, p. 475–488.

Jenny J, LeMarrec A, Monbaron M, 1981, Les couches rouges du Jurassiue moyen du Hant Atlas central, Maroc: corrélations lithostratigraphiques, éléments de datations et cadre tectonosédimentaire: Bulletin de la Société Géologique de France, série 7, v. 23, p. 627–639.

Jesinkey C, Forsyhte RD, Mpodozis C, Davidson J, 1987, Concordant late Paleozoic paleomagnetizations from the Atacama Desert: implications for tectonic models of the Chilean Andes: Earth and Planetary Science Letters, v. 85, p. 461–472.

Ji X, Coney PJ, 1985, Accreted terranes in China: Circum-Pacific Council for Energy and Mineral Resources Earth Sciences Series, v. 1, p. 349–361.

Johnson BD, Powell CM, Veevers JJ, 1976, Spreading history of the eastern Indian Ocean and greater India's northward flight from Antarctica and Australia: Geological Society of America Bulletin, v. 87, p. 1560–1566.

Johnson D, Greene HG, 1988, Modern depositional regimes, offshore Vanuatu: Circum Pacific Council for Energy and Mineral Resources Earth Science Series, v. 8, p. 287–299.

Johnson GD, Raynolds RGH, Burbank DW, 1986, Late Cenozoic tectonics and sedimentation in the north-western Himalayan foredeep: I. Thrust ramping and associated deformation in the Potwar region: International Association of Sedimentologists Special Publication 8, p. 273–291.

Johnson LE, 7 coauthors, 1991, New evidence for crustal accretion in the outer Mariana fore arc: Cretaceous radiolarian cherts and mid-ocean ridge basalt-like lavas: Geology, v. 19, p. 811–814.

Johnson NM, Jordan TE, Johnsson PA, Naeser CW, 1986, Magnetic polarity stratigraphy, age and tectonic setting of fluvial sediments in an eastern Andean foreland basin, San Juan Province, Argentina: International Association of Sedimentologists Special Publication 8, p. 63–75.

Johnson SY, 1984, Stratigraphy, age, and paleogeography of the Eocene Chuckanut Formation, northwest Washington: Canadian Journal of Earth Sciences, v. 21, p. 92–106.

Johnson SY, 1985, Eocene strike-slip faulting and nonmarine basin formation in Washington: Society of Economic Paleontologists and Mineralogists Special Publication 37, p.283–302.

Johnsson MJ, 1990, Tectonic versus chemical-weathering controls on the composition of fluvial sands in tropical environments: Sedimentology, v. 37, p. 713–726.

Johnston CR, Bowin CO, 1981, Crustal reactions resulting from the mid-Pliocene to Recent island arc collision in the Timor region: Bureau of Mineral Resources Journal of Australian Geology and Geophysics, v. 6, p. 223–243.

Jolivet L, Tamaki K, 1992, Neogene kinematics in the Japan Sea region and volcanic activity of the northeast Japan arc: Proceedings of the Ocean Drilling Program, Scientific Results, v. 127/128, Pt. 2, p. 1311–1331.

Jolivet L, Huchon P, Rangin C, 1989, Tectonic setting of western Pacific marginal basins: Tectonophysics, v. 160, p. 23–47.

Jones JG, 1967, Clastic rocks of Espiritu Santo Island, New Hebrides: Geological Society of America Bulletin, v. 78, p. 1281–1287.

Jordan TE, 1981, Thrust loads and foreland basin evolution, Cretaceous, western United States: American Association of Petroleum Geologists Bulletin, v.65, p.2506–2520.

Jordan TE, Allmendinger RW, 1986, The Sierras Pampeanas of Argentina: a modern analogue of Rocky Mountain foreland deformation: American Journal of Science, v. 286, p. 737–764.

Jordan TE, Alonso RN, 1987, Cenozoic stratigraphy and basin tectonics of the Andes Mountains, 20°-28° south latitude: American Association of Petroleum Geologists Bulletin, v. 71, p. 49–64.

Jordan TE, Flemings PB, 1991, Large-scale stratigraphic architecture, eustatic variation, and unsteady tectonism: a theoretical evaluation: Journal of Geophysical Research, v. 96, p. 6681–6699.

Jordan TE, Gardeweg PM, 1989, Tectonic evolution of the late Cenozoic central Andes (20°-33°S), in Ben-Avraham Z (ed.), The evolution of the Pacific Ocean margins: Oxford University Press, New York, p. 193–207.

Jordan TE, Isacks BL, Ramos VA, Allmendinger RW, 1983a, Mountain building in the central Andes: Episodes, v. 1983, p. 20–26.

Jordan TE, 5 coauthors, 1983b, Andean tectonics related to geometry of subducted Nazca plate: Geological Society of America Bulletin, v.94, p.341–361. (Also, see Discussion and Reply, v.95, p.887–880.)

Jordan TE, Flemings PB, Beer JA, 1988, Dating thrust-fault activity by use of foreland basin strata, in Kleinspehn KL, Paola C (eds.), New perspectives in basin analysis: Springer-Verlag, New York, p. 307–330.

Jordan TE, 5 coauthors, 1989, Neogene sedimentary basins in the flat-slab regions, central Andes, Argentina: Abstracts of 28th International Geological Congress, Washington, p. 139.

Jordan TE, 5 coauthors, 1990, Magnetic polarity stratigraphy of the Miocene Rio Azul section, Precordillera thrust belt, San Juan province, Argentina: Journal of Geology, v. 98, p. 519–539.

Jordan TE, Allmendinger RW, Damanti JF, Drake RE, 1993, Chronology of motion in a complete thrust belt: the Precordillera, 30-31°S, Andes Mountains: Journal of Geology, v. 101, p. 137–158.

Jowett EC, Rydzewski A, Jowett RJ, 1987, The Kupferschiefer Cu-Ag ore deposits in Poland: a reappraisal of the evidence of their origin and prsentation of a new genetic model: Canadian Journal of Earth Sciences, v. 24, p. 2016–2037.

Kailasham LN, 1976, Geophysical studies of the major sedimentary basin of the Indian craton, their deep structural features and evolution: Tectonophysics, v. 36, p. 225–245.

Kaeding M, Forsythe RD, Nelson EP, 1990, Geochemistry of the Taitao ophiolite and near-trench intrusions from the Chile margin triple junction: Journal of South American Earth Science, v. 3, p. 161–177.

Kagami H, 20 coauthors, 1986, Initial Reports of the Deep Sea Drilling Project [v. 87]: United States Government Printing Office, Washington, 985 p.

Kaiho K, 1992, Eocene to Quaternary benthic foraminifers and paleobathymetry of the Izu-Bonin arc, Legs 125 and 126: Proceedings of the Ocean Drilling Program, Scientific Results, v. 126, p. 285–310.

Kampunzu AB, Popoff M, 1991, Distribution of the main Phanerozoic African rifts and associated magmatism: introductory notes, in Kampunzu AB, Lubala RT (eds.), Magmatism in extensional structural settings: Springer-Verlag, Berlin, p. 2–10.

Kanamori H, 1971, Great earthquakes at island arcs and the lithosphere: Tectonophysics, v. 12, p. 187–198.

Kanamori H, 1986, Rupture process of subduction-zone earthquakes: Annual Reviews of Earth and Planetary Science, v. 14, p. 293–322.

Kano K, Kosaka K, Murata A, Yanai S, 1990, Intra-arc deformations with vertical rotation axes: the case of the pre-Middle Miocene terranes of southwest Japan: Tectonophysics, v. 176, p. 333–354.

Karig DE, 1970a, Kermadec arc-New Zealand tectonic confluence: New Zealand Journal of Geology and Geophysics, v. 13, p. 21–29.

Karig DE, 1970b, Ridges and basins of the Tonga-Kermadec island arc system: Journal of Geophysical Research, v. 75, p. 239–254.

Karig DE, 1971a, Origin and development of marginal basins in the western Pacific: Journal of Geophysical Research, v.76, p.2542–2561.

Karig DE, 1971b, Structural history of the Mariana Island Arc system: Geological Society of America Bulletin, v. 82, p. 323–344.

Karig DE, 1972, Remnant arcs: Geological Society of America Bulletin, v. 83, p. 1057–1068.

Karig DE, 1974, Evolution of arc systems in the western Pacific: Annual Review of Earth and Planetary Sciences, v. 2, p. 51–75.

Karig DE, 1975, Basin genesis in the Philippine Sea: Initial Reports of the Deep Sea Drilling Project, v. 31, p. 857–879.

Karig DE, 1977, Growth patterns on the upper trench slope, in Talwani M, Pitman WC, III (eds.), Island arcs, deep sea trenches, and back-arc basins [Maurice Ewing Series 1]: American Geophysical Union, Washington, p. 175–185.

Karig DE, 1980, Material transport within accretionary prisms and the "knocker" problem: Journal of Geology, v. 88, p. 27–40.

Karig DE, 1982, Initiation of subduction zones: implications for arc evolution and ophiolite development: Geological Society of London Special Publication 10, p. 563–576.

Karig DE, 1983a, Accreted terranes in the northern part of the Philippine archipelago: Tectonics, v. 2, p. 211–236.

Karig DE, 1983b, Temporal relationships between back arc basin formation and arc volcanism with special reference to the Philippine Sea: American Geophysical Union Geophysical Monograph 27, p. 318–325.

Karig DE, 1985, Kinematics and mechanics of deformation across some accreting forearcs, in Nasu N, Uyeda S, Kushiro I,

Kobayashi K, Kagami H (eds.), Formation of active ocean margins: Terra Scientific Publishing, Tokyo, p. 155–177.

Karig DE, 1986, Physical properties and mechanical state of accreted sediments in the Nankai Trough, S.W. Japan: Geological Society of America Memoir 166, p. 117–133.

Karig DE, Moore GF, 1975, Tectonically controlled sedimentation in marginal basins: Earth and Planetary Science Letters, v.26, p.233–238.

Karig DE, Ranken B, 1983, Marine geology of the forearc region, southern Mariana island arc: American Geophysical Union Geophysical Monograph 27, p. 266–280.

Karig DE, Sharman GF, III, 1975, Subduction and accretion in trenches: Geological Society of America Bulletin, v.86, p.377–389.

Karig DE, 12 coauthors, 1975, Initial Reports of the Deep Sea Drilling Project [v. 31]: United States Government Printing Office, Washington, 927 p.

Karig DE, Caldwell JG, Parmentier EM, 1976, Effects of accretion on the geometry of the descending lithosphere: Journal of Geophysical Research, v. 81, p. 6281–6291.

Karig DE, Anderson RN, Bibee LD, 1978a, Characteristics of back arc spreading in the Mariana Trough: Journal of Geophysical Research, v. 83, p. 1213–1226.

Karig DE, Cardwell RK, Moore GF, Moore DG, 1978b, Late Cenozoic subduction and continental margin truncation along the northern Middle America Trench: Geological Society of America Bulletin, v. 89, p. 265–276.

Karig DE, Suparka S, Moore GF, Hehanussa PE, 1979, Structure and Cenozoic evolution of the Sunda arc in the central Sumatra region: American Association of Petroleum Geologists Memoir 29, p. 223–237.

Karig DE, Lawrence MB, Moore GF, Curray JR, 1980a, Structural framework of the fore-arc basin, NW Sumatra: Journal of the Geological Society of London, v. 137, p. 77–91.

Karig DE, Moore GF, Curray JR, Lawrence MB, 1980b, Morphology and shallow structure of the lower trench slope off Nias Island, Sunda arc: American Geophysical Union Geophysical Monograph 23, p. 179–208.

Karig DE, Barber AJ, Charlton TR, Klemperer S, Hussong DM, 1987, Nature and distribution of deformation across the Banda Arc-Australian collision zone at Timor: Geological Society of America Bulletin, v. 98, p. 18–32.

Karl HA, Cacchione DA, Carlson PR, 1986, Internal-wave currents as a mechanism to account for large sand waves in Navarinsky Canyon head, Bering Sea: Journal of Sedimentary Petrology, v. 56, p. 706–714.

Karl HA, Hampton MA, Kenyon NH, 1989, Lateral migration of Cascadia Channel in response to accretionary tectonics: Geology, v. 17, p. 144–147.

Karner GD, 1986, Effects of lithospheric in-plane stress on sedimentary basin stratigraphy: Tectonics, v. 5, p. 573–588.

Karner GD, Watts AB, 1982, On isostasy at Atlantic-type continental margins: Journal of Geophysical Research, v. 87, p. 2923–2948.

Karner GD, Watts AB, 1983, Gravity anomalies and flexure of the lithosphere at mountain ranges: Journal of Geophysical Research, v. 88, p. 10449–10477.

Karner GD, Steckler MS, Thorne JA, 1983, Long-term thermo-mechanical properties of the continental lithosphere: Nature, v. 304, p. 250–253.

Kasper DC, Larue DK, 1986, Paleogeographic and tectonic implications of quartzose sandstones of Barbados: Tectonics, v. 5, p. 837–854.

Kastens K, 20 coauthors, 1988, ODP Leg 107 in the Tyrrhenian Sea: insights into passive margin and back-arc basin evolution: Geological Society of America Bulletin, v. 100, p. 1140–1156.

Katao H, 1988, Seismic structure and formation of the Yamato Basin: Bulletin of Earthquake Research Institute, University of Tokyo, v. 63, p. 51–86.

Katili JA, 1974, Sumatra, in Spencer AM (ed.), Mesozoic-Cenozoic orogenic belts: Scottish Academic Press, Edinburgh, p. 317–336.

Katz HR, Watters WA, 1965, Geological investigation of the Yahgan Formation and associated igneous rocks of Isla Navarino, southern Chile: New Zealand Journal of Geology and Geophysics., v. 9, p. 323–359.

Kauffman EG, 1977, Geological and biological overview: western interior Cretaceous basin: The Mountain Geologist, v. 14, p. 75–99.

Kauffman EG, 1984, Paleobiogeography and evolutionary response dynamic in the Cretaceous western interior seaway of North America: Geological Association of Canada Special Paper 27, p. 273–306.

Kauffman EG, 1985, Fine-grained deposits and biofacies of the Cretaceous western interior seaway: evidence of cyclic sedimentary processes, in Pratt LM, Kauffman EG, Zelt FB, SEPM Second Annual Midyear Meeting, Field Trip 9: Society of Economic Paleontologists and Mineralogists, Tulsa, p. IV-XIII.

Kay M, 1951, North American geosynclines: Geological Society of America Memoir 48, 143 p.

Kay SM, Kay RW, Citron GP, 1982, Tectonic controls on tholeiitic and calc-alkaline magmatism in the Aleutian arc: Journal of Geophysical Research, v. 87, p. 4051–4072.

Kay SM, Mpodozis C, Romos VA, Munizaga F, 1992, Magma source variations for mid-late Tertiary magmatic rocks associated with a shallowing subduction zone and a thickening crust in the central Andes: Geological Society of America Special Paper 265, p. 113–138.

Kazmin V, 1987, Two types of rifting: dependence on the condition of extension: Tectonophysics, v. 143, p. 85–92.

Keach RW, III, Oliver JE, Brown LD, Kaufman S, 1989, Cenozoic active margin and shallow Cascades structure: COCORP results from western Oregon: Geological Society of America Bulletin, v. 101, p. 783–794.

Keen CE, 1979, Thermal history and subsidence of rifted continental margins–evidence from wells on the Nova Scotian and Labrador shelves: Canadian Journal of Earth Science, v. 16, p. 505–522.

Keen CE, 1985, The dynamics of rifting: deformation of the lithosphere by active and passive driving forces: Geophysical Journal of the Royal Astronomical Society, v. 80, p. 95–120.

Keen CE, Beaumont C, 1990, Geodynamics of rifted continental margins, in Keen MJ, Williams GL (eds.), Geology of the continental margin of eastern Canada [The Geology of North America, v. I-1]: Geological Society of America, Boulder, p. 391–472.

Keen CE, Lewis T, 1982, Measured radiogenic heat production in sediments from continental margin of eastern North America: implications for petroleum generation: American Associa-

tion of Petroleum Geologists Bulletin, v. 66, p. 1402–1407.

Keen CE, Kay WA, Roest WR, 1990, Crustal anatomy of a transform continental margin: Tectonophysics, v. 173, p. 527–544.

Keen MJ, Williams GL, 1990, Geology of the continental margin of eastern Canada, Geology of Canada [The Geology of North America, v. I–1]: Geological Society of America, Boulder, 855 p.

Keith SB, 1978, Paleosubduction geometries inferred from Cretaceous and Tertiary magmatic patterns in southwestern North America: Geology, v.6, p.516–521.

Keller GR, Lidiak EG, Hinze WJ, Braille LW, 1983, The role of rifting in the tectonic development of the midcontinent, USA: Tectonophysics, v. 94, p. 391–412.

Keller GR, 7 coauthors, 1991, A comparative study of the Rio Grande and Kenya rifts: Tectonophysics, v. 197, p. 355–376.

Kelley VC, 1979, Tectonics, middle Rio Grande rift, New Mexico, in Riecker RE (ed.), Rio Grande rift: Tectonics and magmatism: American Geophysical Union, Washington, p. 289–311.

Kelts K, 1981, A comparison of some aspects of sedimentation and translational tectonics from the Gulf of California and the Mesozoic Tethys, northern Penninic margin: Eclogae Geologicae Helvetiae, v. 74, p. 317–338.

Kendall CGSC, Schlager W, 1981, Carbonates and relative changes of sea level: Marine Geology, v. 44, p. 181–212.

Kennett JP, McBirney AR, Thunell RC, 1977, Episodes of Cenozoic volcanism in the circum Pacific region: Journal of Volcanology and Geothermal Research, v. 2, p. 145–163.

Kent DM, 1987, Paleotectonic controls on sedimentation in the northern Williston basin, Saskatchewan, in Longman MW (ed.), Williston basin: anatomy of a cratonic oil province: Rocky Mountain Association of Petroleum Geologists, Denver, p. 45–56.

Keppie JD, 1993, Synthesis of Palaeozoic deformational events and terrane accretion in the Canadian Appalachians: Geologische Rundschau, v. 82, p. 381–431.

Kerrich R, Wyman D, 1990, Geodynamic setting of mesothermal gold deposits: an association with accretionary tectonic regimes: Geology, v. 18, p. 882–885.

Khain VE, 1975, Structure and main stages in tectomagmatic development of the Caucasus: an attempt at geodynamic interpretation: American Journal of Science, v. 275A, p. 131–156.

Khain VE, Michailov AE, 1985, Obshaya geotektonika: Nedra, Moscow, 326 p. (For a German translation, see Chain VE, Michajlov AE, 1989, Allgemeine Geotektonik: VEB Deutscher Verlag für Grundstoffindusrie, Leipzig, 303 p.)

Kidd WSF, Molnar P, 1988, Quaternary and active faulting observed on the 1985 Academia Sinica-Royal Society Geotraverse of Tibet: Philosophical Transactions of the Royal Society of London, v. 327A, p. 337–363.

Kieckhefer RM, Shor GG, Jr, Curray JR, Sugiarta W, Hehuwat F, 1980, Seismic refraction studies of the Sunda Trench and forearc basin: Journal of Geophysical Research, v. 85, p. 863–889.

Kieckhefer RM, Moore GF, Emmel FJ, Sugiarta W, 1981, Crustal structure of the Sunda forearc region west of central Sumatra from gravity data: Journal of Geophysical Research, v. 86, p. 7003–7012.

Kikuchi Y, Tono S, Funayama M, 1991, Petroleum resources in the Japanese island-arc setting: Episodes, v. 14, p. 236–241.

Kilmer FH, 1963, Cretaceous and Cenozoic stratigraphy and paleontology, El Rosario area [PhD thesis]: University of California, Berkeley, 149 p.

Kilmer FH, 1984, Geology of Cedros Island, Baja California, Mexico: Humboldt State University, Arcata, 69p.

Kimbrough DL, 1984, Paleogeographic significance of the Middle Jurassic Gran Canon Formation, Cedros Island, Baja California, in Frizzell VA, Jr (ed.), Geology of the Baja California Peninsula: Pacific Section, Society of Economic Paleontologists and Mineralogists, Los Angeles, p. 107–118.

Kimbrough DL, 1985, Tectonostratigraphic terranes of the Vizcaino Peninsula and Cedros and San Benito Islands, Baja California, Mexico: Circum-Pacific Council for Energy and Mineral Resources Earth Science Series, v. 1, p. 285–298.

Kimbrough DL, Hickey JJ, Tosdal RM, 1987, U-Pb ages of granitoid clasts in upper Mesozoic arc-derived strata of the Vizcaino Peninsula, Baja California, Mexico: Geology, v. 15, p. 26–29.

Kimura G, 1986, Oblique subduction and collision: forearc tectonics of the Kuril arc: Geology, v. 14, p. 404–407.

Kimura G, Mukai A, 1991, Underplated units in an accretionary complex: melange of the Shimanto belt of eastern Shikoku, southwest Japan: Tectonics, v. 10, p. 31–50.

Kimura G, Tamaki K, 1985, Tectonic framework of the Kuril arc since its initiation, in Nasu N, Kobayashi K, Uyeda S, Kushiro I, Kagami H (eds.), Formation of active ocean margins: Terra Scientific Publishing, Tokyo, p. 641–676.

Kimura G, Tamaki K, 1986, Collision, rotation, and back-arc spreading in the region of the Okhotsk and Japan seas: Tectonics, v. 5, p. 389–401.

Kimura M, 1985, Back-arc rifting in the Okinawa Trough: Marine and Petroleum Geology, v. 2, p. 222–240.

Kimura M, 9 coauthors, 1988, Active hydrothermal mounds in the Okinawa Trough backarc basin, Japan: Tectonophysics, v. 145, p. 319–324.

Kimura M, 7 coauthors, 1991, Report on DELP 1988 cruises in the Okinawa Trough—Part 7: geologic investigation of the central rift in the middle to southern Okinawa Trough: Bulletin of the Earthquake Research Institute, University of Tokyo, v. 66, p. 179–209.

King BC, 1970, Vulcanicity and rift tectonics in East Africa, in Clifford TN, Gass IG (eds.), African magmatism and tectonics: Hafner Publishing, Davien, p. 263–283.

King GCP, Stein RS, Rundle JB, 1988, The growth of geological structures by repeated earthquakes 1. Conceptual framework: Journal of Geophysical Research, v. 93, p. 13307–13318.

King PB, 1959, The evolution of North America: Princeton University Press, Princeton, 190 p.

King PB, 1966, The North American Cordillera: Canadian Institute of Mining and Metallurgy Special Volume 8, p. 1–25.

King PB, 1977, The evolution of North America (revised edition): Princeton University Press, Princeton, 197 p.

Kinoshita M, Yamano M, Kasumi Y, Baba H, 1991, Report on DELP 1988 cruises in the Okinawa Trough—Part 8: heat flow measurements: Bulletin of the Earthquake Research Institute, University of Tokyo, v. 66, p. 211–228.

Kinsman DJJ, 1975, Rift valley basins and sedimentary history of trailing continental margins, in Fischer AG, Judson S (eds.), Petroleum and global tectonics: Princeton University Press, Princeton, p.83–126.

Kissel C, Laj C, 1988, The Tertiary geodynamical evolution of the Aegean arc: a palaeomagnetic reconstruction: Tectonophysics, v. 146, p.183–201.

Klaus A, Taylor B, 1991, Submarine canyon development in the Izu-Bonin forearc: a SeaMARC II and seismic survey of Aoga Shima Canyon: Marine Geophysical Researches, v. 13, p. 131–152.

Klaus A, Taylor B, Moore GF, Murakami F, Okamura Y, 1992a, Back-arc rifting in the Izu-Bonnin island arc: structural evolution of Hachijo and Aoga Shima rifts: The Island Arc, v. 1, p. 16–31.

Klaus A, 6 coauthors, 1992b, Structural and stratigraphic evolution of Sumisu Rift, Izu-Bonin Arc: Proceedings of the Ocean Drilling Program, Scientific Results, v. 126, p. 555–574.

Klein GD, 1975, Sedimentary tectonics in southwest Pacific marginal basins based on Leg 30 Deep Sea Drilling Project cores from the South Fiji, Hebrides, and Coral Sea basins: Geological Society of America Bulletin, v. 86, p. 1012–1018.

Klein GD, 1984, Relative rates of tectonic uplift as determined from episodic turbidite deposition in marine basins: Geology, v. 12, p. 48–50.

Klein GD, 1985a, The control of depositional depth, tectonic uplift, and volcanism on sedimentation processes in the back-arc basins of the western Pacific Ocean: Journal of Geology, v. 93, p. 1–25.

Klein GD, 1985b, The frequency and periodicity of preserved turbidites in submarine fans as a quantitative record of tectonic uplift in collision zones: Tectonophysics, v. 119, p. 181–193.

Klein GD, 1987, Current aspects of basin analysis: Sedimentary Geology, v. 50, p. 95–118.

Klein GD, 1991, Origin and evolution of North American cratonic basins: South African Journal of Geology, v. 94, p. 3–18.

Klein GD, 1992a, Depth determination and quantitative distinction of the influence of tectonic subsidence and climate on changing sea level during deposition of midcontinent Pennsylvanian cyclothems: Society of Economic Paleontologists and Mineralogists Contributions to Sedimentology and Paleontology, v. 4, in press.

Klein GD, 1992b, Climatic and tectonic sea-level gauge for midcontinent Pennsylvanian cyclothems: Geology, v. 20, p. 363–366.

Klein GD, Hsui AT, 1987, Origin of intracratonic basins: Geology, v. 15, p. 1094–1098.

Klein GD, Kobayashi K, 1981, Geological summary of the Shikoku Basin and northwestern Philippine Sea, Leg 58, DSDP/IPOD drilling results: Oceanologica Acta SP, p. 181–192.

Klein GD, Kupperman JB, 1992, Pennsylvanian cyclothems: methods of distinguishing tectonically-induced changes in sea level from climatically-induced changes: Geological Society of America Bulletin, v. 104, p. 166–175.

Klein GD, Lee YI, 1984, A preliminary assessment of geodynamic controls on depositional systems and sandstone diagenesis in back-arc basins, western Pacific Ocean: Tectonophysics, v. 102, p. 119–152.

Klein GD, Willard DA, 1989, Origin of the Pennsylvanian coal-bearing cyclothems of North America: Geology, v. 17, p. 152–155.

Klein GD, Okada H, Mitsui K, 1979, Slope sediments in small basins associated with a Neogene active margin, western Hokkaido Island, Japan: Society of Economic Paleontologists

and Mineralogists Special Publication 27, p. 359–374.

Klein GD, 14 coauthors, 1980, Initial Reports of the Deep Sea Drilling Project [v. 58]: United States Government Printing Office, Washington, 1022 p.

Kleinspehn KL, Paola C (eds.), New perspectives in basin analysis: Springer-Verlag, New York, 453 p.

Klemme HD, 1980, Petroleum basins—classifications and characteristics: Journal of Petroleum Geology, v. 3, p. 187–207.

Klemperer SL, Hauge TA, Hauser EC, Oliver JE, Potter CJ, 1986, The Moho in the northern Basin and Range Province, Nevada, along the COCORP 40°N seismic-reflection transect: Geological Society of America Bulletin, v. 97, p. 603–618.

Klimetz MP, 1983, Speculations on the Mesozoic plate tectonic evolution of eastern China: Tectonics, v. 2, p. 139–166.

Klimetz MP, 1985, An outline of the plate tectonics of China: discussion: Geological Society of America Bulletin, v. 96, p. 407.

Klitgord KD, Hutchinson DR, Schouten H, 1988, U.S. Atlantic continental margin: structural and tectonic framework in Sheridan RE, Grow JA (eds.), The Atlantic continental margin U.S. [The Geology of North America, v. I-2]: Geological Society of America, Boulder, p. 19–55.

Klootwijk CT, 1981, The India-Asia collision: a summary of palaeomagnetic constraints, in Geological and ecological studies of Qinghai-Xizang Plateau [Proceedings of symposium on Qinghai-Xizang (Tibet) Plateau (Beijing, China)]: Science Press, Beijing, p. 951–966.

Klootwijk CT, Gee JS, Peirce JW, Smith GM, 1992, Neogene evolution of the Himalayan-Tibetan region: constraints from ODP Site 758, northern Ninetyeast Ridge; bearing on climatic change: Palaeogeography, Palaeoclimatology, Palaeoecology, v. 95, p. 95–110.

Klubov VA, Klevtsova AA, 1981, The Riazan-Saratov trough, in Tectonics of Europe and adjacent areas. Cratons, Baikalides, Caledonides [Explanatory note to the international tectonic map of Europe and adjacent areas, scale 1:2,500,000]: "Nauka", Moscow, p. 146–152.

Kluth CF, 1986, Plate tectonics of the Ancestral Rocky Mountains: American Association of Petroleum Geologists Memoir 42, p. 353–369.

Kluth CF, Coney PJ, 1981, Plate tectonics of the ancestral Rocky Mountains: Geology, v. 9, p. 10–15.

Kneller BC, Edwards DA, McCaffrey WD, Moore R, 1991, Oblique reflection of turbidity currents: Geology, v. 14, p. 250–252.

Knetsch G, 1963, Geologie von Deutschland: Ferdinand Enke Verlag, Stuttgart, 386 p.

Kobayashi K, 1983, Cycles of subduction and Cenozoic arc activity in the northwestern Pacific margin: American Geophysical Union Geodynamics Series 11, p. 287–301.

Kobayashi K, 1984, Subsidence of the Shikoku back-arc basin: Tectonophysics, v. 102, p. 105–117.

Kober L, 1921, Der Bau der Erde: Borntraeger, Berlin, 324 p.

Koizumi I, 1992, Diatom biostratigraphy of the Japan Sea: Leg 127: Proceedings of the Ocean Drilling Program, Scientific Results, v. 127/128, Pt. 1, p. 249–289.

Kokelaar BP, 1988, Tectonic controls of Ordovician arc and marginal basin volcanism in Wales: Journal of the Geological Society of London, v. 145, p. 759–775.

Kokelaar P, 1992, Ordovician marine volcanic and sedimentary record of rifting and volcanotectonism: Snowdon, Wales,

United Kingdom: Geological Society of America Bulletin, v. 104, p. 1433–1455.

Kokelaar P, Busby C, 1992, Subaqueous explosive eruption and welding of pyroclastic deposits: Science, v. 257, p. 196–201.

Kolata DR, Nelson WJ, 1990a, Tectonic history of the Illinois basin: American Association of Petroleum Geologists Memoir 51, p. 263–285.

Kolata DR, Nelson WJ, 1990b, Basin-forming mechanisms of the Illinois basin: American Association of Petroleum Geologists Memoir 51, p. 287–292.

Kolla V, Buffler RT, 1985, Magdalena Fan, Caribbean, in Bouma AH, Normark WR, Barnes NE (eds.), Submarine fans and related turbidite systems: Springer-Verlag, New York, p. 71–78.

Kolla V, Coumes F, 1985, Indus Fan, Indian Ocean, in Bouma AH, Normark WR, Barnes NE (eds.), Submarine fans and related turbidite systems: Springer-Verlag, New York, p. 129–136.

Kolla V, Coumes F, 1987, Morphology, internal structure, seismic stratigraphy, and sedimentation of Indus fan: American Association of Petroleum Geologists Bulletin, v. 71, p. 650–677.

Komar PD, 1969, The channelized flow of turbidity currents with application to Monterey deep-sea fan channel: Journal of Geophysical Research, v. 74, p. 4544–4558.

Komar PD, 1970, The competence of turbidity current flow: Geological Society of America Bulletin, v. 81, p. 1555–1562.

Komar PD, 1977, Computer simulation of turbidity current flow and the study of deep-sea channels and fan sedimentation, in Goldberg ED, McCave IN, O'Brien JJ, Steele JH (eds.), The Sea [v. 6, Marine modeling]: Wiley-Interscience, New York, p. 603–621.

Korsch RJ, 1977, A framework for the Palaeozoic geology of the southern part of the New England geosyncline: Geological Society of Australia Journal, v. 25, p. 339–355.

Korsch RJ, 1984, Sandstone compositions from the New England orogen, eastern Australia: implications for tectonic setting: Journal of Sedimentary Petrology, v. 54, p. 192–211.

Korsch RJ, Lindsay JF, 1989, Relationships between deformation and basin evolution in the intracratonic Amadeus basin, central Australia: Tectonophysics, v. 158, p. 5–22.

Koski RA, Shanks WC, II, Bohrson WA, Oscarson RL, 1988, The composition of massive sulfide deposits from the sediment-covered floor of Escanaba Trough, Gorda Ridge: implications for depositional processes: Canadian Mineralogist, v. 26, p. 655–673.

Koski RA, Lamons RC, Dumoulin JA, Bouse RM, 1993, Massive sulfide metallogenesis at a late Mesozoic sediment-covered spreading axis: evidence from the Franciscan complex and contemporary analogues: Geology, v. 21, p. 137–140.

Kossmat F, 1926, Die Mediterranen Kettengebirge in ihrer Beziehung zum Gleichgewichtszustande der Erdrinde: Abhandlungen der mathematisch-physikalischen klasse der sächsischen Akademie der Wissenschaften, v. 38, p. 1–63.

Kowalik WS, Gold DP, 1976, The use of Landsat-1 imagery in mapping lineaments in Pennsylvania: Utah Geological Association Publication 5, p. 236–249.

Koyama M, Cisowski SM, Gill JB, 1992, Paleomagnetic estimate of emplacement mechanisms of deep basaltic volcaniclastic rocks in the Sumisu Rift, Izu-Bonin Arc: Proceedings of the Ocean Drilling Program, Scientific Results, v. 126, p. 371–379.

Kraus MJ, Bown TM, 1988, Pedofacies analysis; a new approach to reconstructing ancient fluvial sequences: Geological Society of America Special Paper 216, p. 143–152.

Krenkel E, 1922, Die Bruchzonen Ostafrikas: Borntraeger, Berlin, 184 p.

Krieger MH, 1977, Large landslides, composed of megabreccia, interbedded in Miocene basin deposits, southeastern Arizona: United States Geological Survey Professional Paper 1008, 25 p.

Krissek LA, 1984, Continental source area contributions to fine-grained sediments on the Oregon and Washington continental slope, in Stow DAV, Piper DJW (eds.), Fine-grained sediments: deep water processes and facies: Blackwell Scientific, Oxford, p. 363–375.

Kroenke L, 14 coauthors, 1981, Initial Reports of the Deep Sea Drilling Project [v. 59]: United States Government Printing Office, Washington, 820 p.

Krohn MD, 1976, Relations of lineaments to sulfide deposits and fractured zones along Bald Eagle Mountain, Centre, Blair, and Huntingdon counties, Pennsylvania [MS thesis]: Pennsylvania State University, University Park, 104 p.

Krummenacher D, Gastil RG, Bushee J, Doupont J, 1975, K-Ar apparent ages, Peninsular Ranges batholith, southern California and Baja California: Geological Society of America Bulletin, v. 86, p. 760–768.

Kruse S, McNutt M, Phipps-Morgan J, Royden L, Wernicke BP, 1991, Lithosphere extension near Lake Mead, Nevada: a model for ductile flow in the lower crust: Journal of Geophysical Research, v. 96, p. 4435–4456.

Kuehl SA, Hariu TM, Moore WS, 1989, Shelf sedimentation off the Ganges-Brahmaputra river system: evidence for sediment bypassing to the Bengal fan: Geology, v. 17, p. 1132–1135.

Kuenzi WD, Horst OH, McGehee RV, 1979, Effect of volcanic activity on fluvial-deltaic sedimentation in a modern arc-trench gap, southwestern Guatemala: Geological Society of America Bulletin, v. 90, p. 827–838.

Kuhn TS, 1970, The structure of scientific revolutions (Second Edition): University of Chicago Press, Chicago, 210 p.

Kulm LD, Fowler GA, 1974a, Cenozoic sedimentary framework of the Gorda-Juan de Fuca plate and adjacent continental margin—a review: Society of Economic Paleontologists and Mineralogists Special Publication 19, p. 212–229.

Kulm LD, Fowler GA, 1974b, Oregon continental margin structure and stratigraphy: a test of the imbricate thrust model, in Burk CA, Drake CL (eds.), The geology of continental margins: Springer-Verlag, New York, p. 261–283.

Kulm LD, Scheidegger KF, 1979, Quaternary sedimentation on the tectonically active Oregon continental slop: Society of Economic Paleontologists and Mineralogists Special Publication 27, p. 247–263.

Kulm LD, 10 coauthors, 1973, Initial Reports of the Deep Sea Drilling Project [v. 18]: United States Government Printing Office, Washington, 1077 p.

Kulm LD, Resig JM, Moore TC, Jr, Rosato VJ, 1974, Transfer of Nazca Ridge pelagic sediments to the Peru continental margin: Geological Society of America Bulletin, v. 85, p. 769–780.

Kulm LD, Dymond J, Dasch EJ, Hussong DM (eds.), 1981a, Nazca plate: crustal formation and Andean convergence: Geological Society of America Memoir 154, 824 p.

Kulm LD, Prince RA, French W, Johnson S, Masias A, 1981b, Crustal structure and tectonics of the central Peru continental margin and trench: Geological Society of America Memoir 154, p. 445–468.

Kulm LD, Resig JM, Thornburg TM, Shrader H-J, 1982, Cenozoic structure, stratigraphy, and tectonics of the central Peru forearc: Geological Society of London Special Publication 10, p. 151–169.

Kulm LD, Thornburg TM, Suess E, Resig J, Fryer P, 1988, Clastic, diagenetic, and metamorphic lithologies of a subsiding continental block: central Peru forearc: Proceedings of the Ocean Drilling Program, Initial Reports, v. 112, p. 91–107.

Kumpulainen R, Nystuen JP, 1985, Late Proterozoic basin evolution and sedimentation in the westernmost part of Batoscandia, in Gee DG, Sturt BA (eds.), The Caledonide orogen–Scandinavia and related areas, Part I: John Wiley and Sons, London, p. 213–232.

Kusznir NJ, Park RG, 1987, The extensional strength of the continental lithosphere: its dependence on geothermal gradient, and crustal composition and thickness: Geological Society of London Special Publication 28, p. 35–52.

Kusznir NJ, Karner GD, Egan S, 1987, Geometric, thermal and isostatic consequences of detachments in continental lithosphere extension and basin formation: Canadian Society of Petroleum Geologists Memoir 12, p. 185–203.

Kusznir NJ, Marsden G, Egan SS, 1991, A flexural-cantilever simple-shear/pure-shear model of continental lithosphere extension: applications to the Jeanne d'Arc Basin, Grand Banks and Viking graben, North Sea: Geological Society of London Special Publication 56, p. 41–60.

Kvenvolden KA, vonHuene R, 1985, Natural gas generation in sediments of the convergent margin of the eastern Aleutian Trench area: Circum-Pacific Council for Energy and Mineral Resources Earth Science Series, v. 1, p. 31–49.

LaGabrielle Y, 5 coauthors, 1986, The Coast Range ophiolites (northern California): possible arc and back-arc basin remnants; their relations with the Nevadan orogeny: Bulletin Societe Geologique de France, ser. 8, v. 2, p. 981–999.

Laitakari I, 1983, The Jotnian (upper Proterozoic) Sandstone of Stakunta, in Laajoki K, Paakola J (eds.), Exogenic processes and related metallogeny in the Svecokarelian geosynclinal complex: Geological Survey of Finland Guidebook 11, p. 135–139.

Lallemand S, Jolivet L, 1985, Japan Sea: a pull apart basin: Earth and Planetary Science Letters, v. 76, p. 375–389.

Lallemand S, LePichon X, 1987, Coulomb wedge model applied to subduction of seamounts in the Japan Trench: Geology, v. 15, p. 1065–1069.

Lallemand S, Culotta R, vonHuene R, 1989, Subduction of the Daiichi Kashima Seamount in the Japan Trench: Tectonophysics, v. 160, p. 231–247.

Lambeck K, 1983a, The role of compressive forces in intracratonic basin formation and mid-plate orogenies: Geophysical Research Letters, v. 10, p. 845–848.

Lambeck K, 1983b, Structure and evolution of the intracratonic basins of central Australia: Geophysical Journal of the Royal Astronomical Society, v. 74, p. 843–886.

Lambert DD, Unruh DM, Gilbert MC, 1988, Rb-Sr and Sm-Nd-isotopic study of the Glen Mountains layered complex: initiation of rifting within the southern Oklahoma aulacogen: Geology, v. 16, p. 13–17.

Lambiase JJ, 1991, A model for tectonic control of lacustrine stratigraphic sequences in continental rift basins: American Association of Petroleum Geologists Memoir 50, p. 265–276.

Lamerson PR, 1982, The Fossil basin area and its relationship to the Absaroka thrust fault system, in Powers RB (ed.), Geologic studies of the Cordilleran thrust belt: Rocky Mountain Association of Geologists, Denver, p. 279–340.

Landis CA, Bishop DG, 1972, Plate tectonics and regional stratigraphic-metamorphic relations in the southern part of the New Zealand geosyncline: Geological Society of America Bulletin, v. 83, p. 2267–2284.

Landon SM (ed.), 1994, Interior rift basins: American Association of Petroleum Geologists Memoir 59, 276p.

Lanphere MA, Jones DL, 1978, Cretaceous time scale from North America: American Association of Petroleum Geologists Studies in Geology 6, p. 259–268.

LaPierre H, Albarede F, Albers J, Cabanis B, Coulon C, 1985, Early Devonian volcanism in the eastern Klamath Mountains, California: evidence for an immature island arc: Canadian Journal of Earth Sciences, v. 22, p. 214–227.

Larue DK, 1985, Quartzose turbidites of the accretionary complex of Barbados, II: variations in bedding styles, facies and sequences: Sedimentary Geology, v. 42, p. 217–253.

Larue DK, 1991, Organic matter in the Franciscan and Cedros subduction complexes: the problems of 'instantaneous maturation' and 'missing petroleum' in accretionary prisms: Marine and Petroleum Geology, v. 8, p. 468–482.

Larue DK, Hudleston PJ, 1987, Foliated breccias in the active Portuguese Bend landslide complex, California: bearing on melange genesis: Journal of Geology, v. 95, p. 407–422.

Larue DK, Speed RC, 1983, Quartzose turbidites of the accretionary complex of Barbados, I: Chalky Mount succession: Journal of Sedimentary Petrology, v. 53, p. 1337–1352.

Larue DK, 5 coauthors, 1985, Barbados: maturation, source rock potential and burial history within a Cenozoic accretionary complex: Marine and Petroleum Geology, v. 2, p. 96–110.

Larue DK, Smith AL, Schellekens JH, 1991, Oceanic island arc stratigraphy in the Caribbean region: don't take it for granite: Sedimentary Geology, v. 74, p. 289–308.

LASE Study Group, 1986, Deep structure of the U.S. East Coast passive margin from large aperture seismic experiments (LASE): Marine and Petroleum Geology, v. 3, p. 234–242.

Lash GG, 1985, Recognition of trench fill in orogenic flysch sequences: Geology, v. 13, p. 867–870.

Lash GG, 1986a, Anatomy of an early Paleozoic subduction complex in the central Appalachian orogen: Sedimentary Geology, v. 51, p. 75–95.

Lash GG, 1986b, Sedimentology of channelized turbidite deposits in an ancient (early Paleozoic) subduction complex, central Appalachians: Geological Society of America Bulletin, v. 97, p. 703–710.

Lash GG, 1987a, Diverse melanges of an ancient subduction complex: Geology, v. 15, p. 652–655.

Lash GG, 1987b, Geodynamic evolution of the lower Paleozoic central Appalachian foreland basin: Canadian Society of Petroleum Geologists Memoir 12, p. 413–423.

Lash GG, 1988, Along-strike variations in foreland basin evolution: possible evidence for continental collision along an irregular margin: Basin Research, v. 1, p. 71–83.

Lash GG, 1990, The Shochary Ridge sequence, southeastern

Pennsylvania–a possible Ordovician piggyback basin fill: Sedimentary Geology, v. 68, p. 39–53.

Latin DM, Dixon JE, Fitton JG, 1990, Rift-related magmatism in the North Sea basin, in Blundell DJ, Gibbs AD (eds.), Tectonic evolution of the North Sea rifts: Clarendon Press, Oxford, p. 102–144.

Laubscher H-P, 1970, Grundsätzliches zur Tektonik des Rheingrabens, in Illies JH, Mueller S (eds.), Graben problems: E. Schweizerbart'sche Verlagsbuchandlung, Stuttgart, p. 79–87.

Laville E, 1988, A multiple releasing and restraining stepover model for the Jurassic strike-slip basin of the central High Atlas (Morocco), in Manspeizer W (ed.), Triassic-Jurassic rifting. Part A: Elsevier, Amsterdam, p. 499–523.

Lawton TF, 1986a, Compositional trends within a clastic wedge adjacent to a fold-thrust belt: Indianola Group, central Utah, U.S.A.: International Association of Sedimentologists Special Publication 8, p. 411–423.

Lawton TF, 1986b, Fluvial systems of the Upper Cretaceous Mesaverde Group and Paleocene North Horn Formation, central Utah: a record of transition from thin-skinned to thick-skinned deformation in the foreland region: American Association of Petroleum Geologists Memoir 41, p. 423–442.

Lawton TF, Trexler JH, Jr, 1991, Piggyback basin in the Sevier orogenic belt, Utah: implications for development of the thrust wedge: Geology, v. 19, p. 827–830.

Leckie D, 1986, Rates, controls, and sand-body geometries of transgressive-regressive cycles: Cretaceous Moosebar and Gates formations, British Columbia: American Association of Petroleum Geologists Bulletin, v. 70, p. 516–535.

Lee C-S, Shor GG, Jr, Bibee LD, Lee RS, Hilde TWC, 1980, Okinawa Trough: origin of a back-arc basin: Marine Geology, v. 35, p. 219–241.

Lee CW, Burgress CJ, 1978, Sedimentation and tectonic controls in the Early Jurassic, central High Atlas, Morocco: Geological Society of America Bulletin, v. 89, p. 1199–1204.

Lee YI, Klein GD, 1986, Diagenesis of sandstones in the backarc basins of the western Pacific Ocean: Sedimentology, v. 33, p. 651–675.

Leeder MR, 1982, Upper Palaeozoic basins of the British Isles—Caledonide inheritance versus Hercynian plate margin processes: Journal of the Geological Society of London, v. 139, p. 479–491.

Leeder MR, 1991, Denudation, vertical crustal movements and sedimentary basin infill: Geologische Rundschau, v. 80, p. 441–458.

Leeder MR, Alexander J, 1987, The origin and tectonic significance of asymmetrical meander belts: Sedimentology, v. 34, p. 217–226.

Leeder MR, Gawthorpe RL, 1987, Sedimentary models for extensional tilt-block/half-graben basins: Geological Society of London Special Publication 28, p. 139–152.

Leeder MR, Jackson JA, 1993, The interaction between normal faulting and drainage in active extensional basins, with examples from the western United States and central Greece: Basin Research, v. 5, p. 79–102.

Leeder MR, Seger MJ, Stark CP, 1991, Sedimentology and tectonic geomorphology adjacent to active and inactive normal faults in the Megara Basin and Alkyonides Gulf, central Greece: Journal of the Geological Society of London, v. 148, p. 331–343.

Lees GM, 1955, Recent earth movements in the Middle East: Geologische Rundschau, v. 43, p. 221–226.

Leg 135 Scientific Party, 1992, Evolution of backarc basins: ODP Leg 135, Lau Basin: EOS, v. 73, p. 241–247.

Leggett JK, 1980, The sedimentological evolution of a lower Palaeozoic accretionary fore-arc in the Southern Uplands of Scotland: Sedimentology, v. 27, p. 401–417.

Leggett JK (ed.), 1982, Trench-forearc geology: sedimentation and tectonics on modern and ancient active plate margins: Geological Society of London Special Publication 10, 576p.

Leggett JK, McKerrow WS, Casey DM, 1982, The anatomy of a lower Palaeozoic accretionary forearc: the Southern Uplands of Scotland: Geological Society of London Special Publication 10, p. 495–520.

Leggett JK, McKerrow WS,-Soper NJ, 1983, A model for the crustal evolution of southern Scotland: Tectonics, v. 2, p. 187210.

Leggitt SM, Walker RG, Eyles CH, 1990, Control of reservoir geometry and stratigraphic trapping by erosion surface E5 in the Pembina-Carrot Creek area, Upper Cretaceous Cardium Formation, Alberta, Canada: American Association of Petroleum Geologists Bulletin, v. 74, p. 1165-1182

.Leighton MW, Kolata DR, 1990, Selected interior cratonic basins and their place in the scheme of global tectonics: a synthesis: American Association of Petroleum Geologists Memoir 51, p. 729–797.

Leighton MW, Kolata DR, Oltz DF, Eidel JJ (eds.), 1990, Interior cratonic basins: American Association of Petroleum Geologists Memoir 51, 819 p.

Leitch EC, 1975, Plate tectonic interpretation of the Paleozoic history of the New England fold belt: Geological Society of America Bulletin, v. 86, p. 141–144.

Leitch EC, 1984, Marginal basins of the SW Pacific and the preservation and recognition of their ancient analogues: a review: Geological Society of London Special Publication 16, p. 97–109.

Lemcke E, 1937, Der tektonische Bau des Gebiets zwischen Vogelsberg und Rhön: Geotektonische Forschungen, v. 1, p. 28–68.

Lemoine M, 10 coauthors, 1986, The continental margin of the Mesozoic Tethys in the western Alps: Marine and Petroleum Geology, v. 3, p. 179–199.

Leopold LB, Wolman MG, Miller JP, 1964, Fluvial processes in geomorphology: Freeman, San Francisco, 522 p.

LePichon X, Angelier J, 1979, The Hellenic arc and trench system: a key to the neotectonic evolution of the eastern Mediterranean area: Tectonophysics, v. 60, p. 1–42.

LePichon X, Angelier J, 1981, The Aegean Sea: Philosophical Transactions of the Royal Society of London, v. 300A, p. 357–372.

LePichon X, Cochran JR (eds.), 1988, The Gulf of Suez and Red Sea rifting: Tectonophysics, v. 153, 320 p.

LePichon X, Huchon P, 1984, Geoid, Pangea and convection: Earth and Planetary Science Letters, v. 67, p. 123–135.

LePichon X, Sibuet JC, 1981, Passive margins: a model of formation: Journal of Geophysical Research, v. 86, p. 3708–3720.

LePichon X, 16 coauthors, 1987a, Nankai Trough and the fossil Shikoku Ridge: results of box 6 Kaiko survey: Earth and Planetary Science Letters, v. 83, p. 186–198.

LePichon X, 16 coauthors, 1987b, The eastern and western

ends of Nankai Trough: results of box 5 and box 7 Kaiko survey: Earth and Planetary Science Letters, v. 83, p. 199–213.

LePichon X, Bergerat F, Roulet M-J, 1988, Plate kinematics and tectonics leading to the Alpine belt formation; a new analysis: Geological Society of America Special Paper 218, p. 111–131.

Letouzey J, Kimura M, 1985, Okinawa Trough genesis: structure and evolution of a backarc basin developed in a continent: Marine and Petroleum Geology, v. 2, p. 111–130.

Letouzey J, Kimura M, 1986, The Okinawa Trough: genesis of a back-arc basin developing along a continental margin: Tectonophysics, v. 125, p. 209–230.

Levi B, Aguirre L, 1981, Ensialic spreading subsidence in the Mesozoic and Paleogene of central Chile: Journal of the Geological Society of London, v. 138, p.76–81.

Lewis KB, 1980, Quaternary sedimentation on the Hikurangi oblique-subduction and transform margin, New Zealand: International Association of Sedimentologists Special Publication 4, p. 171–189.

Lewis KB, Pantin HM, 1984, Intersection of a marginal basin with a continent: structure and sediments of the Bay of Plenty, New Zealand: Geological Society of London Special Publication 16, p. 121–135.

Lewis SD, Hayes DE, 1984, A geophysical study of the Manila Trench, Luzon, Philippines 2. Fore arc basin structural and stratigraphic evolution: Journal of Geophysical Research, v. 89, p. 9196–9214.

Lewis SD, Ladd JW, Bruns TR, 1988, Structural development of an accretionary prism by thrust and strike-slip faulting: Shumagin region, Aleutian Trench: Geological Society of America Bulletin, v. 100, p. 767–782.

Li CY (ed.), 1982, Tectonic map of Asia: Cartographic Publishing House, Beijing, 1:8,000,000.

Lillegraven JA, Ostresh LM, Jr, 1988, Evolution of Wyoming's early Cenozoic topography and drainage patterns: National Geographic Research, v. 4, p. 303–327.

Lin J, Fuller M, Zhang W, 1985, Preliminary Phanerozoic polar wander paths for the North and South China blocks: Nature, v. 313, p. 444–449.

Lindsay JF, 1987, Sequence stratigraphy and depositional controls in Late Proterozoic–Early Cambrian sediments of Amadeus basin, central Australia: American Association of Petroleum Geologists Bulletin, v. 71, p. 1387–1403.

Lindsay JF, 1990, Forearc basin dynamics and sedimentation controls, Tamworth belt, eastern Australia: Bureau of Mineral Resources Journal of Australian Geology and Geophysics, v. 11, p. 521–528.

Lindsay JF, Korsch RJ, 1989, Interplay of tectonics and sea-level changes in basin evolution: an example from the intracratonic Amadeus basin, central Australia: Basin Research, v. 2, p. 3–25.

Lindsay JF, Korsch RJ, Wilford JR, 1987, Timing the breakup of a Proterozoic supercontinent: evidence from Australian intracratonic basins: Geology, v. 15, p. 1061–1064.

Lindsay JF, Holliday DW, Hulbert AG, 1991, Sequence stratigraphy and the evolution of the Ganges-Brahmaputra delta complex: American Association of Petroleum Geologists Bulletin, v. 75, p. 1233–1254.

Link MH, 1980, Sedimentary facies and mineral deposits of the Miocene Barstow Formation, California, in Fife DL, Brown AR (eds.), Geology and mineral wealth of the California desert: South Coast Geological Society, Santa Ana, p. 191–203.

Link MH, 1982, Provenance, paleocurrents, and paleogeography of Ridge basin, southern California, in Crowell JC, Link MH (eds.), Geologic history of Ridge basin, southern California: Pacific Section, Society of Economic Paleontologists and Mineralogists, Los Angeles, p. 265–276.

Link MH, Osborne RH, 1978, Lacustrine facies in the Pliocene Ridge Basin Group: Ridge basin, California: International Association of Sedimentologists Special Publication 2, p. 169–187.

Link MH, Roberts MJ, 1986, Pennsylvanian paleogeography for the Ozarks, Arkoma, and Ouachita basins in east-central Arkansas, in Stone CG, Haley BR (eds.), Sedimentary and igneous rocks of the Ouachita Mountains of Arkansas, Part 2 [Guidebook 86-3]: Arkansas Geological Commission, Little Rock, p. 37–60.

Linn AM, DePaolo DJ, Ingersoll RV, 1991, Nd-Sr isotopic provenance analysis of Upper Cretaceous Great Valley fore-arc sandstones: Geology, v. 19, p. 803–806.

Linn AM, DePaolo DJ, Ingersoll RV, 1992, Nd-Sr isotopic, geochemical, and petrographic stratigraphy and paleotectonic analysis: Mesozoic Great Valley forearc sedimentary rocks of California: Geological Society of America Bulletin, v. 104, p. 1264–1279.

Lipman PW, 1980, Cenozoic volcanism in the western United States: implications for continental tectonics, in Burchfiel BC, Oliver JE, Silver LT (eds.), Continental tectonics [Studies in Geophysics]: National Research Council, National Academy of Sciences, Washington, p. 161–175.

Lipman PW, 1984, The roots of ash flow calderas in western North America: windows into the tops of granitic batholiths: Journal of Geophysical Research, v. 89, p. 8801–8841.

Lipman PW, 1992, Magmatism in the Cordilleran United States; progress and problems, in Burchfiel BC, Lipman PW, Zoback ML (eds.), The Cordilleran orogen: conterminous U.S. [Geology of North America, v. G3]: Geological Society of America, Boulder, p. 481–514.

Lipman PW, Mullineaux DR (eds.), 1981, The 1980 eruptions of Mount St. Helens, Washington: United States Geological Survey Professional Paper 1250, 844 p.

Lister CRB, 1977, Estimators for heat flow and deep rock properties based on boundary layer theory: Tectonophysics, v. 41, p. 157–171.

Lister GS, Etheridge MA, Symonds PA, 1986, Detachment faulting and the evolution of passive continental margins: Geology, v. 14, p. 246–250. (Also, see Comment and Reply, v. 14, p. 890–892.)

Lister GS, Etheridge MA, Symonds PA, 1991, Detachment models for the formation of passive continental margins: Tectonics, v. 10, p. 1038–1064

Liu H, 1986, Geodynamic scenario and structural styles of Mesozoic and Cenozoic basins in China: American Association of Petroleum Geologists Bulletin, v. 70, p. 377–395.

Liu Y, Gastaldo RA, 1992, Characteristics and provenance of log-transported gravels in a Carboniferous channel deposit: Journal of Sedimentary Petrology, v. 62, p. 1072–1083.

Livelybrooks DW, Clingman WW, Rygh JT, Urquhart SA, Waff HS, 1989, A magnetotelluirc study of the High Cascades graben in central Oregon: Journal of Geophysical Research, v. 94, p. 14173–14184.

Logatchev NA, Rogozhina VA, Solonenko VP, Zorin YA, 1978, Deep structure and evolution of the Baikal rift zone, in Ramberg IB, Neumann ER (eds.), Tectonics and geophysics of continental rifts: D.Reidel, Dordrecht, p. 49–61.

Logatchev NA, Zorin YA, Rogozhina VA, 1983, Baikal rift: active or passive? - comparison of the Baikal and Kenya rift zones: Tectonophysics, v. 94, p. 223–240.

Longwell CR, 1937, Sedimentation in relation to faulting: Geological Society of America Bulletin, v. 48, p. 433–442.

Longwell CR, 1951, Megabreccia developed downslope from large faults: American Journal of Science, v. 249, 343–355.

Lonsdale P, 1975, Sedimentation and tectonic modification of the Samoan archipelagic apron: American Association of Petroleum Geologists Bulletin, v. 59, p. 780–798.

Lonsdale P, 1978, Ecuadorian subduction system: American Association of Petroleum Geologists Bulletin, v. 62, p. 2454–2477.

Lonsdale P, 1989, Geology and tectonic history of the Gulf of California, in Winterer EL, Hussong DM, Decker RW (eds.), The eastern Pacific Ocean and Hawaii [The geology of North America, v. N]: Geological Society of America, Boulder, p. 499–521.

Lorenz JC, 1982, Lithospheric flexure and the history of the Sweetgrass arch, northwestern Montana, in Powers RB (ed.), 1982 symposium: Rocky Mountain Association of Geologists, Denver, p. 77–89.

Lorenz JC, 1988, Triassic-Jurassic rift-basin sedimentology: history and methods: Von Nostrand Reinhold, New York, 315 p.

Lorenz V, Nicholls IA, 1976, The Permocarboniferous basin and range province of Europe. An application of plate tectonics, in Falke H (ed.), The continental Permian in central, west, and south Europe: D. Reidel, Dordrecht, p. 313–342.

Lorenzo JM, Vera EE, 1992, Thermal uplift and erosion across the continent-ocean transform boundary of the southern Exmouth Plateau: Earth and Planetary Science Letters, v. 108, p. 79–92.

Lorenzo JM, Mutter JC, Larson RL, Northwest Australia Study Group, 1991, Development of the continent-ocean transform boundary of the southern Exmouth Plateau: Geology, v. 19, p. 843–846.

Lotze F, 1937, Zur Methodik der Forschungen über saxonische Tektonik: Geotektonische Forschungen, v. 1, p. 6–27.

Lotze F, 1953, Einige Proleme der Osning-Tektonik: Geotektonische Forschungen, v. 9/10, p. 7–12.

Lotze F, 1957, Steinsalz und Kalisalze (2nd edition): Allgemeingeologischer Teil: Borntraeger, Berlin, 465p.

Lotze F, 1971, Geologie Mitteleuropas: E. Schmeizerbart'sche Verlagsbuchhandlung, Stuttgart, 491 p.

Lowe DR, 1985, Ouachita trough, part of a Cambrian failed rift system: Geology, v. 13, p. 790–793.

Lowe DR, 1989, Stratigraphy, sedimentology, and depositional setting of pre-orogenic rocks of the Ouachita Mountains, Arkansas and Oklahoma, in Hatcher RD, Jr, Thomas WA, Viele GW (eds.), The Appalachian-Ouachita orogen in the United States [The Geology of North America, v. F-2]: Geological Society of America, Boulder, p. 575–590.

Lowell JD, 1972, Spitsbergen Tertiary orogenic belt and the Spitsbergen fracture zone: Geological Society of America Bulletin, v. 83, p. 3091–3102.

Lowell JD, Genik, GJ, 1972, Sea-floor spreading and structural evolution of southern Red Sea: American Association of Petroleum Geologists Bulletin, v. 56, p. 247–259.

Lu H, Dong H, Deng X, Li P, Yan J, 1990, Preliminary study of the terranes in west Sichuan: Circum-Pacific Council for Energy and Mineral Resources Earth Sciences Series, v. 13, p. 261–267.

Lu RS, McMillen KJ, 1982, Multichannel seismic survey of the Colombia Basin and adjacent margins: American Association of Petroleum Geologists Memoir 34, p. 395–410.

Lucchitta I, Suneson NH, 1993, Dips and extension: Geological Society of America Bulletin, v. 105, p. 1346–1356.

Ludwig WJ, Windisch CC, Houtz RE, Ewing JI, 1979, Structure of Falkland Plateau and offshore Tierra del Fuego, Argentina: American Association of Petroleum Geologists Memoir 29, p. 125–137.

Luff GC, Pearson BT, 1988, The Marfa basin, in Sloss LL (ed.), Sedimentary cover - North American craton: U.S. [The Geology of North America, v. D-2]: Geological Society of America, Boulder, p. 295–299.

Lundberg N, 1982, Evolution of the slope landward of the Middle America Trench, Nicoya Peninsula, Costa Rica: Geological Society of London Special Publication 10, p. 131–147.

Lundberg N, 1983, Development of forearcs of intraoceanic subduction zones: Tectonics, v. 2, p. 51–61.

Lundberg N, 1988, Present-day sediment transport paths south of the Longitudinal Valley, southeastern Taiwan: Acta Geologica Taiwanica, n. 26, p. 317–332.

Lundberg N, 1991, Detrital record of the early Central American magmatic arc: petrography of intraoceanic forearc sandstones, Nicoya Peninsula, Costa Rica: Geological Society of America Bulletin, v. 103, p. 905–915.

Lundberg N, Dorsey RJ, 1988, Synorogenic sedimentation and subsidence in a Plio-Pleistocene collisional basin, eastern Taiwan, in Kleinspehn KL, Paola C (eds.), New perspectives in basin analysis: Springer-Verlag, New York, p. 265–280.

Lundberg N, Dorsey RJ, 1990, Rapid Quaternary emergence, uplift, and denudation of the Coastal Range, eastern Taiwan: Geology, v. 18, p. 638–641.

Lundberg N, Reed DL, 1991, Continental margin tectonics: forearc processes: Reviews of Geophysics, Supplement, p. 794–806.

Luterbacher HP, Eichenseer H, Betzler C, VandenHurk AM, 1991, Carbonate-siliciclastic depositional systems in the Paleogene of the south Pyrenean foreland basin: a sequence-stratigraphic approach: International Association of Sedimentologists Special Publication 12, p. 391–407.

Luyendyk BP, 1989, Crustal rotation and fault slip in the continental transform zone in southern California, in Kissel C, Laj C (eds.), Paleomagnetic rotations and continental deformation: Kluwer Academic Publishers, Boston, p. 229–246.

Luyendyk BP, 1991, A model for Neogene crustal rotations, transtension, and transpression in southern California: Geological Society of America Bulletin, v. 103, p. 1528–1536.

Luyendyk BP, Hornafius JS, 1987, Neogene crustal rotations, fault slip, and basin development in southern California, in Ingersoll RV, Ernst WG (eds.), Cenozoic basin development of coastal California [Rubey Volume VI]: Prentice-Hall, Englewood Cliffs, p. 259–283.

Luyendyk BP, Kamerling MJ, Terres R, 1980, Geometric model for Neogene crustal rotations in southern California: Geological Society of America Bulletin, v. 91, Part I, p. 211–217.

Luyendyk BP, Kamerling MJ, Terres RR, Hornafius JS, 1985,

Simple shear of southern California during Neogene time suggested by paleomagnetic declinations: Journal of Geophysical Research, v. 90, p. 12454–12466.

Lynch HD, Morgan P, 1987, The tensile strength of the lithosphere and the localization of extension: Geological Society of London Special Publication 28, p. 53–65.

Ma XY, Deng QD, Wang YP, Lin HF, 1982, Cenozoic graben systems in North China: Zeitschrift für Geomorphologie Neue Folge, Supplement, v. 42, p. 99–116.

MacFadden BJ, Swisher CC, III, Opdyke ND, Woodburne MO, 1990, Paleomagnetism, geochronology, and possible tectonic rotation of middle Miocene Barstow Formation, Mojave Desert, southern California: Geological Society of America Bulletin, v. 102, p. 478–493.

Macfarlane A, Carney JN, Crawford AJ, Greene HG, 1988, Vanuatu–a review of the onshore geology: Circum-Pacific Council for Energy and Mineral Resources Earth Science Series, v. 8, p. 45–91.

Mack GH, Jerzykiewicz T, 1989, Provenance of post-Wapiabi sandstones and its implications for Campanian to Paleocene tectonic history of the southern Canadian Cordillera: Canadian Journal of Earth Sciences, v. 26, p. 665–676.

Mack GH, Seager WR, 1990, Tectonic control on facies distribution of the Camp Rice and Palomas formations (Pliocene-Pleistocene) in the southern Rio Grande rift: Geological Society of America Bulletin, v. 102, p. 45–53.

Mack GH, James WC, Thomas WA, 1981, Orogenic provenance of Mississippian sandstones associated with southern Appalachian-Ouachita orogen: American Association of Petroleum Geologists Bulletin, v. 65, p. 1444–1456.

Mack GH, Thomas WA, Horsey CA, 1983, Composition of Carboniferous sandstones and tectonic framework of southern Appalachian-Ouachita orogen: Journal of Sedimentary Petrology, v. 53, p. 931–946.

MacKay ME, Moore GF, 1990, Variation in deformation of the south Panama accretionary prism: response to oblique subduction and trench sediment variation: Tectonics, v. 9, p. 683–698.

MacKay ME, Moore GF, Cochrane GR, Moore JC, Kulm LD, 1992, Landward vergence and oblique structural trends in the Oregon margin accretionary prism: implications and effect on fluid flow: Earth and Planetary Science Letters, v. 109, p. 477–491.

MacKinnon TC, 1983, Origin of the Torlesse terrane and coeval rocks, South Island, New Zealand: Geological Society of America Bulletin, v. 94, p. 967–985.

MacKinnon TC, Howell DG, 1985, Torlesse turbidite system, New Zealand, in Bouma AH, Normark WR, Barnes NE (eds.), Submarine fans and related turbidite systems: Springer-Verlag, New York, p. 223–228.

MacPherson GJ, 1983, The Snow Mountain volcanic complex: an on-land seamount in the Franciscan terrain, California: Journal of Geology, v. 91, p. 73–92.

Makris J, Rihm R, 1991, Shear-controlled evolution of the Red Sea: pull apart model: Tectonophysics, v. 198, p. 441–466.

Makris J, Henke CH, Egloff F, Akamaluk T 1991, The gravity field of the Red Sea and East Africa: Tectonophysics, v. 198, p. 369–382.

Malinverno A, Ryan WBF, 1986, Extension in the Tyrrhenian Sea and shortening in the Apennines as result of arc migration driven by sinking of the lithosphere: Tectonics, v. 5, p. 227–245.

Malizzia DC, 1987, Contribución al conocimiento geológico y estratigráfico de las rocas terciarias del Campo de Talampaya, provincia de La Rioja, Argentina [PhD thesis]: Universidad Nacional de Tucuman, Tucuman, 187 p.

Malizzia DC, Villanueva Garcia A, 1984, Estratigrafía y paleoambiente de sedimentación de la Formación Río Mañero, Provincia de La Rioja, in Actas del IX Congreso Geológico Argentino, v. 5: Asociación Geológica Argentina, Buenos Aires, p. 146–156.

Mann P, Hempton MR, Bradley DC, Burke K, 1983, Development of pull-apart basins: Journal of Geology, v. 91, p. 529–554.

Mansfield CF, 1979, Upper Mesozoic subsea fan deposits in the southern Diablo Range, California: record of the Sierra Nevada magmatic arc: Geological Society of America Bulletin, Part I, v. 90, p. 1025–1046.

Manspeizer W, 1980, Rift tectonics inferred from volcanic and clastic structures, in Manspeizer W (ed.), Field studies of New Jersey geology and guide to field trips [52nd Annual Meeting of the New York State Geological Association]: Rutgers University, New Brunswick, p. 314–350.

Manspeizer W, 1985, The Dead Sea rift: impact of climate and tectonism on Pleistocene and Holocene sedimentation: Society of Economic Paleontologists and Mineralogists Special Publication 37, p. 143–158.

Manspeizer W, 1988, Triassic-Jurassic rifting and opening of the Atlantic: an overview, in Manspeizer W (ed.), Triassic-Jurassic rifting. Part A: Elsevier, Amsterdam, p. 41–79.

Manspeizer W, Cousminer HL, 1988, Late Triassic-Early Jurassic synrift basins of the U.S. Atlantic margin, in Sheridan RE, Grow JA (eds.), The Atlantic continental margin, U.S. [The Geology of North America, v. I-2]: Geological Society of America, Boulder, p. 197–216.

Manspeizer W, Puffer JH, Cousminer HL, 1978, Separation of Morocco and eastern North America: a Triassic-Liassic stratigraphic record: Geological Society of America Bulletin, v. 89, p. 901–920.

Marcher MV, 1961, The Tuscaloosa Gravel in Tennessee and its relation to the structural development of the Mississippi embayment syncline: United States Geological Survey Professional Paper 424, p. B90–B93.

Marjanac T, 1990, Reflected sediment gravity flows and their deposits in flysch of Middle Dalmatia, Yugoslavia: Sedimentology, v. 37, p. 921–929.

Marlow MS, 6 coauthors, 1970, Buldir depression–a late Tertiary graben on the Aleutian Ridge, Alaska: Marine Geology, v. 8, p. 85–108.

Marlow MS, Scholl DW, Buffington EC, Alpha TR, 1973, Tectonic history of the central Aleutian arc: Geological Society of America Bulletin, v. 84, p. 1555–1574.

Marlow MS, Exon NF, Ryan HF, Dadisman SV, 1988, Offshore structure and stratigraphy of New Ireland basin in northern Papua New Guinea: Circum-Pacific Council for Energy and Mineral Resources Earth Science Series, v. 9, p. 137–155.

Marsaglia KM, 1989, The petrology, provenance, and diagenesis of arc-related sands recovered by the Deep Sea Drilling Project on circum-Pacific, Mediterranean, and Caribbean legs [PhD thesis]: University of California, Los Angeles, 384 p.

Marsaglia KM, 1991, Provenance of sands and sandstones from a rifted continental arc, Gulf of California, Mexico: Society

for Sedimentary Geology Special Publication 45, p. 237–248.

Marsaglia KM, 1992, Petrography and provenance of volcaniclastic sands recovered from the Izu-Bonin arc, Leg 126: Proceedings of the Ocean Drilling Program, Scientific Results, v. 126, p. 139–154.

Marsaglia KM, Ingersoll RV, 1992, Compositional trends in arc-related, deep-marine sand and sandstone: a reassessment of magmatic-arc provenance: Geological Society of America Bulletin, v. 104, p. 1637–1649.

Marsaglia KM, Tazaki K, 1992, Diagenetic trends in Leg 126 sandstones: Proceedings of the Ocean Drilling Program, Scientific Results, v. 126, p. 125–138.

Marsaglia KM, Ingersoll RV, Packer BM, 1992, Tectonic evolution of the Japanese islands as reflected in modal compositions of Cenozoic forearc and backarc sand and sandstone: Tectonics, v. 11, p. 1028–1044.

Marsden H, Thorkelson DJ, 1992, Geology of the Hazelton volcanic belt in British Columbia: implications for the Early to Middle Jurassic evolution of Stikinia: Tectonics, v. 11, p. 1266–1287.

Marsh BD, 1982, On the mechanics of igneous diapirism, stoping, and zone melting: American Journal of Science, v. 282, p. 808–855.

Marshak RS, Karig DE, 1977, Triple junctions as a cause for anomalously near-trench igneous activity between the trench and volcanic arc: Geology, v. 5, p. 233–236.

Marshall NF, 1978, Large storm-induced sediment slump reopens an unknown Scripps Canyon tributary, in Stanley DJ, Kelling G (eds.), Sedimentation in submarine canyons, fans, and trenches: Dowden, Hutchinson and Ross, Stroudsburg, p. 73–84.

Martini H-J, 1937, Großschollen und Gräben zwischen Habichtswald und Rheinischem Schiefergebirge: Geotektonische Forschungen, v. 1, p. 69–123.

Maruyama S, Seno T, 1986, Orogeny and relative plate motions: example of the Japanese islands: Tectonophysics, v. 127, p. 305–329.

Mascle A, 22 coauthors, 1988, Proceedings of the Ocean Drilling Program, Initial Reports [v. 110]: Ocean Drilling Program, College Station, 603 p.

Mascle A, Endignoux L, Chennouf T, 1990, Frontal accretion and piggyback basin development at the southern edge of the Barbados Ridge accretionary complex: Proceedings of the Ocean Drilling Program, Scientific Results, v. 110, p. 17–28.

Masson DG, Milsom J, Barber AJ, Sikumbang N, Dwiyanto B, 1991, Recent tectonics around the island of Timor, eastern Indonesia: Marine and Petroleum Geology, v. 8, p. 35–49.

Mathisen ME, McPherson JG, 1991, Volcaniclastic deposits: implications for hydrocarbon exploration: Society for Sedimentary Geology Special Publication 45, p. 27–36.

Matson RG, Moore GF, 1992, Structural influences on Neogene subsidence in the central Sumatra fore-arc basin: American Association of Petroleum Geologists Memoir 53, p. 157–181.

Matsuzawa A, Tamano T, Aoki Y, Ikawa T, 1980, Structure of the Japan Trench subduction zone, from multi-channel seismic-reflection records: Marine Geology, v. 35, p. 171–182.

Mattauer M, 1986, Intracontinental subduction, crust-mantle décollement and crustal-stacking wedge in the Himalayas and other collision belts: Geological Society of London Special Publication 19, p. 37–50.

Mattauer M, 7 coauthors, 1985, Tectonics of the Qinling belt: build-up and evolution of eastern Asia: Nature, v. 317, p. 496–500.

May SR, Ehman KD, Gray GG, Crowell JC, 1993, A new angle on the tectonic evolution of the Ridge basin, a "strike-slip" basin in southern California: Geological Society of America Bulletin, v. 105, p. 1357–1372.

Mayer L, 1991, Central Los Angeles basin, subsidence and thermal implications for tectonic evolution: American Association of Petroleum Geologists Memoir 52, p. 185–195.

Maynard JB, Valloni R, Yu H-S, 1982, Composition of modern deep-sea sands from arc-related basins: Geological Society of London Special Publication 10, p. 551–561.

McBirney AR, 1963, Factors governing the nature of submarine volcanism: Bulletin Volcanologique, v. 26, p. 455–469.

McBirney AR, Williams H, 1965, Volcanic history of Nicaragua: University of California Publications in the Geological Sciences 55, p. 1–65.

McBride EF, 1966, Sedimentary petrology and history of the Haymond Formation (Pennsylvanian), Marathon basin, Texas: University of Texas Bureau of Economic Geology Report of Investigations 57, 101 p.

McBride EF, 1970, Flysch sedimentation in the Marathon region, west Texas: Geological Association of Canada Special Paper 7, p. 67–83.

McBride EF, 1989, Stratigraphy and sedimentary history of pre-Permian Paleozoic rocks of the Marathon uplift, in Hatcher RD, Jr, Thomas WA, Viele GW (eds.), The Appalachian-Ouachita orogen in the United States [The Geology of North America, v. F-2]: Geological Society of America, Boulder, p. 603–620.

McBride JH, 1991, Constraints on the structure and tectonic development of the early Mesozoic south Georgia rift, southeastern United States; seismic reflection data processing and interpretation: Tectonics, v. 10, p. 1065–1083.

McBride JH, Karig DE, 1987, Crustal structure of the outer Banda arc: new free-air gravity evidence: Tectonophysics, v. 140, p. 265–273.

McCaffrey R, 1991, Slip vectors and stretching of the Sumatran fore arc: Geology, v. 19, p. 881–884.

McCaffrey R, 1992, Oblique plate convergence, slip vectors, and forearc deformation: Journal of Geophysical Research, v. 97, p. 8905–8915.

McCaffrey R, Abers GA, 1991, Orogeny in arc-continent collision: the Banda arc and western New Guinea: Geology, v. 19, p. 563–566.

McCallister RH, Boctor NZ, Hinze WJ, 1978, Petrology of the spilitic rocks from the Michigan basin deep drill hole: Journal of Geophysical Research, v. 83, p. 5825–5831.

McCarthy J, Scholl DW, 1985, Mechanisms of subduction accretion along the central Aleutian Trench: Geological Society of America Bulletin, v. 96, p. 691–701.

McCarthy J, Stevenson AJ, Scholl DW, Vallier TL, 1984, Speculations on the petroleum geology of the accretionary body: an example from the central Aleutians: Marine and Petroleum Geology, v. 1, p. 151–167.

McClay KR (ed.), 1992, Thrust tectonics: Chapman and Hall, London, 447 p.

McElhinny MW, Embleton BJJ, Ma XH, Zhang ZK, 1981, Fragmentation of Asia in the Permian: Nature, v. 293, p. 212–216.

McGinnis LD 1970, Tectonics and gravity field in the continental interior: Journal of Geophysical Research, v. 75, p. 317–331.

McGinnis LD, Heigold CP, Ervin CP, Heidi M, 1976, The gravity field and tectonics of Illinois: Illinois State Geological Survey Circular 494, 24 p.

McGookey DP, 1972, Cretaceous System, in Geologic atlas of the Rocky Mountain region: Rocky Mountain Association of Geologists, Denver, p. 190–228.

McGookey DP, 1975, Gulf coast Cenozoic sediments and structure: an excellent example of extra-continental sedimentation: Gulf Coast Association of Geological Societies Transactions, v. 25, p. 104–120.

McIlreath IA, James NP, 1978, Facies models 13: carbonate slopes: Geoscience Canada, v. 5, p. 188–199.

McKenzie DP, 1969, Speculation on the consequences and causes of plate motions: Geophysical Journal of the Royal Astronomical Society, v. 18, p. 1–32.

McKenzie DP, 1978a, Some remarks on the development of sedimentary basins: Earth and Planetary Science Letters, v. 40, p. 25–32.

McKenzie DP, 1978b, Active tectonics of the Alpine-Himalayan belt: the Aegean Sea and surrounding regions: Geophysical Journal of the Royal Astronomical Society, v. 55, p. 217–254.

McKenzie DP, 1981, The variation of temperature with time and hydrocarbon maturation in sedimentary basins formed by extension: Earth and Planetary Science Letters, v. 55, p. 87–98.

McKenzie DP, 1984, A possible mechanism for epeirogenic uplift: Nature, v. 307, p. 616–618.

McKenzie DP, 1986, The geometry of propagating rifts: Earth and Planetary Science Letters, v. 77, p. 176–186.

McKenzie DP, Bickle MJ, 1988, The volume and composition of melt generated by extension of the lithosphere: Journal of Petrology, v. 29, p. 625–679.

McKenzie DP, Jackson JA, 1986, A block model of distributed deformation by faulting: Journal of the Geological Society of London, v. 143, p. 349–353.

McKenzie DP, Sclater JG, 1971, The evolution of the Indian Ocean since the Late Cretaceous: Geophysical Journal of the Royal Astronomical Society, v. 24, p. 437–528.

McKenzie DP, Davies D, Molnar P, 1970, Plate tectonics of the Red Sea and East Africa: Nature, v. 226, p. 243–248.

McKerrow WS, Soper NJ, 1989, The Iapetus suture in the British Isles: Geological Magazine, v. 126, p. 1–8.

McLachlan IR, McMillan IK, 1976, Review and stratigraphic significance of southern Cape Mesozoic palaeontology: Transactions of the Geological Society of South Africa, v. 79, p. 197–212.

McLaughlin RJ, Nilsen TH, 1982, Neogene non-marine sedimentation and tectonics in small pull-apart basins of the San Andreas fault system, Sonoma County, California: Sedimentology, v. 29, p. 865–876.

McLaughlin RJ, Sorg DH, Morton JL, Theodore TG, Meyer CE, 1985, Paragenesis and tectonic significance of base and precious metal occurrences along the San Andreas fault at Point Delgada, California: Economic Geology, v. 80, p. 344–359.

McLaughlin RJ, Blake MC, Jr, Griscom A, Blome CD, Murchey B, 1988, Tectonics of formation, translation, and dispersal of the Coast Range ophiolite of California: Tectonics, v. 7,

p. 1033–1056.

McLean JR, 1977, The Cadomin Formation: stratigraphy, sedimentology, and tectonic implications: Bulletin of Canadian Petroleum Geology, v. 25, p. 792–827.

McMenamin MAS, McMenamin DLS, 1990, The emergence of animals: the Cambrian breakthrough: Columbia University Press, New York, 217 p.

McMillen KJ, Enkeboll RH, Moore JC, Shipley TA, Ladd JW, 1982, Sedimentation in different tectonic environments of the Middle America Trench, southern Mexico and Guatemala: Geological Society of London Special Publication 10, p. 107–119.

McQueen HWS, Beaumont C, 1989, Mechanical models of tilted block basins: American Geophysical Union Geophysical Monograph 48 [IUGG Volume 3], p. 65–71.

McRae LE, 1990, Paleomagnetic isochrons, unsteadiness, and non-uniformity of sedimentation in Miocene fluvial strata of the Siwalik Group, northern Pakistan: Journal of Geology, v. 98, p. 433–456.

Medwedeff DA, 1989, Growth fault-bend folding at southeast Lost Hills, San Joaquin Valley, California: American Association of Petroleum Geologists Bulletin, v. 73, p. 54–67.

Meiburg P, 1982, Saxonische Tektonik und Schollenkinematik am Ostrand des Rheinischen Massifs: Geoktektonische Forschungen, v. 62, p. 1–267.

Melosh HJ, 1990, Mechanical basis for low-angle normal faulting in the Basin and Range province: Nature, v. 343, p. 331–335.

Mercier J-L, 1981, Extensional-compressional tectonics associated with the Aegean arc: comparison with the Andean Cordillera of south Peru–north Bolivia: Philosophical Transactions of the Royal Society of London, v. 300A, p.337–355.

Mercier J-L, Carey-Gailhardis E, Sébrier M, 1991, Palaeostress determinations from fault kinematics: application to the neotectonics of the Himalayas-Tibet and the central Andes: Philosophical Transactions of the Royal Society of London, v. 337A, p. 41–52.

Mercier J-L., Tapponier P, Armijo R, Han T, Zhou J, 1984, Failes normales actives an Tibet: preuves de terrain, in Mercier JL, Li GC (eds), Mission Franco-chinoise an Tibet: Centre National de Recherche Scientifique, Paris, p.393–412.

Merewether EA., Cobban WA, 1986, Biostratigraphic units and tectonism in the mid-Cretaceous foreland of Wyoming, Colorado, and adjoining areas: American Association of Petroleum Geologists Memoir 41, p. 443–468.

Merzer AM., Freund R., 1974, Transcurrent faults, beam theory and the Marlborough fault system. New Zealand: Geophysical Journal of the Royal Astronomical Society, v. 38, p. 553–562.

Merzer AM, Freund R., 1975, Buckling of strike-slip faults in a model and in nature. Geophysical Journal of the Royal Astronomical Society, v. 43, p. 517–530.

Meyerhoff AA, 1982, Hydrocarbon resources in Arctic and subarctic regions, in Embry AF, Blakwill HR (eds), Arctic geology and geophysics: Canadian Society of Petroleum Geologists, Calgary, p. 451–552.

Miall AD, 1978, Tectonic setting and syndepositional deformation of molasse and other nonmarine-paralic sedimentary basins: Canadian Journal of Earth Sciences, v. 15, p. 1613–1632.

Miall AD, 1981, Alluvial sedimentary basins: tectonic setting

and basin architecture: Geological Association of Canada Special Paper 23, p. 1–33.

Miall AD, 1984, Flysch and molasse: the elusive models: Annales Societatis Geologorum Poloniae, v. 54, p. 281–291.

Miall AD, 1986, Eustatic sea-level changes interpreted from seismic stratigraphy: a critique of the methodology with particular reference to the North Sea Jurassic record: American Association of Petroleum Geologists Bulletin, v. 70, p. 131–137.

Miall AD, 1990, Principles of sedimentary basin analysis (2nd edition): Springer-Verlag, New York, 668 p.

Miall AD, 1991, Stratigraphic sequences and their chronostratigraphic correlation: Journal of Sedimentary Petrology, v. 61, p. 497–505.

Miall AD, 1992, The Exxon global cycle chart: an event for every occasion?: Geology, v. 20, p. 787–790.

Miall AD, 1993, The Exxon global cycle chart: an event for every occasion? reply to comments: Geology, v. 21, p. 284.

Middleton GV, Hampton MA, 1976, Subaqueous sediment transport and deposition by sediment gravity flows, in Stanley DJ, Swift DJP (eds.), Marine sediment transport and environmental management: John Wiley, New York, p. 197–218.

Middleton MF, 1980, A model of intracratonic basin formation entailing deep crustal metamorphism: Geophysical Journal of the Royal Astronomical Society, v. 62, p. 1–14.

Milana JP, 1990, Facies y paleohidrología de conglomerados aluviales plio-pleistocenos (San Juan, Argentina): evidencias de fases climáticas en los Andes a los 31° Sur: Universidad de Concepción, II Simposio Sobre el Terciario de Chile, Concepción, p. 215–224.

Milana JP, 1991, Sedimentología y magnetoestratigrafía de formaciones cenozoicas en el area de Mogna y su inserción en el marco tectosedimentario de la Precordillera Oriental [PhD thesis]: Universidad Nacional de San Juan, San Juan, 273 p.

Milana JP, Jordan TE, 1989, Edad del comienzo de la deformación y velocidad de levantamiento del sector norte de la Precordillera oriental, in Actas de la Primera Reunion de Fallas Activas del Noroeste Argentina: Universidad Nacional de San Juan, Facultad de Ciencias Exactas, Físicas Y Naturales, San Juan, p. 63–67.

Milanovsky EE, 1981, Aulacogens of ancient platforms: problems of their origin and tectonic development: Tectonophysics, v. 73, p. 213–248.

Milanovsky EE, 1983, Major stages of rifting evolution in the earth's history: Tectonophysics, v. 94, p. 599–607.

Milanovsky EE, 1987, Riftogenez v Istorii Zemli. Riftogenez v Podvijnih Poyasakh: Nedra, Moscow, 298 p.

Milici RC, deWitt W, Jr, 1988, The Appalachian basin, in Sloss, L.L. (ed.), Sedimentary cover - North American craton: U.S. [The Geology of North America, v. D-2]: Geological Society of America, Boulder, p. 427–469.

Miller EL, Gans PB, Garing J, 1983, The Snake Range décollement: an exhumed mid-Tertiary ductile-brittle transition: Tectonics, v. 2, p. 239–263.

Miller JMG, John B, 1988, Detached strata in a Tertiary low-angle normal fault terrane, SE California: a sedimentary record of unroofing, breaching and continued slip: Geology, v. 16, p. 645–648.

Miller MM, 1989, Intra-arc sedimentation and tectonism: late Paleozoic evolution of the eastern Klamath terrane, California: Geological Society of America Bulletin, v. 101, p. 170–187.

Minster JB, Jordan TH, 1978, Present-day plate motions: Journal of Geophysical Research, v. 83, p. 5331–5354.

Mirtsching A, 1964, Erdöl—und Gaslagerstätten der Sowjetunion und Ihre Geologische Bedeutung: Ferdinand Enke, Stuttgart, 195 p.

Mita N, Maeda M, Tominaga H, 1988, A new rapid prospecting method for submarine hydrothermal activity by determining lipopolysaccharide as bacterial biomass: Geochemical Journal, v. 22, p. 83–88.

Mitchell AHG, 1970, Facies of an early Miocene volcanic arc, Malekula Island, New Hebrides: Sedimentology, v. 14, p. 201–243.

Mitchell AHG, 1978, The Grampian orogeny in Scotland: arc-continent collision and polarity reversal: Journal of Geology, v. 86, p. 643–646.

Mitchell AHG, McKerrow WS, 1975, Analogous evolution of the Burma orogen and the Scottish Caledonides: Geological Society of America Bulletin, v. 86, p. 305–315.

Mitchum RM, Jr, Vail PR, Thompson S, III, 1977, Seismic stratigraphy and global changes of sea level, part 2: the depositional sequence as a basic unit for stratigraphic analysis: American Association of Petroleum Geologists Memoir 26, p. 53–62.

Mitrovica JX, Jarvis GT, 1985, Surface deflections due to transient subduction in a convecting mantle: Tectonophysics, v. 120, p. 211–237.

Mitrovica JX, Beaumont C, Jarvis GT, 1989, Tilting of continental interiors by the dynamical effects of subduction: Tectonics, v. 8, p. 1079–1094.

Miyashiro A, 1973, Metamorphism and metamorphic belts: John Wiley, New York, 492 p.

Miyashiro A, 1986, Hot regions and the origin of marginal basins in the western Pacific: Tectonophysics, v. 122, p. 195–216.

Moberly R, Shepherd GL, Coulburn WT, 1982, Forearc and other basins, continental margin of northern and southern Peru and adjacent Ecuador and Chile: Geological Society of London Special Publication 10, p. 171–189.

Mohr PA, 1971, Outline tectonics of Ethiopia, in Tectonics of Africa: UNESCO Earth Sciences 6, p. 447–458.

Mohr P, 1982, Musings on continental rifts: American Geophysical Union and Geological Society of America Geodynamics Series 8, p. 293–309.

Mohr P, 1991, Structure of Yemeni Miocene dike swarms and emplacement of coeval granite plutons: Tectonophysics, v. 198, p. 203–222.

Moiola RJ, Shanmugam G, 1984, Submarine fan sedimentation, Ouachita Mountains, Arkansas and Oklahoma: Gulf Coast Association of Geological Societies Transactions, v. 34, p. 175–182.

Molnar P, Atwater T, 1978, Interarc spreading and Cordilleran tectonics as alternates related to age of subducted oceanic lithosphere: Earth and Planetary Science Letters, v. 4l, p. 330–340.

Molnar P, Francheteau J, 1975, The relative motion of 'hot spots' in the Atlantic and Indian oceans during the Cenozoic: Geophysical Journal of the Royal Astronomical Society, v. 43, p. 763–774.

Molnar P, Lyon-Caen H, 1988, Some simple physical aspects of the support, structure, and evolution of mountain belts: Geological Society of America Special Paper 218, p. 179–207.

Molnar P, Stock J, 1987, Relative motions of hotspots in the Pacific, Atlantic and Indian oceans since Late Cretaceous time: Nature, v. 327, p. 587–591.

Molnar P, Tapponnier P, 1975, Cenozoic tectonics of Asia: effects of a continental collision: Science, v. 189, p. 419–426.

Molnar P, Tapponnier P, 1977, Relation of the tectonics of eastern China to the India-Eurasia collision: applications of slip-line field theory to large-scale continental tectonics: Geology, v. 5, p. 212–216.

Molnar P, Tapponnier P, 1978, Active tectonics of Tibet: Journal of Geophysical Research, v. 83, p. 5361–5375.

Monger JWH, Price RA, Tempelman-Kluit DJ, 1982, Tectonic accretion and the origin of the two major metamorphic and plutonic welts in the Canadian Cordillera: Geology, v. 10, p. 70–75.

Montradert L, Roberts DG, DeCharpal O, Guennoc P, 1979, Rifting and subsidence of the northern continental margin of the Bay of Biscay: Initial Reports of the Deep Sea Drilling Project, v. 48, 1025–1060.

Moore DG, Curray JR, Emmel FJ, 1976, Large submarine slide (olistostrome) associated with Sunda Arc subduction zone, northeast Indian Ocean: Marine Geology, v. 21, p. 211–226.

Moore GF, 1979, Petrography of subduction zone sandstones from Nias Island, Indonesia: Journal of Sedimentary Petrology, v. 49, p. 71–84.

Moore GF, Karig DE, 1976, Development of sedimentary basins on the lower trench slope: Geology, v. 4, p. 693–697.

Moore GF, Karig DE, 1980, Structural geology of Nias Island, Indonesia: implications for subduction zone tectonics: American Journal of Science, v. 280, p. 193–223.

Moore GF, Shipley TH, 1988, Mechanisms of sediment accretion in the Middle America Trench off Mexico: Journal of Geophysical Research, v. 93, p. 8911–8927.

Moore GF, Silver EA, 1983, Collision processes in the northern Molucca Sea: American Geophysical Monograph Series, v. 27, p. 360–372.

Moore GF, Taylor B, 1988, Structure of the Peru forearc from multichannel seismic-reflection data: Proceedings of the Ocean Drilling Program, Initial Reports, v. 112, p. 71–75.

Moore GF, Curray JR, Moore DG, Karig DE, 1980a, Variations in geologic structure along the Sunda fore arc, northeastern Indian Ocean: American Geophysical Union Geophysical Monograph 23, p. 145–160.

Moore GF, Billman HG, Hehanussa PE, Karig DE, 1980b, Sedimentology and paleobathymetry of Neogene trench-slope deposits, Nias Island, Indonesia: Journal of Geology, v. 88, p. 161–180.

Moore GF, Kadarisman D, Evans CA, Hawkins JW, 1981, Geology of the Talaud Islands, Molucca Sea collision zone, northeast Indonesia: Journal of Structural Geology, v. 3, p. 467–475.

Moore GF, Curray JR, Emmel FJ, 1982, Sedimentation in the Sunda Trench and forearc region: Geological Society of London Special Publication 10, p. 245–258.

Moore GF, Shipley TH, Lonsdale PF, 1986, Subduction erosion versus sediment offscraping at the toe of the Middle America Trench off Guatemala: Tectonics, v. 5, p. 513–523.

Moore GF, 7 coauthors, 1990, Structure of the Nankai Trough accretionary zone from multi-channel seismic reflection data: Journal of Geophysical Research, v. 95, p. 8753–8765.

Moore GT, Woodbury HO, Worzel JL, Watkins JS, Starke GW, 1979, Investigation of Mississippi Fan, Gulf of Mexico: American Association of Petroleum Geologists Memoir 29, p. 383–402P.

Moore GW, 1973, Westward tidal lag as the driving force of plate tectonics: Geology, v. 1, p. 99–100.

Moore JC, 1975, Selective subduction: Geology, v. 3, p. 530–532.

Moore JC, 1989, Tectonics and hydrogeology of accretionary prisms: role of the décollement zone: Journal of Structural Geology, v. 11, p. 95–106.

Moore JC, Allwardt A, 1980, Progressive deformation of a Tertiary trench slope, Kodiak Islands, Alaska: Journal of Geophysical Research, v. 85, p. 4741–4756.

Moore JC, Byrne T, 1987, Thickening of fault zones: a mechanism of mélange formation in accreting sediments: Geology, v. 15, p. 1040–1043.

Moore JC, Karig DE, 1976, Sedimentology, structural geology, and tectonics of the Shikoku subduction zone, southwestern Japan: Geological Society of America Bulletin, v. 87, p. 1259–1268.

Moore JC, Silver EA, 1987, Continental margin tectonics: submarine accretionary prisms: Reviews of Geophysics, v. 25, p. 1305–1312.

Moore JC, 5 coauthors, 1982a, Geology and tectonic evolution of a juvenile accretionary terrane along a truncated convergent margin: synthesis of results from Leg 66 of the Deep Sea Drilling Project, southern Mexico: Geological Society of America Bulletin, v. 93, p. 847–861.

Moore JC, 17 coauthors, 1982b, Offscraping and underthrusting of sediment at the deformation front of the Barbados Ridge: Deep Sea Drilling Project Leg 78A: Geological Society of America Bulletin, v. 93, p. 1065–1077.

Moore JC, 13 coauthors, 1982c, Facies belts of the Middle America Trench and forearc region, southern Mexico: results from Leg 66 DSDP: Geological Society of London Special Publication 10, p. 77–94.

Moore JC, Cowan DS, Karig DE, 1985, Structural styles and deformation fabrics of accretionary complexes: Geology, v. 13, p. 77–79.

Moore JC, 22 coauthors, 1988, Tectonics and hydrogeology of the northern Barbados Ridge: results from Ocean Drilling Program Leg 110: Geological Society of America Bulletin, v. 100, p. 1578–1593.

Moore JC, 11 coauthors, 1991, EDGE deep seismic reflection transect of the eastern Aleutian arc-trench layered lower crust reveals underplating and continental growth: Geology, v. 19, p. 420–424.

Moore JG, 5 coauthors, 1989, Prodigious submarine landslides on the Hawaiian Ridge: Journal of Geophysical Research, v. 94, p. 17465–17484.

Moore TE, 1984, Sedimentary facies and composition of Jurassic volcaniclastic turbidites at Cerro El Calvario, Vizcaino Peninsula, Baja California Sur, Mexico, in Frizzell VA (ed.), Geology of the Baja California Peninsula: Pacific Section, Society of Economic Paleontologists and Mineralogists, Los Angeles, p. 131–148.

Moore TE, 1985, Stratigraphy and tectonic significance of the Mesozoic tectonostratigraphic terranes of the Vizcaino Peninsula, Baja California Sur, Mexico: Circum-Pacific Council for

Energy and Mineral Resources Earth Science Series, v. 1, p. 315–329.

Moores EM, 1991, Southwest U.S.-East Antarctic (SWEAT) connection: a hypothesis: Geology, v. 19, p. 425-428. (Also, see Comment and Reply, v. 20, p. 87–88.)

Moretti I, Froidevaux C, 1986, Thermomechanical models of active rifting: Tectonics, v. 5, p. 501–511.

Morgan P, 1991, A deep look at African rifting: Nature, v. 354, 188–189.

Morgan P, Baker BH (eds.), 1983a, Processes of continental rifting [Developments in Geotectonics 19]: Elsevier, Amsterdam, 680 p.

Morgan P,-Baker BH, 1983b, Introduction—processes of continental rifting: Tectonophysics, v. 94, p. 1–10.

Morgan P, Golombek MP, 1984, Factors controlling the phases and styles of extension in the northern Rio Grande rift: New Mexico Geological Society Guidebook 35, p. 13–19.

Morgan P, Seager WR, Golombek MP, 1986, Cenozoic thermal, mechanical, and tectonic evolution of the Rio Grande Rift: Journal of Geophysical Research, v. 91, p. 6263–6276.

Morgan WJ, 1981, Hotspot tracks and the opening of the Atlantic and Indian Oceans, in Emiliani C (ed.), The sea [v. 7 (The oceanic lithosphere)]: John Wiley, New York, p. 443–487.

Morgenstern NR, 1967, Submarine slumping and the initiation of turbidity currents, in Richards AF (ed.), Marine geotechnique: Illinois University Press, Urbana, p. 189–220.

Morley CK, 1989, Extension, detachments, and sedimentation in continental rifts (with particular reference to East Africa): Tectonics, v. 8, p. 1175–1192.

Morley CK, Nelson RA, Patton TL, Munn SG, 1990, Transfer zones in the East African rift system and their relevance to hydrocarbon exploration in rifts: American Association of Petroleum Geologists Bulletin, v. 74, p. 1234–1253.

Morris RC, 1974a, Carboniferous rocks of the Ouachita Mountains, Arkansas: Geological Society of America Special Paper 148, p. 241–279.

Morris RC, 1974b, Sedimentary and tectonic history of the Ouachita Mountains: Society of Economic Paleontologists and Mineralogists Special Publication 22, p. 120–142.

Morris RC, 1989, Stratigraphy and sedimentary history of post-Arkansas Novaculite Carboniferous rocks of the Ouachita Mountains, in Hatcher RD, Jr, Thomas WA, Viele GW (eds.), The Appalachian-Ouachita orogen in the United States [The Geology of North America, v. F-2]: Geological Society of America, Boulder, p. 591–602.

Morris RC, Proctor KE, Koch MR, 1979, Petrology and diagenesis of deep-water sandstones, Ouachita Mountains, Arkansas and Oklahoma: Society of Economic Paleontologists and Mineralogists Special Publication 26, p. 263–279.

Morris WR, 1992, The depositional framework, paleogeography and tectonic development of the Late Cretaceous through Paleocene Peninsular Range forearc basin in the Rosario embayment, Baja California, Mexico [PhD thesis]: University of California, Santa Barbara, 210 p.

Morris WR, Busby-Spera C, 1987, Relationship between relative sea level fluctuations and petrologic variation in forearc basin sedimentary rocks of Upper Cretaceous Rosario Formation at San Carlos, Baja California del Norte, Mexico: American Association of Petroleum Geologists Bulletin, v. 71, p. 596.

Morris WR, Busby-Spera CJ, 1988, Sedimentologic evolution of a submarine canyon in a forearc basin, Upper Cretaceous Rosario Formation, San Carlos, Mexico: American Association of Petroleum Geologists Bulletin, v. 72, p. 717–737.

Morris W, Busby-Spera C, 1990, A submarine-fan valley-levee complex in the Upper Cretaceous Rosario Formation: implication for turbidite facies models: Geological Society of America Bulletin, v. 102, p. 900–914.

Morris WR, Smith DP, Busby-Spera CJ, 1989, Deep marine conglomerate facies and processes in Cretaceous forearc basins of Baja California, Mexico, in Colburn IP, Abbott PL, Minch J (eds.), Conglomerates in basin analysis: a symposium dedicated to A.O. Woodford [Book 62]: Pacific Section, Society of Economic Paleontologists and Mineralogists, Los Angeles, p. 123–142.

Morrison RR, Brown WR, Edmondson WF, Thomson JN, Young RJ, 1971, Potential of Sacramento Valley gas province, California: American Association of Petroleum Geologists Memoir 15, p. 329–338.

Morton JL, Holmes ML, Koski RA, 1987, Volcanism and massive sulfide formation at a sedimented spreading center, Escanaba Trough, Gorda Ridge, northeast Pacific Ocean: Geophysical Research Letters, v. 14, p. 769–772.

Morton WH, Black R, 1975, Crustal attenuation in Afar, in Pilger A, Rossler A (eds.), Afar depression of Ethiopia: Deutsche Forschauna, Stuttgart, p. 55–65.

Mountain GS, Tucholke BE, 1984, Mesozoic and Cenozoic geology of the U.S. Atlantic continental slope and rise, in Poag CW (ed.), Geologic evolution of the United States Atlantic margin: Van Nostrand-Rheinhold, Stroudsburg, p. 293–341.

Moustafa AR, El-Raey AK, 1993, Structural characteristics of the Suez rift margins: Geologische Rundschau, v. 82, p. 101–109.

Moxon IW, 1988, Sequence stratigraphy of the Great Valley basin in the context of convergent margin tectonics, in Graham SA (ed.), Studies of the geology of the San Joaquin basin [Book 60]: Pacific Section, Society of Economic Paleontologists and Mineralogists, Los Angeles, p. 3–28.

Moxon IW 1990, Stratigraphy and structure of Upper Jurassic–Lower Cretaceous strata, Sacramento Valley, in Ingersoll RV, Nilsen TH (eds.), Sacramento Valley symposium and guidebook [Book 65]: Pacific Section, Society of Economic Paleontologists and Mineralogists, Los Angeles, p. 5–29.

Moxon IW, Graham SA, 1987, History and controls of subsidence in the Late Cretaceous-Tertiary Great Valley forearc basin, California: Geology, v. 15, p. 626–629.

Mrozowski CL, Hayes DE, 1979, The evolution of the Parece Vela Basin, eastern Philippine Sea: Earth and Planetary Science Letters, v. 46, p. 49–67.

Mrozowski CL, Hayes DE, 1980, A seismic reflection study of faulting in the Mariana fore arc: American Geophysical Union Geophysical Monograph 23, p. 223–234.

Muck MM, Underwood MB, 1990, Upslope flow of turbidity currents: a comparison among field observations, theory, and laboratory models: Geology, v. 18, p. 54–57.

Mukhopadhyay M, Dasgupta S, 1988, Deep structure and tectonics of the Burmese arc: constraints from earthquake and gravity data: Tectonophysics, v. 149, p. 299–322.

Mukhopadhyay PK, Rullkotter J, Schaefer RG, Welte DH, 1986, Facies and diagenesis of organic matter in Nankai Trough sediments, Deep Sea Drilling Project Leg 87A: Initial Reports of

the Deep Sea Drilling Project, v. 87, p. 877–890.

Muñoz BJ, Stern CR, 1989, Alkaline magmatism within the segment 38°-39°S of the Plio-Quaternary volcanic belt of the southern South American continental margin: Journal of Geophysical Research, v. 94, p. 4545–4560.

Murakami F, 1988, Structural framework of the Sumisu Rift, Izu-Ogasawara Arc: Bulletin of the Geological Survey of Japan, v. 39, p. 1–21.

Murray CG, Fergusson CL, Flood PG, Whitaker WG, Korsch RJ, 1987, Plate tectonic-model for the Carboniferous evolution of the New England fold belt: Australian Journal of Earth Sciences, v. 34, p. 213–236.

Mutter JC, Talwani M, Stoffa PL, 1982, Origin of seaward dipping reflectors in oceanic crust off the Norwegian margins by "subaerial sea-floor spreading": Geology, v. 10, p. 353–357.

Mutter JC, Buck WR, Zehnder CM, 1988, Convective partial melting 1. A model for the formation of thick basaltic sequences during the initiation of spreading: Journal of Geophysical Research, v. 93, p. 1031–1048.

Mutter JC, Larson RL, Northwest Australia Study Group, 1989, Extension of the Exmouth Plateau, offshore northwestern Australia: deep seismic reflection/refraction evidence for simple and pure shear mechanisms: Geology, v. 17, p. 15–18.

Mutti E, 1985, Turbidite systems and their relations to depositional sequences, in Zuffa GG (ed.), Provenance of arenites: D. Reidel, Dordrecht, p. 65–93.

Mutti E, Normark WR, 1987, Comparing examples of modern and ancient turbidite systems: problems and concepts, in Leggett JK, Zuffa GG (eds.), Marine clastic sedimentology: concepts and case studies: Graham and Trotman, London, p. 1–38.

Mutti E, Ricci Lucchi F, 1972, Le torbiditi dell'Appennino settentrionale: introduzione all' analisi di facies: Memorie della Societa Geologica Italiana 11, p. 161–199.

Mutti E, Ricci Lucchi F, 1978, Turbidites of the northern Apennines: introduction to facies analysis (English translation): International Geology Review, v. 20, p. 125–166.

Muza JP, 1992, Calcareous nannofossil biostratigraphy from the Japan Sea, Sites 798 and 799: evidence for an oscillating Pleistocene oceanographic frontal boundary: Proceedings of the Ocean Drilling Program, Scientific Results, v. 127/128, Pt. 1, p. 155–169.

Myers WB, Hamilton W, 1964, Deformation associated with the Hebgen Lake earthquake of August 17, 1959: United States Geological Survey Professional Paper 435, p. 55–98.

Nábelek J, Chen W-P, Ye H, 1987, The Tangshan earthquake sequence and its implications for the evolution of the North China basin: Journal of Geophysical Research, v. 92, p. 12615–12628.

Nagle HE, Parker ES, 1971, Future oil and gas potential of onshore Ventura basin, California: American Association of Petroleum Geologists Memoir 15, p. 254–297.

Nakamura K, 16 coauthors, 1987, Oblique and near collision subduction, Sagami and Suruga troughs—preliminary results of French-Japanese 1984 Kaiko cruise, Leg 2: Earth and Planetary Science Letters, v. 83, p. 229–242.

Nakao S, Yasa M, Nohara M, Usui A, 1990, Submarine hydrothermal activity in the Izu-Ogasawara arc, western Pacific: Reviews in Aquatic Science, v. 3, p. 95–115.

Namson JS, Davis TL, 1988, Seismically active fold and thrust belt in the San Joaquin Valley, central California: Geological Society of America Bulletin, v. 100, p. 257–273.

Narasimhan TN, Witherspoon PA, Edwards AL, 1978, Numerical model for saturated-unsaturated flow in deformable porous media 2. The algorithm: Water Resources Research, v. 14, p. 255–261.

Natland JH, Tarney J, 1981, Petrologic evolution of the Mariana arc and back-arc basin system–a synthesis of drilling results in the south Philippine Sea: Initial Reports of the Deep Sea Drilling Project, v. 60, p. 877–908.

Naylor MA, Mandl G, Sijpestein CHK, 1986, Fault geometries in basement-induced wrench faulting under different initial stress states: Journal of Structural Geology, v. 7, p. 737–752.

Neev D, Emery KO, 1967, The Dead Sea: depositional processes and environments of evaporites: Israel Ministry of Development Geological Survey Bulletin 41, 147p.

Nelson CH, 1976, Late Pleistocene and Holocene depositional trends, processes, and history of Astoria deep-sea fan, northeast Pacific: Marine Geology, v. 20, p. 129–173.

Nelson CH, Carlson PR, Byrne JV, Alpha TR, 1970, Development of the Astoria canyon-fan physiography and comparison with similar systems: Marine Geology, v. 8, p. 259–291.

Nelson KD, Zhang J, 1991, A COCORP deep reflection profile across the buried Reelfoot rift, south-central United States: Tectonophysics, v. 197, p. 271–293.

Nelson KD, 8 coauthors, 1985a, New COCORP profiling in the southeastern United States. Part I: late Paleozoic suture and Mesozoic rift basin: Geology, v. 13, p. 714–718.

Nelson KD, 5 coauthors, 1985b, New COCORP profiling in the southeastern United States. Part II: Brunswick and east coast magnetic anomalies, opening of the north-central Atlantic Ocean: Geology, v. 13, p. 718–721.

Nelson WJ, 1990a, Comment on "Major Proterozoic basement features of the eastern midcontinent of North America revealed by recent COCORP profiling": Geology, v. 18, p. 378–379.

Nelson WJ, 1990b, Structural styles of the Illinois basin: American Association of Petroleum Geologists Memoir 51, p. 209–243.

Nelson WJ, Krause HF, 1981, The Cottage Grove fault system in southern Illinois: Illinois State Geological Survey Circular 522, 65 p.

Nelson WJ, Lumm DK, 1984, Structural geology of southeastern Illinois and vicinity: Illinois State Geological Survey Contract/Grant Report 1984-2, p. 971–976.

Nesbitt BE, Murowchick JB, Muehlenbachs K, 1986, Dual origin of lode gold deposits in the Canadian Cordillera: Geology, v. 14, p. 506–509.

Ness GE, Kulm LD, 1973, Origin and development of the Surveyor Deep-Sea Channel: Geological Society of America Bulletin, v. 84, p. 3339–3354.

Neugebauer HJ, 1978, Crustal doming and the mechanism of rifting. Part 1: rift formation: Tectonophysics, v. 45, p. 159–186.

Neugebauer HJ, 1983, Mechanical aspects of continental rifting: Tectonophysics, v. 94, p. 91–108.

Neugebauer HJ, Reuther C, 1987, Intrusion of igneous rocks—physical aspects: Geologische Rundschau, v. 76, p. 89–99.

Ni J, York JE, 1978, Cenozoic extensional tectonics of the Tibetan Plateau: Journal of Geophysical Research, v. 83, p. 5277–5384.

Nichols GJ, Hall R, 1991, Basin formation and Neogene sedi-

mentation in a backarc setting, Halmahera, eastern Indonesia: Marine and Petroleum Geology, v. 8, p. 50–61.

Nicholson C, Sorlien CC, Atwater T, Crowell JC, Luyendyk BP, 1994, Microplate capture, rotation of the western Transverse Ranges, and initiation of the San Andreas transform as a low-angle fault system: Geology, v. 22, p. 491–495.

Nicholson PG, 1993, A basin reappraisal of the Proterozoic Torridon Group, northwest Scotland: International Association of Sedimentologists Special Publication 20, p. 183–202.

Nie S, 1991, Paleoclimatic and paleomagnetic constraints on the Paleozoic reconstruction of South China, North China, and Tarim: Tectonophysics, v. 196, p. 279–308.

Nie S, Rowley DB, Yin A, Jin Y, 1993, History of northeastern Tibetan Plateau before India-Asia collision: formation of the Songpan-Ganzi flysch sequence and exhumation of the Dabie Shan ultra-high pressure rocks: Geological Society of America Abstracts with Programs, v. 25, n. 6, p. A-117.

Niem AR, 1976, Patterns of flysch deposition and deep-sea fans in the lower Stanley Group (Mississippian), Ouachita Mountains, Oklahoma and Arkansas: Journal of Sedimentary Petrology, v. 46, p. 633–646.

Niem AR, 1977, Mississippian pyroclastic flow and ash-fall deposits in the deep-marine Ouachita flysch basin, Oklahoma and Arkansas: Geological Society of America Bulletin, v. 88, p. 49–61.

Nilsen TH, 1984, Tectonics and sedimentation of the Upper Cretaceous Hornbrook Formation, Oregon and California, in Crouch JK, Bachman SB (eds.), Tectonics and sedimentation along the California margin [Book 38]: Pacific Section, Society of Economic Paleontologists and Mineralogists, Los Angeles, p. 101–118.

Nilsen TH, 1985a, Stratigraphy, sedimentology, and tectonic framework of the Upper Cretaceous Hornbrook Formation, Oregon and California, in Nilsen TH (ed.), Geology of the Upper Cretaceous Hornbrook Formation, Oregon and California [Book 42]: Pacific Section, Society of Economic Paleontologists and Mineralogists, Los Angeles, p. 51–88.

Nilsen TH, 1985b, Chugach turbidite system, Alaska, in Bouma AH, Normark WR, Barnes NE (eds.), Submarine fans and related turbidite systems: Springer-Verlag, New York, p. 185–192.

Nilsen TH, 1990, Santonian, Campanian, and Maastrichtian depositional systems, Sacramento basin, California, in Ingersoll RV, Nilsen TH (eds.), Sacramento Valley symposium and guidebook [Book 65]: Pacific Section, Society of Economic Paleontologists and Mineralogists, Los Angeles, p. 95–132.

Nilsen TH, Abbott PL, 1981, Paleogeography and sedimentology of Upper Cretaceous turbidites, San Diego, California: American Association of Petroleum Geologists Bulletin, v. 65, p. 1256–1284.

Nilsen TH, McLaughlin RJ, 1985, Comparison of tectonic framework and depositional patterns of the Hornelen strike-slip basin of Norway and the Ridge and Little Sulphur Creek strike-slip basins of California: Society of Economic Paleontologists and Mineralogists Special Publication 37, p. 79–103.

Nilsen TH, Zuffa GG, 1982, The Chugach Terrane, a Cretaceous trench-fill deposit, southern Alaska: Geological Society of London Special Publication 10, p. 213–228.

Nilsen TH, Moore GW, Winkler GR, 1979, Reconnaissance study of Upper Cretaceous to Miocene stratigraphic units and

sedimentary facies, Kodiak and adjacent islands, Alaska: United States Geological Survey Professional Paper 1093, 34 p.

Nishi N, 1988, Structural analysis of the Shimanto accretionary complex, Kyushu, Japan, based on foraminiferal biostratigraphy: Tectonics, v. 7, p. 641–652.

Nishimura A, Murakami F, 1988, Sedimentation of the Sumisu Rift, Izu-Ogasawara arc: Bulletin of the Geological Survey of Japan, v. 39, p. 39–61.

Nishimura A, Yamazaki T, Yuasa M, Mita N, Nakao S, 1988, Bottom sample and heat flow data of Sumisu and Torishima Rifts, Izu-Ogasawara island arc: Geological Survey of Japan Marine Geology Map Series 31, 1:200,000.

Nishimura A, 14 coauthors, 1991, Pliocene-Quaternary submarine pumice deposits in the Sumisu Rift area, Izu-Bonin Arc: Society for Sedimentary Geology Special Publication 45, p. 201–208.

Nishimura A, Rodolfo K, Koizumi A, Gill J, Fujioka K, 1992, Episodic deposition of Pliocene-Pleistocene pumice deposits of Izu-Bonin arc, Leg 126: Proceedings of the Ocean Drilling Program, Scientific Results, v. 126, p. 3–21.

Nohda S, Yoshiyuki T, Yamashita S, Fujii T, 1992, Nd and Sr isotopic study of Leg 127 basalts: implications for the evolution of the Japan Sea backarc basin: Proceedings of the Ocean Drilling Program, Scientific Results, v. 127/128, Pt. 2, p. 899–904.

Norton IO, Sclater JG, 1979, A model for the evolution of the Indian Ocean and the breakup of Gondwanaland: Journal of Geophysical Research, v. 84, p. 6803–6830.

Novikova A, 1964, The Russian plate, in Bogdanoff AA, Muratov MV, Shatsky NS (eds.), Tectonics of Europe: "Nauka", Moscow, p. 54–69.

Nunn JA, 1986, Subsidence and thermal history of the Michigan basin, in Burrus J (ed.), Thermal modeling in sedimentary basins: Editions Technip, Paris, p. 417–436.

Nunn JA, Sleep NH, Moore WE, 1984, Thermal subsidence and generation of hydrocarbons in Michigan basin: American Association of Petroleum Geologists Bulletin, v. 68, p. 296–315.

Nur A, Ben-Avraham Z, 1981, Volcanic gaps and the consumption of aseismic ridges in South America: Geological Society of America Memoir 154, p. 729–740.

Nurmi PA, Haapala I, 1986, The Proterozoic granitoids of Finland: granite types, metallogeny, and relation to crustal evolution: Geological Society of Finland Bulletin, v. 58, p. 203–233.

Oberc J, 1977, The late Alpine epoch in south-west Poland, in Pozaryski W (ed.), Geology of Poland [v. IV (Tectonics)]: Wydawnictwa Geologiczne, Warszawa, p. 451–475.

Obradovich JD, Cobban WA, 1975, A time scale for the Late Cretaceous of the western interior of North America: Geological Association of Canada Special Paper 13, p. 31–54.

Ohmoto H, Skinner BJ (eds.), 1983, The Kuroko and related volcanogenic massive sulfide deposits: Economic Geology Monography 5, 604p.

Oide K, 1968, Geotectonic conditions for the formation of Krakatau-type calderas in Japan: Pacific Geology, v. 1, p. 119–135.

Ojakangas RW, 1968, Cretaceous sedimentation, Sacramento Valley, California: Geological Society of America Bulletin, v. 79, p. 973–1008.

Okada H, 1974, Migration of ancient arc-trench systems: Society of Economic Paleontologists and Mineralogists Special Pub-

lication 19, p. 311–320.

Okada H, 1982, Geological evolution of Hokkaido, Japan: an example of collision orogenesis: Proceedings of the Geologist's Association, v. 93, p. 201–212.

Okada H, 1983, Collision orogenesis and sedimentation in Hokkaido, Japan, in Nasu N, Kobayashi K, Uyeda S, Kushiro I, Kagami H (eds.), Formation of active ocean margins: Terra Scientific Publishing, Tokyo, p. 641–676.

Okamura Y, 1991, Large-scale melange formation due to seamount subduction: an example from the Mesozoic accretionary complex in central Japan: Journal of Geology, v. 99, p. 661–674.

Okay AI, Şengör AMC, 1992, Evidence for intracontinental thrust-related exhumation of the ultra-high-pressure rocks in China: Geology, v. 20, p. 411–414.

Okaya DA, Thompson GA, 1985, Geometry of Cenozoic extensional faulting: Dixie Valley, Nevada: Tectonics, v. 4, p. 107–125.

Okuda Y, 1985, Late Cenozoic evolution of the fore-arc basins off southwest Japan, in Shiki T (ed.), Geology of the northern Philippine Sea: Tokai University Press, Tokyo, p. 247–259.

Okuda Y, Honza E, 1988, Tectonic evolution of the Seinan (SW) Japan fore-arc and accretion in the Nankai Trough: Modern Geology, v. 12, p. 411–434.

Olsen KH, Baldridge WS, Callender JF, 1987, Rio Grande rift: an overview: Tectonophysics, v. 143, p. 119–139.

Olsen PE, 1984, Periodicity of lake-level cycles in the Late Triassic Lockatong Formation of the Newark Basin (Newark Supergroup, New Jersey and Pennsylvania), in Barger A, Imbrie J, Hays J, Kukla G, Saltzman B (eds.), Milankovich and climate [Part I]: D. Riedel, New York, p. 129–146.

Olsen PE, 1986, A 40-million-year lake record of early Mesozoic climatic forcing: Science, v. 234, p. 842–848.

Olsen PE, 1988, Paleontology and paleoecology of the Newark Supergroup (early Mesozoic, eastern North America), in Manspeizer W (ed.), Triassic-Jurassic rifting, Part A: Elsevier, Amsterdam, p. 185–230.

Olsen PE, 1991, Tectonic, climatic, and biotic modulation of lacustrine ecosystems—examples from Newark Supergroup of eastern North America: American Association of Petroleum Geologists Memoir, p. 209–224.

Olsen PE, Kent DV, 1990, Continental coring of the Newark rift: EOS, v. 71, p. 385–394.

Olsen PE, Schlische RW, Gore PJW, 1989, Tectonic, depositional, and palaeoecological history of early Mesozoic rift basins, eastern North America [International Geological Congress Field Trip Guide T-351]: American Geophysical Union, Washington, 174 p.

Olsen PE, Schlische RW, Fedosh MS, 1994, 670 kyr duration of the Early Jurassic flood basalt event in eastern North America estimated using Milankovich stratigraphy: Geology, in press.

Omar GI, Steckler MS, Buck WR, Kohn BP, 1989, Fission track analysis of basement apatites at the western margin of the Suez rift, Egypt: evidence for synchroneity of uplift and subsidence: Earth and Planetary Science Letters, v. 94, p. 316–328.

Orange DL, 1990, Criteria helpful in recognizing shear-zone and diapiric melanges: examples from the Hoh accretionary complex, Olympic Peninsula, Washington: Geological Society of America Bulletin, v. 102, p. 935–951.

Orange DL, Breen NA, 1992, The effects of fluid escape on

accretionary wedges 2. seepage force, slope failure, headless submarine canyons, and vents: Journal of Geophysical Research, v. 97, p. 9277–9295.

Ori GG, 1989, Geologic history of the extensional basin of the Gulf of Corinth (Miocene-Pleistocene), Greece: Geology, v. 17, p. 918–921.

Ori GG, 1993, Continental depositional systems of the Quaternary of the Po Plain (northern Italy): Sedimentary Geology, v. 83, p. 1–14.

Ori GG,-Friend PF, 1984, Sedimentary basins formed and carried piggyback on active thrust sheets: Geology, v. 12, p. 475–478.

Ori GG, Roveri M, Vannoni F, 1986, Plio-Pleistocene sedimentation in the Apenninic-Adriatic foredeep (central Adriatic Sea, Italy): International Association of Sedimentologists Special Publication 8, p. 183–198.

Oriel SS, Tracey JA, Jr, 1970, Uppermost Cretaceous and Tertiary stratigraphy of Fossil basin, southwestern Wyoming: United States Geological Survey Professional Paper 635, 53 p.

Orton GJ, 1991, Emergence of subaqueous depositional environments in advance of a major ignimbrite eruption, Capel Curig Volcanic Formation, Ordovician, North Wales—an example of regional volcanotectonic uplift?: Sedimentary Geology, v. 74, p. 251–288.

Oxburgh ER, Turcotte DL, 1971, Origin of paired metamorphic belts and crustal dilation in island arc regions: Journal of Geophysical Research, v. 76, p. 1315–1327.

Oxburgh ER, Turcotte DL, 1974, Membrane tectonics and the East African Rift: Earth and Planetary Science Letters, v. 22, p. 133–140.

Paakola J, 1983, The Late Cambrian Muhos Formation, in Laajoki K, Paakkola J (eds.), Exogenic processes and related metallogeny in the Svekokarellian geosynclinal complex: Geological Survey of Finland Guidebook 11, Appendix, p. 1–5.

Packer BM, Ingersoll RV, 1986, Provenance and petrology of Deep Sea Drilling Project sands and sandstones from the Japan and Mariana forearc and backarc regions: Sedimentary Geology, v. 51, p. 5–28.

Packham GH, Falvey DA, 1971, An hypothesis for the formation of the marginal seas in the western Pacific: Tectonophysics, v. 11, p. 79–109.

Page BM, Suppe J, 1981, The Pliocene Lichi melange of Taiwan: its plate-tectonic and olistostromal origin: American Journal of Science, v. 281, p. 193–227.

Palmer AR, 1983, The Decade of North American Geology 1983 Time Scale: Geology, v. 11, p. 503–504.

Palmer BA, Neall VE, 1991, Contrasting lithofacies architecture in ring-plain deposits related to edifice construction and destruction, the Quaternary Stratford and Opunake Formations, Egmont Volcano, New Zealand: Sedimentary Geology, v. 74, p. 71–88.

Palmer BA, Walton AW, 1990, Accumulation of volcaniclastic aprons in the Mount Dutton Formation (Oligocene-Miocene), Marysvale volcanic field, Utah: Geological Society of America Bulletin, v. 102, p. 734–748.

Palmer BA, Alloway BV, Neall VE, 1991, Volcanic-debris-avalanche deposits in New Zealand–lithofacies organization in unconfined, wet-avalanche flows: Society for Sedimentary Geology Special Publication 45, p. 89–98.

Pantin HM, Leeder MR, 1987, Reverse flow in turbidity cur-

rents: the role of internal solitons: Sedimentology, v. 34, p. 1143–1155.

Parfenov LM, 1984, Kontinentalniye Okraini i Ostrovniye Dugi Mezozoid Severo-Vostoka Asii: "Nauka", Novosibirsk, 192 p.

Park C, Tamaki K, Kobayashi K, 1990, Age-depth correlation of the Philippine Sea back-arc basins and other marginal basins in the world: Tectonophysics, v. 181, p. 351–371.

Parson LM, 10 coauthors, 1990, Role of ridge jumps and ridge propagation in the tectonic evolution of the Lau backarc basin, southwest Pacific: Geology, v. 18, p. 470–473.

Parson LM, 28 coauthors, 1992, Introduction, background, and principal results of Leg 135, Lau Basin: Proceedings of the Ocean Drilling Program, Initial Reports, v. 135, p. 5–47.

Parsons B, Sclater JG, 1977, An analysis of the variation of ocean floor bathymetry and heat flow with age: Journal of Geophysical Research, v. 82, p. 803–827.

Patacca E, Scandone P, 1989, Post-Tortonian mountain building in the Apennines. The role of the passive sinking of a relic lithospheric slab, in Boriani A, Bonafede M, Piccardo GB, Vai GB (eds.), The lithosphere in Italy: advances in Earth science research: Accademia Nazionale dei Lincei, Rome, p. 157–176.

Patacca E, Sartori R, Scandone P, 1990, Tyrrhenian basin and Apenninic arcs: kinematic relations since late Tortonian times: Memorie della Societa Geologica Italiana, v. 45, p. 425–451.

Patriat P, Achache J, 1984, India-Eurasia collision chronology has implications for crustal shortening and driving mechanism of plates: Nature, v. 311, p. 615–621.

Patterson DL, 1984, Los Chapunes and Valle sandstones: Cretaceous petrofacies of the Vizcaino basin, Baja California, Mexico, in Frizzell VA (ed.), Geology of the Baja California Peninsula: Pacific Section, Society of Economic Paleontologists and Mineralogists, Los Angeles, p. 161–182.

Pavlis TL, Bruhn RL, 1983, Deep-seated flow as a mechanism for the uplift of broad forearc ridges and its role in the exposure of high P/T metamorphic terranes: Tectonics, v. 2, p. 473–497.

Peabody CE, Einaudi MT, 1992, Origin of petroleum and mercury in the Culver-Baer cinnabar deposit, Mayacmas district, California: Economic Geology, v. 87, p. 1078–1103.

Pearce JA, Alabaster T, Shelton AW, Searle MP, 1981, The Oman ophiolite as a Cretaceous arc-basin complex: evidence and implications: Philosophical Transactions of the Royal Society of London, v. A300, p. 299–317.

Pearce JA, Lippard SJ, Roberts S, 1984, Characteristics and tectonic significance of supra-subduction zone ophiolites: Geological Society of London Special Publication 16, p. 77–94.

Pearce JA, 9 coauthors, 1990, Genesis of collision volcanism in eastern Anatolia: Journal of Volcanology and Geothermal Research, v. 44, p. 189–229.

Perfit MR, Heezen BC, 1978, The geology and evolution of the Cayman Trench: Geological Society of America Bulletin, v. 89, p. 1155–1174.

Perman RC, 1990, Depositional history of the Maastrichtian Lewis Shale in south-central Wyoming: deltaic and interdeltaic, marginal marine through deep-water marine, environments: American Association of Petroleum Geologists Bulletin, v. 74, p. 1695–1717.

Perrodon A, Zabek J, 1990, Paris basin: American Association of Petroleum Geologists Memoir 51, p. 633–679.

Perry RB, Nichols H, 1966, Geomorphology of the Amlia basin, Aleutian arc, Alaska: Geographical Review, v. 56, p. 570–576.

Perry SK, Schamel S, 1990, The role of low-angle normal faulting and isostatic response in the evolution of the Suez rift, Egypt: Tectonophysics, v. 174, p. 159–173.

Peters SG, Golding SD, Dowling K, 1990, Mélange- and sediment-hosted gold-bearing quartz veins, Hodgkinson gold field, Queensland, Australia: Economic Geology, v. 85, p. 312–327.

Peters SW, 1979, West African intracratonic stratigraphic sequences: Geology, v. 7, p. 528–531.

Peterson JA, MacCary LM, 1987, Regional stratigraphy and general petroleum geology of the U.S. portion of the Williston basin and adjacent areas, in Peterson JA, Kent DM, Anderson SB, Pilatzke RH, Longman MW (eds.), Williston basin: anatomy of a cratonic oil province: Rocky Mountain Association of Geologists, Denver, p. 9–44.

Pettijohn FJ, Potter PE, Siever R, 1987, Sand and sandstone: Springer-Verlag, New York, 553 p.

Pfiffner OA, 1986, Evolution of the north Alpine foreland basin in the central Alps: International Association of Sedimentologists Special Publication 8, p. 219–228.

Philippson A, 1910, Reisen und Forschungen im westlichen Kleinasien [1. Heft]: Ergäuzungsheft 167 zu Petermanns Mitteilungen, Jena, 104 p.

Phillips JD, 1988, A geophysical study of the northern Hartford basin and vicinity, Massachusetts: United States Geological Survey Bulletin 1776, p. 235–247.

Phipps SP, 1984, Ophiolitic olistostromes in the basal Great Valley sequence, Napa County, northern California Coast Ranges: Geological Society of America Special Paper 198, p. 103–125.

Pichler H, Weyl R 1973, Petrochemical aspects of Central American magmatism: Geologische Rundschau, v. 62, p. 357–396.

Pickering KT, 1984, The Upper Jurassic 'Boulder Beds' and related deposits: a fault-controlled submarine slope, NE Scotland: Journal of the Geological Society of London, v. 141, p. 357–374.

Pickering KT, Hiscott RN, 1985, Contained (reflected) turbidity currents from the Middle Ordovician Cloridorne Formation, Quebec, Canada: an alternative to the antidune hypothesis: Sedimentology, v. 32, p. 373–394.

Pickering KT, Hiscott RN, Hein FJ, 1989, Deep marine environments: clastic sedimentation and tectonics: Unwin Hyman, London, 416 p.

Pickering KT, Underwood MB, Taira A, 1992, Open-ocean to trench turbidity-current flow in the Nankai Trough: flow collapse and reflection: Geology, v. 20, p. 1009–1102.

Pickering KT, Underwood MB, Taira A, 1993, Stratigraphic synthesis of the DSDP–ODP sites in the Shikoku Basin and Nankai Trough and accretionary prism: Proceedings of the Ocean Drilling Program, Scientific Results, v. 131, p. 313–330.

Pigram CJ, Davies PJ, Feary DA, Symonds PA, 1989, Tectonic controls on carbonate platform evolution in southern Papua New Guinea: passive margin to foreland basin: Geology, v. 17, p. 199–202.

Pilger RH, Jr, 1981, Plate reconstruction, aseismic ridges, and low-angle subduction beneath the Andes: Geological Society of America Bulletin, v. 92, p. 448–456.

Pindell JL, 1985, Alleghenian reconstruction and subsequent

evolution of the Gulf of Mexico, Bahamas, and proto-Caribbean: Tectonics, v. 4, p. 1–39.

Pindell J, Dewey JF, 1982, Permo-Triassic reconstruction of western Pangea and the evolution of the Gulf of Mexico/Caribbean region: Tectonics, v. 1, p. 179–211.

Piper DJW, Normark WR, 1983, Turbidite depositional patterns and flow characteristics, Navy Submarine Fan, California Borderland: Sedimentology, v. 30, p. 681–694.

Piper DJW, vonHuene R, Duncan JR, 1973, Late Quaternary sedimentation in the active eastern Aleutian Trench: Geology, v. 1, p. 19–23.

Piper DJW, Shor AN, Farre JA, O'Connell S, Jacobi R, 1985, Sediment slides and turbidity currents on the Laurentian Fan: sidescan sonar investigations near the epicenter of the 1929 Grand Banks earthquake: Geology, v. 13, p. 538–541.

Pisciotto KA, Murray RW, Brumsack H-J, 1992, Thermal history of Japan Sea sediments from isotopic studies of diagenetic silica and associated pore waters: Proceedings of the Ocean Drilling Program, Scientific Results, v. 127/128, Pt. 1, p. 49–56.

Pitman WC, III, 1978, Relationship between eustacy and stratigraphic sequences of passive margins: Geological Society of America Bulletin, v. 89, p. 1389–1403.

Pitman WC, III, Andrews JA, 1985, Subsidence and thermal history of small pull-apart basins: Society of Economic Paleontologists and Mineralogists Special Publication 37, p. 45–49.

Pivnik DA, 1990, Thrust-generated fan-delta deposition: Little Muddy Creek Conglomerate, SW Wyoming: Journal of Sedimentary Petrology, v. 60, p. 489–503.

Platt JP, 1986, Dynamics of orogenic wedges and the uplift of high-pressure metamorphic rocks: Geological Society of America Bulletin, v. 97, p. 1037–1053.

Plint AG, 1988, Sharp-based shoreface sequences and "offshore bars" in the Cardium Formation of Alberta: their relationship to relative changes in sea level: Society of Economic Paleontologists and Mineralogists Special Publication 42, p. 357–370.

Plint AG, 1991, High-frequency relative sea-level oscillations in Upper Cretaceous shelf clastics of the Alberta foreland basin: possible evidence for a glacio-eustatic control?: International Association of Sedimentologists Special Publication 12, p. 409–428.

Plint AG, Walker RG, Bergman KM, 1986, Cardium Formation 6. Stratigraphic framework of the Cardium in subsurface: Bulletin of Canadian Petroleum Geology, v. 34, p. 213–225.

Plint AG, Hart BS, Donaldson WS, 1993, Lithospheric flexure as a control on stratal geometry and facies distribution in Upper Cretaceous rocks of the Alberta foreland basin: Basin Research, v. 5, p. 69–77.

Pollock SG, 1987, Chert formation in an Ordovician volcanic arc: Journal of Sedimentary Petrology, v. 57, p. 75–87.

Poole FG, 1974, Flysch deposits of Antler foreland basin, western United States: Society of Economic Paleontologists and Mineralogists Special Publication 22, p. 58–82.

Posamentier HW, Jervey MT, Vail PR, 1988, Eustatic controls on clastic deposition: Society of Economic Paleontologists and Mineralogists Special Publication 42, p. 109–154.

Posavec M, Taylor D, vanLeeuwen T, Spector A, 1973, Tectonic controls of volcanism and complex movements along the Sumatran fault system: Geological Society of Malaysia Bulletin, v. 6, p. 43–60.

Potter PE, Pryor WA, 1960, Dispersal centers of Paleozoic and later clastics of the upper Mississippi Valley and adjacent areas: Geological Society of America Bulletin, v. 72, p. 1195–1250.

Pouclet A, Bellon H, 1992, Geochemistry and isotopic composition of volcanic rocks from the Yamato Basin: Hole 794D, Sea of Japan: Proceedings of the Ocean Drilling Program, Scientific Results, v. 127/128, Pt. 2, p. 779–789.

Pozaryski W, 1977, The tectonic units of Poland, in Pozaryski W (ed.), Geology of Poland [v. IV (Tectonics)]: Wydawnictwa Geologiczne, Warszawa, p. 50–57.

Pozaryski W, Brochwicz-Lewinski W, 1978, On the Polish trough: Geologie en Mijnbouw, v. 57, p. 545–557.

Pratsch JC, 1982, Wedge tectonics along continental margins: American Association of Petroleum Geologists Memoir 34, p. 221–220.

Pratt T, 7 coauthors, 1989, Major Proterozoic basement features of the eastern midcontinent of North America revealed by COCORP profiling: Geology, v. 17, p. 505–509.

Price RA, 1973, Large-scale gravitational flow of supracrustal rocks, southern Canadian Rockies, in DeJong KA, Scholten R (eds.), Gravity and tectonics: John Wiley and Sons, New York, p. 491–502.

Price RA, 1981, The Cordilleran foreland thrust and fold belt in the southern Canadian Rocky Mountains: Geological Society of London Special Publication 9, p. 427–448.

Price RA, 1986, The southeastern Canadian Cordillera: thrust faulting, tectonic wedging and delamination of the lithosphere: Journal of Structural Geology, v. 8, p. 239–254.

Price RA, Mountjoy EW, 1970, Geologic structure of the Canadian Rocky Mountains between Bow and Athabasca Rivers—a progress report: Geological Association of Canada Special Paper 6, p. 7–25.

Priest GR, 1983, Geothermal exploration in the Oregon Cascades: Oregon Department of Geology and Mineral Industries Special Paper 15, p. 77–88.

Priest GR, 1990, Volcanic and tectonic evolution of the Cascade volcanic arc, central Oregon: Journal of Geophysical Research, v. 95, p. 19583–19599.

Prince RA, Resig JM, Kulm LD, Moore TC, Jr, 1974, Uplifted turbidite basins on the seaward wall of the Peru Trench: Geology, v. 2, p. 607–611.

Proffett JM, 1977, Cenozoic geology of the Yerington district, Nevada, and implications for the nature and origin of Basin and Range faulting: Geological Society of America Bulletin, v. 88, p. 247–266.

Pudsey CJ, Reading HG, 1982, Sedimentology and structure of the Scotland Group, Barbados: Geological Society of London Special Publication 10, p. 291–308.

Puffer JH, Hurtubise DO, Geiger FJ, Leeklar P, 1981, Chemical composition of the Mesozoic basalts of the Newark basin, New Jersey, and the Hartford basin, Connecticut: stratigraphic implications: Geological Society of America Bulletin, v. 92, p. 155–159.

Puigdefabregas C, Munoz JA, Marzo M, 1986, Thrust belt development in the eastern Pyrenees and related depositional sequences in the southern foreland basin: International Association of Sedimentologists Special Publication 8, p. 229–246.

Purser BH, Hotzl H, 1988, The sedimentary evolution of the Red Sea rift: a comparison of the northwest (Egyptian) and northeast (Saudi Arabian) margins: Tectonophysics, v. 153, p. 193–208.

Quade J, Cerling TE, Bowman JR, 1989, Development of Asian monsoon revealed by marked ecological shift during the latest Miocene in northern Pakistan: Nature, v. 342, p. 163–166.

Quinlan GM, 1987, Models of subsidence mechanisms in intracratonic basins, and their applicability to North American examples: Canadian Society of Petroleum Geologists Memoir 12, p. 463–481.

Quinlan GM, Beaumont C, 1984, Appalachian thrusting, lithospheric flexure, and the Paleozoic stratigraphy of the eastern interior of North America: Canadian Journal of Earth Sciences, v. 21, p. 973–996.

Rahmani RA, Lerbekmo JF, 1975, Heavy mineral analysis of Upper Cretaceous and Paleocene sandstones in Alberta and adjacent areas of Saskatchewan: Geological Association of Canada Special Paper 13, p. 607–632.

Ramberg IB, 1976, Gravity interpretation of the Oslo graben and associated igneous rocks: Norges Geologiske Undersfkelse, v. 325, p. 1–194.

Ramsay JG, Huber MI, 1987, The techniques of modern structural geology: Academic Press, New York, 700 p.

Rao R, Xu J, Chen Y, Zou D, 1987, The Triassic System of the Qinghai-Xizang Plateau: People's Republic of China Ministry of Geology and Mineral Resources Geological Memoirs, Series 2, n. 7, 239 p.

Rast N, 1989, The evolution of the Appalachian chain, in Bally AW, Palmer AR (eds.), The geology of North America: an overview [The Geology of North America, v. A]: Geological Society of America, Boulder, p. 323–348.

Rat P, 1975, Structures et phases de structuration dans les plateaux bourguignons et le Nord-ouest du fossé bressan [Geologische Vereinigung 65]: Jahrestagung, Kurzfassungen der Vorträge, p. 33–34.

Raymond LA, 1984, Classification of mélanges: Geological Society of America Special Paper 198, p. 7–20.

Re G, Vilas J, 1990, Análisis de los cambios paleogeográficos ocurridos durante el Cenozoico tardío en la región de Vinchina (Provincia de La Rioja, Argentina), a partir de estudios magnetoestratigráficos [XI Congreso Geológico Argentina, v. 2]: Asociación Geológica Argentina, p. 267–270.

Read JF, 1982, Carbonate platforms of passive (extensional) continental margins: types, characteristics and evolution: Tectonophysics, v. 81, p. 195–212.

Read JF, 1984, Carbonate platform facies models: American Association of Petroleum Geologists Bulletin, v. 69, p. 1–21.

Reading HG, 1980, Characteristics and recognition of strike-slip fault systems: International Association of Sedimentologists Special Publication 4, p.7–26.

Reading HG (ed.), 1986, Sedimentary environments and facies (2nd edtion): Blackwell, Boston, 615 p.

Reading HG, 1991, The classification of deep-sea depositional systems by sediment calibre and feeder system: Journal of the Geological Society of London, v. 148, p. 427–430.

Reck BH, 1987, Implications of measured thermal gradients for water movement through the northeast Japan accretionary prism: Journal of Geophysical Research, v. 92, p. 3683–3690.

Reed DL, Silver EA, Prasetyo H, Meyer AW, 1986, Deformation and sedimentation along a developing terrane suture: eastern Sunda forearc, Indonesia: Geology, v. 14, p. 1000–1003.

Reed DL, Meyer AW, Silver EA, Prasetyo H, 1987, Contourite sedimentation in an intraoceanic forearc system: eastern Sunda arc, Indonesia: Marine Geology, v. 76, p. 223–241.

Reid I, 1989, Effects of lithospheric flow on the formation and evolution of a transform margin: Earth and Planetary Science Letters, v. 95, p. 38–52.

Reimann K-U, 1993, Geology of Bangladesh: Gebruder Borntraeger, Berlin, 160 p.

Reimnitz E, 1971, Surf-beat origin for pulsating bottom currents in the Rio Balsas submarine canyon, Mexico: Geological Society of America Bulletin, v. 82, p. 81–90.

Ren J, Jiang C, Zhang Z, Qin D, 1987, The geotectonic evolution of China: Springer-Verlag, Berlin, 203 p.

Reston TJ, 1988, Evidence for shear zones in the lower crust offshore Britain: Tectonics, v. 7, p. 929–945.

Reston TJ, 1990, The lower crust and the extension of the continental lithophere: kinematic analysis of BIRPS deep seismic data: Tectonics, v. 9, p. 1235–1248.

Reston TJ, 1993, Evidence for extensional shear zones in the mantle, offshore Britain, and their implications for the extension of the continental lithosphere: Tectonics, v. 12, p. 492–506.

Reynolds JH, 1987, Chronology of Neogene tectonics in the central Andes (27-33°S) of western Argentina based on the magnetic polarity stratigraphy of foreland basin sediments [PhD thesis]: Dartmouth College, Hanover, 353 p.

Reynolds JH, Jordan TE, Johnson NM, Damanti JF, Tabbutt KD, 1990, Neogene deformation of the flat-subduction segment of the Argentine-Chilean Andes: magnetostratigraphic constraints from Las Juntas, La Rioja Province, Argentina: Geological Society of America Bulletin, v. 102, p. 1607–1622.

Riba O, 1976, Syntectonic unconformities of the Alto Cardener, Spanish Pyrenees, a genetic interpretation: Sedimentary Geology, v. 15, p. 213–233.

Ricci-Lucchi F, 1986, The Oligocene to Recent foreland basins of the northern Apennines: International Association of Sedimentologists Special Publication 8, p. 105–139.

Richter CF, 1958, Elementary seismology: Freeman, San Francisco, 768 p.

Richter F, McKenzie D, 1978, Simple plate models of mantle convection: Journal of Geophysical Research, v. 44, p. 441–471.

Richter FM, Parsons B, 1975, On the interaction of two scales of convection in the mantle: Journal of Geophysical Research, v. 80, p. 2529–2541.

Ricketts BD, Evenchick CA, Anderson RG, Murphy DC, 1992, Bowser Basin, northern British Columbia: constraints on the timing of initial subsidence and Stickinia—North America terrane interactions: Geology, v. 20, p. 1119–1122.

Ries AC, 1978, The opening of the Bay of Biscay—a review: Earth Science Reviews, v. 14, p. 35–63.

Rigassi DA, Dixon GE, 1972, Cretaceous of the Cape Province, Republic of South Africa, in Dessauvagie TFJ, Whiteman AJ (eds.), African geology: Department of Geology, University of Ibadan, Ibadan, p. 513–517.

Riggs NR, Busby-Spera CJ, 1990, Evolution of a multi-vent volcanic complex within a subsiding arc graben depression: Mount Wrightson Formation, Arizona: Geological Society of America Bulletin, v. 102, p. 1114–1135.

Riggs NR, Busby-Spera CJ, 1991, Facies analysis of an ancient, dismembered, large caldera complex and implications for intra-arc subsidence: Middle Jurassic strata of Cobre Ridge, southern Arizona, USA: Sedimentary Geology, v. 74, p. 39–68.

Riggs NR, Mattinson JM, Busby CJ, 1993, Correlation of Jurassic eolian strata between the magmatic arc and the Colorado Plateau: new U-Pb geochronologic data from southern Arizona: Geological Society of America Bulletin, v. 105, p. 1231–1246.

Riggs SR, Belkamp DF, 1988, Upper Cenozoic processes and environments of continental margin sedimentation: eastern United States, in Sheridan RE, Grow JA (eds.), The geology of the Atlantic continental margin: U.S. [The Geology of North America, v. I-2]: Geological Society of America, Boulder, p. 131–176.

Riggs SR, Manheim FT, 1988, Mineral resources of the U.S. Atlantic continental margin, in Sheridan RE, Grow JA (eds.), The geology of the Atlantic continental margin: U.S. [The Geology of North America, v. I-2]: Geological Society of America, Boulder, p. 501–520.

Rihm R, Makris J, Moller L, 1991, Seismic surveys in the northern Red Sea: asymmetric crustal structure: Tectonophysics, v. 198, p. 279–296.

Roberts D, Grenne T, Ryan PD, 1984, Ordovician marginal basin development in the central Norwegian Caledonides: Geological Society of London Special Publication 16, p. 233–244.

Roberts JM, 1970, Amal field, Libya: American Association of Petroleum Geologists Memoir 14, p. 438–448.

Roberts S, Jackson JA, 1991, Active normal faulting in central Greece: an overview: Geological Society of London Special Publication 56, p. 125–142.

Robertson AHF, 1977, The Kannaviou Formation, Cyprus: volcaniclastic sedimentation of a probably Late Cretaceous volcanic arc: Journal of the Geological Society of London, v. 134, p. 269–292.

Robertson AHF, 1989, Palaeoceanography and tectonic setting of the Jurassic Coast Range ophiolite, central California: evidence from the extrusive rocks and the volcaniclastic sediment cover: Marine and Petroleum Geology, v. 6, p. 194–220.

Robertson AHF, 1990, Sedimentology and tectonic implications of ophiolite-derived clastics overlying the Jurassic Coast Range ophiolite, northern California: American Journal of Science, v. 290, p. 109–163.

Rodgers DA, 1980, Analysis of pull-apart basin development produced by en echelon strike-slip faults: International Association of Sedimentology Special Publication 4, p. 27–41.

Rodgers J, 1987, Chains of basement uplifts within cratons marginal to orogenic belts: American Journal of Science, v. 287, p. 661–692.

Rodolfo KS, 1989, Origin and development of lahar channels at Mabinit, Mayon Volcano, Philippines: Geological Society of America Bulletin, v. 101, p. 414–426.

Rodolfo KS, Solidum RU, Nishimura A, Matsuo Y, Fujioka K, 1992, Major-oxide stratigraphy of glass shards in volcanic ash layers of the Izu-Bonin arc-backarc sites (Sites 788/789 and 790/791): Proceedings of the Ocean Drilling Program, Scientific Results, v. 126, p. 505–517.

Rogers JJW, Novitsky-Evans JM, 1977a, The Clarno Formation of central Oregon, U.S.A.: volcanism on a thin continental margin: Earth and Planetary Science Letters, v. 34, p. 56–66.

Rogers JJW, Novitsky-Evans JM, 1977b, Evolution from oceanic to continental crust in northwestern U.S.A: Geophysical Research Letters, v. 4, p. 347–350.

Romanowicz B, 1982, Constraints on the structure of the Tibet Plateau from pure path phase velocities of Love and Rayleigh Waves: Journal of Geophysical Research, v. 87, p. 6865–6883.

Rose WI, Newhall CG, Bornhorst TJ, Self S, 1987, Quaternary silicic pyroclastic deposits of Atitlan caldera, Guatemala: Journal of Volcanology and Geothermal Research, v. 22, p. 57–80.

Rosencrantz E, Ross MI, Sclater JG, 1988, Age and spreading history of the Cayman trough as determined from depth, heat flow, and magnetic anomalies: Journal of Geophysical Research, v. 93, p. 2141–2157.

Rosendahl BR, 1987, Architecture of continental rifts with special reference to East Africa: Annual Review of Earth and Planetary Sciences, v. 15, p. 445–503.

Rosendahl BR, 7 coauthors, 1986, Structural expressions of rifting: lessons from Lake Tanganyika, Africa: Geological Society of London Special Publication 25, p. 29–44.

Rosenfeld U, Schickor G, 1969, Graben [Deutsches Handwörterbuch der Tektonik, 2 Lieferung: Bundesandstalt für Bodenforschung, Hannover.

Roser BP, Korsch RJ, 1986, Determination of tectonic setting of sandstone-mudstone suites using SiO_2 content and K_2O/Na_2O ratio: Journal of Geology, v. 94, p. 635–650.

Ross CA, 1986, Paleozoic evolution of southern margin of Permian basin: Geological Society of America Bulletin, v. 97, p. 536–554.

Ross DA, 1971, Sediments of the northern Middle America Trench: Geological Society of America Bulletin, v. 82, p. 303–322.

Ross GM, 1986, Eruption style and construction of shallow marine mafic tuff cones in the Narakay volcanic complex (Proterozoic, Hornby Bay Group, Northwest Territories, Canada): Journal of Volcanology and Geothermal Research, v. 27, p. 265–297.

Rothery DA, Drury SA, 1984a, The neotectonics of the Tibetan Plateau deduced from landsat Mss image interpretation [International Symposium on Remote Sensing of Environment, Third Thematic Conference]: Remote Sensing for Exploration Geology, Colorado Springs, p. 321–330.

Rothery DA, Drury SA, 1984b, The neotectonics of the Tibetan Plateau: Tectonics, v. 3, p. 19–26.

Rowley DB, Kidd WSF, 1981, Stratigraphic relationships and detrital composition of the medial Ordovician flysch of western New England: implications for the tectonic evolution of the Taconic orogeny: Journal of Geology, v. 89, p. 199–218. (Also, see v. 90, p. 219–233.)

Royden LH, 1985, The Vienna basin: a thin-skinned pull-apart basin: Society of Economic Paleontologists and Mineralogists Special Publication 37, p. 319–338.

Royden LH, 1993a, The tectonic expression of slab pull at continental convergent boundaries: Tectonics, v. 12, p. 303–325.

Royden LH, 1993b, Evolution of retreating subduction boundaries formed during continental collision: Tectonics, v. 12, p. 629–638.

Royden LH, Karner GD, 1984, Flexure of lithosphere beneath Apennine and Carpathian foredeep basins: evidence for an insufficient topographic load: American Association of Petroleum Geologists Bulletin, v. 68, p. 704–712.

Royden LH, Keen CE, 1980, Rifting process and thermal evolution of the continental margin of eastern Canada determined

from subsidence curves: Earth and Planetary Science Letters, v. 51, p. 343–361.

Royden LH, Keen CE, VonHerzen RP, 1980, Continental margin subsidence and heat flow, important parameters in formation of petroleum hydrocarbons: American Association of Petroleum Geologists Bulletin, v. 64, p. 173–187.

Royden LH, Horvath F, Burchfiel BC, 1982, Transform faulting, extension, and subduction in the Carpathian Pannonian region: Geological Society of America Bulletin, v. 93, p. 717–725.

Royse F, Jr, Warner MA, Reese DL, 1975, Thrust belt structural geometry and related stratigraphic problems, Wyoming-Idaho-northern Utah, in Bolyard DW (ed.), Deep drilling frontiers of the central Rocky Mountains: Rocky Mountain Association of Geologists, Denver, p. 41–54.

Rubin CM, Saleeby JB, 1991, The Gravina sequence: remnants of a mid-Mesozoic oceanic arc in southern southeast Alaska: Journal of Geophysical Research, v. 96, p. 14551–14568.

Ruddiman WF, Kutzbach JE, 1989, Forcing of late Cenozoic northern hemisphere climate by plateau uplift in southern Asia and the American West: Journal of Geophysical Research, v. 94, p. 18409–18427.

Rudkevich MY, 1976, The history and the dynamics of the development of the west Siberian platform: Tectonophysics, v. 36, p. 275–287.

Ryan HF, Coleman PJ, 1992, Composite transform-convergent plate boundaries: description and discussion: Marine and Petroleum Geology, v. 9, p. 89–97.

Ryan HF, Scholl DW, 1989, The evolution of forearc structures along an oblique convergent margin, central Aleutian arc: Tectonics, v. 8, p. 497–516.

Ryder RT, Thomson A, 1989, Tectonically controlled fan delta and submarine fan sedimentation of late Miocene age, southern Temblor Range, California: United States Geological Survey Professional Paper 1442, 59 p.

Ryer TA, 1977, Patterns of Cretaceous shallow-marine sedimentation, Coalville and Rockport area, Utah: Geological Society of America Bulletin, v. 88, p. 177–188.

St Julien P, Hubert C, 1975, Evolution of the Taconian orogen in the Quebec Appalachians: American Journal of Science, v. 275A, p. 337–362.

Sahagian DL, Holland SM, 1991, Eustatic sea-level curve based on a stable frame of reference: preliminary results: Geology, v. 19, p. 1209–1212.

Sahagian D, Jones M, 1993, Quantified Middle Jurassic to Paleocene eustatic variations based on Russian Platform stratigraphy: stage level resolution: Geological Society of America Bulletin, v. 105, p. 1109–1118.

Sakai H, 8 coauthors, 1990, Venting of carbon dioxide-rich fluid and hydrate formation in mid-Okinawa Trough backarc basin: Science, v. 248, p. 1093–1096.

Saleeby JB, 1981, Ocean floor accretion and volcanoplutonic arc evolution of the Mesozoic Sierra Nevada, in Ernst WG (ed.), The geotectonic development of California [Rubey Volume I]: Prentice-Hall, Englewood Cliffs, p. 132–181.

Saleeby JB, Busby-Spera CJ, 1992, Early Mesozoic tectonic evolution of the western U.S. Cordillera, in Burchfiel BC, Lipman PW, Zoback ML (eds.), The Cordilleran orogen: contermi-nous U.S. [The Geology of North America, v. G-3]: Geological Society of America, Boulder, p. 107–168.

Salvador A (ed.), 1991, The Gulf of Mexico basin [The Geology of North America, v. J]: Geological Society of America, Boulder, 568 p.

Sample JC, Fisher DM, 1986, Duplex accretion and underplating in an ancient accretionary complex, Kodiak Islands, Alaska: Geology, v. 14, p. 160–163.

Sample JC, Moore JC, 1987, Structural style and kinematics of an underplated slate belt, Kodiak and adjacent islands, Alaska: Geological Society of America Bulletin, v. 99, p. 7–20.

Sarewitz DR, Lewis SD, 1991, The Marinduque intra-arc basin, Philippines: basin genesis and in situ ophiolite development in a strike-slip setting: Geological Society of America Bulletin, v. 103, p. 597–614.

Sato H, Amano K, 1991, Relationship between tectonics, volcanism, sedimentation and basin development, late Cenozoic, central part of northern Honshu, Japan: Sedimentary Geology, v. 74, p. 323–343.

Saunders AD, Tarney J, 1984, Geochemical characteristics of basaltic volcanism within back-arc basins: Geological Society of London Special Publication 16, p. 59–76.

Savage JC, Hastie LM, 1966, Surface deformation associated with dip-slip faulting: Journal of Geophysical Research, v. 71, p. 4897–4904.

Sawkins FJ, 1976, Massive sulfide deposits in relation to geotectonics: Geological Association of Canada Special Paper 14, p. 221–240.

Sawkins FJ, Burke K, 1980, Extensional tectonics and mid-Paleozoic massive sulfide occurences in Europe: Geologische Rundschau, v. 69, p. 349–360.

Schatski NS, 1961, Vergleichende Tektonik Alter Tafeln [Fortschritte der Sowjetischen Geologie 4]: Akademie Verlag, Berlin, 220 p.

Schermer ER, Busby C, 1994, Jurassic magmatism in the central Mojave Desert: implications for arc paleogeography and preservation of continental volcanic sequences: Geological Society of America Bulletin, v. 106, p. 767–790.

Schlager W, Ginsburg RN, 1981, Bahama carbonate platform—the deep and the past: Marine Geology, v. 44, p. 1–24.

Schlanger SO, Combs J, 1975, Hydrocarbon potential of marginal basins bounded by an island arc: Geology, v. 3, p. 397–400.

Schlee JS, Hinz K, 1987, Seismic stratigraphy and facies of continental slope and rise seaward of Baltimore Canyon trough: American Association of Petroleum Geologists Bulletin, v. 71, p. 1046–1067.

Schlee JS, Dillon WP, Grow JA, 1979, Structure of the continental slope off the eastern United States: Society of Economic Paleontologists and Mineralogists Special Publication 27, p. 95–118.

Schlische RW, 1991, Half-graben basin filling models: new constraints on continental extensional basin development: Basin Research, v. 3, p. 123–141.

Schlische RW, 1992, Structural and stratigraphic development of the Newark extensional basin, eastern North America: evidence for the growth of the basin and its bounding structures: Geological Society of America Bulletin, v. 104, p. 1246–1263.

Schlische RW, 1993, Anatomy and evolution of the Triassic-Jurassic continental rift system, eastern North America: Tectonics, v. 12, p. 1026–1042.

Schlische RW, Olsen PE, 1990, Quantitative filling model for continental extensional basins with applications to early Mesozoic rifts of eastern North America: Journal of Geology, v. 98, v. 135–155.

Schmid R, 1981, Descriptive nomenclature and classification of pyroclastic deposits and fragments: recommendations of the IUGS Subcommission on the Systematics of Igneous Rocks: Geology, v. 9, p. 41–43.

Schmitt JG, Steidtmann JR, 1990, Interior ramp-supported uplifts: implications for sediment provenance in foreland basins: Geological Society of America Bulletin, v. 102, p. 494–501.

Scholl DW, 1974, Sedimentary sequences in the North Pacific trenches, in Burk CA, Drake CL (eds.), The geology of continental margins: Springer-Verlag, New York, p. 493–504.

Scholl DW, Herzer RH, 1992, Geology and resource potential of the southern Tonga Platform: American Association of Petroleum Geologists Memoir 53, p. 139–156.

Scholl DW, Vallier TL, 1983, Subduction and the rock record of Pacific margins, in Carey SW (ed.), Expanding Earth symposium: University of Tasmania, Sydney, p. 235–246.

Scholl DW, Buffington EC, Marlow MS, 1975a, Plate tectonics and the structural evolution of the Aleutian-Bering Sea region: Geological Society of America Special Paper 151, p. 1–31.

Scholl DW, Marlow MS, Buffington EC, 1975b, Summit basins of the Aleutian Ridge, north Pacific: American Association of Petroleum Geologists Bulletin, v. 59, p. 799–816.

Scholl DW, vonHuene R, Vallier TL, Howell DG, 1980, Sedimentary masses and concepts about tectonic processes at underthrust ocean margins: Geology, v. 8, p. 564–568.

Scholl DW, Vallier TL, Stevenson AJ, 1982a, Sedimentation and deformation in the Amlia fracture zone sector of the Aleutian Trench: Marine Geology, v. 48, p. 105–134.

Scholl DW, Vallier TL, Stevenson AJ, 1982b, Arc, forearc, and trench sedimentation and tectonics: Amlia corridor of the Aleutian Ridge: American Association of Petroleum Geologists Memoir 34, p. 413–439.

Scholl DW, Vallier TL, Packham GH, 1985, Framework geology and resource potential of southern Tonga platform and adjacent terranes: Circum-Pacific Council for Energy and Mineral Resources Earth Science Series, v. 2, p. 457–488.

Scholl DW, Grantz A, Vedder JG (eds.), 1987a, Geology and resource potential of the continental margin of western North America and adjacent ocean basins—Beaufort Sea to Baja California: Circum-Pacific Council for Energy and Mineral Resources Earth Science Series, v. 6, 799 p.

Scholl DW, Vallier TL, Stevenson AJ, 1987b, Geologic evolution and petroleum geology of the Aleutian Ridge: Circum-Pacific Council for Energy and Mineral Resources Earth Science Series, v. 6, p. 123–155.

Scholz CA, Rosendahl BR, Scott DL, 1990, Development of coarse-grained facies in lacustrine rift basins: examples from East Africa: Geology, v. 18, p, 140–144.

Schroder E, 1925, Tektonische Studien an niederhessischen Graben [Gottinger Beitrage zur saxonischen Tektonik]: Abhandlungen der Preussischen Geologischen Landesanstalt, Neue Folge, Heft 95, p. 57–82.

Schubert C, 1980, Late-Cenozoic pull-apart basins, Boconó fault zone, Venezuelan Andes: Journal of Structural Geology, v. 2, p. 463–468.

Schubert C, 1984, Basin formation along the Boconó–Morón–El Pilar fault system, Venezuela: Journal of Geophysical Research, v. 89, p. 5711–5718.

Schubert C, 1986, Origin of the Yaracuy basin, Boconó-Moron fault system, Venezuela: Neotectonics, v. 1, p. 39–50.

Schubert C, 1988, Neotectonics of La Victoria fault zone, north-central Venezuela: Annales Tectonicae, v. 2, p. 58–66.

Schubert C 1992, The Bocanó fault, western Venezuela: Annales Tectonicae, v. 6, p. 238–260.

Schuepbach MA, Vail PR, 1980, Evolution of outer highs on divergent continental margins, in National Research Council, Geophysics Study Committee, Continental Tectonics [Studies in Geophysics]: National Academy of Sciences, Washington, p. 50–61.

Schwab FL, 1981, Evolution of the western continental margin, French-Italian Alps: sandstone mineralogy as an index of plate tectonic setting: Journal of Geology, v. 89, p. 349–368.

Schwartz RK, DeCelles PG, 1988, Cordilleran foreland basin evolution in response to interactive Cretaceous thrusting and foreland partitioning, southwestern Montana: Geological Society of America Memoir 171, p. 489–513.

Schweickert RA, Cowan DS, 1975, Early Mesozoic tectonic evolution of the western Sierra Nevada, California: Geological Society of America Bulletin, v. 86, p. 1329–1336.

Schweller WJ, Karig DE, 1982, Emplacement of the Zambales ophiolite into the west Luzon margin: American Association of Petroleum Geologists Memoir 34, p. 441–454.

Schweller WJ, Kulm LD, 1978a, Extensional rupture of oceanic crust in the Chile Trench: Marine Geology, v. 28, p. 271–291.

Schweller WJ, Kulm LD, 1978b, Depositional patterns and channelized sedimentation in active eastern Pacific trenches, in Stanley DJ, Kelling G-(eds.), Sedimentation in submarine canyons, fans, and trenches: Dowden, Hutchinson and Ross, Stroudsburg, p. 311–324.

Schweller WJ, Kulm LD, Prince RA, 1981, Tectonics, structure, and sedimentary framework of the Peru-Chile Trench: Geological Society of America Memoir 154, p. 323–350.

Schweller WJ, Karig DE, Bachman SB, 1983, Original setting and emplacement history of the Zambales ophiolite, Luzon, Philippines, from stratigraphic evidence: American Geophysical Union Geophysical Monograph 27, p. 124–138.

Schweller WJ, Roth PH, Karig DE, Bachman SB, 1984, Sedimentation history and biostratigraphy of ophiolite-related Tertiary sediments, Luzon, Philippines: Geological Society of America Bulletin, v. 95, p. 1333–1342.

Sclater JG, Christie PAF, 1980, Continental stretching: an explanation of the post-mid-Cretaceous subsidence of the central North Sea basin: Journal of Geophysical Research, v. 85, p. 3711–3739.

Scotese CR, Rowley DB, 1985, The orthogonality of subduction: an empirical rule: Tectonophysics, v. 116, p. 173–187.

Scott R, Kroenke L, 1980, Evolution of back arc spreading and arc volcanism in the Philippine Sea: interpretation of Leg 59 DSDP results: American Geophysical Union Geophysical Monograph 23, p. 283–291.

Scott SD, 1980, Geology and structural control of Kuroko-type massive sulphide deposits: Geological Association of Canada Special Paper 20, p. 705–721.

Scrutton RA, 1976, Microcontinents and their significance, in Drake, C.L. (ed.), Geodynamics: progress and prospects: Ameri-

can Geophysical Union, Washington, 238 p.

Scrutton RA, 1979, On sheared passive continental margins: Tectonophysics, v. 59, p. 293–305.

Seager WR, Morgan P, 1979, Rio Grande rift in southern New Mexico, west Texas, and northern Chihuahua, in Riecker RE (ed.), Rio Grande rift: tectonics and magmatism: American Geophysical Union, Washington, p. 87–106.

Searle MP, 10 coauthors, 1987, The closing of Tethys and the tectonics of the Himalaya: Geological Society of America Bulletin, v. 98, p. 678–701.

Sebrier M, Soler P, 1991, Tectonics and magmatism in the Peruvian Andes from late Oligocene time to the present: Geological Society of America Special Paper 265, p. 259–278.

Sebrier M, Mercier JL, Megard F, Laubacher G, Carey-Gailhardis E, 1985, Quaternary normal and reverse faulting and the state of stress in the central Andes of south Peru: Tectonics, v. 4, p. 739–780.

Seely DR, 1977, The significance of landward vergence and oblique structural trends on trench inner slopes, in Talwani M, Pitman WC, III (eds.), Island arcs, deep sea trenches, and backarc basins [Maurice Ewing Series 1]: American Geophysical Union, Washington, p. 187–198.

Seely DR, 1979, The evolution of structural highs bordering major forearc basins: American Association of Petroleum Geologists Memoir 29, p. 245–260.

Seely DR, Vail PR, Walton GG, 1974, Trench slope model, in Burk CA, Drake CL (eds.), The geology of continental margins: Springer-Verlag, New York, p. 249–260.

Segall P, Pollard DD, 1980, Mechanics of discontinuous faults: Journal of Geophysical Research, v. 85, p. 4337–4350.

Sellwood BW, Netherwood RE, 1986, Facies evolution in the Gulf of Suez area: sedimentation history as an indicator of rift initiation and development: Modern Geology, v. 9, p. 43–69.

Şengör AMC, 1976a, Rift valley formation as a combined result of parting and uplift: Geological Society of America Abstracts with Programs, v. 8, p. 65.

Şengör AMC, 1976b, Collision of irregular continental margins: implications for foreland deformation of Alpine-type orogens: Geology, v. 4, p. 779–782.

Şengör AMC, 1978, Über die angebliche primäre Vertikaltektonik im Ägäisraum: Neues Jahrbuch für Geologie und Paläontologie 11, p. 698–703.

Şengör AMC, 1979a, The North Anatolian transform fault: its age, offset and tectonic significance: Journal of the Geological Society of London, v. 136, p. 269–282.

Şengör AMC, 1979b, Mid-Mesozoic closure of Permo-Triassic Tethys and its implication: Nature, v. 279, p. 590–593.

Şengör AMC, 1981, The evolution of Palaeo-Tethys in the Tibetan segment of the Alpides, in Geological and ecological studies of Qinghai-Xizang Plateau [v. 1]: Science Press, Beijing, p. 51–56.

Şengör AMC, 1982, Ege'nin neotektonik evrimini yöneten etkenler, in Erol O, Oygür V (eds.), Batı Anadolu'nun Genç Tektoniğ ve Volkanizması Paneli: Türkiye Jeoloji Kurumu, Ankara, p. 59–71.

Şengör AMC, 1983, Kıt'asal gerilme alanları—Genel, in Canıtez N (ed.), Levha Tektoniği: İTÜ Maden Fakültesi, Ofset Bask+ Atölyesi, Istanbul, p. 461–478.

Şengör AMC, 1984, The Cimmeride orogenic system and the tectonics of Eurasia: Geological Society of America Special

Paper 195, 82 p.

Şengör AMC, 1987a, Aulacogen, in Seyfert CK (ed.), The encyclopedia of structural geology and plate tectonics: Van Nostrand Reinhold, New York, p. 18–25.

Şengör AMC, 1987b, Impactogen, in Seyfert CK (ed.), The encyclopedia of structural geology and plate tectonics: Van Nostrand Reinhold, New York, p. 336–340.

Şengör AMC, 1987c, Cross-faults and differential stretching of hanging walls in regions of low-angle normal faulting: examples from western Turkey: Geological Society of London Special Publication 28, p. 575–589.

Şengör AMC, 1987d, Tectonic subdivisions and evolution of Asia: Bulletin of the Technical University of Istanbul 40, p. 355–435.

Şengör AMC, 1987e, Tectonics of the Tethysides: orogenic collage development in a collisional setting: Annual Reviews of Earth and Planetary Sciences, v. 15, p. 213–244.

Şengör AMC, 1990a, Plate tectonics and orogenic research after 25 years: a Tethyan perspective: Earth Science Reviews, v. 27, p. 1–201.

Şengör AMC, 1990b, A new model for the late Palaeozoic-Mesozoic tectonic evolution of Iran and implications for Oman: Geological Society of London Special Publication 49, p. 797–831.

Şengör AMC, 1990c, Lithotectonic terranes and the plate tectonic theory of orogeny: a critique of the principles of terrane analysis: Circum-Pacific Council for Energy and Mineral Resources Earth Sciences Series, v. 13, p. 9–46.

Şengör AMC, 1991a, Timing of orogenic events: a persistent geological controversy, in Müller DW, McKenzie JA, Weissert H (eds.), Controversies in modern geology: Academic Press, London, p. 405–473.

Şengör AMC, 1991b, Orogenic architecture as a guide to size of ocean lost in collisional mountain belts: Bulletin of the Technical University of Istanbul, v. 44, p. 43–74.

Şengör AMC, 1991c, Late Palaeozoic and Mesozoic tectonic evolution of the Middle Eastern Tethysides: implications for the Palaeozoic geodynamics of the Tethyan realm [Newsletter 2 of IGCP Project 276: Mémoires de Géologie (Lausanne) 10, p. 111–149.

Şengör AMC, 1991d, Plate tectonics and orogenic research after 25 years: synopsis of a Tethyan perspective: Tectonophysics, v. 187, p. 315–344.

Şengör AMC, 1992, The Palaeo-Tethyan suture: a line of demarcation between two fundamentally different architectural styles in the structure of Asia: The Island Arc, v. 1, p. 78–91.

Şengör AMC, 1993, Some current problems on the tectonic evolution of the Mediterranean during the Cainozoic, in Boschi E, Mantovani E, Morelli A (eds.), Recent evolution and Seismicity of the Mediterranean region [NATO ASI Series C402]: Kluwer, Dordrecht, p. 1–51.

Şengör AMC, Burke K, 1978, Relative timing of rifting and volcanism on Earth and its tectonic implications: Geophysical Research Letters, v. 5, p. 419–421.

Şengör AMC, Dewey JF, 1991, Terranology: vice or virtue?, in Dewey JF, 4 coeditors, 1991, Allochthonous terranes: Cambridge University Press, Cambridge, p. 1–21.

Şengör AMC, Hsü KJ, 1984, The Cimmerides of eastern Asia: history of the eastern end of Palaeo-Tethys: Societe Geologique de France Memoires 147, p. 139–167.

Şengör AMC, Kidd WSF, 1979, Post-collisional tectonics of the Turkish-Iranian Plateau and a comparison with Tibet: Tectonophysics, v. 55, p. 361–376.

Şengör AMC, Okuroğulları AH, 1991, The rôle of accretionary wedges in the growth of continents: Asiatic examples from Argand to plate tectonics: Eclogae Geologicae Helvetiae, v. 84, p. 535–597.

Şengör AMC, Burke K, Dewey JF, 1978, Rifts at high angles to orogenic belts: tests for their origin and the upper Rhine graben as an example: American Journal of Science, v. 278, p. 24–40.

Şengör AMC, Yalçın N, Canıtez N, 1980, The origin of the Adana/Cilicia basin: an incompatibility structure arising at the common termination of the East Anatolian and the Dead Sea transform faults: Sedimentary Basins of Mediterranean Margins Abstract Book, Urbino, p.45–46.

Şengör AMC, Yılmaz Y, Sungurlu O, 1984 Tectonics of the Mediterranean Cimmerides: nature and evolution of the western termination of Palaeo-Tethys: Geological Society of London Special Publication 17, p. 77–112.

Şengör AMC, Görür N, Şaroğlu F, 1985, Strike-slip faulting and related basin formation in zones of tectonic escape: Turkey as a case study: Society of Economic Paleontologists and Mineralogists Special Publication 37, p. 227–264.

Şengör AMC, Altıner D, Cin A, Ustaömer T, Hsü KJ, 1988, Origin and assembly of the Tethyside orogenic collage at the expense of Gondwana-land: Geological Society of London Special Publication 37, p. 119–181.

Şengör AMC, Cin A, Rowley DB, Nie S, 1991a, Magmatic evolution of the Tethysides: a guide to reconstruction of collage history: Palaeogeography, Palaeoclimatology, Palaeoecology, v. 87, p. 411–440.

Şengör AMC, Tüysüz O, Serdar H, Erol U, 1991b, The internal structure of the Rhodope-Pontide fragment: implications for the survival of relict oceanic basins in continents: Ozan Sungurlu Symposium Abstracts Books, Ankara, p. 39–40.

Şengör AMC, Natal'in BA, Burtman VS, 1993, Evolution of the Altaid tectonic collage and Palaeozoic crustal growth in Eurasia: Nature, v. 364, p. 299–307.

Seno T, Maruyama S, 1984, Paleogeographic reconstruction and origin of the Philippine Sea: Tectonophysics, v. 102, p. 53–84.

Seyitoğlu G, Scott BC, 1991, Late Cainozoic crustal extension and basin formation in west Turkey: Geological Magazine, v. 128, p. 155–166.

Seyler B, Cluff RM, 1990, Petroleum traps in the Illinois basin: American Association of Petroleum Geologists Memoir 51, p. 361–401.

Shackleton RM, 1955, Pleistocene movements in the Gregory Rift Valley: Geologische Rundschau, v. 43, p. 257–263.

Shackleton RM, Chengfa C, 1988, Cenozoic uplift and deformation of the Tibetan Plateau: the geomorphological evidence: Philosophical Transactions of the Royal Society of London, v. A327, p. 365–378.

Sharp WD, 1980, Ophiolite accretion in the northern Sierra: EOS, v. 61, p. 1122.

Shatsky NS, 1946, Bolshoi Donbass i Sistema Vichita: İzvestija Akademii Nauk SSSR, Serija Geologicheskaya 6, p. 57–90. (Also in Akademik N.S. Shatsky, Izbrannie Trudi, v. 2: "Nauka" Moscow, p. 426–458.)

Shatsky NS, 1955, O proiskhodjdenii Pachelmskogo progiba: Bjulleten Moskovskogo Obschtschestva Ispytatelei Prirody, v. 30, p. 5–26. (Also in Akademik N.S. Shatsky, Izbrannie Trudi, v. 2: "Nauka", Moscow, p. 523–543.)

Shatsky NS, 1964, O progibach Donetskogo typa, in Akademik N.S. Shatsky, Izbrannie Trudi, v. 2: "Nauka", Moscow, p. 544–552.

Shaw HR, 1980, The fracture mechanisms of magma transport from the mantle to the surface, in Hargraves RB (ed.), Physics of magmatic processes: Princeton University Press, Princeton, p. 201–264.

Shaw RD, Etheridge MA, Lambeck K, 1991, Development of the Late Proterozoic to mid-Paleozoic intracratonic Amadeus basin in central Australia: a key to understanding tectonic forces in plate interiors: Tectonics, v. 10, p. 688–721.

Shepard FP, 1981, Submarine canyons; multiple causes and long-time persistence: American Association of Petroleum Geologists Bulletin, v. 65, p. 1062–1077.

Shepard FP, Marshall NF, McLaughlin PA, Sullivan GG, 1979, Currents in submarine canyons and other sea valleys: American Association of Petroleum Geologists Studies in Geology 8, 173 p.

Sheridan RE, 1981, Recent research on passive continental margins: Society of Economic Paleontologists and Mineralogists Special Publication 32, p. 39–55.

Sheridan RE, 1989, The Atlantic passive margin, in Bally AW, Palmer AA (eds.), The geology of North America—an overview [The Geology of North Ameica, v. A]: Geological Society of America, Boulder, p. 81–96.

Sheridan RE, Grow JA (eds.), 1988, The Atlantic continental margin: U.S. [The Geology of North America, v. I–2]: Geological Society of America, Boulder, 610 p.

Sheridan RE, Dill CE, Kraft JC, 1974, Holocene sedimentary environments of the Atlantic inner shelf off Delaware: Geological Society of America Bulletin, v. 85, p. 1319–1328.

Sheridan RE, Enos P, Gradstein FM, Benson WE, 1978, Mesozoic and Cenozoic sedimentary environments of the western North Atlantic: Initial Reports of the Deep Sea Drilling Project, v. 44, p. 971–979.

Sheridan RE, 13 coauthors, 1983, Site 533: Blake Outer Ridge: Initial Reports of the Deep Sea Drilling Project, v. 76, p. 35–140.

Sheridan RE, 8 coauthors, 1993, Deep seismic reflection data of EDGE U.S. mid-Atlantic continental-margin experiment: implications for Appalachian sutures and Mesozoic rifting and magmatic underplating: Geology, v. 21, p. 563–567.

Sherrod DR, Conrey RM, 1988, Geologic setting of the Breitenbush-Austin Hot Springs area, Cascade Range, north-central Oregon: Oregon Department of Geology and Mineral Industries Open-File Report O-88-5, p. 1–14.

Shervais JW, 1990, Island arc and ocean crust ophiolites: contrasts in the petrology, geochemistry and tectonic style of ophiolite assemblages in the California Coast Ranges, in Malpas J, Moores EM, Panayiotou A, Xenophontos C (eds.), Ophiolites, oceanic crustal analogues: Cyprus Geological Survey Department, Nicosia, p. 507–520.

Shervais JW, Kimbrough DL, 1985, Geochemical evidence for the tectonic setting of the Coast Range ophiolite: a composite island arc-oceanic crust terrane in western California: Geology, v. 13, p. 35–38.

Shi Y, Wang C, Langseth MG, Hobart M, vonHuene R, 1988, Heat flow and thermal structure of the Washington-Oregon accretionary prism—a study of the lower slope: Geophysical Research Letters, v. 15, p. 1113–1116.

Shiki T, Misawa Y, 1982, Forearc geological structure of the Japanese Islands: Geological Society of London Special Publication 10, p. 63–73.

Shimamura K, 1989, Topography and sedimentary facies of the Nankai Deep Sea Channel, in Taira A, Masuda F (eds.), Sedimentary facies in the active plate margin: Terra Scientific Publishing, Tokyo, p. 529–556.

Shipboard Party, 1980, Initial Reports of the Deep Sea Drilling Project [v. 56/57, Part 1]: United States Government Printing Office, Washington, 629 p.

Shipley TH, Moore GF, 1985, Sediment accretion and subduction in the Middle America Trench, in Nasu N, Uyeda S, Kushiro I, Kobayashi K, Kagami H (eds.), Formation of active ocean margins: Terra Scientific Publishing, Tokyo, p. 221–255.

Shipley TH, Moore GF, 1986, Sediment accretion, subduction, and dewatering at the base of the trench slope off Costa Rica: a seismic reflection view of the decollement: Journal of Geophysical Research, v. 91, p. 2019–2028.

Shipley TH, Ladd JW, Buffler RT, Watkins JS, 1982, Tectonic processes along the Middle America Trench inner slope: Geological Society of London Special Publication 10, p. 95–106.

Shipley TH, Stoffa PL, Dean DF, 1990, Underthrust sediments, fluid migration paths, and mud volcanoes associated with the accretionary wedge off Costa Rica: Middle America Trench: Journal of Geophysical Research, v. 95, p. 8743–8752.

Shirley K, 1991, Oil still hiding out in Sherwood: American Association of Petroleum Geologists Explorer, August, p. 1, 6–7.

Short PF, Ingersoll RV, 1990, Petrofacies and provenance of the Great Valley Group, southern Klamath Mountains and northern Sacramento Valley, in Ingersoll RV, Nilsen TH (eds.), Sacramento Valley symposium and guidebook [Book 65]: Pacific Section, Society of Economic Paleontologists and Mineralogists, Los Angeles, p. 39–52.

Shouldice DH, 1973, Western Canadian continental shelf: Canadian Society of Petroleum Geologists Memoir 1, p. 7–35.

Shrake DL, 6 authors, 1991, Pre-Mount Simon basin under the CIncinnati arch: Geology, v. 19, p. 139–142.

Shreve RL, Cloos M, 1986, Dynamics of sediment subduction, melange formation, and prism accretion: Journal of Geophysical Research, v. 91, p. 10229–10245.

Siebert L, Glicken HX, Ui T, 1987, Volcanic hazards from Bezmianny- and Bandai-type eruptions: Bulletin of Volcanology, v. 49, p. 435–459.

Siedlecka A, 1975, Late Precambrian stratigraphy and structure of the north-eastern margin of the Fennoscandian shield (east Finnmark-Timan region): Norges Geologiske Undersfkelse 316, p. 313–348.

Siefke JW, 1980, Geology of the Kramer borate deposit, Boron, California, in Fife DL, Brown AR (eds.), Geology and mineral wealth of the California desert: South Coast Geological Society, Santa Ana, p. 260–267.

Sigurdsson H, Sparks RSJ, Carey SN, Huang TC, 1980, Volcanogenic sedimentation in the Lesser Antilles arc: Journal of Geology, v. 88, p. 523–540.

Silver EA, Moore JC, 1978, The Molucca Sea collision zone, Indonesia: Journal of Geophysical Research, v. 83, p. 1681–1691.

Silver EA, Rangin C, 1991, Leg 124 tectonic synthesis: Proceedings of the Ocean Drilling Program, Scientific Results, v. 124, p. 3–9.

Silver EA, Reed DL, 1988, Backthrusting in accretionary wedges: Journal of Geophysical Research, v. 93, p. 3116–3126.

Silver EA, Ellis MJ, Breen NA, Shipley TH, 1985, Comments on the growth of accretionary wedges: Geology, v. 13, p. 6–9.

Silver EA, Abbott LD, Kirchoff-Stein KS, Reed DL, Bernstein-Taylor B, 1991, Collision propagation in Papua New Guinea and the Solomon Sea: Tectonics, v. 10, p. 863–874.

Silver LT, 1986, Observations on the Peninsular Ranges batholith, southern California and Mexico, in space and time: Geological Society of America Abstracts with Programs, v. 18, p. 184.

Silver LT, Stehli FG, Allen CR, 1963, Lower Cretaceous pre-batholithic rocks of northern Baja California, Mexico: American Association of Petroleum Geologists Bulletin, v. 47, p. 2054–2059.

Silver LT, 5 coauthors, 1977, The 1.4 to 1.5 By transcontinental anorogenic plutonic perforation of North America: Geological Society of America Abstracts with Programs, v. 9, p. 1176–1177.

Sinclair HD, Allen PA, 1992, Vertical versus horizontal motions in the Alpine orogenic wedge: stratigraphic response in the foreland basin: Basin Research, v. 4, p. 215–232.

Sinclair HD, Coakley BJ, Allen PA, Watts AB, 1991, Simulation of foreland basin stratigraphy using a diffusion model of mountain belt uplift and erosion: an example from the central Alps, Switzerland: Tectonics, v. 10, p. 599–620.

Singer BS, Myers JD, 1990, Intra-arc extension and magmatic evolution in the central Aleutian arc, Alaska: Geology, v. 18, p. 1050–1053.

Sittler C, 1965, Le Paléogène des Fossés Rhénan et Rhodanien. Etudes sédimentologiques et paléoclimatiques [DS Thèse]: Université de Strasbourg, Strasbourg, 392 p.

Sittler C, 1969, The sedimentary trough of the Rhine graben: Tectonophysics, v. 8, p. 543–560.

Slatt RM, Piper DJW, 1974, Sand-silt petrology and sediment dispersal in the Gulf of Alaska: Journal of Sedimentary Petrology, v. 44, p. 1061–1071.

Sleep NH, 1971, Thermal effects of the formation of Atlantic continental margins by continental breakup: Geophysical Journal of the Royal Astronomical Society, v. 24, p. 325–350.

Sleep NH, 1976, Platform subsidence mechanisms and eustatic sea-level changes: Tectonophysics, v. 36, p. 45–56.

Sleep NH, 1987, Lithospheric heating by mantle plumes: Geophysical Journal of the Royal Astronomical Society, v. 91, p. 1–11.

Sleep NH, Sloss LL, 1978, A deep borehole in the Michigan basin: Journal of Geophysical Research, v. 83, p. 5815–5819.

Sleep NH, Snell NS, 1976, Thermal contraction and flexure of midcontinent and Atlantic marginal basins: Geophysical Journal of the Royal Astronomical Society, v. 45, p. 125–154.

Sleep NH, Toksöz MN, 1971, Evolution of marginal basins: Nature, v. 233, p. 548–551.

Sleep NH, Nunn JA, Chou L, 1980, Platform basins: Annual Review of Earth and Planetary Sciences, v. 8, p. 17–34.

Sloan RJ, Williams BPJ, 1991, Volcano-tectonic control of offshore to tidal-flat regressive cycles from the Dunquin Group

(Silurian) of southwest Ireland: International Association of Sedimentologists Special Publication 12, p. 105–119.

Sloss LL, 1963, Sequences in the intracratonic interior of North America: Geological Society of America Bulletin, v. 74, p. 93–114.

Sloss LL, 1972, Synchrony of Phanerozoic sedimentary-tectonic events in the North American craton and the Russian platform: XXIV International Geological Congress Proceedings, Montreal, Section 6, p. 24–32.

Sloss LL, 1976, Areas and volumes of intracratonic sediments, western North America and eastern Europe: Geology, v. 4, p. 272–276.

Sloss LL, 1979, Global sea level change: a view from the craton: American Association of Petroleum Geologists Memoir 29, p. 461–467.

Sloss LL, 1987, Williston in the family of intracratonic basins, in Longman MW (ed.), Williston basin, anatomy of a cratonic oil province: Rocky Mountain Association of Petroleum Geologists, Denver, p. 1–8.

Sloss LL, 1988a, Tectonic evolution of the craton in Phanerozoic time, in Sloss LL (ed.), Sedimentary cover—North American craton: U.S. [The Geology of North America, v. D-2]: Geological Society of America, Boulder, p. 25–51.

Sloss LL, 1988b, Conclusions, in Sloss LL (ed.), Sedimentary cover—North American craton: US [The Geology of North America, v. D-2]: Geological Society of America, Boulder, p. 493–496.

Sloss LL (ed.), 1988c, Sedimentary cover–North American craton: US [The Geology of North America, v. D-2]: Geological Society of America, Boulder, 506p.

Sloss LL, Scherer W, 1975, Geometry of sedimentary basins: applications to Devonian of North America and Europe: Geological Society of America Memoir 142, p. 71–88.

Sloss LL, Speed RC, 1974, Relationship of Intracratonic and continent-margin tectonic episodes: Society of Economic Paleontologists and Mineralogists Special Publication 22, p. 38–55.

Smellie JL, Davies RES, Thomson MRA, 1980, Geology of a Mesozoic intra-arc sequence on Byers Peninsula, Livingston Island, South Shetland Islands: British Antarctic Survey Bulletin 50, p. 55–76.

Smith AG, Woodcock NH, 1982, Tectonic syntheses of the Alpine-Mediterranean region: a review, in Berckhemer H, Hsü K (eds.), Alpine-Mediterranean geodynamics: American Geophysical Union, Washington, p. 15–38.

Smith DP, Busby CJ, 1993, Mid-Cretaceous crustal extension recorded in deep-marine half-graben fill, Cedros Island, Mexico: Geological Society of America Bulletin, v. 105, p. 547–562.

Smith DP, Busby CJ, 1994, Shallow magnetic inclinations in the Cretaceous Valle Group: remagnetization, compaction, or terrane translation?: Tectonics, v. 12, p. 1258–1266.

Smith GA, 1986, Coarse-grained nonmarine volcaniclastic sediment: terminology and depositional process: Geological Society of America Bulletin, v. 97, p. 1–10.

Smith GA, 1987, The influence of explosive volcanism on fluvial sedimentation: the Deschutes Formation (Neogene) in central Oregon: Journal of Sedimentary Petrology, v. 57, p. 613–629.

Smith GA, 1988, Neogene synvolcanic and syntectonic sedimentation in central Washington: Geological Society of America Bulletin, v. 100, p. 1479–1492.

Smith GA, 1989, Extensional structures within continental-margin arcs: their occurrence in the modern circum-Pacific with emphasis on the central Oregon Cascade Range: EOS, v. 70, p. 1299.

Smith GA, 1991, Facies sequences and geometries in continental volcaniclastic sediments: Society for Sedimentary Geology Special Publication 45, p. 109–121.

Smith GA, Lowe DR, 1991, Lahars: volcano-hydrologic events and deposition in the debris flow—hyperconcentrated flow continuum: Society for Sedimentary Geology Special Publication 45, p. 59–70.

Smith GA, Snee LW, Taylor EM, 1987, Stratigraphic, sedimentologic, and petrologic record of late Miocene subsidence of the central Oregon High Cascades: Geology, v. 15, p. 389–392.

Smith GA, Vincent KR, Snee LW, 1989, An isostatic model for basin formation in and adjacent to the central Oregon High Cascade Range: United States Geological Survey Open-File Report 89-178, p.411–429.

Smith GW, Howell DG, Ingersoll RV, 1979, Late Cretaceous trench-slope basins of central California: Geology, v. 7, p. 303–306.

Smith JG, 1987, New compilation geologic map of the Cascade Range in Washington: Geothermal Resources Council Transactions, v. 40, p. 309–314.

Smith JR, Taylor B, Malahoff A, Petersen L, 1990, Submarine volcanism in the Sumisu Rift, Izu-Bonin arc: Earth and Planetary Science Letters, v. 100, p. 148–160.

Smith RB, 1991, Diagenesis and cementation of lower Miocene pyroclastic sequences in the Sulu Sea, Sites 768, 769, and 771: Proceedings of the Ocean Drilling Program, Scientific Results, v. 124, p. 181–199.

Smythe DK, 6 coauthors, 1982, Deep structure of the Scottish Caledonides revealed by the MOIST reflection profile: Nature, v. 299, p. 338–340.

Snavely PD, Jr, 1987, Tertiary geologic framework, neotectonics, and petroleum potential of the Oregon-Washington continental margin: Circum-Pacific Council for Energy and Mineral Resources Earth Science Series, v. 6, p. 157–189.

Snyder WS, Spinosa C, 1993, Exxon global cycle chart: an event for every occasion: Comment: Geology, v. 21, p. 283–284.

Soares PC, Landim PMBL, Fulfaro VJ, 1978, Tectonic cycles and sedimentary sequences in the Brazilian intracratonic basins: Geological Society of America Bulletin, v. 89, p. 181–191.

Soh W, Pickering KT, Taira A, Tokuyama H, 1991, Basin evolution in the arc-arc Izu collision zone, Mio-Pliocene Miura Group, central Japan: Journal of the Geological Society of London, v. 148, p. 317–330.

Solomon SC, 1987, Secular cooling of the Earth as a source of intraplate stress: Earth and Planetary Science Letters, v. 83, p. 153–158.

Sonder LJ, England PC, Wernicke BP, Christianson RL, 1987, A physical model for Cenozoic extension of western North America: Geological Society of London Special Publication 28, p. 187–201.

Sorochtin OG, 1974, Globalnaya evolutsia zemli: "Nauka", Moscow, 184 p.

Sparks RSJ, Sigurdsson H, Carey SN, 1980a, The entrance of pyroclastic flows into the sea, I. Oceanographic and geologic evidence from Dominica, Lesser Antilles: Journal of Volcanol-

ogy and Geothermal Research, v. 7, p. 87–96.

Sparks RSJ, Sigurdsson H, Carey SN, 1980b, The entrance of pyroclastic flows into the sea. II. Theoretical considerations on subaqueous emplacement and welding: Journal of Volcanology and Geothermal Research, v. 7, p. 97–105.

Speed RC, Larue DK, 1982, Barbados: architecture and implications for accretion: Journal of Geophysical Research, v. 87, p. 3633–3643.

Speed RC, Sleep NH, 1982, Antler orogeny and foreland basin: a model: Geological Society of America Bulletin, v. 93, p. 815–828. (Also, see Discussion and Reply, v. 94, p. 684–686.)

Speed RC, Torrini R, Jr, Smith PL, 1989, Tectonic evolution of the Tobago Trough forearc basin: Journal of Geophysical Research, v. 94, p. 2913–2936.

Speed RC, Barker LH, Payne PLB, 1991, Geologic and hydrocarbon evolution of Barbados: Journal of Petroleum Geology, v. 14, p. 323–342.

Spencer JE, 1984, Role of tectonic denudation in warping and uplift of low-angle faults: Geology, v. 12, p. 95–98.

Spencer JE, Reynolds SJ, 1984, Mid-Tertiary crustal extension in Arizona: Geological Society of America Abstracts with Programs, v. 16, p. 664.

Spieker EM, 1946, Late Mesozoic and early Cenozoic history of central Utah: United States Geological Survey Professional Paper 205-D, p. 117–161.

Spizaharsky TN, Borovikov LI, 1966, Tectonic map of the Soviet Union on a scale of 1:2,500,000, in Scientific communications read to the Commission for the Geological Map of the World: 22nd International Geological Congress, Delhi, p. 111–120.

Spohn T, Schubert G, 1982, Convective thinning of the lithosphere: a mechanism for the initiation of continental rifting: Journal of Geophysical Research, v. 87, p. 4669–4681.

Spohn T, Schubert G, 1983, Convective thinning of the lithosphere: a mechanism for rifting and mid-plate volcanism on Earth, Venus and Mars: Tectonophysics, v. 94, p. 67–90.

Sporli KB, 1978, Mesozoic tectonics, North Island, New Zealand: Geological Society of America Bulletin, v. 89, p. 415–425.

Sporli KB, Ballance PF, 1989, Mesozoic ocean floor/continent interaction and terrane configuration, southwest Pacific area around New Zealand, in Ben-Avraham Z (ed.), The evolution of the Pacific Ocean margins: Oxford Unviersity Press, New York, p. 176–187.

Stabler CL, 1990, Andean hydrocarbon resources–an overview: Circum-Pacific Council for Energy and Mineral Resources Earth Science Series, v. 11, p. 431–438.

Stanley DJ, Kelling G (eds.), 1978, Sedimentation in submarine canyons, fans, and trenches: Dowden, Hutchinson and Ross, Stroudsburg, 395 p.

Stanley RS, Ratcliffe NM, 1985, Tectonic synthesis of the Taconian orogeny in western New England: Geological Society of America Bulletin, v. 96, p. 1227–1250.

Stanley WD, Saad AR, Ohofugi W, 1985, Regional magnetotelluric surveys in hydrocarbon exploration, Parana basin, Brazil: America Association of Petroleum Geologists Bulletin, v. 69, p. 346–360.

Stanley WD, Mooney WD, Fuis GS, 1990, Deep crustal structure of the Cascade Range and surrounding regions from seismic refraction and magnetotelluric data: Journal of Geophysical Research, v. 95, p. 19419–19438.

Steckler MS, 1982, Thermal and mechanical evolution of Atlantic-type margins [PhD thesis]: Columbia University, New York, 261 p.

Steckler MS, 1985, Uplift and extension at the Gulf of Suez: indications of induced mantle convection: Nature, v. 317, p. 135–139.

Steckler MS, Omar GI, 1994, Controls on erosional retreat of the uplifted rift flanks at the Gulf of Suez and northern Red Sea: Journal of Geophysical Research, v. 99, p. 12159–12173.

Steckler MS, tenBrink US, 1986, Lithospheric strength variations as a control on new plate boundaries: examples from the northern Red Sea region: Earth and Planetary Science Letters, v. 79, p. 120–132.

Steckler MS, Watts AB, 1978a, Subsidence of the Atlantic-type continental margin off New York: Earth and Planetary Science Letters, v. 41, p. 1–13.

Steckler MS, Watts AB, 1978b, The Gulf of Lion: subsidence of a young continental margin: Nature, v. 287, p. 425–430.

Steckler MS, Watts AB, 1982, Subsidence history and tectonic evolution of the Atlantic-type continental margins: American Geophysical Union Geodynamics Series, v. 6, p. 184–196.

Steel RJ, 1993, Triassic-Jurassic megasequence stratigraphy in the northern North Sea: rift to post-rift evolution, in Parker JR (ed.), Proceedings of the 4th Conference on Petroleum Geolgy of North West Europe: Geological Society of London Special Publication, in press.

Steel R, Gloppen TG, 1980, Late Caledonian (Devonian) basin formation, western Norway: signs of strike-slip tectonics during infilling: International Association of Sedimentology Special Publication 4, p. 79–103.

Steel RJ, Mæhle S, Nilsen TH, Roe SL, Spinangr A, 1977, Coarsening-upward cycles in the alluvium of Hornelen basin (Devonian), Norway: sedimentary response to tectonic events: Geological Society of America Bulletin, v. 88, p. 1124–1134.

Steel RJ, 5 coauthors, 1985, The Tertiary strike-slip basins and orogenic belt of Spitsbergen: SEPM (Society for Sedimentary Geology) Special Publication 37, p. 339–360.

Steidtmann JR, Schmitt JG, 1988, Provenance and dispersal of tectogenic sediments in thin-skinned thrusted terrains, in Kleinspehn K, Paola C (eds.), New perspectives in basin analysis: Springer-Verlag, New York, p. 353–366.

Stein RS, Barrientos SE, 1985, Planar high-angle faulting in the Basin and Range: geodetic analysis of the 1983 Borah Peak, Idaho, earthquake: Journal of Geophysical Research, v. 90, p. 11355–11366.

Stein RS, King GCP, Rundle JB, 1988, The growth of geological structures by repeated earthquakes 2. Field examples of continental dip-slip faults: Journal of Geophysical Research, v. 93, p. 13319–13331.

Stein RS, Briole P, Ruegy J-C, Tapponnier P, Gasse F, 1991, Contemporary, Holocene and Quaternary deformation of the Asal rift, Djibouti: implications for the mechanics of slow spreading ridges: Journal of Geophysical Research, v. 96, p. 21789–21806.

Stephenson RA, Embry AF, Nakiboglu SM, Hastaoglu MA, 1987, Rift-initiated Permian to Early Cretaceous subsidence of the Sverdrup basin: Canadian Society of Petroleum Geologists Memoir 12, p. 213–231.

Stern RJ, Bloomer SH, 1992, Subduction zone infancy: exam-

ples from the Eocene Izu-Bonin-Mariana and Jurassic California arcs: Geological Society of America Bulletin, v. 104, p. 1621–1636.

Stern RJ, Smoot NC, Rubin M, 1984, Unzipping of the volcano arc, Japan: Tectonophysics, v. 102, p. 153–174.

Stern TA, 1985, A back-arc basin formed within continental ithosphere: the central volcanic region of New Zealand: Tectonophysics, v. 112, p. 385–409.

Stern TA, 1986, Geophysical studies of the upper crust within the central volcanic region, New Zealand: Royal Society of New Zealand Bulletin 23, p. 92–111.

Sterns RG, 1957, Cretaceous, Paleocene, and lower Eocene geologic history of the northern Mississippi embayment: Geological Society of America Bulletin, v. 68, p. 1077–1100.

Stevens SH, Moore GF, 1985, Deformational and sedimentary processes in trench slope basins of the western Sunda Arc, Indonesia: Marine Geology, v. 69, p. 93–112.

Stevenson AJ, Embley R, 1987, Deep-sea fan bodies, terrigenous turbidite sedimentation, and petroleum geology, Gulf of Alaska: Circum-Pacific Council for Energy and Mineral Resources Earth Science Series, v. 6, p. 503–522.

Stevenson AJ, Scholl DW, Vallier TL, 1983, Tectonic and geologic implications of the Zodiac Fan, Aleutian Abyssal Plain, northeast Pacific: Geological Society of America Bulletin, v. 94, p. 259–273.

Stewart AD, 1982, Late Proterozoic rifting in NW Scotland: the genesis of the 'Torridonian': Journal of the Geological Society of London, v. 139, p. 413–420.

Stewart JH, 1972, Initial deposits in the Cordilleran geosyncline: evidence of a late Precambrian (850 m.y.) continental separation: Geological Society of America Bulletin, v. 83, p. 1345–1360.

Stewart JH, 1978, Rift systems in the western United States, in Ramberg IB, Neumann ER (eds.), Tectonics and geophysics of continental rifts: D. Reidel, Dordrecht, p. 89–110.

Stewart RJ, 1976, Turbidites of the Aleutian Abyssal Plain: mineralogy, provenance, and constraints for Cenozoic motion of the Pacific plate: Geological Society of America Bulletin, v. 87, p. 793–808.

Stewart RJ, 1978, Neogene volcaniclastic sediments from the Atka Basin, Aleutian Ridge: American Association of Petroleum Geologists Bulletin, v. 62, p. 87–97.

Stille H (ed.), 1925, Göttinger Beiträge zur saxonischen Tektonik: Abhandlungen der Preußischen Geologischen Landesanstalt, Neue Folge, v. 95, p. 1–207.

Stix J, 1991, Subaqueous, intermediate to silicic-composition explosive volcanism: a review: Earth Science Reviews, v. 31, p. 21–53.

Stockmal GS, 1983, Modeling of large-scale accretionary wedge deformation: Journal of Geophysical Research, v. 88, p. 8271–8287.

Stockmal GS, Beaumont C, 1987, Geodynamic models of convergent margin tectonics: the southern Canadian Cordillera and the Swiss Alps: Canadian Society of Petroleum Geologists Memoir 12, p. 393–411.

Stockmal GS, Beaumont C, Boutilier R, 1986, Geodynamic models of convergent margin tectonics: transition from rifted margin to overthrust belt and consequences for foreland-basin development: American Association of Petroleum Geologists Bulletin, v. 70, p. 181–190.

Stockmal GS, Cant DJ, Bell JS, 1992, Relationship of the stratigraphy of the western Canada foreland basin to Cordilleran tectonics: insights from geodynamic models: American Association of Petroleum Geologists Memoir 55, p. 107–124.

Stow DAV, Lovell JPB, 1979, Contourites: their recognition in modern and ancient sediments: Earth Science Review, v. 14, p. 251–291.

Studer M, 1980, Métamorphisme d'enfouissement dans le Haut Atlas central, Maroc. Essai sur l'évaluation de l'épaisseur des couvertures sédimentaires: Comptes Rendus hébdomadaires de l'Academie des Sciences de Paris, v. 291, p. 457–460.

Suchecki RK, 1984, Facies history of the Upper Jurassic-Lower Cretaceous Great Valley sequence: response to structural development of an outer-arc basin: Journal of Sedimentary Petrology, v. 54, p. 170–191.

Suczek CA, Ingersoll RV, 1985, Petrology and provenance of Cenozoic sand from the Indus Cone and the Arabian Basin, DSDP sites 221, 222, and 224: Journal of Sedimentary Petrology, v. 55, p. 340–346.

Suen CJ, Frey FA, Malpas J, 1979, Bay of Islands ophiolite suite, Newfoundland: petrologic and geochemical characteristics with emphasis on rare earth element geochemistry: Earth and Planetary Science Letters, v. 45, p. 337–348.

Suess E, 1883, Das Antlitz der Erde, v. IA: Tempsky, Prag, 310 p.

Suess E, 26 coauthors, 1988a, Proceedings of the Ocean Drilling Program, Initial Reports [v. 112]: Ocean Drilling Program, College Station, 1015 p.

Suess E, vonHuene R, Leg 112 Shipboard Scientists, 1988b, Ocean Drilling Program Leg 112, Peru continental margin: part 2, sedimentary history and diagenesis in a coastal upwelling environment: Geology, v. 16, p. 939–943.

Sultan M, 6 coauthors, 1993, New constraints on Red Sea rifting from correlations of Arabian and Nubian Neoproterozoic outcrops: Tectonics, v. 12, p. 1303–1319.

Summerhayes CP, Gilbert D, 1982a, Distribution, origin, and hydrocarbon potential of organic matter in sediments from the Pacific margin of southern Mexico: Initial Reports of the Deep Sea Drilling Project, v. 66, p. 541–546.

Summerhayes CP, Gilbert D, 1982b, Distribution, origin, and hydrocarbon potential of organic matter in sediments from the Pacific margin of Guatemala: Initial Reports of the Deep Sea Drilling Project, v. 67, p. 595–599.

Suppe J, 1980a, A retrodeformable cross section of northern Taiwan: Proceedings of the Geological Society of China, v. 23, p. 46–55.

Suppe J, 1980b, Imbricate structure of western foothills belt, south-central Taiwan: Petroleum Geology of Taiwan, n. 17, p. 1–16.

Suppe J, 1984, Kinematics of arc-continent collision, flipping of subduction, and back-arc spreading near Taiwan: Geological Society of China Memoir 6, p. 21–34.

Suppe J, 1985, Principles of structural geology: Prentice-Hall, Englewood Cliffs, 537 p.

Suppe J, 1987, The active Taiwan mountain belt, in Schaer J-P, Rodgers J (eds.), The Anatomy of mountain ranges: Princeton University Press, Princeton, p. 277–293.

Suppe J, Liou JG, Ernst WG, 1981, Paleogeographic origins of the Miocene East Taiwan ophiolite: American Journal of Science, v. 281, p. 228–246.

Suppe J, Chou GT, Hook SC, 1992, Rates of folding and faulting determined from growth strata, in McClay KR (ed.), Thrust tectonics: Chapman and Hall, London, p. 105–121.

Surdam RC, Boles JR, 1979, Diagenesis of volcanic sandstones: Society of Economic Paleontologists and Mineralogists Special Publication 26, p. 227–242.

Surdam RC, Stanley KO, 1979, Lacustrine sedimentation during the culminating phase of Eocene Lake Gosiute, Wyoming (Green River Formation): Geological Society of America Bulletin, v. 90, pt. 1, p. 93–110.

Surkov VS, Djero OG, 1981, Fundament i razvitie platformennogo chekhla Zapadno-Sibirskoi Pliti: "Nedra", Moscow, 143 p.

Surlyk F, 1978, Submarine fan sedimentation along fault scarps on tilted fault blocks (Jurassic-Cretaceous boundary, East Greenland): Gronlands Geologiske Undersogelse Bulletin 128, 108 p.

Surlyk F, 1989, Mid-Mesozoic syn-rift turbidite systems: controls and predictions, in Correlation in hydrocarbon exploration [Norwegian Petroleum Society]: Graham and Trotman, London, p. 231–241.

Surlyk F, 1990, Mid-Mesozoic synrift turbidite systems: controls and predictions, in Collinson ED (ed.), Correlation in hydrocarbon exploration [Norwegian Petroleum Society]: Graham and Trotman, London, p. 231–241.

Sutter JF, 1988, Innovative approaches to dating igneous events in the early Mesozoic basins of the eastern United States: United States Geological Survey Bulletin 1776, p. 194–200.

Svoboda J, 1966, Regional geology of Czechoslovakia, Part I, The Bohemian massif: Geological Survey of Czechoslovakia, Prague, 668 p.

Swanson MT, 1982, Preliminary model for an early transform history in central Atlantic rifting: Geology, v. 10, p. 317–320.

Swift DJP, Hudelson PM, Brenner RL, Thompson P, 1987, Shelf construction in a foreland basin: storm beds, shelf sandbodies, and shelf-slope depositional sequences in the Upper Cretaceous Mesaverde Group, Book Cliffs, Utah: Sedimentology, v. 34, p. 423–457.

Swinden HS, Jenner GA, Fryer BJ, Hertogen J, Roddick JC, 1990, Petrogenesis and paleotectonic history of the Wild Bight Group, an Ordovician rifted island arc in central Newfoundland: Contributions to Mineralogy and Petrology, v. 105, p. 219–241.

Sykes LR, 1978, Intraplate seismicity, reactivation of pre-existing zones of weakness, alkaline magmatism and other tectonic post-continental fragmentation: Reviews of Geophysics and Space Physics, v. 16, p. 621–687.

Sylvester AG (compiler), 1984, Wrench fault tectonics: American Association of Petroleum Geologists Reprint Series 28, 374 p.

Sylvester AG, 1988, Strike-slip faults: Geological Society of America Bulletin, v. 100, p. 1666–1703.

Sylvester AG, Smith RR, 1976, Tectonic transpression and basement-controlled deformation in San Andreas fault zone, Salton Trough, California: American Association of Petroleum Geologists Bulletin, v. 60, p. 2081–2102.

Tabbutt KD, 1990, Temporal constraints on the tectonic evolution of Sierra de Famatina, northwestern Argentina, using the fission-track method to date tuffs interbedded in synorogenic clastic sedimentary strata: Journal of Geology, v. 98, p. 557–566.

Taber S, 1927, Fault troughs: Journal of Geology, v. 35, p. 577–606.

Taira A, 1985, Sedimentary evolution of Shikoku subduction zone: the Shimanto belt and Nankai Trough, in Nasu N (ed.), Formation of active ocean margins: Terra Scientific Publishing, Tokyo, p. 835–851.

Taira A, Niitsuma N, 1986, Turbidite sedimentation in the Nankai Trough as interpreted from magnetic fabric, grain size, and detrital modal analyses: Initial Reports of the Deep Sea Drilling Project, v. 87, p. 611–632.

Taira A, Okada H, Whitaker JHM, Smith AJ, 1982, The Shimanto belt of Japan: Cretaceous-lower Miocene active-margin sedimentation: Geological Society of London Special Publication 10, p. 5–26.

Taira A, Katto J, Tashiro M, Okamura M, Kodama K, 1988, The Shimanto belt in Shikoku, Japan–evolution of Cretaceous to Miocene accretionary prism: Modern Geology, v. 12, p. 5–46.

Taira A, Tokuyama H, Soh W, 1989, Accretion tectonics and evolution of Japan, in Ben-Avraham Z (ed.), The evolution of the Pacific Ocean margins: Oxford University Press, New York, p. 100–123.

Taira A, 28 coauthors, 1991, Proceedings of the Ocean Drilling Program, Initial Reports [v. 131]: Ocean Drilling Program, College Station, 434 p.

Taira A, Hill I, Firth J, the Shipboard Scientific Party, 1992, Sediment deformation and hydrogeology of the Nankai Trough accretionary prism: synthesis of shipboard results of ODP Leg 131: Earth and Planetary Science Letters, v. 109, p. 431–450.

Talwani M, Pitman WC, III (eds.), 1977, Island arcs, deep sea trenches and back-arc basins [Maurice Ewing Series 1]: American Geophysical Union, Washington, 470p.

Tamaki K., 1988, Geological structure of the Sea of Japan on the basis of stratigraphy, basement depth, and heat flow data: Journal of Geomagnetism and Geoelectricity, v. 38, p. 427–446.

Tamaki K, Honza E, 1991, Global tectonics and formation of maginal basins: role of the western Pacific: Episodes, v. 14, p. 224–230.

Tamaki K, 27 coauthors, 1990, Background, objectives, and principal results, ODP Leg 127, Japan Sea: Proceedings of the Ocean Drilling Program, Initial Reports, v. 127, p. 5–33.

Tamaki K, Suyehiro K, Allan J, Ingle JC, Jr, Pisciotto KA, 1992, Tectonic synthesis and implications of Japan Sea ODP drilling: Proceedings of the Ocean Drilling Program, Scientific Results, v. 127/128, Pt. 2, p. 1333–1348.

Tanimura S, Date J, Takahashi T, Ohmoto H, 1983, Geologic setting of the kuroko deposits, Japan: Part II, stratigraphy and structure of the Hokuroku district, in Ohmoto H, Skinner BJ (eds.), The Kuroko and related volcanic massive sulfide deposits: Economic Geology Monograph 5, p. 24–39.

Tankard AJ, 1986, On the depositional response to thrusting and lithospheric flexure: examples from the Appalachian and Rocky Mountain basins: International Association of Sedimentologists Special Publication 8, p. 369–392.

Tankard AJ, 5 coauthors, 1982, Crustal evolution of southern Africa: Springer-Verlag, New York, 523 p.

Tanner PWG, Rex DC, 1979, Timing of events on a Lower Cretaceous island-arc marginal basin system on South Georgia: Geological Magazine, v. 116, p. 167–179.

Tao WC, O'Connell RJ, 1992, Ablative subduction: a two-

sided alternative to the conventional subduction model: Journal of Geophysical Research, v. 97, p. 8877–8904.

Tapponnier P, Molnar P, 1976, Slip-line field theory and large scale continental tectonics: Nature, v. 264, p. 319–324.

Tapponnier P, Mercier J-L, Armijo R, Han T, Zhou J, 1981, Field evidence for active normal faulting in Tibet: Nature, v. 294, p. 410–417.

Tapponnier P, Peltzer G, LeDain AY, Armijo R, Cobbold P, 1982, Propagating extrusion tectonics in Asia: new insights from simple experiments with plasticine: Geology, v. 10, p. 611–616.

Tapponnier P, Peltzer G, Armijo R, 1986, On the mechanics of the collision between India and Asia: Geological Society of London Special Publication 19, p. 115–157.

Tardu T, 6 coauthors, 1987, Akçakale Grabeni'nin yap+sal-stratigrafik özellikleri ve petrol potansiyeli: Türkiye 7 Petrol Kongresi, Bildiriler (Jeoloji), Ankara, p. 36–49.

Tarney J, Saunders AD, Mattey DP, Wood DA, Marsh NG, 1981, Geochemical aspects of back-arc spreading in the Scotia Sea and western Pacific: Philosophical Transactions of the Royal Society of London, v. A300, p. 263–285.

Tatsumi Y, Hamilton DL, Nesbitt RW, 1986, Chemical characteristics of fluid phase released from a subducted lithosphere and origin of arc magmas: evidence from high-pressure experiments and natural rocks: Journal of Volcanology and Geothermal Research, v. 29, p. 293–309.

Taylor B, 1979, Bismark Sea: evolution of a back-arc basin: Geology, v. 7, p. 171–174.

Taylor B, 1992, Rifting and the volcanic-tectonic evolution of the Izu-Bonin-Mariana arc: Proceedings of the Ocean Drilling Program, Scientific Results, v. 126, p. 627–651.

Taylor B, Hayes DE, 1980, The tectonic evolution of the South China Basin: American Geophysical Union Geophysical Monograph 23, p. 89–104.

Taylor B, Hayes DE, 1983, Origin and history of the South China Sea Basin: American Geophysical Union Geophysical Monograph 27, p. 23–56.

Taylor B, Karner GD, 1983, On the evolution of marginal basins: Reviews of Geophysics and Space Physics, v. 21, p. 1727–1741.

Taylor B, 25 coauthors, 1990a, Proceedings of the Ocean Drilling Program, Initial Reports [v. 126]: Ocean Drilling Program, College Station, 1002 p.

Taylor B, 9 coauthors 1990b, ALVIN-Sea Beam studies of the Sumisu Rift, Izu-Bonin Arc: Earth and Planetary Science Letters, v. 100, p. 127–147.

Taylor B, Klaus A, Brown GR, Moore GF, 1991, Structural development of Sumisu Rift, Izu-Bonin Arc: Journal of Geophysical Research, v. 96, p. 16113–16129.

Taylor EM, 1990, Volcanic history and tectonic development of the central High Cascade Range, Oregon: Journal of Geophysical Research, v. 95, p. 19611–19622.

Taymaz T, Jackson JA, McKenzie DP, 1991, Active tectonics of the north and central Aegean Sea: Geophysical Journal International, v. 106, p. 433–490.

Tchalenko JS, 1970, Similarities between shear zones of different magnitudes: Geological Society of America Bulletin, v. 81, p. 1625–1640.

Tchirviniskaya MV, 1981, The Dnieper-Donetz depression, in Tectonics of Europe and adjacent areas. Cratons, Baikalides,

Caledonides. Explanatory note to the international tectonic map of Europe and adjacent areas [1:2,500,000]: "Nauka", Moscow, p. 170–178.

tenBrink US, Ben-Avraham Z, 1989, The anatomy of a pull-apart basin: seismic reflection observations of the Dead Sea Basin: Tectonics, v. 8, p. 333–350.

tenBrink US, 7 coauthors, 1994, Structure of the Dead Sea pull-apart basin from gravity analyses, in press.

Teng LS, 1990, Geotectonic evolution of late Cenozoic arc-continent collision in Taiwan: Tectonophysics, v. 183, p. 57–76.

Teng LS, Chen WS, Wang Y, Song SR, Lo HJ, 1988, Toward a comprehensive stratigraphic system of the Coastal Range, eastern Taiwan: Acta Geologica Taiwanica, n. 26, p. 19–36.

Terres RR, Sylvester AG, 1981, Kinematic analysis of rotated fractures and blocks in simple shear: Seismological Society of America Bulletin, v. 71, p. 1593–1605.

Thiessen R, Burke K, Kidd WSF, 1979, African hotspots and their relation to the underlying mantle: Geology, v. 7, p. 263–266.

Thomas WA, 1976, Evolution of Ouachita-Appalachian continental margin: Journal of Geology, v. 84, p. 323–342.

Thomas WA, 1977, Evolution of Appalachian-Ouachita salients and recesses from reentrants and promontories in the continental margin: American Journal of Science, v. 277, p. 1233–1278.

Thomas WA, 1983, Continental margins, orogenic belts, and intracratonic structures: Geology, v. 11, p. 270–272.

Thomas WA, 1985, The Appalachian-Ouachita connection: Paleozoic orogenic belt at the southern margin of North America: Annual Review of Earth and Planetary Sciences, v. 13, p. 175–199.

Thomas WA, 1988, The Black Warrior basin, in Sloss LL (ed.), Sedimentary cover—North American craton: U.S. [The Geology of North America, v. D-2]: Geological Society of America, Boulder, p. 471–492.

Thomas WA, 1989a, The Appalachian-Ouachita orogen beneath the Gulf Coastal Plain between the outcrops in the Appalachian and Ouachita Mountains, in Hatcher RD, Jr, Thomas WA, Viele GW (eds.), The Appalachian-Ouachita orogen in the United States [The Geology of North America, v. F-2]: Geological Society of America, Boulder, p. 537–553.

Thomas WA, 1989b, Tectonic map of the Ouachita orogen, in Hatcher RD, Jr, Thomas WA, Viele GW (eds.), The Appalachian-Ouachita orogen in the United States [The Geology of North America, v. F-2]: Geological Society of America, Boulder, Plate 9.

Thomas WA, 1991, The Appalachian-Ouachita rifted margin of southeastern North America: Geological Society of America Bulletin, v. 103, p. 415–431.

Thompson TL, 1976, Plate tectonics in oil and gas exploration of continental margins: American Association of Petroleum Geologists Bulletin, v. 60, p. 1463–1501.

Thomson CN, Girty GH, 1994, Early Cretaceous intra-arc ductile strain in Triassic-Jurassic and Cretaceous continental margin arc rocks, Peninsular Ranges, California: Tectonics, v. 13, p. 1108–1119.

Thornburg T, Kulm LD, 1981, Sedimentary basins of the Peru continental margin: structure, stratigraphy, and Cenozoic tectonics from 6°S to 16°S latitude: Geological Society of America Memoir 154, p. 393–422.

Thornburg TM, Kulm LD, 1987a, Sedimentation in the Chile

Trench: depositional morphologies, lithofacies, and stratigraphy: Geological Society of America Bulletin, v. 98, p. 33–52. (Also, see Discussion and Reply, v. 99, p. 598–600.)

Thornburg TM, Kulm LD, 1987b, Sedimentation in the Chile Trench: petrofacies and provenance: Journal of Sedimentary Petrology, v. 57, p. 55–74.

Thornburg TM, Kulm LD, Hussong DM, 1990, Submarine-fan development in the southern Chile Trench: a dynamic interplay of tectonics and sedimentation: Geological Society of America Bulletin, v. 102, p. 1658–1680.

Thorpe RS (ed.), 1982, Andesites: orogenic andesites and related rocks: John Wiley and Sons, Chichester, 725 p.

Thy P, 1992, Petrology of basaltic sills from Ocean Drilling Program Sites 794 and 797 in the Yamato Basin of the Japan Sea: Journal of Geophysical Research, v. 97, p. 9027–9042.

Tiercelin JJ, 1986, The Pliocene Hadar Formation, Afar depression of Ethiopia: Geological Society of London Special Publication 25, p. 221–240.

Tiercelin JJ, 26 coauthors, 1987, Le demi-graben de Baringo-Bagoria, rift Gregory, Kenya: 30,000 ans d'histoire hydrologique et sédimentaire: Bulletin Centres Recherche Explorations Production Elf Aquitaine, v. 11, p. 249–540.

Tiercelin JJ, 5 coauthors, 1988, East African rift system: offset, age and tectonic significance of the Tanganyika-Rukwa-Malawi intracontinental transcurrent fault zone: Tectonophysics, v. 148, p. 241–252.

Tobisch OT, Saleeby JB, Fiske RS, 1986, Structural history of continental volcanic arc rocks, eastern Sierra Nevada, California: a case for extensional tectonics: Tectonics, v. 5, p. 65–94.

Todd BJ, Reid I, Keen CE, 1987, Crustal structure across the southwest Newfoundland transform margin: Canadian Journal of Earth Sciences, v. 25, p. 744–759.

Todd VR, Erskine BG, Morton L, 1988, Metamorphic and tectonic evolution of the northern Peninsular Ranges batholith, southern California in Ernst WG (ed.), Metamorphism and crustal evolution of the western United States [Rubey Volume VII]: Prentice Hall, Englewood Cliffs, p. 895–937.

Toksöz MN, Bird P, 1977, Formation and evolution of marginal basins and continental plateaus, in Talwani M, Pittman WC, III (eds.), Island arcs, deep sea trenches and back-arc basins [Maurice Ewing Series 1]: American Geophysical Union, Washington, p. 379–393.

Toksöz MN, Hsui AT, 1980, Numerical studies of back-arc convection and the formation of marginal basins, in Toksöz MN, Uyeda S, Francheteau J (eds.), Oceanic ridges and arcs: Elsevier, New York, p. 457–476.

Torrini R, Jr, Speed RC, 1989, Tectonic wedging in the forearc basin—accretionary prism transition, Lesser Antilles forearc: Journal of Geophysical Research, v. 94, p. 10549–10584.

Torrini R, Jr, Speed RC, Mattiolli GS, 1985, Tectonic relationships between forearc-basin strata and the accretionary complex at Bath, Barbados: Geological Society of America Bulletin, v. 96, p. 861–874.

Tosdal RM, Haxel G, Wright JE, 1989, Jurassic geology of the Sonoran Desert region, southern Arizona, southeastern California, and northernmost Sonora: construction of a continental margin magmatic arc: Arizona Geological Society Digest 17, p. 397–434.

Tréhu AM, 5 coauthors, 1989, Structure of the lower crust beneath the Carolina trough, U.S. Atlantic continental margin: Journal of Geophysical Research, v. 94, p. 10585–10600.

Treworgy JD, Sargent ML, Kolata DR, 1989, Tectonic subsidence history of the Illinois basin: American Association of Petroleum Geologists Bulletin, v. 73, p. 1040–1041.

Trifonov VG, 1978, Late Quaternary tectonic movements of western and central Asia: Geological Society of America Bulletin, v. 89, p. 1059–1072.

Trincardi F, Zitellini N, 1987, The rifting of the Tyrrhenian Basin: Geo-Marine Letters, v. 7, p. 1–6.

Trümpy R, 1972, Über die Geschwindigkeit der Krustenverkürzung in den Zentralalpen: Geologische Rundschau, v. 61, p. 961–964.

Trümpy R, 1973, The timing of orogenic events in the central Alps, in De Jong KA, Scholten R (eds.), Gravity and tectonics: John Wiley, New York, p. 229–251.

Trümpy R, 1980, Geology of Switzerland. Part A: An outline of the geology of Switzerland: Wepf, Basel, 104 p.

Turcotte DL, 1974, Membrane tectonics: Geophysical Journal of the Royal Astronomical Society, v. 36, p. 33–42.

Turcotte DL, 1979, Flexure [Advances in Geophysics, v. 21]: Academic Press, New York, p. 51–86.

Turcotte DL, Oxburgh ER, 1973, Mid-plate tectonics: Nature, v. 244, p. 337–339.

Turcotte DL, Schubert G, 1982, Geodynamics: applications of continuum mechanics to geological problems: John Wiley, New York, 450 p.

Turner CC, Cohen JM, Connell ER, Cooper DM, 1987, A depositional model for the South Brae oilfield, in Brooks J, Glennie K (eds.), Petroleum geology of northwest Europe: Graham and Trotman, London, p. 853–864.

Uliana MA, Biddle KT, 1987, Permian to late Cenozoic evolution of northern Patagonia: main tectonic events, magmatic activity, and depositional trends: American Geophysical Union Geophysical Monograph 40, p. 271–286.

Uliana MA, Biddle KT, Cerdan J, 1989, Mesozoic extension and the Formation of Argentine sedimentary basins: American Association of Petroleum Geologists Memoir 46, p. 599–614.

Underhill JR, 1991, Implications of Mesozoic-Recent basin development in the western Inner Moray Firth, U.K.: Marine and Petroleum Geology, v. 8, p. 359–369.

Underwood MB, 1984, A sedimentologic perspective on stratal disruption within sandstone-rich melange terranes: Journal of Geology, v. 92, p. 369–385.

Underwood MB, 1985, Sedimentology and hydrocarbon potential of Yager structural complex–possible Paleogene source rocks in Eel River Basin, northern California: American Association of Petroleum Geologists Bulletin, v. 69, p. 1088–1100.

Underwood MB, 1986a, Sediment provenance within subduction complexes–an example from the Aleutian forearc: Sedimentary Geology, v. 51, p. 57–73.

Underwood MB, 1986b, Transverse infilling of the central Aleutian Trench by unconfined turbidity currents: Geo-Marine Letters, v. 6, p. 7–13.

Underwood MB, 1989, Temporal changes in geothermal gradient, Franciscan subduction complex, northern California: Journal of Geophysical Research, v. 94, p. 3111–3125.

Underwood MB, 1991, Submarine canyons, unconfined turbidity currents, and sedimentary bypassing of forearc regions: Critical Reviews in Aquatic Sciences, v. 4, p. 149–200.

Underwood MB, Bachman SB, 1982, Sedimentary facies asso-

ciations within subduction complexes: Geological Society of London Special Publication 10, p. 537–550.

Underwood MB, Bachman SB, 1986, Sandstone petrofacies of the Yager complex and the Franciscan Coastal Belt, Paleogene of northern California: Geological Society of America Bulletin, v. 97, p. 809–817.

Underwood MB, Hathon EG, 1989, Provenance and dispersal of muds south of the Aleutian Arc, North Pacific Ocean: Geo-Marine Letters, v. 8, p. 67–75.

Underwood MB, Howell DG, 1987, Thermal maturity of the Cambria slab, an inferred trench-slope basin in central California: Geology, v. 15, p. 216–219.

Underwood MB, Karig DE, 1980, Role of submarine canyons in trench and trench-slope sedimentation: Geology, v. 8, p. 432–436.

Underwood MB, Norville CR, 1986, Deposition of sand in a trench-slope basin by unconfined turbidity currents: Marine Geology, v. 71, p. 383–392.

Underwood MB, O'Leary JD, Strong RH, 1988, Contrasts in thermal maturity within terranes and across terrane boundaries of the Franciscan Complex, northern California: Journal of Geology, v. 96, p. 399–416.

Underwood MB, Laughland MM, Byrne T, Hibbard JP, DiTullio L, 1992, Thermal evolution of the Tertiary Shimanto belt, Muroto Peninsula, Shikoku, Japan: The Island Arc, v. 1, p. 116–132.

Underwood MB, Orr R, Pickering K, Taira A, 1993, Provenance and dispersal patterns for sediments in the turbidite wedge of Nankai Trough: Proceedings of the Ocean Drilling Program, Scientific Results, v. 131, p. 15–34.

United States Geological Survey, 1987, Recent reverse faulting in the Transverse Ranges, California: U.S. Geological Survey Professional Paper 1339, 203 p.

Unruh JR, Ramirez VR, Phipps SP, Moores EM, 1991, Tectonic wedging beneath fore-arc basins: ancient and modern examples from California and the Lesser Antilles: GSA Today, v. 1, p. 185, 188–190.

Urabe T, Kusakabe M, 1990, Barite silica chimneys from the Sumisu Rift, Izu-Bonin Arc: possible analog to hematitic chert associated with Kuroko deposits: Earth and Planetary Science Letters, v. 100, p. 283–290.

Urabe T, Marumo K, 1991, A new model for Kuroko-type deposits of Japan: Episodes, v. 14, p. 246–251.

Urien CM, Zambrano JJ, 1973, The Geology of the basins of the Argentine continental margin and Malvinas Plateau, in Nairn AEM, Stehli FG (eds.), The ocean basins and margins [v. 1, The South Atlantic]: Plenum Press, New York, p. 135–169.

Uyeda S, 1977, Some basic problems in the trench-arc-back-arc system, in Talwani M, Pitman WC, III (eds.), Island arcs deep sea trenches and back-arc basins [Maurice Ewing Series 1]: American Geophysical Union, Washington, p. 1–14.

Uyeda S, 1982, Subduction zones: an introduction to comparative subductology: Tectonophysics, v. 81, p. 133–159.

Uyeda S, Ben-Avraham Z, 1972, Origin and development of the Philippine Sea: Nature, v. 240, p. 176–178.

Uyeda S, Kanamori H, 1979, Back-arc opening and the mode of subduction: Journal of Geophysical Research, v. 84, p. 1049–1061.

Uyeda S, Miyashiro A, 1974, Plate tectonics and the Japanese Islands: a synthesis: Geological Society of America Bulletin, v. 85, p. 1159–1170.

Vail PR, 7 coauthors, 1977, Seismic stratigraphy and global changes of sea level: American Association of Petroleum Geologists Memoir 26, p. 49–212.

Valloni R, 1985, Reading provenance from modern marine sands, in Zuffa GG (ed.), Provenance of arenites: Reidel Publishing, Dordrecht, p. 309–332.

VanAndel TH, Komar PD, 1969, Ponded sediments of the Mid-Atlantic Ridge between 22 and 23 North latitude: Geological Society of America Bulletin, v. 80, p. 1163–1190.

VanBemmelen RW, 1933, The undation theory of the development of the Earth's crust [Proceedings, v. 2]: 16th International Geological Congress, Washington, p. 965–982.

VanBemmelen RW, 1970, The geology of Indonesia: Martinus Nijoff, The Hague, 732 p.

vanderLingen GJ, 1982, Development of the North Island subduction system, New Zealand: Geological Society of London Special Publication 10, p. 259–274.

vanderLingen GJ, Pettinga JG, 1980, The Makara basin: a Miocene slope-basin along the New Zealand sector of the Australian-Pacific obliquely convergent plate boundary: International Association of Sedimentologists Special Publication 4, p. 191–215.

VanDerVoo R, Watts DR, 1978, Paleomagentic results from igneous and sedimentary rocks from the Michigan basin borehole: Journal of Geophysical Research, v. 83, p. 5844–5848.

VanHouten FB, 1977, Triassic-Liassic deposits of Morocco and east North America; a comparison: American Association of Petroleum Geologists Bulletin, v. 61, p. 79–99.

VanHouten FB, 1981, The odyssey of molasse: Geological Association of Canada Special Paper 23, p. 35–48.

VanHouten FB, 1983, Sirte basin, north-central Libya: Cretaceous rifting above a fixed mantle hotspot?: Geology, v. 11, p. 115–118.

VanSchmus WR, Hinze WJ, 1985, The midcontinent rift system: Annual Review of Earth and Planetary Sciences, v. 13, p. 345–383.

VanWagoner JC, Mitchum RM, Campion KM, Rahmanian VD, 1990, Siliciclastic sequence stratigraphy in well logs, cores, and outcrops: concepts for high-resolution correlation of time and facies: American Association of Petroleum Geologists Methods in Exploration Series 7, 55 p.

vanWeering TCE, 5 coauthors, 1989, The seismic structure of the Lombok and Savu forearc basins, Indonesia: Netherlands Journal of Sea Research, v. 24, p. 251–262.

Vance JA, Clayton GA, Mattinson JM, Naeser CW, 1987, Early and middle Cenzoic stratigraphy of the Mount Rainier-Tieton River area, southern Washington Cascades: Washington Division of Geology and Earth Resources Bulletin 77, p. 269–290.

Vedder JG, Bruns TR, Cooper AK, 1989, Geologic framework of Queen Emma basin, eastern Papua New Guinea: Circum-Pacific Council for Energy and Mineral Resources Earth Science Series, v. 12, p. 59–86.

Veevers JJ, 1981, Morphotectonics of rifted continental margins in embryo (East Africa), youth (Africa-Arabia), and maturity (Australia): Journal of Geology, v. 89, p. 57–82.

Veevers JJ (ed.), 1984, Phanerozoic Earth history of Australia [Oxford Geological Sciences Series 2]: Oxford University Press, New York, 418 p.

Veevers JJ, Falvey DA, Robins S, 1978, Timor Trough and

Australia: facies show topographic wave migrated 80 km during the past 3.-m.y.: Tectonophysics, v. 45, p. 217–227.

Veizer J, Jansen SL, 1979, Basement and sedimentary recycling and continental evolution: Journal of Geology, v. 87, p. 341–370.

Veizer J, Jansen SL, 1985, Basement and sedimentary recycling-2: time dimension to global tectonics: Journal of Geology, v. 93, p. 625–643.

Velbel MA, 1985, Mineralogically mature sandstones in accretionary prisms: Journal of Sedimentary Petrology, v. 55, p. 685–690.

Veltheim A, 1969, On the pre-Quaternary geology of the Bothnian Bay area in the Baltic Sea: Geological Survey of Finland Bulletin 239, 56 p.

Verplanck EP, Duncan RA, 1987, Temporal variations in plate convergence and eruption rates in the western Cascades, Oregon: Tectonics, v. 6, p. 197–209.

Vessel RK, Davies DK, 1981, Nonmarine sedimentation in an active fore arc basin: Society of Economic Paleontologists and Mineralogists Special Publication 31, p. 31–45.

Viallon C, Huchon P, Barrier E, 1986, Opening of the Okinawa basin and collision in Taiwan: a retreating trench model with lateral anchoring: Earth and Planetary Science Letters, v. 80, p. 145–155.

Viele GW, 1989, The Ouachita orogenic belt, in Hatcher RD, Jr, Thomas WA, Viele GW (eds.), The Appalachian-Ouachita orogen in the United States [The Geology of North America, v. F-2]: Geological Society of America, Boulder, p. 555–561.

Viele GW, Thomas WA, 1989, Tectonic synthesis of the Ouachita orogenic belt, in Hatcher RD, Jr, Thomas WA, Viele GW (eds.), The Appalachian-Ouachita orogen in the United States [The Geology of North America, v. F-2]: Geological Society of America, Boulder, p. 695–728.

Vietti JS, 1977, Structural geology of the Ryckman Creek anticline area, Lincoln and Uinta counties, Wyoming: Wyoming Geological Association Guidebook, 29th Annual Field Conference, p. 517–522.

Villemin T, Bergerat F, 1985, Tectonique cassante et paléocontraintes tertiaires de la bordure NE du Fossé Rhénan (R.F.A.): Oberrheinische Geologische Abhandlungen, v. 34, p. 63–87.

Villemin T, Bergerat F, 1987, L'évolution structurale du fossé rhénan au cours du Cénozoique: un bilan de la déformation et des effets thermiques de l'extension: Bulletin de la Société Géologique de France série 8, v. 3, p. 245-255.

Vink GE, Morgan WJ, Zhao W-L, 1984, Preferential rifting of continents: a source of displaced terranes: Journal of Geophysical Research, v. 89, p. 10072–10076.

Vogt PR, 1981, On the applicability of thermal conduction models to mid-plate volcanism: comments on a paper by Gass et al.: Journal of Geophysical Research, v. 86, p. 950–960.

Vogt PR, Avery DE, 1974, Tectonic history of the Arctic basins: partial solutions and unsolved mysteries, in Herman Y (ed.), Marine geology and oceanography of the Arctic seas: Springer Verlag, Berlin, p. 38–117.

vonderBorch CC, 1979, Continent-island arc collision in the Banda arc: Tectonophysics, v. 54, p. 169–193.

vonGärtner HR, 1969, Zur tektonischen und magmatischen Entwicklung der Kratone (Pyrenäen and Kaukasus als extreme Fälle der Aulacogene): Geologisches Jahrbuch, Beihefte, v. 80, p. 117–145.

vonHuene R, 1974, Modern trench sediments, in Burk CA, Drake CL (eds.), The geology of continental margins: Springer-Verlag, New York, p. 207–211.

vonHuene R, 1979, Structure of the outer convergent margin off Kodiak Island, Alaska, from multichannel seismic records: American Association of Petroleum Geologists Memoir 29, p. 261–272.

vonHuene R, 1984, Tectonic processes along the front of modern convergent margins–research of the past decade: Annual Reviews of Earth and Planetary Science, v. 12, p. 359–381.

vonHuene R, Arthur MA, 1982, Sedimentation across the Japan Trench off northern Honshu Island: Geological Society of London Special Publication 10, p. 27–48.

vonHuene R, Culotta R, 1989, Tectonic erosion at the front of the Japan Trench convergent margin: Tectonophysics, v. 160, p. 75–90.

vonHuene R, Kulm LD, 1973, Tectonic summary of Leg 18: Initial Reports of the Deep Sea Drilling Project, v. 18, p. 961–976.

vonHuene R, Lallemand S, 1990, Tectonic erosion along the Japan and Peru convergent margins: Geological Society of America Bulletin, v. 102, p. 704–720.

vonHuene R, Miller J, 1988, Migrated multichannel seismic-reflection records across the Peru continental margin: Proceedings of the Ocean Drilling Program, Initial Reports, v. 112, p. 109–119.

vonHuene R, Scholl DW, 1991, Observations at convergent margins concerning sediment subduction, subduction erosion, and the growth of continental crust: Reviews of Geophysics, v. 29, p. 279–316.

vonHuene R, 17 coauthors, 1980, Leg 67: the Deep Sea Drilling Project Mid-America Trench transect off Guatemala: Geological Society of America Bulletin, v. 91, p. 421–432.

vonHuene R, Langseth M, Nasu N, Okada H, 1982, A summary of Cenozoic tectonic history along the IPOD Japan Trench transect: Geological Society of America Bulletin, v. 93, p. 829–846.

vonHuene R, Kulm LD, Miller J, 1985a, Structure of the frontal part of the Andean convergent margin: Journal of Geophysical Research, v. 90, p. 5429–5442.

vonHuene R, 13 coauthors, 1985b, Initial Reports of the Deep Sea Drilling Project [v. 84]: United States Government Printing Office, Washington, 967 p.

vonHuene R, Fisher MA, Bruns TR, 1987, Geology and evolution of the Kodiak margin, Gulf of Alaska: Circum-Pacific Council for Energy and Mineral Resources Earth Science Series, v. 6, p. 191–212.

vonHuene R, Suess E, Leg 112 Shipboard Scientists, 1988, Ocean Drilling Program Leg 112, Peru continental margin: part 1, tectonic history: Geology, v. 16, p. 934–938.

vonSeidlitz W, 1931, Diskordanz und Orogenese der Gebirge am Mittelmeer: Borntraeger, Berlin, 651 p.

Vrolijk P, Fisher A, Gieskes J, 1991, Geochemical and geothermal evidence for fluid migration in the Barbados accretionary prism (ODP Leg 110): Geophysical Research Letters, v. 18, p. 947–950.

Wadge G, 1986, The dykes and structural setting of the volcanic front in the Lesser Antilles island arc: Bulletin of Volcanology, v. 48, p. 349–372.

Walcott RE-1972, Gravity, flexure, and the growth of sedi-

mentary basins at a continental edge: Geological Society of America Bulletin, v. 83, p. 1845–1848.

Walker GPL, 1984, Downsag calderas, ring faults, caldera sizes, and incremental caldera growth: Journal of Geophysical Research, v. 89, p. 8407–8416.

Walker JA, Moulds TN, Zentilli M, Feigenson MD, 1992, Spatial and temporal variations in volcanics of the Andean central volcanic zone (26 to 28oS): Geological Society of America Special Paper 265, p. 139–158.

Walker RG (ed.), 1984, Facies models (2nd edition): Geoscience Canada Reprint Series 1, 317 p.

Walker RG, 1990, Facies modeling and sequence stratigraphy: Journal of Sedimentary Petrology, v. 60, p. 777–786.

Walker RG, Eyles CH, 1991, Topography and significance of a basinwide sequence-bounding erosion surface in the Cretaceous Cardium Formation, Alberta, Canada: Journal of Sedimentary Petrology, v. 61, p. 473–496.

Wallace RE, 1978, Geometry and rates of change of fault-generated range fronts, north-central Nevada: United States Geological Survey Journal of Research, v. 6, p. 637–650.

Wallace RE, 1984, Faulting related to the 1915 earthquakes in Pleasant Valley, Nevada: United States Geological Survey Professional Paper 1274-A, 33 p.

Wallace RE (ed.), 1990, The San Andreas fault system: United States Geological Survey Professional Paper 1515, 283 p.

Walsh J, Watterson J, 1988, Analysis of the relationship between displacements and dimensions of faults: Journal of Structural Geology, v. 10, p. 239–247.

Walton AW, Palmer BA, 1988, Lahar facies of the Mount Dutton Formation (Oligocene-Miocene) in the Marysvale volcanic field, southwestern Utah: Geological Society of America Bulletin, v. 100, p. 1078–1091.

Wang C, Shi Y, 1984, On the thermal structure of subduction complexes: a preliminary study: Journal of Geophysical Research, v. 89, p. 7709–7718.

Wang HZ (chief compiler) 1985, Atlas of the paleogeography of China: Cartographic Publishing House, Beijing, 85 p.

Wang XM, Liou JG, 1987, The large displacement of the Tanlu fault: evidence from the distribution of coesite-bearing eclogite belt in eastern China: EOS, v. 70, p. 1312–1313.

Wang XM, Liou JG, 1991, Regional ultrahigh-pressure coesite-bearing eclogite terrane in central China: evidence from country rocks, gneiss, marble, and metapelite: Geology, v. 19, p. 933–936.

Ward P, Stanley KO, 1982, The Haslam Formation: a late Santonian-early Campanian forearc basin deposit in the insular belt of southwestern British Columbia and adjacent Washington: Journal of Sedimentary Petrology, v. 52, p. 975–990.

Warner MR, 1990, Basalts, water and shear zones in the lower continental crust?: Tectonophysics, v. 173, p. 163–174.

Warsi WEK, Hilde TWC, Searle RC, 1983, Convergence structures of the Peru Trench between 10oS and 14oS: Tectonophysics, v. 99, p. 313–329.

Waschbusch PJ, Royden LH, 1992a, Episodicity in foredeep basins: Geology, v. 20, p. 915–918.

Waschbusch PJ, Royden LH, 1992b, Spatial and temporal evolution of foredeep basins: lateral strength variations and inelastic yielding in continental lithosphere: Basin Research, v. 4, p. 179–196.

Watanabe T, Langseth MG, Anderson RN, 1977, Heat flow in back-arc basins of the western Pacific, in Talwani M, Pitman WC, III (eds.), Island arcs deep sea trenches and back-arc basins [Maurice Ewing Series 1]: American Geophysical Union, Washington, p. 137–157.

Watkins JS, Drake CL (eds.), Studies in continental margin geology: American Association of Petroleum Geologists Memoir 34, 801 p.

Watkins JS, Montadert L, Dickerson PW (eds.), 1979, Geological and geophysical investigations of continental margins: American Association of Petroleum Geologists Memoir 29, 472 p.

Watkins JS, 15 coauthors, 1982, Initial Reports of the Deep Sea Drilling Project, v. 66, 864 p.

Watkins RW, 1993, Permian carbonate platform development in an island-arc setting, eastern Klamath terrane, California: Journal of Geology, v. 101, p. 659–666.

Watkins RW, Flory RA, 1986, Island arc sedimentation in the Middle Devonian Kennett Formation, eastern Klamath Mountains, California: Journal of Geology, v. 94, p. 753–761.

Watso DC, Klein GD, 1989, Origin of the Cambrian-Ordovician sedimentary cycles of Wisconsin using tectonic subsidence analysis: Geology, v. 17, p. 879–881.

Watson MP, Hayward AB, Parkinson DN, Zhang ZM, 1987, Plate tectonic history, basin development and petroleum source rock deposition onshore China: Marine and Petroleum Geology, v. 4, p. 205–225.

Watterson J, 1986, Fault dimensions, displacements and growth: Pure and Applied Geophysics, v. 124, p. 365–373.

Watts AB, 1978, An analysis of isostasy in the worlds oceans, part 1, Hawaiian-Emperor seamount chain: Journal of Geophysical Research, v. 83, p. 5989–6004.

Watts AB, 1981, The U.S. Atlantic continental margin subsidence history, crustal structure and thermal evolution: American Association of Petroleum Geologists Education Course Notes 19, 75 p.

Watts AB, 1982, Tectonic subsidence, flexure, and global changes of sea level: Nature, v. 297, p. 469–474.

Watts AB, 1988, Gravity anomalies, crustal structure and flexure of the lithosphere at the Baltimore Canyon trough: Earth and Planetary Science Letters, v. 89, p. 221–238.

Watts AB, 1992, The effective elastic thickness of the lithosphere and the evolution of foreland basins: Basin Research, v. 4, p. 169–178.

Watts AB, Thorne J, 1984, Tectonics, global changes in sea level and their relationship to stratigraphical sequences at the U.S. Atlantic continental margin: Marine and Petroleum Geology, v. 1., p. 319–339.

Watts AB, Bodine JH, Ribe NM, 1980, Observations of flexure and the geological evolution of the Pacific Ocean basin: Nature, v. 283, p. 532–537.

Watts AB, Karner GD, Steckler MS, 1982, Lithospheric flexure and the evolution of sedimentary basins: Philosophical Transactions of the Royal Society of London A305, p. 249–281.

Wdowinski S, O'Connell RJ, England P, 1989, A continuum model of continental deformation above subduction zones: application to the Andes and the Aegean: Journal of Geophysical Research, v. 94, p. 10331–10346.

Webb GW, 1981, Stevens and earlier Miocene sandstones, southern San Joaquin Valley, California: American Association of Petroleum Geologists Bulletin, v. 65, p. 438–465.

Weber M, 1921, Zum Problem der Grabenbildung: Zeitschrift der Deutschen Geologische Gesellschaft, v. 73, p. 238–291.

Weber M, 1923, Bemerkungen zur Bruchtektonik: Zeitschrift der Deutschen Geologische Gesellschaft, v. 75, p. 184–192.

Weber M, 1927, Faltengebirge und Vorlandbrüche: Centralblatt für Mineralogie, Geologie und Paläeontologie, v. 5, p. 235–245.

Weimer RJ, 1960, Upper Cretaceous stratigraphy, Rocky Mountain area: American Association of Petroleum Geologists Bulletin, v. 44, p. 1–20.

Weimer RJ, 1970, Rates of deltaic sedimentation and intrabasin deformation, Upper Cretaceous of Rocky Mountain region: Society of Economic Paleontologists and Mineralogists Special Paper 15, p. 270–292.

Weimer RJ, 1986, Relationship of unconformities, tectonics, and sea level changes in the Cretaceous of the Western Interior, United States: American Association of Petroleum Geologists Memoir 41, p. 397–422.

Weinberg RF, 1992, Neotectonic development of western Nicaragua: Tectonics, v. 11, p. 1010–1017.

Weissel JK, 1977, Evolution of the Lau Basin by the growth of small plates, in Talwani M, Pitman WC, III (eds.), Island arcs deep sea trenches and back-arc basins: American Geophysical Union, Washington, p. 429–436.

Weissel JK, Karner GD, 1989, Flexural uplift of rift flanks due to mechanical unloading of the lithosphere during extension: Journal of Geophysical Research, v. 94, p. 13919–13950.

Wernicke BP, 1981, Low angle normal faults in the Basin and Range Province: nappe tectonics in an extending orogen: Nature, v. 192, p. 645–648.

Wernicke BP, 1985, Uniform-sense normal simple shear of the continental lithosphere: Canadian Journal of Earth Sciences, v. 22, p. 108–125.

Wernicke BP, 1991, The fluid crustal layer and its implications for continental dynamics, in Salisbury MH, Fountain DM (eds.), Exposed cross-sections of the continental crust: Kluwer, Dordrecht, p. 509–544.

Wernicke BP, 1992, Cenozoic extensional tectonics of the U.S. Cordillera, in Burchfiel BC, Lipman PW, Zoback ML (eds.), The Cordilleran orogen: conterminus U.S. [The Geology of North America, v. G-3]: Geological Society of America, Boulder, p. 553–581.

Wernicke BP, Axen GJ, 1988, On the role of isostasy in the evolution of normal fault systems: Geology, v. 16, p. 848–851.

Wernicke BP, Burchfiel BC, 1982, Modes of extensional tectonics: Journal of Structural Geology, v. 4, p. 105–115.

Wernicke BP, Tilke PG, 1989, Extensional tectonics framework of the U.S. central Atlantic passive margin: American Association of Petroleum Geologists Memoir 46, p. 7–21.

Wernicke B, Axen GJ, Snow JK, 1988, Basin and Range extensional tectonics at the latitude of Las Vegas, Nevada: Geological Society of America Bulletin, v. 100, p. 1738–1757.

Westbrook GK, 1975, The structure of the crust and upper mantle in the region of Barbados and the Lesser Antilles: Geophysical Journal of the Royal Astronomical Society, v. 43, p. 201–242.

Westbrook GK, 1982, The Barbados Ridge complex: tectonics of a mature forearc system: Geological Society of London Special Publication 10, p. 275–290.

Westbrook GK, Smith MJ, 1983, Long decollements and mud volcanoes: evidence from the Barbados Ridge complex for the role of high porewater pressures in the development of an accretionary complex: Geology, v. 11, p. 279–283.

Westbrook GK, Smith MJ, Peacock JH, Poulter JMJ, 1982, Extensive underthrusting of undeformed sediment beneath the accretionary complex of the Lesser Antilles subduction zone: Nature, v. 300, p. 625–628.

Westbrook GK, Ladd JW, Buhl P, Bangs N, Tiley GJ, 1988, Cross section of an accretionary wedge: Barbados Ridge complex: Geology, v. 16, p. 631–635.

Westerveld J, 1953, Eruptions of acid pumice tuffs and related phenomena along the great Sumatran fault-trough system: Proceedings of the Pacific Science Congress, v. 8, p. 411–438.

Westerveld J, 1963, The tectonic causes of ignimbrite and pumice tuff deposition and of subsequent basalt-andesitic volcanism: Bulletin Volcanologique, v. 125, p. 67–88.

Weyl R, 1980, Geology of Central America: Gebruder Borntraeger, Berlin, 371 p.

Whalen PA, Pessagno EA, Jr, 1984, Lower Jurassic radiolaria, San Hipolito Formation, Vizcaino Peninsula, Baja California Sur, in Frizzell VA, Jr (ed.), Geology of the Baja California Peninsula [Book 39]: Pacific Section, Society of Economic Paleontologists and Mineralogists, Los Angeles, p. 53–65.

Wheeler HE, 1958, Time-stratigraphy: American Association of Petroleum Geologists Bulletin, v. 42, p. 1047–1063.

Wheeler RL, 1989, Persistent segment boundaries on basin-range normal faults: United States Geological Survey Open File Report 89–315, p. 432–444.

Whitaker ST, 1988, Silurian pinnacle reef distribution in Illinois: model for hydrocarbon exploration: Illinois State Geological Survey Petroleum Geology Series 130, 32 p.

White JDL, Busby-Spera CJ, 1987, Deep-marine arc apron deposits and syndepositional magmatism in the Alisitos Group at Punta Cono, Baja California, Mexico: Sedimentology, v. 34, p. 911–927.

White N, 1991, Does the uniform stretching model work in the North Sea?, in Blundell DJ, Gibbs AD (eds.), Tectonic evolution of the North Sea rifts: Clarendon Press, Oxford, p. 217–240.

White N, McKenzie D, 1988, Formation of the "steer's head" geometry of sedimentary basins by differential stretching of the crust and mantle: Geology, v. 16, p. 250–253.

White RS, Klitgord KD, 1976, Sediment deformation and plate tectonics in the Gulf of Oman: Earth and Planetary Science Letters, v. 32, p. 199–209.

White RS, Louden KE, 1982, The Makran continental margin: structure of a thickly sedimented convergent plate boundary: American Association of Petroleum Geologists Memoir 34, p. 499–518.

White RS, McKenzie DP, 1989, Magmatism at rift zones: the generation of volcanic continental margins and flood basalts: Journal of Geophysical Research, v. 94, p. 7685–7729.

White RS, 5 coauthors, 1987, Magmatism at rifted continental margins: Nature, v. 330, 439–444.

White SM, 5 coauthors, 1980, Sediment synthesis: Deep Sea Drilling Project Leg 58, Philippine Sea: Initial Reports of the Deep Sea Drilling Project, v. 58, p. 963–1000.

White TD, Harris JM, 1977, Suid evolution and correlation of African hominid localities: Science, v. 198, p. 13–21.

Whiteman A, 1982, Nigeria: its petroleum geology, resources

and potential, v. 1: Graham and Trotman, London, 166 p.

Whitney RA, 1980, Structural-tectonic analysis of northern Dixie Valley, Nevada [MS thesis]: University of Nevada, Reno, 65 p.

Whittaker A, Bott MHP, Waghorn GD, 1992, Stresses and plate boundary forces associated with subduction plate margins: Journal of Geophysical Research, v. 97, p. 11933–11944.

Wickham J, Pruatt M, Reiter L, Thomson T, 1975, The southern Oklahoma aulacogen: Geological Society of America Abstracts with Programs, v. 7, p. 1332.

Wickham J, Roeder DR, Briggs G, 1976, Plate tectonic models for the Ouachita foldbelt: Geology, v. 4, p. 173–176.

Wilcox RE, Harding TP, Seely DR, 1973, Basic wrench tectonics: American Association of Petroleum Geologists Bulletin, v. 57, p. 74–96.

Wilgus CK, 5 coeditors (eds.), 1988, Sea level changes—an integrated approach: Society of Economic Paleontologists and Mineralogists Special Publication 42, 407 p.

Willard DA, Klein GD, 1990, Tectonic subsidence history of the central Appalachian basin and its influence on Pennsylvanian coal deposition: Southeastern Geology, v. 30, p. 217–239.

Willcox JB, Stagg HMJ, 1990, Australia's southern margin: a product of oblique extension: Tectonophysics, v. 173, p. 269–281.

Williams DM, O'Connor PD, Menuge J, 1992, Silurian turbidite provenance and the closure of Iapetus: Journal of the Geological Society of London, v. 149, p. 349–357.

Williams GD, Stelck CR, 1975, Speculations on the Cretaceous palaeogeography of North America: Geological Association of Canada Special Paper 13, p. 1–20.

Williams H, 1941, Calderas and their origin: University of California Publications in the Geological Sciences Bulletin 21, p. 51–146.

Williams H, 1960, Volcanic history of the Guatemalan Highlands: University of California Publications in the Geological Sciences, v. 38, p. 1–38.

Williams H, McBirney AR, 1979, Volcanology: Freeman, Cooper, San Francisco, 397 p.

Williams MAJ, Assefa G, Adamson DA, 1986, Depositional context of Plio-Pleistocene hominid-bearing formations in the middle Awash valley, southern Afar rift, Ethiopia: Geological Society of London Special Publication 25, p. 241–251.

Williamson PG, Savage RJG, 1986, Early rift sedimentation in the Turkana basin, northern Kenya: Geological Society of London Special Publication 25, p. 267–283.

Wilson CJN, 5 coauthors, 1984, Caldera volcanoes of the Taupo volcanic zone, New Zealand: Journal of Geophysical Research, v. 89, p. 8463–8484.

Wilson G, 1960, The tectonics of the 'great ice chasm', Filchner Ice Shelf, Antarctica: Proceedings of the Geologists' Association, v. 71, part 2, p. 130–138.

Wilson JT, 1954, The development and structure of the crust, in Kuiper G (ed.), The earth as a planet: University of Chicago Press, Chicago, p. 138–214.

Wilson JT, 1968, Static or mobile earth: the current scientific revolution: Proceedings of the American Philosophical Society, v. 112, p. 309–320.

Wilson M, 1989, Igneous petrogenesis: Unwin Hyman, London, 466 p.

Wilson RCL, 1975, Atlantic opening and Mesozoic continental

margin basins of Iberia: Earth and Planetary Science Letters, v. 94, p. 316–328.

Wiltschko DV, Dorr JA, Jr, 1983, Timing of deformation in overthrust belt and foreland of Idaho, Wyoming and Utah: American Association of Petroleum Geologists Bulletin, v. 67, p. 1304–1322.

Wimmenauer W, 1974, The alkaline province of central Europe and France, in Sfrensen H (ed.), The alkaline rocks: John Wiley, Hertfordshire, p. 238–271.

Windley BF, 1993, Uniformitarianism today: plate tectonics is the key to the past: Journal of the Geological Society of London, v. 150, p. 7–19.

Winker CD, Edwards MB, 1983, Nonstable progradational clastic shelf margins: Society of Economic Paleontologists and Mineralogists Special Publication 33, p. 139–157.

Winker CD, Kidwell SM, 1986, Paleocurrent evidence for lateral displacement of the Pliocene Colorado River delta by the San Andreas fault system, southeastern California: Geology, v. 14, p. 788–791.

Winkler W, 1984, Palaeocurrents and petrography of the Gurnigel-Schlieren Flysch: a basin analysis: Sedimentary Geology, v. 40, p. 169–189.

Winn RD, Jr, 1978, Upper Mesozoic flysch of Tierra del Fuego and South Georgia Island: a sedimentologic approach to lithosphere plate restoration: Geological Society of America Bulletin, v. 89, p. 533–547.

Winn RD, Jr, Dott RH, Jr, 1978, Submarine-fan turbidites and resedimented conglomerates in a Mesozoic arc-rear marginal basin in southern South America, in Stanley DJ, Kelling G (eds.), Sedimentation in submarine canyons, fans, and trenches: Dowden, Hutchinson and Ross, Stroudsburg, p. 362–373.

Winn RD, Jr, Stonecipher SA, Bishop MG, 1984, Sorting and wave abrasion: controls on composition and diagenesis in lower Frontier sandstones, southwestern Wyoming: American Association of Petroleum Geologists Bulletin, v. 68, p. 268–284.

Winston D, 1986, Sedimentation and tectonics of the Middle Proterozoic Belt basin and their influence on Phanerozoic compression and extension in western Montana and northern Idaho: American Association of Petroleum Geologists Memoir 41, p. 87–118.

Winterer EL, 1973, Sedimentary facies and plate tectonics of equatorial Pacific: American Association of Petroleum Geologists Bulletin, v. 57, p. 265–282.

Withjack M, 1979, An analytical model of continental rift fault patterns: Tectonophysics, v. 59, p. 59–81.

Witte WK, Kent DV, Olsen PE, 1991, Magnetostratigraphy and paleomagnetic poles from Late Triassic—earliest Jurassic strata of the Newark basin: Geological Society of America Bulletin, v. 103, p. 1648–1662.

Wohletz KH, Heiken G, 1992, Volcanology and geothermal energy: University of California Press, Berkeley, 432 p.

Woodburne MO, Tedford RH, Swisher CC, III, 1990, Lithostratigraphy, biostratigraphy, and geochronology of the Barstow Formation, Mojave Desert, southern California: Geological Society of America Bulletin, v. 102, p. 459–477.

Woodcock NH, 1986, The role of strike-slip fault systems at plate boundaries: Philosophical Transactions of the Royal Society of London, v. A317, p. 13–29.

Woodcock NH, Fisher M, 1986, Strike-slip duplexes: Journal

of Structural Geology, v. 8, p. 725–735.

Worrall DM, 1991, Tectonic history of the Bering Sea and the evolution of Tertiary strike-slip basins of the Bering Shelf: Geological Society of America Special Paper 257, 120 p.

Worrall DM, Snelson S, 1989, Evolution of the northern Gulf of Mexico, with emphasis on Cenozoic growth faulting and the role of salt, in Bally AW, Palmer AR (eds.), The geology of North America–an overview [The Geology of North America, v. A]: Geological Society of America, Boulder, p. 91–138.

Worsley TB, Nance D, Moody JB, 1984, Global tectonics and eustasy for the past 2 billion years: Marine Geology, v. 58, p. 373–400.

Worsley TB, Nance RD, Moody JB, 1986, Tectonic cycles and the history of the earth's biogeochemical and paleoceanographic record: Paleoceoanography, v. 1, p. 233–263.

Wright A, 1984, Sediment distribution and depositional processes operating in the Lesser Antilles intraoceanic arc, eastern Caribbean: Initial Reports of the Deep Sea Drilling Project, v. 78, p. 301–324.

Wright TL, 1987, Geologic summary of the Los Angeles basin, in Wright TL, Heck R (eds.), Petroleum geology of coastal southern California [Guidebook 60]: Pacific Section, American Association of Petroleum Geologists, Bakersfield, p. 21–31.

Wright TL, 1991, Structural geology and tectonic evolution of the Los Angeles basin, California: American Association of Petroleum Geologists Memoir 52, p. 35–134.

Wuellner DE, Lehtonen LR, James WC, 1986, Sedimentary tectonic development of the Marathon and Val Verde basins, west Texas, U.S.A.: a Permo-Carboniferous migrating foredeep: International Association of Sedimentologists Special Publication 8, p. 347–368.

Yamagishi H, 1991, Morphological and sedimentological characteristics of the Neogene submarine coherent lavas and hyaloclastites in southwest Hokkaido, Japan: Sedimentary Geology, v. 74, p. 5–23.

Yamaji A, 1990, Rapid intra-arc rifting in Miocene northeast Japan: Tectonics, v. 9, p. 365–378.

Yamano M, Uyeda S, 1988, Heat flow, in Nairn AEM, Stehli FG, Uyeda S (eds.), The ocean basins and margins [v. 7B]: Plenum Publishing, New York, p. 523–557.

Yamano M, Uyeda S, Aoki Y, Shipley TH, 1982, Estimates of heat flow derived from gas hydrates: Geology, v. 10, p. 339–343.

Yamano M, Honda S, Uyeda S, 1984, Nankai Trough: a hot trench?: Marine Geophysical Researches, v. 6, p. 187–203.

Yamazaki T, 1988, Heat flow in the Sumisu Rift, Izu-Ogasawara arc: Bulletin of the Geological Survey of Japan, v. 39, p. 63–70.

Yamazaki T, Hayashi M, 1976, Geologic background of Otake and other geothermal areas in north-central Kyushu, southwestern Japan: Proceedings [v. 1], Second United Nations Symposium on the Development and Use of Geothermal Resources, Washington, p. 673–684.

Yamazaki T, Okamura Y, 1989, Subducting seamounts and deformation of overriding forearc wedges around Japan: Tectonophysics, v. 160, p. 207–229.

Yamazaki T, Ishihara T, Murakami F, 1991, Magnetic anomalies over the Izu-Ogasawara (Bonin) Arc, Mariana Arc and Mariana Trough: Bulletin of the Geological Survey of Japan, v. 42, p. 655–686.

Yang Z, Cheng Y, Wang H, 1986, The geology of China: Oxford Monographs on Geology and Geophysics 3, 303 p.

Yarnold JC, 1994, Tertiary sedimentary rocks associated with the Harcuvar core complex in Arizona (U.S.A.): insights into paleogeographic evolution during displacement along a major detachment fault system: Sedimentary Geology, v. 89, p. 43–63.

Yarnold JC, Lombard JP, 1989, A facies model for large rock-avalanche deposits formed in dry climates, in Colburn IP, Abbott PL, Minch J (eds.), Conglomerates in basin analysis: a symposium dedicated to A.O. Woodford [Book 62]: Pacific Section, Society of Economic Paleontologists and Mineralogists, Los Angeles, p. 9–31.

Yeats RS, 1973, Newport-Inglewood fault zone, Los Angeles basin, California: American Association of Petroleum Geologists Bulletin, v. 57, p. 117–135.

Yeo RK, 1984, Sedimentology of Upper Cretaceous strata, northern Baja California, Mexico, in Abbott PL (ed.), Upper Cretaceous depositional systems in southern California-northern Baja California [Book 36]: Pacific Section, Society of Economic Paleontologists and Mineralogists, Los Angeles, p. 109–120.

Yılmaz Y, 1985, Türkiye'nin jeolojik tarihinde magmatik etkinlik ve tektonik evrimle ilişkisi [Ketin Simpozyumu Kitabı]: Türkiye Jeoloji Kurumu, Ankara, p. 63–81.

Yılmaz Y, 1990, Comparison of young volcanic associations of western and eastern Anatolia formed under a compressional regime: a review: Journal of Volcanology and Geothermal Research, v. 44, p. 69–87.

Yin A, Ingersoll RV, 1993, Tectonic development of Laramide thrusts and basins in southern U.S. Rocky Mountains: Geological Society of America Abstracts with Programs, v. 25, n. 5, p. 167.

Yin A, Nie S, 1993, An indentation model for the North and South China collision and the development of the Tan-Lu and Honam fault systems, eastern Asia: Tectonics, v. 12, p. 801–813.

Yin A, Paylor ED, II, Norris A, 1992, Analysis of Laramide crustal strain distribution using relative-slip circuits: Geological Society of America Abstracts with Programs, v. 24, n. 6, p. 69.

Yingling V, Heller PL, 1992, Timing and record of foreland sedimentation during the initiation of the Sevier orogenic belt in central Utah: Basin Research, v. 4, p. 279–290.

Zalan PV, 7 coauthors, 1985, Stilos estruturias relacionados a intrusoes magmaticas basicas em rochas sedimentares: PETRO-BRAS Technical Bulletin, v. 28, p. 221–230.

Zalan PV, 8 coauthors, 1990, The Parana basin, Brazil: American Association of Petroleum Geologists Memoir 51, p. 681–708.

Zartman RE, Brock MR, Heyl AV, Thomas HH, 1967, K-Ar and Rb-Sr ages of some alkalic intrusive rocks from central and eastern United States: American Journal of Science, v. 265, p. 858–870.

Zehnder CM, Mutter JC, Buhl P, 1990, Deep seismic and geochemical constraints on the nature of rift-induced magmatism during breakup of the North Atlantic: Tectonophysics, v. 173, p. 545–565.

Zeil W, 1979, The Andes, a geological review: Gebruder Borntraeger, Berlin, 260 p.

Zhang C, 5 coauthors, 1989, Tectonic systems and formation and evolution of the Qinghai-Xizang (Tibet) Plateau: Ministry of Geology and Mineral Resources, Geologic Memoir Series 5,

n. 8, 161 p.

Zhang YK, 1993, The thermal blanketing effect of sediments on the rate and amount of subsidence in sedimentary basins formed by extension: Tectonophysics, v. 218, p. 297–308.

Zhang ZM, Liou JG, Coleman RG, 1984, An outline of the plate tectonics of China: Geological Society of America Bulletin, v. 95, p. 295–312.

Zhang ZM, Liou JG, Coleman RG, 1985, An outline of the plate tectonics of China: reply: Geological Society of America Bulletin, v. 96, p. 408.

Zhao X, Coe R, 1987, Palaeomagnetic constraints on the collision and rotation of North and South China: Nature, v. 327, p. 141–144.

Zhijin Z, 1984, Lower Palaeozoic volcanism of northern Qilianshan, NW China: Geological Society of London Special Publication 16, p. 285–289.

Zhou D, 1987, Triassic deep sea sedimentation in west Qinling Mountians, and its plate tectonic setting [MS thesis]: Chinese Academy of Sciences, Beijing, 128 p.

Zhou D, Graham SA, 1993, Songpan-Ganzi Triassic flysch complex as a remnant ocean basin along diachronous Qinling collisional orogen, central China: Geological Society of America Abstracts with Programs, v. 25, n. 6, p. A118.

Zhu X, 1989. Remarks on the Chinese Mesozoic-Cenozoic sedimentary basins, in Zhu X (ed.), Chinese sedimentary basins: Elsevier, Amsterdam, p. 1–5.

Ziegler PA, 1975, North Sea basin history in the tectonic framework of north-western Europe, in Woodland AW (ed.), Petroleum and the continental shelf of north-west Europe, v. 1: John Wiley, New York, p. 131–149.

Ziegler PA, 1977, Geology and hydrocarbon provinces of the North Sea: Geological Journal, v. 1, p. 7–32.

Ziegler PA, 1978, Northwestern Europe: tectonics and basin development: Geologie en Mijnbouw, v. 57, p. 589–626.

Ziegler PA, 1982, Faulting and graben formation in western and central Europe: Philosophical Transactions of the Royal Society of London, v. A305, p. 114–143.

Ziegler PA, 1987, Late Cretaceous and Cenozoic intraplate compressional deformations in the Alpine foreland–a geodynamic model: Tectonophysics, v. 137, p. 389–420.

Ziegler PA, 1988, Evolution of the Arctic–North Atlantic and the western Tethys: American Association of Petroleum Geologists Memoir 43, 198 p.

Ziegler PA, 1990, Tectonic and palaeogeographic development of the North Sea rift system, in Blundell DJ, Gibbs AD (eds.), Tectonic evolution of the North Sea rifts: Clarendon Press, Oxford, p. 1–36.

Ziegler WH, 1975, Outline of the geological history of the North Sea, in Woodland AW (ed.), Petroleum and the continental shelf of north-west Europe, v. 1: John Wiley, New York, p. 165–190.

Zolnai G, 1991, Continental wrench-tectonics and hydrocarbon habitat [revised]: American Association of Petroleum Geologists Continuing Education Course Note Series 30, variably paginated.

Zonenshain LP, Savostin LA, 1981, Geodynamics of the Baikal rift zone and plate tectonics of Asia: Tectonophysics, v. 76, p. 1–45.

Zonenshain LP, Kuzmin MD, Napov LM, 1990, Geology of the USSR: a plate-tectonic synthesis: American Geophysical Union Geodynamics Series, v. 21, 242 p.

Zou DB, Rao RB, Chen YM, Chen LK, 1984, Triassic turbidites in southern Bayankela Mountains, Tibetan Plateau, in CGQXP Editorial Committee (ed.), Contribution to the Geology of the Tibet Plateau: Geological Publishing House, Beijing, p. 27–47.

Index